Advanced Structured Materials

Volume 60

More information about this series at http://www.springer.com/series/8611

Konstantin Naumenko · Marcus Aßmus
Editors

Advanced Methods of Continuum Mechanics for Materials and Structures

Springer

Editors
Konstantin Naumenko
Institute of Mechanics
Otto von Guericke University
Magdeburg
Germany

Marcus Aßmus
Institute of Mechanics
Otto von Guericke University
Magdeburg
Germany

ISSN 1869-8433 ISSN 1869-8441 (electronic)
Advanced Structured Materials
ISBN 978-981-10-0958-7 ISBN 978-981-10-0959-4 (eBook)
DOI 10.1007/978-981-10-0959-4

Library of Congress Control Number: 2016936648

Printed on acid-free paper

This Springer imprint is published by Springer Nature
The registered company is Springer Science+Business Media Singapore Pte Ltd.

On the occasion of the 60th birthday of Holm Altenbach

One of the favorite discussion topics of Holm is the question how to compute the effective transverse shear stiffness of a plate. This is a tricky task even for plates made of linear-elastic materials. Indeed, to find the transverse shear stiffness within the linear theory of elasticity, a nonlinear equation should be solved (Altenbach 2000; Altenbach et al. 2015). As an example, consider a three-layer laminate plate with skin layers made from the same material and a core layer. Let h be the thickness of the plate and h_c be the thickness of the core layer. Assume that the shear moduli are G_c and G_s for the core and skin layers, respectively. Then according to Altenbach (2000), the effective transverse shear stiffness Γ of the plate can be computed as follows

$$\Gamma = \frac{1}{3} G_s h \lambda^2 \left[1 - \alpha^3 (1 - \mu)\right], \quad \mu = \frac{G_c}{G_s}, \quad \alpha = \frac{h_c}{h}, \tag{1}$$

where λ is the least positive root of the following equation

$$\sin \lambda\alpha \sin \lambda(1 - \alpha) = \mu \cos \lambda\alpha \cos \lambda(1 - \alpha). \tag{2}$$

For homogeneous plates with $\mu = 1$, Eqs. (1) and (2) yield

$$\Gamma = \kappa G h, \quad G = G_s = G_c,$$

where the factor κ takes the value $\pi^2/12$ as originally derived by Mindlin (1951). Sandwich plates applied in lightweight structures are usually composed of relatively thick soft core layer and thin skin layers with a relatively high stiffness. As shown in Altenbach (2000) for sandwich plates, the Reissner's formula (Reissner 1947) follows from Eqs. (1) and (2)

$$\Gamma = G_c h.$$

For laminated glass plates having a thin core layer with the lower shear modulus, the approximate solution of Eq. (2) reads

$$\lambda^2 = \frac{\mu}{\alpha(1 - \alpha)}. \tag{3}$$

With Eq. (1), the approximate value of the transverse shear stiffness is

$$\Gamma = \frac{1}{3} G_c h \frac{1 - \alpha^3 (1 - \mu)}{\alpha(1 - \alpha)}. \tag{4}$$

For thin-walled structures undergoing creep and damage, special iterative procedures are required in order to compute transverse shear deformation (Altenbach and Naumenko 2002).

Beyond the activities on shear correction factors, let us take the opportunity to mention some recent researches by Holm, his students, and collaborators. These include analysis of shells from functionally graded materials (Altenbach and Eremeyev 2008), photovoltaic modules (Schulze et al. 2012; Weps et al. 2013; Eisentrager et al. 2015a, b), nanoscale shell structures, where surface effects have to be taken into account (Altenbach et al. 2009; Altenbach and Eremeyev 2011; Altenbach et al. 2012), inelastic behavior of advanced heat-resistant materials (Altenbach et al. 2008; Langler et al. 2014), micro-mechanics of grain boundary cavitation under creep conditions (Ozhoga-Maslovskaja et al. 2015) and inelastic micro-polar materials (Altenbach and Eremeyev 2014).

It is not straightforward to identify the exact number of books published and/or edited by Holm. We guess that this number would be in the range between 40 and 50. Let us mention some of his textbooks which are used by many professors for teaching applied mechanics and are popular among students as well. These include Engineering Mechanics (Altenbach 2014), Continuum Mechanics (Altenbach 2015), Theory of Plates (Altenbach et al. 1998), and Mechanics of Composite Structural Elements (Altenbach et al. 2004).

Holm has directly supervised over 30 doctoral students, and countless, doctoral and postdoctoral fellows from different countries. For many years, he has been the board member of the interdisciplinary Graduate School 1554 on *Micro-Macro-Interactions in Structured Media and Particle Systems*[5] funded by the German Research Foundation (Deutsche Forschungsgemeinschaft, abbreviated DFG). Many Ph.D. holders from this school and from the former one[6] have managed academic positions in different universities.

Professor Altenbach with assistants and students. Merseburg, 2002

[5]see homepage: www.grk1554.ovgu.de.

[6]Ph.D. schools ("Graduiertenkollegs", abbreviated GRK) are established by German universities to promote young scientists. Their key emphasis is on the qualification of doctoral researchers within the framework of a focused research program and a structured training strategy. The first Ph.D. school at the TU Magdeburg entitled *Modellierung, Berechnung und Identification mechanischer Systeme* was founded in 1992.

This volume contains a collection of contributions on advanced approaches of continuum mechanics written by leading scientists and collaborators, former Ph.D. students, and friends of Holm around the globe. These articles cover not only modern high-impact research areas but also historical essays and fundamentals. We thank all these distinguished authors for sharing this celebration.

We would like to acknowledge the series editor Prof. Andreas Öchsner for giving us the opportunity to publish this volume. We thank Helal Chowdhury, Johanna Eisenträger, Oliver Junge, and Barbara Renner from the Institute of Mechanics, Otto-von-Guericke-University Magdeburg, for their careful readings of parts of the volume. We would like to acknowledge Dr. Christoph Baumann from Springer Publisher for the assistance and support during the preparation of the book.

We wish Holm a wonderful 60th birthday, continued success, many new scientific papers and books, happiness, as well as excellent health for many years to come.

Magdeburg
March 2016

Konstantin Naumenko
Marcus Aßmus

References

Altenbach, H.: Eine direkt formulierte lineare Theorie für viskoelastische Platten und Schalen. Ingenieur-Archiv **58**(3), 215–228 (1988). doi:10.1007/BF00534332

Altenbach, H.: An alternative determination of transverse shear stiffnesses for sandwich and laminated plates. Int. J. Solids Struct. **37**(25), 3503–3520 (2000). doi:10.1016/S0020-7683(99)00057-8

Altenbach, H.: Creep analysis of thin-walled structures. ZAMM—Journal of Applied Mathematics and Mechanics (Zeitschrift für Angewandte Mathematik und Mechanik) **82**(8), 507–533 (2002). doi:10.1002/1521-4001(200208)82:8<507::AID-ZAMM507>3.0.CO;2-Y

Altenbach, H.: Holzmann/Meyer/Schumpich Technische Mechanik Festigkeitslehre. Springer Vieweg, Wiesbaden (2014). doi:10.1007/978-3-658-06041-1

Altenbach, H.: Kontinuumsmechanik: Einführung in die materialunabhängigen und materialabhängigen Gleichungen, 3rd edn. Springer, Berlin Heidelberg (2015). doi:10.1007/978-3-662-47070-1, (originally appeared at Teubner publisher, Leipzig, 1994; published together with his father Johannes Altenbach)

Altenbach, H., Eremeyev, V.: Strain rate tensors and constitutive equations of inelastic micropolar materials. Int. J. Plast **63**, 3–17 (2014). doi:10.1016/j.ijplas.2014.05.009

Altenbach, H., Eremeyev, V.A.: Analysis of the viscoelastic behavior of plates made of functionally graded materials. ZAMM—Journal of Applied Mathematics and Mechanics (Zeitschrift für Angewandte Mathematik und Mechanik) **88**(5), 332–341 (2008). doi:10.1002/zamm.200800001

Altenbach, H., Eremeyev, V.A.: On the shell theory on the nanoscale with surface stresses. Int. J. Eng. Sci. **49**(12), 1294–1301 (2011). doi:10.1016/j.ijengsci.2011.03.011

Altenbach, H., Naumenko, K.: Shear correction factors in creep-damage analysis of beams, plates and shells. JSME International Journal Series A, Solid Mech. Mater. Eng. **45**, 77–83 (2002). doi:10.1299/jsmea.45.77

Altenbach, H., Shilin, P.: Eine nichtlineare Theorie dünner Dreischichtschalen und ihre Anwendung auf die Stabilitätsuntersuchung eines dreischichtigen Streifens. Technische Mechanik **3**(2), 23–30 (1982)

Altenbach, H., Zhilin, P.A.: A general theory of elastic simple shells (in Russian). Adv. in Mech. **11**(4), 107–148 (1988)

Altenbach, H., Morachkovsky, O., Naumenko, K., Sychov, A.: Geometrically nonlinear bending of thin-walled shells and plates under creep-damage conditions. Arch. Appl. Mech. **67**, 339–352 (1997). doi:10.1007/s004190050122

Altenbach, H., Altenbach, J., Naumenko, K.: Ebene Flächentragwerke. Springer, Berlin Heidelberg (1998). doi:10.1007/978-3-642-58721-4

Altenbach, H., Naumenko, K., L'vov, G.I., Pylypenko, S.: Numerical estimation of the elastic properties of thin-walled structures manufactured from short-fiber reinforced thermoplastics. Mech. Compos. Mater. **39**, 221–234 (2003a). doi:10.1023/A:1024566026411

Altenbach, H., Naumenko, K., Zhilin, P.: A micro-polar theory for binary media with application to phase-transitional flow of fiber suspensions. Continuum Mech. Thermodyn. **15**, 539–570 (2003 b). doi:10.1007/s00161-003-0133-5

Altenbach, H., Altenbach, J.W., Kissing, W.: Mechanics of composite structural elements. Springer, Berlin Heidelberg (2004). doi:10.1007/978-3-662-08589-9

Altenbach, H., Naumenko, K., Pylypenko, S., Renner, B.: Influence of rotary inertia on the fiber dynamics in homogeneous creeping flows. ZAMM—Journal of Applied Mathematics and Mechanics (Zeitschrift für Angewandte Mathematik und Mechanik) **87**(2), 81–93 (2007). doi:10.1002/zamm.200610303

Altenbach, H., Naumenko, K., Gorash, Y.: Creep analysis for a wide stress range based on stress relaxation experiments. Int. J. Mod. Phys. B **22**(31–32), 5413–5418 (2008). doi:10.1142/S0217979208050589

Altenbach, H., Eremeyev, V., Morozov, N.: Linear theory of shells taking into account surface stresses. Dokl. Phys. **54**(12), 531–535 (2009). doi:10.1134/S1028335809120039

Altenbach, H., Eremeyev, V.A., Morozov, N.F.: Surface viscoelasticity and effective properties of thin-walled structures at the nanoscale. Int. J. Eng. Sci. **59**, 83–89 (2012). doi:10.1016/j.ijengsci.2012.03.004

Altenbach, H., Eremeyev, V.A., Naumenko, K.: On the use of the first order shear deformation plate theory for the analysis of three-layer plates with thin soft core layer. ZAMM—Journal of Applied Mathematics and Mechanics (Zeitschrift für Angewandte Mathematik und Mechanik) **95**(10), 1004–1011 (2015). doi:10.1002/zamm.201500069

Eisenträger, J., Naumenko, K., Altenbach, H., Köppe, H.: Application of the first-order shear deformation theory to the analysis of laminated glasses and photovoltaic panels. Int. J. Mech. Sci. **96**, 163–171 (2015a). doi:10.1016/j.ijmecsci.2015.03.012

Eisenträger, J., Naumenko, K., Altenbach, H., Meenen, J.: A user-defined finite element for laminated glass panels and photovoltaic modules based on a layer-wise theory. Compos. Struct. **133**, 265–277 (2015b). doi:10.1016/j.compstruct.2015.07.049

Kröner, C., Altenbach, H., Naumenko, K.: Coupling of a structural analysis and flow simulation for short-fiber-reinforced polymers: property prediction and transfer of results. Mech. Compos. Mater. **45**(3), 249–256 (2009). doi:10.1007/s11029-009-9086-5

Längler, F., Naumenko, K., Altenbach, H., Ievdokymov, M.: A constitutive model for inelastic behavior of casting materials under thermo-mechanical loading. J. Strain Anal. Eng. Des. **49**(6), 421–428 (2014). doi:10.1177/0309324714522034

Mindlin, R.D.: Influence of rotatory inertia and shear on flexural motions of isotropic, elastic plates. J. Appl. Mech. **18**(2), 31–38 (1951)

Ozhoga-Maslovskaja, O., Naumenko, K., Altenbach, H., Prygorniev, O.: Micromechanical simulation of grain boundary cavitation in copper considering non-proportional loading. Comput. Mater. Sci. **96**, 178–184 (2015). doi:10.1016/j.commatsci.2014.08.050

Palmow, W.A., Altenbach, H.: Über eine Cosseratsche Theorie für elastische Platten. Technische Mechanik **3**(3), 5–9 (1982)

Reissner, E.: On the bending of elastic plates. Quarterly of Applied Mathematics **5**, 55–68 (1947)
Schulze, S.H., Pander, M., Naumenko, K., Altenbach, H.: Analysis of laminated glass beams for photovoltaic applications. Int. J. Solids Struct. **49**(15), 2027–2036 (2012). doi:10.1016/j.ijsolstr.2012.03.028
Weps, M., Naumenko, K., Altenbach, H.: Unsymmetric three-layer laminate with soft core for photovoltaic modules. Compos. Struct. **105**, 332–339 (2013). doi:10.1016/j.compstruct.2013.05.029

Contents

Part I
Fundamentals and Elasticity

Time Derivatives in Material and Spatial Description—What Are the Differences and Why Do They Concern Us?

Elena A. Ivanova, Elena N. Vilchevskaya and Wolfgang H. Müller

Abstract This paper has many, albeit mostly didactic objectives. It is an attempt toward clarification of several concepts of continuum theory which can lead and have led to confusion. In a way the paper also creates a bridge between the lingo of the solid mechanics and the fluid mechanics communities. More specifically, an attempt will be made, first, to explain and to interpret the subtleties and the relations between the so-called material and spatial description of continuum fields. Second, the concept of time derivatives in material and spatial description will be investigated meticulously. In particular, it will be explained why and how the so-called material and total time derivatives differ and under which circumstances they turn out to be the same. To that end, material and total time derivatives will be defined separately and evaluated in context with local fields as well as during their use in integral formulations, i.e., when applied to balance equations. As a special example the mass balance is considered for closed as well as open bodies. In the same context the concept of a "moving observation point" will be introduced leading to a generalization of the usual material derivative. When the total time derivative is introduced the distinction between the purely mathematical notion of a coordinate system and the intrinsically physics-based concept of a frame of reference will gain particular importance.

E.A. Ivanova · E.N. Vilchevskaya (✉)
Peter the Great Saint-Petersburg Plytechnic Univerity, Politekhnicheskaja 29, 195251 St.-Petersburg, Russia
e-mail: elenaivanova239@gmail.com

E.A. Ivanova · E.N. Vilchevskaya
Institute for Problems in Mechanical Engineering, Russian Academy of Sciences, Bol'shoy Pr. 61, V.O., 199178 St.-Petersburg, Russia
e-mail: vilchevska@gmail.com

W.H. Müller
Chair of Continuum Mechanics and Materials Theory, Institute of Mechanics, Technical University of Berlin, Winsteinufer 5, 10587 Berlin, Germany
e-mail: wolfgang.h.mueller@tu-berlin.de

K. Naumenko and M. Aßmus (eds.), *Advanced Methods of Continuum Mechanics for Materials and Structures*, Advanced Structured Materials 60,
DOI 10.1007/978-981-10-0959-4_1

1 Introduction

There are two fundamentally different approaches for describing the kinematics of continua. One method of observing a structure's motion is based on following individual particles of the body as they move through space and time. It is used for modeling solid matter with various rheological properties (elastic and nonelastic). In the other method motion is described by focusing on a specific location of space through which the structure moves as time passes on. It is mostly used in fluid and gas mechanics as well as in granular media modeling. Following Malvern (1969), we will call the first description *material* and the second one *spatial*.

In order to consider multiphase structures consisting of a solid and fluid phases it is convenient to use the material description for the solid and a modification of the spacial description for fluids. In this case it is very important to understand how the concepts introduced in the different descriptions relate to each other. This holds, in particular, for the time derivatives. The so-called *total derivative in material description* and the *material derivative in spatial description* written in terms of a partial derivative look very similar. Thus it is not surprising that it is a widespread opinion that the total and the material derivatives are different names for the same concept. But as it will be shown further down they are different concepts describing a rate of change of a property of the material point and a rate of change of a property at the space point. Consequently, one aim of the present paper is to give a definition of the total derivative (as an analogue and generalization of the derivative used in the material description), to give a definition of the material derivative (as an analogue and generalization of the derivative used in the spatial description), to make a strict distinction between these concepts, and to investigate the conditions for which they coincide.

In summary, this paper is an attempt to clarify sometimes obscure and confusing statements made in context with the material and spatial description and the associated time derivatives. In this sense it is of didactic value and we only claim to raise awareness of the situation and to provide some comments regarding possible ways out of die-hard conundrums.

2 Total and Material Derivatives in Material and Spatial Descriptions—Literature Review

The main problem with the definition of a *total time derivative* is that the corresponding operator appearing in the laws of continuum theory is not simply a purely mathematical concept. Rather it contains an underlying, definitive physical meaning. In order to clarify the problem we start with a purely mathematical definition. Let $g = \tilde{g}(u(t), v(t), w(t), t)$. The total derivative of $\tilde{g}$ with respect to t (the symbol of the time variable, chosen as a reminder for later purpose) is:

$$\frac{d\tilde{g}}{dt} = \frac{\partial \tilde{g}}{\partial u}\frac{du}{dt} + \frac{\partial \tilde{g}}{\partial v}\frac{dv}{dt} + \frac{\partial \tilde{g}}{\partial w}\frac{dw}{dt} + \frac{\partial \tilde{g}}{\partial t}. \tag{1}$$

Now let the arguments of the function depend on two independent variables, say $u = u_*(s,t)$, $v = v_*(s,t)$, $w = w_*(s,t)$. We can form partial derivatives of $g = g_*(s,t)$ with respect to one of its arguments with the other held constant. Thus in the case of a function of several independent variables (here s and t) every derivative is a partial one. The concept of a total derivative of a function of several independent variables does not exist in mathematics.

In mechanics of continua all quantities characterizing a stress–strain state are functions of several *independent* variables—three spatial coordinates and time. Therefore, introducing the concept of a total derivative with respect to time in a strictly mathematical sense is impossible. An additional physics-based idea postulating what spatial coordinates should be fixed and how to identify them is needed. We will get back to that later.

Within the framework of material description the so-called material points are identified by their position, $\boldsymbol{R}$, in an arbitrary chosen reference configuration. The reference configuration is usually chosen to be fixed in the frame of reference. The total derivative of a vector field, $\boldsymbol{\psi}(\boldsymbol{R}, t)$, is then defined as a partial derivative with respect to t with $\boldsymbol{R}$ held constant, see Dmitrienco (2009):

$$\frac{d\boldsymbol{\psi}}{dt} = \left.\frac{\partial \boldsymbol{\psi}}{\partial t}\right|_{\boldsymbol{R}=\text{const}}. \tag{2}$$

Note that in some cases it makes sense to exclude a rigid body motion from our considerations (e.g., if we are only interested in the (local) deformation, i.e., the displacement, $\boldsymbol{u}$, of the matter of an object, it does not make sense to look at its total motion. Hence, we "take out" the rotation when considering the deformation of a spinning shaft, or the translative/rotative motion of a flying aircraft when bending of its wing becomes an issue, etc.). If a coordinate system comoving with the body is used then the reference position vector depends on time and the definition (2) has to be modified since $\boldsymbol{R} = \hat{\boldsymbol{R}}(t)$.

In a number of books on solid mechanics and nonlinear elasticity the derivative (2) is called material, substitutional, or individual (see, e.g., Ogden 2003; Asaro and Lubarda 2006) and the concept of a total derivative is not introduced. In other books the definition (2) is not given explicitly. Rather the material derivative is defined as a rate of change of a variable, $\boldsymbol{\psi}$, whose arguments are the current position vector of a material particle (the so-called *motion*), $\boldsymbol{r}(\boldsymbol{R}, t)$, and time, t. Then, by virtue of the chain rule of calculus:

$$\dot{\boldsymbol{\psi}} \equiv \frac{D\boldsymbol{\psi}}{Dt} = \left.\frac{\partial \boldsymbol{\psi}}{\partial t}\right|_{\boldsymbol{r}=\text{const}} + \boldsymbol{v}\cdot\nabla\boldsymbol{\psi}, \quad \nabla = \frac{\partial}{\partial \boldsymbol{r}}, \quad \boldsymbol{v} = \left.\frac{\partial \boldsymbol{r}(\boldsymbol{R}, t)}{\partial t}\right|_{\boldsymbol{R}=\text{const}}. \tag{3}$$

In these books the definition of the total derivative is either not given at all or it is said that the material derivative coincides with the total one (Milne-Thomson 1960;

e.g., Batchelor (1970), Lamb (1975), Rouse (1959), Lojtsanskij (1950), Landau and Lifshitz (1959), Prandtl and Tietjens (1929). It should be noted that this definition of a material derivative cannot be regarded as a mathematical definition of the function derivative since an increment of argument $\boldsymbol{v}\delta t$ cannot be expressed in terms of function arguments $\boldsymbol{r}$ and t.

Thus, the formally introduced definition of the material derivative as a derivative of a composite function cannot be used in spatial description. Also, it is worthwhile mentioning that in the velocity of (8) we put $\boldsymbol{v}(\boldsymbol{r}, t)$, whereas in (3) it is a function of the reference position vector, $\boldsymbol{v}(\boldsymbol{R}, t)$.

Now let us consider the definition of the material derivative (2) as a derivative following the motion of the fluid. In the spatial description a given region in space is considered. The material element at position $\boldsymbol{r}$ at time t possesses the velocity $\boldsymbol{v}(\boldsymbol{r}, t)$ and moves to the position $\boldsymbol{r} + \boldsymbol{v}dt$. Thus in order to describe the evolution of matter during the infinitesimal period of time dt one may use the material description taking $\boldsymbol{r}$ and $d\boldsymbol{u}(\boldsymbol{x}, t) = \boldsymbol{v}dt$ as the reference position and the infinitesimal displacement correspondingly. As a result, one may apply mathematical methods developed for the material description to the spatial description, taking at every moment of time the current configuration as the reference one and considering a small vicinity of this configuration, see Ilyushin (1971). Fixing the position vector of the material particle at some moment of time and taking it as the reference position vector contradicts the essence of spatial description, thus the definition (2) cannot be used in it.

It should be noted that since there is only the current configuration in the spatial description it is obvious which coordinates should be fixed. The total derivative is defined as a partial derivative with the observation point, $\boldsymbol{r}$, being held constant,

$$\frac{d\boldsymbol{\psi}}{dt} = \left.\frac{\partial \boldsymbol{\psi}}{\partial t}\right|_{\boldsymbol{r}=\text{const}}. \tag{9}$$

In addition to that the balance equations are formulated for a constant volume containing the observation point. As a result a partial derivative operator appears in front of the integral (Milne-Thomson 1960; Landau and Lifshitz 1959). The total derivative is usually not used in classical hydrodynamics at all and all equations are written in terms of the material and partial derivatives.

However, in order to consider more complicated problems (e.g., fluid flow through a deformable solid or porous media) it is convenient to use a moving observation point fixed within the (elementary) volume jointly traversing space at a given velocity.

An expression for the material derivative in the spatial description with a moving observation point was suggested in Altenbach et al. (2003), Zhilin (2012) and contains the total time derivative:

$$\frac{D\boldsymbol{\psi}}{Dt} = \frac{d\boldsymbol{\psi}(\boldsymbol{r}(t), t)}{dt} + \left(\boldsymbol{v}(\boldsymbol{r}(t), t) - \frac{d\boldsymbol{r}(t)}{dt}\right) \cdot \nabla \boldsymbol{\psi}(\boldsymbol{r}(t), t). \tag{10}$$

A similar expression can be found in the literature on porous media, e.g., Hassanizadeh and Gray (1980). A porous medium is viewed as a body consisting of two coexistent continua. The motion of the solid phase is defined by the current position vector $\boldsymbol{r}^s(\boldsymbol{R}^s, t)$, the motion of the solid so-to-speak. The time rate of change with respect to the solid phase of a quantity ψ is defined as:

$$\frac{D^s\psi}{Dt} = \left.\frac{\partial\psi(\boldsymbol{R}^s, t)}{\partial t}\right|_{\boldsymbol{R}^s=\text{const}} = \left.\frac{\partial\psi(\boldsymbol{r}^s, t)}{\partial t}\right|_{\boldsymbol{r}^s=\text{const}} + \boldsymbol{v}^s \cdot \nabla\psi, \tag{11}$$

where $\boldsymbol{v}^s$ is the solid phase velocity. It is obvious that (11) coincides with the definitions of the total derivative in the material description (2), (3). The time rate of change of the quantity ψ with respect to the fluid phase is given by

$$\frac{D^f\psi}{Dt} = \left.\frac{\partial\psi(\boldsymbol{r}^s, t)}{\partial t}\right|_{\boldsymbol{r}^s=\text{const}} + \boldsymbol{v}^f \cdot \nabla\psi, \tag{12}$$

where $\boldsymbol{v}^f$ is the fluid-phase velocity field.

Subtraction of Eq. (11) from (12) yields the following relation, see Hassanizadeh and Gray (1980):

$$\frac{D^f\psi}{Dt} = \frac{D^s\psi}{Dt} + (\boldsymbol{v}^f - \boldsymbol{v}^s) \cdot \nabla\psi. \tag{13}$$

By taking into account that $d\boldsymbol{r}(t)/dt$ in (10) corresponds to the velocity of the solid phase one can see that the definitions (13) and (10) coincide.

The moving observation point is also considered in the Arbitrary Lagrangian–Eulerian (ALE) technique. ALE is used to account for the deformation of the fluid domain which arises from the displacement and deformation of the solid structure. The material derivative is defined by the fundamental ALE equation (see all of the references immediately below for details):

$$\frac{D\psi}{Dt} = \left.\frac{d\psi}{dt}\right|_{\boldsymbol{R}=\text{const}} + (\boldsymbol{v} - \hat{\boldsymbol{v}}) \cdot \nabla\psi, \tag{14}$$

where $\boldsymbol{v}$ is the velocity of the fluid particle and $\hat{\boldsymbol{v}}$ is referred to indistinctly as the "velocity of the reference point" by Dettmer and Peric (2006) or "velocity of the moving frame" by Del Pin et al. (2007). In fact, it should be called unmistakably "grid velocity" as in Vuong et al. (2015) and Gadala (2004), or "mesh velocity" as in Khoei et al. (2007) or Filipovic et al. (2006). Also note that the concept of a moving grid and of a relative speed inherent to the material derivative was anticipated before ALE became a prominent concept by Müller and Muschik (1983), where $\hat{\boldsymbol{v}}$ was called "mapping velocity." Moreover, d/dt corresponds to the change of the material particle quantity, which is noted by an observer traveling with a point on the reference frame. The definitions (10), (13), and (14) coincide.

However, in some articles, see Dang and Meschke (2014), Preisig and Zimmermann (2011), Sarrate et al. (2001), the material derivative is defined by:

$$\frac{D\psi}{Dt} = \frac{\partial\psi}{\partial t} + (\boldsymbol{v} - \hat{\boldsymbol{v}}) \cdot \nabla\psi, \tag{15}$$

where the symbol $\boldsymbol{v}$ was used for the fluid velocity and $\hat{\boldsymbol{v}}$ for the so-called "velocity of the moving reference," see Dang and Meschke (2014), or "fluid mesh velocity" as termed by Preisig and Zimmermann (2011) or Sarrate et al. (2001). Even if we ignore differences in the linguistic terminology, Eq. (15) coincides with (14), (10), and (13) only if the partial derivative in (15) is defined with the reference position vector of the observation point held constant. At the same time it is written in Surana et al. (2014)[2]:

> ... the Eulerian description with transport
>
> $$\rho\left(\left.\frac{\partial \boldsymbol{v}}{\partial t}\right|_{\boldsymbol{r}=\text{const}} + \boldsymbol{v} \cdot \nabla\boldsymbol{v}\right) - \nabla \cdot \boldsymbol{\sigma} - \rho \boldsymbol{f} = 0,$$
>
> is converted to ALE form by replacing velocity $\boldsymbol{v}$ (velocity at a fixed location $\boldsymbol{r}$) in the convective terms with convective velocity $\bar{\boldsymbol{C}} = \boldsymbol{v} - \hat{\boldsymbol{v}}$.

It is obvious that in this case the material derivative in Surana et al. (2014) differs from the material derivatives defined by (14), (10), and (13). It is seen that there are different formulae for the operator of a material derivative in modern literature. Some authors distinguish between the total and the material derivatives, but sometimes it is used synonymously, namely as "the total time or the material derivative," cf., Milne-Thomson (1960), Lojtsanskij (1950). Sometimes an operator of material derivative is defined through another operator of material derivative (equations analogously to (13)). It conflicts with the classical interpretation of the material derivative as a derivative following the motion of the specific particle.

In many papers the material derivative is written in a form very similar to the one adopted from the classical textbooks. However, it is not specified what is meant by the partial time derivative. As a result, a comparison of material derivatives used by different authors in order to ensure that they coincide or differ is extremely difficult in some cases. Furthermore, definitions of the used notations are not always provided, and only formulae for calculations are listed.

In summary of our review we have to conclude that we are facing, first, the need for a clear distinction between the concept of a material and a spatial description of fields. Second, a distinction of various time derivatives of these fields is required, namely between one unfortunately called material time derivative, despite the fact that it exists in material *and* in spatial description, as well as the other known as total time derivative. Sometimes both coincide in meaning and sometimes they do not. Following this remark, it is the goal of the present paper to give, first, clear mathematical definitions of the material and of the total time derivatives, which can be used for a moving observation point and for a nonconstant reference vector, and, second, to clarify the physical meaning of these operators.

---

[2]This is not a verbal quote. For the convenience of the reader it has been adjusted to the symbols used in this paper.

3 Material Description

3.1 Kinematics of Continua

In material description quantities related to material particles are functions of the reference position, $\boldsymbol{R}$, and of time, t, which we will refer to as "referential variables." Any such function $f = f_*(\boldsymbol{R}, t)$ may be replaced by a function of the spatial variables $\tilde{f}(\boldsymbol{r}, t)$, which has the same value, f, at the corresponding position vector: $f = f_*(\boldsymbol{R}, t) = \tilde{f}(\boldsymbol{r}_*(\boldsymbol{R}, t), t)$.

Suppose that the reference position vector does not depend on time. Then the rate of change of a quantity relevant for characterization of the material particle is

$$\frac{\partial f_*}{\partial t} = \lim_{\Delta t \to 0} \frac{f_*(\boldsymbol{R}, t + \Delta t) - f_*(\boldsymbol{R}, t)}{\Delta t}, \tag{16}$$

if it is a function of the referential variables. Otherwise

$$\begin{aligned} \frac{d_r \tilde{f}}{dt} &= \lim_{\Delta t \to 0} \frac{\tilde{f}(\boldsymbol{r}_*(\boldsymbol{R}, t + \Delta t), t + \Delta t) - \tilde{f}(\boldsymbol{r}_*(\boldsymbol{R}, t), t)}{\Delta t} \\ &= \lim_{\Delta t \to 0} \frac{\tilde{f}(\boldsymbol{r}_*(\boldsymbol{R}, t + \Delta t), t + \Delta t) - \tilde{f}(\boldsymbol{r}_*(\boldsymbol{R}, t + \Delta t), t)}{\Delta t} \\ &\quad + \lim_{\Delta t \to 0} \frac{\tilde{f}(\boldsymbol{r}_*(\boldsymbol{R}, t + \Delta t), t) - \tilde{f}(\boldsymbol{r}_*(\boldsymbol{R}, t), t)}{\Delta t} \\ &= \frac{\partial \tilde{f}(\boldsymbol{r}, t)}{\partial t} + \frac{\partial \boldsymbol{r}_*(\boldsymbol{R}, t)}{\partial t} \cdot \nabla \tilde{f}, \qquad \nabla \equiv \frac{\partial}{\partial \boldsymbol{r}}. \end{aligned} \tag{17}$$

The rate of change of a physical quantity should not depend on the choice of variables. Thus

$$\frac{\partial f_*(\boldsymbol{R}, t)}{\partial t} = \frac{d_r \tilde{f}(\boldsymbol{r}, t)}{dt}, \qquad \frac{d_r}{dt} \equiv \left.\frac{\partial}{\partial t}\right|_{r=\text{const}} + \frac{\partial \boldsymbol{r}_*(\boldsymbol{R}, t)}{\partial t} \cdot \nabla. \tag{18}$$

Equation $(18)_1$ is consistent with the chain rule of calculus. The operator d_r/dt defines the total derivative under the condition that the reference position vector is a constant.

Now let $\boldsymbol{R} = \boldsymbol{R}_{*0}(\boldsymbol{R}_0, t)$, where $\boldsymbol{R}_0$ does not depend on time. The velocity vector is thus defined by

$$\begin{aligned} \boldsymbol{v} &= \lim_{\Delta t \to 0} \frac{\boldsymbol{r}_*(\boldsymbol{R}_{*0}(\boldsymbol{R}_0, t + \Delta t), t + \Delta t) - \boldsymbol{r}_*(\boldsymbol{R}_{*0}(\boldsymbol{R}_0, t), t)}{\Delta t} \\ &= \frac{\partial \boldsymbol{r}_*(\boldsymbol{R}, t)}{\partial t} + \frac{\partial \boldsymbol{R}}{\partial t} \cdot \overset{\circ}{\nabla} \boldsymbol{r} \equiv \frac{d^\circ \boldsymbol{r}}{dt}, \\ \overset{\circ}{\nabla} &\equiv \frac{\partial}{\partial \boldsymbol{R}}, \qquad \frac{d^\circ}{dt} \equiv \left.\frac{\partial}{\partial t}\right|_{\boldsymbol{R}=\text{const}} + \frac{\partial \boldsymbol{R}}{\partial t} \cdot \overset{\circ}{\nabla} \end{aligned} \tag{19}$$

$$\begin{aligned}\frac{d_*^\circ}{dt}\int_{V_0}\rho_*(\boldsymbol{R},t)\,J_*(\boldsymbol{R},t)\,dV_0 &= \frac{\partial}{\partial t}\int_{V_0}\rho_{*0}(\boldsymbol{R}_0,t)\,J_{*0}(\boldsymbol{R}_0,t)\,dV_0\\ &= \int_{V_0}\frac{\partial}{\partial t}\big[\rho_{*0}(\boldsymbol{R}_0,t)\,J_{*0}(\boldsymbol{R}_0,t)\big]dV_0\\ &= \int_{V_0}\frac{d^\circ}{dt}\big[\rho_*(\boldsymbol{R},t)\,J_*(\boldsymbol{R},t)\big]dV_0. \end{aligned} \tag{30}$$

After taking (30) into account and

$$J_*(\boldsymbol{R},t)=\tilde{J}(\boldsymbol{r},t)=\operatorname{Det}\left(\frac{\partial\tilde{\boldsymbol{R}}(\boldsymbol{r},t)}{\partial\boldsymbol{r}}\right)^{-1},\quad \nabla\cdot\tilde{\boldsymbol{v}}(\boldsymbol{r},t)=\tilde{J}^{-1}(\boldsymbol{r},t)\frac{d_r\tilde{J}(\boldsymbol{r},t)}{dt}, \tag{31}$$

we can carry out the differentiation in $(29)_1$:

$$\begin{aligned}&\frac{d_*}{dt}\int_{V(t)}\tilde{\rho}(\boldsymbol{r},t)\,dV\\ &=\frac{d_*^\circ}{dt}\int_{V_0}\rho_*(\boldsymbol{R},t)J_*(\boldsymbol{R},t)\,dV_0\\ &=\int_{V_0}\frac{d^\circ}{dt}\big[\rho_*(\boldsymbol{R},t)J_*(\boldsymbol{R},t)\big]\,dV_0\\ &=\int_{V(t)}\frac{d_r}{dt}\big[\tilde{\rho}(\boldsymbol{r},t)\tilde{J}(\boldsymbol{r},t)\big]\tilde{J}^{-1}(\boldsymbol{r},t)\,dV\\ &=\int_{V(t)}\left[\frac{d_r\tilde{\rho}(\boldsymbol{r},t)}{dt}+\tilde{\rho}(\boldsymbol{r},t)\tilde{J}^{-1}(\boldsymbol{r},t)\frac{d_r\tilde{J}(\boldsymbol{r},t)}{dt}\right]dV\\ &=\int_{V(t)}\left[\frac{d_r\tilde{\rho}(\boldsymbol{r},t)}{dt}+\tilde{\rho}(\boldsymbol{r},t)\nabla\cdot\tilde{\boldsymbol{v}}(\boldsymbol{r},t)\right]dV.\end{aligned} \tag{32}$$

By substituting this result into (29) we obtain the local conservation of mass:

$$\frac{d_r\tilde{\rho}(\boldsymbol{r},t)}{dt}+\tilde{\rho}(\boldsymbol{r},t)\nabla\cdot\tilde{\boldsymbol{v}}(\boldsymbol{r},t)=0. \tag{33}$$

Equation (33) often appears in the literature as:

$$\frac{\partial\tilde{\rho}(\boldsymbol{r},t)}{\partial t}+\nabla\cdot\big[\tilde{\rho}(\boldsymbol{r},t)\tilde{\boldsymbol{v}}(\boldsymbol{r},t)\big]=0. \tag{34}$$

This form is obtained after expanding the total derivative in (33).

In contrast to a partial derivative, the total time derivative is an objective operator, in the sense that it does not depend on the choice of coordinate system. That is why it appears in balance equations in a natural way. A partial derivative can appear in balance equations only after substitution (21).

We will now endeavor to define the total derivative even more stringently.

4 Definition of Total Derivative

Keeping mathematical rigor leads to many different notions of the same physical quality and symbols of the total derivative. In order to facilitate the notation without risking confusion, we shall introduce the general definition of the total time derivative below. However, each formula will be accompanied by verbal remarks, which seem in order, because physics is involved that goes way beyond mathematics. Let us proceed in this spirit.

All quantities in continuum mechanics are functions of spatial coordinates and time. The spatial coordinates may be constants or they may depend upon time. Note that the latter does not imply that we have a function with an argument (i.e., time). The moving coordinates have to depend on other variables that allow us to distinguish different substantial points. Therefore, in order to define a "total time derivative" we have to postulate which spatial coordinates are held constant. In other words, we have to choose a coordinate system with a distinctive feature. The frame of reference could be the one. At this point it is appropriate to introduce the notion "frame of reference" formally.

Imagine in a point O three rigidly connected, perpendicular pointers ("arrows"), $\boldsymbol{e}_1$, $\boldsymbol{e}_2$, and $\boldsymbol{e}_3$. The set $\{O, \boldsymbol{e}_1, \boldsymbol{e}_2, \boldsymbol{e}_3\}$ is called a "frame."

Definition 1 The body of reference is defined by a frame to which a set of points (in space) have been added, whereby a rigid body motion of all the points together with the frame is allowed. The position of the points are labeled relatively to the frame by establishing the reference coordinate system x_1, x_2, x_3 with origin O:

$$\boldsymbol{r}_* = x_1\boldsymbol{e}_1 + x_2\boldsymbol{e}_2 + x_3\boldsymbol{e}_3, \qquad -\infty < (x_1, x_2, x_3) < +\infty. \tag{35}$$

The frame and the reference coordinate system determine the reference body. They are "immutable." This is supposed to mean that once introduced they cannot be changed or this would lead to a different frame of reference (whose definition will come immediately). In order to describe quantitative characteristics of motion we must be able to measure distance *and* time. Hence a "clock" is needed as well:

Definition 2 The reference body with a "clock" is called the "Frame of Reference" (FoR).

Note that a frame of reference is *not* just a mathematical construct. Physics is involved due to the requirement of *measuring* distances in three independent directions and corresponding lengths as well as time.

It is impossible to say anything about the motion of the reference body, because it stands as such alone. However, it is possible to observe and quantify motions of other bodies with respect to the reference body. All physical qualities describing motion, such as velocity, for example, are measured with respect to the frame of reference and do not have any meaning without the reference frame.

In addition to the reference coordinate system one is free to choose any mathematical coordinate system in which the equations are specified. However, the reference coordinate system is a distinctive one since it determines the frame of reference. As an example consider a first coordinate transformation within an FoR, $x_i' = \hat{x}_i'(x_j)$, $i, j \in (1, 2, 3)$. On top of that we now impose a second coordinate transformation $x_i'' = \tilde{x}_i''(x_j') = \tilde{x}_i''(\hat{x}_j'(x_k)) \equiv \hat{x}_i''(x_j)$, $i, j, k \in (1, 2, 3)$. Note that if this operation is applied in context with the spatial dependence of a physical field quantity this would be a purely mathematical operation leading to no change of the meaning or value of that physical quantity. However, if we perform a change of the FoR this could result in a completely different story.

In this context it should be noted that many people do not distinguish between the concepts of frame of reference and coordinate system. Indeed, we read in Cornille (1993), p. 149:

> ... a distinction between mathematical sets of coordinates and physical frames of reference must be made. The ignorance of such distinction is the source of much confusion ...

or in Nerlich (1994), pp. 64–65:

> ... the idea of a reference frame is really quite different from that of a coordinate system. Frames differ just when they define different *spaces* (sets of *rest* points) or times (sets of simultaneous events). So the ideas of a space, a time, of rest and simultaneity, go inextricably together with that of frame. However, a mere shift of origin, or a purely spatial rotation of space coordinates results in a new coordinate system. So frames correspond at best to *classes* of coordinate systems. ...

In order to emphasize it once more: A change of the coordinate system is a purely mathematical operation, where an observer (i.e., the creator and user of the FoR) "sensing" vector quality is not needed. That is why in this case there is no difference how a vector is considered, as a directed segment or as a set of three components. Even more, we can completely exclude base vectors from our considerations and deal only with vector components. We use the notation employed with the coordinate transforms from above in an example. Suppose the coordinates of a vector in the reference coordinate system of the FoR are given by p_i. We would then obtain the corresponding coordinates w.r.t. the two other coordinate systems by $p_j' = \frac{\partial \hat{x}_j'}{\partial x_i} p_i$ and $p_k'' = \frac{\partial \tilde{x}_k''}{\partial x_j'} p_j' = \frac{\partial \tilde{x}_k''}{\partial x_j'} \frac{\partial \hat{x}_j'}{\partial x_i} p_i = \frac{\partial \hat{x}_k''}{\partial x_i} p_i$.

Let $f(x_1, x_2, x_3, t)$ be a function of the reference coordinates and of time. The total time derivative of f is:

$$\frac{df(x_1, x_2, x_3, t)}{dt} = \lim_{\Delta t \to 0} \frac{f(x_1, x_2, x_3, t + \Delta t) - f(x_1, x_2, x_3, t)}{\Delta t}, \tag{36}$$

under the condition that the reference coordinates $x_1,\ x_2,\ x_3$ are held constant and there is an increment in the function only because of the increment in time.

Now let and $g(x(x_1,\ x_2,\ x_3,\ t),\ y(x_1,\ x_2,\ x_3,\ t),\ z(x_1,\ x_2,\ x_3,\ t),\ t)$ be a composite function of several variables, namely $x,\ y$, and z, which are functions like f. Then the total time derivative of g is:

$$\frac{dg}{dt} = \frac{\partial g}{\partial x}\frac{dx}{dt} + \frac{\partial g}{\partial y}\frac{dy}{dt} + \frac{\partial g}{\partial z}\frac{dz}{dt} + \frac{\partial g}{\partial t}. \tag{37}$$

Hence, we arrive at:

Definition 3 The total time derivative is the partial derivative with the reference coordinates held constant.

This definition allows us to drop the function arguments and keep the notation relative to different arguments. In other words we can simply write:

$$\boldsymbol{v} = \frac{d\boldsymbol{r}}{dt}, \qquad \boldsymbol{a} = \frac{d\boldsymbol{v}}{dt}, \qquad \frac{d\rho}{dt} + \rho\nabla\cdot\boldsymbol{v} = 0.$$

Note that in the case of partial derivatives the arguments of the function have to be present. Indeed:

$$\boldsymbol{a}(\boldsymbol{R}, t) = \frac{\partial \boldsymbol{v}_*(\boldsymbol{R}, t)}{\partial t}, \qquad \boldsymbol{a}(\boldsymbol{r}, t) \neq \frac{\partial \tilde{\boldsymbol{v}}(\boldsymbol{r}, t)}{\partial t},$$

$$\frac{\partial \tilde{\rho}(\boldsymbol{r}, t)}{\partial t} + \nabla\cdot\left[\tilde{\rho}(\boldsymbol{r}, t)\tilde{\boldsymbol{v}}(\boldsymbol{r}, t)\right] = 0, \qquad \frac{\partial \rho_*(\boldsymbol{R}, t)}{\partial t} + \nabla\cdot\left[\rho_*(\boldsymbol{R}, t)\boldsymbol{v}_*(\boldsymbol{R}, t)\right] \neq 0.$$

As we shall learn in the next chapter the distinction between various functions, identified by a hat and a tilde, will become obsolete if we turn to the spatial description, where the concepts of a reference and of a current configuration become obsolete and the motion and state of matter is described with respect to an independent grid in space.

5 Spatial Description

5.1 *Body of General Type*

The fundamental laws of mechanics are formulated for a body. Within the material description the body is a material volume. In the spatial description it is not so obvious which object should be considered as the body. In an attempt to make things clearer we start with some definitions.

Definition 4 Consider a closed surface undergoing deformation and motion. A set of material particles located at the present moment within the surface is called the

$$\frac{\delta_r f}{\delta t} = \lim_{\Delta t \to 0} \frac{f(\boldsymbol{r} + \Delta \boldsymbol{s}, t + \Delta t) - f(\boldsymbol{r}, t)}{\Delta t}. \tag{46}$$

The numerator on the right side of (46) describes the change of the property of the given material point in time Δt. Thus, the material derivative determines a rate of change of the property of the material point located at the observation point at time t.

We now show that the formulae (45), (44) are consistent with the definition of the material derivative (46). Indeed, the function $f(\boldsymbol{r} + \Delta \boldsymbol{s}, t + \Delta t)$ can be written as

$$f(\boldsymbol{r} + \Delta \boldsymbol{s}, t + \Delta t) = f(\boldsymbol{r}, t + \Delta t) + \Delta \boldsymbol{s} \cdot \nabla f(\boldsymbol{r}, t + \Delta t), \tag{47}$$

and then it follows from (46):

$$\begin{aligned} \frac{\delta_r f}{\delta t} &= \lim_{\Delta t \to 0} \frac{f(\boldsymbol{r}, t + \Delta t) - f(\boldsymbol{r}, t)}{\Delta t} + \lim_{\Delta t \to 0} \boldsymbol{v}(\boldsymbol{r}, t) \cdot \nabla f(\boldsymbol{r}, t + \Delta t) \\ &= \frac{df(\boldsymbol{r}, t)}{dt} + \boldsymbol{v}(\boldsymbol{r}, t) \cdot \nabla f(\boldsymbol{r}, t). \end{aligned} \tag{48}$$

It should be noted that even though the definition (46) looks like the definitions of the total and partial derivatives, there is a significant difference between them. The material derivative (46) is not a derivative of a function in the mathematical sense. Indeed, the displacement $\Delta \boldsymbol{s} = \boldsymbol{v}(\boldsymbol{r}, t)\Delta t$ on the right side of Eq. (46) cannot be expressed in terms of function arguments. This is due to a peculiarity of the spatial description in which the position vector $\boldsymbol{r}$ is *unrelated* to the evolution of matter and the velocity $\boldsymbol{v}(\boldsymbol{r}, t)$ is an *independent* characteristic. Since $\frac{d\boldsymbol{r}}{dt} = \boldsymbol{0}$ and $\nabla \boldsymbol{r} = \boldsymbol{I}$ ($\boldsymbol{I}$ is the unit tensor), the equation relating the position vector and the velocity,

$$\boldsymbol{v}(\boldsymbol{r}, t) \equiv \frac{\delta_r \boldsymbol{r}}{\delta t} = \frac{d\boldsymbol{r}}{dt} + \boldsymbol{v}(\boldsymbol{r}, t) \cdot \nabla \boldsymbol{r} \tag{49}$$

turns into an identity.

Thus, the velocity $\boldsymbol{v}(\boldsymbol{r}, t)$ is the primary quantity in the spatial description and all other quantities are expressed in terms of $\boldsymbol{v}(\boldsymbol{r}, t)$. For example, the acceleration of a material particle $\boldsymbol{a}(\boldsymbol{r}, t)$ is determined as the material derivative of the velocity:

$$\boldsymbol{a}(\boldsymbol{r}, t) \equiv \frac{\delta_r \boldsymbol{v}(\boldsymbol{r}, t)}{\delta t} = \frac{d\boldsymbol{v}(\boldsymbol{r}, t)}{dt} + \boldsymbol{v}(\boldsymbol{r}, t) \cdot \nabla \boldsymbol{v}(\boldsymbol{r}, t). \tag{50}$$

Note that the first term on the right side of (50) is the local rate of velocity change due to temporal changes at the observation point. It is *not* the acceleration of the material point at position $\boldsymbol{r}$ at time t, because the material point is located at that position only instantaneously.

The spatial description is used mostly in fluid and gas dynamics where the velocity, density, and the pressure are the main unknowns. Due to the complex motion of fluid and gas particles a monitoring of their motion, i.e., of their displacements is hardly feasible. As a result, the displacement vector is usually not considered in classical hydrodynamics. However, from a theoretical point of view, the introduction of this concept is an interesting task. There are different approaches to a formal introduction of the displacement vector. But all of them result in the following differential relation between the velocity $\boldsymbol{v}(\boldsymbol{r},\ t)$ and the displacement vector $\boldsymbol{u}(\boldsymbol{r},\ t)$:

$$\frac{\delta_r \boldsymbol{u}(\boldsymbol{r},\ t)}{\delta t} = \boldsymbol{v}(\boldsymbol{r},\ t). \tag{51}$$

Note that in spatial description this relation is used for determination of the displacement vector provided $\boldsymbol{v}$ is known. Then one can introduce the concept of the reference position vector in the same manner as in the case of the material description $\boldsymbol{R}(\boldsymbol{r},\ t) = \boldsymbol{r} - \boldsymbol{u}(\boldsymbol{r},\ t)$. In contrast to the material description, where the current position vector, $\boldsymbol{r}$, is a function of the reference position vector, $\boldsymbol{R}$, and time, t, the reference position vector, $\boldsymbol{R}$, within the spatial description is a function of the current position vector, $\boldsymbol{r}$, and time, t. This means that we have a different reference configuration for every moment of time. Since

$$\frac{\delta_r \boldsymbol{u}(\boldsymbol{r},\ t)}{\delta t} = \frac{\delta_r (\boldsymbol{R}(\boldsymbol{r},\ t) + \boldsymbol{r})}{\delta t} = \frac{\delta_r \boldsymbol{R}(\boldsymbol{r},\ t)}{\delta t} + \boldsymbol{v}(\boldsymbol{r},\ t) \tag{52}$$

it follows from (51) that

$$\frac{\delta_r \boldsymbol{R}(\boldsymbol{r},\ t)}{\delta t} = \frac{d\boldsymbol{R}}{dt} + \boldsymbol{v} \cdot \nabla \boldsymbol{R} = 0. \tag{53}$$

This differential equation determines the relation between the velocity of the material point, $\boldsymbol{v}$, and its reference position, $\boldsymbol{R}$.

5.4 *Moving Observation Point*

Now consider the closed surface S defined as the boundary of a volume that is no longer fixed in space but moves with a known velocity as a rigid body. The motion of points within the volume is expressed by the field of the position vector $\boldsymbol{r}(x_1, x_2, x_3, t)$, where x_1, x_2, x_3 is the reference coordinate system. The total mass in the volume is determined by Eq. (38). The vector $\boldsymbol{r}^*(t)$ is the position vector of a point fixed with respect to the volume but it moves with respect to the reference coordinate system. The rate of change of the total mass in the volume is the total time derivative of the mass:

$$\frac{dm(\boldsymbol{r}^*(t), t)}{dt} = \left.\frac{\partial m(\boldsymbol{r}^*, t)}{\partial t}\right|_{r^* = const} + \frac{d\boldsymbol{r}^*}{dt} \cdot \nabla_* m(\boldsymbol{r}^*, t), \qquad \nabla_* = \frac{\partial}{\partial \boldsymbol{r}^*}. \tag{54}$$

To clarify the meaning of Eq. (54) we note two special cases. The first of these concerns a volume fixed in space. Then $\boldsymbol{r}^*$ does not depend on time and the rate of change of the total mass is characterized only by the partial derivative. The second special case is relevant when the mass density is inhomogeneously distributed over space and this distribution does not change with time. In this case the first term on the right side of (54) is equal to zero and the change in mass is due to transport of the volume to a different position.

In order to pull the total derivative under the integral sign in (38) a change of variables is required:

$$\boldsymbol{r} = \boldsymbol{r}(\hat{\boldsymbol{r}}, t), \quad \rho(\boldsymbol{r}, t) = \hat{\rho}(\hat{\boldsymbol{r}}, t), \quad \boldsymbol{r}^* = \boldsymbol{r}^*(\hat{\boldsymbol{r}}^*, t), \quad m(\boldsymbol{r}^*, t) = \hat{m}(\hat{\boldsymbol{r}}^*, t), \tag{55}$$

where $\hat{\boldsymbol{r}}$ and $\hat{\boldsymbol{r}}^*$ are fixed in the reference system.

By doing so we can make the following transformations:

$$\frac{dm(\boldsymbol{r}^*, t)}{dt} = \frac{\partial \hat{m}(\hat{\boldsymbol{r}}^*, t)}{\partial t} = \frac{\partial}{\partial t} \int_V \hat{\rho}(\hat{\boldsymbol{r}}, t)\, dV = \int_V \frac{\partial \hat{\rho}(\hat{\boldsymbol{r}}, t)}{\partial t}\, dV = \int_V \frac{d\rho(\boldsymbol{r}, t)}{dt}\, dV, \tag{56}$$

where use has been made of the fact that the volume V is independent of time. Thus, in the case of a moving undeformed volume we obtain

$$\frac{d}{dt} \int_V \rho(\boldsymbol{r}, t)\, dV = \int_V \frac{d\rho(\boldsymbol{r}, t)}{dt}\, dV. \tag{57}$$

Here the total derivative of the mass density is:

$$\frac{d\rho(\boldsymbol{r}, t)}{dt} = \lim_{\Delta t \to 0} \frac{\rho(\boldsymbol{r}(t + \Delta t),\, t + \Delta t) - \rho(\boldsymbol{r}(t),\, t)}{\Delta t} = \frac{\partial \rho(\boldsymbol{r}, t)}{\partial t} + \frac{d\boldsymbol{r}}{dt} \cdot \nabla \rho(\boldsymbol{r}, t). \tag{58}$$

Equation (58) characterizes the rate of change of the mass density at the observation point that moves with velocity $\dfrac{d\boldsymbol{r}}{dt}$.

It should be emphasized that the meaning of the total derivative is the same in both descriptions. The total derivative determines the rate of change of a property related to the matter at the observation point. Within the material description the motion of the observation point coincides with the motion of the material point. This is why, in this particular case, the results of calculation of the material and total derivatives coincide.

The change of mass in the volume is equal to the mass flux through its surface. The rate of mass flow is determined by $\rho(\boldsymbol{r}, t)$ and by the relative velocity of the material points and the surface. Thus the mass balance reads:

$$\frac{d}{dt}\int_V \rho(\boldsymbol{r},t)\,dV = -\int_S \left[\boldsymbol{n}(\boldsymbol{r})\cdot\left(\boldsymbol{v}(\boldsymbol{r},t)-\frac{d\boldsymbol{r}}{dt}\right)\right]\rho(\boldsymbol{r},t)\,dS. \tag{59}$$

After transforming the integral on the right side by means of the divergence theorem and using (57), we arrive at the local mass balance:

$$\frac{d\rho(\boldsymbol{r},t)}{dt} + \nabla\cdot\left[\rho(\boldsymbol{r},t)\left(\boldsymbol{v}(\boldsymbol{r},t)-\frac{d\boldsymbol{r}}{dt}\right)\right] = 0. \tag{60}$$

By taking into the account the following relations:

$$\nabla\cdot\frac{d\boldsymbol{r}}{dt} = \nabla\cdot\frac{\partial\boldsymbol{r}}{\partial t} = \frac{\partial(\nabla\cdot\boldsymbol{r})}{\partial t} = 0, \tag{61}$$

Equation (60) is transformed as follows:

$$\frac{d\rho(\boldsymbol{r},t)}{dt} + \left(\boldsymbol{v}(\boldsymbol{r},t)-\frac{d\boldsymbol{r}}{dt}\right)\cdot\nabla\rho(\boldsymbol{r},t) + \rho(\boldsymbol{r},t)\nabla\cdot\boldsymbol{v}(\boldsymbol{r},t) = 0. \tag{62}$$

Upon introducing the notation

$$\frac{\delta\rho(\boldsymbol{r},t)}{\delta t} = \frac{d\rho(\boldsymbol{r},t)}{dt} + \left(\boldsymbol{v}(\boldsymbol{r},t)-\frac{d\boldsymbol{r}}{dt}\right)\cdot\nabla\rho(\boldsymbol{r},t) \tag{63}$$

the mass balance becomes:

$$\frac{\delta\rho(\boldsymbol{r},t)}{\delta t} + \rho(\boldsymbol{r},t)\nabla\cdot\boldsymbol{v}(\boldsymbol{r},t) = 0. \tag{64}$$

In order to obtain (64) we assume that the volume is not deformed (because of Eq. (57), which hold for an undeformed volume). Rejection of the assumption complicates the derivations but the final Eqs. (62)–(64) remain unchanged.

The operator

$$\frac{\delta}{\delta t} = \frac{d}{dt} + \left(\boldsymbol{v}-\frac{d\boldsymbol{r}}{dt}\right)\cdot\nabla \tag{65}$$

is a generalization of the material-derivative operator (45) for the moving observation point.

Definition 7 If the motion of the observation point $\boldsymbol{r}(t)$ is known then the material derivative of a material point property $f(\boldsymbol{r},t)$ is:

$$\frac{\delta f}{\delta t} = \lim_{\Delta t\to 0}\frac{f(\boldsymbol{r}(t+\Delta t)+\Delta \boldsymbol{s}, t+\Delta t) - f(\boldsymbol{r},t)}{\Delta t}, \qquad \Delta\boldsymbol{s} = \left(\boldsymbol{v}-\frac{d\boldsymbol{r}}{dt}\right)\Delta t, \tag{66}$$

where $\Delta \boldsymbol{s}$ is the displacement with respect to the observation point of a material point that was in the observation point at time t.

The above definition has the same physical meaning as (46). The material derivative characterizes the rate of change a property of the material point that was in the observation point at time t.

It can be shown that Eq. (65) is in agreement with this definition. Since

$$f(\boldsymbol{r}(t+\Delta t)+\Delta \boldsymbol{s}, t+\Delta t)=f(\boldsymbol{r}(t+\Delta t), t+\Delta t)+\Delta \boldsymbol{s} \cdot \nabla f(\boldsymbol{r}(t+\Delta t), t+\Delta t), \tag{67}$$

Equation (66) yields:

$$\frac{\delta f}{\delta t}=\frac{d f(\boldsymbol{r}, t)}{d t}+\left(\boldsymbol{v}(\boldsymbol{r}, t)-\frac{d \boldsymbol{r}}{d t}\right) \cdot \nabla f(\boldsymbol{r}, t). \tag{68}$$

By taking into account

$$\frac{d}{d t}=\left.\frac{\partial}{\partial t}\right|_{\boldsymbol{r}=\text{const}}+\frac{d \boldsymbol{r}}{d t} \cdot \nabla, \tag{69}$$

we rewrite (65) in the form:

$$\frac{\delta}{\delta t}=\left.\frac{\partial}{\partial t}\right|_{\boldsymbol{r}=\text{const}}+\boldsymbol{v} \cdot \nabla. \tag{70}$$

It is easy to see that the expression for the material derivative (70) coincides with the expression for the material derivative with the fixed observation point (44). However, it is impossible to say from these expressions as to whether the observation point is fixed or not. Furthermore, the expression (70) looks similar to the total derivative in the current configuration within the material description (21). Such a coincidence is confusing and obscures the sense and meaning of the total and material derivatives. But from the expressions (65) and (69) the difference between the derivatives is obvious. The material derivative determines a rate of change of a property of the material point located at the observation point at time t, the total derivative determines a rate of change of the property at the observation point. It is true both in the spatial and the material descriptions. Within the material description the observation point is the material point, thus:

$$\boldsymbol{v}=\frac{d \boldsymbol{r}}{d t} \quad \Rightarrow \quad \frac{\delta}{\delta t}=\frac{d}{d t}, \tag{71}$$

and the material derivative coincides with the total derivative. Only in this particular case the statement "the total derivative, it is also the material derivative" can be made. In general, the observation point velocity $\frac{d \boldsymbol{r}}{d t}$ in Eq. (69) does not relate to a material point. If the observation point is fixed or moves independently of the motion of the medium, the total and material derivatives have different meanings and different values. If in a particular case their values coincide it does not mean that the physical meaning of a derivative changes. That is why the difference between the concepts of the total and material derivatives is important. This is particularly relevant when modeling a multicomponent medium, where all components have different velocities with respect to the common observation point.

In conclusion of this section note that the gradient operators have different properties in the material and spatial descriptions. This becomes important if one wants to investigate gradients of displacements, i.e., strains and their time derivatives, i.e., strain rates. Within the material description there are two gradient operators, $\overset{\circ}{\nabla}$ in the reference configuration, and ∇ in the current configuration. It is easy to show that:

$$\overset{\circ}{\nabla}\frac{d}{dt} = \frac{d}{dt}\overset{\circ}{\nabla}, \qquad \nabla\frac{d}{dt} \neq \frac{d}{dt}\nabla. \tag{72}$$

The spatial description deals with the gradient in the current configuration only. In the case of the fixed observation point we have

$$\nabla\frac{d}{dt} = \frac{d}{dt}\nabla, \qquad \nabla\frac{\delta_r}{\delta t} \neq \frac{\delta_r}{\delta t}\nabla. \tag{73}$$

Nevertheless, for a moving observation point the gradient operator is not interchangeable neither with the material nor with the total derivative.

6 Outlook and Conclusions

In Chap. 2 we started by presenting a rather detailed literature review of the various notions of time derivatives for the material and spatial description of continuum fields, which illustrated the confusing, almost desolate state of the subject. This made the need for a rigorous clarification apparent.

For this purpose the concept of material description was carefully analyzed in Chap. 3. The so-called total time derivative was introduced and analyzed for the reference and for the current configuration. Within the material description it may be interpreted as the rate of change of physical field quantities characterizing a material particle. The total time derivative was then examined in context with global balance equations, in particular, the mass balance. The property of the total time derivative being an objective operator independent of the choice of coordinate system was emphasized.

The latter property gave rise for a precise definition and further investigations of the total time derivative in combination with the concepts of Frames of Reference (FoR) and observers in Chap. 4. To this end an FoR was formally defined. The difference between an FoR, being a physics-based concept, and coordinates and transformations thereof, being purely mathematical operations, was pointed out.

Chapter 5 was dedicated to the description of continuum fields in spatial description. Here the considered matter is not necessarily a material volume any more. In order to point out the issue of a possible exchange of matter the notion of a body was introduced. The formulation of balance equations, specifically of the mass balance, was investigated and the operator of a material time derivative in spatial description for a nonmoving position vector, i.e., observation point was introduced. Moreover, an attempt was made to clarify the notion of displacement in spatial description. This culminated in a differential equation between the velocity of a material point and its reference configuration which, under these circumstances, must be viewed as continuously varying. The end of this chapter was devoted to the generalization of the material time derivative for a moving point of observation. It was shown that the material derivative characterizes the rate of change a property of the material point that was in the observation point at the certain moment of time, while the total derivative is the rate of change of property in an observation point. If this point coincides with a material particle (the material description) then (and only then) it is the rate of change of a quantity of the material point. In general, we may conclude that if the observation point is fixed or moves independently of the motion of the medium, the total and material derivatives have different meanings and different values.

Moreover, similarities regarding the mathematical form of the material derivative in spatial description with the total derivative in the current configuration within the material description are nothing else but *amis faux*.

Finally, in context with the mathematical description and the physical interpretation of time derivatives it became expedient to point out the difference between the mathematical concept of a coordinate system representation and the physics-based notion of an FoR. However, the question regarding the indifference of time derivatives w.r.t. changes of an FoR remains an open issue. In particular, an examination of the objectivity of time derivatives in context with the principle of material frame indifference will be presented in future work.

References

Adler, P.: Porous Media: Geometry and Transports. Butterworth-Heinemann, USA (1992)

Altenbach, H., Naumenko, K., Zhilin, P.A.: A micro-polar theory for binary media with application to phase-transitional flow of fiber suspensions. Contin. Mech. Thermodyn. **15**(6), 539–570 (2003)

Asaro, R., Lubarda, V.: Mechanics of Solids and Materials. Cambridge University Press, New York (2006)

Batchelor, G.: An Introduction to Fluid Dynamics. Cambridge University Press, Cambridge (1970)

Cornille, P.: Inhomogeneous waves and Maxwell's equations (Chapter 4). Essays on the Formal Aspects of Electromagnetic Theory. World Scientific, Singapore (1993)

Daily, J., Harleman, D.: Fluid Dynamics. Addison-Wesley, Boston (1966)

Dang, T.S., Meschke, G.: An ALE-PFEM method for the numerical simulation of two-phase mixture flow. Comput. Methods Appl. Mech. Eng. **278**, 599–620 (2014)

Del Pin, F., Idelsohn, S., Onate, E., R A, : The ALE/Lagrangian particle finite element method: A new approach to computation of free-surface flows and fluid object interactions. Comput. Fluids **36**, 27–38 (2007)

Dettmer, W., Peric, D.: A computational framework for free surface fluid flows accounting for surface tension. Comput. Methods Appl. Mech. Eng. **195**, 3038–3071 (2006)

Dmitrienco, U.: Nonlinear mechanics of continus. Physmatlit, Moscow (2009)

Durst, F.: Fluid Mechanics: An Introduction to the Theory of Fluid Flows. Springer, Berlin (1992)

Eringen, C.: Mechanics of Continua. Robert E Krieger Publishing Company, Huntington, New York, (1980)

Filipovic, N., Akira, Mijailovic A.S., Tsuda, Kojic M.: An implicit algorithm within the arbitrary Lagrangian–Eulerian formulation for solving incompressible fluid flow with large boundary motions. Comput. Methods Appl. Mech. Eng. **195**, 6347–6361 (2006)

Fung, Y.: Foundations of Solid Mechanics. Prentice-Hall, Englewood Cliffs (1965)

Gadala, M.: Recent trends in ale formulation and its applications in solid mechanics. Comput. Methods Appl. Mech. Eng. **193**, 4247–4275 (2004)

Granger, R.A.: Fluid Mechanics. Dover Books on Physics (1995)

Gurtin, M.: An Introduction to Continuum Mechanics. Academic Press Inc, London (1981)

Hassanizadeh, M., Gray, W.: General conservation equations for multi-phase systems: 3. constitutive theory for porous media flow. Adv. Water Resour. **3**, 25 (1980)

Ilyushin, A.: Continuum mechanics. Moscow University Press, Moscow (1971)

Khoei, A., Anahid, M., Shahim, K.: An extended arbitrary Lagrangian–Eulerian finite element modeling (X-ALE-FEM) in powder forming processes. J. Mater. Process. Technol. **187–188**, 397–401 (2007)

Lamb, H.: Hydrodynamics. Cambridge University Press, New York (1975)

Landau, L., Lifshitz, E.: Fluid Mechanics, vol. 6, 1st edn. Pergamon Press, Oxford (1959)

Lojtsanskij, L.: Mcchanics of liquid and gas. Moscow (1950)

Malvern, E.: Introduction to the Mechanics of a Continuous Medium. Prentice-Hall Inc, Englewood Cliffs (1969)

Mase, G.: Theory and Problems of Continuum Mechanics. McGraw-Hill Book Company, New York (1970)

Milne-Thomson, L.: Theoretical Hydrodynamics. Martin's Press, Macmillan and Co. LTD, New York (1960)

Müller, W.H., Muschik, W.: Bilanzgleichungen offener mehrkomponentiger systeme. I. massen- und impulsbilanzen. J. Non-Equilib. Thermodyn. **8**(1), 29–46 (1983)

Nerlich, G.: What Spacetime Explains: Metaphysical Essays on Space and Time. Cambridge University Press, Cambridge (1994)

Ogden, R.: Nonlinear Elasticity with Application to Material Modelling. Polish Academy of Sciences, Warsaw (2003)

Pasipoularides, A.: Heart's vortex: intracardiac blood flow phenomena. PMPH-USA (2009)

Petrila, T., Trif, A.: Basics of Fluid Mechanics and Introduction to Computational Fluid Dynamics (Numerical Methods and Algorithms). Springer, Boston (2005)

Prandtl, L., Tietjens, O.: Hydro- und Aeromechanik. Springer, Berlin (1929)

Preisig, M., Zimmermann, T.: Free-surface fluid dynamics on moving domains. Comput. Methods Appl. Mech. Eng. **200**, 372–382 (2011)

Rouse, H.: Advanced Mechanics of Fluids. Wiley, New York (1959)

Sarrate, J., Huerta, A., Donea, J.: Arbitrary Lagrangian–Eulerian formulation for fluid-rigid body interaction. Comput. Methods Appl. Mech. Eng. **190**, 3171–3188 (2001)

Serrin, J.: Mathematical Principles of Classical Fluid Mechanics. Springer, Berlin (1959)

Surana, K., Blackwell, B., Powell, M., Reddy, J.: Mathematical models for fluid-solid interaction and their numerical solutions. J. Fluids Struct. **50**, 184–216 (2014)

Truesdell, C.: A First Course in Rational Continuum Mechanics. John's Hopkins University, Baltimore (1972)

Vuong, A.T., Yoshihara, L., Wall, W.: A general approach for modeling interacting flow through porous media under finite deformations. Comput. Methods Appl. Mech. Eng. **283**, 1240–1259 (2015)

Zhilin, P.A.: Racional'naya mekhanika sploshnykh sred (Rational Continuum Mecanics, in Russian). Politechnic University Publishing House, St. Petersburg (2012)

The Cosserats' Memoir of 1896 on Elasticity

Gérard A. Maugin

Abstract Nowadays the Cosserat brothers are mostly cited for their work on so-called "Cosserat continua" of 1909 that practically initiated the theory of "oriented media" as generalized continua. But in 1896 they had already published a lengthy well-structured memoir on the theory of elasticity. This memoir is often considered as a foundational work on the modern approach to elasticity as it beautifully summarizes what was achieved in the nineteenth century but with original traits that will permeate further the twentieth century developments with an emphasis on finite deformations, the interest for applying the thermodynamic laws, the allied formulation of the notion of stress (internal forces), questions of stability, and the use of curvilinear coordinates, though still without using vector and/or tensor analysis. The present contribution examines in detail the contents of this epoch-making work of 1896, its main sources (e.g. Kirchhoff, Kelvin, Saint-Venant, Boussinesq, and Poincaré) and its insertion in the then current technical literature. We try to appraise its importance and its legacy in the modern developments of continuum mechanics, especially after the revival of the field by Truesdell and others.

1 Introduction

At the time of writing of this contribution, the most cited work of the Cosserat brothers, Eugène and François, certainly is their book of 1909 (Cosserat and Cosserat 1909). This is due to a justified renewal of interest in continua endowed with a microstructure (in particular, so-called "micropolar continua" also rightly named "Cosserat continua"). These are not classical in the sense that such media exhibit nonsymmetric stress tensors and so-called moment (or couple) stresses. Year 2009 marked with some emphasis the hundredth anniversary of the publication of this

G.A. Maugin (✉)
Institut Jean le Rond d'Alembert UMR CNRS 7190, Sorbonne Universités,
Université Pierre et Marie Curie – Paris 6, Tour 55, 4 place Jussieu,
75252 Paris Cedex 05, France
e-mail: gerard.maugin@upmc.fr

K. Naumenko and M. Aßmus (eds.), *Advanced Methods of Continuum Mechanics for Materials and Structures*, Advanced Structured Materials 60,
DOI 10.1007/978-981-10-0959-4_2

famous but rarely read opus (cf. Maugin and Metrikine 2010). In the period 1896–1914, the Cosserats in fact published together no less than 21 works in the field of theoretical mechanics. Out of these, 14 were short notes—of three or four pages—to the Paris Academy of Sciences. Apart from their book of 1909, the only long original memoir they published was a long paper in a true serial scientific journal in Toulouse in 1896 (Cosserat and Cosserat 1896), while their other publications in the field are scattered in odd places, often as supplements to lecture notes or books by other authors [Koenigs, Chwolson (in French translation), Appell, Voss (also in French translation)]. This paper of 1896 is the object of the present perusal. The originality of its contents is a discussed matter, whether the paper provides a nice overview of nineteenth century continuum mechanics or it does bring a new enriching viewpoint with specific traits of the brothers' talents and rigour, a positive appraisal certainly expressed by Truesdell on different occasions (cf. Truesdell 1952a; Truesdell and Toupin 1960). Our own opinion is that the Cosserats demonstrated a deep understanding of the bases of continuum mechanics and thus clarified many points, and they exhibited a style and ideas that were to bear fruits during the following 60 years or about.

2 About the Cosserats and Their Scientific Environment

In order to grasp the essentials of the Cosserats' personalities and achievements, we need to comprehend their scientific formation and to appraise the scientific environment they shared at a time that may schematically be called the "Belle Epoque" (roughly, 1880–1914). In that period the two most prestigious schools in France were the *Ecole Polytechnique* and the *Ecole Normale Supérieure* (ENS), both in Paris, and accessible only after a difficult competitive entrance examination. The former was destined to form engineers essentially for the needs of the State although the programme in mathematics was the highest possible with the best available teachers. To be fully trained in more engineering matters the best alumni from *Ecole Polytechnique* had to follow an "Ecole d'application" of which the most well-known one was the *Ecole Nationale des Ponts et Chaussées* (ENPC). Students who successfully completed their study in the two schools would become members of the elitist "Corps of Engineers of *Ponts et Chaussées*", one of the most desired titles in the French Third Republic. This opened the way to both technical and managerial positions at the highest level in the State or in private companies (e.g. the newly expanding railway companies). Notice that not much was said about universities (or rather faculties) which fell under the unique directorship of the Ministry of Education. Famous French scientists, physicists and mathematicians of the nineteenth century belonged to the Corps of *Ponts et Chaussées*, among them, Cauchy, Navier, Lamé, Duhamel, Coriolis, Clapeyron, Poncelet, Liouville, Arago and Barré de Saint-Venant. "Poor" Boussisnesq who "modestly" graduated from the University of Montpellier had a much harder work to achieve to reach the same stratospheric medium. Another prestigious school of application of *Polytechnique* was the National School of Mines in Paris. Henri Poincaré thus belonged to the "Corps of Engineers of Mines"—which in time became even more prestigious than the one of *Ponts et Chaussées*—although he devoted his whole life to mathematics and mathematical physics.

The *Ecole Normale Supérieure* was initially destined to form teachers for *Lycées*, i.e. secondary high schools educating students from age 12 to 18 with a final diploma called the "Baccalauréat" with a strong emphasis on classics. Then they could attempt a university or continue to prepare for the difficult examination entrance to *Polytechnique* and ENS. Very good students were admitted to both schools and selected the one that pleased them most. Under the influence of Louis Pasteur the ENS also became a "fish tank" for creative scientists who would soon join and then surpassed the polytechnicians.

François Cosserat (1852–1914), the elder of the two brothers, graduated from the Ecole Polytechnique and became a member of the Corps of *Ponts et Chaussées*. He had a professional career in the fast growing development of railways with the North and then the East companies of Railways in France. Eugène Cosserat (1866–1931), his cadet by 14 years, was educated in mathematics at the *Ecole Normale Supérieure* in Paris and became a professional (mathematical) astronomer with a career spent almost entirely in Toulouse in the south-west of France. As such he had to teach courses in analysis, astronomy and celestial mechanics, but he also had a marked interest in differential geometry already exhibited in his doctoral thesis.

We do not know what prompted the interest of the Cosserat brothers for rational mechanics and the theory of elasticity in particular. It may be the lectures received by François at both *Polytechnique* and ENPC and then the influence of this older brother on his cadet. The cooperation of the two brothers lasted from 1896 to the death of François in 1914. Anyway, they must have been bright students to start with and endowed with some easiness to grasp fundamental concepts and a gift to expand them as neither François nor Eugène were officially professional mathematicians in the field of mechanics. But they were enlightened amateurs with all technical abilities and a background of true professionals. Both became members of the Paris Academy of Sciences (François in 1896, and Eugène in 1919). François was even elected President of the French Society of Mathematics (*Société Mathématique de France*) in 1913 one year before his death. François was certainly confronted to the works of Adhémar Barré de Saint-Venant (1797–1886) and Joseph V. Boussinesq (1842–1929) at the ENPC. In his engineering curriculum he met with the works of his great predecessors, namely, Gabriel Lamé (1795–1870) and Alfred Clebsch (1833–1872), both authors of the first comprehensive treatises on elasticity (cf. Lamé 1852; with a tremendous expansion by Barré de Saint-Venant 1883 for the latter in French translation), and also Gustav Kirchhoff (1824–1887) in Kirchhoff (1852) and James Clerk Maxwell (1831–1873) (cf. Maxwell 1853). Eugène Cosserat defended his Sorbonne thesis in mathematics before a committee formed by Gaston Darboux (1842–1917), Paul Appell (1855–1930) and Gabriel Koenigs (1858–1931)—see Lebon (1910). This thesis on geometry was published in the *Annales* of the Faculty of Sciences of Toulouse in 1885. Darboux was the author of a formidable work—in four volumes—on the theory of surfaces and an ardent propagandist of the notion of *mobile triad* that was readily adopted by the Cosserats. Paul Appell became professor of rational mechanics at the Sorbonne in 1885 and, among many creative works, published an influential encyclopaedic treatise on rational mechanics (starting in 1893 with many augmented editions) and practically became the godfather of all mechanicians in France in the period of interest. Koenigs, a student of Darboux, became professor of mechanics at the Sorbonne while publishing a successful treatise on kinematics (Lessons of 1895–1897, Koenigs 1895, see also Lovett 1900). Both Darboux and Koenigs left a strong print on the Cosserats' work of 1896 as witnessed by the large number of citations to their books. Other French contemporaries of the two brothers were Henri Poincaré (1854–1912), Pierre Duhem (1861–1916), Marcel Brillouin (1854–1948), Emile Picard (1856–1941), Emile Jouguet (1871–1943), Jacques Hadamard (1865–1963), and Paul Painlevé (1863–1933), all educated at the ENS save Poincaré. Eugène Cosserat was very close to Hadamard and Painlevé. Contemporaries outside France were Woldemar Voigt (1850–1919), August Föppl (1854–1924), Hermann von Helmholtz (1821–1894), Georg Hamel (1887–1954), and Ludwig Boltzmann (1844–1906) in Germany, Josiah Willard Gibbs (1838–1905) in the USA, and William Thomson (1824–1907; *aka* Lord Kelvin), A.E.H. Love (1863–1940), and Lord Rayleigh (1842–1919) in the UK. What are really missing in the interactions with foreign scientists are any contacts with, and citations to, Italian mechanical engineers and mathematicians. The strangest fact is the lack of connection with Gabrio Piola (1794–1850), apparently eclipsed by Kirchhoff. In all, the scientific environment of the Cosserat brothers in Paris was stupendous, and they dutifully cited all scientists—that they studied in detail—at the proper place of their works with high accuracy. Hard working in such a rich environment and equipped with knowledge of the most influential foreign languages, the Cosserats were in a most favourable frame to develop their original views although their activity in rational mechanics was only an aside to their professional occupations. The result is all the more remarkable.

3 The Cosserats' Paper of 1896

Preliminary remark: In their general kinematic description the Cosserat brothers note the direct deformation $(x, y, z) \rightarrow (x_1, y_1, z_1)$ that we note $(X^K,\ K = 1, 2, 3) \rightarrow (x^i,\ i = 1, 2, 3)$ in modern indicial notation. They note (u, v, w) the components of the displacement that we would note $(u^i, i = 1, 2, 3)$. The initial density they note ρ and the final one ρ_1 while we shall use ρ_0 and ρ for these two, respectively. We repeatedly use the convention of summation over repeated indices. The Cosserats do not use any vector or tensor notation and thus have to give all components explicitly, but we rewrote the main cited equations in the modern outlook to help the reader. We hope that this does not create any confusion, still always referring to the original equations of the Cosserats in their text where necessary (i.e. page number and equation number). They do not treat the dynamical case.

3.1 Deformations

The Cosserats define finite deformation just like Green in his celebrated memoir of 1839 (Green 1839), but they emphasize the intimate link with the use of the theory of curvilinear coordinates. That is, to be unambiguous, their formulas (3) and (4) are none other than the modern formulas for the Cauchy–Green strain of material components E_{KL} and for the finite deformation gradient $\boldsymbol{F}$ of components $F^i_{.K}$ such that

$$\begin{aligned} \mathrm{d}\boldsymbol{x} &= \boldsymbol{F}\,\mathrm{d}\boldsymbol{X}, \quad F = \left\{ F^i_{.K} = \frac{\partial x^i}{\partial X^K} \right\}, \\ \boldsymbol{E} &= \tfrac{1}{2}\left(\boldsymbol{F}^T\boldsymbol{F} - \boldsymbol{1}\right) = \left\{ E_{KL} = \tfrac{1}{2}\left(F^i_{.K} F^j_{.L} g_{ij} - \delta_{KL} \right) \right\}. \end{aligned} \tag{1}$$

The (relative) strain measures here introduced have also been considered by Barré de Saint-Venant, Kirchhoff, Lord Kelvin (William Thomson) and Boussinesq. The six functions given by the elements of E_{KL} cannot be completely arbitrary as they must verify a system of second-order partial differential equations (known as compatibility conditions; cf. Barré de Saint-Venant 1864). Whenever all E_{KL}'s vanish it means that the deformed configuration is deduced form the original one by a displacement "en bloc", combined or not combined with a symmetry transformation (cf. p. I.12). This is of fundamental importance because it defines what is understood by a *rigid-body motion*. After Lord Kelvin, a homogeneous deformation is one in which the E_{KL}'s are all constants or they vanish identically. This allows one to introduce analytically the simple form (homographic transformation) of homogeneous deformations (Eq. (5), p. I.13), particular cases being those of linear dilatations and angular dilatations. For a sufficiently small portion of the undeformed body about a point P, one can substitute a homogeneous deformation to the actual deformation at P. In this they essentially adopt the viewpoint of W. Thomson (Lord Kelvin) with so-called linear dilatations and angular dilatations as main constructive elements. Following the original work of Cauchy (1827) they pay special attention to the notions of ellipsoids of deformation, rotation at a point, and pure deformation. The first ellipsoid E of deformation clearly corresponds to a transformation of an initially spherical form into an ellipsoid. Reciprocally, the second ellipsoid E_1 relates to the inverse relation between a sphere in the final configuration and an ellipsoid in the initial configuration. The three axes of E can be brought parallel to those of E_1 by an appropriate *rotation*. The vanishing of such rotation corresponds to what Thomson and Tait (1867), p. 132, call a *pure* deformation. This combination of pure deformation and rotation (pp. I.19–I.25) materializes in what is called the *polar decomposition* (attributed to, but not proved by, Cauchy) of the deformation gradient—noted $\boldsymbol{F} = \boldsymbol{RU} = \boldsymbol{VR}$ in modern treatises (e.g. Truesdell and Toupin 1960). As noted in the modern formula the rotation can be effected first and pure deformation next, or in the other order but with a different pure deformation (in fact in a different space; cf. Footnote in p. I.20). In their geometric proof the Cosserats exploit the transformation of quadratic forms and the notion of principal axes of the involved ellipsoids. They also have to introduce the cubic dilatation and the Jacobian determinant—that they note Δ—of the deformation, i.e. $J = \det \boldsymbol{F}$ in modern notation. These considerations lead them directly to introduce the invariants of deformation (p. I.26) and the useful notions of simple extension (stretch) and simple shear (pp. I.25–I.28). The usefulness of the notion of *simple shear*, e.g. (with coefficient γ characterizing the amount of shear)

$$x = X + \gamma Y, \qquad y = Y, \qquad z = Z, \tag{2}$$

had particularly been emphasized by a certain Louis Vicat (1786–1861)—cf. Vicat (1833)—and above all Barré de Saint-Venant in his lectures of 1837–1838 (Barré de Saint-Venant 1837, 1838) at the School of *Ponts et Chaussées*—see also Brillouin (1891). As we know now, the notion of simple shear is often used as a test deformation in the characterization of nonlinear elastic responses for various materials with a priori prescribed strain energy.

Infinitesimally small deformations are correctly introduced (p. I.29) by the Cosserats with the help of an ordering parameter (that they note t) and expansion of the displacement components in integer powers of this parameter, assuming uniform convergence of the corresponding series and of those defined by their derivatives with respect to the initial coordinates. In this context, special cases are those of linear and angular "dilatations" (following the vocabulary introduced by Cauchy). Principal dilatations (or stretches) are those expanded along the axes of the second ellipsoid of deformation. In the case where both linear dilatations and relative shears vanish, then it is shown, following a method due to Darboux, that the displacement field is one of the rigid-body types that we can write in direct notation as

$$\boldsymbol{u} = \boldsymbol{u}_0 + \boldsymbol{\omega} \times \boldsymbol{X} \tag{3}$$

where both $\boldsymbol{u}_0$ and $\boldsymbol{\omega}$ are translation and rotation of constant values.

Finally, the Cosserats (pp. I.35–I.37) recall the necessary and sufficient conditions that a system of six functions of coordinates must satisfy to be that of a symmetric deformation associated with an existing displacement. These conditions form a set of six second-order partial differential equations, an auxiliary system, now called the compatibility condition of Navier and Saint-Venant, but in fact introduced by Barré de Saint-Venant (1864) in his commented edition of Navier's lectures (cf. Navier 1864). Related works by Boussinesq (1871), Beltrami (1889), Love (1892) and Cesàro (1894) are cited in this context.

Globally, in this introduction to the deformation theory of continua, the Cosserats do not innovate so much but they faithfully incorporate all progress made since Cauchy till the work of their contemporaries (Poincaré, Darboux, Koenigs, Kelvin,...). Still, they cultivate this fruitful view that general deformations must be considered first, leaving infinitesimal deformations as infinitesimally small limits in a strict mathematical vision.

Truesdell (1952a), p. 53, however, notes that the Cosserats missed the long innovative paper of Finger (1894)—obviously very recent at the time of the Cosserats' publication—where Finger introduced the spatial strain measure named after him, i.e. $(\boldsymbol{c}^{-1})^i_{.j} = (\boldsymbol{F}\boldsymbol{F}^T)^i_{.j} = F^i_{.K} F^k_{.L} \delta_{kj} \delta^{KL}$, which would have made simple the formulation of elasticity constitutive equations for finite strain in isotropic bodies.

3.2 Internal Forces (stresses) in a Continuum

The Cosserats do not elaborate much about the original introduction of the notion of *stresses* (according to the coinage of Rankine), i.e. more traditionally, *internal forces* in a continuum. They skip Cauchy's classical argument to introduce (p. I.39) the stress notion at a cut at a point in a body, simply remarking in passing that the cut is tangent to an infinity of curved surfaces, so that only the *normal* to the cut at a point is involved, and stresses (as we shall call them now) are forces per unit area in contrast to body forces that are *mass* forces. The Cosserats do not refer to these internal forces as "tensors" (following Voigt 1898 or, as if they had followed Gibbs 1881–1884) and others, "linear vector functions". But here, to facilitate the reading by modern students, we denote by the Cartesian tensor components $t^{j}_{.i}$ or t^{ij} the stress in the *actual* (after deformation) configuration, and will avoid any direct (no indices) notation that could create some confusion.

Cauchy's equilibrium equations are stated as (cf. Eqs. (24) and (23) in pp. I.39–I.40) in the following traditional form:

$$\frac{\partial}{\partial x_i} t^{i}_{.j} + \rho f_j = 0 \tag{4}$$

at internal points in the body and

$$t_j = n_i \, t^{i}_{.j} \tag{5}$$

at its regular boundary of unit outward normal of components n_i. But the Cosserats have formulated the deformation theory essentially in the undeformed reference configuration (see preceding section). They thus want to reformulate Eqs. (4) and (5) in the appropriate framework, that is, per unit undeformed volume and unit undeformed area. They rightly think that the required manoeuvre must be analogous to what is done in hydrodynamics in passing from Euler to Lagrange equations. This is called a "pull back operation" in modern treatises, and this is in fact defined by the celebrated Piola transformation (Piola 1836), but the Cosserats refer only to Kirchhoff (1852) for this operation which they achieved astutely by associating to Eqs. (4) and (5) a form of the principle of virtual work and then effecting the required transformation in this global formulation (pp. I.42–I.48). Noting δu^j the virtual displacement, one obtains thus the global expression (Eq. I.26)

$$\int_V \rho f_j \delta u^j \, \mathrm{d}V + \int_{\partial V} t_j \delta u^j da - \int_V t^{i}_{.j} \frac{\partial}{\partial x_i} \left(\delta u^j\right) \, \mathrm{d}V = 0. \tag{6}$$

On this occasion, the Cosserats remark on the definition of a virtual "rigidifying" deformation which cancels out the last expression in the left-hand side of Eq. (6). The lengthy transformation of (6) that we do not repeat yields the following expression of the principle of virtual work:

$$\int_{V_0} \rho_0 f_j \delta u^j \, \mathrm{d}V_0 + \int_{\partial V_0} T_j \, \delta u^j da_0 - \int_{V_0} S^{KL} \delta E_{KL} \, \mathrm{d}V_0 = 0, \tag{7}$$

assuming that $J = \det \boldsymbol{F}$ is everywhere positive and the continuity equation reads $\rho_0 = \rho J$. Here δE_{KL} is the variation of the Cauchy–Green strain measure resulting from the variation δu^j, and S^{KL} is the conjugate stress (now called the second Piola–Kirchhoff stress). The Cosserats are then able to transform (7) in the form

$$\int_{\partial V_0} (T_j - N_K T^K_{.j}) \, \delta u^j da_0 + \int_{V_0} \left(\frac{\partial}{\partial X^K} T^K_{.j} - \rho_0 f_j \right) \delta u^j \, \mathrm{d}V_0 = 0 \,, \tag{8}$$

with the definition of the object $T^K_{.j}$ (now called the first Piola–Kirchhoff stress) given by (in our notation; cf. Eq. (36) in p. I.48)

$$T^K_{.j} = \frac{\partial J}{\partial F^i_{.K}} t^i_{.j} = J X^K_{,i} t^i_{.j} \,, \tag{9}$$

and N_K denotes the components of the unit outward normal to the surface body in the undeformed configuration. The localisation of (8) provides the two equations (cf. Eqs. (34) and (35) in p. I.46)

$$\frac{\partial}{\partial X^K} T^K_{.j} + \rho_0 f_j = 0 \qquad \text{in } V_0, \tag{10}$$

$$T_j = N_K T^K_{.j} \qquad \text{at } \partial V_0. \tag{11}$$

Here, as emphasized by the Cosserats (top of p. I.47), the hybrid geometrical object $T^K_{.j}$ represents a force in the direction of the actual axis noted i, but per unit area in the undeformed configuration. Equations (10)–(11) were given by Marcel Brillouin (1884, 1885).

On using an identity established by Carl Neumann (1860),

$$\frac{\partial}{\partial X^K} \left(\frac{\partial J}{\partial F^i_{.K}} \right) = 0 \,, \tag{12}$$

one can revert to the actual (Eulerian form of the) equation of equilibrium as proved by Boussinesq (1869) since with (9) and (10) one has

$$\frac{\partial}{\partial X^K} \left(t^i_{.j} \frac{\partial J}{\partial F^i_{.K}} \right) + \rho_0 f_j = 0. \tag{13}$$

But (see p. I.49)

$$\frac{\partial}{\partial X^K}\left(t^i_{.j}\frac{\partial J}{\partial F^i_{.K}}\right) = \frac{\partial J}{\partial F^i_{.K}}\frac{\partial}{\partial X^K}t^i_{.j}$$

$$= J\frac{\partial X^K}{\partial x^i}\frac{\partial}{\partial X^K}t^i_{.j} = J\frac{\partial}{\partial x^i}t^i_{.j} = \frac{\rho_0}{\rho}\frac{\partial}{\partial x^i}t^i_{.j}, \tag{14}$$

whence Eq. (4). Note that the virtual work of external forces can be written as (cf. Eqs. (6) and (7) above):

$$\delta T_e = \int_V t^j_{.i}\frac{\partial}{\partial x^j}\delta u^i \,\mathrm{d}V = \int_{V_0} T^K_{.i}\frac{\partial}{\partial X^K}\delta x^i \,\mathrm{d}V_0. \tag{15}$$

The Cosserats then discuss the notion of isostatic surfaces after Lamé, Boussinesq and Weingarten (1881), a subject matter that we skip here. In concluding their chapter II the Cosserats evoke the equilibrium equations (cf. p. I.58) in a straight cylinder (before deformation), i.e. a thin rod, and mention those that would be obtained in plates of any thickness loaded on their edge. These are the equations expanded by Clebsch and Barré de Saint-Venant (1883) in the French translation of the book of Clebsch (1883).

One has to wait for the next chapter to witness an introduction of elasticity constitutive equations on thermodynamic bases.

3.3 Equations of Equilibrium

In their formulation of the equations of equilibrium for *elastic* bodies (Chapter III), the Cosserat brothers are strongly influenced by the thermodynamic works of Kelvin (Thomson 1855, 1856, 1857); also (Thomson and Tait 1867), and the recent considerations brought to the field by Pierre Duhem (1887, 1894). That means that they exploit the formulation of the first and second laws of thermodynamics, respectively, then called the *principle of equivalence of heat and work* (with the symbol $E > 0$ standing for the so-called mechanical equivalent of heat, and ignored in modern texts with appropriate physical units) and the *principle of Carnot and Clausius*. For a body in its natural state (homogeneous and without deformation), one then considers homogeneous deformations from this natural state with the same absolute temperature T for all material points. The state of this body after deformation from the natural state is defined by six strains and the temperature. It is assumed (p. I.60) that these seven parameters remain within acceptable limits so that any alteration of the body may be viewed as a continuous sequence of equilibrium states and it corresponds to a reversible evolution (following Duhem). In the sequence of these states the body is maintained in such states by the application of a unique system of external forces with the external bodies kept at the same temperature as the considered body. The two principles of thermodynamics then read:

stress in the deformed configuration is then given in a form attributed to Boussinesq (1869) that we can rewrite in condensed form as

$$t^i_{.j} = J^{-1} F^i_{.K} T^K_{.j} = J^{-1} x^i_{,K} S^{KL} x^p_{,L} g_{jp} = J^{-1} x^i_{,K} \frac{\partial W}{\partial E_{KL}} x^p_{,L} g_{pj}. \tag{28}$$

This, obviously, is not reported in this tensorial form, but the Cosserats give only the form taken in full by two of the components of $t^i_{.j}$ (cf. Eq. (61), or (62), p. I.65) indicating that other components are easily deduced.

In the above-specified conditions the mechanical equilibrium equations are obtained as (cf. Eq. (63)–(64), p. I.66)

$$\frac{\partial}{\partial X^K}\left(\frac{\partial W}{\partial F^i_{.K}}\right) + \rho_0 f_i = 0 \tag{29}$$

at internal points in the body and

$$T_i = N_K \frac{\partial W}{\partial F^i_{.K}} \tag{30}$$

at its regular boundary of unit outward pointing normal of components N_K in the undeformed configuration.

In the rest of this chapter, the Cosserats deal with various matters that include a "paradox" previously dealt with by Poincaré, Kirchhoff and others, notions on stability, the choice of a natural state, the question of material symmetry, and the case of infinitesimal deformations. The paradox referred to by the Cosserats concerns the possible a priori existence of a function Φ of the gradient components $F^i_{.K}$ such that

$$\int_V \delta\Phi \, \mathrm{d}V - \delta T_e = 0. \tag{31}$$

This means that for an equilibrium position one must have Eqs. (27) and (30) with W replaced by Φ, so that, for any part of the body, Eq. (27) must be written with W replaced by Φ. But the quantities $F^i_{.K}$ cannot be taken arbitrarily as they must obey a set of three partial differential equations (Eq. (37) in p. I.49) of which the general integral is an arbitrary function of the six components E_{KL} of the finite deformation. This is the requirement for Eq. (31) to be compatible with the existence of internal forces. This was noticed by Poincaré in his lectures on elasticity (Poincaré 1892, p. 77) but also by Kirchhoff (1852), C. Neumann (1860), and closer to the Cosserats by Cellérier (1893).

The second remark relates to the stability of equilibrium and the notions of "bifurcation" equilibrium and "limit" equilibrium of Poincaré. We must recall that the years 1890s are fruitful as regards questions of stability. This is particularly true of the works of Henri Poincaré with his marked interest in the stability of liquid masses in rotation, a subject also of interest to Paul Appell (1888) in his treatise on rational mechanics, and the original work by Aleksandr Lyapunov (1857–1918) with his Doctoral thesis (in Russian) on "The general problem of stability of motion" at Kharkov, Ukraine (Love 1892). Although the Cosserats had some knowledge of Russian, Lyapunov's work came too late to influence them, but will influence Pierre Duhem when the latter will have identified a potential akin to a Lyapunov function. Thus, the Cosserats are mostly influenced by Poincaré and his considerations on stability in his lectures on elasticity (Poincaré 1892), Chapters III and IV). Along this line, one first notes that in the absence of external forces, the function $W - T_e$ reduces to W. If the latter is minimum at the natural state, then one can only say that the corresponding equilibrium is stable only in so far as deformations are concerned (but it is not stable in a general way). But, now, if the externally applied forces vary in a continuous way depending on a parameter y, assuming that T_e exits for all values of y, then one is led to a situation identical to that envisaged by Poincaré in his study of the equilibrium of a fluid mass in rotation (Poincaré 1885), so that one has to consider Poincaré's notions of "bifurcation" and "limit" equilibria (cf. Cosserats, p.I.69) and to imagine a linear series of equilibrium forms that correspond to a series of real values of y related to the critical points of y functions defined by the system of equilibrium equations. That is all for this remark.

The next remark relates to the choice of a *natural* state. In the absence of external loading, one admits the existence of a natural state that corresponds to a vanishing of the derivatives of the function W with respect to the strain components. We can assume that W can be expanded in the positive entire powers of the strain components, providing thus an expression of the type (Cosserats, Eq. (67), p. I.70)

$$W = W_2 + W_3 + \cdots , \tag{32}$$

where W_k denotes a homogeneous polynomial of degree k, assuming that the constant term has been set equal to zero without loss in generality. For a natural state corresponding to a stable equilibrium from the point of view of deformations (see above), it is sufficient that W be positive for all infinitesimally small components of the strain. This classically yields the definite positiveness of the quadratic form W_2. Following more generally Poincaré (1892), Sects. 27 through 30, one can assume that there exists a first-order contribution W_1 so that 27 elasticity coefficients—at most—will be defined from W_1 and W_2 in the absence of any specific symmetry. The Cosserats then turn to the special case of *isotropy* for a homogeneous body. Invoking the traditional three invariants of strains, W_1 contains only one coefficient ν while W_2 contains the famous two Lamé coefficients, λ and μ, which have to satisfy the inequality

$$3\lambda + 2\mu > 0, \quad \mu > 0, \tag{33}$$

to warrant stability about a natural state (for which $\nu = 0$).

The chapter concludes with the formulation of equations in the case of infinitesimal deformations. This brief analysis (pp. I.72–I.74) introduces an order parameter noted t by the Cosserats; this leads to an expansion of the displacement field in successive positive integer powers of t. This is also the case of the function W. This follows considerations of Darboux in his theory of surfaces (Darboux 1887–1896, Vol. 4, p. 65 on) and Poincaré in his general approach to problems of mathematical physics (cf. Poincaré 1894). The study of W_2 in fact follows the developments offered by Poincaré (1892), pp. 46–58, that we shall not repeat. A theorem due to Kirchhoff (1852) applies when forces vanish. Finally, the standard equilibrium equations are deduced for infinitesimal strains (Cosserats, Eq. (80), p.I.77) rewritten in modern intrinsic notation for the isotropic case as

$$(\lambda + \mu)\,\nabla\theta + \mu\,\Delta\boldsymbol{u} + \rho\boldsymbol{f} = \mathbf{0}, \tag{34}$$

where $\theta = \nabla \cdot \boldsymbol{u}$ denotes the dilatation.

3.4 On Curvilinear Coordinates

The long and final chapter IV must have been welcomed by most readers when the paper was published. It deals with the basic problem of the formulation of the equations of elasticity in curvilinear coordinates. This was approached by pioneers such as Lamé and Beltrami and other scientists before the advent of tensor calculus. But the Cosserat brothers are still living in a period where vector calculus still is in development and is rarely applied (see Crowe 1967, for a historical perspective) and tensor analysis is still in infancy with no clear-cut application but for the notion of tensor introduced by W. Voigt (1898) and that of dyadic by J.W. Gibbs (1881–1884). What the Cosserats propose is to implement the theory of the *mobile triad* introduced by G. Darboux in his general studies of surfaces (Darboux 1887–1896). This is not so surprising since this theory is in full blossom and "é la mode" in these years 1890s. Furthermore, Eugène Cosserat was a disciple of Darboux, who in fact belonged to his Doctoral thesis committee. The main point in this approach is the consideration of a displacement field that depends on three independent parameters (noted ρ_i , $i = 1, 2, 3$ by the Cosserats,) and the important role played by rotations. Then one first envisages the case where the mobile system has a fixed point (it can only rotate). But the interesting case for continuum mechanics is one where the mobile triad of three rectangular axes moves in any way through space so that nine new entities (related to translation) have to be adjoined to the nine rotation parameters (director cosines). In all this is equivalent to a single motion but observed in different systems of axes. Advance in the theory (p. I.83) is made by following Gauss (1827) (in 2D) and Lamé (1859) (in 3D) in exploiting the geometric representation of a system of curvilinear coordinates by considering three families of surfaces and looking for the expression of an arc of any curve traced in space in terms of the ρ_i's, with a drastic simplification if the curvilinear coordinates are rectangular. The change in surface element is evaluated in the same conditions. The consideration of a *reference* mobile triad is emphasized (cf. p. I.65). This allows one to deal with geometric questions related to surfaces and curves traced in space (problem of conjugated tangents on one of the surfaces ρ_i = const. with i fixed, or the problem of establishing the differential equation for curvature lines). This looks very much like exercises given in the past to students in competition for admission to *Grandes Ecoles*. But for applications to continuum mechanics in 3D, one must focus on kinematic formulas where parameters ρ_i are none other than the original orthogonal coordinates (noted x, y, z by the Cosserats, but simply $X^K,\ K = 1, 2, 3$ in modern indicial notation). Then translations are given by the displacement. The latter has to be projected on the mobile triad, and the strain components can be expressed in terms of this projection (cf. Eq. (98) in p. I.91). External applied forces also are reported to the mobile axes. In the end one can write down the equations of equilibrium in this framework (See Eq. (100), p. I.92). The result is a set of partial differential equations satisfied by both translations and rotations, the knowledge of which is intimately related to the triple system of surfaces in which the primitive rectangular coordinate planes have been transformed. Then the Cosserats specialize to the case of infinitesimally small deformations with corresponding expansions of various quantities in the already introduced ordering parameter t, resulting in fact in expressions already given by Beltrami and Barré de Saint-Venant. These are given by Eq. (I.103) for a simple natural state and an isotropic elastic body. The Cosserats mention that the case of thin straight rods and thin plates would be treated in the like manner, but the corresponding elaboration is postponed to further works. The more general case where the body before deformation is reported to an arbitrary triple system of surfaces (with parameters ρ_i) is then lengthily exposed in the rest of the chapter together with equilibrium equations relative to the deformed body. This is achieved with the help of the principle of virtual work (Eq. (32) in p. I.101) for both finite and infinitesimally small deformations with, we must say, rather atrocious equations in terms of the ρ_i 's (for illustration, see Eqs. (116)–(117) in p. I.103 on).

In all, the contents of this chapter IV seem a bit obsolete to our modern eyes used to reasoning with tensors. But in the circumstances of the period where both vector and tensor analyses are not yet sufficiently developed and/or applied, the Cosserats' efforts are certainly justified in spite of the obvious laborious feeling that we gather from them and the somewhat old-fashioned geometric character that permeates them. These may not have been felt as such by the contemporaries of the Cosserats.

4 Summary and Conclusion

This brief perusal of the long paper published by the Cosserat brothers in 1896 brings us to the following general remarks and conclusion. First, the very length and detail of the paper lean towards an interpretation of this paper as an aborted series of lectures on a field of marked interest at the period. Indeed, the first chapters of the opus support this interpretation, especially in the theory of deformations. However, the detailed and accurate description with appropriate references reveals a typical trait of the brothers' style. They are clearly mathematical and, in spite of their professions,

models of interphases (Benveniste and Miloh 2001; Hashin 1962, 1990, 1991; Rubin and Benveniste 2004; Dong et al. 2014, among other). The most practical and popular of them are (arguably) the Gurtin–Murdoch (Gurtin and Murdoch 1975) model and the spring layer model (e.g., Benveniste and Miloh 2001 or Hashin 1962, 1990, 1991). However, with those simplified models the analysis of the effective properties of composite materials is also complicated, even with the use of computational techniques. To make the problem more manageable analytically, various simplified methods have been devised, one of which is based on the notion of *equivalent inhomogeneity*. The idea behind that notion is to replace the original inhomogeneities and their interphases (however complex they may be) with uniform equivalent inhomogeneities with properly defined properties, embed them in the matrix material and analyze the resulting composite as if there were no interphases.

The concept of equivalent inhomogeneity has a long history. Hashin (1962, 1990, 1991), in application of his composite sphere assemblage to analysis of the effective bulk modulus, introduced a similar concept and discussed its possible extension to multilayer systems. That idea has been subsequently followed within the so-called differential scheme (Shen and Li 2005; Sevostianov and Kachanov 2007). In this approach layers of infinitesimal thickness were added to the original spherical inhomogeneity to form an interphase with properties varying across its thickness. With addition of each layer the properties of the system were defined either by the Mori–Tanaka scheme or Hashin–Shtrikman (Hashin and Shtrikman 1963) upper bound estimate. Equivalent inhomogeneities have also been presented in the contributions of Duan et al. (2007) and Gu et al. (2014) in which three different models of interphases were considered: the Gurtin–Murdoch material surface model, the spring layer, and the continuum models. Their definitions were identical and based on equality of the energy changes (introduced in Eshelby 1957) caused by insertion of a spherical inhomogeneity together with its interphase and the changes caused by insertion of the equivalent inhomogeneity. While the bulk modulus of the equivalent inhomogeneity obtained that way was identical as that of Hashin (1991) and depended only on the properties of the original inhomogeneity and of the interphase, its shear modulus, however, turned out to also include the moduli of the matrix. The formula defining equivalent shear modulus expectedly reflects the properties of the entire system, not just those of the inhomogeneity and its interphase that it is supposed to replace. Thus, for a specific inhomogeneity and its specific interphase, the criterion adopted by Duan et al. (2007) and Gu et al. (2014) leads to infinite number of "equivalent inhomogeneities" which seems nonphysical, and it is unlike any of the previously presented equivalent inhomogeneities (Hashin 1991; Shen and Li 2005; Sevostianov and Kachanov 2007).

To the best knowledge of the Authors only the differential scheme of Shen and Li (2005) and Sevostianov and Kachanov (2007) could lead to the equivalent inhomogeneity for interphase with properties varying piecewise continuously across its thickness. Also Duan et al. (2007) considered interphase assumed to be a three-dimensional continuum. All other effective inhomogeneities have been defined either for the Gurtin–Murdoch interphase model or for the spring layer model. There are however situations, such as coated particle composites for example, where there are three materials involved. Depending on the properties of those materials, a Gurtin–Murdoch model may be appropriate on the boundaries between any pair of the involved materials while the coating itself, being often soft, may be adequately modeled by a spring layer. Consequently, there are three interphase mechanisms in the space between the original inhomogeneity and the matrix: two (possibly different) Gurtin–Murdoch models and one spring layer model, and the equivalent inhomogeneity for such interphase problem must include all three of them. That case is not covered by any of the existing developments.

Another example of a situation where inclusion of multiple mechanisms in the description of interphases is appropriate relates to the Carbon Nanotube (CNT) reinforced materials. In that case the CNT can be modeled as high-stiffness surface, and Gurtin–Murdoch model is very adequate for its mathematical description. Outside of CNT, however, there exists a zone of rather weak bonding with the matrix (typically polymer) and various techniques used to improve that bonding (so-called functionalization, Ma and Kim 2011; Tserpes and Silvestre 2014) have been only partially successful. Such a weak (or soft) zone surrounding the CNT can be modeled by a spring layer of some thickness h. In this case one Gurtin–Murdoch and one spring layer model would need to be combined.

There is a definition of equivalent inhomogeneity that is easily adaptable to the problems whose interphase may involve a combination of the Gurtin–Murdoch and spring layer models. It is called the *energy-equivalent inhomogeneity* that has been originally introduced by Nazarenko et al. (2015) for the Gurtin–Murdoch model and, subsequently, by Nazarenko and Stolarski (2016) for the spring layer model. Presentation of how these two developments can be exploited in defining the equivalent inhomogeneity for the problems whose interphases involve a combination of Gurtin–Murdoch and spring layer models is the main goal of this contribution. To make this presentation self-contained, however, the basic assumptions and results obtained in the above two original contributions are outlined in Sects. 2 and 3. Section 4 contains the effective properties for the interphase composed of two Gurtin–Murdoch models and one spring layer model, along with an explanation of how these results have been arrived at. In Sect. 5 the developments presented herein are discussed in the context of other definitions of equivalent inhomogeneities. Several conclusions are presented in Sect. 6.

2 General Formulations of Equivalence

In principle, the overall methodology adopted in this work follows the idea pursued in Duan et al. (2007), Gu et al. (2014), Hashin (1962, 1990, 1991), Nazarenko et al. (2015), Nazarenko and Stolarski (2016), Sevostianov and Kachanov (2007), Shen and Li (2005), and it consists of replacing the original inhomogeneity and the surrounding interface (irrespectively of the model used to describe it) by uniform equivalent inhomogeneity which is then perfectly bonded to the matrix. That way the composites including interphases can be analyzed using techniques available for composites without interphases. What is substantially different in the presented approach is the way the properties of the equivalent inhomogeneity are developed. In this regard the typical method of homogenization is used as described below.

Consider the original inhomogeneity and any surrounding domain, which differs from the matrix, as a single system. The boundary Γ_2 of that system with the matrix, Fig. 1, is subjected to displacements $\boldsymbol{u}_2 = \boldsymbol{\varepsilon}_{\mathrm{eq}} \cdot \boldsymbol{r}_2$, consistent with a constant strain $\boldsymbol{\varepsilon}_{\mathrm{eq}}$. The solution of that problem can be obtained by minimization of its potential energy

$$E = \frac{1}{2}\int_{V_1} \boldsymbol{\varepsilon}_1 : \boldsymbol{C}_1 : \boldsymbol{\varepsilon}_1 \, \mathrm{d}V_1 + E_{\mathrm{int}}, \quad E_{\mathrm{int}} = \frac{1}{2}\int_{\Omega_{\mathrm{int}}} \boldsymbol{\varepsilon}_{\mathrm{int}} : \boldsymbol{C}_{\mathrm{int}} : \boldsymbol{\varepsilon}_{\mathrm{int}} \, \mathrm{d}\Omega_{\mathrm{int}}, \tag{1}$$

where V_1 is the domain of the inhomogeneity, Ω_{int} is the surrounding domain and

$$\boldsymbol{\varepsilon} = \boldsymbol{\varepsilon}(\boldsymbol{u}) = \nabla^{\mathrm{sym}} \boldsymbol{u}, \tag{2}$$

with continuous and piecewise differentiable displacements $\boldsymbol{u}$. The domain Ω_{int} may represent a volume or a surface, depending on its physical interpretation and the associated mathematical description. The interpretation of the strains $\boldsymbol{\varepsilon}_{\mathrm{int}}$ and of the stiffness tensor $\boldsymbol{C}_{\mathrm{int}}$ also needs to be adjusted accordingly.

The vanishing first variation of potential energy, $\delta_{\boldsymbol{u}} E = 0$, allows to determine displacements $\boldsymbol{u}$, if not exactly, then approximately, and—consequently—the strains in Eq. (2) can be expressed in the form:

$$\boldsymbol{u} \Rightarrow \boldsymbol{\varepsilon}_1 = \boldsymbol{\mathcal{T}}_1 : \boldsymbol{\varepsilon}_{\mathrm{eq}}, \quad \boldsymbol{\varepsilon}_{\mathrm{int}} = \boldsymbol{\mathcal{T}}_{\mathrm{int}} : \boldsymbol{\varepsilon}_{\mathrm{eq}}, \tag{3}$$

where $\boldsymbol{\mathcal{T}}_1$ and $\boldsymbol{\mathcal{T}}_{\mathrm{int}}$ are fourth order tensors. Inserting the above formulas in Eq. (1) yields the equilibrium energy of the system to be

$$E = \frac{1}{2}\boldsymbol{\varepsilon}_{\mathrm{eq}} : \left(\int_{V_1} \boldsymbol{\mathcal{T}}_1^T : C_1 : \boldsymbol{\mathcal{T}}_1 \, \mathrm{d}V_1 + \int_{\Omega_{\mathrm{int}}} \boldsymbol{\mathcal{T}}_{\mathrm{int}}^T : \boldsymbol{C}_{\mathrm{int}} : \boldsymbol{\mathcal{T}}_{\mathrm{int}} \, \mathrm{d}\Omega_{\mathrm{int}} \right) : \boldsymbol{\varepsilon}_{\mathrm{eq}}, \tag{4}$$

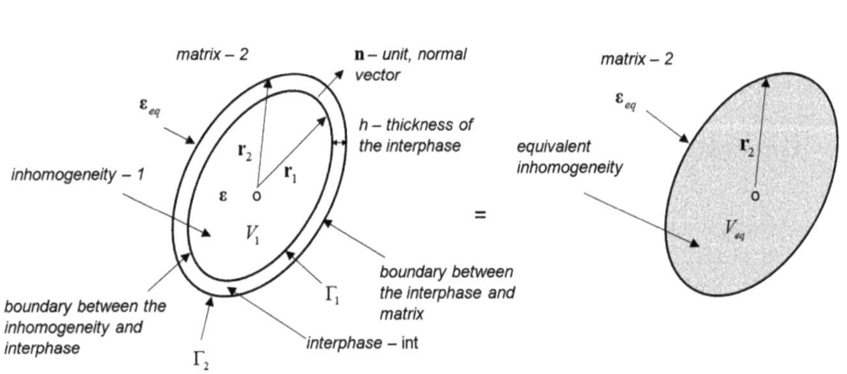

Fig. 1 Schematic illustration of inhomogeneity with interphase and equivalent inhomogeneity

were $\mathcal{T}^T$ denotes the "major" transposition, i.e., for a tensor $\mathcal{T}$ whose components (relative to an orthogonal coordinate system) are T_{ijkl}, the components of $\mathcal{T}^T$ are T_{klij}. The above equation implies that the properties of equivalent inhomogeneity can be described by the formula

$$\mathcal{C}_{\text{eq}} = \int_{V_1} \mathcal{T}_1^T : \mathcal{C}_1 : \mathcal{T}_1 \, \mathrm{d}V_1 + \int_{\Omega_{\text{int}}} \mathcal{T}_{\text{int}}^T : \mathcal{C}_{\text{int}} : \mathcal{T}_{\text{int}} \, \mathrm{d}\Omega_{\text{int}}. \tag{5}$$

The above general process covers all possible shapes of the inhomogeneities and all possible types of interphases, including the continuous interphase. It will be subsequently specified for the most commonly used types: the Gurtin–Murdoch model of the interphase Gurtin and Murdoch (1975, 1978) and a simplified analysis of the spring layer model (Duan et al. 2007; Gu et al. 2014; Hashin 1962, 1990, 1991; Nazarenko et al. 2016; Nazarenko and Stolarski 2016). It will also be argued that only because of the proposed formulation of those problems, a very elementary process allows to consider simultaneous presence of both of those interphase models, which in some applications is necessary. It also allows to apply the differential scheme for more complex inhomogeneous interphases.

3 Gurtin–Murdoch and Spring-Layer Interphase Models

One possible approximate solution of the above problem may be based on the assumption that the displacements prescribed on Γ_2 cause the original inhomogeneity to undergo constant straining combined with rigid body motion. This approximation is particularly suitable for small inhomogeneities and it has been shown in Nazarenko et al. (2015) to render the effective properties of composites whose accuracy is on par with those obtained by other approximate, or even formally exact, methods. Mathematically, that assumption is described by the following equation:

$$\boldsymbol{u}(\boldsymbol{r}) = \boldsymbol{u}_0 + (\boldsymbol{\omega}_1 + \boldsymbol{\varepsilon}_1) \cdot \boldsymbol{r}, \tag{6}$$

where $\boldsymbol{\varepsilon}_1 = const$, $\boldsymbol{\omega}_1$ and $\boldsymbol{u}_0$ are the parameters describing displacement field within the original inhomogeneity, and $\boldsymbol{r}$ is the position vector relative to a reference point O, Fig. 1. Consequently, the first part of Eq. (1) may be written in the form

$$\frac{1}{2} \int_{V_1} \boldsymbol{\varepsilon}_1 : \mathcal{C}_1 : \boldsymbol{\varepsilon}_1 \, \mathrm{d}V_1 = \frac{V_1}{2} \boldsymbol{\varepsilon}_1 : \mathcal{C}_1 : \boldsymbol{\varepsilon}_1. \tag{7}$$

The second part of Eq. (1) has the form dependent on the interphase model used.

For the Gurtin–Murdoch interphase model $\Omega_{\text{int}} = \Gamma_1 = \Gamma_2$, $\boldsymbol{\varepsilon}_1 = \boldsymbol{\varepsilon}_{\text{eq}}$. Physically, the model can be interpreted as vanishing thickness isotropic membrane separating two different materials while preserving continuity of their displacements.

Consequently, (neglecting the effects of the surface displacement gradient, Nazarenko et al. (2015)) the interphase part of the energy is

$$E_{\mathrm{int}} = \frac{1}{2} \oint_{\Gamma_1} \left[2\left(\mu_{\mathrm{s}} - \tau_0\right) \boldsymbol{\varepsilon}_{\mathrm{s}} : \boldsymbol{\varepsilon}_{\mathrm{s}} + \left(\lambda_{\mathrm{s}} + \tau_0\right) \operatorname{tr}\left(\boldsymbol{\varepsilon}_{\mathrm{s}}\right)^2 \right] \mathrm{d}\Gamma_1, \tag{8}$$

where λ_{s} and μ_{s} are the surface Lame parameters, τ_0 is the deformation-independent (residual) surface stress and $\boldsymbol{\varepsilon}_{\mathrm{s}}$ represents the surface strains on Γ_1 dependent on $\boldsymbol{\varepsilon}_1$ via the relation Nazarenko et al. (2015)

$$\boldsymbol{\varepsilon}_{\mathrm{s}} = \boldsymbol{\mathcal{I}}_{\mathrm{s}} : \boldsymbol{\varepsilon}_1, \tag{9}$$

where $\boldsymbol{\mathcal{I}}_{\mathrm{s}}$ is the fourth order surface identity tensor.

The spring-layer model of the interphase has been widely used to describe the so-called "soft interphase" (Duan et al. 2007; Gu et al. 2014; Hashin 1962, 1990, 1991; Nazarenko et al. 2016; Nazarenko and Stolarski 2016). For that model the interphase energy is expressed by the formula

$$E_{\mathrm{int}} = \oint_{\Gamma_1} \frac{1}{2} \left\{ K_{\mathrm{n}} \left| \Delta \boldsymbol{u}_{\mathrm{n}} \right|^2 + K_{\mathrm{t}} \left| \Delta \boldsymbol{u}_{\mathrm{t}} \right|^2 \right\} \mathrm{d}\Gamma_1 \tag{10}$$

where K_{n} and K_{t} are the stiffnesses of the normal and tangential linear spring layers and

$$\Delta \boldsymbol{u} = \boldsymbol{u}_2 - \boldsymbol{u}_1. \tag{11}$$

In the above equation $\boldsymbol{u}_1$ is the displacement on the surface Γ_1 defined by Eq. (6) with $\boldsymbol{r} = \boldsymbol{r}_1$ and $\boldsymbol{u}_2 = \boldsymbol{\varepsilon}_{\mathrm{eq}} \cdot [\boldsymbol{r}_1 + h\boldsymbol{n}]$ the displacement on Γ_2, expressed in terms of the interphase thickness h and $\boldsymbol{n}$, the unit normal to Γ_1.

Whereas the above presentation is valid for arbitrary original inhomogeneities, the involved integrals may be difficult to evaluate analytically if their shapes are complex. However, to illustrate the basic features of the proposed approach and its properties it is sufficient to consider their simple geometry. Thus, spherical inhomogeneities are discussed in the following sections.

3.1 Equivalent Inhomogeneity for the Gurtin–Murdoch Interphase Model

The details of the development for this model of interphase (and the numerical examples illustrating the accuracy of the approach) are presented in Nazarenko et al. (2015). As shown there $\boldsymbol{\varepsilon}_1 = \boldsymbol{\varepsilon}_{\mathrm{eq}}$, the absence of $\boldsymbol{\omega}_1$ and $\boldsymbol{u}_0$ eliminates the need for minimization of energy E in this case, and the properties of equivalent inhomogeneity are obtained simply by inserting Eq. (9) into Eq. (8), and subsequently into Eq. (1). The result is

$$\mathcal{C}_{\text{eq}} = \mathcal{C}_1 + \frac{1}{2V_I} \oint_{\Gamma_1} \left[2\bar{\mu}_s \mathcal{I}_s + \bar{\lambda}_s \boldsymbol{I}_s \otimes \boldsymbol{I}_s \right] \mathrm{d}\Gamma_1. \tag{12}$$

Further evaluation of the above formula (Nazarenko et al. 2015) shows that for spherical and isotropic inhomogeneities, and for isotropic surface properties tensor $\mathcal{C}_{\text{eq}}$ is also isotropic and its bulk and shear moduli are

$$K_{\text{eq}} = K_1 + \hat{K}_s, \ \mu_{\text{eq}} = \mu_1 + \hat{\mu}_s, \tag{13}$$

where K_1 and μ_1 are the original bulk and shear moduli of the inhomogeneity, while

$$\hat{K}_s = 2\frac{\left[2\bar{\mu}_s + 2\bar{\lambda}_s\right]}{3r_1}, \ \hat{\mu}_s = \frac{7\bar{\mu}_s + \bar{\lambda}_s}{5r_1}, \tag{14}$$

with $\bar{\lambda}_s = \lambda_s + \tau_0$, $\bar{\mu}_s = \mu_s - \tau_0$ appearing as a result of the surface contribution in Gurtin–Murdoch surface model, and r_1 being the radius of spherical inhomogeneity.

Via comparisons with the results obtained by other approaches it is also shown in Nazarenko et al. (2015) that use of the energy-equivalent inhomogeneity leads to remarkably accurate effective properties of composites with Gurtin–Murdoch surface model.

3.2 Equivalent Inhomogeneity for the Spring Layer Interphase Model

The energy-based formulation of the spring layer interphase model has been introduced in Nazarenko and Stolarski (2016) where, in contrast with all previous discussions of the spring layer model, its thickness h was assumed to be finite. Finite thickness of the interphase is also assumed here.

In view of Eq. (11), in this case the potential energy depends on all parameters involved in the description of the problem. The discretised form of energy is an algebraic quadratic form and its minimization leads to a system of linear algebraic equations. Their solution shows that $\boldsymbol{\omega}_1 = \mathbf{0}$ and $\boldsymbol{u}_0 = \mathbf{0}$. The strain $\boldsymbol{\varepsilon}_1$ is, obviously, linearly related to $\boldsymbol{\varepsilon}_{\text{eq}}$ which, as indicated in Eq. (4), leads to equilibrium potential energy being the quadratic with respect to $\boldsymbol{\varepsilon}_{\text{eq}}$.

Under the assumptions adopted here, the properties for the equivalent inhomogeneity for a spherical shape are also isotropic and the associated bulk and shear moduli are (Nazarenko and Stolarski 2016)

$$K_{\text{eq}} = \frac{r_1^2}{[r_1 + h]} \frac{K_n K_1}{[r_1 K_n + 3K_1]}, \ 2\mu_{\text{eq}} = \frac{2r_1^2}{[r_1 + h]} \frac{\mu_1 [2K_n + 3K_t]}{[2r_1 K_n + 3r_1 K_t + 10\mu_1]}. \tag{15}$$

In the above equations r_1 is the radius of spherical inhomogeneity and h is the interphase thickness.

In Nazarenko et al. (2016) a different equivalent inhomogeneity approach was persecuted, whereby the properties of equivalent inhomogeneity were evaluated on the basis of Lurie solution (Lurie 2005). This resulted in

$$K_{\mathrm{eq}} = \frac{r_1^2}{[r_1 + h]} \frac{K_{\mathrm{n}} K_1}{[r_1 K_{\mathrm{n}} + 3K_1]}, \tag{16}$$

$$2\mu_{\mathrm{eq}} = \frac{2\mu_1 r_1^2}{5[r_1 + h]} \times \frac{4\mu_1[7 + 5\nu_1][2K_{\mathrm{n}} + 3K_t] + 5[7 - 10\nu_1]K_{\mathrm{n}}K_{\mathrm{t}}r_1}{\left[4\mu_1^2(7 + 5\nu_1) + 2\mu_1((7 - 4\nu_1)K_n + (7 - \nu_1)K_t) + (7 - 10\nu_1)K_n K_t r_1\right]}. \tag{17}$$

It is seen from Eqs. (15)–(17) that bulk moduli for equivalent inhomogeneity obtained on the basis of both approaches are identical. It should also be noted that expressions for bulk modulus obtained in Duan et al. (2007) and Hashin (1991) are identical as well. However, the expressions for shear moduli of equivalent inhomogeneity for both the model proposed herein and the one based on Lurie solution are different.

4 Coated Particles and Complex Interphases

In some problems, such as coated particle composites, there are clearly identifiable boundaries Γ_1 and Γ_2 between pairs of the three phases present in those cases. Each of those boundaries may be associated with a different Gurtin–Murdoch model. Assuming that the coating is soft and can be modeled by the spring layer of thickness h, the development of the equivalent inhomogeneity for such problems may involve variable number of interphase models. It may involve only one model (the spring layer model, if the influence of the Gurtin–Murdoch models associated with both Γ_1 and Γ_2 is negligible), two models (if the influence of the Gurtin–Murdoch model on only one surface is negligible) or three models (if none is negligible).

The energy-based approach described above, allows for an easy combination of those models in one expression defining the properties of the equivalent inclusion, irrespectively of how many (one, two or three) interphase mechanism are included. That ease is crucially associated with the assumptions made in development of equivalent inclusion for each of the interphase models separately. In particular, it is critical to assume that in the spring layer model presented in Sect. 2 displacements on the boundary Γ_1 are associated with $\boldsymbol{\varepsilon}_1 = const$. That allows for sequential (or hierarchical) inclusion of the models to obtain the final result. This can be done very formally but an informal, descriptive explanation of the process leading to the final result is outlined below.

According to the development presented in Sect. 1 the original inhomogeneity and the Gurtin–Murdoch model on Γ_2 can be equivalently replaced by a homogeneous inhomogeneity with the properties given by Eqs. (13) and (14). In the next step the spring layer model is added to the previous system. The new problem consists of the previously defined equivalent inhomogeneity and the new model of interphase, namely the spring layer model. Thus, the properties of the subsequent equivalent inhomogeneity describing equivalently this problem are defined by Eq. (15) in which K_1 and μ_1 are replaced by the bulk and shear modulus for the previously defined equivalent inhomogeneity. Inclusion of the Gurtin–Murdoch model on Γ_1 involves one more step, following identical logic. This leads to the following formula involving all three mechanisms possibly present in coated particle composites.

$$K_{\mathrm{eq}} = \frac{r_1^2}{[r_1 + h]} \frac{K_{\mathrm{n}} \left[3 r_1 K_1 + 2\left(2\bar{\mu}_{\mathrm{s}[1]} + 2\bar{\lambda}_{\mathrm{s}[1]}\right)\right]}{\left[3 (r_1)^2 K_{\mathrm{n}} + 3\left(3 r_1 K_1 + 2\left(2\bar{\mu}_{\mathrm{s}[1]} + 2\bar{\lambda}_{\mathrm{s}[1]}\right)\right)\right]} + 2\frac{\left[2\bar{\mu}_{\mathrm{s}[2]} + 2\bar{\lambda}_{\mathrm{s}[2]}\right]}{3\,[r_1 + h]}, \tag{18}$$

$$2\mu_{\mathrm{eq}} = \frac{2r_1^2}{[r_1 + h]} \frac{[2K_{\mathrm{n}} + 3K_{\mathrm{t}}]\left[5 r_1 \mu_1 + 7\bar{\mu}_{\mathrm{s}[1]} + \bar{\lambda}_{\mathrm{s}[1]}\right]}{\left[10 (r_1)^2 K_{\mathrm{n}} + 15 (r_1)^2 K_{\mathrm{t}} + 10\left(5 r_1 \mu_1 + 7\bar{\mu}_{\mathrm{s}[1]} + \bar{\lambda}_{\mathrm{s}[1]}\right)\right]} + \frac{7\bar{\mu}_{\mathrm{s}[2]} + \bar{\lambda}_{\mathrm{s}[2]}}{5\,[r_1 + h]}. \tag{19}$$

where $\bar{\lambda}_{\mathrm{s}[1]}$, $\bar{\mu}_{\mathrm{s}[1]}$ and $\bar{\lambda}_{\mathrm{s}[2]}$, $\bar{\mu}_{\mathrm{s}[2]}$ are surface Lame parameters on the surfaces Γ_1 and Γ_2, correspondently.

It is obvious that the Gurtin–Murdoch model on either or both surfaces Γ_1 and Γ_2 can be excluded simply by assuming that the associated material parameters vanish.

5 Discussion

One can, of course, define the equivalent properties for the original inhomogeneity and an interphase of any complexity differently. For example, Duan et al. (2007) and Gu et al. (2014) use for that purpose the Eshelby formula for the change of energy when an inhomogeneity is inserted in a uniformly strained infinite medium (Eshelby 1957). In their work the properties of equivalent inhomogeneity are obtained by equating such energy change associated with insertion of the original inhomogeneity and its single interphase and the one associated with insertion of the uniform equivalent inhomogeneity. The equivalent bulk modulus they obtained was dependent only on the properties of the original inhomogeneity and its interphase, and identical with that developed by Hashin (1991) or Nazarenko and Stolarski (2016), which is also reproduced in Eqs. $(15)_1$ and (16). However, the shear moduls (quite expectedly) was also dependent on the properties of the infinite medium (matrix), in addition to be very complex. For multiple mechanisms describing the interphase, such an approach would clearly necessitate repetition of the entire analysis included in Duan et al. (2007) and Gu et al. (2014). Given the complexity of that analysis for a single interphase, it is difficult to anticipate its outcome in the case of multiple interphase mechanisms.

In general, when the properties of equivalent inhomogeneity depend on the properties of the material outside of its domain, it is difficult to envision a process (if it exists at all) allowing combination of the formulas obtained for individual interphase models in order to develop properties for the equivalent inhomogeneity for problems with multiple interphase mechanisms. In fact, even if the properties of equivalent inhomogeneity depend only on the properties of the material within its domain such combination may not be possible. For example, the equivalent properties for the spring layer model based on the Lurie solution developed by Nazarenko et al. (2016) and quoted in Eqs. (16) and (17) cannot be combined with the Gurtin–Murdoch model quoted in Eqs. (13) and (14) if such a model was to be appropriate to include between the spring layer and the original interphase (i.e., boundary Γ_1). The reason is that, in the Lurie approach, the strains within the original inhomogeneity are not constant, while the contribution of the Gurtin–Murdoch surface to the properties of equivalent inhomogeneity in Eqs. (13) and (14) are based on the assumption of constant strains However, Gurtin–Murdoch model on Surface Γ_2 can be (additively) combined with the spring layer model, as done in Eqs. (18) and (19).

The approach and the assumptions discussed in this work are not the only ones allowing for easy combination of various models to develop properties of equivalent inhomogeneities. For example, somewhat similar process is used in the work of Sevostianov and Kachanov (2007), who employ Hashin–Shtrikman (Hashin and Shtrikman 1963) upper bound estimate to add consecutive layer of different material within the so-called differential scheme to find properties of equivalent inclusion for a more complex interphase. However, in that approach the equivalent properties at any stage of the process depend only on the material within the equivalent inhomogeneity at that stage too.

The ability of the sequential addition of various interphase models in the evaluation of the final equivalent inhomogeneity discussed herein, while advantageous, cannot possibly be considered as the determining factor in evaluating the validity of the approach. A balance between its accuracy and ability is of paramount importance. Although examination of that balance requires additional investigations, a glimpse into accuracy of the assumptions facilitating such sequential addition (energy-based approach and properties $\boldsymbol{\varepsilon}_1 = const$ in Sect. 2) can be gained from their application to the cases of a single interphase model presented by Nazarenko et al. (2015), Nazarenko et al. (2016) and Nazarenko and Stolarski (2016). It was documented there that the effective properties of composites based on those assumptions were of accuracy comparable with that of other popular methods, including formally exact (series) numerical solutions.

6 Concluding Remarks

It has been argued in this work that in some cases, such as coated particle composites, modeling of interphases may necessitate inclusion of multiple known, and practically useful, mechanisms. In this context it has been demonstrated that the approach of Nazarenko et al. (2015), Nazarenko and Stolarski (2016) based on the concept of equivalent inclusion can easily accommodate such need. For that case a closed-form formula describing the effective properties of the equivalent inhomogeneity involving an interphase combining the spring layer model and two Gurtin–Murdoch models is presented.

In addition to the development of the closed-form expressions for a specific problem, general conditions that need to be fulfilled in order to be able to use the results obtained for individual models in construction of some more complex models are discussed. Those conditions happen to be fulfilled by the energy-based approach of Nazarenko et al. (2015), Nazarenko and Stolarski (2016), even though the only goal in those contributions was to obtain a simple and sufficiently accurate solution for the effective properties of composites with simple interphases.

It is to be underscored that inclusion of multiple mechanisms in description of interphases is not reserved only for coated particle composites. It is, for example, very relevant in modeling Carbon Nanotube (CNT) reinforced materials. In that case the CNT is a small radius, high-stiffness surface which can be very appropriately described by Gurtin–Murdoch model. Even though it is not coated, the bonding between CNT and the matrix material (e.g., polymer) is often weaker than the matrix itself Ma and Kim (2011), Tserpes and Silvestre (2014) and can be adequately modeled by a spring layer of some thickness h. These two distinct mechanisms can be combined and the formula given in Eqs. (18) and (19) are very relevant in this case.

Acknowledgments LN and HS gratefully acknowledge the financial support by the German Research Foundation (DFG) via Project NA1203/1-1.

References

Benveniste, Y., Miloh, T.: Imperfect soft and stiff interfaces in two-dimensional elasticity. Mech. Mater. **33**(6), 309–323 (2001). doi:10.1016/S0167-6636(01)00055-2

Dong, H., Wang, J., Rubin, M.: Cosserat interphase models for elasticity with application to the interphase bonding a spherical inclusion to an infinite matrix. Int. J. Solid. Struct. **51**(2), 462–477 (2014). doi:10.1016/j.ijsolstr.2013.10.020

Duan, H., Yi, X., Huang, Z., Wang, J.: A unified scheme for prediction of effective moduli of multiphase composites with interface effects. part i: theoretical framework. Mech. Mater. **39**(1), 81–93 (2007). doi:10.1016/j.mechmat.2006.02.009

Eshelby, J.D.: The determination of the elastic field of an ellipsoidal inclusion, and related problems. Proc. R. Soc. Lond. A: Math. Phys. Eng. Sci. **241**(1226), 376–396 (1957). doi:10.1098/rspa.1957.0133

Gu, S.T., Liu, J.T., He, Q.C.: Size-dependent effective elastic moduli of particulate composites with interfacial displacement and traction discontinuities. Int. J. Solid. Struct. **51**(13), 2283–2296 (2014). doi:10.1016/j.ijsolstr.2014.02.033

Gurtin, M.E., Murdoch, A.I.: A continuum theory of elastic material surfaces. Arch. Ration. Mech. Anal. **57**(4), 291–323 (1975). doi:10.1007/BF00261375

Gurtin, M.E., Murdoch, A.I.: Surface stress in solids. Int. J. Solid. Struct. **14**(6), 431–440 (1978). doi:10.1016/0020-7683(78)90008-2

Hashin, Z.: The elastic moduli of heterogeneous materials. J. Appl. Mech. **29**(1), 143–150 (1962). doi:10.1115/1.3636446

Hashin, Z.: Thermoelastic properties of fiber composites with imperfect interface. Mech. Mater. **8**(4), 333–348 (1990). doi:10.1016/0167-6636(90)90051-G

Hashin, Z.: The spherical inclusion with imperfect interface. J. Appl. Mech. **58**(2), 444–449 (1991). doi:10.1115/1.2897205

Hashin, Z., Shtrikman, S.: A variational approach to the theory of the elastic behaviour of multiphase materials. J. Mech. Phys. Solid. **11**(2), 127–140 (1963). doi:10.1016/0022-5096(63)90060-7

Lurie, A.I.: Theory of Elasticity. Springer, Heidelberg (2005). doi:10.1007/978-3-540-26455-2

Ma, P.C., Kim, J.K.: Carbon Nanotubes for Polymer Reinforcement. CRC Press (2011). ISBN 9781439826218 - CAT# K11260

Nazarenko, L., Stolarski, H.: Energy-based definition of equivalent inhomogeneity for various interphase models and analysis of effective properties of particulate composites. Composites Part B. **94**, 82–94 (2016). doi:10.1016/j.compositesb.2016.03.015

Nazarenko, L., Bargmann, S., Stolarski, H.: Energy-equivalent inhomogeneity approach to analysis of effective properties of nanomaterials with stochastic structure. Int. J. Solid. Struct. **59**, 183–197 (2015). doi:10.1016/j.ijsolstr.2015.01.026

Nazarenko, L., Bargmann, S., Stolarski, H.: Lurie solution for spherical particle and spring layer model of interphases: its application in analysis of effective properties of composites. Mech. Mater. **96**, 39–52 (2016). doi:10.1016/j.mechmat.2016.01.011

Rubin, M., Benveniste, Y.: A cosserat shell model for interphases in elastic media. J. Mech. Phys. Solid. **52**(5), 1023–1052 (2004). doi:10.1016/j.jmps.2003.09.030

Sevostianov, I., Kachanov, M.: Effect of interphase layers on the overall elastic and conductive properties of matrix composites. applications to nanosize inclusion. Int. J. Solid. Struct. **44**(3–4), 1304–1315 (2007). doi:10.1016/j.ijsolstr.2006.06.020

Shen, L., Li, J.: Homogenization of a fibre/sphere with an inhomogeneous interphase for the effective elastic moduli of composites. Proc. R. Soc. Lond. A: Math. Phys. Eng. Sci. **461**(2057), 1475–1504 (2005). doi:10.1098/rspa.2005.1447

Tserpes, K., Silvestre, N. (eds.): Modeling of Carbon Nanotubes, Graphene and Their Composites. Springer Series in Materials Science. Springer, Cham (2014). doi:10.1007/978-3-319-01201-8

The Influence of Distributed Dislocations on Large Deformations of an Elastic Sphere

Evgeniya V. Zhbanova and Leonid M. Zubov

Abstract From the point of view of nonlinear elasticity theory the equilibrium problem for elastic sphere was considered taking into account distributed edge dislocations. We used the system of equations that consists of the incompatibility equations with a given dislocation density tensor, equilibrium equations, and constitutive equations of the material. For the isotropic material and spherically symmetric distribution of the edge dislocations, the problem was reduced to the second-order ordinary differential equation. In the framework of harmonic (semi-linear) material, the exact solution of this equation was found for any function which defines the edge dislocation density. In particular, we studied the case of dislocations concentrated on a spherical surface within a body. It was established that this surface was the discontinuity surface of strains and stresses. In addition to eigenstress problem, we solved a problem of the loading of a hollow sphere with external or internal hydrostatic pressure. Influence of dislocations on resistance of the sphere to the compression or blowing was investigated.

1 Introduction

Dislocations are the important and widespread element of the solids microstructure. The dislocation models make it possible to describe various properties of modern materials (Gutkin and Ovid'ko 2004; Besson et al. 2010; Clayton 2011; Maugin 2012). Moreover, they are used for the description of effects such as inelasticity, inner friction, plastic flow, fatigue, destruction, crystal growth, and so on. In many cases, the number of dislocations in a limited volume is large. For these cases, it is appropriated to transit from the discrete set of dislocations to their continuous

E.V. Zhbanova (✉) · L.M. Zubov
Institute of Mathematics, Mechanics and Computer Science of Southern Federal University, Milchakova str. 8a, 344090 Rostov on Don, Russia
e-mail: jane.zhbanova@gmail.com

L.M. Zubov
e-mail: zubovl@yandex.ru

K. Naumenko and M. Aßmus (eds.), *Advanced Methods of Continuum Mechanics for Materials and Structures*, Advanced Structured Materials 60,
DOI 10.1007/978-981-10-0959-4_4

distribution, and to use the continual theory of continuously distributed dislocations. Nowadays the theory of continuously distributed dislocations plays an important role among other generalized models of continuum such as micropolar, micromorphic, and strain-gradient models (Forest and Sedláček 2003; Forest 2008; Eremeyev et al. 2013; Clayton et al. 2006; Altenbach et al. 2011; Altenbach and Eremeyev 2013, 2014). Using the dislocation distribution, one can model various defects in crystal and nanostructural materials. Although the equations of the nonlinear continual dislocation theory are well known (Kondo 1952; Bilby et al. 1955; Kröner 1960; Zubov 1997; Derezin and Zubov 1999, 2011), there is just a small list (Yavary and Goriely 2012; Zelenina and Zubov 2013) of the known exact solutions of the boundary problems of the spatial nonlinear elasticity theory for bodies with dislocations distributed with a special density. The importance of such problems is pointed out, in particular, in the paper Kröner (1960). The exact solutions of the nonlinear continual dislocation theory allow, in particular, to discover a new quantitative and qualitative effects of the deformation of bodies with dislocations.

2 Input Relations

Let $\boldsymbol{r} = x_s \boldsymbol{i}_s$ and $\boldsymbol{R} = X_k \boldsymbol{i}_k$ be the radius vectors of a point in the elastic medium in the reference and deformed configurations, respectively, x_s and X_k $(k, s = 1, 2, 3)$ the Cartesian coordinates of the reference and final states of the body, $\boldsymbol{i}_k$ the fixed coordinate orts. Later on, the operators of gradient, rotor, and divergence in the reference configuration will be used:

$$\mathrm{grad}\boldsymbol{\Psi} = \boldsymbol{r}^s \otimes \frac{\partial \boldsymbol{\Psi}}{\partial q^s}, \qquad \mathrm{rot}\boldsymbol{\Psi} = \boldsymbol{r}^s \times \frac{\partial \boldsymbol{\Psi}}{\partial q^s},$$
$$\mathrm{div}\boldsymbol{\Psi} = \boldsymbol{r}^s \cdot \frac{\partial \boldsymbol{\Psi}}{\partial q^s}, \qquad \boldsymbol{r}^s = \boldsymbol{i}_m \frac{\partial q^s}{\partial x_m},$$

where q^s are the curvilinear coordinates, $\boldsymbol{\Psi}$ is the arbitrary differentiable tensor field. We consider the problem of finding the position $\boldsymbol{R}(\boldsymbol{r})$ of a point of the deformed elastic body by the differentiable and single-valued field $\mathbf{C} = \mathrm{grad}\boldsymbol{R}$ given in the domain v. In the case of the multi-connected domain, the vector field $\boldsymbol{R}(\boldsymbol{r})$ in general cannot be uniquely defined, that implies the presence in the body of translational dislocations, each of which is characterized by the Burgers vector. Following Nye (1953), Eshelby (1956), Landau and Lifshitz (1975), let us come from the discrete dislocations to their continuous distribution and define the dislocation density as the second-rank tensor field $\boldsymbol{\alpha}$, whose flux across any surface yields the total Burgers vector of the dislocations crossing this surface. This definition leads to the following equation for the distortion tensor:

$$\mathrm{rot}\mathbf{C} = \boldsymbol{\alpha}. \tag{1}$$

If the tensor field of the dislocation density $\boldsymbol{\alpha}(\boldsymbol{r})$, which should satisfy the solenoidality condition $\text{div}\boldsymbol{\alpha} = 0$, is considered as specified, then (1), called the incompatibility equation, with the equilibrium equation (Lurie 1990; Ogden 1997)

$$\text{div}\mathbf{D} = 0, \tag{2}$$

the constitutive equation for the elastic material

$$\mathbf{D}(\mathbf{C}) = \mathrm{d}W(\mathbf{G})/\mathrm{d}\mathbf{C}, \qquad \mathbf{G} = \mathbf{C}\cdot\mathbf{C}^T, \tag{3}$$

and the boundary conditions, makes it possible to find the distortion tensor field, and, consequently, the stress field in the body. In (2), (3), $\mathbf{D}$ is the Piola's asymmetric stress tensor, $\mathbf{G}$ is the metric tensor, W is the specific energy of deformation. Mass forces are not taken into account. Note that there is no vector field $\boldsymbol{R}(\boldsymbol{r})$ for $\boldsymbol{\alpha} \neq 0$.

Incompatibility equation (1) of the first order can be transformed (Derezin and Zubov 1999, 2011) to nonlinear incompatibility equations of the second order with respect to components of the metric tensor $\mathbf{G}$. These equations compose the basis of the classical continuum theory of dislocations (Kondo 1952; Bilby et al. 1955; Kröner 1960) and, from the differential geometry point of view, represent the equality to zero of the Riemann–Cartan curvature tensor of the metrically connected space with torsion. The torsion tensor of the specified space is expressed through the dislocation density tensor. The incompatibility equations of the second order are quite complex, and their use for solving the boundary value problems of the nonlinear theory of dislocations is related to great difficulties. For this reason, we use further the incompatibility equations in the form of (1).

3 Formulation of the Equilibrium Problem for the Sphere with Continuously Distributed Dislocations

We consider an elastic body as a spherical shell with the outer radius r_0 and the inner radius r_1. Let us introduce the spherical coordinates r, ϕ, θ by the following formulas:

$$x_1 = r\cos\phi\cos\theta, \qquad x_2 = r\sin\phi\cos\theta, \qquad x_3 = r\sin\theta.$$

We denote the unit vectors tangent to the coordinate lines by $\boldsymbol{e}_r$, $\boldsymbol{e}_\phi$, $\boldsymbol{e}_\theta$. Suppose we have the dislocation density tensor as follows:

$$\boldsymbol{\alpha} = \beta(r)\left(\boldsymbol{e}_\phi \otimes \boldsymbol{e}_\theta - \boldsymbol{e}_\theta \otimes \boldsymbol{e}_\phi\right). \tag{4}$$

Spherically symmetric distribution of edge dislocations (4) satisfies the solenoidality condition for any function $\beta(r)$. We search the distortion tensor field in the form of

$$\mathbf{C} = C_1(r)\boldsymbol{e}_r \otimes \boldsymbol{e}_r + C_2(r)\left(\boldsymbol{e}_\phi \otimes \boldsymbol{e}_\phi + \boldsymbol{e}_\theta \otimes \boldsymbol{e}_\theta\right). \tag{5}$$

Substituting (4) and (5) into the incompatibility tensor Eq. (1), one can make sure it reduces to a single scalar differential equation

$$\frac{C_1 - C_2}{r} - \frac{\mathrm{d}C_2}{\mathrm{d}r} = \beta(r). \tag{6}$$

Considering nonlinear elastic material of the sphere be isotropic, we write the general constitutive equation for the isotropic elastic material (Lurie 1990; Ogden 1997):

$$\begin{aligned}
&\mathbf{D} = a_1\mathbf{C} + a_2\mathbf{G}\cdot\mathbf{C} + a_3\mathbf{C}^{-T},\\
&a_k = a_k\left(I_1, I_2, I_3\right), \qquad k = 1, 2, 3,\\
&I_1 = \mathrm{tr}\mathbf{G}, \qquad I_2 = \frac{1}{2}\left(\mathrm{tr}^2\mathbf{G} - \mathrm{tr}\mathbf{G}^2\right), \qquad I_3 = \det\mathbf{G}.
\end{aligned} \tag{7}$$

Here, I_1, I_2, I_3 are the deformation invariants, a_k are the certain functions of invariants. From (5), (7) it follows that the Piola's stress tensor has the representation similar to (5):

$$\mathbf{D} = D_1(r)\boldsymbol{e}_r \otimes \boldsymbol{e}_r + D_2(r)\left(\boldsymbol{e}_\phi \otimes \boldsymbol{e}_\phi + \boldsymbol{e}_\theta \otimes \boldsymbol{e}_\theta\right). \tag{8}$$

In view of (8), the vector equation of equilibrium (2) can be brought into a form of a single scalar equation

$$\frac{\mathrm{d}D_1}{\mathrm{d}r} + \frac{2(D_1 - D_2)}{r} = 0. \tag{9}$$

For the arbitrary isotropic material, the system of Eqs. (6), (9) can be easily reduced to a single nonlinear second-order ordinary differential equation for $C_2(r)$ by the constitutive equations (3). For some material models, it is possible to reduce the above system to the equation defining D_1 stress.

From this point on, we will use the semi-linear (harmonic) compressible material (John 1960; Lurie 1990; Ogden 1997), the state equation of which is given by

$$\mathbf{D} = \frac{2\mu}{1-2\nu}\left(\nu\mathrm{tr}\mathbf{U} - 1 - \nu\right)\mathbf{A} + 2\mu\mathbf{C}, \tag{10}$$

where μ, ν are the constants, $\mathbf{U}$ and $\mathbf{A}$ are the elements of polar decomposition of the distortion tensor: $\mathbf{C} = \mathbf{U}\cdot\mathbf{A}$, at what $\mathbf{U}$ is the symmetric positive defined stretch tensor, $\mathbf{A}$ is the properly orthogonal rotation tensor. In the area of the small strains there is no difference between the constitutive equation (10) and the Hooke's law, with μ and ν being the shear modulus and the Poisson's ratio.

4 Eigenstress Problem

Physically, deformation of the medium must satisfy the condition $\det\mathbf{C} > 0$, from which, due to (5), we obtain the inequality $C_1(r) > 0$. Distortion component $C_2(r)$ may be both positive and negative. On this basis, Piola's stress tensor (10) can be written componentwise:

$$D_1 = \frac{2\mu}{1-2\nu}\left(2\nu\,|C_2| + (1-\nu)\,C_1 - 1 - \nu\right), \tag{11}$$

$$D_2 = \frac{2\mu(\operatorname{sgn}C_2)}{1-2\nu}\left(2\nu\,|C_2| + \nu C_1 - 1 - \nu\right) + 2\mu C_2. \tag{12}$$

Eigenstress problem has two solutions. In the first solution $C_2(r) > 0$, and in the second one $C_2(r) < 0$. We consider cases of the sign of C_2 component separately.

Solving (11), (12) for C_1, C_2 in the case of $C_2 > 0$, we obtain:

$$C_1 = \frac{D_1 - 2\nu D_2}{2\mu(1+\nu)} + 1, \qquad C_2 = \frac{D_2(1-\nu) - \nu D_1}{2\mu(1+\nu)} + 1. \tag{13}$$

Substituting (13) into (6), and taking into account (9), we have the second-order differential equation for the D_1 stress:

$$r^2\frac{\mathrm{d}^2 D_1}{\mathrm{d}r^2} + 4r\frac{\mathrm{d}D_1}{\mathrm{d}r} = \frac{4\mu(\nu+1)}{\nu-1} r\beta(r). \tag{14}$$

Next we consider a general case of spherically symmetric distribution of edge dislocations, i.e., when $\beta(r)$ is an arbitrary function. Equation (14) can be written in the following form:

$$D_1'' + \frac{4}{r}D_1' = \gamma(r), \qquad \gamma(r) \equiv \frac{4\mu(\nu+1)}{\nu-1}\frac{\beta(r)}{r}. \tag{15}$$

Solution is represented by

$$D_1(r) = -\frac{1}{3r^3}\int_{r_1}^{r} \rho^4\gamma(\rho)d\rho + \frac{1}{3}\int_{r_1}^{r} \rho\gamma(\rho)d\rho + A_1 + \frac{A_2}{r^3}. \tag{16}$$

The boundary conditions for (25) follow from the requirement of non-loading surfaces of the sphere $r = r_1, r = r_0$

$$D_1(r_1) = 0, \qquad D_1(r_0) = 0 \tag{17}$$

With the help of (17) we find the constants of integration:

$$A_1 = -\frac{A_2}{r_1^3}, \qquad A_2 = \frac{r_0^3 r_1^3}{3\left(r_0^3 - r_1^3\right)} \left(\int_{r_1}^{r_0} \rho\gamma(\rho)d\rho - \frac{1}{r_0^3} \int_{r_1}^{r_0} \rho^4\gamma(\rho)d\rho \right).$$

Let us consider the case of dislocations concentrated on the surfaces of the sphere with the radius r_*. In this case, we define the function $\beta(r)$ as

$$\beta(r) = \beta_0\delta(r - r_*), \qquad r_1 \le r_* \le r_0. \tag{18}$$

where $\delta(r - r_*)$ is the Dirac delta function. Substituting (18) into (16), we obtain solution in its final form:

$$D_1 = \frac{4\mu\beta_0(\nu+1)}{3(\nu-1)} \frac{r_0^3\left(r_1^3 - r^3\right)}{\left(r_0^3 - r_1^3\right)r^3} \left(1 - \left(\frac{r_*}{r_0}\right)^3\right), \quad r_1 \le r < r_*; \tag{19}$$

$$D_1 = \frac{4\mu\beta_0(\nu+1)}{3(\nu-1)} \frac{\left(r_*^3 - r_1^3\right)\left(r^3 - r_0^3\right)}{\left(r_0^3 - r_1^3\right)r^3}, \quad r_* < r \le r_0. \tag{20}$$

Based on (19) and (20), (11)–(13), one can find D_2, C_1, C_2.

$$D_2 = \frac{2\mu\beta_0(\nu+1)}{3(\nu-1)} \frac{r_0^3\left(r_1^3 + 2r^3\right)}{\left(r_1^3 - r_0^3\right)r^3} \left(1 - \left(\frac{r_*}{r_0}\right)^3\right), \quad r_1 \le r < r_*; \tag{21}$$

$$D_2 = \frac{2\mu\beta_0(\nu+1)}{3(\nu-1)} \frac{\left(r_*^3 - r_1^3\right)\left(r_0^3 + 2r^3\right)}{\left(r_0^3 - r_1^3\right)r^3}, \quad r_* < r \le r_0;$$

$$C_1 = 1 + \frac{2\beta_0}{3(\nu-1)} \frac{r_0^3\left((1+\nu)r_1^3 - (1-2\nu)r^3\right)}{\left(r_0^3 - r_1^3\right)r^3} \left(1 - \left(\frac{r_*}{r_0}\right)^3\right), \quad r_1 \le r < r_*;$$

$$C_1 = 1 + \frac{2\beta_0}{3(\nu-1)} \frac{\left(r_1^3 - r_*^3\right)\left((1+\nu)r_0^3 - (1-2\nu)r^3\right)}{\left(r_0^3 - r_1^3\right)r^3}, \quad r_* < r \le r_0;$$

$$C_2 = 1 + \frac{\beta_0}{3(\nu-1)} \frac{r_0^3\left(-(1+\nu)r_1^3 - 2(1-2\nu)r^3\right)}{\left(r_0^3 - r_1^3\right)r^3} \left(1 - \left(\frac{r_*}{r_0}\right)^3\right), \quad r_1 \le r < r_*;$$

$$C_2 = 1 + \frac{\beta_0}{3(\nu-1)} \frac{\left(r_1^3 - r_*^3\right)\left(2(2\nu-1)r^3 - (1+\nu)r_0^3\right)}{\left(r_0^3 - r_1^3\right)r^3}, \quad r_* < r \le r_0.$$

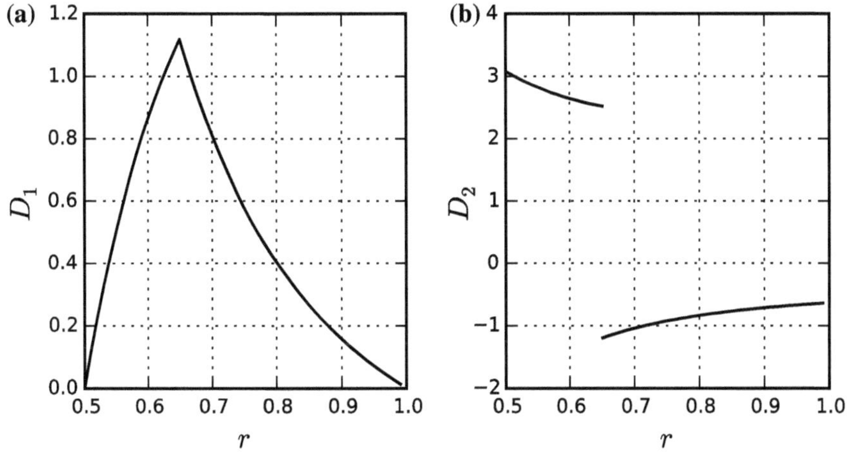

Fig. 1 Components of the Piola's stress tensor in the case of $\beta_0 = 1$

For plotting we will use $\mu = 1$, $\nu = 0.3$. Inspection of Fig. 1a reveals that the D_1 stress is continuous. It follows from Figs. 1b–2b that the sphere of radius r_* tends to be the discontinuity surface of D_2, C_1, C_2. Thus, dislocations concentrated on the surface, which is inside a body, produce the discontinuous fields of stresses and strains. The discontinuity surface is a surface, on which dislocations are concentrated. This means that with the help of dislocations distributed on a surface, we can model the discontinuous fields of strains in elastic bodies.

Note that the jump amount of the function $C_2(r)$ on the spherical surface $r = r_*$ is independent of material properties and expressed in terms of β_0. Indeed, we integrate the incompatibility equation (6) for $\beta(r) = \beta_0\delta(r - r_*)$ over the interval

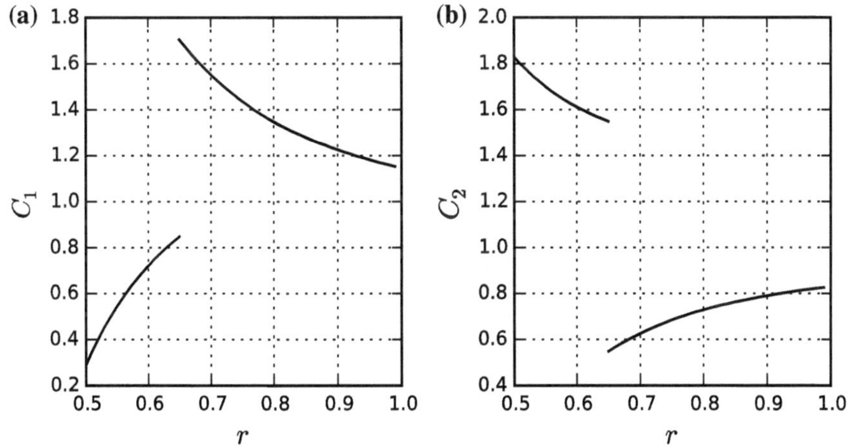

Fig. 2 Components of the distortion tensor in the case of $\beta_0 = 1$

$(r_* - \varepsilon, r_* + \varepsilon)$, where $\varepsilon > 0$. Taking into account the boundedness of the functions $C_1(r)$, $C_2(r)$, and letting ε tend to zero, we obtain

$$C_2^- - C_2^+ = \beta_0, \qquad C_2^- = \lim_{\varepsilon\to 0} C_2(r_* - \varepsilon), \quad C_2^+ = \lim_{\varepsilon\to 0} C_2(r_* + \varepsilon).$$

Moreover, in virtue of (19), (21), if dislocations are concentrated on the outer surface of the sphere ($r_* = r_0$) or on the inner surfaces of the sphere ($r_* = r_1$), then $D_1 \equiv 0$, $D_2 \equiv 0$. This means that dislocations distributed on the boundary of body do not produce a stressed state in it.

For the solid sphere, i.e., for $r_1 = 0$, we obtain:

$$D_1 = -\frac{4}{3}\frac{\nu+1}{\nu-1}\mu\beta_0\frac{r_0^3 - r_*^3}{r_0^3}, \quad 0 \le r < r_*; \tag{22}$$

$$D_1 = \frac{4}{3}\frac{\nu+1}{\nu-1}\mu\beta_0\frac{r_*^3(r^3 - r_0^3)}{r_0^3 r^3}, \quad r_* < r \le r_0;$$

$$D_2 = -\frac{4}{3}\frac{\nu+1}{\nu-1}\mu\beta_0\frac{r_0^3 - r_*^3}{r_0^3}, \quad 0 \le r < r_*; \tag{23}$$

$$D_2 = \frac{2}{3}\frac{\nu+1}{\nu-1}\mu\beta_0\frac{r_*^3(2r^3 + r_0^3)}{r_0^3 r^3}, \quad r_* < r \le r_0;$$

$$C_1 = \frac{4\nu\beta_0\left(r_0^3 - r_*^3\right) - 2\beta_0\left(r_0^3 - r_*^3\right) - 3(1-\nu)r_0^3}{3(\nu-1)r_0^3}, \quad 0 \le r < r_*;$$

$$C_1 = -\frac{2\beta_0\nu\left(r_0^3 + 2r^3\right)r_*^3 + 2\beta_0\left(r_0^3 - r^3\right)r_*^3 - 3(\nu-1)r_0^3 r^3}{3(\nu-1)r_0^3 r^3}, \quad r_* < r \le r_0;$$

$$C_2 = \frac{4\nu\beta_0\left(r_0^3 - r_*^3\right) - 2\beta_0\left(r_0^3 - r_*^3\right) - 3(1-\nu)r_0^3}{3(\nu-1)r_0^3}, \quad 0 \le r < r_*;$$

$$C_2 = -\frac{2\beta_0(2\nu-1)r^3 r_*^3 - \beta_0(1+\nu)r_0^3 r_*^3 - 3(\nu-1)r^3 r_0^3}{3(\nu-1)r_0^3 r^3}, \quad r_* < r \le r_0;$$

According to (22), (23), if $\beta_0 > 0, 0 < \nu < 1$, the sphere $0 \le r \le r_*$ is in the state of a full stretch, and a spherical stress tensor is constant.

In the case of unbounded medium ($r_1 \to 0$, $r_0 \to \infty$), we obtain:

$$D_1 = -\frac{4}{3}\frac{\nu+1}{\nu-1}\mu\beta_0, \quad 0 \le r < r_*,$$

$$D_1 = -\frac{4}{3}\frac{\nu+1}{\nu-1}\mu\beta_0\left(\frac{r_*}{r}\right)^3, \quad r_* < r \le r_0;$$

$$D_2 = -\frac{4}{3}\frac{\nu+1}{\nu-1}\mu\beta_0, \quad 0 \le r < r_*,$$

$$D_2 = \frac{2}{3}\frac{\nu+1}{\nu-1}\mu\beta_0\left(\frac{r_*}{r}\right)^3, \quad r_* < r \le r_0;$$

$$C_1 = 1 + \frac{2\beta_0(2\nu-1)}{3(\nu-1)}, \quad 0 \le r < r_*,$$
$$C_1 = 1 - \frac{2}{3}\frac{\nu+1}{\nu-1}\beta_0\left(\frac{r_*}{r}\right)^3, \quad r_* < r \le r_0;$$
$$C_2 = 1 + \frac{2\beta_0(2\nu-1)}{3(\nu-1)}, \quad 0 \le r < r_*,$$
$$C_2 = 1 + \frac{\beta_0}{3}\frac{\nu+1}{\nu-1}\left(\frac{r_*}{r}\right)^3, \quad r_* < r \le r_0.$$

Now, consider the second solution of the eigenstress problem in which $C_2 < 0$.

Solving (11), (12) for C_1, C_2 in the case of $C_2 < 0$, we obtain:

$$C_1 = \frac{D_1 + 2\nu D_2}{2\mu(1+\nu)} + 1, \qquad C_2 = \frac{D_2(1-\nu) + \nu D_1}{2\mu(1+\nu)} - 1. \tag{24}$$

Substituting (24) into (6), and taking into account (9), we have the second-order differential equation for the D_1 stress:

$$r^2\frac{d^2 D_1}{dr^2} + 4r\frac{dD_1}{dr} + \frac{4\nu}{\nu-1}D_1 = 4\mu\frac{\nu+1}{\nu-1}\left(r\beta\left(r\right) - 2\right). \tag{25}$$

The boundary conditions for (25) are (17).

For $\beta(r) \equiv 0$, the unstressed state $D_1 \equiv 0$ is obviously not a solution of the boundary value problem (25), (17). This means that in an elastic spherical shell, stresses exist when no dislocations and external forces are present. It is well known that this phenomenon may be possible for eversion of a spherical shell. We shall show that for $\beta(r) \equiv 0$, the boundary value problem (25), (17) describes a state of the sphere made of semi-linear material and turned inside out indeed. In addition to the Lagrangian spherical coordinates r, ϕ, θ, let us introduce Eulerian spherical coordinates R, Φ, Θ

$$X_1 = R\cos\Phi\cos\Theta, \qquad X_2 = R\sin\Phi\cos\Theta, \qquad X_3 = R\sin\Theta, \tag{26}$$

where X_1, X_2, X_3 are the Cartesian coordinates of body's points in the deformed state. Deformation of turning inside out is given by Lurie (1990), Ogden (1997), Zubov and Moiseyenko (1983)

$$R = R(r), \qquad \Phi = \phi, \qquad \Theta = -\theta. \tag{27}$$

Instead of generally accepted expression (27), we will describe the sphere's turning inside out by the following coordinate transformation

$$R = R(r), \qquad \Phi = \phi + \pi, \qquad \Theta = -\theta. \tag{28}$$

Deformation (28) differs from the deformation (27) by the fact of the additional rotation of the sphere as a perfectly rigid body by 180° about the X_3-axis. It is clear that this rotation does not affect the stress state of the sphere. Representation (28) has a substantial advantage over (27), in that it leads the distortion tensor $\mathbf{C}$ to particularly simple representation in the case of the sphere's turning inside out. Based on $\mathbf{C} = \mathrm{grad}\boldsymbol{R}$, (26), (27), we have

$$\mathbf{C} = \frac{\partial X_k}{\partial r}\boldsymbol{e}_r \otimes \boldsymbol{i}_k + \frac{1}{r\cos\theta}\frac{\partial X_k}{\partial \phi}\boldsymbol{e}_\phi \otimes \boldsymbol{i}_k + \frac{1}{r}\frac{\partial X_k}{\partial \theta}\boldsymbol{e}_\theta \otimes \boldsymbol{i}_k$$
$$= -\frac{\mathrm{d}R}{\mathrm{d}r}\boldsymbol{e}_r \otimes \boldsymbol{e}_r - \frac{R}{r}\boldsymbol{e}_\phi \otimes \boldsymbol{e}_\phi - \frac{R}{r}\boldsymbol{e}_\theta \otimes \boldsymbol{e}_\theta. \quad (29)$$

Comparing (29) with the representation (5), one can obtain

$$C_1 = -\frac{\mathrm{d}R}{\mathrm{d}r}, \qquad C_2 = -\frac{R}{r}. \quad (30)$$

Elimination of the function $R(r)$ from the relation (30) leads to the compatibility equation

$$\frac{\mathrm{d}C_2}{\mathrm{d}r} - \frac{C_1 - C_2}{r} = 0. \quad (31)$$

which coincides with (6) for $\beta(r) \equiv 0$. The Eqs. (9), (8) and the constitutive equations (24) give the following equation defining $D_1(r)$:

$$r^2\frac{d^2D_1}{dr^2} + 4r\frac{dD_1}{dr} + \frac{4\nu}{\nu - 1}D_1 = -8\mu\frac{\nu + 1}{\nu - 1}. \quad (32)$$

The above equation coincides with (25) when $\beta(r) \equiv 0$. Thus, the case of $\beta(r) \not\equiv 0$, $C_2 < 0$ can be interpreted as a spherically symmetric distribution of edge disloca-

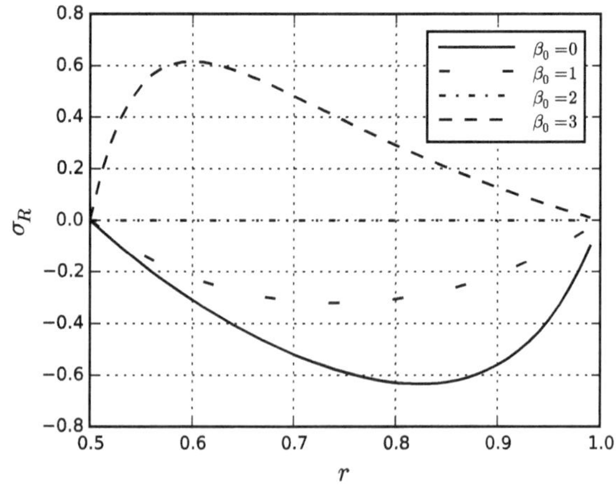

Fig. 3 The σ_R component of the Cauchy stress tensor

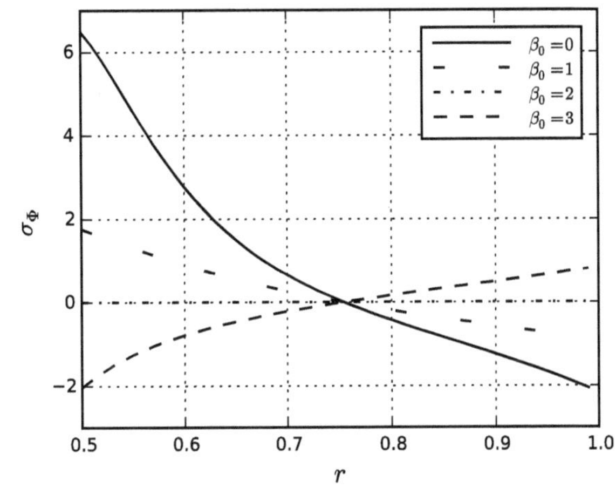

Fig. 4 The σ_Φ component of the Cauchy stress tensor

tions in an elastic sphere turned inside out. Note that the problem of equilibrium and stability of an elastic sphere made of semi-linear material and turned inside out, without taking into account the dislocations, was solved earlier (Zubov and Moiseyenko 1983).

The solution of the boundary value problem (17), (25), in the case of $\beta(r) = \beta_0 r^{-1}$, $\beta_0 = const$, $\mu = 1$, $\nu = 0.3$, is shown in Figs. 3 and 4, where the distribution of the Cauchy stress components $\sigma_R = C_2^{-2} D_1$, $\sigma_\Phi = (C_1 C_2)^{-1} D_2$ is represented.

5 Lame's Problem

We consider the Lame's problem for a sphere for the case of $C_2 > 0$. The sphere is supposed to be loaded by a constant pressure on the inside and from the outside. Consequently, the boundary conditions can be written as follows:

$$D_1(r_0) = -p_0 C_2^2(r_0), \tag{33}$$

$$D_1(r_1) = -p_1 C_2^2(r_1). \tag{34}$$

Then, constants of integration in representation of D_1 (16), that has been derived in previous section, are

$$A_1 = \frac{1}{1 - r_0^{-3} r_1^3} \left(\frac{1}{3 r_0^3} \int_{r_1}^{r_0} \rho^4 \gamma(\rho) d\rho - \frac{1}{3} \int_{r_1}^{r_0} \rho \gamma(\rho) d\rho + \frac{r_1^3}{r_0^3} p_1 C_2^2(r_1) - p_0 C_2^2(r_0) \right), \tag{35}$$

$$A_2 = -A_1 r_1^3 - p_1 C_2^2(r_1) r_1^3 \tag{36}$$

Let dislocation density have the form of

$$\beta(r) = \beta_0 r^{\kappa}, \qquad \beta_0 = const, \qquad \kappa = const.$$

Then, having in mind (11)–(13), (16), we obtain

$$D_1 = \frac{4\mu\beta_0(\nu+1)}{(\kappa+1)(\kappa+4)(\nu-1)} r^{\kappa+1} + A_1 + A_2 r^{-3}, \tag{37}$$

$$D_2 = \frac{2\mu\beta_0(\nu+1)(\kappa+3)}{(\kappa+1)(\kappa+4)(\nu-1)} r^{\kappa+1} + A_1 - \frac{A_2}{2} r^{-3}, \tag{38}$$

$$C_1 = 1 + \frac{2\beta_0\left(1-2\nu-\nu(\kappa+1)\right)}{(\kappa+1)(\kappa+4)(\nu-1)} r^{\kappa+1} + \frac{A_1(1-2\nu)+A_2(1+\nu)r^{-3}}{2\mu(1+\nu)}, \tag{39}$$

$$C_2 = 1 + \frac{(1-2\nu)A_1}{2\mu(1+\nu)} - \frac{A_2}{4\mu} r^{-3} - \frac{\beta_0}{(\kappa+4)(\kappa+1)}\left(\kappa+3+\frac{2\nu}{\nu-1}\right) r^{\kappa+1}. \tag{40}$$

Cases of loading on the inside and from the outside are separately considered.

If $p_0 = 0$, $p_1 \neq 0$, then we have the nonlinear boundary condition (34), which corresponds to blowing up the sphere.

If we solve a problem for all p_1 values selected from a certain range, we will be able to construct the relationship $K_1(p_1)$, where $K_1 = C_2(r_1)$. This relationship characterizes a nonlinear resistance of the sphere to application of internal pressure. The construction of the relationship $K_1(p_1)$ can be simplified by finding the inverse function $p_1(K_1)$. Finding of the relationship $p_1(K_1)$ requires the solving of the linear boundary value problem with the next boundary conditions

$$D_1(r_0) = 0, \tag{41}$$

$$C_2(r_1) = K_1, \tag{42}$$

where K_1 is a given value.

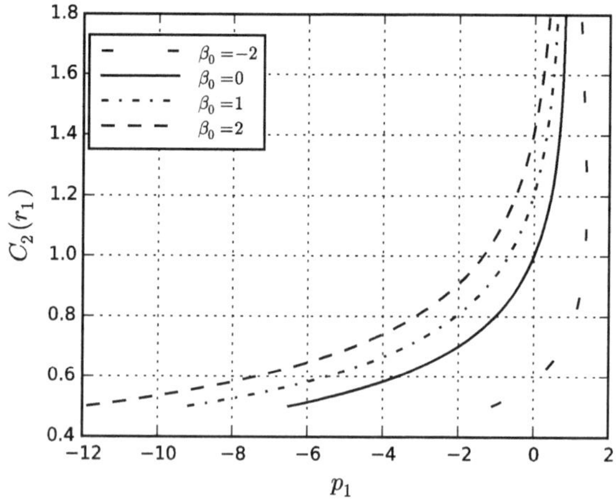

Fig. 5 Dependence of $C_2(r_1)$ on pressure p_1

From Fig. 5, one may observe that positive dislocation density decreases resistance of the sphere to blowing up, and the negative one increases it.

We have no difficulty to find components of the Cauchy stress tensor, which are illustrated in Figs. 6 and 7. It is clear from Fig. 7 that there is a spherical surface, circular stresses on which is independent of the dislocation density. The presence of dislocations intensifies the unevenness of distribution of the σ_Φ stress along the vessel thickness. Figures 6 and 7 show that the negative dislocations increase stresses σ_R, σ_Φ for its absolute value.

If $p_1 = 0$, $p_0 \neq 0$, then we have the nonlinear boundary condition (33), which corresponds to hydrostatic pressure on the sphere. As before, we first consider the linear boundary condition for $D_1(r_1)$. The boundary conditions take the form:

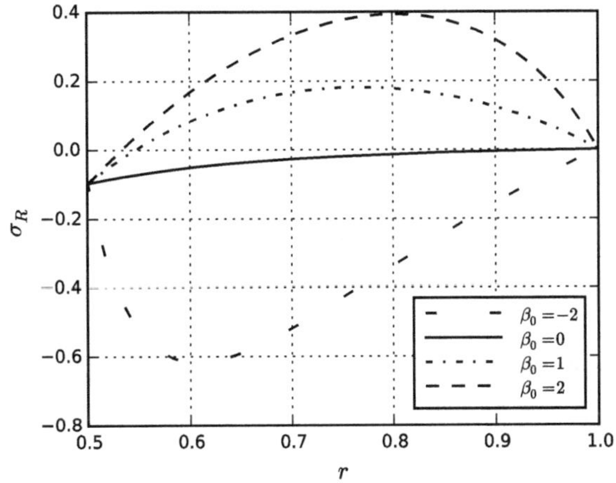

Fig. 6 The σ_R component of the Cauchy stress tensor; $p_1 = 0.1, \kappa = 1$

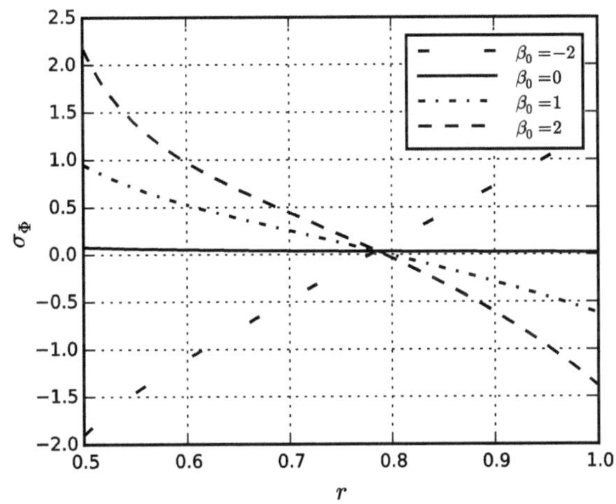

Fig. 7 The σ_Φ component of the Cauchy stress tensor; $p_1 = 0.1, \kappa = 1$

Fig. 8 Dependence of $C_2(r_0)$ on pressure p_0

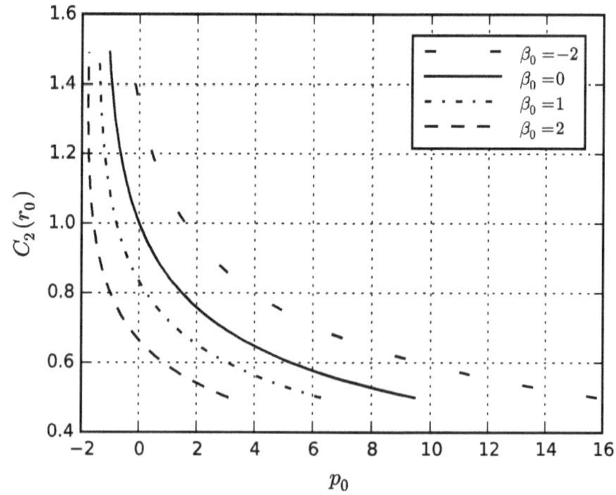

$$D_1(r_1) = 0, \tag{43}$$

$$C_2(r_0) = K_0, \tag{44}$$

where K_0 is a given value.

From Fig. 8, one may observe that positive dislocation density decreases resistance of the sphere to hydrostatic pressure, and the negative one increases it.

It is clear from Fig. 10 that, as in the case of blowing up, there is a spherical surface, circular stresses on which is independent of the dislocation density. The presence of dislocations intensifies the unevenness of distribution of the σ_Φ stress along the vessel thickness. Figures 9, 10 show that the negative dislocations increase stresses σ_R, σ_Φ for its absolute value.

Fig. 9 The σ_R component of the Cauchy stress tensor; $p_0 = 0.1$, $\kappa = 1$

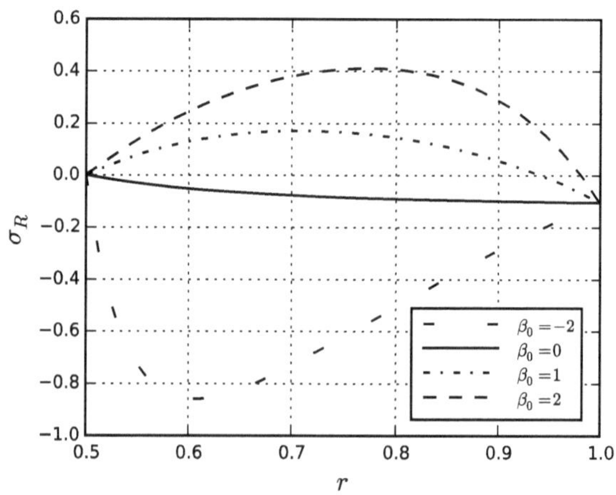

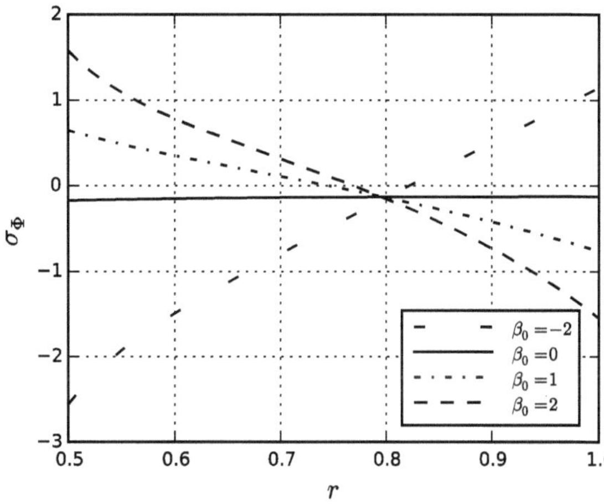

Fig. 10 The σ_Φ component of the Cauchy stress tensor; $p_0 = 0.1, \kappa = 1$

6 Conclusion

In this paper we have investigated a spherically symmetric problem of nonlinear continual theory of dislocations. For the harmonic compressible material we have found exact solutions, which describe eigenstresses in an elastic hollow sphere caused by spherically symmetric distribution of edge dislocations. We ascertain that the eigenstress problem has two spherically symmetric solutions for the same dislocation density. We have shown that dislocations concentrated on a spherical surface within a body determine the discontinuous fields of strains and stresses. At the same time, dislocations distributed over the boundary of the elastic body do not produce eigenstresses in a body. Also we have examined the influence of distributed dislocations on large deformations of an elastic sphere loaded with internal or external pressure.

Acknowledgments The authors acknowledge support by the Russian Foundation of Basic Research (15-01-01492).

References

Altenbach, H., Eremeyev, V.A. (eds.): Generalized Continua: From the Theory to Engineering Applications. CISM Courses and Lectures. Springer, Wien (2013)

Altenbach, H., Eremeyev, V.: Strain rate tensors and constitutive equations of inelastic micropolar materials. Int. J. Plast. **63**, 3–17 (2014)

Altenbach, H., Maugin, G.A., Erofeev, V. (eds.): Mechanics of Generalized Continua, Advanced Structured Materials, vol. 7. Springer, Berlin (2011)

Besson, J., Cailletaud, G., Chaboche, J.L., Forest, S., Blétry, M.: Non-linear Mechanics of Materials. Springer, Dordrecht (2010)

Bilby, B.A., Bullough, R., Smith, E.: Continuous distributions of dislocations: a new application of the methods of non-Riemannian geometry. Proc. Roy. Soc. London **A231**, 263–273 (1955)

Clayton, J.D.: Nonlinear Mechanics of Crystals. Springer, Dordrecht (2011)
Clayton, J.D., McDowell, D.L., Bammann, D.J.: Modeling dislocations and disclinations with finite micropolar elastoplasticity. Int. J. Plast. **22**(2), 210–256 (2006)
Derezin, S.V., Zubov, L.M.: Equations of a nonlinear elastic medium with continuously distributed dislocations and disclinations. Dokl. Phys. **44**(6), 391–394 (1999)
Derezin, S.V., Zubov, L.M.: Disclinations in nonlinear elasticity. Ztsch Angew Math und Mech **91**, 433–442 (2011)
Eremeyev, V.A., Lebedev, L.P., Altenbach, H.: Foundations of Micropolar Mechanics. SpringerBriefs in Applied Sciences and Technologies. Springer, Heidelberg (2013)
Eshelby, J.D.: The continuum theory of lattice defects. In: Seitz, F., Turnbull, D. (eds.) Solid State Physics, vol. 3, pp 79–144. Academic Press, New York (1956)
Forest, S.: Some links between Cosserat, strain gradient crystal plasticity and the statistical theory of dislocations. Phil. Mag. **88**(30–32), 3549–3563 (2008)
Forest, S., Sedláček, R.: Plastic slip distribution in two-phase laminate microstructures: dislocation-based versus generalized continuum approaches. Phil. Mag. **83**(2), 245–276 (2003)
Gutkin, M.Y., Ovid'ko, I.A.: Plastic Deformation in Nanocrystalline Materials. Springer, Berlin (2004)
John, F.: Plane strain problems for a perfectly elastic material of harmonic type. Commun. Pure Appl. Math. **XIII**, 239–296 (1960)
Kondo, K.: On the geometrical and physical foundations in the theory of yielding. In: Proceeding of 2nd Japan National Congress of Applied Mechanics, Tokyo, pp 41–47 (1952)
Kröner, E.: Allgemeine Kontinuumstheorie der Versetzungen und Eigenspannungen. Arch. Ration. Mech. Anal. **4**, 273–334 (1960)
Landau, L.D., Lifshitz, E.M.: Theory of Elasticity, Theoretical Physics, vol. 7. Pergamon, Oxford (1975)
Lurie, A.I.: Nonlinear Theory of Elasticity. North-Holland, Amsterdam (1990)
Maugin, G.A.: Defects, dislocations and the general theory of material inhomogeneity. In: Sansour C, Skatulla S (eds) Generalized continua and dislocation theory. CISM courses and lectures, vol. 537, pp. 1–83. Springer, Vienna (2012)
Nye, J.F.: Some geometrical relations in dislocated crystals. Acta Metall. **1**(2), 153–162 (1953)
Ogden, R.W.: Non-linear Elastic Deformations. Dover, New York (1997)
Yavary, A., Goriely, A.: Riemann-Cartan geometry of nonlinear dislocation mechanics. Arch. Ration. Mech. Anal. **205**, 59–118 (2012)
Zelenina, A.A., Zubov, L.M.: Nonlinear effects during the tension, bend, and torsion of elastic bodies with distributed dislocations. Dokl. Phys. **58**(8), 354–357 (2013)
Zubov, L.M.: Nonlinear Theory of Dislocations and Disclinations in Elastic Bodies. Springer, Berlin (1997)
Zubov, L.M., Moiseyenko, S.I.: Stability of equilibrium of an elastic sphere turned inside out (in russian). Izv Akad Nauk SSSR Mekh Tverd Tela **5**, 148–155 (1983)

Part II
Visco-Elastic Media and Coupled Phenomena

A Closed-Form Solution for a Linear Viscoelastic Self-gravitating Sphere

Wolfgang H. Müller and Elena N. Vilchevskaya

Abstract Following up on the classical solutions by Love for a linear-elastic self-gravitating sphere, this paper presents the corresponding extension to a linear viscoelastic body of the Kelvin–Voigt type. The solution is expressed in closed form by making use of Laplace transforms. Applications to the genesis of terrestrial planets are sought and the evolution of the Love radius and possible extensions to large deformations are discussed. As a new result, it turns out that in the early days of planet formation there is no Love radius and that it takes time for the Love radius to develop.

1 What This Paper Is About and What It Is Not

Viscoelasticity is of current interest to geology. Typically geologists investigate the *temporal* evolution of deformation within the Earth's outer crust caused by earthquakes or other gravitational load shifts such as melting ice, using viscoelastic material models, e.g., Campbell (1974), Ragazzo and Ruiz (2015) and Tanaka et al. (2009).

The present paper is not going in this direction at all. Rather it is a idealistic continuum approach toward an understanding of the genesis of terrestrial planets and the subsequent state of deformation in a large self-gravitating object. However, there is also a certain esthetic aspect in the solution we are about to present, and in order to quote Keats we may say that "beauty is truth, truth beauty." More specifi-

W.H. Müller (✉)
Chair of Continuum Mechanics and Materials Theory, Institute of Mechanics,
Technical University of Berlin, Einsteinufer 5, 10587 Berlin, Germany
e-mail: wolfgang.h.mueller@tu-berlin.de

E.N. Vilchevskaya
Institute for Problems in Mechanical Engineering of the Russian Academy of Sciences,
Bol'shoy Pr. 61, V.O., 199178 St.-Petersburg, Russia

E.N. Vilchevskaya
Peter the Great Saint-Petersburg Polytechnic University,
Politekhnicheskaja 29, 195251 St.-Petersburg, Russia
e-mail: vilchevska@gmail.com

K. Naumenko and M. Aßmus (eds.), *Advanced Methods of Continuum Mechanics for Materials and Structures*, Advanced Structured Materials 60,
DOI 10.1007/978-981-10-0959-4_5

cally, our result is a follow-up on the classical solutions found by the great A.E.H. Love for a self-gravitating *linear-elastic* sphere, see Love (1892, 1906, 1927). We shall extend his beautiful formulae to a linear-elastic model of the Kelvin–Voigt type. In other words, we will explore the *temporal development of the static linear-elasticity solution* of a self-gravitating terrestrial planet. In particular, we shall look at the temporal evolution of the Love radius, i.e., the position of the transition zone between compression and tension within a self-gravitating "solid" sphere. This may even be of practical use, since it is related to damage during the early stages of a developing terrestrial planet. However, we shall not endeavor to investigate this in full quantitative detail, at least not here. Surprisingly, our results will be of closed form, thanks to the efforts of one of the authors in a completely different field of research, cf. Frelova (2016). This shows the power of continuum theory: Everything is connected, a maxim we chose to start our salute to our esteemed colleague Holm Altenbach!

2 Literature Review and Putting the Problem into Perspective

Today it is a commonly accepted opinion that terrestrial planets, such as Mercury, Venus, Earth, and Mars, but also other huge solid celestial objects, specifically the Moon,[1] are the result of a coagulation process of "rocky" matter, a.k.a. "planetesimals," to form so-called "protoplanets" during the early stages of the developing solar system, cf., Wetherill (1990). In order to quote Lissauer (1993), pg. 134: "... in this picture, planet formation is fundamentally different from star formation in that planetary growth begins with the accumulation of solid bodies, with the accretion of substantial amounts of gas occurring after a planet becomes sufficiently massive ..." and, pg. 136, "... These planetesimals continue to agglomerate via pairwise mergers. ... Growth via binary collisions proceeds until the protoplanets become dynamically isolated from each other."

Hearing all this, we might conclude that the mathematical modeling of the genesis of a planet is exclusively numerical and within the field of discrete mechanics or (better) discrete systems, since there will be thermodynamics aspects involved, see, e.g., Kenyon (2006). However, it is always wise to look at a problem from different angles and, consequently, we promote the continuum perspective in what follows.

Let us consider the following scenario: A spherical, initially homogeneous, unstressed sphere (the planet in *statu nascendi*) undergoes self-gravity. We must ask as to whether static equilibrium is possible and how it is reached? Two rather idealized scenarios come to mind.

First, imagine that gravity is "suddenly switched on." Then, we will essentially face a situation similar to that of a moving masspoint connected to a linear-elastic

---

[1] It is said that also the gas giants initially need a rocky core of sufficient size which is then able to attract gas, if available in the region of its formation.

Hookean spring: Due to the inertial terms in the equation of motion and due to the potential of a linear oscillator pertinent to a radially symmetric, self-gravitating sphere, this sphere will begin to shrink below the radius determined by static equilibrium of forces. While doing so, stress-related forces will build up so that the sphere will finally start to rebounce. Provided there is no dissipation it will reach its initial radius again. This will happen over and over if we assume the material of the sphere to be perfectly elastic without internal friction and without heat conduction, so that isothermal conditions prevail. In other words, without dissipation there will be a constant exchange between the elastic energy, the gravitational potential, and the kinetic energies: The motion of the self-gravitating matter would never come to a standstill. Of course, in the real world there are dissipative processes acting. The shrinking will be accompanied by dissipation in terms of viscoelastic or viscoplastic deformation, and there will be heat conduction. All of this will, in the end, bring motion to a standstill, and the sphere will arrange itself in thermomechanical equilibrium, i.e., there will be equilibrium of gravitational and inner, stress-related forces in a state of homogeneous temperature. It should be mentioned that the final equilibrium state of a self-gravitating sphere has been modeled in closed mathematical form at the end of the nineteenth century by Love (1892), who used linear elasticity at small deformations for this purpose. The interested reader will find detailed information about Love's solution and interpretation in the Appendix.

In conclusion, we shall not attempt to model the dynamic transition toward that equilibrium for various reasons. First of all, its treatment would be fully numerical based on large deformations expressed in terms of velocities. This makes it difficult if not impossible to compare it to Love's analysis of equilibrium, which was based on small strains. Moreover, choosing an adequate numerical technique would be required. Surely there will be more than one, all of them with certain pros and cons. Finally, the question which initial conditions are appropriate is difficult to answer, since gravity is not simply "switched on" but always present. Hence in terms of capturing reality our dynamic continuum model could not seriously compete with the discrete mechanics approach of planetesimal masses bouncing into each other, sticking together, and finally forming a primordial planet that relaxes stress- and displacement-wise under the influence of their mutual gravitational attraction. In short, the fully dynamic continuum model requires too much effort for little gain.

For all these reasons, we shall eat humble pie and turn alternatively to a quasistatic treatment instead. This way inertial forces in the equations of motion can be neglected and the self-gravitating sphere will quasistatically and isothermally move into its final state of deformation. Such a situation is frequently conjured up in so-called $p\,\mathrm{d}V$-thermodynamics, for example, if we allow the pressure on a piston to change very slowly so that the gas which is trapped in the corresponding container has time to accommodate pressure- and temperature-wise. However, in our approach the time parameter will enter through a *viscoelastic* model used to connect stresses, strains, and their corresponding *rates*. More specifically, in order to be able to study the temporal development of the solution for the displacements, strains, and stresses toward Love's closed-form solutions we will make use of a linear viscoelastic model of the Kelvin–Voigt type, i.e., small deformation theory will reappear.

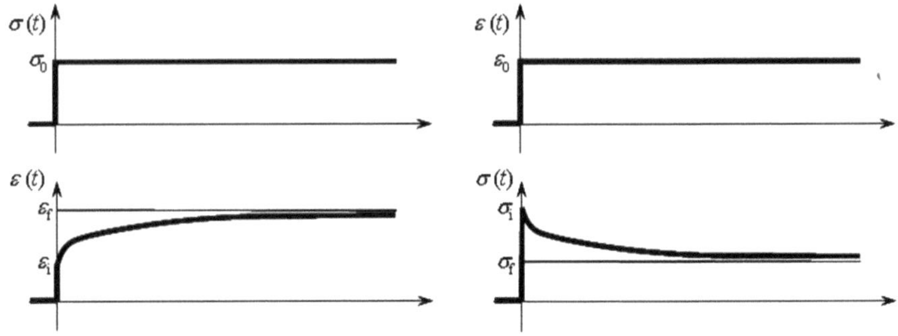

Fig. 1 Stress- versus displacement-controlled viscoelastic experiments (see text)

In this context recall the two fundamental types of quasistatic experiments always mentioned in combination with quasistatic, linear viscoelasticity (see Fig. 1)[2]: In the first one a linear-viscoelastic strip is suddenly subjected to a constant "dead load," i.e., a constant uniaxial tensile stress (the "cause"), σ_0, is prescribed. Under such circumstances we also speak of load-control. The "effect" consists of an elastic strain, ϵ_{i}, instantaneously built up. After that the strip gradually creeps quasistatically toward its final total strain, ϵ_{f}. The counterpart to this experiment consists of prescribing a strain of a fixed amount (the "cause"), ϵ_0, and to observe the stress response (the "effect"). This is what we call a displacement-controlled test. It turns out that the stress response immediately overshoots to a high level, σ_{i}, and is then reduced by creeping quasistatically toward a final lower value, σ_{f}. This time we speak of stress-relaxation.

In the following section we shall state and solve the linear-viscoelastic problem for a self-gravitating sphere mathematically and study the behavior of the corresponding solution which, surprisingly, will also be of closed-form. Moreover, we shall also investigate as to whether this fits into the traditional pattern of stress or strain controlled experiments.

3 A Viscoelastic Model of Self-gravitation

3.1 Viscoelasticity of the Kelvin–Voigt Type

Recall the 1D representation of the so-called Kelvin–Voigt model: A Hookean spring and a dashpot are arranged in parallel: Fig. 2. If we apply a displacement, δ, at the outer points of this rheological model it will be transferred equally to the spring and to the dashpot, $\delta = \delta_1 = \delta_2$, whereas the resulting force is the sum of the forces due

[2]Some readers may want to consult Lakes (2009), pg. 4 or Müller and Müller (2009), pg. 370 for more information.

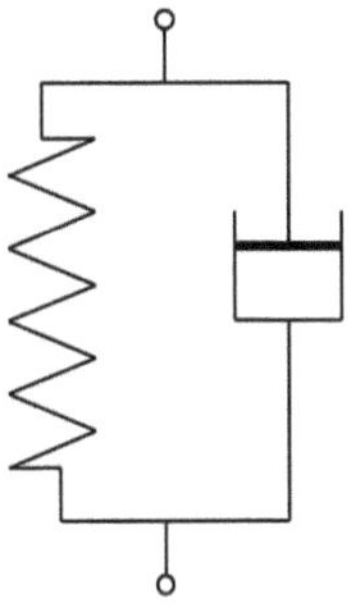

Fig. 2 The Kelvin–Voigt model of 1D linear viscoelasticity

to both elements, $F = F_1 + F_2$. In strength-of-materials-terminology we may say that the strains and, hence, the strain rates are equal, $\epsilon = \epsilon_1 = \epsilon_2 \Rightarrow \dot{\epsilon} = \dot{\epsilon}_1 = \dot{\epsilon}_2$, whereas the stresses are additive, $\sigma = \sigma_1 + \sigma_2$, and where the dot refers to a time derivative. The spring is now modeled by Hooke's law, $\sigma_1 = E\epsilon_1$, and the dashpot by a Newton–Navier–Stokes relationship, $\sigma_2 = \eta\,\dot{\epsilon}_2$. If we combine these equations, we arrive at:

$$\sigma = E\left(\frac{\eta}{E}\dot{\epsilon} + \epsilon\right). \tag{1}$$

We can then introduce a strain-based relaxation time, $\tau_\epsilon = {}^{\eta}/_{E}$, which will come in handy once we turn to dimensionless equations.

One-dimensional rheological models allow us to find three-dimensional analogues if we accept the daring idea of replacing the 1D stress and strain simply by stress and strain *tensors*. We proceed to illustrate this idea for the case of the Kelvin–Voigt model, Eq. (1), and write:

$$\boldsymbol{\sigma} = \boldsymbol{\sigma}_1 + \boldsymbol{\sigma}_2\,, \quad \boldsymbol{\epsilon} = \boldsymbol{\epsilon}_1 = \boldsymbol{\epsilon}_2. \tag{2}$$

Now we assume that the planet can be modeled as a linear, isotropic medium. We therefore consider the following customary constitutive equations for linear elasticity and viscosity during further analysis (the acronyms "dil" and "dev" refer to dilatoric and deviatoric parts of the strain (rate) tensors, respectively):

$$\boldsymbol{\sigma}_1 = 3k\,\boldsymbol{\epsilon}_1^{\text{dil}} + 2\mu\,\boldsymbol{\epsilon}_1^{\text{dev}}\,, \quad \boldsymbol{\sigma}_2 = 3\eta'\,\dot{\boldsymbol{\epsilon}}_2{}^{\text{dil}} + 2\eta\,\dot{\boldsymbol{\epsilon}}_2{}^{\text{dev}}\,, \tag{3}$$

where k and μ refer to the bulk modulus and the shear modulus, respectively. Moreover, η' and η are known as coefficients of bulk and shear viscosity.

If we now combine Eqs. (2) and (3) suitably we finally arrive at the following relation:

$$\boldsymbol{\sigma} = 3k\left(\frac{\eta'}{k}\dot{\boldsymbol{\epsilon}}^{\text{dil}} + \boldsymbol{\epsilon}^{\text{dil}}\right) + 2\mu\left(\frac{\eta}{\mu}\,\dot{\boldsymbol{\epsilon}}^{\text{dev}} + \boldsymbol{\epsilon}^{\text{dev}}\right). \tag{4}$$

Hence, in principle, we must distinguish between two different relaxation times:

$$\tau_{\mathrm{v},\epsilon} = \frac{\eta'}{k}, \quad \tau_{\mathrm{s},\epsilon} = \frac{\eta}{\mu}, \tag{5}$$

where the indices v and s are supposed to remind us of the dilatoric (**v**olumetric) and deviatoric (**s**hear) parts, and the index ϵ of the strain-related relaxation process.

However, it is known that the bulk viscosity is a rather elusive parameter and very difficult to measure, see Gad-el Hak and Bandyopadhyay (1995). Therefore, we will neglect it in what follows and obtain from the previous equations because of $\boldsymbol{\epsilon}^{\mathrm{dil}} := \frac{1}{3}\,\mathrm{Tr}\,\boldsymbol{\epsilon}\mathbf{I}$ and $\boldsymbol{\epsilon}^{\mathrm{dev}} := \boldsymbol{\epsilon} - \boldsymbol{\epsilon}^{\mathrm{dil}}$:

$$\boldsymbol{\sigma} = 3k\,\boldsymbol{\epsilon}^{\mathrm{dil}} + 2\mu\,\boldsymbol{\epsilon}^{\mathrm{dev}} + 2\eta\,\dot{\boldsymbol{\epsilon}}^{\mathrm{dev}} \equiv \lambda\,\mathrm{Tr}\,\boldsymbol{\epsilon}\,\mathrm{I} + 2\mu\,\boldsymbol{\epsilon} + 2\eta\left(\dot{\boldsymbol{\epsilon}} - \frac{1}{3}\,\mathrm{Tr}\,\dot{\boldsymbol{\epsilon}}\,\mathbf{I}\right), \tag{6}$$

I being the unit tensor.

Our main objective is to determine the displacement, $\boldsymbol{u}$, in spherical coordinates (for obvious reasons). We assume perfect spherical symmetry, hence $\boldsymbol{u} = u_r(r)\boldsymbol{e}_r$, u_r being its radial component and $\boldsymbol{e}_r$ being the radial unit vector. All necessary equations will be written in spherical coordinates. Specifically, we recall Eqs. (37) and (38) from the Appendix, which we complement by:

$$\dot{\epsilon}_{rr} = \dot{u}_r', \qquad \dot{\epsilon}_{\vartheta\vartheta} \equiv \dot{\epsilon}_{\varphi\varphi} = \frac{\dot{u}_r}{r}, \qquad \dot{\epsilon}_{r\vartheta} = \dot{\epsilon}_{r\varphi} = \dot{\epsilon}_{\vartheta\varphi} \equiv 0\,. \tag{7}$$

Then we obtain analogously to Eq. (39):

$$\begin{aligned}
\sigma_{rr} &= (\lambda + 2\mu)u_r' + 2\lambda\frac{u_r}{r} + \frac{4}{3}\eta\left(\dot{u}_r' - \frac{\dot{u}_r}{r}\right),\\
\sigma_{\vartheta\vartheta} \equiv \sigma_{\varphi\varphi} &= \lambda u_r' + 2(\lambda + \mu)\frac{u_r}{r} + \frac{2}{3}\eta\left(\frac{\dot{u}_r}{r} - \dot{u}_r'\right),\\
\sigma_{r\vartheta} &= \sigma_{r\varphi} = \sigma_{\vartheta\varphi} \equiv 0.
\end{aligned} \tag{8}$$

The equilibrium conditions (34) hold and we arrive similarly to Eq. (40) at the following Partial Differential Equation (PDE) for $u_r(r,t)$:

$$u_r'' + 2\frac{u_r'}{r} - 2\frac{u_r}{r^2} + \frac{4}{3}\frac{\eta}{\lambda + 2\mu}\left(\dot{u}_r'' + 2\frac{\dot{u}_r'}{r} - 2\frac{\dot{u}_r}{r^2}\right) = \frac{4\pi\rho_0^2 G}{3(\lambda + 2\mu)}r, \tag{9}$$

where the dot means differentiation w.r.t time, t, and the dash differentiation w.r.t. position, r.

This PDE must be solved in combination with two boundary conditions and one initial condition. We will study the case of a viscoelastic sphere of outer radius, r_{o}. The boundary conditions state that the displacement stays finite and vanishes in the center and that there is no traction at the outer boundary, r_{o}:

$$u_r(r = 0, t) = 0 \,, \quad \sigma_{rr}(r = r_o, t) = 0. \tag{10}$$

3.2 Solution in Dimensionless Form

Analogously to Eqs. (47) and $(49)_1$ we define

$$u \equiv u(x, \tau) = \frac{u_r}{r_o} \,, \quad x = \frac{r}{r_o} \,, \quad \tau = \frac{\lambda + 2\mu}{\eta} t \,, \quad \alpha = \frac{8\pi G \rho_0^2 r_o^2}{3(\lambda + 2\mu)}. \tag{11}$$

Then the PDE (9) assumes the form:

$$u'' + 2\frac{u'}{x} - 2\frac{u}{x^2} + \frac{4}{3}\left(\dot{u}'' + 2\frac{\dot{u}'}{x} - 2\frac{\dot{u}}{x^2}\right) = \frac{\alpha}{2} x, \tag{12}$$

where the dot now refers to differentiation w.r.t. dimensionless time, τ, and the dash means differentiation w.r.t. dimensionless position, x.

The nonvanishing stresses are normalized by $\lambda + 2\mu$ (identified by a tilde) and read:

$$\tilde{\sigma}_{rr} = u' + \frac{2\nu}{1-\nu}\frac{u}{x} + \frac{4}{3}\left(\dot{u}' - \frac{\dot{u}}{x}\right), \tag{13}$$

$$\tilde{\sigma}_{\vartheta\vartheta} \equiv \tilde{\sigma}_{\varphi\varphi} = \frac{\nu}{1-\nu} u' + \frac{1}{1-\nu}\frac{u}{x} - \frac{2}{3}\left(\dot{u}' - \frac{\dot{u}}{x}\right).$$

The boundary conditions (10) take the following form:

$$u(0, \tau) = 0 \,, \tag{14}$$

$$\tilde{\sigma}_{rr}(1, \tau) \equiv u'(1, \tau) + \frac{2\nu}{1-\nu} u(1, \tau) + \frac{4}{3}\Big[\dot{u}'(1, \tau) - \dot{u}(1, \tau)\Big] = 0,$$

and the initial condition reads:

$$u(x, 0) = 0 \,, \quad x \in [0, 1], \tag{15}$$

this is to say that we expect no displacements initially, because "gravitation has just been switched on at $\tau = 0$."

We solve the PDE (12) by mapping it onto Laplace space w.r.t. time $\tau \leftrightarrow s$ and then finding a solution of the corresponding Ordinary Differential Equation (ODE). The Laplace transform of the displacement will be identified by a bar, $\bar{u} = \bar{u}(x, s)$, and we may write according to the usual rules of Laplace transforms:

$$\left(1+\frac{4}{3}s\right)\left(\bar{u}''(x,s)+2\frac{\bar{u}'(x,s)}{x}-2\frac{\bar{u}(x,s)}{x^2}\right)-$$
$$\frac{4}{3}\left[u''(x,0)+2\frac{u'(x,0)}{x}-2\frac{u(x,0)}{x^2}\right]=\frac{\alpha x}{2}\frac{1}{s}. \tag{16}$$

The term in brackets drops out. We can give two reasons for that. First, there is the initial condition (15), according to which the displacement (and all its derivatives) shall vanish initially. Second, we note that this very term represents the (stationary) ODE of the gravitational problem provided gravitation is not present, see Eq. (40), which is zero to begin with. The solution of the remaining ODE for $\bar{u}(x,s)$ is completely analogous to the one presented in Eq. (41). We may write:

$$\bar{u}(x,s)=Ax+\frac{B}{x^2}+\frac{\alpha}{20}x^3\frac{1}{s\left(1+\frac{4}{3}s\right)},\quad 0\le x\le 1, \tag{17}$$

In order to determine the constants of integration we have to transform the boundary conditions (14) into Laplace space as follows:

$$\bar{u}(0,s)=0\,, \tag{18}$$

$$\left(1+\frac{4}{3}s\right)\bar{u}'(1,s)+\left(\frac{2\nu}{1-\nu}-\frac{4}{3}s\right)\bar{u}(1,s)-\frac{4}{3}\left[u'(1,0)-u(1,0)\right]=0.$$

For the same reasons as before the term in parentheses in the second equation drops out. The first equation requires us to put $B=0$. Moreover, the remaining linear equation for A in Eq. $(17)_2$ can be solved and, after back transform into real time space, the final result reads as follows:

$$u(x,\tau=0)=0\,, \tag{19}$$

$$u(x,\tau>0)=-\frac{\alpha}{20}x\left[\frac{3-\nu}{1+\nu}-x^2\right]\left[1-\exp\left(-\tfrac{3}{4}\tau\right)\right]-\frac{\alpha}{10}\frac{1-\nu}{1+\nu}x\exp\left(-\tfrac{3}{4}\tau\right).$$

Note that special attention has been given to the case $\tau=0$: If we consider the limit case $\tau\to 0$ we find a nonvanishing initial displacement. Moreover, it can be seen that the initial and boundary conditions from Eqs. (14), (15) are indeed satisfied. For $\tau\to\infty$ the stationary relation shown in Eq. (48) is obtained.

We are now in a position to determine the dimensionless stresses from Eq. (13):

$$\tilde{\sigma}_{rr}=-\frac{\alpha}{20}\left(1-x^2\right)\left[\frac{3-\nu}{1+\nu}-\frac{1+\nu}{1-\nu}\exp\left(-\frac{3}{4}\tau\right)\right], \tag{20}$$
$$\tilde{\sigma}_{\vartheta\vartheta}\equiv\tilde{\sigma}_{\varphi\varphi}=-\frac{\alpha}{20}\frac{3-\nu}{1-\nu}\left[1-\frac{1+3\nu}{3-\nu}x^2-\frac{1+\nu}{3-\nu}\left(1-2x^2\right)\exp\left(-\frac{3}{4}\tau\right)\right].$$

It is easy to see that in the limit $\tau \to \infty$ the stresses of the stationary solution from Eq. (52) result.

3.3 Evaluation and Discussion of the Results

Figure 3 depicts the temporal evolution of the displacement as a function of radial distance in dimensionless form as predicted by Eq. (19) for the choice $\nu = 0.3$. Note that immediately after "gravity has been switched on" the dependence is nearly linear.[3] Consequently, the minimum is located at the outer radius $x = 1$. It is an edge minimum and not a "true" minimum with vanishing derivative.

In this context recall the notion of the "Love radius." It indicates the position where the radial strains within a self-gravitating sphere changes sign and it was first discovered by A.E.H. Love. In equilibrium this (normalized) position is given by:

$$x_{\text{Love}} = \sqrt{\frac{3 - \nu}{3(1 + \nu)}}, \tag{21}$$

and the details of the derivation of the formula can be found in the Appendix.

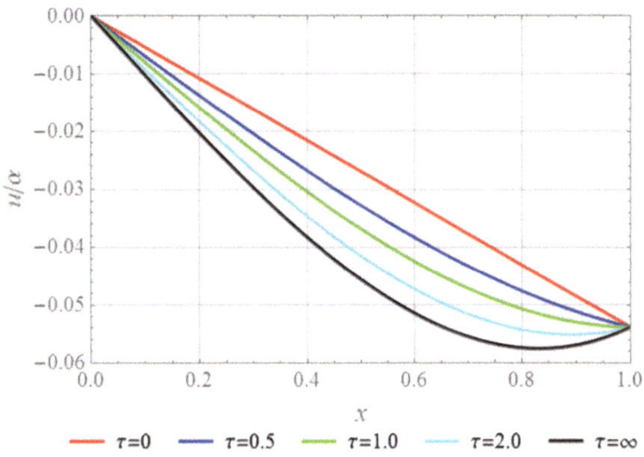

Fig. 3 Temporal development of the displacement as a function of radial position (see text)

[3]The label $\tau = 0$ in Figs. 3, 4 is to be understood in the sense $\tau \approx 0$ (i.e., very small but not equal to zero). A more detailed discussion of this degenerated case can be found in Müller and Weiss (2016).

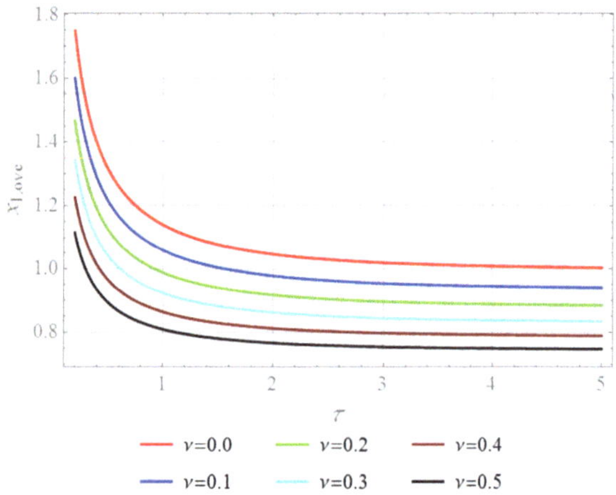

Fig. 4 Temporal development of the Love radius (see text)

Also recall that in the present case the radial strain is nothing else but the derivative of the radial displacement w.r.t. position, i.e., the slope to that curve. Moreover, the Love radius is defined by a true minimum of the radial displacement with zero slope. Consequently, in the transient case, a tensile region does not exist initially. It takes a certain while until the prominent feature of a true minimum corresponding to the location of the Love radius evolves.

We can obtain the location of the Love radius by (formal) differentiation of the displacement shown in Eq. (19) w.r.t. x. The result is:

$$x_{\text{Love}} = \frac{1}{\sqrt{3}}\sqrt{\frac{3-\nu}{1+\nu} + \frac{1-\nu}{1+\nu}\frac{1}{\exp\left(\frac{3}{4}\tau\right) - 1}}. \tag{22}$$

It is easily seen that this expression tends to the "elastic" Love radius shown in Eq. (21) if τ goes to infinity (Fig. 4).

Finally, Fig. 5 presents the (dimensionless) stresses. It is noteworthy that the radial stress component relaxes monotonically without a qualitative change in the shape of the curve. This is not so for the angular stresses, whose minima switch from $x = 1$ to $x = 0$ as time goes on.

Based on the results presented in Figs. 3 and 5, we must conclude that the process of relaxation in a self-gravitating terrestrial planet is not as simple as in the textbook example of a load and displacement-controlled viscoelastic strip shown in Fig. 1. This is due to the fact that we face a three-dimensional state of stress after "switching on" a spatially varying body force.

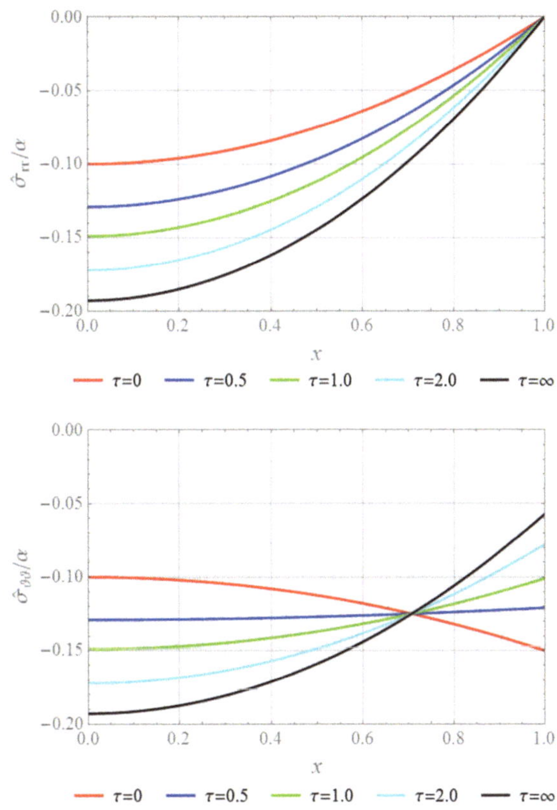

Fig. 5 Temporal development of the stresses (see text)

4 Outlook and Conclusions

The main objective of this paper was to present an analysis of the temporal development of the displacements, strains, and stresses in a self-gravitating sphere. The model was based on a radially symmetric linear viscoelastic constitutive model of the Kelvin–Voigt type. An analytical solution was found based on Laplace transforms. It was shown how the displacement and stresses relax to the stationary linear-elastic solution, originally due to Love, which was also reviewed in an appendix. In particular it was shown that the so-called Love radius, which marks the transition between the regions of compressive and tensile strain, does not exist in the early stages. It takes some time to develop.

In future work we will investigate alternative viscoelastic models, for example a generalization of the Zener type. We will also attempt to predict the relaxation time scales based on recent measurements of the viscosity of (liquid) iron and igneous rock.

Appendix: Love's Solution—Its History in Modern Form

The following passages were primarily written for the benefit of readers who do not solve linear-elastic problems on a daily basis. However, they also contain interpretations not originally provided by Love, for example intuitive explanations of the meaning of the normalizing coefficients for the displacements and for the stresses.

A.1 The Primary Assumptions of Love and Their Limitations

Love's solution for the state of deformation in a self-gravitating sphere is a static one. Hence the balance of momentum degenerates to the following equation:

$$\nabla \cdot \boldsymbol{\sigma} = -\rho \boldsymbol{f}, \tag{23}$$

where ρ denotes the local current mass density, and $\boldsymbol{\sigma}$ the Cauchy stress tensor. The specific body force, $\boldsymbol{f}$, i.e., the gravitational acceleration, is conservative and originates from self-gravity. Hence a gravitational potential $U^{\text{grav}}(\boldsymbol{x})$ exists, where $\boldsymbol{x}$ denotes an arbitrary (current) position within the body, and we may write:

$$\boldsymbol{f}(\boldsymbol{x}) = -\nabla U^{\text{grav}}(\boldsymbol{x}). \tag{24}$$

The gravitational potential obeys Poisson's equation:

$$\Delta U^{\text{grav}}(\boldsymbol{x}) = 4\pi G \rho(\boldsymbol{x}). \tag{25}$$

For the stress tensor we initially assume that Hooke's law holds, so that there are no rate effects:

$$\boldsymbol{\sigma} = \lambda \,\text{Tr}\boldsymbol{\epsilon}\, \mathbf{I} + 2\mu\, \boldsymbol{\epsilon}, \tag{26}$$

where the *linear* strain tensor has been used:

$$\boldsymbol{\epsilon} = \tfrac{1}{2}(\nabla \boldsymbol{u} + \nabla \boldsymbol{u}^{\top}). \tag{27}$$

$\boldsymbol{u}$ refers to the displacement vector, i.e., to $\boldsymbol{u} = \boldsymbol{x} - \boldsymbol{X}$, $\boldsymbol{X}$ being the reference position of a material point of the sphere. λ and μ are Lamé's elastic constants.

At this point three remarks for putting Love's approach in perspective are in order, which are all related somehow. They all circle around the question "What happens if the deformations prove to be large?" We shall see that they can be large, we want to point out possible remedies, we will provide some citations for further reading, but we shall not endeavor to work it all out in this paper.

The first comment concerns the nabla operator used in the aforementioned equations: Note that *all* nabla operators above indicate differentiation w.r.t. the *current* spatial position, $\boldsymbol{x}$. This is how the theory works for "linear elasticians:" There is only one gradient in linear theory of elasticity at small deformations, namely that one. Hence for them putting an emphasis on it sounds trivial. However, there is a world outside of linear elasticity as understood by Sokolnikoff (1956) or Timoshenko and Goodier (1951), in order to quote just two references of that denomination. Indeed, it is possible to understand linear elasticity as a limit case of nonlinear materials theory. Then Hooke's law results in a natural way written in terms of gradients with respect to the reference configuration, $\boldsymbol{X}$, see Wang and Truesdell (1973), pp. 170, or Müller (1973), p. 72. However, in the next breath it is said, see Truesdell and Toupin (1960), Sects. 57 and 301, that it does not really matter, and these gradients can be replaced by derivatives w.r.t. the current position, since the deformations are so small. What they do not say, though, is that it *does matter* from a principal, didactic point-of-view.

Second, from the standpoint of linear elasticity, the set of Eqs. (23)–(27) serves only one purpose: It allows us to calculate the displacement $\boldsymbol{u}(\boldsymbol{x})$. To this end the mass density $\rho(\boldsymbol{x})$ must be considered as known, and for linear-elasticians it is, in the simplest case, a space-independent constant, $\rho = \rho_0$. If we relate it to our problem we may consider it to be the mass density of the homogeneous sphere before gravity has been switched on. However, recall that this does not mean that the current mass density is also a constant, even if it is one in the reference state. It is dependent on deformation and it can be determined from mass conservation. In general, now turning back to nonlinear theory for a moment, it is well known that we may write:

$$\rho(\boldsymbol{x}) = \frac{\rho_0}{\det \boldsymbol{F}(\boldsymbol{x})}, \tag{28}$$

where $\boldsymbol{F}(\boldsymbol{x}) \equiv \nabla_X \boldsymbol{x}$ is the deformation gradient pertinent to a material point. Recall once more that Eq. (28) is the result of the physical principle of local mass conservation and geometry, i.e., nonlinear kinematics, and, as such, it holds for arbitrary deformations. If we insist on studying small deformations, we must replace Eq. (28) by:

$$\rho(\boldsymbol{x}) \approx \rho_0 \left[1 - \operatorname{Tr} \boldsymbol{\epsilon}(\boldsymbol{x})\right]. \tag{29}$$

Consequently, the argument now runs as follows: Once the linear strain $\boldsymbol{\epsilon}(\boldsymbol{x})$ is known from a linear-elastic analysis based on the (static) balance of momentum in combination with Hooke's law, during which the mass density is assumed to be spatially constant, the spatial distribution of the current mass density in the strained body can be calculated from Eq. (29). In other words, one does *not* solve a coupled problem and does *not* make use of the balances of mass and momentum simultaneously. Indeed, in our problem we calculate the strains or rather the displacements from Eqs. (23)–(27) *after* the current mass density in the body force has been replaced by a constant reference mass density, hence mimicking homogeneous initial conditions for the mass distribution of a terrestrial planet. For conciseness of this paper, the question as to how the current mass density will look like and how it compares with today's knowledge of the inner mass distribution of Earth will be discussed elsewhere.[4]

Third, the use of a constant reference mass density in linear elasticity turns into a very subtle point when applied to problems of self-gravity. Observe that in the current local balance of momentum the current mass density appears explicitly and linearly in *three* locations: (i) in the inertial term (not shown in Eq. (23), because we restrict ourselves to quasistatic conditions); (ii) in the product of the term for the body force density, and (iii) in the acceleration part of the body force density, if we consider the case of full self-gravitational interaction. The latter will be demonstrated explicitly in Subsection (A.2.1). Now recall once more that all of this is ignored in linear elasticity where the current mass density is simply replaced by a constant value ρ_0, everywhere. We may rephrase it in the jargon of technical mechanics by saying that the forces are applied to the undeformed structure and a first order theory is used to calculate the resulting deformation. Thus we would like to reemphasize that the model "linear elasticity" is defined by three prerequisites (also see, for example, the beginning of Kienzler and Schröder (2009), namely, first, a linear relationship between stress and strain, second, strains and displacements to be small and, third, equilibrium of an undeformed element.

However, the use of linear elasticity in self-gravitational problems remains questionable. Indeed, we shall see that for certain celestial objects, in particular the Earth, the strains we are about to predict from the linear theory of elasticity can become very large. Probably the first to notice was A.E.H. Love after applying linear elasticity in the way defined above and discovering what is known as the Love radius, a radial transition point within a self-gravitating spherical body, where radial strains switch from compression to tension and, consequently, may result in damage of the body. Love muses in sudden attacks of self-doubt about his approach, namely in Love (1892), Article 127: "There is another difficulty in the application of the result [for the strains and for the breaking stress] to the case of the Earth. The necessary limitation to the mathematical theory is that the strain found from it must always be "small". ..." and in Love (1927), Article 75: "The Earth is an example of a body which must be regarded as being in a state of initial stress, for the stress that must exist in the interior is much too great to permit of the calculation, by the ordinary methods, of strains reckoned from the unstressed state as unstrained state."

Consequently, we could come to the conclusion to abandon linear elasticity completely and to use a deformation-wise nonlinear theory instead. Indeed, this was done in a series of papers from the school of Seth, who was one of the first to study and use nonlinear deformation measures for elasticity problems, see, e.g., Chattarji (1953) or Bose and Chattarji (1963). In principle, this requires solving the coupled problem, namely the balances of mass *and* momentum, unless an empirical expression for the (current) mass density distribution is assumed. The latter was the case in the aforementioned papers from the school of Seth. One of their main conclusions was that the position of the Love radius as predicted by linear elasticity at small deformations does not change much when switching to large deformation theory.

---

[4]The interested reader is referred to Müller and Weiss (2016).

This in mind, do we now feel completely reconciled by using the linear theory elasticity in context with problems of self-gravity? The answer is unfortunately "no." Indeed, a detailed study of the nonlinear problem shows that there are many open questions, ranging from numerical issues to the use of the proper nonlinear stress-strain relationship.[5] Nevertheless, this paper is not the place to explore this issue completely. We must and will assume the position of Galileo's Simplicio, who would use the linear-elastic solution anyway, no matter how large the self-gravitating mass really is. To quote Churchill: "Now this is not the end. It is not even the beginning of the end. But it is, perhaps, the end of the beginning" or shall we say the dawn of awareness?

In the next section we shall briefly summarize Love's linear elasticity results at small deformations and provide some additional comments for better explanation and clarification of the problem. For example we shall introduce and interpret various parameters in terms of their physical meaning, which can be used for normalization of the solution. Moreover, we will show that for certain celestial objects, such as Mercury, the linear elasticity solution with small deformations can be considered as valid. It will also serve as a starting point as well as for comparison with the results from linear viscoelasticity in Sect. 3.

A.2 Review of Love's Linear-Elastic Model of Self-gravitation

A.2.1 Analysis of the Strictly Radially Symmetric Case

We start from the Poisson equation describing Newtonian gravity as shown in Eq. (24) and assume purely radial dependencies:

$$\frac{1}{r^2}\frac{\mathrm{d}}{\mathrm{d}r}\left(r^2\frac{\mathrm{d}U^{\mathrm{grav}}(r)}{\mathrm{d}r}\right) = 4\pi G\rho(r) \quad \Rightarrow \quad \frac{\mathrm{d}U^{\mathrm{grav}}}{\mathrm{d}r} = G\frac{m(r)}{r^2}, \tag{30}$$

where $m(r)$ denotes the total mass within a spherical region of radial extension r:

$$m(r) = 4\pi\int_{\tilde{r}=0}^{\tilde{r}=r}\rho(\tilde{r})\tilde{r}^2\mathrm{d}\tilde{r}\,, \quad 0 \le r \le r_{\mathrm{o}}, \tag{31}$$

and r_{o} stands for the current outer radius of the spherical body. Consequently, according to Eq. (24), under these circumstances the volume density of body force is given by:

$$\rho(r)\boldsymbol{f}(r) = -G\frac{\rho(r)m(r)}{r^2}\boldsymbol{e}_r. \tag{32}$$

[5]The interested reader may want to consult Müller and Weiss (2016) for further information.

This includes the well-known high school result according to which the gravitational force at a distance r within a *homogeneous* sphere is given by Newton's law of gravity for point masses: The attracting mass is given by all the matter below the position r, i.e., $m(r)$, and can be thought of as being concentrated in the origin of the sphere, i.e., $r = 0$. The to-be-attracted mass at the radial position r is given by $\mathrm{d}m = \rho(r)\,\mathrm{d}V$, $\mathrm{d}V$ being the corresponding volume element to be used for multiplication in Eq. (24). Moreover, the gravitational force is attractive, as indicated by the negative direction of the current radial unit vector, $\boldsymbol{e}_r$. It should be pointed out that in this equation the mass density within the sphere does not necessarily have to be homogeneous. Rather it can be a function of the current radius, $\rho(r)$, and the high-school result still holds. This is very often not clearly stated in textbooks, especially if use of the Poisson equation is avoided for mathematical simplicity.

However, as outlined before, it is customary in linear elasticity to use the body force of an undeformed structure in Eq. (23). More specifically, we pretend everything is initially homogeneous and use a *constant* mass density, ρ_0, such that:

$$\rho(r)\boldsymbol{f}(r) \approx -G\frac{\rho_0 m(r)}{r^2}\,\boldsymbol{e}_r \approx -\frac{4\pi G\rho_0^2}{3} r\,\boldsymbol{e}_r. \tag{33}$$

Note that a two-step approximation was involved here. First, the current mass density, $\rho(r)$, in Eq. (23) or in (32) was replaced by the reference mass density, ρ_0. Second, no distinction is made between the current and the reference radius on the right hand side of Eq. (32). We will now use the approximation (33) in Eq. (23), which reads in spherical coordinates as follows:

$$\begin{aligned}
\frac{\partial\sigma_{rr}}{\partial r} + \frac{1}{r}\frac{\partial\sigma_{r\vartheta}}{\partial\vartheta} + \frac{1}{r\sin\vartheta}\frac{\partial\sigma_{r\varphi}}{\partial\varphi} + \frac{2\sigma_{rr} - \sigma_{\vartheta\vartheta} - \sigma_{\varphi\varphi} + \sigma_{r\vartheta}\cot\vartheta}{r} &= \frac{4\pi G\rho_0^2}{3} r,\\
\frac{\partial\sigma_{r\vartheta}}{\partial r} + \frac{1}{r}\frac{\partial\sigma_{\vartheta\vartheta}}{\partial\vartheta} + \frac{1}{r\sin\vartheta}\frac{\partial\sigma_{\vartheta\varphi}}{\partial\varphi} + \frac{3\sigma_{r\vartheta} + \left(\sigma_{\vartheta\vartheta} - \sigma_{\varphi\varphi}\right)\cot\vartheta}{r} &= 0,\\
\frac{\partial\sigma_{r\varphi}}{\partial r} + \frac{1}{r}\frac{\partial\sigma_{\vartheta\varphi}}{\partial\vartheta} + \frac{1}{r\sin\vartheta}\frac{\partial\sigma_{\varphi\varphi}}{\partial\varphi} + \frac{3\sigma_{r\varphi} + 2\sigma_{\vartheta\varphi}\cot\vartheta}{r} &= 0.
\end{aligned} \tag{34}$$

Moreover, Hooke's law reads in spherical coordinates as follows:

$$\begin{aligned}
&\sigma_{rr} = \lambda(\epsilon_{\vartheta\vartheta} + \epsilon_{\varphi\varphi}) + (\lambda + 2\mu)\epsilon_{rr}\ ,\ \sigma_{\vartheta\vartheta} = \lambda(\epsilon_{rr} + \epsilon_{\varphi\varphi}) + (\lambda + 2\mu)\epsilon_{\vartheta\vartheta},\\
&\sigma_{\varphi\varphi} = \lambda(\epsilon_{rr} + \epsilon_{\vartheta\vartheta}) + (\lambda + 2\mu)\epsilon_{\varphi\varphi}\ ,\\
&\sigma_{r\vartheta} = 2\mu\,\epsilon_{r\vartheta}\ ,\sigma_{r\varphi} = 2\mu\,\epsilon_{r\varphi}\ ,\sigma_{\vartheta\varphi} = 2\mu\,\epsilon_{\vartheta\varphi}.
\end{aligned} \tag{35}$$

And finally the linear strain tensor is linked with spatial derivatives of the displacements by:

$$\epsilon_{rr} = \frac{\partial u_r}{\partial r}\ ,\ \epsilon_{\vartheta\vartheta} = \frac{1}{r}\frac{\partial u_\vartheta}{\partial\vartheta} + \frac{u_r}{r}\ ,\ \epsilon_{\varphi\varphi} = \frac{1}{r\sin\vartheta}\frac{\partial u_\varphi}{\partial\varphi} + \frac{u_r}{r} + \frac{\cot\vartheta}{r}u_\vartheta,$$

$$\epsilon_{r\varphi} = \tfrac{1}{2}\left(\frac{1}{r\sin\vartheta}\frac{\partial u_r}{\partial \varphi} + \frac{\partial u_\varphi}{\partial r} - \frac{u_\varphi}{r}\right), \tag{36}$$

$$\epsilon_{r\vartheta} = \tfrac{1}{2}\left(\frac{1}{r}\frac{\partial u_r}{\partial \vartheta} + \frac{\partial u_\vartheta}{\partial r} - \frac{u_\vartheta}{r}\right), \quad \epsilon_{\vartheta\varphi} = \tfrac{1}{2}\left(\frac{1}{r\sin\vartheta}\frac{\partial u_\vartheta}{\partial \varphi} + \frac{1}{r}\frac{\partial u_\varphi}{\partial \vartheta} - \frac{\cot\vartheta}{r}u_\varphi\right).$$

We now proceed to solve these equations. To this end we make use of the semi-inverse method. Because of symmetry it seems reasonable to seek for solutions with the following *ansatz*:

$$u_r = u_r(r)\,, \quad u_\vartheta = 0\,, \quad u_\varphi = 0. \tag{37}$$

Consequently, we find for the linear strains:

$$\epsilon_{rr} = u_r'(r)\,, \qquad \epsilon_{\vartheta\vartheta} \equiv \epsilon_{\varphi\varphi} = \frac{u_r}{r}\,, \qquad \epsilon_{r\vartheta} = \epsilon_{r\varphi} = \epsilon_{\vartheta\varphi} \equiv 0\,, \tag{38}$$

where the dash refers to a differentiation w.r.t. r. Because of that Hooke's law (35) reduces to:

$$\sigma_{rr} = (\lambda + 2\mu)u_r' + 2\lambda\frac{u_r}{r}, \qquad \sigma_{\vartheta\vartheta} \equiv \sigma_{\varphi\varphi} = \lambda u_r' + 2(\lambda + \mu)\frac{u_r}{r}, \tag{39}$$

$$\sigma_{r\vartheta} = \sigma_{r\varphi} = \sigma_{\vartheta\varphi} \equiv 0\,.$$

Thus, the angular components of the balance of momentum shown in Eq. (34) are identically satisfied and the first one results in an ordinary differential equation of second order (a dash refers to differentiation with respect to the radius, r):

$$u_r'' + 2\frac{u_r'}{r} - 2\frac{u_r}{r^2} = \frac{4\pi\rho_0^2 G}{3(\lambda + 2\mu)}r. \tag{40}$$

The general solution consists of the full solution to the homogeneous part and one particular solution of the inhomogeneous case. It reads with two constants of integration, A and B, respectively, as follows:

$$u_r = Ar + \frac{B}{r^2} + \frac{4\pi\rho_0^2 G}{30(\lambda + 2\mu)}r^3. \tag{41}$$

Two conditions are needed in order to determine the two constants of integration. First, we require that the solution does not become singular at $r = 0$ and, second, the traction must be continuous at the outer radius, r_o, of the sphere. Hence, $\sigma_{rr}\big|_{r=r_o} = 0$, because the influence of the outer atmospheric pressure of roughly 1bar is negligibly small when it comes to the deformation of a solid. With Eq. $(39)_1$ we find that:

$$B = 0\,, \qquad A = -\frac{4\pi G\rho_0^2}{30(\lambda + 2\mu)}\frac{3-\nu}{1+\nu}r_o^2 \equiv -\frac{4\pi G\rho_o^2}{90\,k}\frac{3-\nu}{1-\nu}r_o^2, \tag{42}$$

because $\lambda = \frac{E\nu}{(1-2\nu)(1+\nu)}$ and $\mu = \frac{E}{2(1+\nu)}$, E being Young's modulus and ν Poisson's ratio, respectively. Hence the radial displacement reads:

$$u_r = -\frac{2\pi G\rho_0^2 r_o^2}{15(\lambda + 2\mu)}\left(\frac{3-\nu}{1+\nu} - \frac{r^2}{r_o^2}\right) r \equiv -\frac{2\pi G\rho_0^2 r_o^2}{45\,k}\frac{1+\nu}{1-\nu}\left(\frac{3-\nu}{1+\nu} - \frac{r^2}{r_o^2}\right) r. \tag{43}$$

For completeness, the nonvanishing stresses then follow from Eq. (39) as:

$$\sigma_{rr} = -\frac{2\pi G\rho_0^2 r_o^2}{15}\frac{3-\nu}{1-\nu}\left(1 - \frac{r^2}{r_o^2}\right), \tag{44}$$
$$\sigma_{\vartheta\vartheta} \equiv \sigma_{\varphi\varphi} = -\frac{2\pi G\rho_0^2 r_o^2}{15}\frac{3-\nu}{1-\nu}\left(1 - \frac{1+3\nu}{3-\nu}\frac{r^2}{r_o^2}\right).$$

Note the common factor $\frac{2\pi G\rho_0^2 r_o^2}{15}$ in front of all these expressions. On first glance it does not allow for an easy intuitive interpretation. However, on second thought, note that (within the approximations made) the total mass of the gravitating sphere, m_0, the gravitational acceleration on its surface, g, and its surface area, A_o, are given by:

$$m_0 = \frac{4\pi}{3}\rho_0 r_o^3\,, \quad g = \frac{Gm_0}{r_o^2}\,, \quad A_o = 4\pi r_o^2. \tag{45}$$

Hence, we may write the ominous factor as:

$$\frac{2\pi G\rho_0^2 r_o^2}{15} \equiv \frac{3m_0 g}{10A_o}, \tag{46}$$

and interpret it, with the exception of the fraction $\frac{3}{10}$, as an average "pressure," namely the ratio between the "total gravitational force," $m_0 g$, distributed over the total surface area, A_o.

A.2.2 Dimensionless Formulation

For a numerical analysis it is best to work with dimensionless quantities. Since the outer radius, r_o, is the only length parameter in the problem, there is no other choice for a dimensionless distance and a dimensionless displacement but to define:

$$x = \frac{r}{r_o}\,, \quad u = \frac{u_r}{r_o}. \tag{47}$$

This allows us to rewrite Eq. (43) as follows:

$$u = -\frac{\alpha}{20}\left(\frac{3-\nu}{1+\nu} - x^2\right)x \equiv -\frac{\alpha_k}{30}\frac{1+\nu}{1-\nu}\left(\frac{3-\nu}{1+\nu} - x^2\right)x, \tag{48}$$

with two dimensionless factors:

$$\alpha = \frac{8\pi G\rho_0^2 r_o^2}{3(\lambda + 2\mu)}, \quad \alpha_k = \frac{4\pi G\rho_0^2 r_o^2}{3k}, \tag{49}$$

because the bulk modulus is given by $k = \frac{E}{3(1-2\nu)}$.

Whilst the appearance of α is a straightforward consequence of Eq. (43) the need for α_k must be explained. To this end note that the dimensionless expressions in the parentheses of Eq. (48) still contain Poisson's ratio. However, Poisson's ratio of a terrestrial planet is *not* an immediately accessible parameter. A homogenization technique has to be applied in order to find out which effective elastic parameters such an object has. On the other hand, if we evaluate the parentheses in this equation at the surface of the planet, i.e., at $x = 1$, we obtain twice the fraction $\frac{1-\nu}{1+\nu}$. Hence Poisson's ratio disappears completely in the expression for the normalized displacement if we use the dimensionless factor α_k. In this case one only needs to know the effective compressibility of the planet, a parameter that can vary within certain physically reasonable bounds. And what is more, since $u(x = 1)$ can be interpreted as an average strain characterizing the state of deformation of the planet, which we wish to access numerically, it is very useful to have one elastic parameter less to worry about. A final comment is in order in context with Eq. (48). It obviously provides a restriction to the size of α so that small strain theory applies. However, as it was pointed out above, we will discuss this issue in detail in Müller and Weiss (2016) and not here.

Later we shall be interested in a strain-based failure criterion. Hence it is useful to know the strains explicitly:

$$\epsilon_{rr} = -\frac{\alpha}{20}\left(\frac{3-\nu}{1+\nu} - 3x^2\right), \quad \epsilon_{\vartheta\vartheta} \equiv \epsilon_{\varphi\varphi} = -\frac{\alpha}{20}\left(\frac{3-\nu}{1+\nu} - x^2\right). \tag{50}$$

We now turn to the stresses given by Eq. (44). Differently from the case of length related quantities there are *various* possibilities for normalization. First, as Eq. (44) suggestively seems to indicate, we can use the factor $\frac{2\pi G\rho_0^2 r_o^2}{15}$, which we have interpreted as an "average gravitational pressure" before. However, alternatively, we may use combinations of (effective) elastic constants. There is $\lambda + 2\mu$, which is related to the velocity of P-waves $\left(v_P = \sqrt{\frac{\lambda+2\mu}{\rho_0}}\right)$, and, hence, a physically accessible quantity. Moreover, we may turn to the compressibility k, which is a direct measure of the resistance of a planet's response to its own self-gravity. However, in this paper we restrict ourselves for simplicity to the choice:

$$\tilde{\boldsymbol{\sigma}} = \frac{\boldsymbol{\sigma}}{\lambda + 2\mu} \tag{51}$$

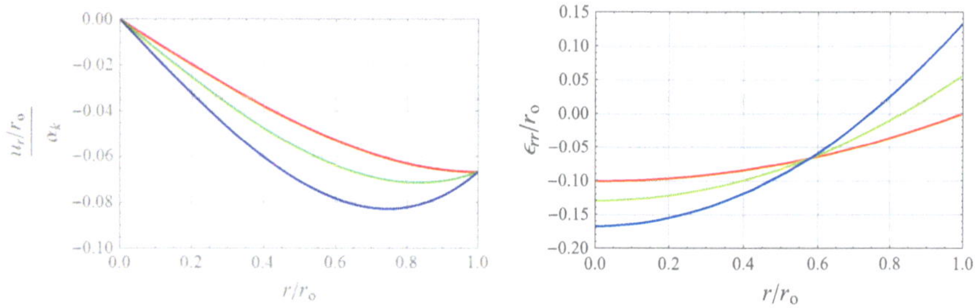

Fig. 6 Normalized displacement versus dimensionless radius (see text)

and obtain:

$$\tilde{\sigma}_{rr} = -\frac{\alpha}{20}\frac{3-\nu}{1-\nu}\left(1-x^2\right), \qquad \tilde{\sigma}_{\vartheta\vartheta} \equiv \tilde{\sigma}_{\varphi\varphi} = -\frac{\alpha}{20}\frac{3-\nu}{1-\nu}\left(1-\frac{1+3\nu}{3-\nu}x^2\right). \tag{52}$$

A.2.3 Numerical Evaluation and Graphical Representation

Figure 6 (left) illustrates the dependence of the normalized displacement, $u \equiv \frac{u_r}{r_o}$ per α_k, on $x \equiv \frac{r}{r_o}$ for three different choices of Poisson's ratio, $\nu = 0$ (red), $\nu = 0.3$ (green), and $\nu = 0.5$ (blue). Note that, as it should be, the radial displacement is negative and that the curves show a minimum. Because of Eq. $(36)_1$ we may interpret the slope of the curves as radial strain multiplied by α_k. Hence the only positive radial strains can be found to the right of the minimum. The transition point between positive and negative strains (see Fig. 6, right), identifiable by locating the minimum of the displacement, is a.k.a. the *Love radius* and given by:

$$r_{\text{Love}} = r_0\sqrt{\frac{3-\nu}{3(1+\nu)}}. \tag{53}$$

This result was first mentioned by Love in his books on linear elasticity, namely in Article 127 of Love (1892) and later in Article 98 of Love (1906) or Love (1927). An intuitive explanation for the necessity of its occurrence is as follows: Unlike a homogeneous, isotropic sphere subjected to a constant external pressure, the state of strain in our case is not homogeneous and not isotropic. We face a nonconstant "external hydrostatic pressure," so-to-speak, given by an effective gravitational force depending linearly on the distance from the center. This in combination with Poisson's effect, i.e., the ability of a radial strain "making up" for the lateral contractions, $\epsilon_{\vartheta\vartheta}$ and $\epsilon_{\varphi\varphi}$, which are purely compressive in nature everywhere: The contractive force is proportional to r, the stretching stress (compensation of the lateral contraction) is proportional to r^2. Hence from some r onward the force is not strong enough for compression, resulting in a transition from the negative to the positive, in other words in the existence of a Love radius. In his later editions Love does no longer comment on the physical significance of his radius. However, the first edition makes it perfectly clear that he was aiming at a failure criterion, namely specifically at what is known today as *maximum principal strain theory*. Love realized that the angular strains are always negative, whereas the radial strains may become positive above the Love radius, cf. Eq. (50) and Fig. 6 (right), and he provided an expression for the maximum radial strain, which is the one at the surface:

$$\epsilon_{rr}\Big|_{x=1} = \frac{\alpha}{5}\frac{\nu}{1+\nu}. \tag{54}$$

According to Love the corresponding tensile breaking stress is then given by:

$$T_0 = E\epsilon_{rr}\Big|_{x=1} = \frac{E\alpha}{5}\frac{\nu}{1+\nu}. \tag{55}$$

This is quite a daring concept, because a planet like Earth is heterogeneous, and surely not perfectly linear elastic, and the materials it is made of might not be susceptible to strain-based failure, and so on, and so on. But even if we accept his idea in principle, what is the proper Young's modulus to be used for a heterogeneous object like Earth? On second glance, however, note that the factor α also contains Young's modulus in its denominator (cf. Eq. $(49)_1$) since $\lambda + 2\mu = \frac{(1-\nu)E}{(1-2\nu)(1+\nu)}$. Thus we do not need to know E but only Poisson's ratio, ν, which runs within well-known bounds, namely $0 \leq \nu \leq 0.5$. Hence, we may rewrite Love's result as follows:

$$T_0 = \frac{\nu(1-2\nu)}{1-\nu}\frac{8\pi G\rho_0^2 r_o^2}{15}. \tag{56}$$

If we assume Poisson's ratio to be that of iron, $\nu = 0.3$, and the mean mass density of Earth $\rho_0 = 5500\,\frac{\text{kg}}{\text{m}^3}$ together with its (average) outer radius $r_o = 6370\,\text{km}$, we obtain $T_0 > 30{,}000\,\text{MPa}$, which is a value for a breaking stress far beyond physical

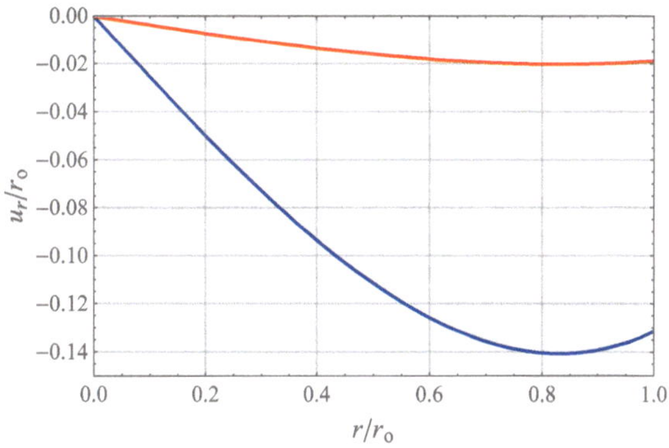

Fig. 7 Normalized displacement versus dimensionless radius for Mercury and Earth (see text)

Constitutive Modelling of the Glass Transition and Related Phenomena: Relaxation of Shear Stress Under Pressure

Alexander Lion, Michael Johlitz and Christoph Mittermeier

Abstract In industrial fabrication processes as well as in many applications of polymer parts, the glass transition plays a significant role. This is due to high mechanical processing speeds, high temperatures or large cooling rates. The mechanical, the thermomechanical and the caloric properties of polymers differ below and above the glass transition which is a thermoviscoelastic phenomenon. It depends on the ratio between the intrinsic time scale of the polymer and that of the thermomechanical loading process. If both scales are comparable, the material is in the glass transition region. Otherwise it is in the equilibrium or in the glassy region. In the industry, there are increasing demands to simulate fabrication processes in order to estimate the resulting behaviour of the polymer parts before they are manufactured. To this end, constitutive models of finite thermoviscoelasticity are needed which can represent the volumetric as well as the isochoric mechanical behaviour of the polymer in combination with the caloric and the thermomechanical properties. In a recent paper of the authors, the concept of a hybrid free energy has been developed. This approach will be applied in the current essay where the pressure-dependent relaxation behaviour under shear deformations is of interest.

1 Introduction

In industrial production processes of polymer parts, under extreme temperature conditions or under dynamic loadings in broad frequency ranges, the glass transition plays an important role. It influences the mechanical properties of the polymer, its

---

A. Lion (✉) · M. Johlitz · C. Mittermeier
Faculty of Aerospace Engineering, Institute of Mechanics,
Universität der Bundeswehr München, 85577 Neubiberg, Germany
e-mail: alexander.lion@unibw.de

M. Johlitz
e-mail: michael.johlitz@unibw.de

C. Mittermeier
e-mail: christoph.mittermeier@unibw.de

K. Naumenko and M. Aßmus (eds.), *Advanced Methods of Continuum Mechanics for Materials and Structures*, Advanced Structured Materials 60,
DOI 10.1007/978-981-10-0959-4_6

thermal expansion behaviour as well as its caloric properties (see e.g. Altenbach 2015; Jackle 1986 or Greiner and Schwarzl 1984). In order to understand the glass transition it is generally accepted to introduce two timescales. The first one is denoted as $\tau_{\mathrm{m}}(\theta)$, corresponds to the typical relaxation time of the polymer and depends on the temperature θ. The viscosity of the polymer is more or less proportional to $\tau_{\mathrm{m}}(\theta)$ and the reciprocal value $1/\tau_{\mathrm{m}}(\theta)$ can be linked to the molecular mobility. With decreasing temperature, the characteristic timescale of the material $\tau_{\mathrm{m}}(\theta)$ increases progressively. A well-known model to describe this behaviour is the WLF-equation which has been developed by Williams et al. (1955). In another approach which is also discussed in Williams et al. (1955) and can be related to the WLF-equation, the relaxation time depends exponentially on the reciprocal value of the linearly temperature-dependent free volume of the polymer. The free volume is the difference between the total volume of the polymer and that which is occupied by its molecules. The second timescale τ_{l} is determined by the external thermomechanical loading process. Based on its temporal characteristics, it can be the reciprocal value of the frequency, the duration of a ramp or the holding time under constant mechanical or thermal input. In the case of $\tau_{\mathrm{l}} >> \tau_{\mathrm{m}}(\theta)$, the external loading process is slow in comparison with the response time of the material. Hence, the molecules respond to changes in the load without any delay such that the polymer is in an equilibrium state. In the opposite case of $\tau_{\mathrm{l}} << \tau_{\mathrm{m}}(\theta)$, the external loading is fast in comparison with the response time of the material such that the molecules cannot follow. In other words, the molecules are frozen and the polymer is in the glassy state. In the case of comparable timescales $\tau_{\mathrm{l}} \approx \tau_{\mathrm{m}}(\theta)$ the polymer is in the glass transition regime. Since the characteristic time $\tau_{\mathrm{m}}(\theta)$ of the material is temperature-dependent, the glass transition can be reached either by changing the temperature or, when the temperature is constant, by the dynamics of the external loading process. Sometimes, the terms "mechanical glass transition" and "thermal glass transition" are used.

In order to simulate glass transition-induced phenomena of polymer parts under external thermomechanical loadings or in industrial production processes, constitutive models in combination with the finite element method are required. Since stress- or temperature-induced volume changes of polymers are demonstrably small, their mechanical behaviour is slightly compressible. In order to implement such a model into finite element software, formulations in which the stress tensor is the dependent mechanical state variable are needed. Accordingly, the specific Helmholtz free energy is a convenient thermodynamic potential. On the other hand, calorimetric material properties and the thermal expansion behaviour are commonly considered under constant hydrostatic pressure. In order to model such a material behaviour, the specific Gibbs free energy is the naturally related thermodynamic potential (see Lion et al. 2010; Peters et al. 2011 or Lion et al. 2011 and the citations therein). In order to combine both approaches, a so-called hybrid free energy has been proposed in Lion et al. (2014). It is a function of the hydrostatic pressure, the isochoric part of the deformation gradient and the absolute temperature. To represent inelastic effects, it can depend on additional internal state variables. This type of thermodynamic potential is applied in the current article.

In Sect. 2, the most essential fundamentals are introduced. In Sect. 3, a constitutive model of finite nonlinear thermoviscoelasticity is proposed. Section 4 shows some simulations and theoretical considerations. Section 5 closes with a short discussion.

2 Fundamentals

For the fundamentals of continuum mechanics, the readers are referred, for example, to the textbooks of Altenbach (2015), Haupt (2002), Altenbach and Altenbach (1994) or Malvern (1969) and for details with regard to the constitutive representation of the glass transition with thermoviscoelastic methods to Lion et al. (2011), Peters et al. (2011) or Lion et al. (2014) and the citations therein.

The deformation gradient is denoted as $\boldsymbol{F}$, its determinant as $J = \det(\boldsymbol{F})$ and the volumetric strain as $\varepsilon_{\mathrm{vol}} = J - 1$. Following Flory (1961), the isochoric and volumetric parts of the deformation gradient $\hat{\boldsymbol{F}}$ and $\bar{\boldsymbol{F}}$ are introduced such that $\boldsymbol{F}$ can be multiplicatively split as follows:

$$\boldsymbol{F} = \bar{\boldsymbol{F}} \cdot \hat{\boldsymbol{F}}, \qquad \hat{\boldsymbol{F}} = J^{-1/3}\boldsymbol{F}, \qquad \bar{\boldsymbol{F}} = J^{1/3}\mathbf{1}. \tag{1}$$

Based on this approach, the classical right Cauchy–Green tensor $\boldsymbol{C} = \boldsymbol{F}^{\mathrm{T}} \cdot \boldsymbol{F}$ and an isochoric right Cauchy–Green tensor $\hat{\boldsymbol{C}} = \hat{\boldsymbol{F}}^{\mathrm{T}} \cdot \hat{\boldsymbol{F}}$ as well as their related Green strain tensors are defined:

$$\boldsymbol{E} = \frac{1}{2}(\boldsymbol{C} - \mathbf{1}), \qquad \hat{\boldsymbol{E}} = \frac{1}{2}\left(\hat{\boldsymbol{C}} - \mathbf{1}\right). \tag{2}$$

Between these two Green strain tensors, the following relation holds:

$$\boldsymbol{E} = J^{2/3}\hat{\boldsymbol{E}} + \frac{1}{2}\left(J^{2/3} - 1\right)\mathbf{1}. \tag{3}$$

The Cauchy stress tensor $\boldsymbol{T}$ is expressed as the sum of a spherical and a deviatoric part:

$$\boldsymbol{T} = -p\mathbf{1} + \boldsymbol{T}^{\mathrm{D}}, \qquad p = -\frac{1}{3}\mathrm{tr}(\boldsymbol{T}), \qquad \boldsymbol{T}^{\mathrm{D}} = \boldsymbol{T} - \frac{1}{3}\mathrm{tr}(\boldsymbol{T})\mathbf{1}. \tag{4}$$

The scalar p is the hydrostatic pressure. Taking these decompositions into account, two stress tensors of the second Piola–Kirchhoff type are introduced:

$$\tilde{\boldsymbol{T}} = J\boldsymbol{F}^{-1} \cdot \boldsymbol{T} \cdot \boldsymbol{F}^{\mathrm{T}-1}, \qquad \hat{\tilde{\boldsymbol{T}}} = J\hat{\boldsymbol{F}}^{-1} \cdot \boldsymbol{T}^{\mathrm{D}} \cdot \hat{\boldsymbol{F}}^{\mathrm{T}-1}. \tag{5}$$

A short calculation based on (4) and (5) leads to the following relation between them:

$$\tilde{\boldsymbol{T}} = -pJ^{1/3}\hat{\boldsymbol{C}}^{-1} + J^{-2/3}\hat{\tilde{\boldsymbol{T}}}\,. \tag{6}$$

Considering (3) and (6), the total stress power decomposes additively into a volumetric and an isochoric term (cf. Lion et al. (2014)):

$$\tilde{\boldsymbol{T}} : \dot{\boldsymbol{E}} = -p\dot{\varepsilon}_{\text{vol}} + \hat{\tilde{\boldsymbol{T}}} : \dot{\hat{\boldsymbol{E}}}\,. \tag{7}$$

Following Lion et al. (2014), the specific hybrid free energy ϕ is introduced by a Legendre transformation in which ψ is the Helmholtz free energy per unit mass and ρ_{R} the density:

$$\phi = \psi + \frac{1}{\rho_{\text{R}}} p\varepsilon_{\text{vol}}\,. \tag{8}$$

As a result of the previous discussion, (7) and (8) are inserted into the classical formulation of the Clausius–Duhem inequality (cf. e.g. Malvern 1969 or Haupt 2002) which has to be nonnegative for arbitrary thermomechanical processes:

$$-\rho_{\text{R}}\dot{\psi} + \tilde{\boldsymbol{T}} : \dot{\boldsymbol{E}} - \rho_{\text{R}}s\dot{\theta} - \frac{\boldsymbol{q}_{\text{R}} \cdot \boldsymbol{g}_{\text{R}}}{\theta} \geq 0. \tag{9}$$

In this inequality, s is the specific entropy, the vectors $\boldsymbol{q}_{\text{R}}$ and $\boldsymbol{g}_{\text{R}}$ are the heat flux and the temperature gradient and θ is the absolute temperature. Considering (7), (8) and (9) the following result is obtained:

$$-\rho_{\text{R}}\dot{\varphi} + \hat{\tilde{\boldsymbol{T}}} : \dot{\hat{\boldsymbol{E}}} + \varepsilon_{\text{vol}}\dot{p} - \rho_{\text{R}}s\dot{\theta} - \frac{\boldsymbol{q}_{\text{R}} \cdot \boldsymbol{g}_{\text{R}}}{\theta} \geq 0. \tag{10}$$

It substantiates that the isochoric Green strain tensor $\hat{\boldsymbol{E}}$, the pressure p and the temperature θ are the natural independent variables of the specific hybrid free energy. In order to model the thermoviscoelastic behaviour of polymers in the vicinity of the glass transition, additional variables are introduced:

$$\varphi = \varphi\left(p, \hat{\boldsymbol{E}}, \theta, \boldsymbol{Q}_1, \ldots, \boldsymbol{Q}_n, q_1, \ldots, q_m\right). \tag{11}$$

The internal variables $\boldsymbol{Q}_k$ and q_k are second-order tensors and scalars. They are introduced to represent history-dependent phenomena like the glass transition and viscoelasticity. Inserting the time rate of (11) into (10) the following result is found after a short calculation:

$$\begin{aligned}&\left(\hat{\tilde{\boldsymbol{T}}} - \rho_{\text{R}}\frac{\partial\varphi}{\partial\hat{\boldsymbol{E}}}\right) : \dot{\hat{\boldsymbol{E}}} + \left(\varepsilon_{\text{vol}} - \rho_{\text{R}}\frac{\partial\varphi}{\partial p}\right)\dot{p} - \rho_{\text{R}}\left(s + \frac{\partial\varphi}{\partial\theta}\right)\dot{\theta}\\ &\quad -\rho_{\text{R}}\left(\frac{\partial\varphi}{\partial\boldsymbol{Q}_k} : \dot{\boldsymbol{Q}}_k + \frac{\partial\varphi}{\partial q_k}\dot{q}_k\right) - \frac{\boldsymbol{q}_{\text{R}} \cdot \boldsymbol{g}_{\text{R}}}{\theta} \geq 0.\end{aligned} \tag{12}$$

Applying the standard argumentation (see, e.g. Haupt 2002) for the evaluation of (12) under consideration of the relation $\hat{\boldsymbol{C}}^{-1} : \dot{\hat{\boldsymbol{C}}} = 0$ which holds for unimodular tensors and is equivalent to $\hat{\boldsymbol{C}}^{-1} : \dot{\hat{\boldsymbol{E}}} = 0$, the following potential relations are obtained:

$$\hat{\tilde{\boldsymbol{T}}} = \rho_R \frac{\partial \varphi}{\partial \hat{\boldsymbol{E}}} + \Phi \hat{\boldsymbol{C}}^{-1}, \qquad \varepsilon_{\text{vol}} = \rho_R \frac{\partial \varphi}{\partial p}, \qquad s = -\frac{\partial \varphi}{\partial \theta}. \tag{13}$$

The residual inequality which has to be satisfied by the evolution equations for the internal variables and the heat flux vector reads as follows:

$$-\rho_R \left(\frac{\partial \phi}{\partial \boldsymbol{Q}_k} : \dot{\boldsymbol{Q}}_k + \frac{\partial \phi}{\partial q_k} \dot{q}_k \right) - \frac{\boldsymbol{q}_R \cdot \boldsymbol{g}_R}{\theta} \geq 0. \tag{14}$$

The parameter Φ in the potential relation for the stress tensor in (13) can be computed. To this end, it has to be demanded that the tensor $\boldsymbol{T}^D = \hat{\boldsymbol{F}} \cdot \hat{\tilde{\boldsymbol{T}}} \cdot \hat{\boldsymbol{F}}^T / J$ is deviatoric. This leads to the following result:

$$\Phi = -\frac{\rho_R}{3} \frac{\partial \phi}{\partial \hat{\boldsymbol{E}}} : \hat{\boldsymbol{C}}, \qquad \hat{\tilde{\boldsymbol{T}}} = \rho_R \left(\frac{\partial \varphi}{\partial \hat{\boldsymbol{E}}} - \frac{1}{3} \left(\frac{\partial \varphi}{\partial \hat{\boldsymbol{E}}} : \hat{\boldsymbol{C}} \right) \hat{\boldsymbol{C}}^{-1} \right). \tag{15}$$

The proposed approach which is based on the hybrid free energy leads to potential relations for the isochoric part of the second Piola–Kirchhoff stress tensor, the volumetric strain and the specific entropy. In order to compute the total second Piola–Kirchhoff stress tensor (6), the potential relation for the volumetric strain has to be inverted such that a relation in the form of $p = f(\varepsilon_{\text{vol}}, \theta, \ldots)$ is obtained. This rearrangement can easily be done under the assumption of linearity with respect to the dependence of the volumetric strain on the pressure.

3 A Simple Model for the Hybrid Free Energy in Finite Thermoviscoelasticity

In order to model the thermoviscoelastic properties of polymers under large strains isotropic material behaviour is assumed such that thermal expansion is a pure volumetric effect. Since caloric phenomena are usually studied under prescribed pressure and temperature, the hybrid free energy density is additively split into the sum of two contributions:

$$\varphi = \varphi_{\text{vol}}(p, \theta, \delta) + \varphi_{\text{iso}}\left(\hat{\boldsymbol{E}}, \theta, \boldsymbol{Q}_1, \ldots, \boldsymbol{Q}_n, \delta\right). \tag{16}$$

In the case of calorimetric simulations, the strains are purely volumetric and the stress tensor is spherical such that $\hat{\boldsymbol{E}} \equiv \boldsymbol{0}$ holds and $\phi_{\text{iso}}\left(\boldsymbol{0}, \theta, \boldsymbol{Q}_1, \ldots, \boldsymbol{Q}_n, \delta\right) = 0$ can be assumed. For the volumetric part of (16) the model proposed in Lion et al. (2014) is applied:

Accordingly, the factors of pressure and temperature in (41) can be physically interpreted as equilibrium values of the reciprocal bulk modulus and the thermal expansion coefficient. If the material is in its equilibrium state and the pressure or the temperature is changed with an infinitely high rate, the internal variable δ is frozen and does not change. In this case, the material is in the glassy state and the spontaneous response of the volumetric strain is as follows:

$$\dot{\varepsilon}_{\text{vol\_0}} = \alpha_0 \dot{\theta} - \frac{1}{\kappa_0} \dot{p}. \tag{42}$$

The factors of the temperature rate and the pressure rate are the glassy thermal expansion coefficient and the related value of the reciprocal bulk modulus. Comparing (41) with (42) and assuming nonnegative material constants, the equilibrium values of the two coefficients are larger than their values in the glassy state. For the computation as well as for numerical and theoretical investigations of the isobaric specific heat of this model, the reader is referred to Lion et al. (2014) and the citations therein. In this article, the behaviour of specific heat and volumetric strain is simulated for temperature excitations with piecewise constant rates below and above the glass transition.

For the simulations in the present work, the constitutive Eqs. (33)–(36) were solved with the mathematical software MATLAB. To this end, the following constants were used:

$$\rho_R = 1050 \frac{\text{kg}}{\text{m}^3} \qquad B = 0.1$$

$$f_g = 0.1 \qquad \alpha_f = 4.8 \times 10^{-4} \frac{1}{\text{K}}$$

$$A_0 = 10^{-3} \frac{1}{\text{s}} \qquad e = 7.9 \frac{\text{J}}{\text{kgK}}$$

$$d = 79.5 \frac{\text{J}}{\text{kg}} \qquad \eta = 10^8 \text{Ns/m}^2$$

$$\beta_0 = 3.6 \frac{1}{\text{K}} \qquad \alpha_0 = 6 \times 10^{-4} \frac{1}{\text{K}}$$

$$w = 1.1 \times 10^{-5} \frac{\text{m}^2\text{J}}{\text{kgN}} \qquad \mu = 10^3 \text{Nm/kg}$$

$$c_{p0} = 10^3 \frac{\text{J}}{\text{kgK}} \qquad \theta_0 = 333\text{K}$$

Their values were not experimentally identified but roughly estimated such that the model can reproduce the typical behaviour of polymers in the vicinity of the glass transition.

In order to provide explanatory simulations, deformation-controlled shear processes under isobaric and isothermal conditions are studied. The tangent of the shear angle is described by the time-dependent shear ratio $\gamma(t)$, the related shear stress is $\tau(t)$, the pressure is p and θ is the absolute temperature. In order to express the isochoric part of the deformation gradient, the related right Cauchy-Green tensor and the Cauchy stress the dyads of the Cartesian unit vectors are used:

$$\hat{\boldsymbol{F}} = \boldsymbol{e}_1 \otimes \boldsymbol{e}_1 + \boldsymbol{e}_2 \otimes \boldsymbol{e}_2 + \boldsymbol{e}_3 \otimes \boldsymbol{e}_3 + \gamma \boldsymbol{e}_1 \otimes \boldsymbol{e}_2, \tag{43}$$

$$\hat{\boldsymbol{C}} = \boldsymbol{e}_1 \otimes \boldsymbol{e}_1 + \left(1+\gamma^2\right) \boldsymbol{e}_2 \otimes \boldsymbol{e}_2 + \boldsymbol{e}_3 \otimes \boldsymbol{e}_3 + \gamma \left(\boldsymbol{e}_1 \otimes \boldsymbol{e}_2 + \boldsymbol{e}_2 \otimes \boldsymbol{e}_1\right), \tag{44}$$

$$\boldsymbol{T} = -p \left(\boldsymbol{e}_1 \otimes \boldsymbol{e}_1 + \boldsymbol{e}_2 \otimes \boldsymbol{e}_2 + \boldsymbol{e}_3 \otimes \boldsymbol{e}_3\right) + \tau \left(\boldsymbol{e}_1 \otimes \boldsymbol{e}_2 + \boldsymbol{e}_2 \otimes \boldsymbol{e}_1\right). \tag{45}$$

Based on the constitutive relation for the second Piola–Kirchhoff stress tensor (36), the Cauchy stress (45) can be straightforwardly computed under consideration of (1) and (5):

$$\boldsymbol{T} = \left(\varepsilon_{\mathrm{vol}} + 1\right)^{-1/3} \hat{\boldsymbol{F}} \cdot \tilde{\boldsymbol{T}} \cdot \hat{\boldsymbol{F}}^{\mathrm{T}}. \tag{46}$$

The inelastic isochoric right Cauchy-Green tensor is an internal state variable of the model, has to satisfy the initial condition $\hat{\boldsymbol{C}}_{\mathrm{i}}(0) = \mathbf{1}$ and is expressed as follows:

$$\hat{\boldsymbol{C}}_{\mathrm{i}} = \alpha_{\mathrm{i}} \boldsymbol{e}_1 \otimes \boldsymbol{e}_1 + \beta_{\mathrm{i}} \boldsymbol{e}_2 \otimes \boldsymbol{e}_2 + \lambda_{\mathrm{i}} \boldsymbol{e}_3 \otimes \boldsymbol{e}_3 + \gamma_{\mathrm{i}} \left(\boldsymbol{e}_1 \otimes \boldsymbol{e}_2 + \boldsymbol{e}_2 \otimes \boldsymbol{e}_1\right). \tag{47}$$

In order to implement the condition $\det\left(\hat{\boldsymbol{C}}_{\mathrm{i}}\right) = 1$ the four scalar functions in (47) depend on each other. If three of them a known, the fourth one can be calculated:

$$\lambda_{\mathrm{i}} = \frac{1}{\alpha_{\mathrm{i}} \beta_{\mathrm{i}} - \gamma_{\mathrm{i}}^2}. \tag{48}$$

In the following simulations, the evolution Eq. (34) for the internal variable δ is solved for two different types of initial conditions:
Equilibrium w.r.t. θ and p:

$$\delta(0) = \frac{wp - e\left(\theta - \theta_0\right)}{d}, \tag{49}$$

Equilibrium only w.r.t. θ:

$$\delta(0) = -\frac{e\left(\theta - \theta_0\right)}{d}. \tag{50}$$

The initial condition (49) makes sure that the simulation starts in an equilibrium state with regard to both pressure and temperature. During the application of the time-dependent shear deformation both the internal variable δ and the volumetric strain $\varepsilon_{\mathrm{vol}}$ remain temporally constant.

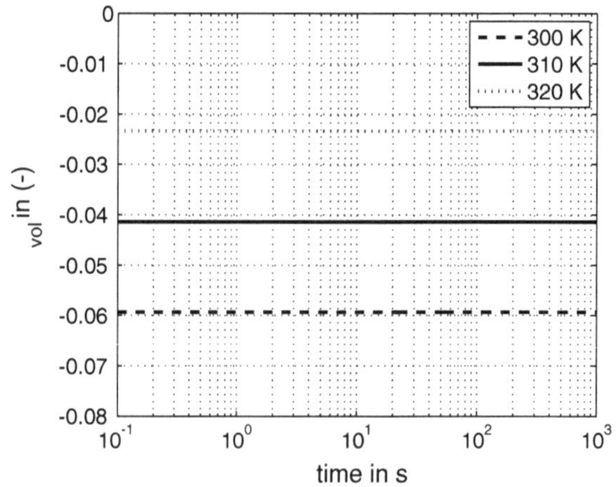

Fig. 1 Volumetric strain as function of time and temperature for $p = 0$ and start in equilibrium

The initial condition (50) provokes that the simulation starts in an equilibrium state with regard to temperature and vanishing pressure. If, at the very beginning of such a simulation, the pressure is instantaneously changed from zero to a constant value, the internal variable δ and the volumetric strain are evolving with time. Since the volumetric strain alters the viscosity function η in the evolution Eq. (33) for the inelastic right Cauchy-Green tensor, the viscoelastic response behaviour of the shear stress is influenced. Such effects are known as physical ageing.

In all simulations, the shear ratio is increased with a constant rate of $\dot{\gamma} = 1/\text{s}$ to the final value of $\gamma_{\max} = 1$. Then, it is kept constant for 10^3s. In Figs. 1, 3 and 5, the simulated volumetric strain is plotted as function of the time for different values of temperature and pressure and for the different initial conditions. The dependence

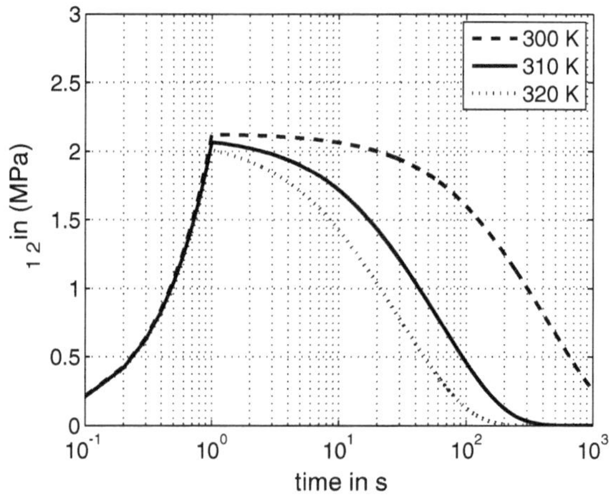

Fig. 2 Shear stress as function of time and temperature for $p = 0$ and start in equilibrium

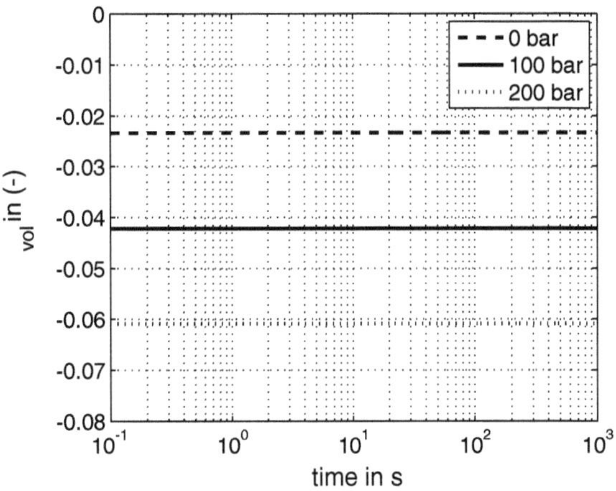

Fig. 3 Volumetric strain as function of time and pressure for $\theta = 320$K and start in equilibrium

of ε_{vol} on temperature and pressure which is depicted in Figs. 1 and 3 is physically reasonable. In Fig. 5, the simulations started in a non-equilibrium state with regard to the pressure such that ε_{vol} exhibits time-dependent creep behaviour just until the related equilibrium values are reached. Since the viscosity function is influenced by the free volume as described in (37) and (39) the stress relaxation and the evolution of the inelastic strains depend on pressure too. Figures 4 and 6 demonstrate that the stress relaxation becomes slower with increasing hydrostatic pressure. Experimental investigations of the pressure dependence of the creep modulus of three different thermoplastics can be found in Naumann and Stommel (2011). The numerical simulations depicted in Fig. 4 reproduce qualitatively the behaviour observed by Naumann and Stommel (2011) (Fig. 2).

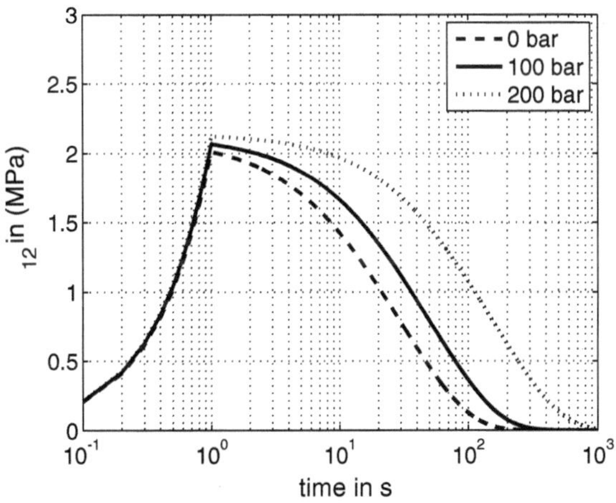

Fig. 4 Shear stress as function of time and pressure for $\theta = 320$K and start in equilibrium

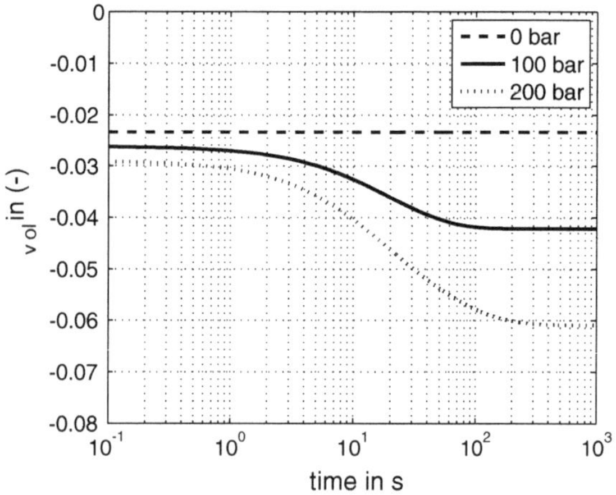

Fig. 5 Volumetric strain as function of time and pressure for $\theta = 320\text{K}$ and start in non-equilibrium

5 Discussion and Conclusion

In this investigation, a simple approach of finite nonlinear viscoelasticity which is based on the concept of the hybrid free energy function is proposed. In Lion et al. (2014), this concept was developed and applied to describe the thermal expansion and the isobaric specific heat of glass-forming materials under temperature histories with piecewise constant rates. In the current article, it is demonstrated that this concept is also applicable to represent the isochoric behaviour of glass-forming polymers under mechanical shear loads. In a recent approach developed by Naumann and Stommel (2011), the stiffness parameters of their Maxwell series depend on the free volume. In the present model, the viscosity depends on the free volume. The approach of

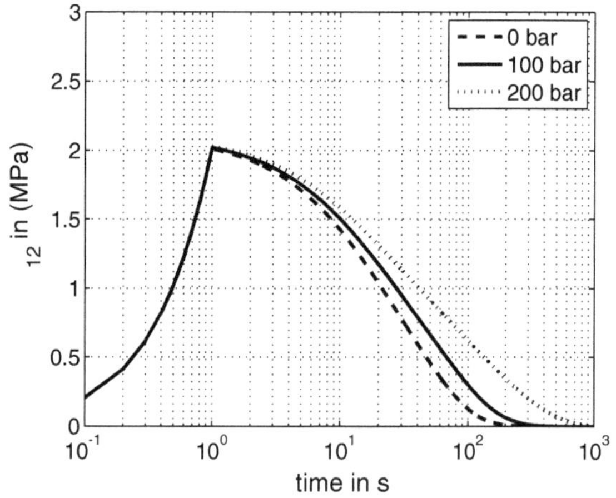

Fig. 6 Shear stress as function of time and pressure for $\theta = 320\text{K}$ and start in non-equilibrium

finite multiplicative viscoelasticity which has been used here to formulate the model can easily be extended by introducing a larger set of multiplicative decompositions of the deformation gradient in parallel. Due to more material parameters, such a generalisation enables the quantitative representation of the thermoviscoelastic material behaviour of real polymers in the vicinity of the glass transition.

As a first practical conclusion, it is mentioned that the free volume as independent variable of the viscosity (or the relaxation time) in the evolution equations of the internal variables allows the representation of pressure- and temperature-dependent viscoelastic effects in parallel. The second conclusion is that the constitutive approach which is based on the hybrid free energy allows modelling the calorimetric material behaviour (isobaric specific heat) under thermal loadings as shown in Lion et al. (2014) and the isochoric mechanical behaviour under mechanical loads as shown in this study.

References

Altenbach, H.: Kontinuumsmechanik: Einführung in die Materialunabhängigen und Materialabhängigen Gleichungen, 3rd edn. Springer, Heidelberg (2015). doi:10.1007/978-3-662-47070-1

Altenbach, .H., Altenbach, J.: Einfürung in die Kontinuumsmechanik. Teubner (1994)

Doolittle, A.K.: Studies in newtonian flow ii. the dependence of the viscosity of liquids on free-space. J. Appl. Phys. **22**(12), 1471–1475 (1951)

Flory, P.: Thermodynamic relations for high elastic materials. Trans. Faraday Soc. **57**, 829–838 (1961)

Greiner, R., Schwarzl, F.: Thermal contraction and volume relaxation of amorphous polymers. Rheol. Acta **23**(4), 378–395 (1984)

Haupt, P.: Continuum Mechanics and Theory of Materials. Springer, Heidelberg (2002)

Jackle, J.: Models of the glass transition. Rep. Prog. Phys. **49**(2), 171 (1986)

Johlitz, M., Scharding, D., Diebels, S., Retka, J., Lion, A.: Modelling of thermo-viscoelastic material behaviour of polyurethane close to the glass transition temperature. ZAMM-J. Appl. Math. Mech./Zeitschrift für Angewandte Mathematik und Mechanik **90**(5), 387–398 (2010)

Lion, A.: On the large deformation behaviour of reinforced rubber at different temperatures. J. Mech. Phys. Solid **45**(11), 1805–1834 (1997)

Lion, A., Liebl, C., Kolmeder, S., Peters, J.: Representation of the glass-transition in mechanical and thermal properties of glass-forming materials: a three-dimensional theory based on thermodynamics with internal state variables. J. Mech. Phys. Solid **58**(9), 1338–1360 (2010)

Lion, A., Peters, J., Kolmeder, S.: Simulation of temperature history-dependent phenomena of glass-forming materials based on thermodynamics with internal state variables. Thermochim. Acta **522**(1), 182–193 (2011)

Lion, A., Dippel, B., Liebl, C.: Thermomechanical material modelling based on a hybrid free energy density depending on pressure, isochoric deformation and temperature. Int. J. Solid. Struct. **51**(3), 729–739 (2014)

Lubliner, J.: A model of rubber viscoelasticity. Mech. Res. Commun. **12**, 93–99 (1985)

Malvern, L.E.: Introduction to the Mechanics of a Continuous Medium. Prentice-Hall, Upper Saddle River (1969)

Naumann, T., Stommel, M.: Simulation of the long term behavior of polymers on the basis of short term data. Tech. Mech. **32**, 446–462 (2011)

field proportional to the magnitude of the deformation, cf. Landau et al. (1984), Kittel (2004), McMeeking and Landis (2005). Both effects can be described in linearized kinematics so that the responses of the mechanical and the electrical field can simply be added.

In contrast, EAPs typically exploit significant displacement levels to enable movement and to generate force. When stimulated by an electric field, the polymer undergoes large elongations which require to sustain high stresses and voltages. As a consequence limiting effects like material stability and dielectric breakdown also play a role. The electromechanical interaction is evidently complex, and a model of electroactive materials requires a fully nonlinear coupling of the electrical and the mechanical field.

In this contribution, we propose a constitutive model for electroactive materials based on a multiplicative decomposition of the deformation gradient into passive and active components. Our model extends superposition approaches popular in recent literature, cf. Ask et al. (2012), Klassen et al. (2012), Suo (2010), Zwecker and Klinkel (2011); it is deduced from a consistent thermodynamic framework and relies on an additive decomposition of the underlying Helmholtz free energy density of the material. In addition, we suggest an evolution equation for the active deformation and present a numerical study to illustrate its effect on the overall electroactive response.

2 Formulation of the Problem

We consider an EAP and denote a point in its material configuration $\mathcal{B}_0 \subset \mathbb{R}^3$ by $\boldsymbol{X} = (X_1, X_2, X_3)^T$. An electric field applied during a time interval $[0, \bar{t}]$ causes a deformation and the material point is now located at position $\boldsymbol{x}(\boldsymbol{X}, t)$ of the current configuration $\mathcal{B}_t \subset \mathbb{R}^3$. The corresponding deformation mapping $\boldsymbol{\chi}(\boldsymbol{X}, t) : \mathcal{B}_0 \times [0, \bar{t}\,] \rightarrow \mathcal{B}_t$ is uniquely described by its material gradient,

$$\boldsymbol{F}(\boldsymbol{X}, t) = \frac{\partial \boldsymbol{\chi}}{\partial \boldsymbol{X}} = \nabla_X \boldsymbol{x}(\boldsymbol{X}, t), \tag{1}$$

whereas the electric field is given by the spacial gradient of a potential ϕ,

$$\boldsymbol{e}(\boldsymbol{x}, t) = -\frac{\partial \phi}{\partial \boldsymbol{x}} = -\nabla_x \phi(\boldsymbol{x}, t) . \tag{2}$$

2.1 *Mechanical Fields and Equilibrium*

The local equilibrium reads with respect to the material configuration

$$\nabla_X \cdot \boldsymbol{P} + \boldsymbol{f}_0^V = 0 \qquad \text{in } \mathcal{B}_0 \times [0, \bar{t}] , \tag{3}$$

where $\boldsymbol{P}$ is the first Piola–Kirchhoff stress tensor and $\boldsymbol{f}_0^V, \boldsymbol{f}_0^A$ are body forces and traction. The boundary of $\mathcal{B}_0$ with outward normal $\boldsymbol{N}$ is denoted by $\partial\mathcal{B}_0$. Displacement and traction boundary conditions are

$$\boldsymbol{x} = \bar{\boldsymbol{x}} \qquad \text{on } \partial\mathcal{B}_0^1 \times [0, \bar{t}], \tag{4}$$

$$\boldsymbol{PN} = \bar{\boldsymbol{f}}_0^A \qquad \text{on } \partial\mathcal{B}_0^2 \times [0, \bar{t}]. \tag{5}$$

Note that (3) is the static form of the linear momentum balance and the angular momentum balance is satisfied through the symmetry of the product $\boldsymbol{PF}^T = \boldsymbol{FP}^T$. The weak form of the momentum balance follows with admissible variations $\delta\boldsymbol{x}$ in its standard form:

$$\int_{\mathcal{B}_0} \boldsymbol{P}\delta\boldsymbol{F}\,\mathrm{d}V = \int_{\mathcal{B}_0} \boldsymbol{f}_0^V \delta\boldsymbol{x}\,\mathrm{d}V + \int_{\partial\mathcal{B}_{0,2}} \boldsymbol{f}_0^A \delta\boldsymbol{x}\,\mathrm{d}V\,. \tag{6}$$

For completeness, we state here the right Cauchy-Green deformation tensor $\boldsymbol{C} = \boldsymbol{F}^T\boldsymbol{F}$ and remind that the material's volume change is measured by $J = \det \boldsymbol{F} = \mathrm{d}\varrho_0/\,\mathrm{d}\varrho$, where ϱ_0 and ϱ are the mass density per unit reference and deformed volume, respectively.

2.2 *Electric Fields and Equilibrium*

The electric induction (or electric displacement) $\boldsymbol{d}$ is proportional to the electric field $\boldsymbol{e}$ through the vacuum permittivity ε_0,

$$\boldsymbol{d} = \varepsilon_0\boldsymbol{e} + \boldsymbol{\pi}\,, \tag{7}$$

where $\boldsymbol{\pi}$ is the polarization tensor which depends on the dielectric material. For a local volumetric charge density ρ_0^V, the electric field (2) and the electric induction must satisfy the equilibrium equations of electrostatics,

$$\nabla_x \times \boldsymbol{e} = 0 \quad \text{and} \quad \nabla_x \cdot \boldsymbol{d} + \rho_0^V = 0 \qquad \text{in } \mathcal{B}_0 \times [0, \bar{t}]. \tag{8}$$

Surface charges are assumed to be absent. Let now $\boldsymbol{E}$, $\boldsymbol{D}$, and $\boldsymbol{\Pi}$ denote the corresponding material fields. Then we deduce from the fields (1) and (2),

$$\boldsymbol{e} = -\nabla_x\phi = \left(-\nabla_X^T\phi\,\boldsymbol{F}^{-1}\right)^T = -\boldsymbol{F}^{-T}\nabla_X^T\phi = \boldsymbol{F}^{-T}\boldsymbol{E}\,,$$

which gives $\boldsymbol{E} = \boldsymbol{F}^T\boldsymbol{e}$. Correspondingly follows for the material electric induction the transformation $\boldsymbol{D} = J\boldsymbol{F}^{-1}\boldsymbol{d}$ and for the material polarization $\boldsymbol{\Pi} = J\boldsymbol{F}^{-1}\boldsymbol{\pi}$. Finally, the material electric induction is expressed as

$$D = J\varepsilon_0 C^{-1} E + \Pi \,. \tag{9}$$

Here the first term accounts for the distortion of the electric field due to material deformation. From (8), we derive the weak form which—after a Legendre transform of variables—becomes in the material configuration:

$$\int_{\mathcal{B}_0} D \cdot \delta E \, \mathrm{d}V = \int_{\mathcal{B}_0} \rho_0^V \delta\phi \, \mathrm{d}V \,. \tag{10}$$

2.3 Energy Balance and Dissipation Inequality

The specific internal energy U of the system is composed from contribution of the mechanical and electric energy. The local form of the rate reads

$$\dot{U} = P : \dot{F} + E \cdot D + \varrho_0 Q - \nabla_X \cdot H_T \tag{11}$$

where Q is the heat supply per unit mass and H_T is the material energy flux vector. The local Clausius–Duhem form of the second law of thermodynamics states the non-negativeness of the total entropyproduction $\dot{\Pi}^S$,

$$\dot{\Pi}^S = \dot{N} + \frac{1}{T}\left(\nabla_X \cdot H_T - \varrho_0 Q - \frac{1}{T} H_T \nabla_X T\right) \geq 0 \,. \tag{12}$$

After multiplication with absolute temperature T and with (11), we state

$$T\dot{\Pi}^S = T\dot{N} - \dot{U} + P : \dot{F} + E \cdot D - \frac{1}{T} H_T \cdot \nabla_X T \geq 0.$$

Following Landau et al. (1984), we now use a transformation of the equation of state $\tilde{U} = U - D \cdot E$ to write the energy balance (11)

$$\dot{\tilde{U}} = P : \dot{F} - D \cdot E + \varrho_0 Q - \nabla_X H_T \tag{13}$$

and the entropy production

$$T\dot{\Pi}^S = T\dot{N} - \dot{\tilde{U}} + P : \dot{F} - D \cdot \dot{E} - \frac{1}{T} H_T \cdot \nabla_X T \geq 0 \,. \tag{14}$$

3 Constitutive Model for the Electromechanical Problem

The equations of mechanical equilibrium (3) and of electrical equilibrium (8)—as well as their variational counterparts (6) and (10)—are formulated independently and presented in a decoupled form. Coupling between mechanics and electricity arises from the constitutive relations which will be derived here from thermodynamical considerations.

3.1 Equations of State

Following the classical approach of Coleman and Noll (1963), we assume the thermodynamic state of a material point at $\boldsymbol{X}$ and its infinitesimal neighborhood being completely defined by the variables of state: deformation gradient $\boldsymbol{F}$, electric field $\boldsymbol{E}$, entropy density N, and a set of internal variables $\boldsymbol{Z}$. The latter may account for dissipative phenomena such as damage or phase change. Then it holds for the local energy $U = U(\boldsymbol{F}, N, \boldsymbol{E}, \boldsymbol{Z})$ and the for temperature $T = T(\boldsymbol{F}, N, \boldsymbol{E}, \boldsymbol{Z})$. Taking the differential gives

$$\dot{U} = \frac{\partial U}{\partial \boldsymbol{F}} : \dot{\boldsymbol{F}} + \frac{\partial U}{\partial N}\dot{N} + \frac{\partial U}{\partial \boldsymbol{E}}\dot{\boldsymbol{E}} - \boldsymbol{Y} : \dot{\boldsymbol{Z}} \tag{15}$$

where we introduced the thermodynamic forces conjugated to $\mathbf{Z}$, $\boldsymbol{Y} = -\partial U/\partial \mathbf{Z}$. The first Piola–Kirchhoff decomposes into the equilibrium stress $\boldsymbol{P}^e$ and the rate dependent viscous stress $\boldsymbol{P}^v$,

$$\boldsymbol{P} = \boldsymbol{P}^e(\boldsymbol{F}, N, \boldsymbol{E}, \boldsymbol{Z}) + \boldsymbol{P}^v(\boldsymbol{F}, N, \boldsymbol{E}, \boldsymbol{Z}, \dot{\boldsymbol{F}}). \tag{16}$$

Plugging these relations into (14) gives

$$T\dot{\Pi}^S = \left(\boldsymbol{P}^e - \frac{\partial U}{\partial \boldsymbol{F}}\right) : \dot{\boldsymbol{F}} + \boldsymbol{P}^v : \dot{\boldsymbol{F}} + \left(T - \frac{\partial U}{\partial N}\right)\dot{N} - \left(\boldsymbol{D} + \frac{\partial U}{\partial \boldsymbol{E}}\right)\dot{\boldsymbol{E}}$$
$$- \frac{1}{T}\boldsymbol{H}_T \nabla_X T + \boldsymbol{Y} : \dot{\boldsymbol{Z}} \geq 0. \tag{17}$$

Since inequality (17) must hold for any admissible process, the terms in brackets need to be zero. This defines equations of state for the equilibrium stress, the temperature and the electric induction. The dissipation remains as

$$T\dot{\Pi}^S = \boldsymbol{P}^v : \dot{\boldsymbol{F}} - \frac{1}{T}\boldsymbol{H}_T \nabla_X T + \boldsymbol{Y} : \dot{\boldsymbol{Z}} \geq 0. \tag{18}$$

Here, as thermodynamic potential is alternative to (15), we use the Helmholtz free energy density A, obtained through a Legendre transform: $A(\boldsymbol{F}, T, \boldsymbol{E}, \boldsymbol{Z}) = \inf_N \{U(\boldsymbol{F}, N, \boldsymbol{E}, \boldsymbol{Z}) - TN\}$. This results in the equations of state

$$\boldsymbol{P}^e = \frac{\partial A}{\partial \boldsymbol{F}}, \qquad N = \frac{\partial A}{\partial T}, \qquad \boldsymbol{D} = -\frac{\partial A}{\partial \boldsymbol{E}}, \tag{19}$$

and the thermodynamic force conjugate to $\boldsymbol{Z}$ becomes $\boldsymbol{Y} = -\partial A/\partial \boldsymbol{Z}$. The general framework is completed with the definition of the kinetic relations that enable to determine $\boldsymbol{P}^v$, $\boldsymbol{Z}$ and $\boldsymbol{H}_T$.

In the following we are interested in isothermal and time-independent processes. We simplify the expression of the Helmholtz free energy by dropping the dependence on the internal variables and on the temperature; the equations of state will reduce to $(19)_1$ and $(19)_3$.

3.2 *Kinematics of Electromechanics*

To define the constitutive equations of a coupled process, we assume a multiplicative decomposition of the deformation gradient (1) into elastic and inelastic (i.e. active) components,

$$\boldsymbol{F}(\boldsymbol{X}, t) = \boldsymbol{F}^e \boldsymbol{F}^a. \tag{20}$$

The elastic part of the deformation gradient is related to the passive response of the material, while the active part is introduced to describe the geometrical changes of the initial configuration induced by the action of the electric potential. This decomposition refers to a non-compatible intermediate configuration, where the inelastic phenomena locally take place without inducing a stress state. The compatibility requirement will relax the continuum from the intermediate configuration to the current deformed configuration. Here equilibrium and compatibility conditions are fully satisfied, see Fig. 1. We assume the active component to be volume preserving and enhance the stretch in the direction of the electric field. Specifically, it follows a finite evolution equation of the form

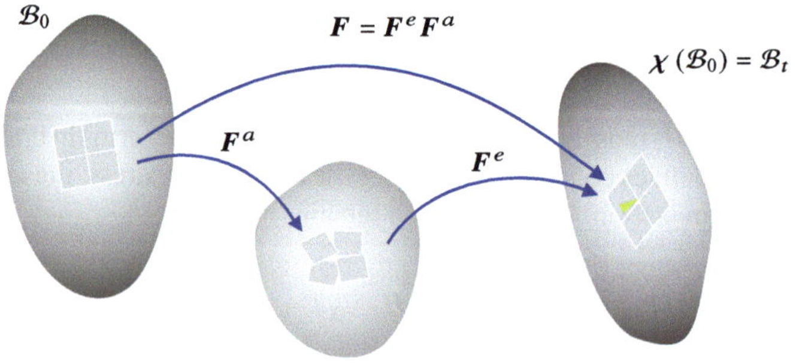

Fig. 1 Reference, current and non-compatible intermediate configuration of the body

$$F^a = I + \kappa \frac{E \otimes E}{|E|^2}, \tag{21}$$

where κ is a dimensionless material parameter. The multiplicative decomposition of the deformation gradient is a convenient mathematical representation of the change of configuration of a system undergoing multiphysics processes in large deformations Eckart (1948). The formal introduction in nonlinear continuum mechanics of the multiplicative decomposition can be attributed to [70] in the case of thermoelasticity, and to [39] in the case of phenomenological elastoplasticity. Also for active biomechanical interactions the approach has seen successful applications recently, Ambrosi et al. (2011), Gizzi et al. (2015), Nardinocchi and Teresi (2007).

3.3 Helmholtz Free Energy

We assume a separation of the arguments in the sense of relation (20) and decompose the Helmholtz free energy additively,

$$A(F, F^e, E) = A^e(F^e) + A^a(F, E).$$

Here the first term represents the classical strain energy density of hyperelastic materials and the second term is an inelastic free energy density that accounts for the effects of the electric field. Additional energy contributions, e.g. due to eigendeformations, may be added.

With $(19)_1$ the equilibrium stress P follows as the sum of two terms, the passive stresses P^p and active stresses P^a,

$$P = P^p + P^a = \frac{\partial A^e}{\partial F^e}\frac{\partial F^e}{\partial F} + \frac{\partial A^a}{\partial F} = P^e F^{a-T} + P^a \tag{22}$$

The passive stress is the pulled-back elastic stress and the active stress is derived from the active part of the free energy A^a; it is the Maxwell stress associated with the presence of an electric field plus an additional contribution of the polarization.

Correspondingly, the electric induction follows from $(19)_3$ as

$$D = -\frac{\partial A}{\partial E} = -\frac{\partial A^M(F, E)}{\partial E} - \frac{\partial A^\Pi(F, F^a(E), E)}{\partial F^i}\frac{\partial F^a(E)}{\partial E}, \tag{23}$$

where we implicitly state $A^a(F, F^a, E) = A^M(F, E) + A^\Pi(F, F^a, E)$. Here the first term accounts for the electric induction in the absence of electric distortions carried by the material, i.e. the one observed in the vacuum, and the second term is related to the polarization Π. It accounts for the effects due to the presence of a material through the electric distortions F^a, see Fig. 2 for an illustration.

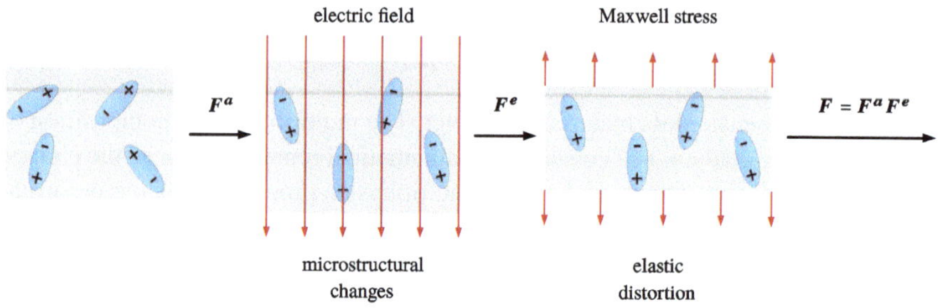

Fig. 2 Active and passive deformation in a dielectric

3.4 Neo-Hookean Dielectric EAP

Let us consider a sample material of neo-Hookean type with classical dielectric properties. The latter corresponds to a polarization $\boldsymbol{\pi} = \varepsilon_r \boldsymbol{e}$ with constant relative permittivity $\varepsilon_r = 1 + \chi_\mathrm{e}$. The elastic strain energy density is

$$A^e(\boldsymbol{F}^e) = \frac{1}{4}(J^2 - 1 - 2\log J) + \frac{\mu}{2}\left(\mathrm{tr}\,\bar{\boldsymbol{C}}^e - 3\right) , \tag{24}$$

where K and μ are bulk and shear modulus and $\bar{\boldsymbol{C}} = J^{-2/3}\boldsymbol{F}^e\boldsymbol{F}^e$. It defines the equilibrium stress in the intermediate configuration as

$$\boldsymbol{P}^e = \frac{\partial A^e}{\partial \boldsymbol{F}^e} = \frac{K}{2}(J^2 - 1)\boldsymbol{F}^{e-1} + \frac{2}{3}J^{-2/3}\mu\boldsymbol{F}^e, \tag{25}$$

and the passive first Piola–Kirchhoff stress tensor is its pull-back to the reference configuration.

The active part of the free energy for a classical dielectric is

$$\begin{aligned} A^a(\boldsymbol{F}, \boldsymbol{E}) &= -\frac{1}{2}\varepsilon_0\varepsilon_r J\, \boldsymbol{e} \cdot \boldsymbol{e} \\ &= -\frac{1}{2}\varepsilon_0(1 + \chi_\mathrm{e})J\, \boldsymbol{E}\boldsymbol{F}^{-1} \cdot \boldsymbol{F}^{-T}\boldsymbol{E}, \end{aligned} \tag{26}$$

It defines the active first Piola–Kirchhoff stress

$$\boldsymbol{P}^a = -\varepsilon_0(1 + \chi_\mathrm{e})J\left[\boldsymbol{E}\boldsymbol{F}^{-1} \otimes \boldsymbol{F}^{-T}\boldsymbol{E} - \frac{1}{2}\left(\boldsymbol{E}\boldsymbol{F}^{-1} \cdot \boldsymbol{F}^{-T}\boldsymbol{E}\right)\boldsymbol{I}\right]\boldsymbol{F}^{-T}, \tag{27}$$

and with $\varepsilon_r = 1 + \chi_\mathrm{e}$, we obtain the mechanical stresses (22) in the form:

$$\boldsymbol{P} = \boldsymbol{P}^e\boldsymbol{F}^{a,-T} + \varepsilon_0\varepsilon_r J\left[\boldsymbol{E}\boldsymbol{F}^{-1}\otimes \boldsymbol{F}^{-T}\boldsymbol{E} - \frac{1}{2}\left(\boldsymbol{E}\boldsymbol{F}^{-1} \cdot \boldsymbol{F}^{-T}\boldsymbol{E}\right)\boldsymbol{I}\right]\boldsymbol{F}^{-T}. \tag{28}$$

Note that here we assume a deformation independent permittivity, typical of classic dielectrics. More generally, a dielectric may be characterized by a permittivity of the type $\chi_e = \chi_e(\boldsymbol{F}, \boldsymbol{F}^a)$, which supports an additive density $A^a = A^M + A^\Pi$, and admits a derivative of the form (23).

Let us have a closer look at the true stresses in the material. By considering purely deviatoric active effects, i.e. a purely elastic volumetric deformation $J = J^e$, standard transformation rules provide the passive Cauchy stress tensor:

$$\boldsymbol{\sigma}^p = J^{-1}\boldsymbol{P}^p\boldsymbol{F}^T = J^{-1}J^e\boldsymbol{\sigma}^e\boldsymbol{F}^{-T}\boldsymbol{F}^T = \boldsymbol{\sigma}^e. \tag{29}$$

The active stress follows correspondingly from (27) as

$$\boldsymbol{\sigma}^a = J^{-1}\boldsymbol{P}^a\boldsymbol{F}^T = \varepsilon_0(1+\chi_e)\left[\boldsymbol{e}\otimes\boldsymbol{e} - \frac{1}{2}(\boldsymbol{e}\cdot\boldsymbol{e})\boldsymbol{I}\right] \tag{30}$$

which for $\chi_e = 0$ coincides with the well-known Maxwell stresses, Suo (2010), Dorfmann and Ogden (2005).

Finally, we derive the electric induction from the free energy density (26).

$$\boldsymbol{D} = -\frac{\partial A}{\partial \boldsymbol{E}} = \frac{\partial}{\partial \boldsymbol{E}}\left(\frac{1}{2}\varepsilon_0\varepsilon_r J\boldsymbol{E}\boldsymbol{F}^{-1}\cdot\boldsymbol{F}^{-T}\boldsymbol{E}\right) = J\varepsilon_0\varepsilon_r\boldsymbol{C}^{-1}\boldsymbol{E} \tag{31}$$

Linearization of the relations (28) and (31) with respects to $\boldsymbol{F}$ and $\boldsymbol{E}$ gives the consistent tangent of the coupled problem.

4 Parametric study

With the aim to adapt and optimize polymeric actuators, we performed several experimental investigations on EAPs recently, see Gaida et al. (2013), Gaida et al. (2014). Our observations motivated the derivation of the presented material model. In particular, the decomposition (20) of the deformation gradient, $\boldsymbol{F} = \boldsymbol{F}^e\boldsymbol{F}^a$ has been introduced to account for the coupling of elastic and active effects. Equation (21) determines the evolution of the inelastic part of deformation.

$$\boldsymbol{F}^a = \boldsymbol{I} + \kappa\frac{\boldsymbol{E}\otimes\boldsymbol{E}}{|\boldsymbol{E}|^2}$$

The material-specific parameter $\kappa \in (-1, \infty)$ is introduced to control the coupling intensity. In the limit case $\kappa = -1$, the tensor $\boldsymbol{F}^a$ may have undefined entries. The case $\kappa = 0$ corresponds to an uncoupled model, where the reference configuration and the intermediate configuration coincide, see also Fig. 1. This may be an appropriate model for a material without inelastic deformations, cf. Dorfmann and Ogden (2005),

Suo (2010), Zwecker and Klinkel (2011). In practice, such a model may suffice if a homogenous electric field is applied on a uniform isotropic EAP of simple geometry.

This corresponds well to our experimental observations where we used a silicon film kit provide by *Danfoss Polypower A/S* to prepare dielectric EAPs, Bien (2015). The specimen have been slightly prestretched at 4.3 % and then exposed to an electric field with a voltage in the range of 500–1500 V. For a rectangular specimen, the difference of measured and theoretical stretch is small, with relative error $\leq 0.05\,\%$. When the used geometry, however, causes a non-homogenous deformation, we observe a deviation from the ideal state which grows with the applied voltage. This may be explained with non-uniform changes in the microstructure of the polymer which locally enhance the deformation caused by the electric field via Maxwell stress. Such microstructural changes may become permanent as well as induced electric dipole moments in the material, see Fig. 2. The active part of the deformation gradient $\boldsymbol{F}^a$ has been introduced to account for these local effects.

In the following, we study the proposed framework analytically and show the effect of parameter κ on the stretch at given voltage. To this end, we assume the material to be an incompressible neo-Hookean with strain energy density (24) and $J = 1$. The deformation gradient is formulated in principle stretches λ_a, $a = 1, 2, 3$, for the resulting equations see also Reppel et al. (2013).

Uniaxial State

The applied electric field is constant in one direction.

$$\boldsymbol{E} = \begin{bmatrix} 0 \\ 0 \\ E \end{bmatrix} \tag{32}$$

With prescribed elongation $\lambda = l/L$, the elastic component of the deformation gradient $\boldsymbol{F}^e$ has a diagonal form, and with field (32) in (21) the active component follows accordingly.

$$\boldsymbol{F}^e = \begin{bmatrix} \lambda^{-1/2} & 0 & 0 \\ 0 & \lambda^{-1/2} & 0 \\ 0 & 0 & \lambda \end{bmatrix} \qquad \boldsymbol{F}^a = \begin{bmatrix} 1 & 0 & 0 \\ 0 & 1 & 0 \\ 0 & 0 & 1+\kappa \end{bmatrix} \tag{33}$$

The equilibrium stress (25) results in one component $P_3^e = \mu(\lambda - 1/\lambda^2)$, the corresponding passive stresses follow as $P_1^p = P_2^p = 0$ and $P_3^p = \mu(\lambda - \lambda^{-2})/F_3$. The active stresses (27) evaluate to $P_1^a = P_2^a = 0$ and $P_3^a = \varepsilon_0 \varepsilon_r E^2/F_3^3$. We reformulate in terms of true stresses to obtain the Cauchy stress tensor:

$$\boldsymbol{\sigma} = \boldsymbol{\sigma}^p + \boldsymbol{\sigma}^a = \mu \begin{bmatrix} 0 & 0 & 0 \\ 0 & 0 & 0 \\ 0 & 0 & \lambda^2 - \frac{1}{\lambda} \end{bmatrix} + \frac{1}{2}\varepsilon_0 \varepsilon_r \begin{bmatrix} 0 & 0 & 0 \\ 0 & 0 & 0 \\ 0 & 0 & \frac{E^2}{\lambda^2(1+\kappa)^2} \end{bmatrix} \tag{34}$$

Presuming equilibrium, i.e. $|\sigma_{33}^p| = |\sigma_{33}^a|$, leads to:

$$\left(\lambda^4 - \lambda\right)(1+\kappa)^2 = \varepsilon_0 \varepsilon_r \frac{E^2}{\mu}. \tag{35}$$

With different voltages, this equation has iteratively been solved for λ; the results are displayed in Figs. 3 and 4. The chosen material parameters are: $\mu = 3.3\,\text{MPa}$, $\varepsilon_0 = 8.854 \times 10^{-12}\,\text{As/Vm}$, $\varepsilon_r = 8$ and an initial length of $L_3 = 40\,\mu\text{m}$.

From Eq. (35) and the plotted results, it is clearly visible, that for $\kappa < 0$ the electrically induced deformation of the strip is enlarged. Negative κ-values enhance the electromechanical effect, $\lambda(\kappa < 0) > \lambda(\kappa = 0)$. In our experiments, we also observed a raise of stretch for the non-homogenous specimen; these results, however, are preliminary at this point. On the opposite side, for $\kappa > 0$ the electrically induced deformation reduces, $\lambda(\kappa > 0) < \lambda(\kappa = 0)$. Positive κ-values in the evolution equation (21) work against an active deformation. For large values of κ, we observe a convergence against $\lambda = 1$. This means, the deformation is completely compensated, there is no electroactive effect. For $\kappa = -1$ function (35) is not defined, the material would be completely squeezed. We consider this to be the lower limit for κ. Meaningful values, however, are much smaller and we specify the suggested range of material parameters to $-1 \ll \kappa \leq 1$.

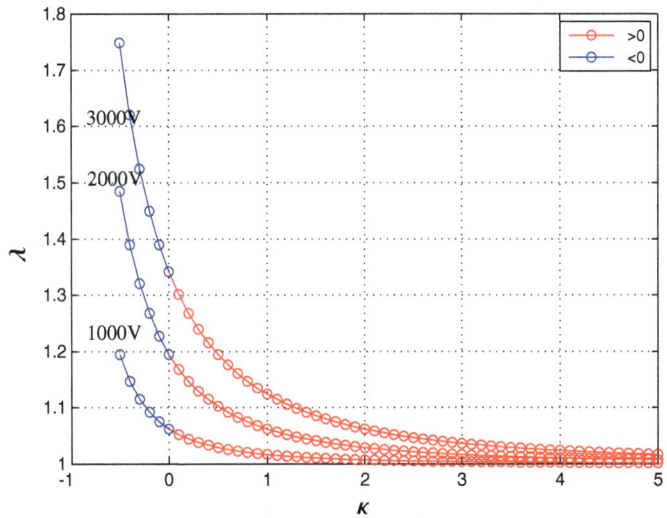

Fig. 3 Principle stretch λ_3 versus parameter κ at a voltage of 1000 V, 2000 V and 3000 V; negative values of κ are plotted in *blue*, positive values in *red*

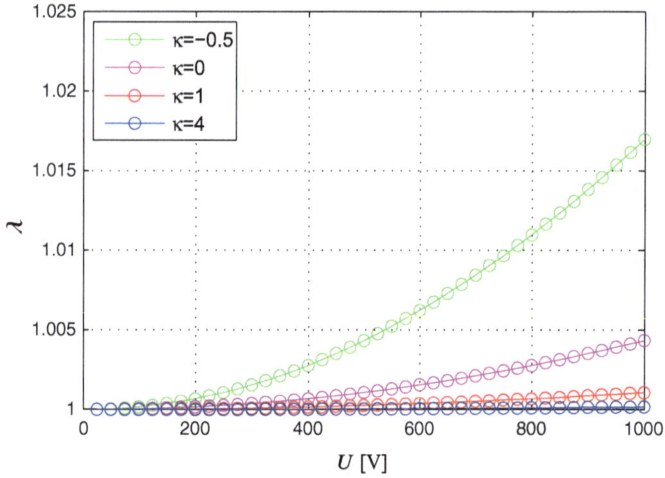

Fig. 4 Effect of variable κ on the stretch λ at in the uniaxial case

References

Ambrosi, D., Arioli, G., Nobile, F., Quarteroni, A.: Electromechanical coupling in cardiac dynamics: the active strain approach. SIAM J. Appl. Math. **71**(2), 605–621 (2011)

Ask, A., Menzel, A., Ristinmaa, M.: Electrostriction in electro-viscoelastic polymers. Mech. Mater. **50**, 9–21 (2012)

Bien, W.: Verformungsberechnungen hyperelastischer DEAP-Aktoren anhand analytischer und numerischer Methoden sowie Validierung mittels einer Hochspannungsvorrichtung. Diplomarbeit, Universität Siegen (2015)

Coleman, B.D., Noll, W.: The thermodynamics of elastic materials with heat conduction and viscosity. Arch. Ration. Mech. Anal. **13**(1), 167–178 (1963)

Dorfmann, A., Ogden, R.W.: Nonlinear electroelasticity. Acta Mechanica **174**, 167–183 (2005)

Eckart, C.: The thermodynamics of irreversible processes, IV. The theory of elasticity and anelasticity. Phys. Rev. **73**, 373–380 (1948)

Gaida, P., Weinberg, K., Jahr, A.: Preliminary survey of a diaphragm EAP stack for the use as an actuator in a water valve. In: Proceedings of the 14th International Workshop on Research and Education in Mechatronics (REM 2013) (2013)

Gaida, P., Weinberg, K., Jahr, A.: Viscoelastic modelling and experimental results of a dielectric electro-active polymer diaphragm actuator. PAMM **14**, 1–2 (2014)

Gizzi, A., Cherubini, C., Filippi, S., Pandolfi, A.: Theoretical and numerical modeling of nonlinear electromechanics with applications to biological active media. Commun. Comput. Phys. **17**(1), 93–126 (2015)

Karsten, R., Ibis, M., Groche, P., Schlaak, H.F.: Investigation of an applicability of dielectric elastomer actuators (DEA) for active vibration attenuation on planar structures. In: Proceedings of the AIA-DAGA 2013 Conference on Acoustics (2013)

Kittel, C.: Introduction to Solid State Physics. Wiley, New York (2004)

Klassen, M., Xu, B., Klinkel, S., Müller, R.: A solid shell finite element formulation to simulate the behaviour of thin dielectric elastomer structures. Technische Mechanik **32**(1), 38–52 (2012)

Landau, L.D., Lifshitz, E.M., Pitaevskii, L.P.: Electrodynamics of Continuous Media, vol. 8. Butterworth-Heinemann, Oxford (1984)

McMeeking, R.M., Landis, C.M.: Electrostatic forces and stored energy for deformable dielectric materials. J. Appl. Mech. **72**, 581–590 (2005)

Miehe, C., Kiefer, B., Rosato, D.: An incremental variational formulation of dissipative magnetostriction at the macroscopic continuum level. Int. J. Solids Struct. **48**, 1846–1866 (2011)

Nardinocchi, P., Teresi, L.: On the active response of soft living tissues. J. Elast. **88**, 27–39 (2007)

Ogden, R.W., Steigman, D.: Mechanics and Electrodynamics of Magneto- and Electro-elastic Materials. CISM International Centre for Mechanical Sciences. Springer, Udine (2011)

Reppel, T., Dally, T., Weinberg, K.: On the elastic modeling of highly extensible polyurea. Technische Mechanik **33**(1), 19–33 (2013)

Rosenthal, M., Weaber, C., Polyakov, I., Zarrabi, A., Gise, P.: Designing components using smartMOVE electroactive polymer technology. In: Proceedings of the SPIE 6927:692,704–692,704–12 (2008)

Suo, Z.: Theory of dielectric elastomers. Acta Mechanica Solida Sinica **23**(6), 549–578 (2010)

Thiele, C.: Electroactive polymers and devices 2013-2018: Forecasts, technologies, players electronic & ionic eaps and their applications (2013)

Vogel, F., Bustamante, R., Steinmann, P.: On some mixed variational principles in electroelastostatics. Int. J. Non-Linear Mech. **47**, 341–354 (2012)

Zwecker, S., Klinkel, S.: A solid shell finite element formulation to simulate the behaviour of thin dielectric elastomer structures. PAMM **11**(1), 499–500 (2011)

Part III
Thermo-Elasto-Plasticity

Finite Elastoplasticity with Thermal Effects—Some Unexpected Phenomena

Otto T. Bruhns

Abstract A thermomechanical cyclic process of heating and cooling with fixed stresses and fixed plastic strains, respectively, at the corresponding half-cycles is discussed within the frame of large deformations and a coupled theory of Eulerian thermo-elastoplasticity. This discussion was stimulated by some comments of unknown reviewers during an in parts annoying peer-reviewing process. Within these comments, some effects observed in thermoplasticity were characterised as "against physics" and thus were used to discredit the submitted paper. In this paper, the main issues of these reviewers are re-examined for a simple uniaxial case. Accordingly, the effects are named as "some unexpected phenomena". It could be shown that these phenomena are in accord with physics and, moreover, that comparable experimental results underline this position. The underlying Eulerian theory of thermo-elastoplasticity is associated with the recently introduced objective logarithmic rate. This paper also discloses a hitherto not published motivation for introducing this specific time derivative.

1 Introduction

In the past decades, the thermodynamic foundation of elastoplasticity was studied by several researchers. For small deformations, e.g. by Ziegler (1958, 1963), Naghdi (1960), Kestin and Rice (1970), Perzyna (1971), Lubliner (1972) and recently by Maugin (1992), Houlsby and Puzrin (2000), Collins (2005) and many others. In the presence of finite strain and large rotation, these endeavours were continued and extended, e.g. by Green and Naghdi (1965, 1968), Kratochvíl and Dillon (1969), Lee (1970), Lehmann (1972, 1984, 1989), Mandel (1972, 1974), Rice (1975), Haupt (1977), Anand (1985), Anand and Lush (1987), Casey (1998), Rajagopal and Srinivasa (1998), Bertram (2003), Anand et al. (2009), Bertram and Krawietz (2012).

O.T. Bruhns (✉)
Institute of Mechanics, Ruhr-Universität Bochum, 44780 Bochum, Germany
e-mail: otto.bruhns@rub.de

K. Naumenko and M. Aßmus (eds.), *Advanced Methods of Continuum Mechanics for Materials and Structures*, Advanced Structured Materials 60,
DOI 10.1007/978-981-10-0959-4_8

2 Basic Relations

2.1 Energy Balance and Entropy Inequality

For a deformable body in the purely mechanical sense, as basic field variables the deformation gradient $\boldsymbol{F}$ and the Cauchy stress $\boldsymbol{\sigma}$ at each particle characterise the local deformation state relative to a reference configuration and the local stressed state, respectively. Let $\boldsymbol{X}$ and $\boldsymbol{x}$ be the reference and the current position vector of a material particle, respectively. Then the deformation gradient $\boldsymbol{F}$ is given by

$$\boldsymbol{F} = \frac{\partial \boldsymbol{x}}{\partial \boldsymbol{X}}. \tag{1}$$

In a general case with thermal effects, a further basic field variable is the temperature. In addition, two quantities are introduced to describe the efflux or influx of heat as well as the heat absorption or release. They are the heat flux vector $\boldsymbol{q}$ and the heat supply r. Throughout, $\boldsymbol{q}$ is measured per unit area in the current configuration, and r per unit reference volume.[3] Locally, these two quantities together specify the heat entering an infinitesimal volume element.

In a material body, there may exist complicated interactions among the different fields of deformation, stress and temperature. In fact, a thermodynamic process occurring in a material body is a rather involved process in which the mechanical work and the heat are constantly converted into each other. In general, there may be a strong coupling between deformation and heat. For various kinds of thermomechanical behaviour of deformable material bodies, certain universal laws or principles may be formulated for the description of common physical features. In the purely mechanical case, there are the material objectivity or frame-indifference principle and the laws of conservation for momentum and mass. The first requires that objective quantities should be employed in constitutive formulations, while the latter two are represented by Cauchy's law of motion and the continuity equation, respectively. For these, refer, e.g. to Truesdell and Noll (1965) for details. In addition, further universal laws or principles, namely, the three thermodynamic laws, are formulated in a general case with thermal effects. Details may be found, e.g. in Truesdell and Toupin (1960), Truesdell (1984), Müller (1985), Šilhavý (1997), Haupt (2002).

First, it is assumed that there exists a temperature quantity T, called the absolute temperature, which is always positive,

$$T > 0. \tag{2}$$

The above statement is sometimes referred to as the zeroth law of thermodynamics.

Second, from the observation on the interconvertibility of heat and mechanical work, it is assumed that the mechanical work done by the external loadings and the

[3] Note that, unlike the usual treatment, here the so-called intensive quantities, such as the r above and those introduced later, are not defined as quantities measured per unit mass.

internal stress, and the non-mechanical work contributed by the heat flux and the heat supply, are converted into the kinetic energy and the internal energy of the material body. According to this fact, a new physical quantity, namely, the internal energy, is introduced. Let φ represent the internal energy per unit reference volume. Then from the foregoing energy balance, together with Cauchy's law of motion and the continuity equation, the following relation[4] may be derived:

$$\dot{\varphi} = \boldsymbol{\tau} : \boldsymbol{D} - J\nabla \cdot \boldsymbol{q} + r, \tag{3}$$

for all possible thermodynamic processes. Here and henceforth, J and $\boldsymbol{\tau}$ are used to denote the volume ratio (Jacobian) and the Kirchhoff stress, given by

$$J = \det \boldsymbol{F}, \quad \boldsymbol{\tau} = J\boldsymbol{\sigma}. \tag{4}$$

Moreover, the notation ∇ is used to designate the formal differentiation vector with respect to the current position vector $\boldsymbol{x}$, namely,

$$\nabla = \frac{\partial}{\partial \boldsymbol{x}},$$

and $\boldsymbol{D}$ is the natural deformation rate (stretching), given by the symmetric part of the velocity gradient,

$$\boldsymbol{D} = \frac{1}{2}\left(\nabla \dot{\boldsymbol{x}} + (\nabla \dot{\boldsymbol{x}})^T\right) = \frac{1}{2}\left(\dot{\boldsymbol{F}} \cdot \boldsymbol{F}^{-1} + (\dot{\boldsymbol{F}} \cdot \boldsymbol{F}^{-1})^T\right). \tag{5}$$

As indicated by Eq. (3), known as the first law of thermodynamics, the changing rate of internal energy is furnished by the sum of the (internal) stress power and the (internal) non-mechanical power due to heat flux and heat supply.

Next, on the basis of the observation on the irreversibility property of macroscopic physical phenomena, it is assumed that there exists a scalar field quantity η measured per unit reference volume, called specific entropy, so that the following inequality holds

$$\dot{\eta} \geq -J\nabla \cdot \left(\frac{\boldsymbol{q}}{T}\right) + \frac{r}{T} \tag{6}$$

for all possible thermodynamic processes. Physically, the quantity η represents the total local dissipation.

The entropy η and the internal energy φ are related to each other through the first and second laws, as expressed by Eq. (3) and inequality (6). To render this relation clear and direct, alternate forms of Eq. (3) and the inequality (6) are derivable by introducing the Helmholtz free energy per unit reference volume:

---

[4]Let $\boldsymbol{n}$ be the outward normal at a point on a closed material surface. We assume that $\boldsymbol{q} \cdot \boldsymbol{n} > 0$ means the efflux of the heat through the material surface. Thus follows the minus sign in Eq. (3) for the term contributed by the heat flux $\boldsymbol{q}$.

$$\psi = \varphi - T\eta. \tag{7}$$

Then the energy balance (3) may be reformulated as

$$\dot{\psi} = \boldsymbol{\tau} : \boldsymbol{D} - J\nabla \cdot \boldsymbol{q} + r - \eta\dot{T} - T\dot{\eta}. \tag{8}$$

Using this relation, we may recast inequality (6) in the form:

$$\boldsymbol{\tau} : \boldsymbol{D} - (\dot{\psi} + \eta\dot{T}) - \frac{J}{T}\boldsymbol{q} \cdot \nabla T \geq 0. \tag{9}$$

Inequality (6) or its alternate form (9), known as the second law of thermodynamics or Clausius–Duhem inequality, expresses the entropy principle for the irreversibility of macroscopic physical phenomena in deformable material bodies.

Furthermore, we introduce a quantity $\mathcal{D}$ by

$$\mathcal{D} = T\dot{\eta} - (r - J\nabla \cdot \boldsymbol{q}) = \boldsymbol{\tau} : \boldsymbol{D} - (\dot{\psi} + \eta\dot{T}). \tag{10}$$

This quantity, known as *internal dissipation*, provides the remaining part after the deduction of the total local heating from the total local entropy rate times the temperature. The Planck inequality (cf., e.g. Truesdell 1984)

$$\mathcal{D} \geq 0 \tag{11}$$

requires that the internal dissipation should always be non-negative. It should be positive for a process of dissipative deformations such as elastoplastic deformations.

For a thermomechanical behaviour of a material body, there will be a strong coupling between the different fields of deformation, stress and temperature, respectively. The material behaviour will be characterised by constitutive equations relating these field quantities and their histories. The thermodynamic laws as expressed by (2), (3) and (6), or their alternate forms (2), (8) and (9), place restrictions on various kinds of material behaviour.

We note here that the stress power per unit volume in Eq. (3) plays an essential role in continuum mechanics, especially when deriving constitutive relations. Since $\boldsymbol{\tau}$ and $\boldsymbol{D}$ constitute the stress power, the Kirchhoff stress and the deformation rate may be regarded as pair of energetically conjugated quantities (refer to Hill 1968, 1978 and Macvean 1968). This assignment is, however, not unique, as the Kirchhoff stress may be replaced by other stress measures and thus the stress power may be formulated in different configurations (see Macvean 1968). For instance, $\boldsymbol{\tau}$ and $\boldsymbol{D}$ in this expression may be replaced by other stress measures and kinematical quantities. Thus, we may arrive, e.g. at the following equivalent expressions:

$$\boldsymbol{\tau} : \boldsymbol{D} = \boldsymbol{P} : \dot{\boldsymbol{F}} = \boldsymbol{S} : \dot{\boldsymbol{E}}, \tag{12}$$

providing additional pairs of conjugated quantities, where

$$\boldsymbol{P} = \boldsymbol{\tau} \cdot \boldsymbol{F}^{-T}, \quad \boldsymbol{S} = \boldsymbol{F}^{-1} \cdot \boldsymbol{\tau} \cdot \boldsymbol{F}^{-T} \tag{13}$$

are first and second Piola-Kirchhoff stress tensors, respectively, and

$$\boldsymbol{E} = \frac{1}{2}(\boldsymbol{F}^T \cdot \boldsymbol{F} - \mathbf{1}) \tag{14}$$

is the Lagrangean Green strain.

2.2 Some Additional Remarks About Deriving the Logarithmic Rate

In the remaining part of this section, the motivation for deriving the logarithmic rate is elucidated[5] from the aforementioned thermodynamical considerations.

Similar as Zhilin did, we begin with the stress power $\boldsymbol{\tau} : \boldsymbol{D}$ and the equivalence expressions (12). Kinematical tensors such as $\boldsymbol{D}$, $\dot{\boldsymbol{F}}$ and $\dot{\boldsymbol{E}}$ herein express the rate of the local shape change of the body. If one of these quantities equals the rate of a local deformation tensor, this tensor is said to be *work conjugated* to the respective conjugated stress tensor.

If now the thermodynamical consistency of a constitutive relation is analysed—or in other words—if the bounds of such consistency with thermodynamics are examined, the internal dissipation $\mathcal{D}$ should fulfil Planck's inequality (11).

In a Lagrangean configuration, for which the material time derivative is objective, the assignment of conjugated quantities is unique, such that $\boldsymbol{P}$ and $\boldsymbol{F}$ as well as $\boldsymbol{S}$ and $\boldsymbol{E}$ form work conjugated pairs. For an isotropic thermoelastic material, e.g. with state variables $\boldsymbol{E}$ and T, and the Helmholtz free energy $\psi = \psi(\boldsymbol{E}, T)$, inequality (11) becomes

$$\mathcal{D} = \left\{\boldsymbol{S} - \frac{\partial \psi}{\partial \boldsymbol{E}}\right\} : \dot{\boldsymbol{E}} - \left\{\eta + \frac{\partial \psi}{\partial T}\right\} \dot{T} \geq 0. \tag{15}$$

Since this inequality is assumed to hold true for all admissible processes, two equations of state can be deduced as

$$\boldsymbol{S} = \frac{\partial \psi}{\partial \boldsymbol{E}} \quad \text{and} \quad \eta = -\frac{\partial \psi}{\partial T}. \tag{16}$$

[5]It should be noted that in a recent book by Altenbach et al. (2013) an interesting and regrettably until then unpublished idea of the late P.A. Zhilin is explained and commented in a particular Chapter on "Material Strain Tensor". In this contribution (pp. 321–331) a method is suggested to introduce a so-called material strain tensor for inelastic continua, which will be conjugated to the Cauchy stress $\boldsymbol{\sigma}$. Starting from the definition of the stress power $\boldsymbol{\sigma} : \boldsymbol{D}$, a condition is derived which has to be fulfilled by this strain. For some particular cases, e.g. isotropic elastic material, it is demonstrated that the Hencky strain will fulfil this condition.

For an Eulerian configuration, on the other hand, this procedure[6] turns out to be more complicated and ambiguous. The main reason for this more complex situation is due to the natural deformation rate $\boldsymbol{D}$ incorporated in the definition of the stress power $\boldsymbol{\tau} : \boldsymbol{D}$, which should be expressible as a rate of an Eulerian strain tensor. Moreover, in an Eulerian description the material time derivative loses its objectivity and must be replaced by an objective one. Thus, a dual unknown problem has to be solved: *An unknown objective time derivative* $\overset{\circ}{(\bullet)}$ *(say) of an unknown Eulerian strain measure* $\bar{\boldsymbol{e}}$ *should give the deformation rate, i.e.*

$$\overset{\circ}{\bar{\boldsymbol{e}}} = \boldsymbol{D}. \tag{17}$$

This problem came up to the present author in the early eighties of last century, and from the very beginning there was a strong conjecture that the solution of this problem—if it would exist—should be related with the Hencky strain $\boldsymbol{h}$.[7]

The infinite number of Eulerian strain measures and the many objective time derivatives, which have been discussed in those days may underline the apparent ambiguity of the requirement (17). Two classical solutions may be given with the Eulerian Finger strain $\boldsymbol{a}$ and Almansi strain $\boldsymbol{A}$, respectively,

$$\boldsymbol{a} = \frac{1}{2}(\boldsymbol{B} - \boldsymbol{I}), \quad \boldsymbol{A} = \frac{1}{2}(\boldsymbol{I} - \boldsymbol{B}^{-1}), \tag{18}$$

where $\boldsymbol{B} = \boldsymbol{F} \cdot \boldsymbol{F}^T$ is the left Cauchy–Green tensor. Applying here, e.g. the non-corotational lower and upper (convected) Oldroyd rates,[8] respectively, to these strain measures, leads us to the two solutions (see, e.g. Haupt 2002; Bruhns et al. 2004)

$$\begin{aligned} \overset{\circ}{\boldsymbol{A}}{}^{\mathrm{Ol}} &= \dot{\boldsymbol{A}} + \boldsymbol{A} \cdot \boldsymbol{L} + \boldsymbol{L}^T \cdot \boldsymbol{A} = \boldsymbol{D}, \\ \overset{\circ}{\boldsymbol{a}}{}^{\mathrm{Ou}} &= \dot{\boldsymbol{a}} - \boldsymbol{a} \cdot \boldsymbol{L}^T - \boldsymbol{L} \cdot \boldsymbol{a} \quad = \boldsymbol{D}, \end{aligned} \tag{19}$$

where as usual $\boldsymbol{L} = \nabla\dot{\boldsymbol{x}} = \dot{\boldsymbol{F}} \cdot \boldsymbol{F}^{-1}$ is the velocity gradient.

This result has been generalised in Bruhns et al. (2004), with the following conclusion: *The deformation rate* $\mathbf{D}$ *is expressible as a Hill-type non-corotational rate of any given Eulerian Seth-Hill strain* $\mathbf{e}^{(m)}$. Both equations (19) are two particular cases of this general result, where $m = \pm 1$.

Next, in 1960, Prager made an interesting observation, which significantly should affect the description of Eulerian elastoplasticity. He noted that the yield function should be stationary with the vanishing of the stress rates. This requirement, called the yielding stationarity condition, has been shown to be necessary for the consistency of

[6] This procedure is sometimes attributed to Coleman and Noll (1963).

[7] See the definition of the Hencky strain at the end of Sect. 3.3. Moreover, reference is made to the historical notes in Bruhns (2014a, b) and the remarks in the aforementioned Zhilin article.

[8] The former of these rates is sometimes called Cotter–Rivlin rate, refer to, e.g. Bruhns (2014b) and the further remarks and references therein.

the composite structure of Eulerian elastoplasticity (Prager 1960, Xiao et al. 2000). One of the consequences of this criterion is that the use of non-corotational rates in finite elastoplasticity should be excluded.[9]

For corotational rates, on the other hand, the requirement (17) can be reduced to

$$\overset{\circ}{\bar{e}} = \dot{\bar{e}} + \bar{e} \cdot \boldsymbol{\Omega} - \boldsymbol{\Omega} \cdot \bar{e} = \boldsymbol{D}. \tag{20}$$

Here, $\boldsymbol{\Omega}$ is a general skew-symmetric second-order Eulerian (or spin) tensor. The well-known Zaremba–Jaumann rate provides an example of corotational rates with $\boldsymbol{\Omega} = \boldsymbol{W}$, where $\boldsymbol{W}$ is the vorticity tensor, given by the skew-symmetric part of the velocity gradient $\boldsymbol{W} = \frac{1}{2}(\boldsymbol{L} - \boldsymbol{L}^T)$.

In a series of papers, starting from 1997,[10] it has been shown that there exists a unique solution for the requirement (20), namely the strain must be the Hencky strain $\bar{e} = \boldsymbol{h}$ and the rate must be the logarithmic rate with the spin $\boldsymbol{\Omega} = \boldsymbol{\Omega}^{\log}$.[11]

It has been demonstrated further that this logarithmic rate has a remarkable property, namely that a hypoelastic relation of grade zero with a logarithmic rate applied to the stress and with the deformation rate can be integrated to give an elastic relation between this stress and the Hencky strain (see also Xiao et al. 1997c, 2007a and Bruhns et al. 1999). Thus, with reference to the discussion initiated by Simó and Pister (1984), contrary to the situation with the other then discussed objective rates, a hypoelastic relation of this kind exhibits no dissipation (refer to Meyers et al. 2000). This means that in an Eulerian description of elastoplasticity, where an additive split of the deformation rate is used with a hypoelastic relation representing the elastic part, there exists only one solution, where this elastic part runs without dissipation, namely when the logarithmic rate is applied to the stress and the Hencky strain is used. And, contrary to what is observed in numerous articles, the often used Zaremba–Jaumann rate fails to have this property.

For an Eulerian thermoelastic material,[12] e.g. with state variables $\boldsymbol{h}$ and T, and the Helmholtz free energy $\psi = \hat{\psi}(\boldsymbol{h}, T)$, inequality (11) can be written as

$$\mathcal{D} = \left\{ \boldsymbol{\tau} - \frac{\partial \hat{\psi}}{\partial \boldsymbol{h}} \right\} : \overset{\circ}{\boldsymbol{h}}{}^{\log} - \left\{ \eta + \frac{\partial \hat{\psi}}{\partial T} \right\} \dot{T} \geq 0, \tag{21}$$

where $\overset{\circ}{\boldsymbol{h}}{}^{\log} = \boldsymbol{D}$, and the two equations of state become

---

[9] We must note that the Zaremba–Jaumann rate was favoured in the discussion by Prager (1960), since this was the only corotational rate in the four classical rates then known.

[10] Reference is made, e.g. to Xiao et al. (1997a, b, 1998b, a, 1999, 2000) and Bruhns et al. (2004). Some historical remarks are also contained in Bruhns (2014b), especially in footnotes 44 and 46 therein.

[11] For the definition of the logarithmic spin $\boldsymbol{\Omega}^{\log}$, we refer to Eq. (36).

[12] We still confine our considerations here to an isotropic material.

$$\boldsymbol{\tau} = \frac{\partial \hat{\psi}}{\partial \boldsymbol{h}} \quad \text{and} \quad \eta = -\frac{\partial \hat{\psi}}{\partial T}. \tag{22}$$

As usual from these results, we may deduce the Gibbs relation

$$\dot{\psi} = \boldsymbol{\tau} : \boldsymbol{D} - \eta \dot{T}. \tag{23}$$

Introducing this with the help of the transformation (7) into the energy balance (3), eliminates the stress power:

$$T\dot{\eta} = -J\nabla \cdot \boldsymbol{q} + r. \tag{24}$$

Finally, utilising the equations of state converts this expression to an evolution equation for the temperature (refer, e.g. to Haupt 2002)

$$c\dot{T} = T\,\frac{\partial \boldsymbol{\tau}}{\partial T} : \boldsymbol{D} - J\nabla \cdot \boldsymbol{q} + r, \tag{25}$$

where Eq. $(22)_1$ has been applied, and

$$c = -T\,\frac{\partial^2 \hat{\psi}}{\partial T^2}$$

is the specific heat at constant strain. The first term of this heat conduction equation (25) is representing the thermoelastic coupling effect. And this general result even holds true for an elastoplastic behaviour, e.g. with a Helmholtz free energy form $\psi = \hat{\psi}(\boldsymbol{h}^e, T, \boldsymbol{\alpha}, \kappa)$.[13] Then merely the $\boldsymbol{D}$ in Eq. (25) is replaced by a $\boldsymbol{D}^e$. Thus, it is absolutely not clear why and where this thermoelastic coupling effect, which sometimes is also called elastic heat, should be "against the fundamental physical law" (refer to Remark 1).

3 Elastoplastic Materials with Thermal Effects

3.1 Basic Features of Thermoelastoplastic Materials

Elastoplastic material behaviour is a combination or coupling of two disparate idealised behaviours, namely, purely elastic and purely plastic. The former is spring- or solid-like, whereas the latter is fluid-like. Although constitutive formulations may be established for these two idealised behaviours in definite ways, it turns out that a realistic, consistent constitutive formulation for finite elastoplasticity as inextricable combination of these two has been a challenging task. Thermal effects would further

[13] See the Appendix.

introduce considerable complexity. Certain typical features of elastoplasticity with thermal effects may be summarised below.

1. There exists a *yield limit* expressed by a yield function f. The stress and the temperature are always bounded by this limit, i.e.

$$f \leq 0. \tag{26}$$

   The closed hypersurface in stress-temperature space determined by the equation $f = 0$ is referred to as the yield surface.
2. If the stress and the temperature are kept within the region $f < 0$, a thermoelastic behaviour is expected. In this case, including *unloading*, the stress work is recoverable. Whenever the yield limit is attained and maintained, coupled thermo-elastoplastic behaviour is expected and irreversible plastic flow is induced. In this *loading* case, the stress work is partly recoverable and partly irrecoverable (refer to Farren and Taylor 1925), but usually the dissipated behaviour of fluid-like may play a dominant role.
3. Except for some idealised considerations, the yield surface is changing and moving with plastic flow in the loading case. This phenomenon, known as hardening behaviour, results from the micro-structural rearrangement due to the movement and interactions of vast number of dislocations and other defects.

On account of the above features, a complete constitutive formulation of finite elastoplasticity with thermal effects is a composition of constitutive ingredients in several aspects, including the characterisation of the yield function, the loading–unloading condition, the constitutive relations between stress and deformation rates in both loading and unloading cases, as well as the description of hardening behaviour.

As mentioned earlier, many proposals for elastoplastic formulations have been made based on different starting-points and assumptions. We here propose an Eulerian rate formulation by extending the main ideas as explained in Xiao et al. (2006). With direct physical relevance to the fluid-like characteristics of elastoplastic deformations, this formulation establishes direct relations between the Kirchhoff stress $\boldsymbol{\tau}$ and the stretching $\boldsymbol{D}$.

3.2 Eulerian Rate Constitutive Formulations

We begin with the experimental result that the stress power $\boldsymbol{\tau} : \boldsymbol{D}$ is a composition of two parts, namely a recoverable and an irrecoverable part. This leads us to the following decomposition of the natural deformation rate $\boldsymbol{D}$:

$$\boldsymbol{D} = \boldsymbol{D}^e + \boldsymbol{D}^p. \tag{27}$$

With this intention, the part $\boldsymbol{D}^e$ is related to thermoelastic behaviour in the sense that the quantity $\boldsymbol{\tau} : \boldsymbol{D}^e$ furnishes the recoverable part of the total stress power, whereas the part $\boldsymbol{D}^p$ is associated with plastic flow and $\boldsymbol{\tau} : \boldsymbol{D}^p$ gives the irrecoverable part of the total stress power.[14]

In a fully general sense to describe the hardening behaviour, the yield function f may be regarded as function of the current values of stress and temperature and the deformation history. However, such a general form might hardly be tractable for realistic purposes. Usually it is assumed that a set of internal variables may be introduced to characterise the hardening behaviour with adequate accuracy. As is widely done, we here introduce one scalar variable κ and a tensorial variable $\boldsymbol{\alpha}$. The former is usually identified with the plastic work[15]:

$$\dot{\kappa} = \boldsymbol{\tau} : \boldsymbol{D}^p. \tag{28}$$

This variable describes the *isotropic hardening*. The hardening variable $\boldsymbol{\alpha}$ is an objective symmetric Eulerian tensor and usually called the *back stress*, which characterises the *kinematic hardening*. The two hardening variables[16] together describe combined isotropic–kinematic hardening behaviour. They are changing with plastic flow in a process of elastoplastic deformations. Constitutive relations governing their changes, known as evolution equations, will be discussed later.

With these hardening variables, the yield function f for a thermo-elastoplastic material is formulated as function of the set $(\boldsymbol{\tau}, T, \boldsymbol{\alpha}, \kappa)$. As a result, the yield limit (26) becomes[17]

$$f = f(\boldsymbol{\tau}, T, \boldsymbol{\alpha}, \kappa) \leq 0. \tag{29}$$

Now we are in the position to formulate the rate-type constitutive relation for the two parts $\boldsymbol{D}^e$ and $\boldsymbol{D}^p$ in (27). First, we propose an objective Eulerian rate formulation for the part $\boldsymbol{D}^e$ for thermoelastic behaviour. This leads to an objective Eulerian rate equation of hypoelastic type:

$$\boldsymbol{D}^e = \frac{\partial^2 \bar{W}}{\partial \boldsymbol{\tau}^2} : \overset{\circ}{\boldsymbol{\tau}} + \frac{\partial^2 \bar{W}}{\partial \boldsymbol{\tau} \partial T} \dot{T}. \tag{30}$$

where the scalar function

$$\bar{W} = \bar{W}(\boldsymbol{\tau}, T) \tag{31}$$

---

[14] We note that herein also use has been made of the so-called iso-stress assumption that the stresses $\boldsymbol{\tau}$ within these two products remain the same, which is common for most isotropic materials.

[15] The so-called equivalent plastic strain is also widely used. Unlike the plastic work, however, such a quantity might not possess direct physical meaning related to the concepts of work and energy in thermodynamics.

[16] More internal variables may be used and can be treated by following a way similar to the subsequent development.

[17] A usually treated form is, e.g.

$$f(\boldsymbol{\tau} - \boldsymbol{\alpha}) - y(T, \kappa) \leq 0.$$

In this case, the yield surface moves about in stress-temperature space with the moving of the centre represented by the back stress $\boldsymbol{\alpha}$, but changes its shape and size only in a uniform manner.

is a complementary thermoelastic potential.[18]

Next, we formulate the flow rule for the plastic stretching $\boldsymbol{D}^p$. With the rate-independence property, a general form may be given by (refer, e.g. to de Boer 1977)

$$\boldsymbol{D}^p = \xi \boldsymbol{\Phi}(\boldsymbol{\tau}, T, \boldsymbol{\alpha}, \kappa; \overset{\circ}{\boldsymbol{\tau}}, \dot{T}) = \xi \boldsymbol{\mathcal{A}}(\boldsymbol{\tau}, T, \boldsymbol{\alpha}, \kappa) :\overset{\circ}{\boldsymbol{\tau}} + \xi \boldsymbol{G}(\boldsymbol{\tau}, T, \boldsymbol{\alpha}, \kappa)\dot{T}, \tag{32}$$

where $\boldsymbol{\mathcal{A}}$ and $\boldsymbol{G}$ are fourth- and second-order constitutive tensors relying on the variable set $(\boldsymbol{\tau}, T, \boldsymbol{\alpha}, \kappa)$. Throughout, ξ is used to represent the *plastic indicator*, taking values 1 and 0 for loading and unloading, respectively (details will be given later).

Moreover, we formulate the evolution equations for the hardening variables κ and $\boldsymbol{\alpha}$. The widely used evolution for κ has been presented with equation (28). A reasonable generalisation of this relation may be given with:

$$\dot{\kappa} = \boldsymbol{C}(\boldsymbol{\tau}, T, \boldsymbol{\alpha}, \kappa) : \boldsymbol{D}^p + \xi D(\boldsymbol{\tau}, T, \boldsymbol{\alpha}, \kappa)\dot{T}. \tag{33}$$

For the back stress $\boldsymbol{\alpha}$, a general form of the evolution equation is as follows:

$$\overset{\circ}{\boldsymbol{\alpha}} = \boldsymbol{\mathcal{H}}(\boldsymbol{\tau}, T, \boldsymbol{\alpha}, \kappa) : \boldsymbol{D}^p + \xi \boldsymbol{\Theta}(\boldsymbol{\tau}, T, \boldsymbol{\alpha}, \kappa)\, \dot{T}. \tag{34}$$

Here again, ξ is the plastic indicator, and $\boldsymbol{C}$ and D as well as $\boldsymbol{\mathcal{H}}$ and $\boldsymbol{\Theta}$ are constitutive tensors relying on the whole variable set $(\boldsymbol{\tau}, T, \boldsymbol{\alpha}, \kappa)$.

3.3 Consistency Criteria and Logarithmic Rate

The specific rates just introduced in Eqs. (30), (32) and (34) have to fulfil the objectivity requirement and, moreover, prior to the occurrence of yielding the rate formulation (30) with $\boldsymbol{D}^p = \boldsymbol{0}$, i.e. $\boldsymbol{D}^e = \boldsymbol{D}$ should be exactly integrable to deliver a thermoelastic relation. This requirement, known as the exact integrability criterion, is introduced in Bruhns et al. (1999) and Xiao et al. (1999) in the purely mechanical case. Certain integrability conditions have to be treated, as exemplified in Bernstein (1960) and Truesdell and Noll (1965) for the case of the Zaremba–Jaumann rate. With integrability conditions, Simó and Pister (1984) demonstrated that none of the then well-known classical rates could fulfil the exact integrability criterion even for the simplest case, namely, the widely used hypoelastic equation of grade zero.

On the other hand, it is demonstrated by Bruhns et al. (1999, 2004) and Xiao et al. (1999, 2005) that of all possible corotational rates there exists one and only

[18] As constitutive functions of objective Eulerian variables, the potential function $\bar{W}$ and the yield function f in the foregoing, as well as the constitutive functions introduced in the subsequent development should be isotropic in accord with the requirement from the objectivity principle; refer to, e.g. Simó and Hughes (1998). In this case, only initial isotropy would be permissible and, accordingly, any kind of initial anisotropy would be excluded. We note further that this potential may be one part of a more general complementary energy function $g = g(\boldsymbol{\tau}, T, \boldsymbol{\alpha}, \kappa)$ for thermo-elastoplastic behaviour (see the Appendix).

one which can fulfil the exact integrability criterion. This rate is the logarithmic rate introduced in Lehmann et al. (1991), Reinhardt and Dubey (1996), Xiao et al. (1997b, 1998a, b).

Combining the above results and applying them to the general case with thermal effects, it may be concluded that the above objective rates should be logarithmic rates

$$\begin{aligned} \overset{\circ}{\boldsymbol{\tau}}{}^{\log} &= \dot{\boldsymbol{\tau}} + \boldsymbol{\tau} \cdot \boldsymbol{\Omega}^{\log} - \boldsymbol{\Omega}^{\log} \cdot \boldsymbol{\tau}, \\ \overset{\circ}{\boldsymbol{\alpha}}{}^{\log} &= \dot{\boldsymbol{\alpha}} + \boldsymbol{\alpha} \cdot \boldsymbol{\Omega}^{\log} - \boldsymbol{\Omega}^{\log} \cdot \boldsymbol{\alpha}, \end{aligned} \tag{35}$$

where $\boldsymbol{\Omega}^{\log}$ is the logarithmic spin

$$\boldsymbol{\Omega}^{\log} = \boldsymbol{W} + \sum_{r \neq s}^{m} \left(\frac{1 + (b_r/b_s)}{1 - (b_r/b_s)} + \frac{2}{\ln(b_r/b_s)} \right) \boldsymbol{B}_r \cdot \boldsymbol{D} \cdot \boldsymbol{B}_s. \tag{36}$$

In the above, b_t and $\boldsymbol{B}_t$ are the m distinct eigenvalues and the corresponding eigenprojections of the left Cauchy–Green tensor.

Thus, with the logarithmic rate, the exactly integrable Eulerian rate equation for thermoelastic behaviour is reformulated as

$$\boldsymbol{D}^e = \frac{\partial^2 \bar{W}}{\partial \boldsymbol{\tau}^2} : \overset{\circ}{\boldsymbol{\tau}}{}^{\log} + \frac{\partial^2 \bar{W}}{\partial \boldsymbol{\tau} \partial T} \dot{T}. \tag{37}$$

Prior to yielding, this rate equation with a natural state is integrable to deliver the following finite thermoelastic relation:

$$\boldsymbol{h} = \frac{\partial \bar{W}}{\partial \boldsymbol{\tau}}. \tag{38}$$

Here,

$$\boldsymbol{h} = \frac{1}{2} \ln(\boldsymbol{F} \cdot \boldsymbol{F}^T) = \frac{1}{2} \sum_{r=1}^{m} (\ln b_r) \boldsymbol{B}_r$$

is Hencky's logarithmic strain.

3.4 Continuity for Plastic Flow and Unified Loading–Unloading Conditions

For plastic flow, two consistency conditions must be considered. First, the flow rule (32) must be continuous for the transition between loading and neutral loading as expressed by (47) and (48) below throughout the stress-temperature space. That is also the case for the evolution equations (33) and (34). From these and with the logarithmic rate, we reduce Eqs. (32)–(34) to (see, e.g. Green and Naghdi 1965 and Appendix A of Xiao et al. 2007b)

$$\boldsymbol{D}^p = \xi \hat{f}\, \boldsymbol{P}, \tag{39}$$

$$\dot{\kappa} = \boldsymbol{C} : \boldsymbol{D}^p, \tag{40}$$

$$\overset{\circ}{\boldsymbol{\alpha}}{}^{\log} = \boldsymbol{\mathcal{H}}(\boldsymbol{\tau}, T, \boldsymbol{\alpha}, \kappa) : \boldsymbol{D}^p. \tag{41}$$

Here, the symmetric second-order constitutive tensor $\boldsymbol{P}$ relies on the variable set $(\boldsymbol{\tau}, T, \boldsymbol{\alpha}, \kappa)$. Throughout, $\hat{f}$ is a scalar rate-type quantity given by

$$\hat{f} \equiv \frac{\partial f}{\partial \boldsymbol{\tau}} : \overset{\circ}{\boldsymbol{\tau}}{}^{\log} + \frac{\partial f}{\partial T}\, \dot{T}. \tag{42}$$

Furthermore, in terms of a plastic potential (see, e.g. von Mises 1928 and Hill 1950), the flow rule (39) may be further reduced to

$$\boldsymbol{D}^p = \xi\, \frac{\hat{f}}{h}\, \frac{\partial p}{\partial \boldsymbol{\tau}}, \tag{43}$$

where h, the plastic modulus, will be given later, and the scalar function

$$p = p(\boldsymbol{\tau}, T, \boldsymbol{\alpha}, \kappa), \tag{44}$$

is the plastic potential.

The plastic flow $\boldsymbol{D}^p$, governed by the flow rule (43), is directed towards the outward normal $\partial p / \partial \boldsymbol{\tau}$ of the plastic potential. This requires

$$\boldsymbol{D}^p : \frac{\partial p}{\partial \boldsymbol{\tau}} > 0$$

in the loading case ($\xi = 1$). Hence we infer

$$\frac{\hat{f}}{h} > 0 \tag{45}$$

for loading.

Next, the second consistency condition for plastic flow is

$$\dot{f} = 0.$$

Formulating this condition in a log-rotating frame (refer to Xiao et al. 2007b) and using Eqs. (40), (41) and (43), we derive the plastic modulus h as follows[19]:

[19]The plastic modulus here should not be mistaken with the Hencky strain, introduced before.

$$h = -\left(\frac{\partial f}{\partial \kappa}\left(\boldsymbol{C} : \frac{\partial p}{\partial \boldsymbol{\tau}}\right) + \frac{\partial f}{\partial \boldsymbol{\alpha}} : \mathcal{H} : \frac{\partial p}{\partial \boldsymbol{\tau}}\right). \tag{46}$$

Now we study the loading–unloading conditions for a judgement about the occurrence of plastic flow. As usual, three cases may be distinguished for elastoplasticity with strain hardening and strain softening as well as perfect elastoplasticity, which correspond to positive, negative and vanishing plastic modulus, namely, $h > 0$, $h < 0$, $h = 0$, respectively. The classical conditions, formulated in terms of the quantity $\hat{f}$ given by (42), are known to apply only to the strain-hardening case and fail to combine these three cases. A unified form of loading–unloading conditions in purely mechanical case was presented earlier by Hill (1958, 1959). Its definite form extended to the general case with thermal effects is suggested in Xiao et al. (2007b)

$$\boldsymbol{D}^p \neq \boldsymbol{0} \quad (\xi = 1) \text{ if } f = 0 \text{ and } \frac{\hat{f}}{h} > 0, \tag{47}$$

$$\boldsymbol{D}^p = \boldsymbol{0} \quad (\xi = 1) \text{ if } f = 0 \text{ and } \hat{f} = 0, \tag{48}$$

$$\boldsymbol{D}^p = \boldsymbol{0} \quad (\xi = 0) \text{ if } f = 0 \text{ and } \hat{f} < 0, \tag{49}$$

$$\boldsymbol{D}^p = \boldsymbol{0} \quad (\xi = 0) \text{ if } f < 0. \tag{50}$$

Related with conditions (47)–(50), we speak of loading, neutral loading, unloading from an elastoplastic state, and a thermoelastic state, respectively.

It may be evident that the above conditions yield the classical conditions for the strain-hardening case with $h > 0$, while both loading and unloading are possible whenever the stress rate and temperature rate are such that $\hat{f} > 0$ for the strain-softening case. For the case of perfect elastoplasticity with $h = 0$, the above conditions lead to $\hat{f} = 0$ for loading.

The loading–unloading conditions in the foregoing are in terms of the stress rate. An alternate form is derivable in terms of the natural deformation rate. We therefore refer to Xiao et al. (2007b).

Finally, combining Eqs. (37) and (43), we obtain the relation between the total deformation rate and the total stress rate as follows:

$$\boldsymbol{D} = \frac{\partial^2 \bar{W}}{\partial \boldsymbol{\tau}^2} : \overset{\circ}{\boldsymbol{\tau}}{}^{\log} + \frac{\partial^2 \bar{W}}{\partial \boldsymbol{\tau} \partial T}\dot{T} + \xi \frac{\hat{f}}{h}\frac{\partial p}{\partial \boldsymbol{\tau}}. \tag{51}$$

Now the plastic indicator may be given by

$$\xi = \begin{cases} 1 \text{ if } f = 0 \text{ and } \dfrac{\hat{f}}{h} \geq 0, \\ 0 \text{ if } f < 0 \text{ or } (f = 0,\ \hat{f} < 0). \end{cases} \tag{52}$$

3.5 The Essential Structure with Associated Flow Rule

Just for simplicity, let us assume a description within the frame of classical thermoplasticity. If the plastic potential p is given in accord with the yield function f, which for a von Mises type description and with reference to footnote 17 may be introduced in the form:

$$f(\boldsymbol{\tau}, T, \boldsymbol{\alpha}, \kappa) = (\boldsymbol{\tau}' - \boldsymbol{\alpha}) : (\boldsymbol{\tau}' - \boldsymbol{\alpha}) - y^2(T, \kappa) = 0, \tag{53}$$

the flow rule (43) is just the associated normality rule

$$\boldsymbol{D}^p = \xi \frac{\hat{f}}{h} \frac{\partial f}{\partial \boldsymbol{\tau}}. \tag{54}$$

Here a $(\bullet)'$ designates the deviator of a second-order tensor $(\bullet)$, e.g.

$$\boldsymbol{\tau}' = \boldsymbol{\tau} - \frac{1}{3}\mathrm{tr}(\boldsymbol{\tau})\boldsymbol{I},$$

$\boldsymbol{I}$ is the second-order identity tensor, and the back stress $\boldsymbol{\alpha}$ is introduced as a deviatoric quantity, namely $\mathrm{tr}(\boldsymbol{\alpha}) = 0$.

If the yield surface is convex, we have

$$(\boldsymbol{\tau}' - \boldsymbol{\tau}_0') : \frac{\partial f}{\partial \boldsymbol{\tau}} > 0, \tag{55}$$

for a regular point $\boldsymbol{\tau}'$ on the yield surface and for any point $\boldsymbol{\tau}_0'$ inside the yield surface. In particular, we have

$$(\boldsymbol{\tau}' - \boldsymbol{\alpha}) : \frac{\partial f}{\partial \boldsymbol{\tau}} > 0, \tag{56}$$

since the back stress $\boldsymbol{\alpha}$ assigning the centre of the yield surface should be included in the yield surface, namely,

$$f(\boldsymbol{\tau}, T, \boldsymbol{\alpha}, \kappa) < 0.$$

The evolution equation (40) of the scalar valued internal variable κ is here with

$$\boldsymbol{C} = \frac{1}{2} \frac{\partial f}{\partial \boldsymbol{\tau}} \tag{57}$$

analogously reduced to

$$\dot{\kappa} = \frac{1}{2} \frac{\partial f}{\partial \boldsymbol{\tau}} : \boldsymbol{D}^p = (\boldsymbol{\tau}' - \boldsymbol{\alpha}) : \boldsymbol{D}^p. \tag{58}$$

$$s : \overset{\circ}{\boldsymbol{\tau}}{}^{\log} = \frac{s : \mathcal{C} : \boldsymbol{D} + s : \boldsymbol{M}\dot{T} - s : \mathcal{C} : s\frac{4}{h}\left(\frac{1}{2}y_0^2 + B\kappa\right)\omega\dot{T}}{1 + \frac{4}{h}s : \mathcal{C} : s}, \tag{72}$$

and finally introducing this expression into Eq. (70), the inverted form of the total constitutive relation becomes

$$\overset{\circ}{\boldsymbol{\tau}}{}^{\log} = \mathcal{C} : \boldsymbol{D} + \boldsymbol{M}\dot{T} - 4\mathcal{C} : s\,\frac{s : \mathcal{C} : \boldsymbol{D} + s : \boldsymbol{M}\dot{T} + \left(\frac{1}{2}y_0^2 + B\kappa\right)\omega\dot{T}}{h + 4s : \mathcal{C} : s}. \tag{73}$$

For an isotropic material $\boldsymbol{M}$ takes the form $\boldsymbol{M} = m\boldsymbol{I}$, where m is the stress-temperature modulus. Thus, expression (73) may be further reduced to

$$\overset{\circ}{\boldsymbol{\tau}}{}^{\log} = \mathcal{C} : \boldsymbol{D} + m\boldsymbol{I}\dot{T} - 4\mathcal{C} : s\,\frac{s : \mathcal{C} : \boldsymbol{D} + \left(\frac{1}{2}y_0^2 + B\kappa\right)\omega\dot{T}}{h + 4s : \mathcal{C} : s}. \tag{74}$$

since the product $s : \boldsymbol{M}$ vanishes.

It clearly turns out that a change in the temperature due to two different effects may cause a change in the stress or strain. These effects are the thermal expansion, or alternatively, the thermal stresses on the one side, and the influence of the thermal softening during plastic flow on the other side. The latter is deviatoric and thus related with an isochoric motion, whereas the former is associated to a purely elastic volumetric motion.

4 Application to Simple Examples

Let us now discuss a simple example.[21] A rod subject to a uniaxial tension with stress τ and a fixed temperature T_1 is loaded beyond the yield point and then the applied load at τ^* is kept constant, whereas now the temperature is increased to reach a value $T_2 > T_1$. This thermomechanical process is compared with the result of a direct loading of the same rod at temperature T_2.

Moreover, the influence of a cyclic process of heating and cooling on the behaviour of the rod is touched in short.

[21] For the sake of simplicity, here a purely thermoplastic behaviour is assumed, i.e. the elastic parts of the deformation are neglected. Moreover, the hardening of the material may be represented by a linear isotropic model according to Eq. (61). Even for this simplified model, it turns out from the subsequent calculations that the reviewer's statement that: "coupled temperature and plastic deformation may not be a correct concept in physics" (refer to Remark 2) seems to be disproved.

4.1 Step 1, Isothermal Loading

The first step is described by an isothermal process with $\dot{T} = 0$. For the sake of simplicity, we adopt isotropic hardening with $c = 0$. In this specific case we have

$$\frac{\hat{f}}{h} = \frac{\tau\dot{\tau}}{3y^2 B(1 - \omega(T_1 - T_0))} > 0,$$

and thus loading, provided the plastic modulus does not change its sign.

From Eq. (65), we find

$$\dot{h}^p = \frac{\dot{\tau}}{B(1 - \omega(T_1 - T_0))},$$

which can be integrated to give

$$\tau - \tau_0(T_1) = B(1 - \omega(T_1 - T_0))h^p,$$

where h^p is the plastic part of the (uniaxial) logarithmic strain and $\tau_0(T_1)$ is the yield stress at T_1, i.e.

$$\tau_0(T_1) = y_0\sqrt{\frac{3}{2}}(1 - \omega(T_1 - T_0)).$$

The first step is terminated at $\tau = \tau^*$, which allows us to calculate h_1^p, the plastic strain at τ^*

$$h_1^p = \frac{\tau^* - \tau_0(T_1)}{B(1 - \omega(T_1 - T_0))}.$$

Moreover, the plastic work κ_1 at τ^* is then calculated from Eq. (63)

$$\kappa_1 = \frac{\tau^{*2} - \tau_0^2(T_1)}{3B(1 - \omega(T_1 - T_0))}.$$

An equivalent result can be given for loading at temperature T_2, namely

$$\tau - \tau_0(T_2) = B(1 - \omega(T_2 - T_0))h^p,$$

with

$$\tau_0(T_2) = y_0\sqrt{\frac{3}{2}}(1 - \omega(T_2 - T_0)).$$

The plastic strain and the plastic work at stress τ^* become

$$h_2^p = \frac{\tau^* - \tau_0(T_2)}{B(1 - \omega(T_2 - T_0))}, \quad \kappa_2 = \frac{\tau^{*2} - \tau_0^2(T_2)}{3B(1 - \omega(T_2 - T_0))}.$$

4.2 Step 2, Heating at Fixed Stress

In the following two steps of the process, we are merely interested in a solution for the underlying constitutive relations, rather than in the solution of a possible initial-boundary value problem. Accordingly, the system will be treated as adiabatic.

We now keep the load τ^* fixed and increase the temperature from T_1 to T_2. Thus, with $\dot{\tau} = 0$,

$$\frac{\hat{f}}{h} = \frac{(y_0^2 + 2B\kappa)\,\omega\,\dot{T}}{3y^2 B(1 - \omega(T - T_0))} > 0,$$

the system is in loading for $\dot{T} > 0$.

From Eq. (65), we find

$$\dot{h}^p = \frac{\tau^* \omega\,\dot{T}}{2B(1 - \omega(T - T_0))^2},$$

and with the initial condition at h_1^p this can be integrated to give

$$h^p = \frac{\tau^*}{2B(1 - \omega(T - T_0))} + \frac{\tau^* - 2\tau_0(T_1)}{2B(1 - \omega(T_1 - T_0))}.$$

and finally

$$h_3^p = \frac{\tau^*}{2B(1 - \omega(T_2 - T_0))} + \frac{\tau^* - 2\tau_0(T_1)}{2B(1 - \omega(T_1 - T_0))},$$

wherein h_3^p is the plastic strain at $T = T_2$.

Using the initial condition κ_1, the plastic work κ_3 at $T = T_2$ is determined

$$\kappa_3 = \frac{\tau^{*2}}{3B(1 - \omega(T_2 - T_0))} - \frac{\tau_0^2(T_1)}{3B(1 - \omega(T_1 - T_0))}.$$

4.3 Step 3, Cooling at Fixed Plastic Strain

In this third step, we keep the plastic strain constant (at the value h_3^p) and decrease the temperature back to the starting value, i.e. $T_2 \to T_1$. Thus, with $\boldsymbol{D}^p = 0$, i.e. $h^p = h_3^p$ and $\kappa = \kappa_3$, this subsequent cooling process appears as neutral loading:

The conditions $f = 0$ and $\hat{f} = 0$ are fulfilled for

$$\frac{2}{3}\tau^2 = (y_0^2 + 2B\kappa_3)(1 - \omega(T - T_0)), \quad \dot{\tau} > 0.$$

Introducing here $T = T_1$, the respective stress becomes

$$\tau^{**} = \sqrt{\frac{3}{2}(y_0^2 + 2B\kappa_3)(1 - \omega(T_1 - T_0))}.$$

To illustrate these results, we take the following material constants for thermoplastic behaviour:

$$y_0 = 100\,\text{MPa}, \quad B = 3000\,\text{MPa},$$
$$T_1 - T_0 = 200\,\text{K}, \quad T_2 - T_0 = 400\,\text{K}, \quad \omega = \frac{1}{6} \times 10^{-2}\,\text{K}^{-1}.$$

With these parameters, we get from the above-derived results:

$$\tau_0(T_1) = y_0, \quad \tau_0(T_2) = \frac{1}{\sqrt{2}} y_0, \quad \tau^* = 120\,\text{MPa}, \quad \tau^{**} = 169.71\,\text{MPa},$$
$$h_1^p = 1.0\,\%, \quad h_2^p = 4.93\,\%, \quad h_3^p = 4.0\,\%,$$
$$\kappa_1 = 0.73\,\text{MPa}, \quad \kappa_2 = \kappa_3 = 3.13\,\text{MPa}.$$

The stress-plastic strain behaviour of the rod during the different processes is depicted in Fig. 1.

This Figure shows the two isothermal stress-plastic strain curves at temperatures T_1 and $T_2 > T_1$. The functions of the underlying model have been chosen such that a linear isotropic hardening is reproduced. Moreover, it is shown that the initial yield

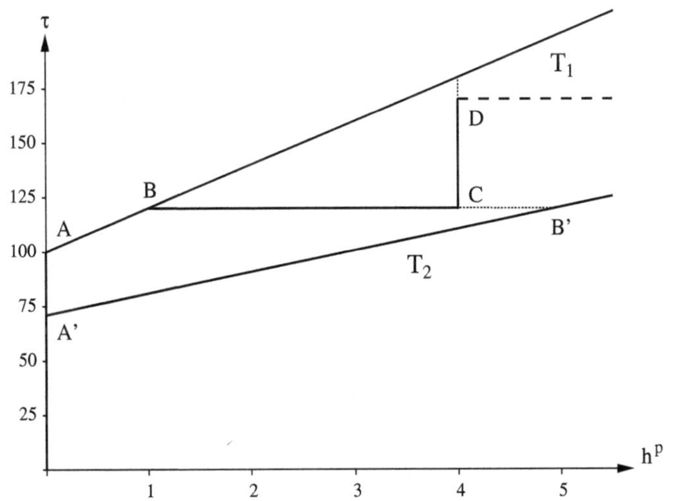

Fig. 1 Isothermal stress-plastic strain diagrams for T_1 and $T_2 > T_1$, and non-isothermal loading paths

stress as well as the slope of the stress-plastic strain curve decrease as the temperature increases.

Several different isothermal and non-isothermal loading paths have been calculated and compared in this Figure.

We first start an isothermal tension process at T_1 increasing the load (from the origin to point A and then from A to B). At point B the stress has reached the value of $\tau = \tau^* = 120\,\text{MPa}$, and plastic strain and plastic work are $h^p = h_1^p = 1.0\,\%$ and $\kappa = \kappa_1 = 0.73\,\text{MPa}$, respectively.

We now change the process to a non-isothermal heating, where as in a classical creep test $\tau = \tau^*$ is kept fixed during this process. The stress-plastic strain point is moving from B to C, with plastic strain $h_3^p = 4.0\,\%$ and plastic work $\kappa_3 = 3.13\,\text{MPa}$, respectively, at point C. It should be noted, however, that point C although here the temperature T_2 is reached does not reach point B' on the isothermal stress-plastic strain curve for temperature T_2.

This process is compared with the isothermal tension test at T_2. Again, the rod is loaded up to $\tau = \tau^* = 120\,\text{MPa}$. Here the respective point moves from A' with initial yield stress $\tau_0(T_2) = 70.71\,\text{MPa}$ to B', where plastic strain and plastic work have reached the values $h_2^p = 4.93\,\%$ and $\kappa_2 = \kappa_3 = 3.13\,\text{MPa}$, respectively. This result will not be altered if prior to this process a neutral loading (from point A to A') is added, i.e. a heating from $T_1 \to T_2$, while $\boldsymbol{D}^p = 0$, $h^p = 0$ and $\kappa = 0$.

The question may arise, what would happen in point C if subsequently the loading is reversed to cooling. A direct reversal with $\tau = \tau^* = \text{const}$ in the direction of point B would be excluded[22] since in this case $\hat{f} < 0$ and unloading would occur. The same would happen for any other process with a negative plastic strain rate. There is only one solution to this problem when the plastic strain and the plastic work remain fixed. Then $h^p = h_3^p$ and $\kappa = \kappa_3$ and a neutral loading moves point C to point D with $\tau = \tau^{**} = 169.71\,\text{MPa}$. Again, the point does not reach the isothermal curve for T_1.

This sequence of processes may be continued as it is marked with the dashed line. We then will observe a stepwise continuation of curves B-C-D towards larger values of stress and plastic strain, bounded by the two stress-plastic strain curves for T_1 and T_2. Thus, a sequence of heating (with fixed stress) and cooling (with fixed plastic strain) between two temperatures T_1 and T_2 will produce a creep-like behaviour during the first part of this cycle and a (necessary) increase of the stress during the second part. In both cases, however, the values of stresses and plastic strains will not reach the corresponding values of the isothermal curves. In other words, a process of cyclic heating and cooling—subject to the above-mentioned conditions of stress and plastic strain, respectively—will produce ratchetting, and stress ratchetting as well, within the bounds of the isothermal stress-plastic strain curves.

[22]Unloading, indeed, could be a solution to this problem. However, here we are primarily interested in plastic solutions, where the plastic indicator remains $\xi = 1$. Moreover, any process with further increasing plastic strain should also be excluded.

With reference to the remarks of the two unknown reviewers, we simply note that as demonstrated with the equation of heat production[23] three different effects may contribute to the evolution of temperature: First, the heat flux as well as the heat source may change the temperature. The second contribution is due to the thermoelastic coupling—which frequently is neglected. And the third contribution is a consequence of the dissipation during plastic work and the work done by the internal variables as representatives of micro-structural changes. Obviously, a change in temperature also may affect these quantities. Thus, a sentence like: "A change in temperature itself (through heating or cooling) should not produce stress and/or plastic deformation in a material" does not reflect the behaviour of a thermo-elastoplastic model.

In the present thermoplastic case, where no heat flux is acting and the elastic strain and rate of deformation are neglected, the right-hand side of equation (80) reduces to the inelastic stress power or rate of plastic work—if further no additional internal variables are used. This means that the plastic work increases as the temperature increases (heating). If further during this process the stress is kept constant, the plastic strain must increase. Comparable results have also been observed experimentally during creep tests performed at low carbon steels under temperature variation (refer, e.g. to Taira and Ohnami 1960 and Iwasaki et al. 1987). For creep additional effects like thermal activation have to be considered. These phenomena are not discussed in this short report. It should, however, be mentioned that possible recovery terms introduced in the evolution equations of the back stress (59) would underline the ratchetting of the material (refer, e.g. to Zhu et al. 2016).

Appendix

We introduce a free energy function $\psi = \psi(\boldsymbol{h}^e, T, \boldsymbol{\alpha}, \kappa)$ per unit reference volume, where $\boldsymbol{h}^e$ may be understood as a formal Hencky strain.[24]

Then from the dissipation inequality (11), we may derive

$$\mathcal{D} = \left\{\boldsymbol{\tau} - \frac{\partial \psi}{\partial \boldsymbol{h}^e}\right\} : \boldsymbol{D}^e - \left\{\eta + \frac{\partial \psi}{\partial T}\right\} \dot{T} + \boldsymbol{\tau} : \boldsymbol{D}^p - \frac{\partial \psi}{\partial \boldsymbol{\alpha}} : \overset{\circ}{\boldsymbol{\alpha}}{}^{\log} - \frac{\partial \psi}{\partial \kappa} \dot{\kappa} \geq 0, \tag{75}$$

and the corresponding equations of state

$$\boldsymbol{\tau} = \frac{\partial \psi}{\partial \boldsymbol{h}^e} \quad \text{and} \quad \eta = -\frac{\partial \psi}{\partial T}. \tag{76}$$

With these results, the production of free energy turns out to be

$$\dot{\psi} = \boldsymbol{\tau} : \boldsymbol{D}^e - \eta \dot{T} - \mathcal{D}_i, \tag{77}$$

---

[23] See Eqs. (80) or (82) in the Appendix.

[24] See, e.g. Xiao et al. (2007b), Appendix B and footnote 9 therein.

Bertram, A., Krawietz, A.: On the introduction of thermoplasticity. Acta Mechanica **223**, 2257–2268 (2012)

de Boer, R.: On non-isothermal elastic-plastic and elastic-viscoplastic deformations. Int. J. Solids Struct. **13**, 1203–1217 (1977)

Bruhns, O.T., Meyers, A., Xiao, H.: On non-corotational rates of Oldroyd's type and relevant issues in rate constitutive formulations. Proc. Roy. Soc. Lond. A **460**, 909–928 (2004)

Bruhns, O.T., Xiao, H., Meyers, A.: Self-consistent Eulerian rate type elasto-plasticity models based upon the logarithmic stress rate. Int. J. Plast. **15**, 479–520 (1999)

Bruhns, O.T.: Some remarks on the history of plasticity—Heinrich Hencky, a pioneer of the early years. In: Stein, E. (ed.) The History of Theoretical, Material and Computational Mechanics, pp. 133–152. Springer, Heidelberg (2014a)

Bruhns, O.T.: The Prandtl-Reuss equations revisited. Z. Angew. Math. Mech. **94**, 187–202 (2014b)

Casey, J.: On elastic-thermo-plastic materials at finite deformations. Int. J. Plast. **14**, 173–191 (1998)

Coleman, B.D., Noll, W.: The thermodynamics of elastic materials with heat conduction and viscosity. Arch. Rat. Mech. Anal. **13**, 167–178 (1963)

Collins, I.F.: Elastic/plastic models for soils and sands. Int. J. Mech. Sci. **47**, 493–508 (2005)

Farren, W.S., Taylor, G.I.: The heat developed during plastic extension of metals. Proc. Roy. Soc. Lond. A **107**, 422–451 (1925)

Green, A.E., Naghdi, P.M.: A general theory of an elastic-plastic continuum. Arch. Rat. Mech. Anal. **18**, 251–281, (1965). (corrigenda **19**, 408)

Green, A.E., Naghdi, P.M.: A thermodynamic development of elastic-plastic continua. In: Parkus, H., Sedov, L.I. (eds.) Procedings of the IUTAM Symposium Irreversible Aspects of Continuum Mechanics and Transfer of Physical Characteristics in Moving Fluids, pp. 117–131. Springer, Heidelberg (1968)

Haupt, P.: Viskoelastizität und Plastizität. Springer, Berlin (1977)

Haupt, P.: Continuum Mechanics and Theory of Materials, 2nd edn. Springer, Berlin (2002)

Hill, R.: The Mathematical Theory of Plasticity. Clarendon Press, Oxford (1950)

Hill, R.: A general theory of uniqueness and stability in elastic-plastic solids. J. Mech. Phys. Solids **6**, 236–249 (1958)

Hill, R.: Some basic principles in the mechanics of solids without a natural time. J. Mech. Phys. Solids **7**, 209–225 (1959)

Hill, R.: On constitutive inequalities for simple materials - I. J. Mech. Phys. Solids **16**, 229–242 (1968)

Hill, R.: Aspects of invariance in solid mechanics. Adv. Appl. Mech. **18**, 1–75 (1978)

Houlsby, G.T., Puzrin, A.M.: A thermomechanical framework for constitutive models for rate-independent dissipative materials. Int. J. Plast. **16**, 1017–1047 (2000)

Iwasaki, Y., Hiroe, T., Igari, T.: An application of the viscoplasticity theory to the inelastic analysis at elevated temperature. Trans. Jpn. Soc. Mech. Eng. Ser. A **493**, 1838–1843 (1987)

Kestin, J., Rice, J.R.: Paradoxes in the applications of thermodynamics to strained solids. In: Stuart, E.B., Cal'Or, B., Brainard, A.J. (eds.) A Critical Review of Thermodynamics, pp. 275–298. Mono Book Corporation, Baltimore (1970)

Kratochvíl, J., Dillon, O.W.: Thermodynamics of elastic-plastic materials as a theory with internal state variables. J. Appl. Phys. **40**, 3207–3218 (1969)

Lee, E.H.: Thermo-elastic-plastic analysis at finite strain. In: Boley, B.A. (ed.) Thermoinelasticity, pp. 156–169. Springer, Heidelberg (1970)

Lehmann, T.: Some thermodynamic considerations of phenomenological theory of non-isothermal elastic-plastic deformations. Arch. Mech. **24**, 975–989 (1972)

Lehmann, T. (ed.): The constitutive law in thermoplasticity. CISM Courses and Lectures, 281st edn. Springer, New York (1984)

Lehmann, T.: On the balance of energy and entropy at inelastic deformations of solid bodies. Eur. J. Mech. A/Sol. **8**, 235–251 (1989)

Lehmann, T., Guo, Z.H., Liang, H.Y.: The conjugacy between Cauchy stress and logarithm of the left stretch tensor. Eur. J. Mech. A/Solids **10**, 395–404 (1991)

Lubliner, J.: On the thermodynamic foundations of non-linear solid mechanics. Int. J. Non-Linear Mech. **7**, 237–254 (1972)
Macvean, D.B.: Die Elementararbeit in einem Kontinuum und die Zuordnung von Spannungs- und Verzerrungstensoren. ZAMP **18**, 157–184 (1968)
Mandel, J.P.: Plasticité Classique et Viscoplasticité. In: CISM Courses and Lectures, 97 edn. Springer, Wien (1972)
Mandel, J.P.: Thermodynamics and plasticity. In: Domingos, J.J., Nina, M.N.R., Whitelaw, J.H. (eds.) Foundations of Continuum Thermodynamics, pp. 283–304. The MacMillan Press, London (1974)
Maugin, G.A.: The Thermomechanics of Plasticity and Fracture. Cambridge University Press, Cambridge (1992)
Meyers, A., Xiao, H., Bruhns, O.T.: Elastic stress ratchetting and corotational stress rates. Techn. Mechanik **23**, 92–102 (2003)
Miehe, C.: A theory of large-strain isotropic thermoplasticity based on metric transformation tensors. Arch. Appl. Mech. **66**, 45–64
von Mises, R.: Mechanik der plastischen Formänderung von Kristallen. Z. Angew. Math. Mech. **8**, 161–185 (1928)
Müller, I.: Thermodynamics. Pitman, London (1985)
Naghdi, P.M.: Stress-strain relations in plasticity and thermoplasticity. In: Lee, E.H., Symonds, P.S. (eds.), Plasticity, Proceedings of the 2nd Symposium Naval Structural Mechanics, pp. 121–169. Pergamon Press, New York (1960)
Naghdi, P.M.: A critical review of the state of finite plasticity. Z. Angew. Math. Phys. **41**, 315–394 (1990)
Perzyna, P.: Thermodynamic theory of viscoplasticity. Adv. Appl. Mech. **11**, 313–354 (1971)
Prager, W.: An elementary discussion of definitions of stress rate. Quart. Appl. Math. **18**, 403–407 (1960)
Rajagopal, K.R., Srinivasa, A.R.: Mechanics of the inelastic behaviour of materials. Parts I and II. Int. J. Plast. **14**, 945–967, 969–995, (1998)
Reinhardt, W.D., Dubey, R.N.: Coordinate-independent representation of spin tensors in continuum mechanics. J. Elast. **42**, 133–144 (1996)
Rice, J.R.: Continuum mechanics and thermodynamics of plasticity in relation to micro-scale deformation mechanism. In: Argon, A.S. (ed.) Constitutive Equations in Plasticity, pp. 21–79. MIT Press, Cambridge (1975)
Šilhavý, M.: Mechanics and Thermodynamics of Continuous Media. Springer, Berlin (1997)
Simó, J.C., Hughes, T.J.R.: Computational Inelasticity. Springer, New York (1998)
Simó, J.C., Pister, K.S.: Remarks on rate constitutive equations for finite deformation problems: computational implications. Compt. Meth. Appl. Mech. Eng. **46**, 201–215 (1984)
Taira, S., Ohnami, M.: Creep and creep rupture under varying temperatures. J. Jpn. Soc. Test. Mater. **9**, 89–103 (1960)
Truesdell, C., Noll, W.: The non-linear field theories of mechanics. In: Flügge, S. (ed.), Handbuch der Physik. vol. III/3. Springer, Berlin (1965)
Truesdell, C., Toupin, R.A.: The classical field theories. In: Flügge, S. (ed.), Handbuch der Physik. vol. III/1, pp. 226–793. Springer, Berlin (1960)
Truesdell, C.A. (ed.): Rational Thermodynamics, 2nd edn. Springer, New York (1984)
Xiao, H., Bruhns, O.T., Meyers, A.: A new aspect in the kinematics of large deformations. In: Gupta, N.K. (ed.) Plasticity and Impact Mechanics, pp. 100–109. New Age International Ltd., New Delhi (1997a)
Xiao, H., Bruhns, O.T., Meyers, A.: Logarithmic strain, logarithmic spin and logarithmic rate. Acta Mechanica **124**, 89–105 (1997b)
Xiao, H., Bruhns, O.T., Meyers, A.: Hypo-elasticity model based upon the logarithmic stress rate. J. Elast. **47**, 51–68 (1997c)
Xiao, H., Bruhns, O.T., Meyers, A.: Strain rates and material spins. J. Elast. **52**, 1–41 (1998a)

Xiao, H., Bruhns, O.T., Meyers, A.: On objective corotational rates and their defining spin tensors. Int. J. Solids Struct. **35**, 4001–4014 (1998b)

Xiao, H., Bruhns, O.T., Meyers, A.: Existence and uniqueness of the integrable-exactly hypoelastic equation $\boldsymbol{\tau}^{\circ *} = \lambda(\text{tr}\mathbf{D})\mathbf{I} + 2\mu\mathbf{D}$ and its significance to finite inelasticity. Acta Mechanica **138**, 31–50 (1999)

Xiao, H., Bruhns, O.T., Meyers, A.: The choice of objective rates in finite elastoplasticity: general results on the uniqueness of the logarithmic rate. Proc. Roy. Soc. Lond. A **456**, 1865–1882 (2000)

Xiao, H., Bruhns, O.T., Meyers, A.: Objective rates, path-dependence properties and non-integrability problems. Acta Mechanica **176**, 135–151 (2005)

Xiao, H., Bruhns, O.T., Meyers, A.: Elastoplasticity beyond small deformations. Acta Mechanica **182**, 31–111 (2006)

Xiao, H., Bruhns, O.T., Meyers, A.: The integrability criterion in finite elastoplasticity and its constitutive implications. Acta Mechanica **188**, 227–244 (2007a)

Xiao, H., Bruhns, O.T., Meyers, A.: Thermodynamic laws and consistent Eulerian formulation of finite elastoplasticity with thermal effects. J. Mech. Phys. Solids **55**, 338–365 (2007b)

Zhu, Y., Kang, G., Kan, Q., Bruhns, O.T.: Logarithmic stress rate based constitutive model for cyclic loading in finite plasticity. Int. J. Plast. **54**, 34–55 (2014)

Zhu, Y., Kang, G., Kan, Q., Bruhns, O.T., Liu, Y.: Thermo-mechanically coupled cyclic elasto-viscoplastic constitutive model of metals: Theory and application. Int. J, Plasticity (to appear) (2016)

Ziegler, H.: An attempt to generalize Onsager's principle, and its significance for rheological problems. Z. Angew. Math. Phys. **9**, 748–763 (1958)

Ziegler, H.: Some extremum principles in irreversible thermodynamics with application to continuum mechanics. In: Sneddon, I., Hill, R. (eds.) Progress in Solid Mechanics, vol. 4. North-Holland, Amsterdam (1963)

Crystal-Plasticity Simulation of Micromachining of Single-Crystal Metal: Methodology and Analysis

Qiang Liu, Srihari Dodla, Anish Roy and Vadim V. Silberschmidt

Abstract A crystal-plasticity modelling framework was implemented to simulate micromachining of a single-crystal metal. A new shear strain-based criterion was proposed to control material removal. This criterion was implemented in three different modelling techniques: element deletion, arbitrary Lagrangian–Eulerian (ALE) adaptive remeshing and smooth particle hydrodynamics (SPH) in a general-purpose finite-element software package ABAQUS. The three different modelling approaches were compared in terms of their computational accuracy and efficiency. Based on these studies, an optimized modelling strategy was proposed to simulate micro-scratching of single-crystal copper. The validity of the suggested methodology was corroborated through comparison between FE simulations and experimental data in terms of cutting forces, chip morphology and pile-up patterns in the work-piece.

1 Introduction

The production of small-scaled components with complex features is gaining increasing importance due to the trend of miniaturization of products. As a result there is a growing need for fast, reliable, mass micromachining of functional components (Engel and Eckstein 2002). In contrast to conventional macro-scale machining, a process zone in micromachining is usually limited to one or several grains of a metallic work-piece material. A component in the microscale exhibits different material

Q. Liu · S. Dodla · A. Roy (✉) · V.V. Silberschmidt
Wolfson School of Mechanical and Manufacturing Engineering,
Loughborough University, Leicestershire LE11 3TU, UK
e-mail: A.Roy3@lboro.ac.uk

Q. Liu
e-mail: Q.Liu2@lboro.ac.uk

S. Dodla
e-mail: S.Dodla@lboro.ac.uk

V.V. Silberschmidt
e-mail: V.Silberschmidt@lboro.ac.uk

K. Naumenko and M. Aßmus (eds.), *Advanced Methods of Continuum Mechanics for Materials and Structures*, Advanced Structured Materials 60,
DOI 10.1007/978-981-10-0959-4_9

behaviour when compared to its macro-scale counterpart (Greer and De Hosson 2011; Kraft et al. 2010). Consequently, a cutting response in the microscale differs significantly from that of its macro-scale counterpart. For example, a cutting force and chip morphology were found to depend on crystal orientation in several experimental studies on micromachining of single-crystal metals (Lee et al. 2000; Mian et al. 2011). To better understand local deformation processes at a tool–work–piece interface in a micromachining process, a thorough analysis of deformation mechanisms at grain level is required (Mian et al. 2011; Shi and Attia 2010).

In recent years, a finite-element (FE) method has been widely used to model macro-scale machining of metallic materials, including analysis of chip morphology and temperature effects (Buchkremer et al. 2014; Babitsky et al. 2004), influence of cutting conditions on structure of machining subsurface (Arısoy and Özel 2015; Jafarian et al. 2014) as well as optimization of machining parameters (Hokka et al. 2012; Elhami et al. 2015). Similar research was also conducted in the field of micromachining; for example, in the work of Jin and Altintas (2012), a relationship between cutting forces and chip thickness and a tool edge radius was identified through series of FE simulation results for micromilling. FE simulations could provide an in-depth understanding of the underpinning micromechanisms during the micromachining process that are difficult to capture experimentally. Liu and Melkote (2007) adopted FE simulations to reveal a micromechanism of a size effect in metal microcutting, i.e. a nonlinear increase in specific cutting energy with a decrease in the depth-of-cut. The simulation results indicated that a strain-gradient effect in a deformation zone was a dominating reason of the size effect. Parle et al. (2014) studied microcrack formation in orthogonal machining of carbon steel and evaluated a contribution of microcracks to the total specific energy of a shear zone under different machining conditions. In the work of Zong et al. (2016), the FE method was used to capture crack propagation in a process of chip formation in ZnS crystal, and then used to predict critical cutting parameters.

However, in these FE simulations, anisotropy at the grain level, induced by crystallographic structure, was not considered. To overcome this drawback, a single-crystal-plasticity (SCP) theory, incorporating the effect of crystal orientation and activated slip systems, was developed (Roters et al. 2010) to study inherently anisotropic deformation behaviour at a smallest practical length scale of metallic materials. SCP-based studies by Zahedi et al. (2013) and Tajalli et al. (2014) showed that chip formation and cutting forces in single-crystal copper depend on the initial crystal orientations in the work-piece with respect to a cutting direction. Moreover, the works by Demiral et al. (2014) and Pal and Stucker (2013), based on strain-gradient crystal-plasticity simulations, indicated that inhomogeneous plastic deformation could affect machinability of a work-piece. In contrast to the extensive numerical studies, these crystal-plasticity-based simulations still lack comprehensive experimental validations.

In FE simulations of micromachining, the modelling of material removal or chip separation is an important issue in addition to anisotropic deformation behaviour at the grain level. To our knowledge, three modelling methods are usually employed to simulate the material removal: element deletion (Hibbitt et al. 2012), arbitrary Lagrangian Eulerian (ALE) adaptive remeshing (Hibbitt et al. 2012) and smooth particle hydrodynamics (SPH) (Hibbitt et al. 2012; Monaghan 2005). For example, element deletion was adopted by Demiral et al. (2014) and Arisoy and Ozel (2015), the ALE adaptive remeshing was employed by Amini et al. (2008), Jin and Altintas (2012) and Parle et al. (2014), while SPH was utilized in simulations of Zahedi et al. (2013; 2015) and Zhang and Dong (2015). However, there is still a lack of direct comparison among these modelling methods, which is one of the primary motivations of this study.

This chapter is organized as follows: a theoretical framework of the SCP theory and its numerical implementation are summarized in Sect. 2, followed by description of the suggested modelling procedure in Sect. 3. Simulation results from different modelling methodologies are analyzed in Sect. 4. An optimized modelling strategy was chosen to predict the microscratching of single-crystal copper, and the comparison of simulations with experiment is discussed in the subsequent section. We conclude with some remarks in last Sect. 6.

2 Constitutive Relations

2.1 Theory

In this section, a classical crystal-plasticity theory adopted in this study is reviewed. A deformation gradient, $\boldsymbol{F}$, can be decomposed into its elastic and plastic parts:

$$\boldsymbol{F} = \boldsymbol{F}_e \boldsymbol{F}_p, \tag{1}$$

where the subscripts 'e' and 'p' denote the elastic and plastic parameters, respectively. By applying the product rule of differentiation, one can obtain the rate of the total deformation gradient, $\dot{\boldsymbol{F}}$, as

$$\dot{\boldsymbol{F}} = \boldsymbol{F}_e \dot{\boldsymbol{F}}_p + \boldsymbol{F}_e \dot{\boldsymbol{F}}_p. \tag{2}$$

Therefore, the velocity gradient, $\boldsymbol{L}$, can be introduced following its definition $\boldsymbol{L} = \dot{\boldsymbol{F}}\boldsymbol{F}^{-1}$ as

$$\boldsymbol{L} = \dot{\boldsymbol{F}}_e \boldsymbol{F}_e^{-1} + \boldsymbol{F}_e (\dot{\boldsymbol{F}}_p \boldsymbol{F}_p^{-1}) \boldsymbol{F}_e^{-1} = \boldsymbol{L}_e + \boldsymbol{L}_p. \tag{3}$$

It is assumed that the plastic velocity gradient, $\boldsymbol{L}_p$, is induced by shearing on each slip system in a crystal. Hence, $\boldsymbol{L}_p$ is formulated as the sum of shear rates on all the slip systems, i.e.

$$\boldsymbol{L}_p = \sum_{i=1}^{N} \dot{\gamma}^{\alpha} \boldsymbol{s}^{\alpha} \otimes \boldsymbol{m}^{\alpha}, \tag{4}$$

where $\dot{\gamma}^{\alpha}$ is the shear slip rate on the slip system α, N is the total number of slip systems, and unit vectors $\boldsymbol{s}^{\alpha}$ and $\boldsymbol{m}^{\alpha}$ define the slip direction and the normal to the slip plane in the deformed configuration, respectively. Furthermore, the velocity gradient can be expressed in terms of a symmetrically rate of stretching, $\boldsymbol{D}$, and an antisymmetric rate of spin, $\boldsymbol{W}$,

$$\boldsymbol{L} = \boldsymbol{D} + \boldsymbol{W} = (\boldsymbol{D}_e + \boldsymbol{W}_e) + (\boldsymbol{D}_p + \boldsymbol{W}_p). \tag{5}$$

Using Eqs. (3) and (4), it can be deduced

$$\boldsymbol{D}_e + \boldsymbol{W}_e = \dot{\boldsymbol{F}}_e \boldsymbol{F}_e^{-1}, \qquad \boldsymbol{D}_p + \boldsymbol{W}_p = \sum_{i=1}^{N} \dot{\gamma}^{\alpha} \boldsymbol{s}^{\alpha} \otimes \boldsymbol{m}^{\alpha}. \tag{6}$$

Following the work of Huang (1991), a constitutive law is expressed as the relationship between the elastic part of the symmetric rate of stretching, $\boldsymbol{D}_e$, and the Jaumann rate of Cauchy stress, $\overset{\triangledown}{\boldsymbol{\Sigma}}$, i.e.

$$\overset{\triangledown}{\boldsymbol{\Sigma}} + \boldsymbol{\Sigma}(\boldsymbol{I} : \boldsymbol{D}_e) = \boldsymbol{C} : (\boldsymbol{D} - \boldsymbol{D}_p), \tag{7}$$

where $\boldsymbol{I}$ is the second-order unit tensor, $\boldsymbol{C}$ is the fourth-order, possibly anisotropic, elastic stiffness tensor. The Jaumann stress rate is expressed as

$$\overset{\triangledown}{\boldsymbol{\Sigma}} = \dot{\boldsymbol{\Sigma}} - \boldsymbol{W}_e \boldsymbol{\Sigma} + \boldsymbol{\Sigma} \boldsymbol{W}_e. \tag{8}$$

On each slip system, the resolved shear stress, τ^{α}, is expressed by a Schmid law,

$$\tau^{\alpha} = \mathrm{sym}(\boldsymbol{s}^{\alpha} \otimes \boldsymbol{m}^{\alpha}) : \boldsymbol{\Sigma}. \tag{9}$$

The relationship between the shear rate, $\dot{\gamma}^{\alpha}$, and the resolved shear stress, τ^{α}, on the slip system α is expressed by a power law proposed by Hutchinson (1976),

$$\dot{\gamma}^{(\alpha)} = \dot{\gamma}_0 \left| \frac{\tau^{\alpha}}{g^{\alpha}} \right|^n \mathrm{sgn}(\tau^{\alpha}), \tag{10}$$

where $\dot{\gamma}_0$ is the reference shear rate, g^{α} is the slip resistance and n is the rate-sensitivity parameter. The evolution of g^{α} is given by

$$\dot{g}^{\alpha} = \sum_{i=1}^{n} h_{\alpha\beta} |\dot{\gamma}^{\beta}|, \tag{11}$$

where $h_{\alpha\beta}$ is the hardening modulus that is calculated from

$$h_{\alpha\alpha} = (h_0 - h_s)\,\mathrm{sech}^2\left(\frac{(h_0 - h_s)\gamma}{\tau_s - \tau_0}\right) + h_s,$$
$$h_{\alpha\beta} = q h_{\alpha\alpha}(\alpha \neq \beta),\ \gamma = \sum_{\alpha}\int_0^t |\dot{\gamma}^{\alpha}|dt. \tag{12}$$

The proposed hardening relation is modified from the original relation proposed by Asaro (1983). Here, h_0 and h_s are the initial and saturated hardening modulus, q is the latent hardening ratio, τ_0 and τ_s are the shear stresses at the onset of yield and the saturation of hardening, respectively. γ is the accumulative shear strain over all the slip systems.

2.2 *Implementation in Explicit ABAQUS Environment*

Implementation of the SCP theory in implicit FE environment, by means of a so-called UMAT subroutine, was introduced in the work of Huang (1991), where a time integration scheme and a stress update algorithm were presented as

$$\dot{\gamma}^{\alpha}\Delta t = \Delta\gamma^{\alpha} = \Delta t[(1-\theta)\dot{\gamma}^{\alpha}|_t + \theta\dot{\gamma}^{\alpha}|_{t+\Delta t}], \tag{13}$$

$$\Delta\boldsymbol{\Sigma} = \overset{\nabla}{\boldsymbol{\Sigma}}\,\Delta t. \tag{14}$$

Here, Δt is the time increment in the FE calculation; θ ranges from 0 to 1, representing different time integration schemes (as an example, setting $\theta = 0$ yields a simple Euler time integration scheme); the Jaumann stress rate, $\overset{\nabla}{\boldsymbol{\Sigma}}$, was defined in Eq. (8). In this chapter, the SCP theory is implemented employing a VUMAT subroutine in the explicit ABAQUS environment. The time integration scheme was identical to the one implemented in the UMAT; however, the stress update algorithm had to be modified due to the difference of the defined stress rate for ABAQUS/standard and ABAQUS/explicit formulations. The former employed the Jaumann stress rate, but the latter was based on the Green–Naghdi stress rate (Hibbitt et al. 2012).

In contrast to the Jaumann stress rate defined in Eq. (8), the Green–Naghdi stress rate was defined as

$$\overset{\triangle}{\boldsymbol{\Sigma}} = \dot{\boldsymbol{\Sigma}} - \boldsymbol{\Omega}_e\boldsymbol{\Sigma} + \boldsymbol{\Sigma}\boldsymbol{\Omega}_e. \tag{15}$$

Here, Ω was found from a right polar decomposition of the total deformation gradient, $\boldsymbol{F}$, as

$$\boldsymbol{\Omega} = \dot{\boldsymbol{R}}\boldsymbol{R}^T, \qquad \boldsymbol{F} = \boldsymbol{V}\boldsymbol{R}, \tag{16}$$

where $\boldsymbol{R}$ and $\boldsymbol{V}$ are the right rotation and stretch tensors, respectively. To evaluate the stress update defined by the Green–Naghdi stress rate using the Jaumann rate, one can use the Hughes–Winget algorithm (Hughes and Winget 1980), as

$$\boldsymbol{\Sigma}|_{t+\triangle t} = \triangle \boldsymbol{R} \boldsymbol{\Sigma}|_t \triangle \boldsymbol{R}^T + \triangle \boldsymbol{\Sigma}, \tag{17}$$

$$\triangle \boldsymbol{R} = \left[\boldsymbol{I} - \frac{1}{2} (\triangle \boldsymbol{W} - \triangle \boldsymbol{\Omega}) \right]^{-1} \left[\boldsymbol{I} + \frac{1}{2} (\triangle \boldsymbol{W} - \triangle \boldsymbol{\Omega}) \right], \tag{18}$$

$$\triangle \boldsymbol{W} = \boldsymbol{W} \triangle t, \triangle \boldsymbol{\Omega} = \boldsymbol{\Omega} \Delta t, \tag{19}$$

where $\triangle \boldsymbol{R}$ is the relative spin increment tensor and $\boldsymbol{I}$ is the second-order unit tensor, $\triangle \boldsymbol{\Sigma}$ is the stress increment obtained with the Jaumann stress rate (Eq. (14)). Another essential difference in ABAQUS/Explicit is that the stress and strain tensors are defined based on the spatial coordinate system (i.e. with respect to the local coordinate system rotating with the volume), in contrast to the material coordinate system (i.e. a fixed global coordinate system) used in ABAQUS/Standard. Therefore, during the conversion of UMAT to VUMAT, the stress update algorithm in VUMAT should be rewritten as

$$\boldsymbol{\Sigma}|_{t+\triangle t} = \triangle \boldsymbol{R} \boldsymbol{\Sigma}|_t \triangle \boldsymbol{R}^T + \boldsymbol{R} \triangle \boldsymbol{\Sigma} \boldsymbol{R}^T. \tag{20}$$

3 Modelling Procedure

3.1 Finite-Element Model

Without loss of generality, a single-crystal copper with a face-centred cubic (FCC) crystallographic structure was used in the study. A FE model was developed to simulate a microscratching processes as shown in Fig. 1a. A cutting tool had a wedge angle of 60° and a clearance angle of 6.25°. Deformation behaviour of single-crystal copper was described with the SCP model introduced in Sect. 2.1, implemented in the subroutine VUMAT formulated in Sect. 2.2. For FCC single-crystal copper, slip was assumed to occur on the usual twelve [111]<110> slip systems, in which [111] represents the normal to slip plane and <110> is the slip direction. In our simulations, the cutting tool was assumed to be rigid, and the contact condition between the cutting tool and the work-piece was assumed to be frictionless. A cutting direction was in the negative X direction (Fig. 1a), and a cutting velocity was 10 mm/min. The groove produced by microscratching is shown in Fig. 1b, which is a result of two stages of microscratching: In Stage 1, a linearly increasing cutting depth (ranging from 0 to 18 μm) was set for the cutting tool, while in Stage 2 a constant cutting depth (18 μm) was retained. To reduce a computational cost, the two stages were modelled using two FE models. In the first model, the length, height and width of the work-piece were 340, 200 and 120 μm, respectively. In the second model, the respective magnitudes were 240, 200 and 120 μm.

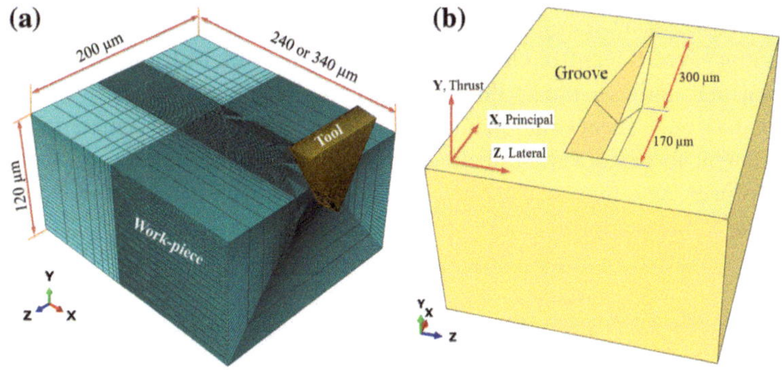

Fig. 1 Finite-element model (**a**) for simulation of microscratching and groove geometry (**b**)

The work-piece was meshed using eight-node brick elements with reduced integration (C3D8R) in ABAQUS. For comparison of different modelling techniques of material removal, a mesh with 2×10^5 elements was used for computational efficiency. Here, Stage 2 of microscratching was considered, with a fixed cutting depth of 18 μm. For a full FE simulation replicating a physical experiment, a mesh with 1×10^6 brick elements was used after an extensive mesh-sensitivity study. To improve accuracy, a finer local mesh was used in regions near the cutting zone with a height of 27 μm. In this study, a normal to the top surface of the work-piece has the [110] crystal orientation. Two orientations were chosen as cutting directions, [1-10], and another rotated by an angle of 45° with respect to it. The cutting forces were defined as follows: a force along the cutting direction (X direction) is called the principal force, and those along the Y and Z directions are the thrust and lateral forces, respectively (Fig. 1b).

3.2 *Modelling of Material Removal*

Three modelling techniques to simulate material removal in a machining scenario were considered in our study. To determine chip separation in a FE simulation, a criterion for material removal is required (Parle et al. 2014). In macro-scale machining studies, fracture-energy-based damage criteria were typically used to control material removal (Jin and Altintas 2012; Zhang et al. 2012), such as a widely used Johnson–Cook model (Johnson and Cook 1985) (though its applicability came under some noteworthy criticism (Zhang et al. 2015)). Modelling damage in single-crystal metal remains a significant challenge due to its inherently anisotropic character of evolution in these materials. Typically, a simplified isotropic damage model was used (Kim and Yoon 2015), claiming that a shear strain-based damage model provides the most accurate prediction based on experimental data. As extreme shear deformation occurs at the interface between a cutting tool and a work-piece during machining, a shear strain-based criterion (i.e. the accumulative shear strain over all the slip systems, γ) was employed to control element deletion in the work of Demiral et al. (2014). However, our studies indicate that using accumulated shear strain (Demiral et al. 2014) or maximum shear strain (Kim and Yoon 2015) as criterion for material removal is insufficient to capture the effects of crystallographic orientation on the cutting force. To overcome this drawback, a new shear strain-based criterion is proposed here in the following form:

$$\max(\gamma - \gamma_{cr}, \gamma_{sl,min} - \gamma_{sl,cr}) \geq 0, \tag{21}$$

$$\gamma_{sl,min} = \min(\gamma^{\alpha}), \quad \alpha = 1, 2, \ldots N \tag{22}$$

where $\gamma_{sl,cr}$ and γ_{cr} are the critical values of shear strain on a single slip system and the accumulated shear strain on all the slip systems, respectively. That is, damage in a single crystal is considered based on two scenarios of partial and full activation of slip systems. In other words, both shear on individual slip systems as well as the overall slip due to all slip systems are monitored. If either the critical value for an individual slip system ($\gamma_{sl,cr}$) or the accumulated slip (γ_{cr}) is attained, the element is removed. In our study, the values of γ_{cr} and $\gamma_{sl,cr}$ were 6.0 and 0.068, respectively; these parameters were chosen based on initial calibration tests. In a nominal sense, these values imply that failure in single-crystal copper occurs easier for the full activation of slip systems than for that of a partial one.

The proposed criterion was implemented to control element deletion or conversion of finite-elements into SPH particles. Such a criterion of material removal was not required when the ALE adaptive remeshing technique was employed. In this case, severely deformed elements in a shear zone of the work-piece could be re-meshed with regular elements and simultaneously separated from the latter as a chip (Parle et al. 2014; Hibbitt et al. 2012). In FE simulation of microscratching, a contact interface between the cutting tool and the work-piece was more irregular compared to that of microcutting. As a result, with acceptable levels of computational cost, the process of chip separation in microscratching could not be well simulated when only ALE adaptive remeshing was employed. Instead, material removal modelling using both ALE adaptive remeshing and element deletion was considered as a case study. The latter was compared to that with only element deletion so that the roles of ALE adaptive remeshing in FE simulation could be elucidated.

4 Analysis and Discussion

4.1 *Comparison of Implicit and Explicit Implementations of SCP Theory*

To validate the conversion of UMAT into VUMAT introduced in Sect. 2.2, several numerical tests were carried out in ABAQUS/Standard and ABAQUS/Explicit,

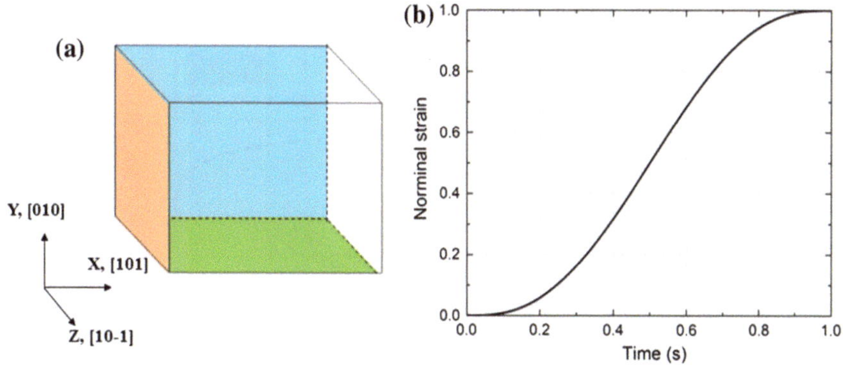

Fig. 2 Schematic of implicit and explicit FE simulations using SCP theory: **a** crystal orientation of numerical sample; **b** deformation regime

respectively. Figure 2 showed the schematic of numerical test, including a crystal orientation of numerical sample (Fig. 2a) and a deformation regime (Fig. 2b). These numerical tests were divided into three cases of deformation of the numerical sample: tension, shear in the XY plane and shear in the XZ plane. The loading was realized by employing a displacement boundary condition in ABAQUS, and the applied nominal strain was up to 100 % in order to check implementation of the SCP theory for large deformations. In the explicit FE method, an inertial force was introduced into the equation of force balance, which was one of the most important differences from the implicit formulation. To eliminate a possible inertial effect, the increase in the nominal strain, ϵ_{norm}, with time, t, followed a smooth curve, introduced as

$$\epsilon_{\text{norm}}(t) = \epsilon_{\text{f}}\xi^3(10 - 15\xi + 6\xi^2), \quad \xi = \frac{t}{t_{\text{f}} - t} \tag{23}$$

where ϵ_{f} and t_{f} are the final nominal strain and time, respectively. The variation of with time is illustrated in Fig. 2b.

For the case of tension, stress–strain curves obtained with the implicit and explicit FE simulations are compared in Fig. 3 (the inset shows shapes of the modelled sample before and after deformation). To quantitatively demonstrate a difference between the Jaumann stress rate (JSR) and the Green–Naghdi stress rate (GSR), the numerical results obtained with VUMAT with account for the JSR are also provided for comparison. Apparently, the two used VUMAT formulations yield the same tensile response as UMAT. Therefore, for samples under tension (or compression), there was no difference between the JSR and GSR due to the absence of rotation deformation (here $\boldsymbol{R} = \boldsymbol{I}$).

A similar comparison for shearing in the XY plane (Fig. 4a) and XZ plane (Fig. 4b) was carried out. As shown in the insets in Fig. 4, an elongation effect of the sample became significant under large shear deformation. As a result, the magnitude of direct stress becomes comparable to the level of shear stresses induced in shearing. Shear

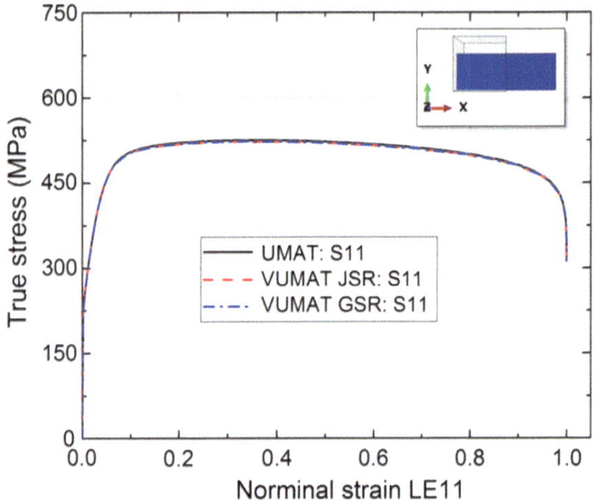

Fig. 3 Comparison of implicit and explicit implementations of SCP theory for tension

deformation are accompanied by inherent rotation in the computations; consequently, a difference between the JSR and GSR is clearly observed on stress–strain curves for shear deformation. Such a difference was significant only when shear strain was relatively high, e.g. larger than 0.5. Thus, a special attention should be paid to the difference for different FE formulations when the finite-deformation-based constitutive theory is to be implemented. The stress–strain curves obtained with VUMAT with the GSR matched perfectly those from UMAT under shearing deformation, thus validating implementation of the explicit formulation of the SCP theory described in Sect. 2.2.

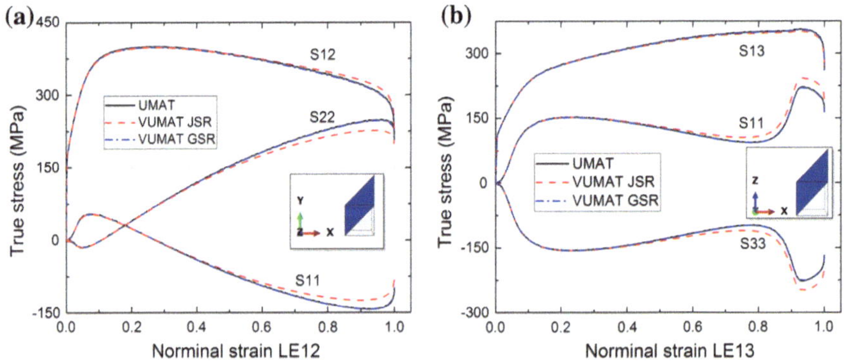

Fig. 4 Comparison of implicit and explicit FE simulations in shearing in XY plane (**a**) and XZ plane (**b**)

4.2 Comparison of Material Removal Method

To compare different modelling techniques of material removal, a cutting force and a computational time were used as primary parameter. All numerical tests were performed in ABAQUS v.6.14. The ALE adaptive remeshing scheme was implemented after every 50 increments, and 3 mesh sweeps were created to reduce element distortion each time. In SPH-based simulations, a continuum finite-element was converted into an SPH particle once the criterion of material removal was satisfied for it.

The cutting forces obtained with the three material removal methods are shown in Fig. 5 for both 0° and 45° directions; both the principal and thrust forces are presented. For all the modelling techniques of material removal, the cutting forces initially increased with the engagement of the cutting tool, and approached a constant level with the full tool engagement in the work-piece. For the studied directions, the principal forces were much larger than the thrust forces. Compared to the case of element deletion, the introduced ALE adaptive remeshing scheme had no significant effect on the cutting forces in both 0° and 45° directions. Some differences were observed between the cases of element deletion and use of SPH. Especially in the engagement stage, the cutting forces obtained in the simulation with SPH were larger than those from modelling incorporating only element deletion. A possible reason was a continuing contribution of SPH particles to the overall cutting/reaction forces even after the material removal criterion was satisfied. However, this difference became less significant at the full-engagement stage with relatively constant cutting forces for the two cutting directions studied. The average cutting forces and the relative computational times obtained with the three modelling techniques are summarized in Table 1.

The obtained simulation results indicate that the difference in average principal and thrust forces is negligible. For comparability, the computational times for different studied cases were normalized with that for the 0° direction simulation with the element deletion method. Apparently, significant differences in computation time for

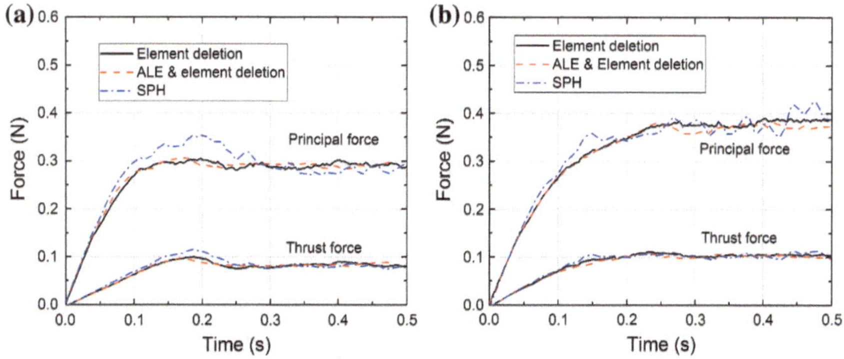

Fig. 5 Comparison of cutting forces obtained with different material removal methods for 0° (**a**) and 45° (**b**) directions

Table 1 Comparison of average cutting forces and computational times for different modelling techniques of material removal

Parameter	Element deletion		ALE and element deletion		SPH	
	0°	45°	0°	45°	0°	45°
Average principal force (N)	0.269	0.329	0.265	0.323	0.277	0.327
Average thrust force (N)	0.074	0.089	0.073	0.087	0.076	0.090
Relative computational time	1	1.25	6.89	9.48	12.3	12.5

different techniques were observed. Among the three used methods, element deletion resulted in the least computational time whilst SPH was the most expensive. The computational times for the ALE adaptive remeshing and SPH were about 7 and 10 times of that for element deletion. The computational cost estimation was based on numerical experiments with 2×10^5 elements used to mesh the work-piece. Clearly, the quantitative estimates with regard to computational costs are expected to vary for meshes with different element densities; still, qualitatively, the element deletion technique is computationally cheaper in comparison to two other used approaches. Interestingly, the simulation for the 45° cutting direction was consistently more computationally expensive than that for the 0° cutting direction.

In summary, the three modelling techniques of material removal produced almost identical results for the cutting force, but element deletion was by far the most computationally efficient.

5 Application of Crystal-Plasticity Simulation to Experiment

In this section, experimental results from microscratching of single-crystal copper were employed to validate current modelling methodology presented before. Here, the Green–Naghdi stress rate and the element deletion method were chosen based on the analysis in Sect. 4. First, experimental data on deformation in single-crystal copper were employed to calibrate material parameters of the work-piece. The experimental data for single-crystal copper under compression was reported by Takeuchi (1975). As shown in Fig. 6, numerical results obtained after calibration show an excellent match with the experimental data for both [100] and [111] crystal orientations. The calibrated model parameters are listed in Table 2, and were used to simulate the microscratching process of single-crystal copper.

The variation of cutting forces with time during a full microscratching process, including its two stages-with increasing depth and constant depth, is shown for the 0° and 45° directions (Fig. 7). In the first stage, both principal and thrust forces increased with time due to the continuing increase in the cutting depth. In the second stage, cutting forces approached constant levels. In our FE simulation, the cutting process was carried out for 1 s in the second stage. As the two stages in the microscratching

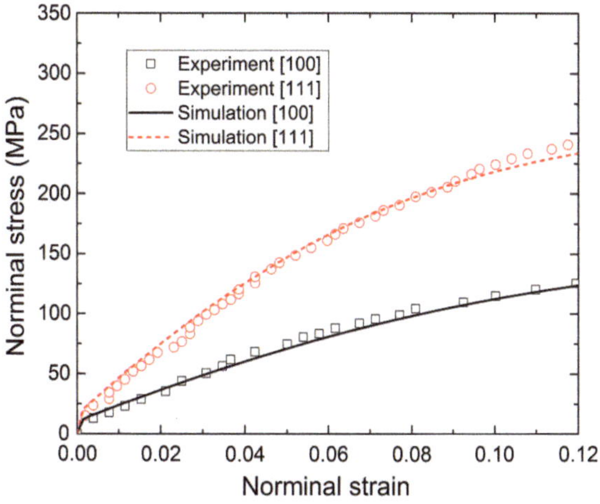

Fig. 6 Comparison of experimental and numerical stress–strain behaviour under compression for single-crystal copper

Table 2 Material parameters of single-crystal copper

Parameter	Definition	Value	Unit
$\dot{\gamma}_0$	Reference shear rate	0.001	s^{-1}
n	Rate-sensitivity parameter	50	–
τ_0	Initial slip resistance	4.0	MPa
τ_s	Saturated slip resistance	52.0	MPa
h_0	Initial hardening modulus	180.0	MPa
h_s	Saturated hardening modulus	24.0	MPa
q	Latent hardening ratio	1.2	–

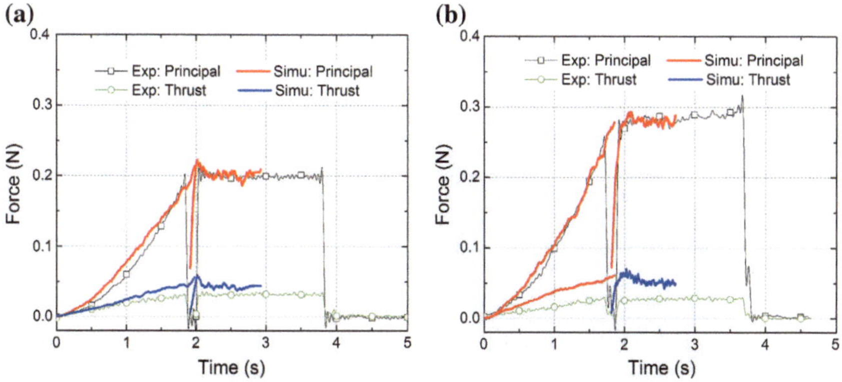

Fig. 7 Comparison of cutting forces for FE simulations and experimental data in microscratching of single-crystal copper for 0° (**a**) and 45° (**b**) directions

were simulated using two different FE runs, an apparent discontinuity was observed on the cutting force-time curves (due to concatenation of two curves). The experimental curves also show a clear discontinuity at the point of transition from Stage 1 to 2; this is due to the nature of the experiment, in which at the completion of Stage 1 the tool movement was stopped before the start of Stage 2. As shown in Fig. 7, the simulation results correlate well with the experimental data for the principal forces (with the exception of the discontinuous region between the stages). A significant dependence of the principal force on crystal orientation was observed both in the experiments and simulations. In Stage 2, the principal force for the 45° direction was about 40 % higher than that for the 0° direction. This anisotropy in the principal force was accurately captured in the FE simulations, thus validating the suggested criterion of material removal. Thrust force predictions were comparatively poorer. A possible reason for this could be relatively low magnitudes of the thrust force. The thrust forces showed a lower dependence on the cutting direction both in the simulation results and the experimental data for the normal to work-piece surface fixed at the [110] crystal orientation.

Another important aspect in FE simulations of metal machining is their capability of capturing chip morphology. The chip morphology obtained with our FE simulations and experiment is compared in Fig. 8. When cutting was along the 0°

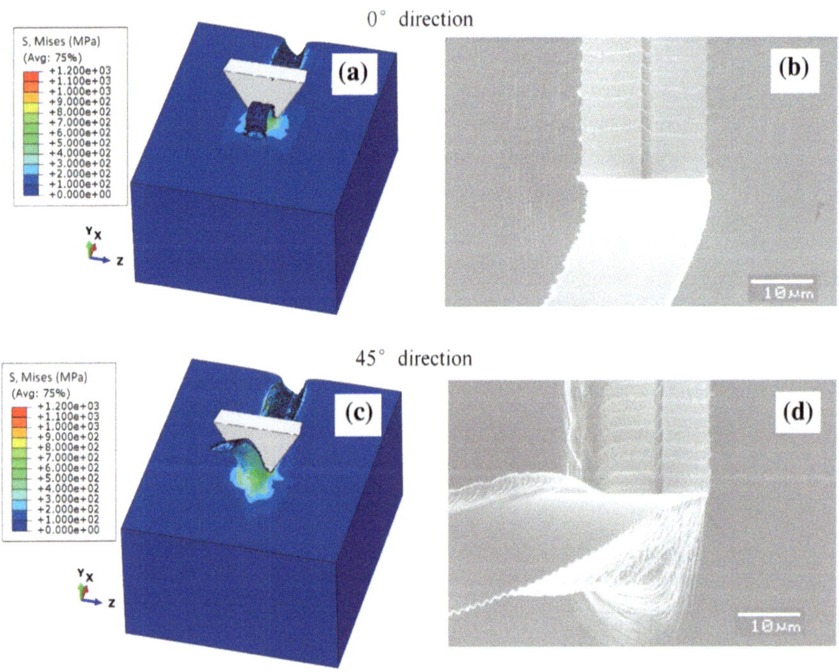

Fig. 8 Comparison of chip morphology for FE simulations and experiments in microscratching of single-crystal copper: **a** simulation in 0° direction; **b** experiment in 0° direction; **c** simulation in 45° direction; **d** experiment in 45° direction

direction (i.e. [1-10] orientation), activated slip systems in the FCC crystallographic structure were symmetrical with respect to it. Therefore, in the FE simulation, the chip separated from work-piece symmetrically. The stress field was also symmetric relative to the cutting plane (Fig. 8a). The simulated chip morphology was verified by the experimental results shown in Fig. 8b. In contrast, microscratching along the 45° direction yielded an asymmetrical stress field and chip formation (Fig. 8c, d) as the cutting direction did not coincide with main axes of symmetry of the FCC structure.

Furthermore, distributions of displacement normal to the work-piece's surface are shown in Fig. 9 for both 0° and 45° directions (here, the chip was removed for clarity). Again, symmetrical and asymmetrical deformation fields in the work-piece were observed for the microscratching in the 0° and 45° directions, respectively. Ahead of the chip, higher deformation was observed for directions at about ±45° with respect to the cutting direction for microscratching in the 0° direction; however, larger deformations were found to the right of cutting plane for microscratching in the 45° direction. Consequently, the pile-up height to the right of groove was larger than that to the left in the FE simulation for microscratching in the 45° direction. The surface

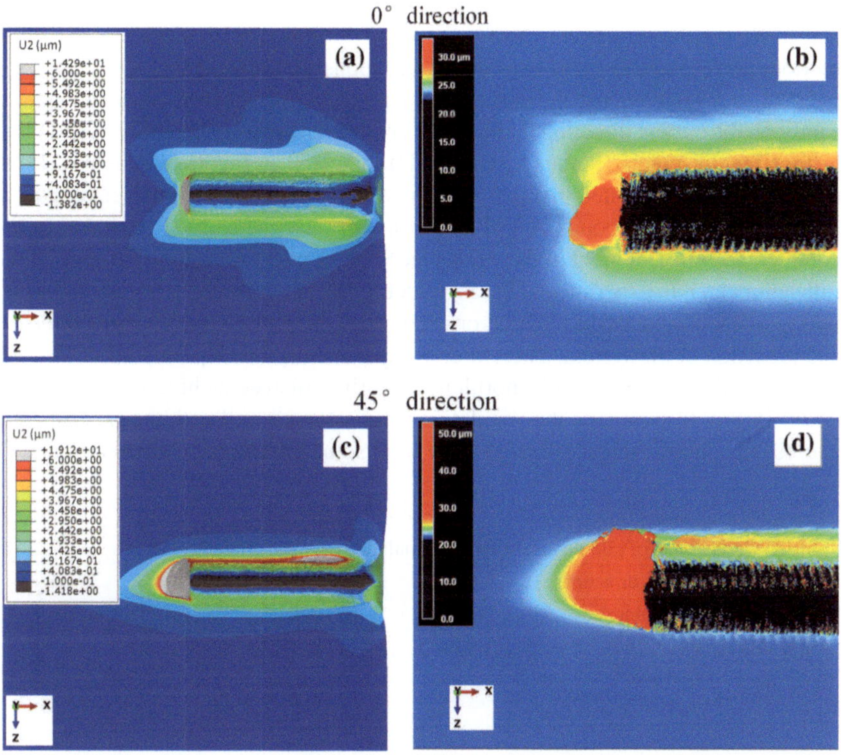

Fig. 9 Comparison of displacement fields in work-piece for FE simulation and experimental data on microscratching of single-crystal copper: **a** simulation in 0° direction; **b** experiment in 0° direction; **c** simulation in 45° direction; **d** experiment in 45° direction

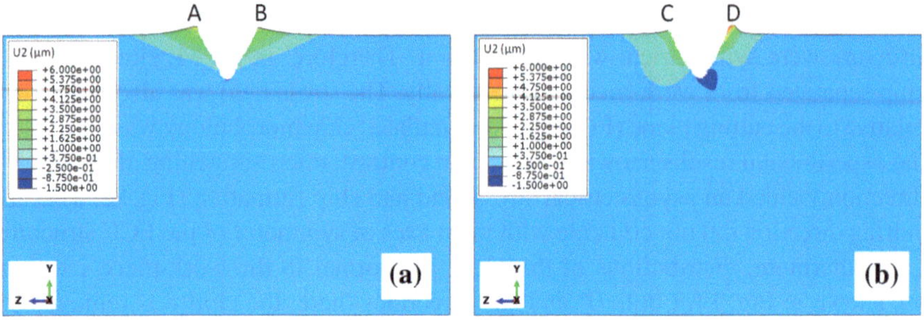

Fig. 10 Pile-up morphology in cross section of work-piece in FE simulation for 0° (**a**) and 45° (**b**) directions

morphology of the deformed work-piece was measured in the experiments, and the surface height is shown by contour plots in Fig. 9b, d. Clearly, the surface morphology in the FE simulations was qualitatively consistent with that in experiment.

Based on the obtained simulation results, a representative cross section perpendicular to the cutting direction was chosen for the deformed work-piece; the two cross sections corresponding to the two cutting directions are shown in Fig. 10. Clearly, the residual groove geometry and surface integrity of the work-piece depended significantly on the cutting direction in the single crystal. The pile-up height corresponding to four feature points (Points A through D) indicated in Fig. 10 was assessed; the obtained results are listed in Table 3. For both simulation and experimental results, the pile-up height of points A and B was not exactly the same. It was due to the complex deformation field in the work-piece during microscratching (in spite of the full symmetry). In experiments, the asymmetry was more pronounced due to the difficulty in ensuring a perfect spatial orientation of the process due to inherent errors and precision of the system. Therefore, in experiments, the chip separation was less symmetrical than the groove's morphology during microscratching in the 0° direction (Fig. 8a). In summary, the simulation results agreed well with the experimental data for the pile-up height although there were some differences for Points A and C.

Table 3 Comparison of pile-up height for FE simulations and experiments

	0° direction		45° direction	
	Point A	Point B	Point C	Point D
Simulation (μm)	3.75	3.30	1.26	4.20
Experiment (μm)	4.00	3.40	1.0	4.25

6 Conclusions

This chapter presented a methodology for crystal-plasticity-based simulations of micromachining of single-crystal metals. Numerical implementation of the crystal-plasticity theory in the explicit finite-element environment was introduced, and the user subroutine incorporating the Green–Naghdi stress rate was shown to match accurately the theoretical predictions. A new chip separation criterion, accounting simultaneously for partial and full activation of slip systems, was proposed to control material removal. This criterion was implemented using three types of modelling techniques: element deletion, ALE adaptive remeshing and SPH. These techniques showed no significant effect on the predicted level of cutting forces; however, computationally the element deletion was much less expensive compared to others. Results of numerical simulations of microscratching tests on single-crystal copper incorporating the Green–Naghdi stress rate and the element deletion technique were found to correlate well with the experimental data. Furthermore, a quantitative agreement in pile-up patterns of the machined work-piece surface was obtained between the simulations and the experiments. The proposed modelling methodology could be extended to simulate micromachining of other single-crystal metals.

Acknowledgments Funding from the Engineering and Physical Sciences Research Council (UK) through grant EP/K028316/1 and Department of Science and Technology (India), project MAST, is gratefully acknowledged. The authors are grateful to Prof. Takashi Matsumura and Dr. Shoichi Tamura (Japan) for experimental results on microscratching of single-crystal copper.

References

Amini, S., Soleimanimehr, H., Nategh, M., Abudollah, A., Sadeghi, M.: FEM analysis of ultrasonic-vibration-assisted turning and the vibratory tool. J. Mater. Process. Technol. **201**(1), 43–47 (2008)

Arısoy, Y.M., Özel, T.: Prediction of machining induced microstructure in Ti-6Al-4V alloy using 3-D FE-based simulations: effects of tool micro-geometry, coating and cutting conditions. J. Mater. Process. Technol. **220**, 1–26 (2015)

Asaro, R.J.: Crystal plasticity. J. Appl. Mech. **50**(4b), 921–934 (1983)

Babitsky, V., Mitrofanov, A., Silberschmidt, V.: Ultrasonically assisted turning of aviation materials: simulations and experimental study. Ultrasonics **42**(1), 81–86 (2004)

Buchkremer, S., Wu, B., Lung, D., Münstermann, S., Klocke, F., Bleck, W.: FE-simulation of machining processes with a new material model. J. Mater. Process. Technol. **214**(3), 599–611 (2014)

Demiral, M., Roy, A., El Sayed, T., Silberschmidt, V.V.: Numerical modelling of micro-machining of f.c.c single crystal: influence of strain gradients. Comput. Mater. Sci. **94**, 273–278 (2014)

Elhami, S., Razfar, M., Farahnakian, M.: Analytical, numerical and experimental study of cutting force during thermally enhanced ultrasonic assisted milling of hardened AISI 4140. Int. J. Mech. Sci. **103**, 158–171 (2015)

Engel, U., Eckstein, R.: Microforming-from basic research to its realization. J. Mater. Process. Technol. **125**, 35–44 (2002)

Greer, J.R., De Hosson, J.T.M.: Plasticity in small-sized metallic systems: intrinsic versus extrinsic size effect. Prog. Mater. Sci. **56**(6), 654–724 (2011)

Hibbitt, D., Karlsson, B., Sorensen, P.: Abaqus 6.12. 3 manual. Abaqus 612 3 Manual (2012)

Hokka, M., Leemet, T., Shrot, A., Baeker, M., Kuokkala, V.T.: Characterization and numerical modeling of high strain rate mechanical behavior of Ti-15-3 alloy for machining simulations. Mater. Sci. Eng.: A **550**, 350–357 (2012)

Huang, Y.: A user-material subroutine incroporating single crystal plasticity in the ABAQUS finite element program. Harvard University (1991)

Hughes, T.J., Winget, J.: Finite rotation effects in numerical integration of rate constitutive equations arising in large-deformation analysis. Int. J. Numer. Methods Eng. **15**(12), 1862–1867 (1980)

Hutchinson, J.: Bounds and self-consistent estimates for creep of polycrystalline materials. Proc. R. Soc. Lond. A: Math., Phys. Eng. Sci., R. Soc. **348**, 101–127 (1976)

Jafarian, F., Ciaran, M.I., Umbrello, D., Arrazola, P., Filice, L., Amirabadi, H.: Finite element simulation of machining Inconel 718 alloy including microstructure changes. Int. J. Mech. Sci. **88**, 110–121 (2014)

Jin, X., Altintas, Y.: Prediction of micro-milling forces with finite element method. J. Mater. Process. Technol. **212**(3), 542–552 (2012)

Johnson, G.R., Cook, W.H.: Fracture characteristics of three metals subjected to various strains, strain rates, temperatures and pressures. Eng. Fract. Mech. **21**(1), 31–48 (1985)

Kim, J.B., Yoon, J.W.: Necking behavior of AA 6022–T4 based on the crystal plasticity and damage models. Int. J. Plast. **73**, 3–23 (2015)

Kraft, O., Gruber, P.A., Mönig, R., Weygand, D.: Plasticity in confined dimensions. Ann. Rev. Mater. Res. **40**, 293–317 (2010)

Lee, W., To, S., Cheung, C.: Effect of crystallographic orientation in diamond turning of copper single crystals. Scripta Materialia **42**(10), 937–945 (2000)

Liu, K., Melkote, S.N.: Finite element analysis of the influence of tool edge radius on size effect in orthogonal micro-cutting process. Int. J. Mech. Sci. **49**(5), 650–660 (2007)

Mian, A., Driver, N., Mativenga, P.: Identification of factors that dominate size effect in micro-machining. Int. J. Mach. Tools Manuf. **51**(5), 383–394 (2011)

Monaghan, J.J.: Smoothed particle hydrodynamics. Rep. Prog. Phys. **68**(8), 1703 (2005)

Pal, D., Stucker, B.: A study of subgrain formation in Al 3003 H-18 foils undergoing ultrasonic additive manufacturing using a dislocation density based crystal plasticity finite element framework. J. Appl. Phys. **113**(20), 203517 (2013)

Parle, D., Singh, R.K., Joshi, S.S., Ravikumar, G.: Modeling of microcrack formation in orthogonal machining. Int. J. Mach. Tools Manuf. **80**, 18–29 (2014)

Roters, F., Eisenlohr, P., Hantcherli, L., Tjahjanto, D., Bieler, T., Raabe, D.: Overview of constitutive laws, kinematics, homogenization and multiscale methods in crystal plasticity finite-element modeling: theory, experiments, applications. Acta Materialia **58**(4), 1152–1211 (2010)

Shi, B., Attia, H.: Current status and future direction in the numerical modeling and simulation of machining processes: a critical literature review. Mach. Sci. Technol. **14**(2), 149–188 (2010)

Tajalli, S., Movahhedy, M., Akbari, J.: Simulation of orthogonal micro-cutting of FCC materials based on rate-dependent crystal plasticity finite element model. Comput. Mater. Sci. **86**, 79–87 (2014)

Takeuchi, T.: Work hardening of copper single crystals with multiple glide orientations. Trans. Jpn. Inst. Met. **16**(10), 629–640 (1975)

Zahedi, S.A., Demiral, M., Roy, A., Silberschmidt, V.V.: FE/SPH modelling of orthogonal micro-machining of f.c.c single crystal. Comput. Mater. Sci. **78**, 104–109 (2013)

Zahedi, S.A., Roy, A., Silberschmidt, V.V.: Modelling of vibration assisted machining f.c.c single crystal. Procedia CIRP **31**, 393–398 (2015)

Zhang, H., Dong, X.: Physically based crystal plasticity FEM including geometrically necessary dislocations: numerical implementation and applications in micro-forming. Comput. Mater. Sci. **110**, 308–320 (2015)

Zhang, Y., Mabrouki, T., Nelias, D., Courbon, C., Rech, J., Gong, Y.: Cutting simulation capabilities based on crystal plasticity theory and discrete cohesive elements. J. Mater. Process. Technol. **212**(4), 936–953 (2012)

Zhang, Y., Outeiro, J., Mabrouki, T.: On the selection of Johnson-Cook constitutive model parameters for Ti-6Al-4V using three types of numerical models of orthogonal cutting. Procedia CIRP **31**, 112–117 (2015)

Zong, W., Cao, Z., He, C., Xue, C.: Theoretical modelling and FE simulation on the oblique diamond turning of ZnS crystal. Int. J. Mach. Tools Manuf. **100**, 55–71 (2016)

1 Introduction

Inverter power modules developed for hybrid and electric cars have multilayered structures to meet the requirements for high power density, heat dissipation, and reliability (Ishiko 2004; Wada et al. 2010; Kimura et al. 2014). Layers such as Si dies, insulators, brazing metals, solders, and metallic baseplates are mounted in this type of power modules. Since these layers have considerably different CTEs (coefficients of thermal expansion), warpage can occur under soldering and thermal loading. Warpage may affect the heat radiation from power modules to heat extractors. Warpage is, therefore, regarded as one of the important issues in designing inverter power modules for hybrid and electric cars.

Tanie et al. (2011) studied a fundamental subject for the warpage issue mentioned above. In that study, cyclic thermal tests of Si/solder/OFHC-Cu layered plates were performed to confirm the computational findings of Nakane et al. (2008, 2010). The Cu layer was either as-received or fully annealed before soldering. Cyclic growth/recovery of warpage was thus experimentally observed to occur with an increase in the number of thermal cycles. The cyclic growth occurred when the Cu layer was fully annealed before soldering and was relatively thick in comparison with the Si and solder layers. Tanie et al. (2011) also performed finite element analysis to simulate the cyclic thermal tests. In the analysis, however, no strain hardening in the solder was assumed to simplify the analysis, although the solder exhibited noticeable strain hardening at low temperatures under cyclic thermal loading in the tests. Moreover, the modes of ratcheting in the solder and copper layers were not shown as results of the analysis, although the validity of a material model for ratcheting depends on the mode to be simulated.

Material modeling for ratcheting has greatly advanced especially in the 1990s and 2000s, leading to the following understanding, as reviewed by Ohno (1997), Kang (2008) and Chaboche (2008). The nonlinear kinematic hardening model of Armstrong and Frederick (1966) usually overpredicts ratcheting. This deficiency is improved to a considerable extent by sophisticating the dynamic recovery of back stress evolution (e.g., Chaboche 1991; Ohno and Wang 1993a, b, 1994; Jiang and Sehitoglu 1996a, b; Abdel-Karim and Ohno 2000; Bari and Hassan 2002; Kang 2004). The multi-mechanism models proposed by Cailletaud and Saï (1995) can be effective for simulating ratcheting (Taleb and Cailletaud 2010; Saï 2011). The multiplicative AF (Armstrong–Frederick) model proposed by Dafalias et al. (2008) can be also effective for simulating ratcheting (Dafalias and Feigenbaum 2011). Hence, one can now say that ratcheting is simulated well, or fairly well, using a proper material model though its validity may be limited to a certain type and range of loading conditions.

Ratcheting of solders has been studied by Chen et al. (2007), Bai and Chen (2009), and Kobayashi and Sasaki (2009). It has thus been observed that, even at room temperature, solders exhibit more significant ratcheting under lower rate loading, as observed in the ratcheting tests of SS304 at 700 °C by Kang et al. (2006). To simulate such rate-dependent ratcheting of solders, it was proposed to take into account the rate-dependence in the dynamic recovery of back stress (Chen et al. 2007; Bai and Chen 2009). Chen et al. (2007) thus showed that the multiaxial ratcheting of Pb-63Sn solder under lower rate loading was more properly simulated using the Armstrong and Frederick model. However, they used a rate-independent material model despite the rate-dependence of ratcheting. It is therefore worthwhile to examine their finding using a rate-dependent material model for Pb-63Sn and also for other solders. It is noted that, to the best of the authors' knowledge, no data of multiaxial ratcheting have been published with respect to solders other than those of Pb-63Sn reported by Chen et al. (2007).

In this study, the cyclic thermal tests of Si/solder/Cu layered plates (Tanie et al. 2011) are analyzed by assuming three material models for the solder and two material models for the copper. Strain hardening and rate-dependence in the solder are taken into account in the three material models using the test data of Yamada and Ogawa (1996). The material parameters in the two models for the copper are identified in as-received and annealed states using the test data of Tanie et al. (2011), Wada et al. (2010), and Zhang and Jiang (2008). It is thus demonstrated that the cyclic growth and cyclic recovery of warpage observed in the cyclic thermal tests are simulated well if proper material models are used for the solder and copper in finite element analysis. It is shown that multiaxial ratcheting occurred in the solder in contrast to the uniaxial type of ratcheting or cyclic strain recovery in the copper. It is also shown that the Armstrong and Frederick model unexpectedly works well for the multiaxial ratcheting in the solder. To verify this unexpected finding, the multiaxial ratcheting of Pb-63Sn (Chen et al. 2007) is simulated using a rate-dependent material model with the Armstrong and Frederick model.

2 Cyclic Thermal Tests of Layered Plates (Tanie et al. 2011)

Figure 1 illustrates the Si/solder/Cu layered plates tested by Tanie et al. (2011). Each layered plate consisted of a Si chip and an OFHC-Cu plate that were bonded using a solder sheet of 0.1 mm thickness. The Si chips had their (111) surfaces parallel to the Cu plates. The Si chips had a constant thickness of 0.19 mm. The following three thicknesses were chosen for the Cu plates: $H_{Cu} = 0.2$, 0.5, and 1.0 mm. The Cu plates were either as-received or fully annealed, and had 5 μm thick Ni coating before soldering.

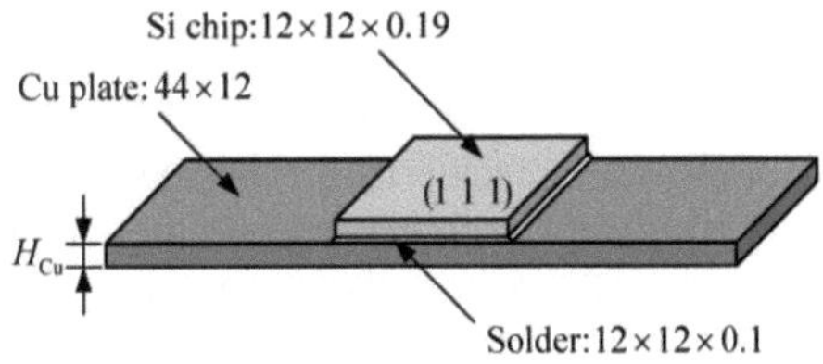

Fig. 1 Shape of the Si/solder/Cu layered plates tested by Tanie et al. (2011); size in mm

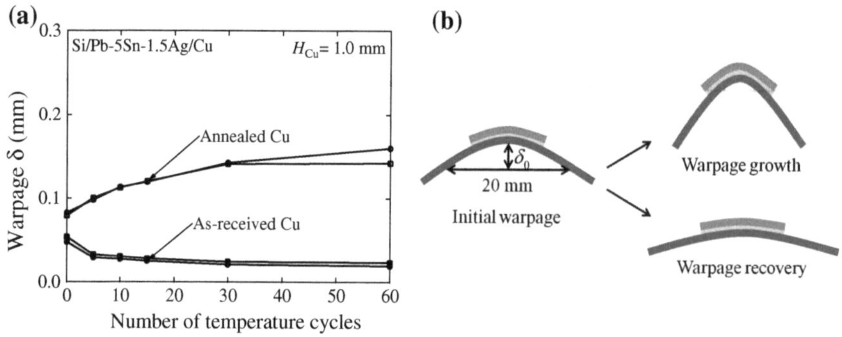

Fig. 2 Cyclic variation in warpage δ of Si/Pb-5Sn-1.5Ag/OFHC-Cu with $H_{Cu} = 1.0$ mm; **a** experimental data (Tanie et al. 2011) and **b** schematic illustration

The layered plates were subjected to cyclic thermal loading in a test chamber with a temperature controller. The temperature input to the controller was set so as to induce temperature cycling between $T_{min} = -40\,°C$ and $T_{max} = 150\,°C$ with a period of 3000 s, which consisted of temperature dropping for 600 s, holding at T_{min} for 1200 s, rising for 600 s, and holding at T_{max} for 600 s. The temperature cycling started from room temperature. The warpage δ of each layered plate was measured before and during testing. To measure δ at room temperature, the temperature cycling was interrupted to take out the layered plates from the test chamber at each end of the 5, 10, 15, 30, and 60th temperature cycles. A laser displacement sensor with a resolution of 0.2 μm was used to measure δ along a longitudinal center line 20 mm long in the central region on the back surface of the Cu layer at room temperature. This led to defining δ to be the maximum deflection along this 20-mm-long center line (Fig. 2b). When the back surface of the Cu layer was concave, δ was taken to be positive.

Figure 2a shows the cyclic variations in δ measured for the Si/Pb-5Sn-1.5Ag/Cu plates with $H_{Cu} = 1.0$ mm. As depicted in the figure, the growth and recovery of δ occurred with an increase in the number of temperature cycles when the Cu plates were annealed and as-received, respectively. Figure 2b exaggeratedly illustrates the cyclic growth and recovery of δ.

3 Material Models

This section describes the material models that were used to analyze the cyclic variations in warpage δ shown in Fig. 2a. In the analysis, three and two material models were considered for the solder and Cu layers, respectively, and the strain ε was additively decomposed into thermal, elastic, and (visco)plastic parts based on the assumption of small deformation:

$$\varepsilon = \varepsilon^{t} + \varepsilon^{e} + \varepsilon^{p}. \tag{1}$$

Table 1 Material parameters of Si layer; stress in MPa

Young's modulus	Poisson's ratio	CTE
$E = 1.69 \times 10^5$	$\nu = 0.262$	$\alpha = 3.0 \times 10^{-6}$

In this section, direct notations are used for tensors, and inner products between them are indicated by colons (e.g., $\boldsymbol{\sigma} : \boldsymbol{\varepsilon} = \sigma_{ij}\varepsilon_{ij}$). A superposed dot denotes differentiation with respect to time t, and $\| \ \|$ signifies the Frobenius norm of second rank tensors (e.g., $\|\boldsymbol{\sigma}\|^2 = \sigma_{ij}\sigma_{ij}$).

3.1 Si Chips

The Si chips were presumed to be thermalelastic (Table 1). It is noted that the Si chips had in-plane elastic isotropy because the Si chips had their (111) surfaces parallel to the Cu plates (Sect. 2). It is also noted that a plane stress state almost prevailed in the Si chips in the cyclic thermal tests. Accordingly, the (111) in-plane isotropic elastic constants were used for the elastic behavior of the Si chips.

3.2 Solder Layer

Solders usually exhibit significant rate-dependence and negligible cyclic hardening under cyclic inelastic loading (Busso et al. 1992; Kanchanomai et al. 2002; Bai and Chen 2009). For the solder layer shown in Fig. 1, therefore, only kinematic hardening was taken into account through a power-law viscoplastic equation:

$$\dot{\boldsymbol{\varepsilon}}^{\mathrm{p}} = \frac{3}{2}\dot{\varepsilon}_0 \left(\frac{\sigma_{\mathrm{e}}}{\sigma_0(T)}\right)^{n(T)-1} \frac{\boldsymbol{s} - \boldsymbol{a}}{\sigma_0(T)}, \tag{2}$$

where $\dot{\varepsilon}_0$ is a reference strain rate, $\sigma_0(T)$ and $n(T)$ are temperature-dependent material parameters, $\boldsymbol{s}$ and $\boldsymbol{a}$ denote the deviatoric parts of stress and back stress, respectively, and σ_{e} indicates the effective stress defined as

$$\sigma_{\mathrm{e}} = \sqrt{\frac{3}{2}}\|\boldsymbol{s} - \boldsymbol{a}\|. \tag{3}$$

The following three kinematic hardening models were assumed for the evolution of $\boldsymbol{a}$ in the solder: the linear kinematic hardening (LK) model, the Armstrong and Frederick (AF) model, and the multilinear kinematic hardening (MLK) model proposed by Ohno and Wang (1993a). For all three models, $\boldsymbol{a}$ was supposed to be expressed as[1]

$$\boldsymbol{a} = h(T)\boldsymbol{b}, \tag{4}$$

where $h(T)$ is a temperature-dependent material parameter, and $\boldsymbol{b}$ is the nondimensional back stress. Then the LK and AF models, respectively, have

$$\dot{\boldsymbol{b}} = \frac{2}{3}\dot{\boldsymbol{\varepsilon}}^{\mathrm{p}}, \tag{5}$$

$$\dot{\boldsymbol{b}} = \frac{2}{3}\dot{\boldsymbol{\varepsilon}}^{\mathrm{p}} - \zeta \boldsymbol{b}\dot{p}. \tag{6}$$

Here, ζ is a temperature-independent material parameter, and $\dot{p}$ denotes the accumulated (visco)plastic strain rate expressed as

$$\dot{p} = \sqrt{\frac{2}{3}}\|\dot{\boldsymbol{\varepsilon}}^{\mathrm{p}}\|. \tag{7}$$

The MLK model of Ohno and Wang (1993a) can have

$$\boldsymbol{b} = \sum_{i=1}^{M} \boldsymbol{b}^{(i)}, \tag{8}$$

$$\dot{\boldsymbol{b}}^{(i)} = \frac{2}{3}\dot{\boldsymbol{\varepsilon}}^{\mathrm{p}} - \zeta^{(i)} H(f^{(i)}) \left\langle \dot{\boldsymbol{\varepsilon}}^{\mathrm{p}} : \frac{\boldsymbol{b}^{(i)}}{b_{\mathrm{e}}^{(i)}} \right\rangle \boldsymbol{b}^{(i)}, \tag{9}$$

where M and $\zeta^{(i)}$ are temperature-independent material parameters, H denotes Heaviside's step function, $\langle\ \rangle$ indicates Macaulay's brackets, and $f^{(i)}$ and $b_{\mathrm{e}}^{(i)}$ are defined as

$$f^{(i)} = b_{\mathrm{e}}^{(i)} - \frac{1}{\zeta^{(i)}}, \quad b_{\mathrm{e}}^{(i)} = \sqrt{\frac{3}{2}}\|\boldsymbol{b}^{(i)}\|. \tag{10}$$

Table 2 gives the material parameters used for the solder layer in this study. These parameters were identified by fitting the tensile stress–strain curves of Pb-5Sn at $\dot{\varepsilon} = 0.01$ and 0.001 %/s at −40, 25, 80, and 120 °C (Fig. 3). All stresses and strains are nominal in this study based on the assumption of small deformation. It is noted

[1]For the nonisothermal uniaxial AF model expressed as $\dot{\alpha} = (\partial\alpha/\partial\varepsilon^{\mathrm{p}})\dot{\varepsilon}^{\mathrm{p}} + (\partial\alpha/\partial T)\dot{T}$ with $h = h(p, T)$ and $\zeta = \zeta(p, T)$, the temperature-history-independent evolution condition of back stress α under thermal-mechanical tensile loading is expressed as $\partial^2(\ln h)/\partial\varepsilon^{\mathrm{p}}\partial T = -\partial\zeta/\partial T$ (Ohno and Wang 1991). This condition is satisfied if $h = h_1(p)h_2(T)$ and $\zeta = \zeta(p)$. Consequently, when Eqs. (4) and (6) are assumed, the evolution of back stress depends on T but not on the history of T.

Table 2 Material parameters of solder layer; stress in MPa, strain in mm/mm, time in s, and T in K

Elastic constants	$E(T) = 31.1 - 0.049T, \quad \nu = 0.35$
CTE	$\alpha = 29.2 \times 10^{-6}$
AF model	$\dot{\varepsilon}_0 = 1.00 \times 10^{-4}$
	$\sigma_0(T) = -0.0533T + 31.4, \quad n(T) = -0.0303T + 19.1$
	$h(T) = 50353\ \exp(-0.0153T), \quad \zeta = 75$
LK model	$\dot{\varepsilon}_0 = 1.00 \times 10^{-4}$
	$\sigma_0(T) = 69.9\ \exp(-0.00470T), \quad n(T) = 46.1\ \exp(-0.00469T)$
	$h(T) = 18388\ \exp(-0.0157T)$
MLK model	$\dot{\varepsilon}_0 = 1.00 \times 10^{-4}$
	$\sigma_0(T) = -0.0511T + 30.1, \quad n(T) = -0.0355T + 20.6$
	$h^{(1)}(T) = 11917\ \exp(-0.0130T), \quad \zeta^{(1)} = 4.00 \times 10^2$
	$h^{(2)}(T) = 4974.0\ \exp(-0.0130T), \quad \zeta^{(2)} = 2.00 \times 10^2$
	$h^{(3)}(T) = 6217.5\ \exp(-0.0130T), \quad \zeta^{(3)} = 1.00 \times 10^2$
	$h^{(4)}(T) = 3005.1\ \exp(-0.0130T), \quad \zeta^{(4)} = 5.00 \times 10$
	$h^{(5)}(T) = 5699.3\ \exp(-0.0130T), \quad \zeta^{(5)} = 0.00$

that Pb-5Sn is regarded as having an equivalent chemical composition to Pb-5Sn-1.5Ag because Pb-10Sn and Pb-5Sn-2.5Ag provided almost completely the same tensile curves as each other at room temperature and 0.7 T_m (melting temperature in K) (Schoeller et al. 2009).

The LK, AF, and MLK models expressed as Eqs. (5), (6), and (9) have no material parameter to control ratcheting. These models were chosen because no experimental data have been available with respect to the ratcheting of Pb-5Sn. It is emphasized that, before starting the finite element analysis described in Sect. 4, the AF model and the LK and MLK models were supposed to be the extreme models to overpredict and underpredict ratcheting of Pb-5Sn. This was because the AF model usually

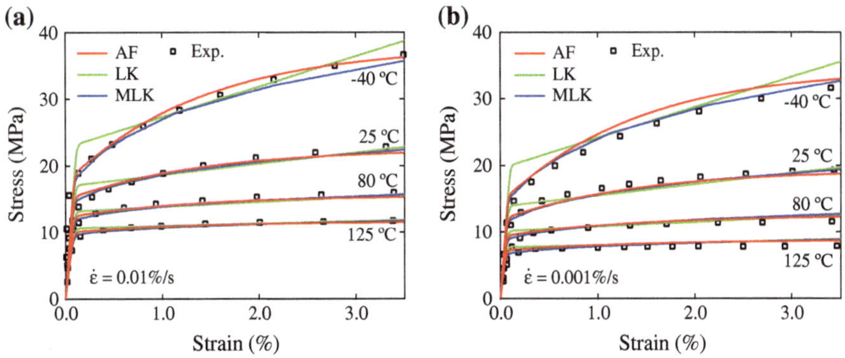

Fig. 3 Tensile stress–strain curves of Pb-5Sn; experiment (Yamada and Ogawa 1996) and simulation; **a** $\dot{\varepsilon} = 0.01$ %/s and **b** $\dot{\varepsilon} = 0.001$ %/s

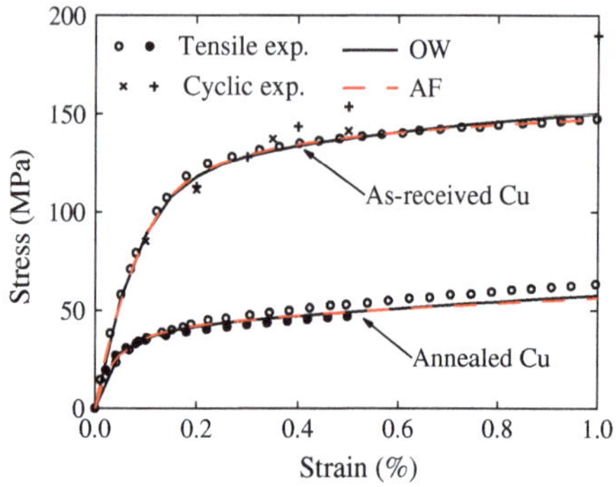

Fig. 4 Tensile and cyclic stress–strain data of as-received and annealed OFHC-Cu; •, × and + indicate the tensile and cyclic data of annealed OFHC-Cu by Wada et al. (2010) and Zhang and Jiang (2008)

over-predicts ratcheting, while the LK and MLK models underpredict it (Ohno 1997; Kang 2008; Chaboche 2008).

3.3 Cu Plates

Tanie et al. (2011) performed tensile tests of the as-received and fully annealed, 1.0 mm thick OFHC-Cu plates used in the cyclic thermal tests, and found that the copper was almost identical to those tested by Wada et al. (2010) and Zhang and Jiang (2008), as illustrated in Fig. 4. The plastic flow stress of the as-received copper was suggested to have much smaller dependence on temperature than the viscoplastic flow stress of the solder.[2] For the Cu plates, therefore, we assumed temperature-independent plasticity based on the normality to the Mises-type yield surface with its center $\boldsymbol{a}$ and radius Y:

$$\dot{\boldsymbol{\varepsilon}}^{\mathrm{p}} = \frac{3}{2}\dot{p}\frac{\boldsymbol{s} - \boldsymbol{a}}{Y}, \tag{11}$$

$$F = \frac{3}{2}(\boldsymbol{s} - \boldsymbol{a}) : (\boldsymbol{s} - \boldsymbol{a}) - Y^2, \tag{12}$$

where F denotes the yield function, and $\dot{p}$ is defined by Eq. (7). It was further assumed that $\boldsymbol{a}$ is decomposed into M parts (Chaboche 1989):

$$\boldsymbol{a} = \sum_{i=1}^{M} \boldsymbol{a}^{(i)}. \tag{13}$$

[2]The saturated uniaxial hysteresis loops of OFHC-Cu at $\Delta\varepsilon = 1.0\,\%$ and $\dot{\varepsilon} = 0.1\,\%/\mathrm{s}$ suggest that the peak stress at 125 °C is about 15 % smaller than that at room temperature (Freed 1988). Almost the same temperature dependence was found by performing tensile tests of the as-received copper at 140 °C in addition to that at room temperature in this study. This temperature dependence is much smaller than that of the solder shown in Fig. 3.

By expressing each $\boldsymbol{a}^{(i)}$ as

$$\boldsymbol{a}^{(i)} = h^{(i)}\boldsymbol{b}^{(i)}, \tag{14}$$

the cyclic thermal tests described in Sect. 2 were analyzed using the following two evolution equations of $\boldsymbol{b}^{(i)}$ based on Armstrong and Frederick (1966) and Ohno and Wang (1993a):

$$\dot{\boldsymbol{b}}^{(i)} = \frac{2}{3}\dot{\boldsymbol{\varepsilon}}^{\mathrm{p}} - \zeta^{(i)}\boldsymbol{b}^{(i)}\dot{p}, \tag{15}$$

$$\dot{\boldsymbol{b}}^{(i)} = \frac{2}{3}\dot{\boldsymbol{\varepsilon}}^{\mathrm{p}} - \zeta^{(i)}(\zeta^{(i)}b_{\mathrm{e}}^{(i)})^{k^{(i)}}\left\langle \dot{\boldsymbol{\varepsilon}}^{\mathrm{p}} : \frac{\boldsymbol{b}^{(i)}}{b_{\mathrm{e}}^{(i)}} \right\rangle \boldsymbol{b}^{(i)}, \tag{16}$$

where $h^{(i)}$, $\zeta^{(i)}$, and $k^{(i)}$ are material parameters, and $b_{\mathrm{e}}^{(i)}$ is defined by Eq. $(10)_2$. Equations (15) and (16) are referred to as the AF (Armstrong–Frederick) and OW (Ohno–Wang) models hereafter.

Table 3 provides the material parameters used for the Cu plates in this study. The material parameters except $k^{(i)}$ were identified by fitting the tensile and cyclic data shown in Figs. 4, 5, and 6. A procedure to identify the material parameters, except

Table 3 Material parameters of Cu layer; stress in MPa

Elastic constants	$E = 1.23 \times 10^5, \quad \nu = 0.34$
CTE	$\alpha = 17.0 \times 10^{-6}$
Annealed (AF)	$Y_0 = 12.55$
	$h_0^{(1)} = 9.60 \times 10^4, \quad \zeta^{(1)} = 5.00 \times 10^3$
	$h_0^{(2)} = 9.33 \times 10^3, \quad \zeta^{(2)} = 1.00 \times 10^3$
	$h_0^{(3)} = 3.85 \times 10^2, \quad \zeta^{(3)} = 1.67 \times 10^2$
	$h_0^{(4)} = 7.77 \times 10^2, \quad \zeta^{(4)} = 1.00 \times 10^2$
	$\rho(p) = 1.0 + 2.11[1.0 - \exp(-9.41p)]$
Annealed (OW)	$Y_0 = 12.55$
	$h_0^{(1)} = 6.91 \times 10^4, \quad \zeta^{(1)} = 1.10 \times 10^4, \quad k^{(1)} = 3.0$
	$h_0^{(2)} = 2.68 \times 10^4, \quad \zeta^{(2)} = 3.61 \times 10^3, \quad k^{(2)} = 3.0$
	$h_0^{(3)} = 1.00 \times 10^4, \quad \zeta^{(3)} = 1.49 \times 10^3, \quad k^{(3)} = 3.0$
	$h_0^{(4)} = 3.57 \times 10^3, \quad \zeta^{(4)} = 7.50 \times 10^2, \quad k^{(4)} = 3.0$
	$h_0^{(5)} = 1.09 \times 10^3, \quad \zeta^{(5)} = 3.60 \times 10^2, \quad k^{(5)} = 3.0$
	$h_0^{(6)} = 9.38 \times 10^2, \quad \zeta^{(6)} = 1.00 \times 10^2, \quad k^{(6)} = 3.0$
	$\rho(p) = 1.0 + 2.11[1.0 - \exp(-9.41p)]$
As-received (AF, OW)	Y_0 and $h_0^{(i)}$ in the above columns are replaced by $Y = 3.11Y_0$ and $h^{(i)} = 3.11h_0^{(i)}$, and $\rho(p) = 1.0$

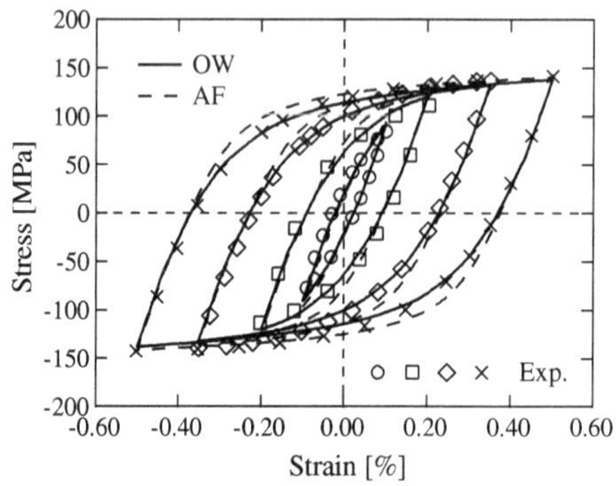

Fig. 5 Hysteresis loops of annealed OFHC-Cu in saturated state of cyclic hardening; experiment (Wada et al. 2010) and simulation

for $k^{(i)}$, in the OW model was described in the appendix by Nakane et al. (2008). It is noted that, for the annealed copper, the material parameters Y and $h^{(i)}$ in Eqs. (12) and (14) were taken to be dependent on accumulated plastic strain p to represent the cyclic hardening depicted in Fig. 6:

$$Y = \rho(p)Y_0, \quad h^{(i)} = \rho(p)h_0^{(i)}. \tag{17}$$

The AF and OW models provided almost the same results in Fig. 6. For the as-received copper, on the other hand, no cyclic hardening was assumed, because the tensile curve of the as-received copper was close to the saturated cyclic data on stress and strain amplitudes ($\Delta\sigma_{\mathrm{sat}}/2$ and $\Delta\varepsilon/2$) of the annealed copper as far as we considered $\Delta\varepsilon/2 \leq 0.5\,\%$ (Fig. 4).

The OW model has $k^{(i)}$ to control ratcheting. This material parameter was identified by fitting the ratcheting data shown in Fig. 7. Since these data were obtained by performing stress-controlled tests after the saturation of cyclic hardening in strain-controlled cyclic tests (Zhang and Jiang 2008), the ratcheting in the figure was simulated using the material parameters in the saturated state of cyclic hardening (i.e., the material parameters of the as-received copper).

It is worthwhile to point out that, as seen from Fig. 6, the strain-range-dependent cyclic hardening of the annealed copper at $\Delta\varepsilon/2 \leq 0.5\,\%$ was simulated well without

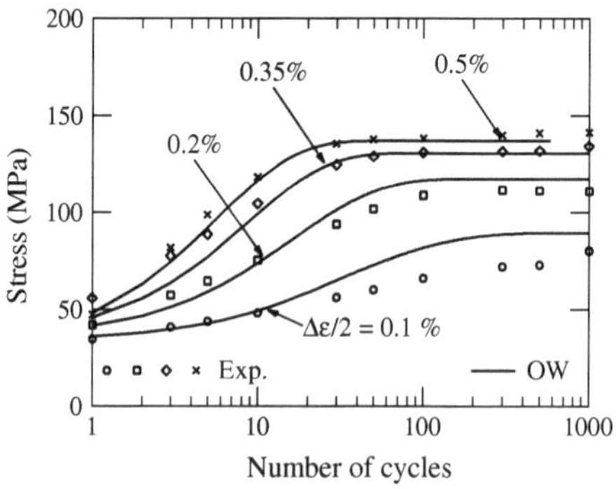

Fig. 6 Variation in tensile peak stress of annealed OFHC-Cu under uniaxial cyclic loading of constant strain range; experiment (Wada et al. 2010) and simulation; almost identical results by AF and OW models

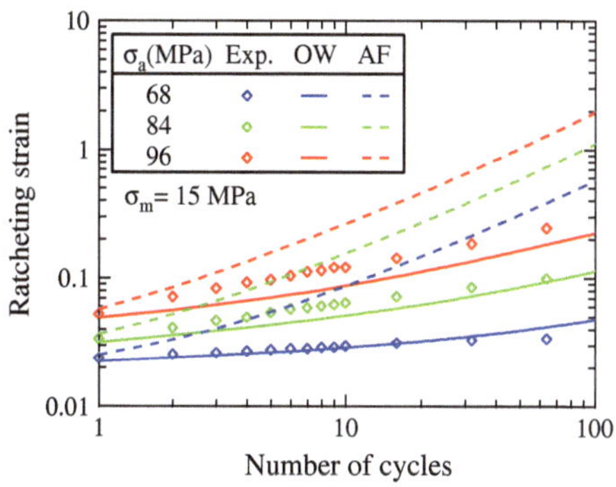

Fig. 7 Uniaxial ratcheting strain of OFHC-Cu; experiment after saturation of cyclic hardening (Zhang and Jiang 2008), and simulation for as-received OFHC-Cu; experimental data corrected by subtracting the initial scatters

introducing any memory surface such as those proposed by Chaboche et al. (1979) and Ohno (1982), which were reviewed by Ohno (2015). The saturated stress–strain hysteresis loops at $\Delta\varepsilon/2 \leq 0.5\,\%$ were also reproduced well without any memory surface (Fig. 5). The strain-range-dependent cyclic hardening at $\Delta\varepsilon/2 \leq 0.5\,\%$ is explained by the fact that the saturated amplitude of back stress, $\Delta\alpha_{sat}/2$, depends on $\Delta\varepsilon/2$ though the saturated value Y_{sat} of Y has no dependence on $\Delta\varepsilon/2$.

4 Analysis Method

The tested layered plates with $H_{Cu} = 1.0\,\text{mm}$ were divided into finite elements in a quarter model based on the geometric symmetries. The finite element analysis was performed using Abaqus. Twenty-node quadratic brick elements with reduced integration (C3D20R) were used for the quarter model (Fig. 8). The number of elements was 2300. The material models of the solder and copper, described in Sects. 3.2 and 3.3, were fully implicitly implemented in Abaqus with a user subroutine UMAT (Ohno et al. 2013, 2016). Hereafter, we use the abbreviations such as AF-sol and OW-Cu which, respectively, indicate "the AF model used for the solder" and "the OW model used for the copper".

The temperature variation shown in Fig. 9 was uniformly applied to the quarter model under the following boundary conditions: $u_x = u_y = u_z = 0$ at the origin

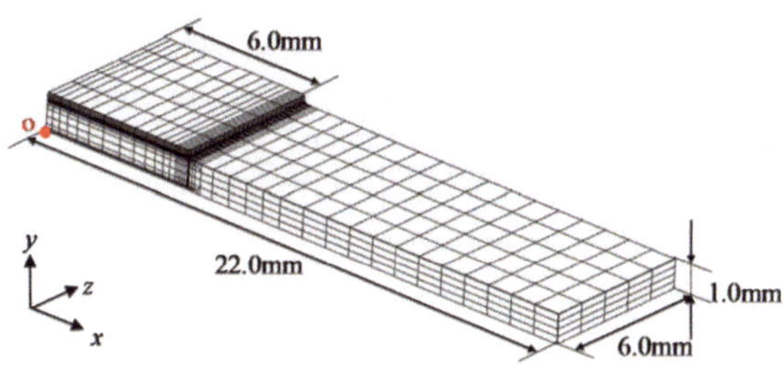

Fig. 8 Finite element division of a quarter of layered plate with $H_{Cu} = 1.0\,\text{mm}$

Fig. 9 Temperature variation

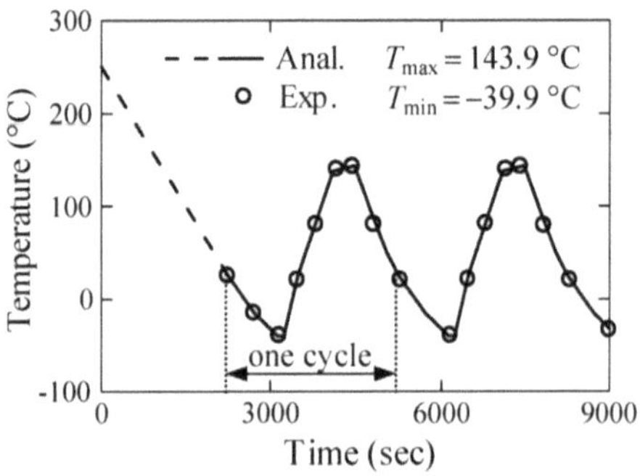

O, $u_x = 0$ at $x = 0$, and $u_z = 0$ at $z = 0$. This temperature variation consisted of an initial drop and a subsequent cyclic variation, which are indicated by dashed and solid lines in Fig. 9. The initial drop was introduced in the analysis for simulating the initial warpage induced by soldering, although no actual temperature data were available for the initial drop. The subsequent cyclic variation was determined by piecewise linearly approximating the actual temperature variation that was measured with an accuracy of ±0.6 °C using a thermocouple attached to one of the layered plates in the test chamber.

The finite element analysis described above was performed in the following 12 cases: the three solder models, two copper models, and two heat treatment conditions of the copper. An initial time increment of 10 s was used in all analysis cases, and was found successful under the default convergence condition in Abaqus.

5 Analysis Results

5.1 *Cyclic Variations in Warpage*

In each analysis case, the warpage δ along the 20-mm-long longitudinal center line on the back surface of the Cu layer was outputted at 25 °C at the end of every temperature cycle in accordance with the experimental measurement described in Sect. 2. The cyclic variations in δ thus computed are plotted in Fig. 10 to show the following.

- The cyclic variations in δ significantly depended on the material models assumed for the copper and solder layers. In other words, both the copper and solder layers need to have proper material models for predicting the cyclic variations in δ.
- Using the pair of OW-Cu and AF-sol, the observed cyclic growth and cyclic recovery of δ depicted in Fig. 2a were reproduced well except for the initial values of δ (Fig. 10a).

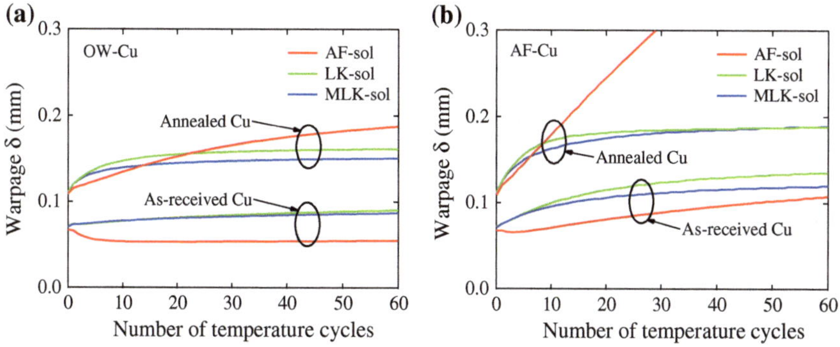

Fig. 10 Warpage δ of Si/Pb-5Sn/OFHC-Cu with $H_{Cu} = 1.0$ mm; **a** OW model and **b** AF model for OFHC-Cu

It is noted that it was hard to predict the initial values of δ correctly, because no actual data were available with respect to the temperature change in the soldering process; nevertheless, the initial values of δ were predicted passably well.

As described above, the cyclic variations in δ were predicted well by assuming the pair of OW-Cu and AF-sol in the analysis. Hence, the computational results obtained using the pair of OW-Cu and AF-sol are used below to show ratcheting in the copper and solder layers.

5.2 *Ratcheting in the Copper Layer*

Figure 11 illustrates the deformed configurations of the longitudinal, middle cross section at the end of the 60th cycle in the analysis. As seen from the figure, the annealed copper plate had larger deflection as a result of the cyclic growth of δ than the as-received copper plate in which the cyclic recovery of δ occurred. Figure 12 shows the hysteresis loops of σ_{xx} and ε_{xx}^{m} at point A in the copper layer. Point A is close to the top surface of the copper layer (Fig. 11). Here, ε_{xx}^{m} denotes the mechanical part of ε_{xx}, i.e., $\varepsilon_{xx}^{m} = \varepsilon_{xx}^{e} + \varepsilon_{xx}^{p}$. Point A also had the hysteresis loops of σ_{zz} and ε_{zz}^{m} that were almost the same as those of σ_{xx} and ε_{xx}^{m}.

It is seen from Fig. 12 that significant ratcheting occurred in the annealed copper plate while the mean values of σ_{xx} and ε_{xx}^{m} decreased to some extent in the

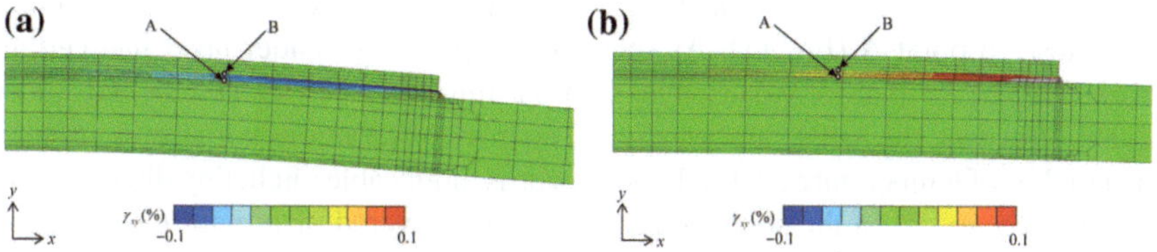

Fig. 11 Deformed configuration and γ_{xy} contour map at the end of 60th temperature cycle; layered plate with **a** annealed and **b** as-received OFHC-Cu; displacement multiplied by three

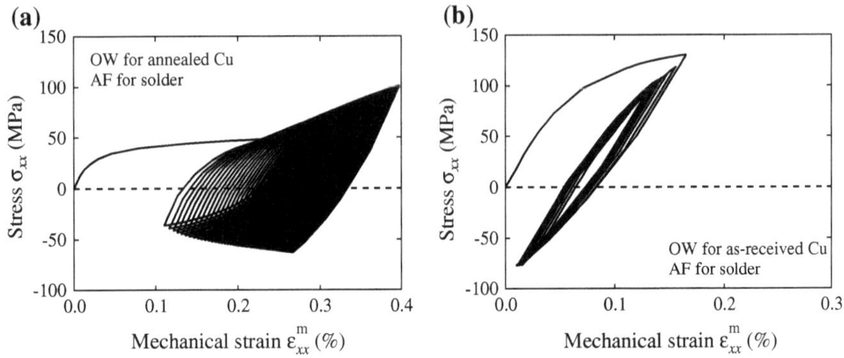

Fig. 12 Hysteresis loops of σ_{xx} and $\varepsilon^{\mathrm{m}}_{xx}$ at point A in copper in layered plates with **a** annealed and **b** as-received OFHC-Cu; analysis using OW and AF models for Cu and solder

as-received copper plate with the increase in the number of temperature cycles. This ratcheting was caused by the positive mean values of σ_{xx} that were induced by the CTE mismatch and the temperature dependence of viscoplastic flow stress in the solder layer under cyclic thermal loading, as described in detail in an axisymmetric case by Nakane et al. (2008). The ratcheting can be classified as uniaxial ratcheting, though mean stress is not constant. On the other hand, the cyclical decrease in σ_{xx} and $\varepsilon^{\mathrm{m}}_{xx}$ was brought by the CTE-mismatch-induced elastic bending energies in the Si and Cu layers; these energies allowed the warpage to recover cyclically as viscoplastic deformation cyclically proceeded in the solder layer (Nakane et al. 2010). Obviously, the elastic bending energy in the as-received copper plate was much larger than that in the annealed copper plate, because the as-received copper had much higher plastic flow stress than the annealed copper (Fig. 4). It is thus concluded that the uniaxial-type of ratcheting and cyclic strain recovery occurred in the copper layer to induce the cyclic growth and cyclic recovery of δ, respectively, under cyclic thermal loading.

5.3 *Ratcheting in the Solder Layer*

The above-mentioned strain variations at point A in the copper layer suggested checking the hysteresis loops of σ_{xx} and $\varepsilon^{\mathrm{m}}_{xx}$ at point B in the solder layer (Fig. 13). Point B is close to point A (Fig. 11). As seen from Fig. 13, the solder layer had certain amplitudes of $\varepsilon^{\mathrm{m}}_{xx}$ that were caused by the CTE mismatch between the Si and solder layers. However, the mean value of $\varepsilon^{\mathrm{m}}_{xx}$ had no marked variations with the increase in the number of temperature cycles. In other words, noticeable ratcheting did not occur with respect to $\varepsilon^{\mathrm{m}}_{xx}$ in the solder layer even in the layered plate with the annealed copper layer.

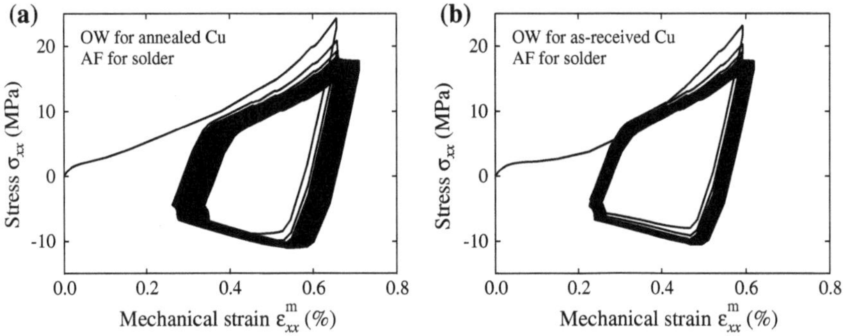

Fig. 13 Hysteresis loops of σ_{xx} and ε_{xx}^{m} at point B in solder in layered plates with **a** annealed and **b** as-received OFHC-Cu; analysis using OW and AF models for Cu and solder

The γ_{xy} contour maps on the longitudinal, middle cross section at the end of the 60th cycle are also sown in Fig. 11. Here, γ_{xy} denotes engineering shear strain, i.e., $\gamma_{xy} = 2\varepsilon_{xy}$. As seen from the figure, γ_{xy} was negative in the solder layer when the cyclic growth of δ occurred in the analysis with the annealed copper plate; on the other hand, γ_{xy} was positive in the solder layer when the cyclic recovery of δ was properly simulated by considering the as-received copper for the copper layer. The opposite signs of γ_{xy} in the solder layer are thus considered to be a key point to explain the cyclic growth and cyclic recovery of δ.

Figure 14 shows the hysteresis loops of σ_{xy} and γ_{xy} at point B in the solder layer. It is seen that ratcheting occurred in the solder layer to result in the opposite signs of γ_{xy} depending on the heat treatment conditions of the copper layer. Needless to say, the cyclic variations in σ_{xy} and γ_{xy} in the solder layer were caused by the CTE mismatch between the silicon and copper layers under cyclic thermal loading, and the opposite signs of γ_{xy} are regarded as consequences of the cyclic growth and cyclic recovery

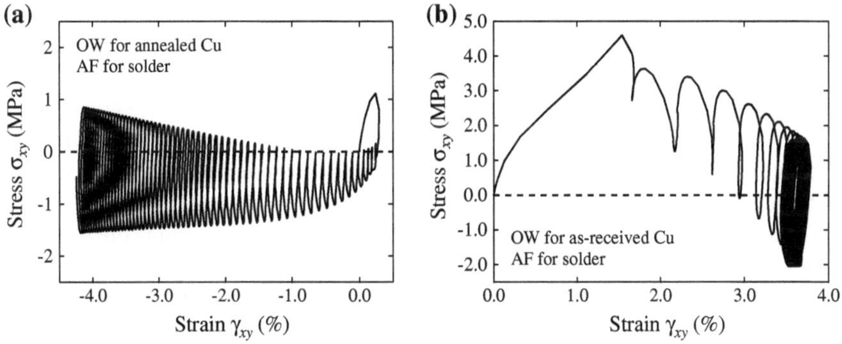

Fig. 14 Hysteresis loops of σ_{xy} and γ_{xy} at point B in solder in layered plates with **a** annealed and **b** as-received OFHC-Cu; analysis using OW and AF models for Cu and solder

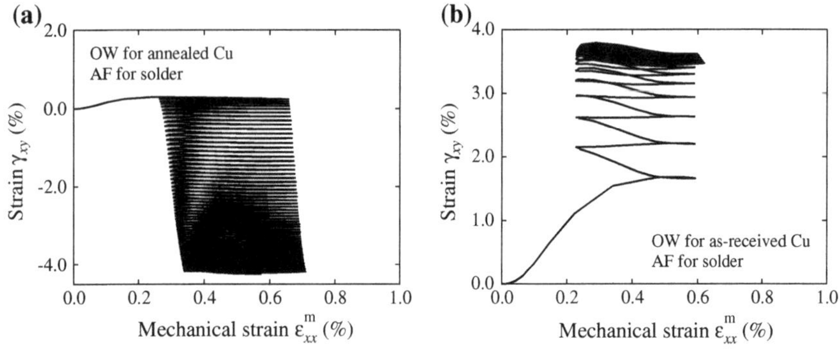

Fig. 15 Hysteresis loops of $\varepsilon^{\mathrm{m}}_{xx}$ and γ_{xy} at point B in solder in layered plates with **a** annealed and **b** as-received OFHC-Cu; analysis using OW and AF models for Cu and solder

of δ predicted by the pair of OW-Cu and AF-sol. However, it must be pointed out that the magnitudes of σ_{xy} in the figure, which are only about 1.0 MPa at the end of the 60th cycle, are considerably small in comparison with the tensile viscoplastic flow stresses of the solder shown in Fig. 3. Therefore, it cannot be said that the cyclic variations in σ_{xy} solely governed the ratcheting in the solder layer, although the mean values of σ_{xy} were obviously nonzero (Fig. 14). In short, the ratcheting in the solder layer is not classified as the uniaxial type.

Figure 15 depicts the strain trajectories concerning $\varepsilon^{\mathrm{m}}_{xx}$ and γ_{xy} at point B in the solder. It is clear from the figure that multiaxial ratcheting occurred in the solder; i.e., γ_{xy} developed either negatively or positively with the increase in the number of cyclic changes in $\varepsilon^{\mathrm{m}}_{xx}$. This was because the mean values of σ_{xy}, which were combined with the cyclic changes in $\varepsilon^{\mathrm{m}}_{xx}$, were either negative or positive depending on the heat treatment conditions of the copper layer (Fig. 14). It is emphasized that the ratcheting in the solder was multiaxial in contrast to the uniaxial type of ratcheting in the copper.

6 Multiaxial Ratcheting of Pb-63Sn Solder

It was described in Sect. 5 that the pair of OW-Cu and AF-sol provided good predictions for the cyclic variations in warpage δ observed by Tanie et al. (2011). However, good performance of the AF model for the ratcheting in the solder was not expected, because the AF model usually overpredicts ratcheting (Ohno 1997; Kang 2008; Chaboche 2008). This section is, therefore, devoted to examining the AF model using the multiaxial ratcheting data of Pb-63Sn solder reported by Chen et al. (2007). It is noted that, to the best of the authors' knowledge, only their data were published with respect to multiaxial ratcheting of solders.

Figure 16 depicts the shear stress–strain hysteresis loops obtained in the multiaxial ratcheting tests of Pb-63Sn in which constant tensile stress of $\sigma = 5\,\mathrm{MPa}$ was

Fig. 16 Shear hysteresis loops in multiaxial ratcheting tests of Pb-63Sn; $\dot{\gamma} = 1.0$, 0.1, 0.01 %/s, $\Delta\gamma = 0.52\,\%$, $\sigma = 5\,\text{MPa}$; experiment (Chen et al. 2007) and simulation

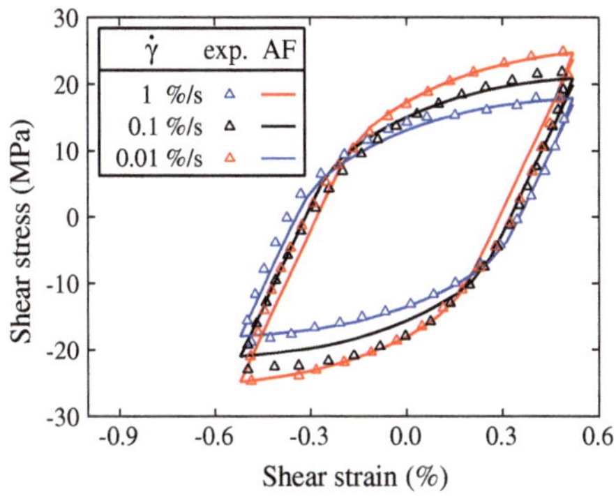

Table 4 Material parameters of AF model for Pb-63Sn solder; stress in MPa, strain in mm/mm, and time in s

Elastic constants	$E = 2.80 \times 10^4$, $\nu = 0.37$
AF model	$\dot{\varepsilon}_0 = 1.00 \times 10^{-4}$
	$\sigma_0 = 21.3$, $n = 9.30$
	$h = 9.80 \times 10^3$, $\zeta = 8.00 \times 10^2$

combined with cyclic shear straining at $\dot{\gamma} = 1.0$, 0.1, and 0.01 %/s at room temperature (Chen et al. 2007). As shown in the figure, the shear stress–strain hysteresis loops were simulated well using the AF model represented in Sect. 3.2. The material parameters used for the simulation are given in Table 4.

Figures 17 and 18 compare the predicted and experimental increases in tensile strain ε in the multiaxial ratcheting at $\dot{\gamma} = 1.0$, 0.1 and 0.01 %/s under $\sigma = 5\,\text{MPa}$. Obviously, the AF model gave better predictions for the multiaxial ratcheting when $\dot{\gamma}$ was smaller, as clearly seen from Fig. 18; i.e., the difference of experiments and predictions became about three times smaller as $\dot{\gamma}$ decreased from 1.0 to 0.01 %/s. The data in the figure imply that the difference could be negligible if $\dot{\gamma} \approx 0.001$ %/s.

Fig. 17 Multiaxial ratcheting of Pb-63Sn; $\dot{\gamma} =$ 1.0, 0.1, 0.01 %/s, $\Delta\gamma = 0.52\,\%$, $\sigma = 5\,\text{MPa}$; experiment (Chen et al. 2007) and simulation

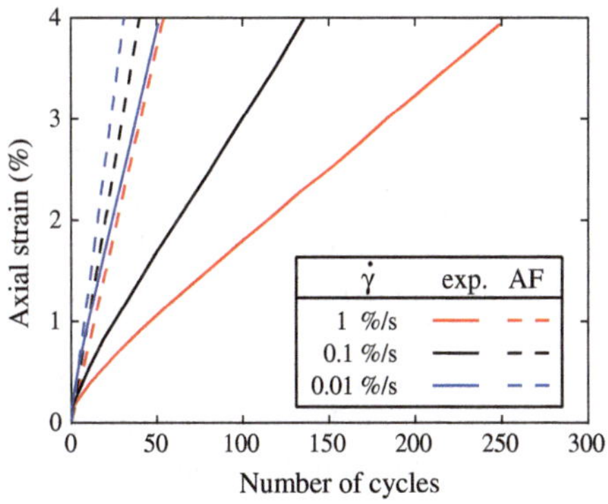

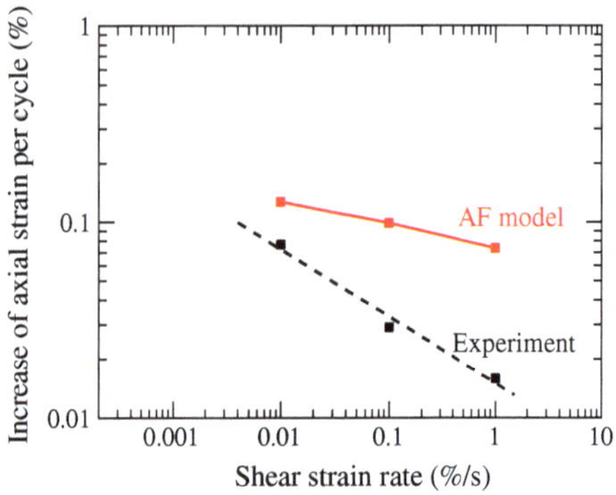

Fig. 18 Effect of shear strain rate on multiaxial ratcheting rate of Pb-63Sn; $\Delta\gamma = 0.52\,\%$, $\sigma = 5\,\text{MPa}$; experiment (Chen et al. 2007) and simulation

Point B in the solder in the Si/solder/Cu layered plates had the equivalent strain range equal to about 0.75 % in the analysis results described in Sect. 5, and the temperature cycling had the period of 3000 s (Fig. 9). Consequently, the equivalent strain rate at point B was about 5.0×10^{-4} %/s. This equivalent strain rate corresponds to $\dot{\gamma} \approx 0.001$ %/s. Therefore, the multiaxial ratcheting data shown in Fig. 18 imply that the AF model could be appropriate for the multiaxial ratcheting in the solder in in the Si/solder/Cu layered plates.

7 Conclusions

In this study, the cyclic thermal tests of Si/Pb-5Sn-1.5Ag/OFHC-Cu layered plates (Tanie et al. 2011) were analyzed by assuming three material models for the solder and two material models for the copper. Strain hardening and rate-dependence in the solder were taken into account in the three material models using the tensile test data of Yamada and Ogawa (1996) on Pb-5Sn, noting that Pb-10Sn and Pb-5Sn-2.5Ag provided almost completely the same tensile curves as each other at room temperature and $0.7T_{\mathrm{m}}$ (Schoeller et al. 2009). The material parameters in the two models for the copper were identified in as-received and fully annealed states using the tensile and cyclic test data of Tanie et al. (2011), Wada et al. (2010), and Zhang and Jiang (2008). The following findings were thus obtained:

- The cyclic growth and cyclic recovery of warpage observed in the cyclic thermal tests were reproduced well when the Armstrong and Frederick (1966) model and the Ohno and Wang (1993a) model were assumed for the solder and copper, respectively.
- Multiaxial ratcheting occurred in the solder in contrast to the uniaxial type of ratcheting or cyclic strain recovery in the copper. For this multiaxial ratcheting in the solder, the Armstrong and Frederick model unexpectedly worked well.

To confirm the above-mentioned unexpected finding, the Armstrong and Frederick model was applied to the multiaxial ratcheting of Pb-63Sn (Chen et al. 2007). It was thus suggested that the Armstrong and Frederick model could be appropriate for the multiaxial ratcheting at such slow strain rates as in the solder in the cyclic thermal tests of Tanie et al. (2011).

References

Abdel-Karim, M., Ohno, N.: Kinematic hardening model suitable for ratcheting with steady-state. Int. J. Plast. **16**, 225–240 (2000)

Armstrong, P.J., Frederick, C.O.: A mathematical representation of the multiaxial Bauschinger effect. CEGB Report RD/BN/731, Berkeley Nuclear Laboratories, UK (1966)

Bai, N., Chen, X.: A new unified constitutive model with short- and long-range back stress for lead-free solders of Sn-3Ag-0.5Cu and Sn-0.7Cu. Int. J. Plast. **25**, 2181–2203 (2009)

Bari, S., Hassan, T.: An advancement in cyclic plasticity modeling for multiaxial ratcheting simulation. Int. J. Plast. **18**, 873–894 (2002)

Busso, E.P., Kitano, M., Kumazawa, T.: A visco-plastic constitutive model for 60/40 tin-lead solder used in IC package joints. J. Eng. Mater. Technol. **114**, 331–337 (1992)

Cailletaud, G., Saï, K.: Study of plastic/viscoplastic models with various inelastic mechanisms. Int. J. Plast. **11**, 991–1005 (1995)

Chaboche, J.L.: Constitutive equations for cyclic plasticity and cyclic viscoplasticity. Int. J. Plast. **5**, 247–302 (1989)

Chaboche, J.L.: On some modifications of kinematic hardening to improve the description of ratcheting effects. Int. J. Plast. **7**, 661–678 (1991)

Chaboche, J.L.: A review of some plasticity and viscoplasticity constitutive theories. Int. J. Plast. **24**, 1642–1693 (2008)

Chaboche, J.L., Dang Van, K., Cordier, G.: Modelization of the strain memory effect on the cyclic hardening of 316 stainless steel. In: Proceedings of the 5th International Conference on Structural Mechanics in Reactor Technology, vol. L, Paper No. L11/3 (1979)

Chen, X., Kim, K.S., Abdel-Karim, M., Sakane, M.: Strain rate dependent constitutive model of multiaxial ratcheting of 63Sn-37Pb solder. J. Electron. Packag. **129**, 278–286 (2007)

Dafalias, Y.F., Feigenbaum, H.P.: Biaxial ratcheting with novel variations of kinematic hardening. Int. J. Plast. **27**, 479–491 (2011)

Dafalias, Y.F., Kourousis, K.I., Saridis, G.J.: Multiplicative AF kinematic hardening in plasticity. Int. J. Solids Struct. **45**, 2861–2880 (2008)

Freed, A.D.: Thermoviscoplastic model with application to copper. NASA Technical Paper 2845 (1988)

Ishiko, M.: Recent R&D activities of power devices for hybrid electric vehicles. R&D Rev. Toyota CRDL **39**(4), 1–6 (2004)

Jiang, Y., Sehitoglu, H.: Modeling of cyclic ratcheting plasticity, part I: development of constitutive relations. J. Appl. Mech. **63**, 720–725 (1996a)

Jiang, Y., Sehitoglu, H.: Modeling of cyclic ratcheting plasticity, part II: comparison of model simulations with experiments. J. Appl. Mech. **63**, 726–733 (1996b)

Kanchanomai, C., Miyashita, Y., Mutoh, Y.: Low-cycle fatigue behavior of Sn-Ag, Sn-Ag-Cu, and Sn-Ag-Cu-Bi lead-free solders. J. Electron. Mater. **31**, 456–465 (2002)

Kang, G.Z.: A visco-plastic constitutive model for ratcheting of cyclically stable materials and its finite element implementation. Mech. Mater. **36**, 299–312 (2004)

Kang, G.Z.: Ratcheting: recent progresses in phenomenon observation, constitutive modeling and application. Int. J. Fatigue **30**, 1448–1472 (2008)

Kang, G.Z., Kan, Q.H., Zhang, J., Sun, Y.F.: Time-dependent ratchetting experiments of SS304 stainless steel. Int. J. Plast. **22**, 858–894 (2006)

Kimura, T., Saitou, R., Kubo, K., Nakatsu, K., Ishikawa, H., Sasaki, K.: High-power-density inverter technology for hybrid and electric vehicle applications. Hitachi Rev. **63**, 96–102 (2014)

Kobayashi, T., Sasaki, K.: Experiments and simulations of uniaxial ratcheting deformation of Sn-3Ag-0.5Cu and Sn-37Pb solder alloys. J. Mater. Sci.: Mater. Electron. **20**, 343–353 (2009)

Nakane, K., Ohno, N., Tsuda, M., Yagi, Y., Nakagawa, I., Atsumi, T.: Thermal ratcheting of solder-bonded elastic and elastoplastic layers. Int. J. Plast. **24**, 1819–1836 (2008)

Nakane, K., Ohno, N., Tanie, H.: Thermal ratcheting of solder-bonded layered plates: cyclic recovery and growth of deflection. Comput. Mech. **46**, 259–268 (2010)

Ohno, N.: A constitutive model of cyclic plasticity with a nonhardening strain region. J. Appl. Mech. **49**, 721–727 (1982)

Ohno, N.: Recent progress in constitutive modeling for ratcheting. Mater. Sci. Res. Int. **3**, 1–9 (1997)

Ohno, N.: Material models of cyclic plasticity with extended isotropic hardening: a review. Mech. Eng. Rev. **2**, Paper No. 14-00425 (2015)

Ohno, N., Wang, J.D.: Transformation of a nonlinear kinematic hardening rule to a multisurface form under isothermal and nonisothermal conditions. Int. J. Plast. **7**, 879–891 (1991)

Ohno, N., Wang, J.D.: Kinematic hardening rules with critical state of dynamic recovery, part I: formulation and basic features for ratcheting behavior. Int. J. Plast. **9**, 375–390 (1993a)

Ohno, N., Wang, J.D.: Kinematic hardening rules with critical state of dynamic recovery, part II: application to experiments of ratcheting behavior. Int. J. Plast. **9**, 391–403 (1993b)

Ohno, N., Wang, J.D.: Kinematic hardening rules for simulation of ratcheting behavior. Eur. J. Mech. A-Solids **13**, 519–531 (1994)

Ohno, N., Tsuda, M., Kamei, T.: Elastoplastic implicit integration algorithm applicable to both plane stress and three-dimensional stress states. Finite Elem. Anal. Des. **66**, 1–11 (2013)

Ohno, N., Tsuda, M., Sugiyama, H., Okumura, D.: Elastic-viscoplastic implicit integration algorithm applicable to both plane stress and three-dimensional stress states. Finite Elem. Anal. Des. **109**, 54–64 (2016)

Saï, K.: Multi-mechanism models: present state and future trends. Int. J. Plast. **27**, 250–281 (2011)

Schoeller, H., Bansal, S., Knobloch, A., Shaddock, D., Cho, J.H.: Microstructure evolution and the constitutive relations of high-temperature solders. J. Electron. Mater. **38**, 802–809 (2009)

Taleb, L., Cailletaud, G.: An updated version of the multimechanism model for cyclic plasticity. Int. J. Plast. **26**, 859–874 (2010)

Tanie, H., Nakane, K., Urata, Y., Tsuda, M., Ohno, N.: Warpage variations of Si/solder/OFHC-Cu layered plates subjected to cyclic thermal loading. Microelectron. Reliab. **51**, 1840–1844 (2011)

Wada, K., Yagi, Y., Nakagawa, I., Atsumi, T., Ohno, N.: Analysis of thermal expansion in elastic and elatoplastic layers subjected to cyclic thermal loading. Microelectron. Reliab. **50**, 1626–1630 (2010)

Yamada, H., Ogawa, K.: Evaluation of thermal fatigue life of solder joints in electronic components. R&D Rev. Toyota CRDL **31**, 43–52 (1996)

Zhang, J.X., Jiang, Y.Y.: Constitutive modeling of cyclic plasticity deformation of a pure polycrystalline copper. Int. J. Plast. **24**, 1890–1915 (2008)

A Note on Plasticity with Additional Internal Variables

Michael Wolff and Michael Böhm

Abstract This note deals with plastic behavior, more precisely with the role of possible additional internal variables for the description of observable effects. Based on the experimental experience, models in use are sometimes modified in order to capture new effects which cannot be described by the original version of the model. This modification can be done with additional internal variables. In doing so, one has to take care that the modified models remain thermodynamically consistent. We discuss this issue on the basis of two phenomena in cyclic plasticity. Moreover, we show how the concept of two-mechanism models can be used for the description of interactions between the arising phenomena.

1 Introduction

In many cases, complex material behavior is modeled with the aid of internal variables. An alternative concept uses functionals (see Haupt 2002 for a detailed discussion of both approaches). Sometimes, existing models of material behavior are modified to take special observable effects into account. For instance, in cyclic plasticity the strain-memory effect (see Nouailhas et al. 1985, e.g.) and an additional hardening due to nonproportional loading (see Benallal and Marquis 1987, e.g.) may occur. In Taleb and Cailletaud (2010), a current modeling of both phenomena was presented in order to achieve a good agreement between simulations and ratcheting experiments. However, a strong thermodynamic foundation of the approach was not in the focus there. This note aims to fill the gap.

A further important tool for modeling complex material behavior are multimechanism models (abbreviated as MM-models) which have been investigated for about 25 years. In applications, one encounters mostly two-mechanism models. In the case

M. Wolff (✉) · M. Böhm
Zentrum für Technomathematik, Universität Bremen, Bremen, Germany
e-mail: mwolff@math.uni-bremen.de

M. Böhm
e-mail: mbohm@math.uni-bremen.de

K. Naumenko and M. Aßmus (eds.), *Advanced Methods of Continuum Mechanics for Materials and Structures*, Advanced Structured Materials 60,
DOI 10.1007/978-981-10-0959-4_11

of small deformations, the characteristic trait of serial multimechanism models (with m mechanisms) is the decomposition of the inelastic strain into a weighted sum of m parts (plastic or viscoplastic, e.g.). These parts are sometimes called mechanisms. We refer to Saï (2011), Wolff et al. (2011, 2015), Taleb and Cailletaud (2011), Taleb (2013), Kröger (2013) for current overviews and further references. In comparison with rheological models (see Palmov 1998, 2008, e.g.), there can be an interaction among the inelastic mechanisms. This leads to new possibilities in modeling. The interaction among mechanisms allows a good description of important observable effects (inverse strain-rate sensibility, e.g.). We refer to Contesti and Cailletaud (1989) and Cailletaud and Saï (1995). As a consequence of the coupling among the mechanisms, the thermodynamic consistency of an MM-model generally does not follow from the thermodynamic consistency of the single components. It can be ensured by sufficient conditions involving the material functions (see Wolff and Taleb 2008; Wolff et al. 2011, 2015).

In this theoretical note, we model two phenomena in the cyclic plasticity mentioned above in a thermodynamically consistent way with the aid of additional variables which do not occur directly in the free energy (in Sect. 2). In Sect. 3, we demonstrate how a possible interaction of these phenomena can be described within the framework of two-mechanism models.

2 Modeling of Elastoplastic Behavior and Some Extensions

After providing some basics concerning plasticity with hardening and damage in Sect. 2.1, we consider two extensions to the phenomena in cyclic plasticity in Sect. 2.2.

2.1 A Model with One Mechanism Including Damage

Considering a serial material behavior in small-deformation setting, we assume an additive split of the full strain $\boldsymbol{\varepsilon}$ as well as the free energy ψ into their corresponding thermoelastic and inelastic parts $\boldsymbol{\varepsilon}_{te}$, $\boldsymbol{\varepsilon}_{in}$ and ψ_{te}, ψ_{in}, i.e.,

$$\boldsymbol{\varepsilon} = \boldsymbol{\varepsilon}_{te} + \boldsymbol{\varepsilon}_{in}, \qquad \psi = \psi_{te} + \psi_{in}. \tag{1}$$

Here, plastic behavior is of our focus. Thus, instead of "inelastic" we mostly employ "plastic" abbreviated by the index "p". A clear distinction between both partial strains will follow from (6). The thermoelastic and inelastic parts of the free energy are assumed to have the following structure (for comments and references see Remark 1).

$$\begin{aligned}\psi_{te}(\boldsymbol{\varepsilon}_{te},\theta,d_E) :&= \frac{(1-d_E)}{2\varrho_0}\boldsymbol{\mathcal{E}}(\theta)\boldsymbol{\varepsilon}_e:\boldsymbol{\varepsilon}_e+\psi_{cal}(\theta)\\ &= \frac{(1-d_E)}{2\varrho_0}\boldsymbol{\mathcal{E}}\big(\boldsymbol{\varepsilon}_{te}-(\theta-\theta_0)\boldsymbol{G}(\theta)\big):\big(\boldsymbol{\varepsilon}_{te}-(\theta-\theta_0)\boldsymbol{G}(\theta)\big)\\ &\quad+\psi_{cal}(\theta), \end{aligned} \tag{2}$$

$$\psi_{in}(\boldsymbol{\alpha},q,\theta,d_\alpha,d_q) := \frac{(1-d_\alpha)}{2\varrho_0}\,C(\theta)\,\boldsymbol{\alpha}:\boldsymbol{\alpha}+\frac{(1-d_q)}{2\varrho_0}\,Q(\theta)\,q^2. \tag{3}$$

ψ_{cal} is the purely temperature-dependent part of the free energy (calorimetric function, e.g., see Helm and Haupt 2003), not of further interest here, $\boldsymbol{\mathcal{E}}$-fourth-order elasticity tensor, $\boldsymbol{\varepsilon}_e$-pure elastic strain, $\varrho_0 > 0$-mass density in the reference configuration, θ, θ_0-temperature, reference temperature, $\boldsymbol{G}$-(symmetric) second-order heat-dilatation tensor, $\boldsymbol{\alpha}$-tensorial internal variable related to kinematic hardening, q-scalar internal variable related to isotropic hardening, d_E, d_α, d_q scalar internal damage variables related to elasticity, kinematic and isotropic hardening, respectively, $Q > 0$, $C > 0$-kinematic and isotropic hardening moduli. (We exclude the simple cases $C = 0$ and $Q = 0$.) Q expresses the initial sloop of the curve of isotropic hardening. The parameter C allows an analogous interpretation for kinematic hardening in the case of uniaxial direction. The fourth-order elasticity tensor $\boldsymbol{\mathcal{E}} = (\mathcal{E}_{ijkl})$ has the following properties for all admissible temperatures θ as well as in all body points x (see Haupt 2002; Altenbach 2012; Bertram and Glüge 2015, e.g.).

$$\mathcal{E}_{ijkl}(\theta,x)=\mathcal{E}_{jikl}(\theta,x)=\mathcal{E}_{klij}(\theta,x)=\mathcal{E}_{ijlk}(\theta,x)\quad \forall\, i,j,k,l\in\{1,2,3\}, \tag{4}$$

$$\exists\, c_0>0 \;:\; \boldsymbol{\mathcal{E}}(\theta,x)\boldsymbol{\alpha}:\boldsymbol{\alpha}\geq c_0\|\boldsymbol{\alpha}\|^2 \quad \text{for all symmetric second-order tensors } \boldsymbol{\alpha}. \tag{5}$$

The current stress tensor $\boldsymbol{\sigma}$ is assumed to be a function of $\boldsymbol{\varepsilon}_{te}$, θ, and d_E, but *not* of $\boldsymbol{\varepsilon}_{in}$, $\boldsymbol{\alpha}$, q, d_α, d_q (see Remark 1 (iii) for reasoning and references):

$$\boldsymbol{\sigma}=\boldsymbol{\sigma}(\boldsymbol{\varepsilon}_{te},\theta,d_E) \tag{6}$$

Moreover, it is assumed that for a given θ and d_E this function is invertible with respect to $\boldsymbol{\varepsilon}_{te}$. Here we exclude the case of $\boldsymbol{\varepsilon}_{te}\equiv 0$ like for pure creep without an elastic domain.

In a standard way, we define the thermodynamic forces associated with hardening and damage:

$$\boldsymbol{X}=\varrho_0\frac{\partial\psi_{in}}{\partial\boldsymbol{\alpha}}=(1-d_\alpha)C(\theta)\,\boldsymbol{\alpha},\quad R=\varrho_0\frac{\partial\psi_{in}}{\partial q}=(1-d_q)Q(\theta)\,q, \tag{7}$$

$$Y_\alpha=\varrho_0\frac{\partial\psi_{in}}{\partial d_\alpha}=-\frac{1}{2}\,C(\theta)\,\boldsymbol{\alpha}:\boldsymbol{\alpha},\quad Y_q=\varrho_0\frac{\partial\psi_{in}}{\partial d_q}=-\frac{1}{2}\,Q(\theta)\,q^2, \tag{8}$$

$$Y_E=\varrho_0\frac{\partial\psi_{te}}{\partial d_E}=-\frac{1}{2}\,\boldsymbol{\mathcal{E}}(\theta)\,\boldsymbol{\varepsilon}_e:\boldsymbol{\varepsilon}_e. \tag{9}$$

Usually, X is called back stress, R—isotropic hardening. In a standard way, we formulate evolution equations for $\boldsymbol{\alpha}$, q, d_E, d_α, d_q, as well as initial conditions and restrictions for the damage variables:

$$\dot{\boldsymbol{\alpha}} = \dot{\boldsymbol{\varepsilon}}_p - \frac{D}{C}\boldsymbol{X}\dot{s}_p, \qquad \boldsymbol{\alpha}(0) = 0, \qquad \dot{q} = \dot{s}_p - \frac{b}{Q}R\dot{s}_p, \qquad q(0) = 0, \tag{10}$$

$$\dot{d}_i = f_{di}, \qquad d_i(0) = 0, \qquad 0 \le d_i \le 1, \qquad i = E, \alpha, q. \tag{11}$$

The generally temperature-depending material functions $D \ge 0$, $b \ge 0$ and $f_d \ge 0$ may depend on internal variables and on stress. The plastic (generally inelastic) accumulation s_p is defined by

$$s_{in}(t) = s_p(t) = \int_0^t \sqrt{\frac{2}{3}\dot{\boldsymbol{\varepsilon}}_p(s) : \dot{\boldsymbol{\varepsilon}}_p(s)}\,\mathrm{d}s. \tag{12}$$

Thus, s_p can be considered as a further internal variable. Differentiation of (12) yields its evolution equation. In the case $b \equiv 0$, s_p coincides with q, and one gets linear isotropic hardening. Otherwise, the free energy does not depend directly on s_p (see Remark 1 (iv)).

The standard evolution equation for the plastic strain $\boldsymbol{\varepsilon}_p$ reads as

$$\dot{\boldsymbol{\varepsilon}}_p = \lambda\sqrt{\frac{3}{2}}\,\frac{\boldsymbol{\sigma}^* - \boldsymbol{X}^*}{\|\boldsymbol{\sigma}^* - \boldsymbol{X}^*\|} =: \lambda\,\mathbf{n}, \qquad \boldsymbol{\varepsilon}_p(0) = 0. \tag{13}$$

$\boldsymbol{\sigma}^*$ is the deviator of $\boldsymbol{\sigma}$, $\mathbf{n}$—outer normal to the stress sphere. The plastic multiplier λ is defined via the flow rule in plasticity (see (16)), or it is directly defined in viscoplasticity or creep. Due to the von Mises approach, the flow function is defined by

$$f(\boldsymbol{\sigma}, \boldsymbol{X}, R, \sigma_0) = \sqrt{\frac{3}{2}(\boldsymbol{\sigma}^* - \boldsymbol{X}^*) : (\boldsymbol{\sigma}^* - \boldsymbol{X}^*)} - (\sigma_0 + R), \tag{14}$$

with the initial yield stress $\sigma_0(\theta) > 0$. Note that there may be an implicit dependence on damage via $\boldsymbol{\sigma}$, $\boldsymbol{X}$, R. In plasticity, a general constraint occurs (see Remark 1 (v)),

$$f(\boldsymbol{\sigma}, \boldsymbol{X}, R, \sigma_0) \le 0, \tag{15}$$

and the plastic multiplier is defined by

$$\begin{aligned} \lambda &\ge 0, \qquad \text{if } f(\boldsymbol{\sigma}, \boldsymbol{X}, R, \sigma_0) = 0,\\ \lambda &= 0, \qquad \text{if } f(\boldsymbol{\sigma}, \boldsymbol{X}, R, \sigma_0) < 0. \end{aligned} \tag{16}$$

Here we assume the usual Fourier law of heat conduction (see Altenbach 2012, e.g.). Since heat conduction is not in our direct focus here, we omit formulas and details. Performing the standard approach of evaluation of the entropy inequality (Clausius–Duhem inequality) (see Haupt 2002; Besson et al. 2010, e.g.), one obtains the usual stress relation

$$\boldsymbol{\sigma} = \varrho_0 \frac{\partial \psi_{te}}{\partial \boldsymbol{\varepsilon}_{te}} = (1-d)\boldsymbol{\mathcal{E}}\big(\boldsymbol{\varepsilon}_{te} - (\theta - \theta_0)\boldsymbol{G}(\theta)\big), \tag{17}$$

the usual entropy relation which we omit as well as the Clausius–Planck inequality (i.e., the remaining inequality without heat dissipation)

$$\begin{aligned} \boldsymbol{\sigma} : \dot{\boldsymbol{\varepsilon}}_p - \Big(\boldsymbol{X} : \dot{\boldsymbol{\varepsilon}}_p - \frac{D}{C}\boldsymbol{X} : \boldsymbol{X}\dot{s}_p\Big) - \Big(R\dot{s}_p - \frac{b}{Q}R^2\dot{s}_p\Big) + -Y_\alpha f_{d\alpha} - Y_q f_{dq} - Y_E f_{dE} \\ = \sigma_0\dot{s}_p + \frac{D}{C}\boldsymbol{X} : \boldsymbol{X}\dot{s}_p + \frac{b}{Q}R^2\dot{s}_p + \frac{f_{d\alpha}}{2}C\boldsymbol{\alpha} : \boldsymbol{\alpha} + \frac{f_{dq}}{2}Q\,q^2 + \frac{f_{dE}}{2}\boldsymbol{\mathcal{E}}\boldsymbol{\varepsilon}_e : \boldsymbol{\varepsilon}_e \geq 0 \end{aligned} \tag{18}$$

In the first line in (18), the first part represents the dissipated power of plastic deformation, the brackets (without minus sign) represent the rate of the stored energy due to kinematic and isotropic hardening, respectively, and, finally, the remaining parts stand for the dissipated power due to damage. Obviously, under the given restrictions to the material functions, the inequality is always fulfilled. Or, in other words, the sum of the dissipated power is not smaller than the rate of stored energy. Thus, the model presented is thermodynamically consistent.

Remark 1

1. In this note, the mechanical part of the thermoelastic free energy ψ_{te} in (2) is assumed to be a quadratic form of the pure elastic strain $\boldsymbol{\varepsilon}_e = \boldsymbol{\varepsilon}_{te} - (\theta - \theta_0)\boldsymbol{G}$ as in Lemaitre and Chaboche (1990). As a consequence, the force Y_E defined in (9) is not positive, and the thermodynamic consistency is ensured for a nondecreasing damage variable d_E (corresponding to a nonnegative f_{dE}). For a widely spread alternative, we refer to Haupt (2002). In Wolff et al. (2015), these aspects have been discussed in more detail.
2. For simplicity only scalar damage variables are considered here. For more complex approaches, we refer to Naumenko and Altenbach (2007) and Murakami (2012), e.g.
3. Based on experimental experience, the stress tensor $\boldsymbol{\sigma}$ is usually assumed to depend only on purely elastic quantities like $\boldsymbol{\varepsilon}_{te}$ (and possible on θ and d_E), but not on $\boldsymbol{\varepsilon}_{in}$ as well as on internal variables like $\boldsymbol{\alpha}$ and q associated with inelastic phenomena (hardening, e.g.) (see Haupt 2002; Bertram 2012; Bertram and Glüge 2015, e.g.). An exception is pure inelastic behavior like pure creep without an elastic domain or some kinds of viscoelasticity, which requires a special consideration.
4. Note: Actually, besides $\boldsymbol{\alpha}$, q, d_α, d_q, d_e in the model above, the two additional internal variables s_p and $\boldsymbol{\varepsilon}_p$ arise. However, the free energy does not depend

directly on them. (An exclusion is the special case of linear hardening (Prager's model, see Besson et al. 2010) with $D \equiv 0$ and $b \equiv 0$, yielding $\boldsymbol{\alpha} = \boldsymbol{\varepsilon}_p$ and $q = s_p$.) The occurrence of internal variables "outside of the free energy" formally contradicts the rule of equipresence (see Truesdell and Noll 1965; Coleman and Gurtin 1967). We think that this is not a real drawback, since we could define the free energy as a function which is constant with respect to these additional internal variables. Clearly, these additional internal variables do not change the structure of the remaining inequality.
In this context, the thermodynamical forces defined in (7)–(9) are not internal variables, they can be regarded as abbreviations.

5. Plastic behavior expressed by (14)–(16) can be equivalently described with the aid of variational inequalities. We refer exemplarily to Duvaut and Lions (1976), Han and Reddy (1999), Simo and Hughes (1998) for theory and to Boettcher et al. (2015) for an application to complex material behavior.

Remark 2 (*An estimate of the isotropic hardening*) Using $(7)_2$ and $(10)_2$, we can solve the Cauchy problem for q (assuming the remaining quantities known) and obtain

$$q(t) = \int_0^t \dot{s}_p(s) e^{-\int_s^t (1-d_q(\tau)) b(\tau) \dot{s}_p(\tau)\, \mathrm{d}\tau}\, \mathrm{d}s. \tag{19}$$

Assuming

$$b \geq b_1 > 0 \qquad \text{for all admissible arguments,} \tag{20}$$

$$0 \leq d_q(t) \leq d_1 < 1, \tag{21}$$

one obtains from (19) the global estimate for q:

$$0 \leq q(t) \leq \frac{1}{b_1(1-d_1)}(1 - e^{-b_1(1-d_1)s_p(t)}). \tag{22}$$

Using $(7)_2$, we get the estimate:

$$0 \leq R(t) \leq R_{max} := \frac{Q_1}{b_1(1-d_1)}, \tag{23}$$

where

$$0 < Q_0 \leq Q(\theta) \leq Q_1 \qquad \text{for all admissible } \theta. \tag{24}$$

In particular, if b depends only on θ, for *isothermal* processes without damage one obtains

$$R(t) = R_\theta(t) = \frac{Q(\theta)}{b(\theta)}(1 - e^{-b(\theta)s_p(t)}). \tag{25}$$

That means, R is a function of the plastic accumulation s_p alone, and its curve has the typical form with the initial sloop of Q and the asymptotic value of $\frac{Q}{b}$.

2.2 *An Extension of the Model Presented*

In cyclic plasticity, special phenomena are observed which require extensions of the above approaches. In the usual model presented above, the parameter $Q(\theta)$ (=initial sloop of the isotropic hardening curve) does not undergo an increase due to cyclic hardening. In order to model this and other phenomena within the framework developed above, one can use additional internal variables which do not occur in the free energy. See Remark 1 (iv) for discussion.

2.2.1 Strain-Memory Effect

Experiments show that the maximum value of the isotropic hardening R may increase under a strain-controlled regime (see Nouailhas et al. 1985 for details as well as Fig. 1 for an illustration). To realize this increase and motivated by (23) and (25), we propose a decrease of the parameter b via

$$\dot{b} = (b_1 - b)\mu\,\dot{\tau}, \qquad b(0) = b_0 > b_1 > 0. \tag{26}$$

Thus, b turns into an internal variable, τ is a further internal variable which will be dealt with later, $\mu > 0$ is a material parameter possibly depending on θ, $b_0 = b_0(\theta)$ is the initial value of b (for fixed θ). It is maximal and can be obtained from a monotone tension test with the original material. The quantity $b_1(\theta)$ is an assumed minimal value of b. The approach (26) is a slight modification of the one in Nouailhas et al. (1985)

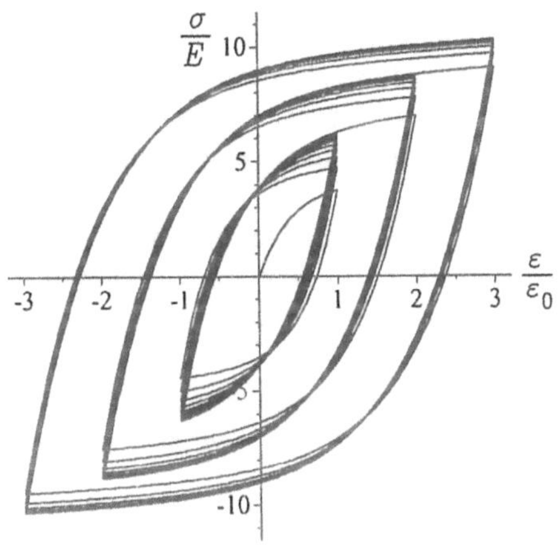

Fig. 1 Illustration of the strain-memory effect: Cyclic hardening with step-wise increasing strain amplitudes in a strain-controlled regime (taken from Haupt 2002, Fig. 11.4, p. 447).

Remark 3 1. In our setting (in contrast to Taleb and Cailletaud 2010), the Heaviside function H is used. The quotient in (34) is bounded, however not defined if one of its arguments becomes zero. Using the Heaviside function in this case, $\cos(\alpha)$ is zero. In doing so, a drawback mentioned in Taleb and Cailletaud (2010) can be overcome.
2. In the case of proportional loading, the rates of the deviatoric stress $\dot{\boldsymbol{\sigma}}^*$ and of the plastic strain $\dot{\boldsymbol{\varepsilon}}_p$ are collinear. Thus, $\cos(\alpha) = 1$ and $A = 0$, yielding the constant solution $b = b_s = b_0$ to (31). Clearly, there is no hardening due to nonproportionality. On the contrary, the case of $\cos(\alpha) = 0$ and $A = 1$ ("full nonproportionality") causes the fastest decrease of b, corresponding to the best hardening.
3. The dependence of the internal variable b on the derivative of the deviatoric stress $\dot{\boldsymbol{\sigma}}^*$ is nonstandard. Although the free energy does not depend on b, the internal variable q depends via (11) on b. Thus, when exploiting the entropy inequality to obtain sufficient conditions for thermodynamic consistency, there may arise some difficulties. In this case, the derivative $\dot{\boldsymbol{\sigma}}^*$ can be temporarily changed by a difference quotient. At the end, a limiting process can be performed returning the derivative. This work is in progress.

3 An Application of Multimechanism Models

After providing some basics of multimechanism models in Sects. 3.1, 3.2, we will consider a possible coupling between the two cyclic-plastic-phenomena presented above in Sect. 2.2. This coupling will be realized with the aid of a two-mechanism approach.

3.1 Preparations

In a small deformation setting, the characteristic trait of multimechanism models (in series) is the additive decomposition of the inelastic strain $\boldsymbol{\varepsilon}_{in}$ into a weighted sum. For 2M-models this means

$$\boldsymbol{\varepsilon}_{in} = A_1\,\boldsymbol{\varepsilon}_1 + A_2\,\boldsymbol{\varepsilon}_2 \tag{38}$$

where $A_1 > 0$, $A_2 > 0$ are real numbers (see Wolff et al. 2015 for extensions). The full strain $\boldsymbol{\varepsilon}$ and the free energy ψ are additively decomposed as in (1). After that, partial stresses are defined:

$$\boldsymbol{\sigma}_i = A_i\boldsymbol{\sigma} \qquad i = 1, 2. \tag{39}$$

For theory and applications of multimechanism models we refer to Saï (2011), Wolff et al. (2011, 2015), Taleb and Cailletaud (2011), Taleb (2013), Kröger (2013), as well as to the references therein.

In the sequel, we deal with a special two-mechanism model with one common flow criterion (abbreviated by 2M1C) and with two isotropic variables (see Remark 4 (i)), considering only plastic behavior. Thus, instead of the index *in* we use *p*. In Taleb and Cailletaud (2010), a 2M1C-model with two isotropic hardenings R_1 and R_2 has been considered to take the two special effects mentioned above into account. However, no possible coupling between R_1 and R_2 was dealt with. It is now the aim to extend the approach in Taleb and Cailletaud (2010), bringing it into in a clear thermodynamic setting. The thermoelastic part of the free energy ψ_{te} is defined as in (2). The inelastic part of the free energy is assumed as follows:

$$\psi_{in} = \psi_{kin} + \psi_{iso} = \frac{(1-d_\alpha)}{2\varrho_0}\Big(C_{11}(\theta)\,\boldsymbol{\alpha}_1 : \boldsymbol{\alpha}_1 + 2C_{12}(\theta)\,\boldsymbol{\alpha}_1 : \boldsymbol{\alpha}_2 + C_{22}(\theta)\,\boldsymbol{\alpha}_2 : \boldsymbol{\alpha}_2\Big)$$
$$+ \frac{(1-d_q)}{2\varrho_0}\Big(Q_{11}(\theta)\,q_1^2 + 2Q_{12}(\theta)\,q_1 q_2 + Q_{22}(\theta)\,q_2^2\Big). \tag{40}$$

ψ_{kin} stands for kinematic, ψ_{iso} for isotropic hardening. Now, one has two kinematic internal tensorial variables $\boldsymbol{\alpha}_1$ and $\boldsymbol{\alpha}_2$ as well as two isotropic scalar internal variables q_1 and q_2. Besides, there are three kinematic and three isotropic hardening moduli C_{11}, C_{12}, C_{22} and Q_{11}, Q_{12}, Q_{22}, respectively, all are possibly temperature-dependent. We assume for all admissible temperatures θ

$$C_{11}(\theta) > 0, \qquad C_{22}(\theta) > 0, \qquad C_{12}^2(\theta) \le C_{11}(\theta)C_{22}(\theta), \tag{41}$$
$$Q_{11}(\theta) > 0, \qquad Q_{22}(\theta) > 0, \qquad Q_{12}^2(\theta) \le Q_{11}(\theta)Q_{22}(\theta) \tag{42}$$

Therefore, the symmetric matrices

$$\begin{pmatrix} C_{11} & C_{12} \\ C_{12} & C_{22} \end{pmatrix}, \qquad \begin{pmatrix} Q_{11} & Q_{12} \\ Q_{12} & Q_{22} \end{pmatrix}$$

are positive semidefinite. However, they may be singular. The singularity of the first matrix $\mathbf{C}$ causes ratcheting phenomena in plasticity (see Cailletaud and Saï 1995) and in special sorts of viscoelasticity (see Wolff et al. 2012). The conditions (41) and (42), respectively, mean that ψ_{kin} and ψ_{iso} are nonnegative and convex. Thus during hardening, energy cannot be received from the material structure. This would have no physical sense (see Wolff and Taleb 2008 for further discussion and numerical examples).

The thermodynamic force Y_E is defined in (9). In analogy with (7), (8) the back stresses, isotropic hardenings and the remaining thermodynamic force related to damage are defined by

$$\boldsymbol{X}_i = \varrho_0 \frac{\partial \psi_{kin}}{\partial \boldsymbol{\alpha}_i} = (1 - d_\alpha)\Big(C_{i1}(\theta)\,\boldsymbol{\alpha}_1 + C_{i2}(\theta)\,\boldsymbol{\alpha}_2\Big), \qquad i = 1, 2, \tag{43}$$

$$R_i = \varrho_0 \frac{\partial \psi_{iso}}{\partial q_i} = (1 - d_q)\Big(Q_{i1}(\theta)\,q_1 + Q_{i2}(\theta)\,q_2\Big), \qquad i = 1, 2, \tag{44}$$

$$Y_\alpha = \varrho_0 \frac{\partial \psi_{in}}{\partial d_\alpha} = -\frac{1}{2}\Big\{C_{11}(\theta)\,\boldsymbol{\alpha}_1 : \boldsymbol{\alpha}_1 + 2C_{12}(\theta)\,\boldsymbol{\alpha}_1 : \boldsymbol{\alpha}_2 + C_{22}(\theta)\,\boldsymbol{\alpha}_2 : \boldsymbol{\alpha}_2\Big\}, \tag{45}$$

$$Y_q = \varrho_0 \frac{\partial \psi_{in}}{\partial d_q} = -\frac{1}{2}\Big\{Q_{11}(\theta)\,q_1^2 + 2Q_{12}(\theta)\,q_1 q_2 + Q_{22}(\theta)\,q_2^2\Big\}. \tag{46}$$

Based on the von Mises plasticity, we introduce further quantities:

$$J_i = \sqrt{\frac{3}{2}(\boldsymbol{\sigma}_i^* - \boldsymbol{X}_i^*) : (\boldsymbol{\sigma}_i^* - \boldsymbol{X}_i^*)} = \sigma_V(\boldsymbol{\sigma}_i - \boldsymbol{X}_i), \qquad i = 1, 2, \tag{47}$$

$$J = \Big(J_1^\beta + J_2^\beta\Big)^{\frac{1}{\beta}}$$

where $\beta > 1$ is a further parameter (see Wolff et al. 2015). The flow function for the 2M1C-model under consideration is defined by (see Remark 4 (i))

$$f(\boldsymbol{\sigma}_i, \boldsymbol{X}_i, R_i, R_0) = J - (R_1 + R_2 + R_0), \tag{48}$$

where the quantity R_0 is given by

$$R_0 = 2^{\frac{1}{\beta}}\sigma_0, \tag{49}$$

where $\sigma_0 = \sigma_0(\theta)$ is the initial yield stress (see Taleb and Cailletaud 2010; Wolff et al. 2015). In the case under consideration, the general constraint reads as

$$f(\boldsymbol{\sigma}_i, \boldsymbol{X}_i, R_i, R_0) \leq 0 \tag{50}$$

Defining the direction vectors $\mathbf{n}_i$ of the partial strains via

$$\mathbf{n}_i = \frac{3}{2}\,\frac{\boldsymbol{\sigma}_i^* - \boldsymbol{X}_i^*}{J_i}\left(\frac{J_i}{J}\right)^{\beta-1}, \qquad i = 1, 2, \tag{51}$$

one supposes the following evolution equations for the partial strains by

$$\boldsymbol{\varepsilon}_i = \lambda\,\mathbf{n}_i, \qquad i = 1, 2. \tag{52}$$

λ is the common plastic multiplier fulfilling

$$\begin{aligned} \lambda &\geq 0, \qquad \text{if } f(\boldsymbol{\sigma}_i, \boldsymbol{X}_i, R_i, R_0) = 0, \\ \lambda &= 0, \qquad \text{if } f(\boldsymbol{\sigma}_i, \boldsymbol{X}_i, R_i, R_0) < 0. \end{aligned} \tag{53}$$

Clearly, one has the relation

$$\lambda = \left(\dot{s}_1^{\frac{\beta}{\beta-1}} + \dot{s}_2^{\frac{\beta}{\beta-1}}\right)^{\frac{\beta-1}{\beta}}, \tag{54}$$

where the rates of accumulation of the partial strains are given by

$$\dot{s}_i = \sqrt{\frac{2}{3}\dot{\boldsymbol{\varepsilon}}_i : \dot{\boldsymbol{\varepsilon}}_i} = \varepsilon_V(\dot{\boldsymbol{\varepsilon}}_i), \qquad i = 1, 2. \tag{55}$$

The evolution equations for the kinematic internal variables $\boldsymbol{\alpha}_1, \boldsymbol{\alpha}_2$ are given by (see Wolff et al. 2015)

$$\begin{aligned} \dot{\boldsymbol{\alpha}}_1 &= \dot{\boldsymbol{\varepsilon}}_1 - \frac{D_{11}}{C_{11}}\boldsymbol{X}_1\lambda - \frac{D_{12}}{\sqrt{C_{11}C_{22}}}\boldsymbol{X}_2\lambda, \\ \dot{\boldsymbol{\alpha}}_2 &= \dot{\boldsymbol{\varepsilon}}_1 - \frac{D_{21}}{\sqrt{C_{11}C_{22}}}\boldsymbol{X}_1\lambda - \frac{D_{22}}{C_{22}}\boldsymbol{X}_2\lambda, \end{aligned} \tag{56}$$

For several extensions, we refer to Cailletaud and Saï (1995), Taleb et al. (2006), and Wolff et al. (2015). We assume the following evolution equations for the isotropic internal variables: q_1, q_2.

$$\begin{aligned} \dot{q}_1 &= \lambda - \frac{b_{11}}{Q_{11}}R_1\lambda - \frac{b_{12}}{\sqrt{Q_{11}Q_{22}}}R_2\lambda, \\ \dot{q}_2 &= \lambda - \frac{b_{21}}{\sqrt{Q_{11}Q_{22}}}R_1\lambda - \frac{b_{22}}{Q_{22}}R_2\lambda, \end{aligned} \tag{57}$$

There arise two new matrices **D** and **B** of parameter functions

$$\begin{pmatrix} D_{11} & D_{12} \\ D_{21} & D_{22} \end{pmatrix}, \quad \begin{pmatrix} b_{11} & b_{12} \\ b_{21} & b_{22} \end{pmatrix}.$$

These matrices are generally nonsymmetric; their entries may be functions of temperature and further quantities. More general, in analogy to the considerations in Sects. 2.2.1 and 2.2.2, these functions may play the role of additional variables having own evolution equations. Up to now, diagonal matrices are mostly in use (see Taleb and Cailletaud 2010; Saï 2011). However, further even "nonsymmetric" couplings between the mechanisms are possible, using nonsymmetric matrices (see Wolff et al. 2011 for an example). In order to keep a good analogy to the case of models with one mechanism (1M-models) like in Sect. 2.1, the evolution equations (56) and (57) are written in this special form, using quotients like $\frac{b_{11}}{Q_{11}}$ (see Wolff et al. 2015). The evolution equation for the damage variables d_i $(i = E, \alpha, q)$ is assumed as above as shown in (11) for the case of a 1M-model. The difference may be in the concrete form of the nonnegative functions f_{di}.

Assuming the usual Fourier law of heat conduction and performing the standard approach again, the Clausius–Planck inequality reads as

$$
\begin{aligned}
&\boldsymbol{\sigma}_1 : \dot{\boldsymbol{\varepsilon}}_1 + \boldsymbol{\sigma}_2 : \dot{\boldsymbol{\varepsilon}}_2 - \boldsymbol{X}_1 : \dot{\boldsymbol{\varepsilon}}_1 - \boldsymbol{X}_2 : \dot{\boldsymbol{\varepsilon}}_2 + \sum_{i,j=1}^{2} \frac{D_{ij}}{\sqrt{C_{ii}C_{jj}}} \boldsymbol{X}_j : \boldsymbol{X}_i \lambda + \\
&- R_1\lambda - R_2\lambda + \sum_{i,j=1}^{2} \frac{b_{ij}}{\sqrt{Q_{ii}Q_{jj}}} R_i R_j \lambda - Y_E f_{dE} - Y_\alpha f_{d\alpha} - Y_q f_{dq} \\
&= R_0\lambda + \sum_{i,j=1}^{2} \frac{D_{ij}}{\sqrt{C_{ii}C_{jj}}} \boldsymbol{X}_j : \boldsymbol{X}_i \lambda + \sum_{i,j=1}^{2} \frac{b_{ij}}{\sqrt{Q_{ii}Q_{jj}}} R_i R_j \lambda + \\
&- Y_E f_{dE} - Y_\alpha f_{d\alpha} - Y_q f_{dq} \geq 0. \qquad (58)
\end{aligned}
$$

In order to ensure thermodynamic consistency, the matrices **D** and **B** must be a positive semidefinite for all admissible arguments. Obviously, this is equivalent to

$$
(D_{12} + D_{21})^2 \leq 4D_{11}^2 D_{22}^2, \qquad (b_{12} + b_{21})^2 \leq 4b_{11}^2 b_{22}^2 \qquad (59)
$$

for all admissible arguments. Using $(43)_2$ and (57), one obtains evolution equations for the isotropic hardenings R_1 and R_2:

$$
\begin{aligned}
\dot{R}_1 &= Q_{11}\lambda - b_{11}R_1\lambda - \sqrt{\frac{Q_{11}}{Q_{22}}} b_{12}R_2\lambda + Q_{12}\lambda - \frac{Q_{12}}{\sqrt{Q_{11}Q_{22}}} b_{21}R_1\lambda + \\
&- \frac{Q_{12}}{Q_{22}} b_{22}R_2\lambda + \dot{\theta}\frac{\partial Q_{11}}{\partial \theta} q_1 + \dot{\theta}\frac{\partial Q_{12}}{\partial \theta} q_2, \qquad (60)
\end{aligned}
$$

$$
\begin{aligned}
\dot{R}_2 &= Q_{12}\lambda - \frac{Q_{12}}{Q_{11}} b_{11}R_1\lambda - \frac{Q_{12}}{\sqrt{Q_{11}Q_{22}}} b_{12}R_2\lambda + Q_{22}\lambda - \sqrt{\frac{Q_{22}}{Q_{11}}} b_{21}R_1\lambda + \\
&- b_{22}R_2\lambda + \dot{\theta}\frac{\partial Q_{12}}{\partial \theta} q_1 + \dot{\theta}\frac{\partial Q_{22}}{\partial \theta} q_2. \qquad (61)
\end{aligned}
$$

Remark 4 1. The 2M1C-model introduced in this subsection has one common flow criterion and two associated isotropic internal variables. This extension of previous MM-models allowing several isotropic variables for one criterion was introduced in Wolff et al. (2015). In this context, the hardenings R_i $(i = 1, 2)$ can be regarded as partial ones, and their sum $R_1 + R_2$ can be understood as hardening related to the flow criterion. In an analogous manner (Wolff et al. 2015), a mechanism in MM-models may have several kinematic variables as in the so-called Chaboche models with one mechanism (see Chaboche 2008).

2. The concept of two- and multimechanisms can be also applied to viscoplastic behavior (e.g., Wolff et al. 2015) as well as to some sorts of creep (e.g., Wolff and Böhm 2010; Bökenheide 2015) and viscoelasticity (e.g., Wolff et al. 2012). More-

over, if mechanisms do not belong to the same criterion, they may be different, plastic and viscoplastic, e.g.,

3.2 *Modeling of Special Effects in Cyclic Plasticity*

Now it is the aim to use the two-mechanism approach for a simultaneous modeling of the strain-memory effect and of the nonproportional loading effect like in Taleb and Cailletaud (2010). In addition, we propose a coupling between the two effects, using the possibilities due to a generally nondiagonal matrix $\mathbf{B}$. For this reason, we relate the isotropic hardening R_1 to the strain-memory effect and R_2 to the nonproportional loading effect. The material function b_{11} is assumed to be an internal variable satisfying an evolution equation as in (26). After that, we proceed as above in Sect. 2.2.1. The only difference is that now the direction vector $\mathbf{n}$ of the full inelastic strain $\boldsymbol{\varepsilon}_p$ is given by

$$\mathbf{n} = A_1 \mathbf{n}_1 + A_2 \mathbf{n}_2. \tag{62}$$

Analogously, the material function b_{22} is assumed to be an internal variable having the evolution equation as in (31). If one sets $Q_{12} \equiv 0$ and $b_{12} = b_{21} \equiv 0$, the situation in Taleb and Cailletaud (2010) is reproduced in a different setting. However, choosing suitable quantities Q_{12} and b_{12}, b_{21}, possible coupling phenomena can be taken into account.

Setting $Q_{12} \equiv 0$ from (60) and (61), we get

$$\dot{R}_1 = Q_{11}\lambda - b_{11}R_1\lambda - \sqrt{\frac{Q_{11}}{Q_{22}}}b_{12}R_2\lambda + \dot{\theta}\frac{\partial Q_{11}}{\partial \theta}q_1, \tag{63}$$

$$\dot{R}_2 = Q_{22}\lambda - \sqrt{\frac{Q_{22}}{Q_{11}}}b_{21}R_1\lambda - b_{22}R_2\lambda + \dot{\theta}\frac{\partial Q_{22}}{\partial \theta}q_2. \tag{64}$$

Now the roles of b_{12} and b_{21} become clear. b_{12} expresses the influence of the isotropic hardening due to nonproportional loading on the evolution of the isotropic hardening due to the strain-memory effect. Analogously, b_{12} expresses the reverse effect. Note that due to the possible nonsymmetry of the matrix $\mathbf{B}$, these both influences may be different. The only restriction (within the frame of this modeling) is that this matrix is positive semidefinite. Thus, b_{12} and b_{21} can be chosen as (possibly temperature-dependent) parameters or as functions with own evolution equations so long as there holds

$$(b_{12} + b_{21})^2 \leq 4b_{11}^2 b_{22}^2 \qquad \text{for all admissible arguments.} \tag{65}$$

Clearly, a simple case consists in choosing $b_{12} = b_{12}(\theta)$ and $b_{21} = b_{21}(\theta)$ in accordance with

$$(b_{12}(\theta) + b_{21}(\theta))^2 \leq 4b^2_{1,s-m}(\theta)b^2_{1,n-p}(\theta) \qquad \text{for all admissible } \theta, \tag{66}$$

where $b_{1,s-m}$ is the minimum of b_{11} associated with the strain-memory effect (see (26)), and $b_{1,n-p}$ is the minimal b_1 in (35).

4 Conclusion

In this note, we have dealt with a thermodynamically consistent modeling of observable effects in cyclic plasticity within the framework of two-mechanism models, using additional internal variables. These effects are the strain-memory effect (see Nouailhas et al. 1985, e.g.) and an additional hardening due to nonproportional loading (see Benallal and Marquis 1987, e.g.). In Taleb and Cailletaud (2010), a current modeling of both phenomena was presented in order to achieve a good agreement between simulations and ratcheting experiments. However, a strong thermodynamic foundation of the approach was not in the focus there but this note aims to fill the gap.

A detailed verification based on experimental data as well as a detailed comparison with the results by Taleb and Cailletaud (2010) remains for further work.

References

Altenbach, H.: Kontinuumsmechanik - Einführung in die materialabhängigen Gleichungen - 2. Springer Vieweg, Berlin (2012)

Benallal, A., Marquis, D.: Constitutive equations for nonproportional cyclic elasto-viscoplasticity. J. Eng. Mater. Technol. **109**(4), 326–336 (1987)

Bertram, A.: Elasticity and Plasticity of Large Deformations: An Introduction. Springer, Berlin (2012)

Bertram, A., Glüge, R.: Festkörpermechanik (überarbeitete Ausgabe). Otto-von-Guericke-Universität Magdeburg (2015)

Besson, J., Cailletaud, G., Chaboche, J.L., Samuel, F., Blétry, M.: Non-linear Mechanics of Materials, Solid Mechanics and its Applications, vol. 167. Springer, New York (2010)

Boettcher, S., Böhm, M., Wolff, M.: Well-posedness of a problem of thermo-elasto-plasticity with phase transitions in steel. ZAMM (2015)

Bökenheide, S.: Modelling and simulation of inelastic phenomena in the material behaviour of steel during heat treatment processes. Ph.D. thesis, University of Bremen, Germany (2015)

Cailletaud, G., Saï, K.: Study of plastic/viscoplastic models with various inelastic mechanisms. Int. J. Plast. **11**, 991–1005 (1995). doi:10.1016/S0749-6419(95)00040-2

Chaboche, J.: A review of some plasticity and viscoplasticity constitutive theories. Int. J. Plast. **24**, 1642–1693 (2008). doi:10.1016/j.ijplas.2008.03.009

Chaboche, J.L., Dang-Van, K., Cordier, G.: Modelization of the strain memory effect on the cyclic hardening of 316 stainless steel. In: Proceedings of the 5th International Conference on SMiRT, Div. L, Berlin, Germany (1979)
Coleman, B.D., Gurtin, M.E.: Thermodynamics with internal state variables. J. Chem. Phys. **47**(2), 597–613 (1967). doi:10.1063/1.1711937
Contesti, E., Cailletaud, G.: Description of creep-plasticity interaction with non-unified constitutive equations: application to an austenitic steel. Nucl. Eng. Des. **116**, 265–280 (1989). doi:10.1016/0029-5493(89)90087-3
Duvaut, G., Lions, J.L.: Inequalities in Mechanics and Physics. A Series of Comprehensive Studies in Mathematics. Springer, New York (1976)
Han, W., Reddy, D.B.: Plasticity - Mathematical Theory and Numerical Analysis. Springer, New York (1999)
Haupt, P.: Continuum Mechanics and Theory of Materials. Springer, New York (2002)
Helm, D., Haupt, P.: Shape memory behavior: modeling within continuum thermomechanics. Int. J. Solids Struct. **40**, 827 (2003). doi:10.1016/S0020-7683(02)00621-2
Kröger, N.H.: Multi-mechanism models - theory and applications. Ph.D. thesis, Universität Bremen, Germany (2013)
Lemaitre, J., Chaboche, J.L.: Mechanics of Solid Materials. Cambridge University Press, Cambridge (1990)
Murakami, S.: Continuum Damage Mechanics: A Continuum Mechanics Approach to the Analysis of Damage and Fracture, vol. 185. Springer, New York (2012)
Naumenko, K., Altenbach, H.: Modeling of Creep for Structural Analysis. Springer, New York (2007)
Nouailhas, D., Cailletaud, G., Policella, H., Marquis, D., Dufailly, J., Lieurade, H., Ribes, A., Bollinger, E.: On the description of cyclic hardening and initial cold working. Eng. Fract. Mech. **21**(4), 887–895 (1985)
Palmov, V.A.: Vibrations of Elasto-Plastic Bodies. Springer, New York (1998)
Palmov, V.A.: Theory of defining equations in nonlinear thermodynamics of deformable bodies (in Russian). St. Petersburg State Polytechnical University (2008)
Saï, K.: Multi-mechanism models: present state and future trends. int. j. plast. **27**, 250–281 (2011). doi:10.1016/j.ijplas.2010.05.003
Simo, J.C., Hughes, T.J.R.: Computational Inelasticity. Springer, New York (1998)
Taleb, L.: About the cyclic accumulation of the inelastic strain observed in metals subjected to cyclic stress control. Int. J. Plast. **43**, 1–19 (2013)
Taleb, L., Cailletaud, G.: An updated version of the multimechanism model for cyclic plasticity. Int. J. Plast. **26**(6), 859–874 (2010). doi:10.1016/j.ijplas.2009.11.002
Taleb, L., Cailletaud, G.: Cyclic accumulation of the inelastic strain in the 304L SS under stress control at room temperature: Ratcheting or creep? Int. J. Plast. **27**, 1936–1958 (2011). doi:10.1016/j.ijplas.2011.02.001
Taleb, L., Cailletaud, G., Blaj, L.: Numerical simulation of complex ratcheting tests with a multi-mechanism model type. Int. J. Plast. **22**, 724–753 (2006). doi:10.1016/j.ijplas.2005.05.003
Truesdell, C., Noll, W.: The Non-linear Field Theories of Mechanics. Encyclopedia of Physics. Springer, New York (1965)
Wolff, M., Böhm, M.: Two-mechanism models and modelling of creep. In: Proceedings of the 3rd International Conference on Nonlinear Dynamics, Charkiw, Ukraine (2010)
Wolff, M., Taleb, L.: Consistency for two multi-mechanism models in isothermal plasticity. Int. J. Plast. **24**, 2059–2083 (2008). doi:10.1016/j.ijplas.2008.03.001
Wolff, M., Böhm, M., Taleb, L.: Thermodynamic consistency of two-mechanism models in the non-isothermal case. Technische Mechanik **31**, 58–80 (2011)

Wolff, M., Böhm, M., Bökenheide, S., Kröger, N.: Two-mechanism approach in thermo-viscoelasticity with internal variables. Technische Mechanik **32**(2–5), 608–621 (2012)
Wolff, M., Bökenheide, S., Böhm, M.: Some new extensions to multi-mechanism models for plastic and viscoplastic material behavior under small strains. Contin. Mech. Thermodyn. (2015). doi:10.1007/s00161-015-0418-5

Finite Inelastic Deformations of Compressible Soft Solids with the Mullins Effect

Heng Xiao, Hao Li, Zhao-Ling Wang and Zheng-Nan Yin

Abstract It is shown that finite inelastic behavior of soft solids with the Mullins effect may be directly simulated by establishing finite elastoplastic J_2-flow models. New results in three respects are presented, including (i) general compressible deformations are taken into account for the purpose of bypassing limitations of the usual incompressibility constraint; (ii) any damage-like variables and associated evolution equations are not involved; and (iii) any given number of unloading curves of any given shapes in the Mullins effect may be simulated by direct, explicit procedures.

1 Motivation and Introduction

Soft solid materials, such as biological tissues and polymeric materials, are widely used in engineering. Unlike usual solids with small elastic deformations, such materials exhibit very complex inelastic deformation behavior coupled with large elastic deformations, such as the well-known Mullins effect associated with the stress softening at unloading and the permanent set after unloading, etc. In the past decades, many efforts have been made to establish constitutive models for soft solids. Results have been given based on damage-like variables, e.g., earlier in Simo (1987) and later in Miehe (1995), Lion (1996), Ogden and Roxburg (1999), Miehe and Keck

H. Xiao (✉) · H. Li · Z.-L. Wang · Z.-N. Yin
State Key Laboratory of Advanced Special Steels and Shanghai Institute
for Applied Mathematics and Mechanics, Shanghai University,
Yanchang Road 149, Shanghai 200072, China
e-mail: xiaoheng@shu.edu.cn

H. Li
e-mail: haoli@shu.edu.cn

Z.-N. Yin
e-mail: zhengnanyin@shu.edu.cn

Z.-L. Wang
Weifang University, School of Mathematics and Information Science,
Shandong, Weifang 261061, China

K. Naumenko and M. Aßmus (eds.), *Advanced Methods of Continuum Mechanics for Materials and Structures*, Advanced Structured Materials 60,
DOI 10.1007/978-981-10-0959-4_12

(2000), Beatty and Krishnaswamy (2000), Laiarinandrasana and Piques (2003), Lin and Schomburg (2003), Chagnon et al. (2004), Dorfman and Ogden (2004), Diani et al. (2006a), and Li et al. (2008). Results based on similar approaches may be found in Besdo and Ihlemann (2003a), Qi and Boyce (2004), De Tommasi et al. (2006), etc. On the other hand, micro-mechanical models based on macromolecular mechanisms were proposed earlier in Govindjee and Simo (1991) and later by Marckmann et al. (2002) , Besdo and Ihlemann (2003b), Drozdov and Dorfman (2003), Göktepe and Miehe (2005), Diani et al. (2006b), Ayoub et al. (2010), Itskov et al. (2010), Zaïri et al. (2011), Dargazany et al. (2014), and many others.

Irrespective of numerous results, there are issues that need to be treated (cf., Diani 2009). In fact, a large number of parameters should be introduced in each existing model. For a given material sample, complicated numerical procedures should be performed to find out such parameters, in a sense of fitting test data given for this sample as closely as possible. Whenever the model applicability is validated for a material sample, it may not be clear that it will be the case for another material sample. On the other hand, the incompressibility constraint is usually assumed in order to reduce undue complexities involved. Although this usual assumption may provide meaningful results, it leaves two issues outstanding. The one is that particular procedures treating this constraint have to be carried out in numerical implementation, which may give rise to possible errors and instabilities (see, e.g., Simo and Hughes 1998). The other is that soft solid materials may display appreciable compressibility. In this case, the incompressibility constraint may not be suitable.

In this study we are going to address the above issues. Toward this objective, we are going to demonstrate that finite inelastic behavior of soft solids may be simulated by establishing finite elastoplastic J_2-flow models with nonlinear hardening. New results in three respects are presented, including (i) general compressible deformations are taken into account for the purpose of bypassing limitations of the usual incompressibility constraint; (ii) any damage-like variables and associated evolution equations are not involved; and (iii) any given number of unloading curves of any given shapes in the Mullins effect may be simulated by means of direct, explicit procedures.

The main content is arranged as follows. In Sect. 2, a rate-independent J_2-flow model will be proposed, with compressible elastic behavior evolving with the plastic work. In Sect. 3, the evolving elastic potential and the yield limit incorporated in the proposed model will be determined by means of direct, explicit procedures, so that any given sets of data for the stress softening in the uniaxial unloading case may be automatically, accurately fitted and, in the meantime, the proposed model with identified multiaxial constitutive quantities is ready in use for general case of multiaxial deformation. In Sect. 4, numerical examples in fitting test data will be presented for the purpose of model validation. Finally, remarks for further studies will be given in Sect. 5.

To conclude this introduction, some notations that will be used are explained as follows. The notations $(\cdot)$ and $(:)$ placed between two tensors mean the single dot product and the double dot product, respectively, and tr$\boldsymbol{S}$ represents the trace of second-order tensor $\boldsymbol{S}$. Moreover, the superimposed dot $\dot{\overline{(\ )}}$ represents the material time rate. Finally, sgn(x) denotes the sign function of x that takes values $1, 0, -1$ for $x > 0$, $x = 0$, $x < 0$, respectively.

2 Finite Strain J_2-flow Model with Evolving Rubberlike Elasticity

The starting point is to establish a new J_2-flow model of finite elastoplastic deformations with evolving elastic behavior. The consistent Eulerian rate formulation of finite elastoplasticity based on the corotational logarithmic rate will be applied. Detail in this respect may be found in Bruhns et al. (1999) and Xiao et al. (2006, 2007). Toward our objective, we are going to extend this formulation to a broader case with evolving rubberlike elasticity. Such a formulation in the incompressibility case has been suggested in a most recent study (Xiao 2015). In what follows, a broader case for general compressible deformations will be treated.

Let $\boldsymbol{F}$ and $\boldsymbol{F}^T$ be the deformation gradient and its transposition and let $\boldsymbol{h}$ and $\boldsymbol{D}$ be Hencky's logarithmic strain and the stretching. Moreover, let $\boldsymbol{\tau}$ and $\tilde{\boldsymbol{\tau}}$ be the Kirchhoff stress and its deviatoric part, namely, $\boldsymbol{\tau} = J\boldsymbol{\sigma}$ where $\boldsymbol{\sigma}$ is the Cauchy stress (true stress) and $J = det\boldsymbol{F}$ is the volumetric ratio..

2.1 *Separation of the Stretching*

It is assumed that the stretching $\boldsymbol{D}$ is composed of an elastic part, $\boldsymbol{D}^e$, and a plastic part, $\boldsymbol{D}^p$, as shown below:

$$\boldsymbol{D} = \boldsymbol{D}^e + \boldsymbol{D}^p. \tag{1}$$

An objective rate equation should be given for the elastic part $\boldsymbol{D}^e$, which governs finite hyperelastic behavior of soft solids. Moreover, a flow rule should be given for the plastic part $\boldsymbol{D}^p$, which governs dissipated inelastic behavior of soft solids.

Below the two Eulerian rate type constitutive equations indicated above will be presented, separately.

2.2 *Elastic Rate Equation with Evolving Elastic Potential*

To determine the elastic part $\boldsymbol{D}^e$, a Eulerian rate equation is given to establish a linear relationship between the elastic part $\boldsymbol{D}^e$ and an objective stress rate:

$$\boldsymbol{D}^e = \frac{\partial^2 \overline{W}}{\partial \boldsymbol{\tau}^2} : \overset{\circ}{\boldsymbol{\tau}}{}^{\log} + \frac{\partial^2 \overline{W}}{\partial \boldsymbol{\tau} \partial \kappa} \dot{\kappa}, \tag{2}$$

where $\overset{\circ}{\boldsymbol{\tau}}{}^{\log}$ is the logarithmic rate of the Kirchhoff stress $\boldsymbol{\tau}$ and, besides,

$$\overline{W} = \overline{W}(\boldsymbol{\tau}, \kappa), \tag{3}$$

is a complementary elastic potential characterizing compressible hyperelastic behavior of soft solids.

As a departure from usual treatment, here the potential $\overline{W}$ (cf. Eq. (3)) will evolve with the plastic work κ (see Eq. (8) given slightly later). This idea will prove essential in simulating the stress softening behavior.

It may be demonstrated (Xiao et al. 2007) that, with the logarithmic stress rate $\overset{\circ}{\boldsymbol{\tau}}{}^{\log}$, the elastic rate equation Eq. (2) prior to the initial yielding, i.e., at $\boldsymbol{D}^e = \boldsymbol{D}$, is exactly integrable to deliver a finite hyperelastic relation, namely,

$$\boldsymbol{h} = \frac{\partial \overline{W}}{\partial \boldsymbol{\tau}}. \tag{4}$$

This renders the meaning of the complementary potential $\overline{W}$ clear. Since the latter relies on the plastic work κ, it may be also clear that the elastic behavior will be different for different cases of unloading, as is the case in the Mullins effect.

2.3 Normality Flow Rule

The plastic part $\boldsymbol{D}^p$ is governed by the following normality rule (Bruhns et al. 2003, 2005 and Xiao et al. 2007):

$$\boldsymbol{D}^p = \frac{\xi}{\hat{h}} \left(\frac{\partial f}{\partial \boldsymbol{\tau}} : \overset{\circ}{\boldsymbol{\tau}}{}^{\log} \right) \frac{\partial f}{\partial \boldsymbol{\tau}}. \tag{5}$$

Here, the yield function f is of von Mises type:

$$f = \frac{1}{2} \mathrm{tr} \tilde{\boldsymbol{\tau}}^2 - \frac{1}{3} r^2. \tag{6}$$

A simple, direct approach will be proposed to treat tension–compression asymmetry, as will be shown in the next section. In the above, the yield limit r relies on the plastic work κ and a well-designed stress invariant, χ, and is taken to be of the following form:

$$r = r(\kappa, \chi) = \frac{1}{2}(r_+ + r^-) + \frac{1}{2} \mathrm{sgn}(\chi)(r^+ - r^-), \tag{7}$$

where

$$\dot{\kappa} = \boldsymbol{\tau} : \boldsymbol{D}^p. \tag{8}$$

In the above, the stress invariant χ is introduced for the purpose of characterizing the tension–compression asymmetry and in this sense the r^+ and r^- are the yield limits in the cases of tension and compression, respectively, and both are dependent merely on the plastic work κ, namely,

$$\chi = \sqrt{6}\frac{J_3}{J_2^{1.5}}, \tag{9}$$

and

$$r^+ = r^+(\kappa), \quad r^- = r^-(\kappa). \tag{10}$$

Here and henceforward, J_2 and J_3 are used to signify the second and third basic invariants of the deviatoric Kirchhoff stress $\tilde{\boldsymbol{\tau}}$, viz.,

$$J_2 = \mathrm{tr}\tilde{\boldsymbol{\tau}}^2, \quad J_3 = \mathrm{tr}\tilde{\boldsymbol{\tau}}^3. \tag{11}$$

It may readily be shown that $-1 \le \chi \le 1$. In particular,

$$\chi = \begin{cases} +1 & \text{for uniaxial tension,} \\ -1 & \text{for uniaxial compression.} \end{cases} \tag{12}$$

Moreover, the ξ in Eq. (5) is the plastic indicator taking values 1 and 0 for the loading and unloading cases (cf. Bruhns et al. 2003; Xiao et al. 2007), viz.,

$$\xi = \begin{cases} 1 & \text{for} \quad \left[f = 0, \ \frac{1}{\hat{h}}\frac{\partial f}{\partial \boldsymbol{\tau}} : \overset{\circ}{\boldsymbol{\tau}}{}^{\log} \ge 0 \right], \\ 0 & \text{for} \quad f < 0 \quad \text{or} \quad \left[f = 0, \ \frac{1}{\hat{h}}\frac{\partial f}{\partial \boldsymbol{\tau}} : \overset{\circ}{\boldsymbol{\tau}}{}^{\log} \le 0 \right]. \end{cases} \tag{13}$$

The plastic modulus $\hat{h}$ may be derived from the plastic consistency condition $\dot{f} = 0$ and given by

$$\hat{h} = \frac{4}{9} r^3 r' \tag{14}$$

with

$$r' = \frac{\partial r}{\partial \kappa}. \tag{15}$$

2.4 On the Thermodynamic Consistency and Others

Since pronounced intrinsic dissipation is a typical feature of inelastic deformations, a physically sound model for inelastic behavior should be placed on the ground of thermodynamic principles. In continuum thermodynamics, the second law is formulated by the Clausius–Duhem inequality with restrictions on the entropy production in every possible thermodynamic process. In a sense of thermodynamic consistency, a constitutive model should identically fulfill the Clausius–Duhem inequality. Most recently, the thermodynamic consistency of Eulerian elastoplastic rate models has been demonstrated (Xiao et al. 2007 and Xiao 2013, 2014a, b, 2015) in an explicit sense of presenting both the specific free energy function and the specific entropy function which identically meet the Clausius–Duhem inequality. Following the same procedures proposed in these recent studies, results may be derived for the model proposed here. Detail is not pursued here and may be referred to the aforementioned references.

A J_2-flow model with nonlinear combined hardening has been established most recently (Xiao 2015) for incompressible deformations. The model proposed here is for general compressible deformations but simpler in the sense that nonlinear anisotropic hardening is not incorporated. It will be demonstrated that this simplified model in a broader case may provide good simulation of the complicated stress softening behavior of soft solids displaying the Mullins effects.

The deformation features of the proposed model are characterized by two constitutive quantities, namely, the evolving elastic potential $\overline{W}$ (cf., Eq. (3)) and the yield limit r (cf., Eq. (7)). The form of the potential $\overline{W}$ should be well chosen, so that large rubberlike elastic deformation behavior may be characterized at unloading. Furthermore, the yield limit r should be such that it joins the potential $\overline{W}$ to characterize the inelastic deformation behavior. Detail will be given in the next section.

3 Explicit Procedures for Determining Constitutive Quantities

When the two constitutive quantities, namely, the potential $\overline{W}$ and the yield limit r, are given for general multiaxial cases, the proposed model is ready for use in treating various multiaxial deformation problems. Except for some degenerate cases, however, it might be far from being trivial to obtain suitable forms of the constitutive functions in a model. Usually, specific forms of constitutive functions with a number of unknown parameters have to be chosen on an ad hoc basis. To validate the applicability of chosen forms of constitutive functions, it should be known whether or not suitable values of these unknown parameters may be found by means of trial-and-error procedures, in a sense of fitting adequate test data as closely as possible. For given trial values of the parameters at each step, the response for each process of

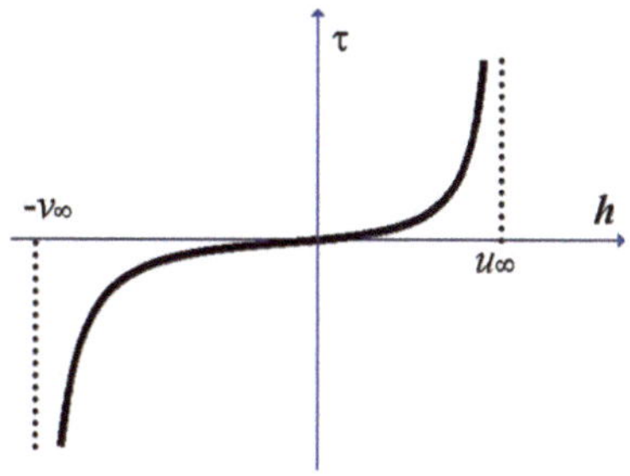

Fig. 1 Uniaxial strain-stiffening curve at monotone loading

deformation should be determined from the model by means of complicated numerical procedures.

For soft solids, usually test data for uniaxial responses are given both for the monotone loading case and for various cases of unloading, as shown in Fig. 1 for the loading curve and in Fig. 2 for unloading curves. Responses from model predictions for given trial values of parameters should be compared with test data for the purpose of model validation. For a reasonable model, by means of trial-and-error procedures of searching for suitable forms of the constitutive functions and parameters, reasonable accord with test data may be eventually achieved after iterative numerical procedures. Since both the loading curve and each unloading curve display complicated nonlinearities coupled with large strain and, in particular, a number of unloading curves should be taken into consideration and any two of them may be of different shapes, usual implicit procedures for both searching for reasonable forms of constitutive functions and identifying a set of unknown parameters would be unduly complicated.

It will be demonstrated in this section that, for the purpose of representing the stress softening behavior of a soft solid material, explicit procedures may be introduced to obtain the two constitutive functions in the proposed model, in the sense that the predictions from the proposed model can automatically, accurately fit any given sets of data both for the uniaxial monotone loading case and for various unloading cases.

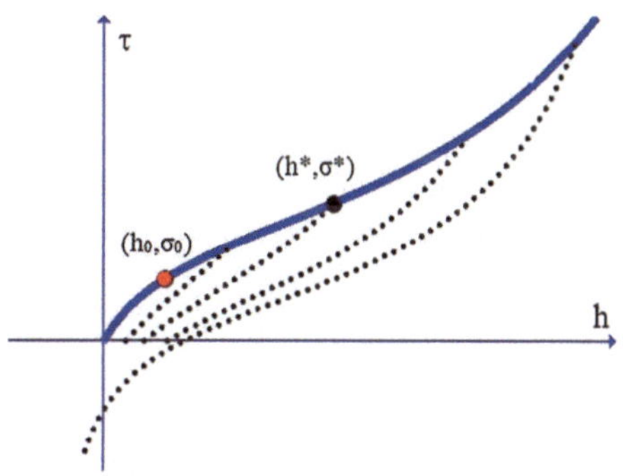

Fig. 2 Uniaxial unloading curves displaying the stress softening effect

3.1 Uniaxial Stress–Strain Curves at Loading and Unloading

As shown in Fig. 1, the uniaxial stress–strain curve of a soft solid material under monotone loading is of S-shape. Here and henceforward, τ and h are used to represent the axial Kirchhoff stress and the axial Hencky strain, respectively. Two noticeable features of such a curve may be noticed, namely, (i) the stress τ may grow very rapidly as the strain h is approaching either a *compression limit*, $-v_\infty$, or an *extension limit*, u_∞, and (ii) the stress response may display pronounced tension–compression asymmetry. The former arises from limiting extensibility of long chain-like macromolecules and is known as the strain-stiffening effect.

On the other hand, the well-known Mullins effect is in association with the stress softening at unloading. When a sample is loaded to a certain point (h^*, τ^*) and then unloaded, there emerges the stress softening effect as described below (cf., e.g., Xiao 2015):

(i) After the unloading is completed, nearly the whole of the strain may be recovered, but, nevertheless, a very small residual strain (permanent set) may be induced. The stress–strain curve in an unloading process may appreciably deviate from the monotone loading curve up to the unloading point (h^*, τ^*) at the onset of unloading;

(ii) for each strain $h \in [0, h^*)$, the stress on the unloading curve is always lower than the stress on the monotone loading curve, referred to as stress softening at unloading;

(iii) when reloading is introduced, the part of the reloading curve till the point (h^*, τ^*) is nearly coincident with the unloading curve in the foregoing, but the part of the reloading curve after the point (h^*, τ^*) follows the monotone loading curve after a transition at the point (h^*, τ^*);

(iv) and, finally, the larger the strain h^* at the unloading point is, the more pronounced the stress softening effect becomes.

As shown in Fig. 2, a family of unloading curves is associated with different unloading points (h^*, τ^*) (the loading curve is shown only for the tension case). An unloading curve covering both compression and tension is sketchily shown in Fig. 2. It should be noted that the stress softening behavior under tension may differ in nature from that under compression, namely, the unloading curve need not be symmetric under tension and compression.

For a soft material sample of interest, test data are given for the loading curve and for a certain number of unloading curves. For the latter, consider M data sets for the unloading curves associated with M unloading stresses $\tau_1^* < \tau_2^* < \cdots < \tau_M^*$, which correspond to M values of the plastic work below: $\kappa_1 < \kappa_2 < \cdots < \kappa_M$. Specifically, for the unloading curve associated with the unloading stress τ_s^* or the plastic work κ_s for each $s \in \{0, 1, \ldots, M\}$, a set of strain–stress data is given below:

$$(h_0^{(s)}, \tau_0^{(s)}), (h_1^{(s)}, \tau_1^{(s)}), \ldots, (h_{N_s}^{(s)}, \tau_{N_s}^{(s)}). \tag{16}$$

In the above, the first point $(h_0^{(s)}, \tau_0^{(s)})$ (cf., Fig. 2) is just given by the strain h_s^* and the stress τ_s^* at the onset of unloading, and, moreover, the residual strain (permanent set) at vanishing stress, i.e., the point $(h_s^p, 0)$, should be incorporated in the above set of data.

Given a data set for the loading curve and data sets for a certain number of unloading curves, as described in the above, the main effort is to find out the evolving elastic potential $\overline{W}$ (cf., Eq. (3)) and the yield limit r (cf., Eq. (7)) such that the model established in Sect. 2 fits these data as closely as possible. In what follows, we are going to show that explicit procedures may be introduced to obtain these two constitutive quantities in a sense of accurately matching test data for any given shapes of loading and unloading curves.

3.2 Multiaxial Elastic Potential in Explicit Form

We first present the evolving elastic potential $\overline{W}$. The main procedures will follow those suggested in a most recent study (Xiao 2015).

A single-variable function for the unloading curve associated with the unloading point (h_s^*, τ_s^*) or the plastic work κ_s may be obtained in a sense of accurately fitting any given date set as shown in Eq. (16). Indeed, it may be given by an interpolating function, such as a Lagrange interpolant and an other form of function. Let such a function be designated by $h = g_s(\tau)$. Then, we have

$$g_s(0) = h_s^p, \quad g_s(\tau_\alpha) = h_\alpha, \quad \alpha = 0, 1, \ldots, N_s. \tag{17}$$

It may readily be shown that in the uniaxial case, the evolving potential Eq. (3) reduces to a function of the axial Kirchhoff stress τ and the plastic work κ, denoted as

$$\overline{w} = \overline{w}(\tau, \kappa). \tag{18}$$

For the unloading curve associated with each unloading point (h_s^*, τ_s^*) or each plastic work κ_s, the elastic rate equation (2) yields

$$\dot{h} = \overline{\left(\frac{\partial \overline{w}}{\partial \tau}\right)}^{\bullet}.$$

For the unloading curve with number s, the plastic work is held fixed and given by κ_s. Then, the integration of this equation from the point $(h_s^p, 0)$ to the point (h, τ) in an unloading process gives rise to

$$h - h_s^p = \frac{d\overline{w}_s}{d\tau}. \tag{19}$$

In the above, we denote $\overline{w}_s = \overline{w}(\tau, \kappa_s)$. Moreover, the following property is assumed:

$$\left.\frac{\partial \overline{w}}{\partial \tau}\right|_{\tau=0} = 0. \tag{20}$$

From Eqs. (19) and (27) we deduce

$$\frac{d\overline{w}_s}{d\tau} = g_s(\tau) - h_s^p. \tag{21}$$

Hence we obtain

$$\overline{w}_s = \overline{w}_s(\tau) = \int_0^\tau \left(g_s(\tau) - h_s^p\right) d\tau. \tag{22}$$

From the above procedures we obtain M uniaxial potentials $\overline{w}_1, \ldots, \overline{w}_M$ for the M unloading curves at issue. However, each of them is restricted to the uniaxial case. Now we show that, by means of the multiaxial invariants χ (cf. Eq. (9)) and the three basic stress invariants, the sought multiaxial potential $\overline{W}$ (cf., Eq. (3)) is obtainable from the foregoing M uniaxial potentials. To this end, a crucial step is to extend each uniaxial potential $w_s(\tau)$ to a multiaxial potential below:

$$\overline{W}_s = \frac{1}{3}(1-2\nu)\overline{w}_s(I) + \frac{2}{3}(1+\nu)\overline{w}_s\left(\text{sgn}(\chi)\sqrt{1.5J_2}\right), \tag{23}$$

where

$$I = \text{tr}\boldsymbol{\tau} \tag{24}$$

is the first stress invariant. In the above, $\nu > 0$ is a parameter characterizing the compressibility effect and may be referred to as generalized Poisson ratio. Detail in this respect may be found in Yuan et al. (2015). In particular, the incompressibility case is incorporated as a special case when $\nu = 0.5$. It may be verified that, in the uniaxial case, each function $\overline{W}_s$ for general multiaxial compressible deformations, given by Eq. (23), exactly supplies its uniaxial counterpart $\overline{w}_s$.

We then present the sought multiaxial potential $\overline{W}(\boldsymbol{\tau}, \kappa)$ in unified form which exactly gives the M multiaxial potentials $\overline{W}_1, \ldots, \overline{W}_M$ for the M plastic works $\kappa_1, \ldots, \kappa_M$. Such a potential may be given by an interpolating function with the M interpolating nodes $(\kappa_s, \overline{W}_s)$ for $s = 1, \ldots, M$. For instance, it may be given by a Lagrange interpolant below:

$$\overline{W} = \overline{W}(\boldsymbol{\tau}, \kappa) = \sum_{s=1}^{M} \overline{W}_s\, L_s(\kappa) \tag{25}$$

with the Lagrangian base functions

$$L_s(\kappa) = \frac{P(\kappa)}{(\kappa - \kappa_s)P'(\kappa_s)}, \quad s = 1, \ldots, M, \tag{26}$$

$$P(\kappa) = \prod_{s=1}^{M} (\kappa - \kappa_s). \tag{27}$$

From Eqs. (12) and (23)–(27) it may be concluded that, in the uniaxial case, the evolving potential Eq. (25) with Eqs. (22)–(44), (26) and (27) exactly supplies the M uniaxial potentials $\overline{w}_1(\tau), \ldots, \overline{w}_M(\tau)$ given by Eq. (22) and, therefore, can exactly fit any given M sets of data for M unloading curves (cf. Eq. (16), (17)). In fact, for the uniaxial case with the uniaxial stress τ, by utilizing Eq. (12) and $1.5J_2 = \tau^2$ we deduce

$$\overline{W}_s = \begin{cases} \overline{w}_s(|\tau|) = \overline{w}_s(\tau) & \text{for uniaxial tension } \tau \geq 0, \\ \overline{w}_s(-|\tau|) = \overline{w}_s(\tau) & \text{for uniaxial compression } \tau \leq 0. \end{cases} \tag{28}$$

From this and Eqs. (25)–(27), we then infer that $\overline{W} = \overline{w}_s(\tau)$ for the uniaxial case at $\kappa = \kappa_s$. Thus, it follows from Eqs. (22) to (27) that the multiaxial potential $\overline{W}$ given can automatically reproduce each unloading curve.

The above results show that, with single-variable functions for unloading curves, as shown by Eq. (17), the multiaxial potential $\overline{W}$ may be determined by the explicit, direct procedures proposed. With sufficient test data for a certain number of unloading curves, each such single-variable function may be given directly by an interpolating function or by choosing an other suitable form of function. It is worthwhile to point out that no adjustable parameters need be identified in the former case, while parameters of physical meanings may be introduced and readily determined in the latter case, as will be shown in subsequent development.

3.3 *Determination of the Yield Limit*

Next we are going to explain how to determine the stress limit r. According to Eq. (7), it needs to determine the yield limits r^+ and r^- for the tension and compression cases. To this end, we first treat the uniaxial extension case for the loading case. In this case, the yield limit is just r^+ (cf., Eqs. (7) and (10)). Let a set of data be given for a monotone uniaxial stress–strain curve starting at the initial yield point (h_0^+, τ_0^+) at extension (cf. Fig. 2). Then, a single-variable function for this curve may be given also by an interpolating function or an other form of function, denoted as

$$\begin{cases} h = f^+(\tau), & \tau \geq \tau_0^+, \\ h_0 = f^+(\tau_0^+), \end{cases} \tag{29}$$

In a process of monotone loading in the uniaxial extension, the following relations may be derived (cf., Xiao 2015):

$$\begin{cases} \tau = r^{+} = r^{+}(\kappa), \\ \frac{d}{d\kappa}\left(f^{+}(\tau) - \frac{\partial \overline{w}}{\partial \tau}\right) = \frac{1}{\tau}, \end{cases} \tag{30}$$

where $\overline{w} = \overline{w}(\tau, \kappa)$ is the elastic potential for uniaxial extension and given by

$$\overline{w} = \sum_{s=1}^{M} \overline{w}_s(\tau) L_s(\tilde{\kappa}). \tag{31}$$

where $\overline{w}_s(\tau)$ and $L_s(\kappa)$ are given by Eqs. (22) and (26), (27). Thus, by working out Eq. $(30)_2$ we infer that the stress limit $r^{+} = \tau$ is determined by the following first-order differential equation:

$$\frac{d\tau}{d\kappa} = \frac{\frac{df^{+}}{d\tau} - \frac{\partial^2 \overline{w}}{\partial \tau^2}}{\frac{1}{\tau} + \frac{\partial^2 \overline{w}}{\partial \tau \partial \kappa}} \tag{32}$$

with $\tau|_{\kappa=0} = \tau_0^{+}$.

Results for the compression case may be derived following the same procedures in the above. The only difference is that Eq. $(30)_1$ should be replaced by $\tau = -r^{-}$. Let $f^{-}(\tau)$ be a function representing the monotone loading curve in the uniaxial compression case, namely,

$$\begin{cases} h = f^{-}(\tau), \quad \tau \leq \tau_0^{-} < 0, \\ h_0^{-} = f^{-}(\tau_0^{-}), \end{cases} \tag{33}$$

with the initial yield point (h_0^{-}, τ_0^{-}) under compression. Given test data, the above function may be given by an interpolating function or an other form of function, as in the monotone tension case. Then, the yield limit $r^{-} = -\tau$ in the compression case may be determined by

$$\frac{d\tau}{d\kappa} = \frac{\frac{df^{-}}{d\tau} - \frac{\partial^2 \overline{w}}{\partial \tau^2}}{\frac{1}{\tau} + \frac{\partial^2 \overline{w}}{\partial \tau \partial \kappa}} \tag{34}$$

with $\tau|_{\kappa=0} = \tau_0^{-}$.

3.4 Shape Functions for Uniaxial Loading and Unloading

In the above results in explicit forms, it needs to input the two shape functions $f^{\pm}(\tau)$ for the two monotone loading curves in both cases of extension and compression. Either of them may be directly given by interpolating functions, which is also the case for the shape functions $g_s(\tau)$ (cf., Eq. (17)) for each unloading curve. As such, any given data may be exactly fitted without involving usual procedures of identifying unknown parameters. As mentioned before, these shape functions may also be given by other forms of functions. Sometimes, simple forms of shape functions with parameters of physical meanings may be instrumental.

Soft solid materials display two noticeable deformation features, namely, the strain-stiffening effect and the tension–compression asymmetry, as indicated in Sect. 3.1. It has been found (Xiao 2012) that a simple form of shape function with strain limits may accurately characterize complex nonlinearities of rubberlike elasticity with tension–compression asymmetry. Here, we are going to show that a new, simpler form may be introduced to represent each elastic unloading curve and given below:

$$\tau = \frac{(1-\gamma^+)E(h-h^p)}{1-\frac{h-h^p}{u}} + \gamma^+ E(h-h^p) \quad \text{for} \quad h-h^p \geq 0 \ (\tau \geq 0), \tag{35}$$

and

$$\tau = \frac{(1-\gamma^-)E(h-h^p)}{1+\frac{h-h^p}{v}} + \gamma^- E(h-h^p) \quad \text{for} \quad h-h^p \leq 0 \ (\tau \leq 0). \tag{36}$$

Hence, we havc

$$h - h^p = g(\tau) = \frac{1}{2}(g^+ + g^-) + \frac{1}{2}\text{sgn}(\tau)(g^+ - g^-), \tag{37}$$

where g^+ and g^- are given by inverting the two functions given by Eqs. (35)–(36) and of the forms below:

$$\begin{cases} g^+ = \dfrac{2u\tau}{\tau + Eu + \sqrt{(\tau+Eu)^2 - 4\gamma^+ Eu\tau}}, \\ g^- = \dfrac{-2v\tau}{\tau - Ev - \sqrt{(\tau-Ev)^2 + 4\gamma^- Ev\tau}}. \end{cases} \tag{38}$$

In the above, h^p is the plastic strain after unloading, E is the Young's modulus at infinitesimal strain, u and v are strain limits in extension and compression, as shown in Fig. 1, and $\gamma^+ < 1$ and $\gamma^- < 1$ are dimensionless factors characterizing the growth degree of stress at the strain limits. Each of them may rely on the plastic work κ, viz.,

$$E = E(\kappa), \quad u = u(\kappa), \quad v = v(\kappa), \quad \gamma^{\pm} = \gamma^{\pm}(\kappa). \tag{39}$$

For each unloading curve associated with a value of the plastic work, values of the above parameters may be determined by fitting the two functions given by Eqs. (37)–(38) to test data. As such, a set of parameter values may be obtained for a family of unloading curves. Then, each of the foregoing κ-dependent parameter may also be given directly by an interpolating function. It may be noted that the parameters introduced directly represent the very features of each elastic unloading curve at issue, as indicated in Fig. 1, and, accordingly, are determinable from each unloading curve in an uncoupled manner.

In particular, the unloading curve starting at the initial yield points $(h_0^\pm, \tau_0^\pm)$ is associated with $\kappa = 0$ and coincident exactly with the initial elastic curve linking the just mentioned two initial yield points. This curve may be given simply by a linear relation between the axial stress τ and the axial Hencky strain h, namely,

$$\begin{cases} h = g_0(\tau) = \tau/E_0, \ \tau_0^- \le \tau \le \tau_0^+, \\ \kappa = 0, \end{cases} \tag{40}$$

where $E_0 = E(0)$ is the initial Young's modulus.

On the other hand, the two monotone loading curves (cf. Eqs. (29) and (33)) under tension and compression start at the initial yield points $(h_0^\pm, \tau_0^\pm)$ and may also be given by the same form of function, as shown below:

$$\tau = \tau_0^+ + E_0^+(h - h_0^+)\left(\frac{1-m}{1-\frac{h-h_0^+}{u_\infty}} + m\right), \tag{41}$$

for the monotone loading curve under extension, and

$$\tau = \tau_0^- + E_0^-(h - h_0^-)\left(\frac{1-n}{1+\frac{h-h_0^-}{v_\infty}} + n\right), \tag{42}$$

for the monotone loading curve under compression. Here, $E_0^\pm$ are the two slopes at the two initial yield points $(h_0^\pm, \tau_0^\pm)$ (cf. Fig. 2) under tension and compression, $u_\infty > 0$ and $v_\infty > 0$ are the extension limit and the compression limit under monotone loading, as shown in Fig. 1, and $m < 1$ and $n < 1$ are two dimensionless factors.

From the inverted forms of the last two functions we obtain

$$h = f^+(\tau) = h_0^+ + \frac{2(\tau - \tau_0^+)}{E_0^+ + \frac{\tau-\tau_0^+}{u_\infty} + \sqrt{(E_0^+ + \frac{\tau-\tau_0^+}{u_\infty})^2 - 4mE_0^+\frac{\tau-\tau_0^+}{u_\infty}}} \tag{43}$$

and

$$h = f^-(\tau) = h_0^- + \frac{2(\tau - \tau_0^-)}{E_0^- - \frac{\tau-\tau_0^-}{v_\infty} + \sqrt{(E_0^- - \frac{\tau-\tau_0^-}{v_\infty})^2 + 4nE_0^-\frac{\tau-\tau_0^-}{v_\infty}}}. \tag{44}$$

4 Numerical Examples

With the proposed model in which the constitutive functions are given in explicit forms, as established in Sects. 2 and 3, an explicit, accurate simulation of the stress softening effect for any given test data for both monotone loading behavior and unloading behavior may be achieved. Toward this objective, it is sufficient to specify the two shape functions $f^{\pm}(\tau)$ (cf. Eqs. (43)–(44)) for the monotone loading case and a family of shape functions, $g_s^{\pm}(\tau)$ (cf. Eq. (38)), for various unloading cases. As indicated before, it may be straightforward to present these shape functions by usual interpolating procedures. However, they may be given by other forms of functions. As illustrative example, in this section the simple forms of shape functions given in Sect. 3.4 will be used to fit test data in literature.

In what follows, test data will be fitted using the initial elastic curve equation (40) and the monotone loading curve equation (41) for the tension case, as well as the unloading curve equation (35) in the tension case. Here the compression case will not be involved, since no test data for this case are given in the treated examples. The values of the parameters incorporated in the equations just mentioned will be determined following the scheme explained in Xiao (2015).

Model predictions will be compared with two sets of data in literature. In the two examples below, the generalized Poisson ratio is taken to be as follows: $\nu = 0.4999$. Since now the model is free of the incompressibility constraint, the incompressibility condition is rendered irrelevant.

4.1 *Simulation of Data from Dorfman and Ogden (2004)*

We first consider the data provided by Dorfman and Ogden (2004). These data are concerned with a particle-reinforced compound with 60 phr of carbon black for the

Fig. 3 Comparing model predictions with test data by Dorfman and Ogden (2004) (the initial yield stress is indicated by *circle* o)

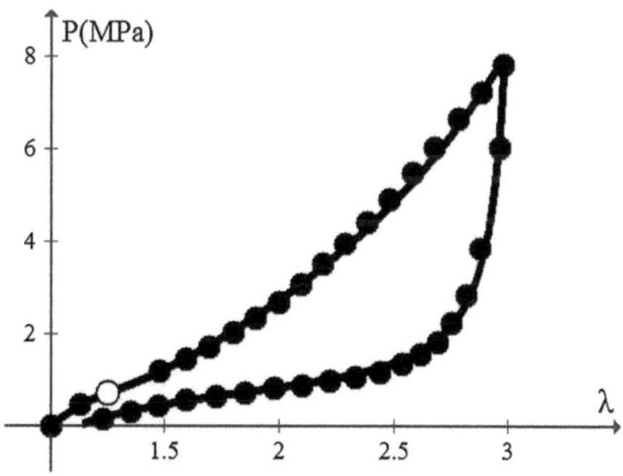

monotone loading case till the stretch $\lambda = 3$ and then for the unloading case starting at this stretch.

For the monotone loading case, results are almost the same as those in Xiao (2015). The shape function for the monotone loading curve is given by Eq. (41) with the parameter values below: $h_0^+ = \ln 1.248$, $\tau_0^+ = 0.905\,\text{MPa}$, $E_0^+ = 3.3\,\text{MPa}$, $m = -2.3$, $u_\infty = \ln 3.6$. For the unloading case, the unloading curve is presented by Eq. (35) with $h^p = \ln 1.12$, $E = 2.2\,\text{MPa}$, $u = \ln(3.05/1.12)$, $\gamma^+ = 0.77$. Moreover, the initial elastic curve till the initial yield point (h_0^+, τ_0^+) is given by Eq. (40) with $h_0^+ = \ln 1.128$, $\tau_0^+ = 0.905\,\text{MPa}$, $E_0 = 4.08\,\text{MPa}$. Results are shown in Fig. 3 by plotting the curves of the axial nominal stress P against the axial stretch λ. Accurate simulation is achieved.

4.2 *Simulation of the Data from Mullins and Tobin (1957)*

We next consider the data from Mullins and Tobin (1957) for a vulcanizate sample, which are concerned with the monotone loading case till the stretch $\lambda = 5$ and the unloading cases starting at $\lambda = 3, 4, 5$, separately.

For the initial elastic curve (cf., Eq. (40)) till the initial yield point (h_0^+, τ_0^+), the relevant parameter values are given by $E_0 = 38.086\,\text{MPa}$, $h_0^+ = \ln 1.541$, $\tau_0^+ = 16.470\,\text{MPa}$, Moreover, for the monotone loading curve starting at the foregoing initial yield point is simulated by Eq. (41) with $h_0^+ = \ln 1.541$, $\tau_0^+ = 16.470\,\text{MPa}$, $E_0^+ = 32.8\,\text{MPa}$, $m = -3.66$, $u_\infty = \ln 5.18$. Finally, the three unloading curves are simulated by applying Eq. (35) with the parameter values listed in Table 1.

Results are shown in Fig. 4 by plotting curves of the axial nominal stress P against the axial engineering strain $\lambda - 1$. Again, accurate simulation is achieved.

Table 1 Parameter values for three unloading curves

Parameters	E(MPa)	u	h_p	γ
1st curve	25.00	$\ln \dfrac{3.566}{1.14}$	1.14	0.42
2nd curve	23.00	$\ln \dfrac{4.5}{1.24}$	1.24	0.24
3nd curve	20.00	$\ln \dfrac{5.35}{1.3}$	1.3	0.1

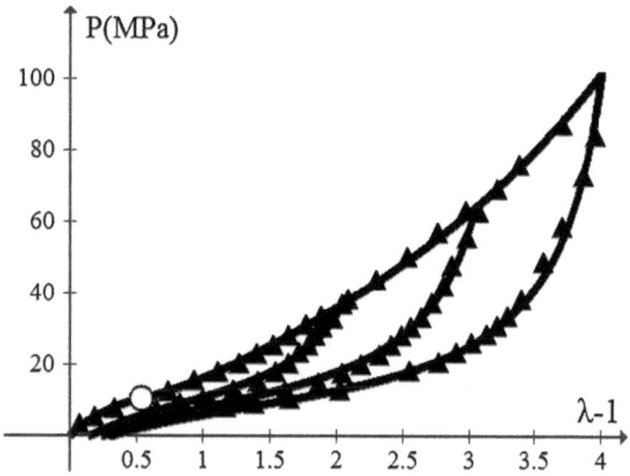

Fig. 4 Comparing model predictions with test data by Mullins and Tobin (1957) (the initial yield point is indicated by *circle* o)

5 Concluding Remarks

New J_2-flow elastoplasticity models with evolving rubberlike elasticity has been proposed to achieve an explicit, accurate simulation of complicated inelastic behavior of soft solid materials displaying the Mullins effect. Explicit, direct procedures arc proposed to obtain the constitutive quantities incorporated in the proposed model from suitable uniaxial data. It has been demonstrated that the constitutive quantities may be determined for general multiaxial deformations and, in the meantime, extensive test data for any given shapes of loading and unloading curves may be accurately, automatically fitted for the first time. Results are applicable for a broad case of tension–compression asymmetry. In particular, it has been demonstrated that all may be done in the scope of classical elastoplasticity and, therefore, that no additional variables need be introduced and treated. As a consequence, not only usual implicit, complicated procedures both in searching for suitable forms of constitutive functions and in identifying values of unknown parameters may be bypassed, but also, in the meantime, an explicit, accurate simulation may be achieved for any given sets of test data.

This study represents a development of the model suggested in a most recent study (Xiao 2015), in the sense that simpler results are derived in a sense free of the usually assumed incompressibility constraint. However, rate effects and fatigue effects under cyclic loading are not treated and will be studied by developing the free, smooth elastoplasticity models in (Xiao et al. 2014) and (Xiao 2014b). These respects will be studied in elsewhere.

Acknowledgments This work was carried out under joint support of the funds from Natural Science Foundation of China (No.: 11372172; No.: 11542020) and from the 211-project launched by the Education Committee of China through Shanghai University (No.: S.15-0303-15-208) and from the science and technology development project lauched by Weifang municipal government (No.: 2015GX018).

References

Ayoub, G., Zaïri, F., Naït-Abdelaziz, M., Gloaguen, J.M.: Modelling large deformation behaviour under loading-unloading of semi-crystalline polymers: application to a high density polyethylene. Int. J. Plast. **26**, 329–347 (2010)

Beatty, M.F., Krishnaswamy, S.: A theory of stress-softening in incompressible isotropic materials. J. Mech. Phys. Solids **48**, 1931–1965 (2000)

Besdo, D., Ihlemann, J.: Properties of rubberlike materials under large deformations explained by self organizing linkage patterns. Int. J. Plast. **19**, 1001–1018 (2003a)

Besdo, D., Ihlemann, J.: A phenomenological constitutive model for rubberlike materials and its numerical applications. Int. J. Plast. **19**, 1019–1036 (2003b)

Bruhns, O.T., Xiao, H., Meyers, A.: Self-consistent Eulerian rate type elastoplasticity models based on the logarithmic stress rate. Int. J. Plast. **15**, 479–520 (1999)

Bruhns, O.T., Xiao, H., Meyers, A.: Some basic issues in traditional Eulerian formulations of finite elastoplasticity. Int. J. Plast. **19**, 2007–2026 (2003)

Bruhns, O.T., Xiao, H., Meyers, A.: A weakened form of Ilyushin's postulate and the structure of self-consistent Eulerian finite elastoplasticity. Int. J. Plast. **21**, 199–219 (2005)

Chagnon, G., Verron, E., Gornet, L., Marckmann, G., Charrier, P.: On the relevance of continuum damage mechanics as applied to the Mullins effect in elastomers. J. Mech. Phys. Solids **52**, 1627–1650 (2004)

Dargazany, R., Khiêm, V.N., Itskov, M.: A generalized network decomposition model for the quasi-static inelastic behavior of filled elastomers. Int. J. Plast. **63**, 94–109 (2014)

De Tommasi, D., Puglisi, G., Saccomandi, G.: A micromechanics based model for the Mullins effect. J. Rheol. **50**, 495–512 (2006)

Diani, J.: A review on the Mullins effect. Eur. Polym. J. **45**, 601–612 (2009)

Diani, J., Brieu, M., Gilormini, P.: Observation and modeling of anisotropic visco-hyperelastic behavior of a rubberlike material. Int. J. Solids Struct. **43**, 3044–3056 (2006a)

Diani, J., Brieu, M., Vacherand, J.M.: A damage directional constitutive model for Mullins effect with permanent set and induced anisotropy. Eur. J. Mech. A/Solids **25**, 483–496 (2006b)

Dorfman, A., Ogden, R.W.: A constitutive model for the Mullins effect with permanent set in particle-reinforced rubber. Int. J. Solids Struct. **41**, 1855–1878 (2004)

Drozdov, A.D., Dorfman, A.: A micro-mechanical model for the response of filled elastomers at finite strain. Int. J. Plast. **19**, 1037–1067 (2003)

Göktepe, S., Miehe, Ch.: A micro-macro approach to rubber-like materials. Part III: the micro-sphere model of anisotropic Mullins type damage. J. Mech. Phys. Solids **53**, 2259–2283 (2005)

Govindjee, S., Simo, J.: A micro-mechanical continuum model for carbon black filled rubbers incorporating Mullins effect. J. Mech. Phys. Solids **39**, 87–112 (1991)

Itskov, M., Ehret, A., Kazakevičiute-Makovska, R., Weinhold, G.: A thermodynamically consistent phenomenological model of the anisotropic Mullins effect. ZAMM-J. Appl. Math. Mech. **90**, 370–386 (2010)

Laiarinandrasana, L., Piques, R.: Visco-hyperelastic model with internal state variable coupled with discontinuous damage concept under total Lagrangian formulation. Int. J. Plast. **19**, 977–1000 (2003)

Li, J., Mayau, D., Lagarrigue, V.: A constitutive model dealing with damage due to cavity growth and the Mullins effect in rubber-like materials under triaxial loading. J. Mech. Phys. Solids **56**, 953–973 (2008)

Lin, R.C., Schomburg, U.: A finite elastic-viscoelastic-elastoplastic material law with damage: theoretical and numerical aspects. Comput. Methods Appl. Mech. Eng. **192**, 1591–1627 (2003)

Lion, A.: A constitutive model for carbon black filled rubber: experimental investigations and mathematical representation. Contin. Mech. Thermodyn. **8**, 153–169 (1996)

Marckmann, G., Verron, E., Goornet, L., Chagnon, G., Charrier, P., Fort, P.: A theory of network alteration for the Mullins effect. J. Mech. Phys. Solids **50**, 2911–2028 (2002)

Miehe, Ch.: Discontinuous and continuous damage evolution in Ogden-type large-strain elastic materials. Eur. J. Mech. A/Solids **14**, 697–720 (1995)

Miehe, Ch., Keck, J.: Superimposed finite elastic-viscoelastic-plastoelastic stress response with damage in filled rubbery polymers. Experiments, modeling and algorithmic implementation. J. Mech. Phys. Solids **48**, 323–365 (2000)

Mullins, L., Tobin, N.R.: Theoretical model for the elastic behaviour of filler-reinforced vulcanized rubbers. Rubber Chem. Technol. **30**, 551–571 (1957)

Ogden, R.W., Roxburg, D.G.: A pseudo-elastic model for the Mullins effect in filled rubber. Proc. R. Soc. Lond. A **455**, 2861–2877 (1999)

Qi, H.J., Boyce, M.C.: Constitutive model for stretch-induced softening of the stress-stretch behavior of elastomeric materials. J. Mech. Phys. Solids **52**, 2187–2205 (2004)

Simo, J.: On a fully three-dimensional finite-strain viscoelastic damage model: formulation and computational aspects. Comput. Methods Appl. Mech. Eng. **60**, 153–173 (1987)

Simo, J., Hughes, T.J.R.: Computational Inelasticity. Academic Press, Boston (1998)

Xiao, H.: An explicit, direct approach to obtaining multi-axial elastic potentials that exactly match data of four benchmark tests for rubberlike materials-part 1: incompressible deformations. Acta Mechanica **223**, 2039–2063 (2012)

Xiao, H.: Pseudo-elastic hysteresis out of finite recoverable elastoplastic flows. Int. J. Plast. **41**, 82–96 (2013)

Xiao, H.: An explicit, straightforward approach to modeling SMA pseudo-elastic hysteresis. Int. J. Plast. **53**, 228–240 (2014a)

Xiao, H.: Thermo-coupled elastoplasticity model with asymptotic loss of the material strength. Int J. Plast. **63**, 211–228 (2014b)

Xiao, H.: A direct, explicit simulation of finite strain multiaxial inelastic behavior of polymeric solids. Int. J. Plast. **71**, 146–169 (2015)

Xiao, H., Bruhns, O.T., Meyers, A.: Elastoplasticity beyond small deformations. Acta Mechanica **182**, 31–111 (2006)

Xiao, H., Bruhns, O.T., Meyers, A.: Thermodynamic laws and consistent Eulerian formulations of finite elastoplasticity with thermal effects. J. Mech. Phys. Solids **55**, 338–365 (2007)

Xiao, H., Bruhns, O.T., Meyers, A.: Free rate-independent elastoplastic equations. ZAMM-J. Appl. Math. Mech. **94**, 461–476 (2014)

Yuan, L., Gu, Z.X., Yin, Z.N.: New compressible hyperelastic models for rubberlike matereials. Acta Mechanica **226**, 1182–1196 (2015)

Zaïri, F., Naït-Abdelaziz, M., Gloaguen, J.M., Lefebvre, J.M.: A physically-based constitutive model for anisotropic damage in rubber-toughened glassy polymers during finite deformation. Int. J. Plast. **27**, 25–51 (2011)

Effect of Geometric Dimension on the Dissipative Property of the Structures Consisting of NiTi Shape Memory Alloy Wires

Chao Yu, Guozheng Kang and Qianhua Kan

Abstract Effect of the geometric dimension of NiTi shape memory alloy (SMA) wires on the dissipative property of their structural components is predicted by a physical mechanism-based thermo-mechanically coupled constitutive model in this work. Two types of NiTi SMA structural components, i.e., the single-wire and multi-wire ones, are considered. The dissipative property of the component is measured by its accumulated dissipation energy obtained during cyclic deformation. The calculated results show that at low (lower than 1×10^{-5}/s), moderate (from 5×10^{-5}/s to 1.5×10^{-4}/s), and high strain rates (higher than 5×10^{-4}/s), the accumulated dissipation energy decreases, changes non-monotonically, and increases with the increasing number of wires, respectively.

1 Introduction

Owing to the solid-to-solid diffusionless thermo-elastic martensite transformation, NiTi SMA can exhibit super-elasticity, shape memory effect, high damping capacity, and excellent biocompatibility. Recently, the structures made by NiTi SMA wires had been often used as the energy absorber (Jani et al. 2014). To evaluate the design of structures, the dissipative property is one of the most important factors, which is determined by the deformation characteristics of NiTi SMA.

C. Yu · G. Kang (✉) · Q. Kan
Applied Mechanics and Structure Safety Key Laboratory of Sichuan Province,
School of Mechanics and Engineering, Southwest Jiaotong University, Chengdu 610031
, Sichuan, People's Republic of China
e-mail: guozhengkang@home.swjtu.edu.cn

C. Yu
e-mail: yuchao5045@163.com

Q. Kan
e-mail: qianhuakan@home.swjtu.edu.cn

K. Naumenko and M. Aßmus (eds.), *Advanced Methods of Continuum Mechanics for Materials and Structures*, Advanced Structured Materials 60,
DOI 10.1007/978-981-10-0959-4_13

where $\boldsymbol{\alpha}$ and ϑ_{r} are the second-ordered thermal expansion tensor and the reference temperature (hereafter, which is set as room temperature), respectively. $\boldsymbol{\mathcal{C}}$ is the isotropic fourth-ordered elastic tensor, and can be written as

$$\boldsymbol{\mathcal{C}} = \frac{E\nu}{(1-2\nu)(1+\nu)}\boldsymbol{\delta}\otimes\boldsymbol{\delta} + \frac{E}{(1+\nu)}\boldsymbol{\mathcal{I}} \tag{4}$$

$$I_{ijkl} = \frac{1}{2}\left(\delta_{il}\delta_{jk} + \delta_{ik}\delta_{jl}\right) \tag{5}$$

test, where E and ν are the elastic modulus and Poisson's ratio, respectively. $\boldsymbol{\delta}$ and $\boldsymbol{\mathcal{I}}$ are the second-ordered delta and fourth-ordered identity tensors, respectively.

The transformation strain rate and plastic strain rate can be written as

$$\dot{\boldsymbol{\varepsilon}}^{\mathrm{tr}} = g^{\mathrm{tr}}\boldsymbol{N}_{\mathrm{tr}}\,\dot{\xi} \tag{6}$$

$$\dot{\boldsymbol{\varepsilon}}^{p} = \boldsymbol{N}_{\mathrm{p}}\dot{\gamma} \tag{7}$$

where $\boldsymbol{N}_{\mathrm{tr}}$ and $\boldsymbol{N}_{\mathrm{p}}$ are the direction tensors of martensite transformation and plastic deformation, respectively. g^{tr} is the magnitude of transformation strain generated in full forward transformation. $\dot{\xi}$ and $\dot{\gamma}$ are the rate of martensite volume fraction and dislocation slipping, respectively. Referring to Lagoudas and Entchev (2004), $\boldsymbol{N}_{\mathrm{tr}}$ and $\boldsymbol{N}_{\mathrm{p}}$ can be given as follows:

$$\boldsymbol{N}_{\mathrm{tr}} = \begin{cases} \sqrt{\dfrac{3}{2}}\dfrac{\boldsymbol{\sigma}^{\mathrm{dev}} + \boldsymbol{B}}{\left\|\boldsymbol{\sigma}^{\mathrm{dev}} + \boldsymbol{B}\right\|} & \dot{\xi} > 0 \\ \dfrac{\boldsymbol{\varepsilon}^{\mathrm{tr}}_{\mathrm{recent}}}{g^{\mathrm{tr}}\xi_{\mathrm{recent}}} & \dot{\xi} < 0 \end{cases} \tag{8}$$

$$\boldsymbol{N}_{\mathrm{p}} = \sqrt{\frac{3}{2}}\frac{\boldsymbol{\sigma}^{\mathrm{dev}}}{\left\|\boldsymbol{\sigma}^{\mathrm{dev}}\right\|} \tag{9}$$

where $\boldsymbol{\sigma}^{\mathrm{dev}}$ is the deviator of stress tensor $\boldsymbol{\sigma}$. $\boldsymbol{\varepsilon}^{\mathrm{tr}}_{\mathrm{recent}}$ denotes the transformation strain at the current transformation reversal (i.e., the point at which the current forward transformation is ended) and the scalar ξ_{recent} is the martensite volume fraction at the current transformation reversal. $\boldsymbol{B}$ is the internal stress tensor, and can be written as

$$\boldsymbol{B} = \boldsymbol{B}_{\mathrm{n}}\frac{\boldsymbol{N}_{\mathrm{tr}}}{\|\boldsymbol{N}_{\mathrm{tr}}\|} \tag{10}$$

where $\|\boldsymbol{N}_{\mathrm{tr}}\|$ is the norm of N_{tr}, i.e., $\|\boldsymbol{N}_{\mathrm{tr}}\| = \sqrt{\boldsymbol{N}_{\mathrm{tr}} : \boldsymbol{N}_{\mathrm{tr}}}$. It is seen that the direction of $\boldsymbol{B}$ is assumed to be identical to that of martensite transformation, where the scalar $\boldsymbol{B}_{\mathrm{n}}$ is the norm of $\boldsymbol{B}$.

By the second law of thermodynamics and the constructed Helmholtz's energy (Yu et al. 2014, 2015), the driving forces of martensite transformation and dislocation slipping can be defined as

$$\pi_{\mathrm{tr}} = g^{\mathrm{tr}}\left(\boldsymbol{\sigma} + \boldsymbol{B}\right) : \boldsymbol{N}_{\mathrm{tr}} + \Delta\eta^{0}\left(\vartheta - \vartheta_{0}\right) - X \tag{11}$$

$$\pi_{p} = \boldsymbol{\sigma} : \boldsymbol{N}_{\mathrm{p}} - \tau_{\mathrm{c}} \tag{12}$$

X and τ_c are the transformation and slipping resistances, respectively. $\Delta\eta^0$ is a material parameter reflecting the entropy difference between austenite and martensite phases.

The evolution equations for martensite transformation can be written as

$$\dot{\xi} = \left(\frac{\pi_{\mathrm{tr}}}{Y}\right)^{m_{\mathrm{tr}}} \qquad \text{if } \pi_{\mathrm{tr}} > 0 \text{ and } \xi < 1 - \xi_{\mathrm{d}} \tag{13}$$

$$\dot{\xi} = -\left(\frac{-\pi_{\mathrm{tr}}}{Y}\right)^{m_{\mathrm{tr}}} \qquad \text{if } \pi_{\mathrm{tr}} < 0 \text{ and } \xi > \xi_{\mathrm{r}} \tag{14}$$

$$\dot{\xi} = 0 \qquad \text{other conditions} \tag{15}$$

where Y is a positive variable controlling the width of stress–strain hysteresis loop, and evolves during the cyclic deformation. Under the isothermal condition, m_{tr} represents the viscosity of martensite transformation. Here, m_{tr} is set as a large number (e.g., $m_{\mathrm{tr}} = 100$) to reflect the weak viscosity of super-elastic NiTi SMA discussed by Grabe and Bruhns (2008). ξ_{d} is the volume fraction of remained austenite phase which cannot be transformed into the martensite one due to the restraint of dislocation slipping occurred in the austenite phase near the interface between the austenite and induced-martensite phases. Thus, the term $1-\xi_{\mathrm{d}}$ in Eq. (13) represents the maximum volume fraction of austenite phase which can be transformed into the martensite one. ξ_{r} is the residual volume fraction of martensite phase.

Based on the experimental observations, the transformation resistance X is introduced into the constitutive model to describe the transformation hardening presented during the cyclic deformation of super-elastic NiTi SMA, and can be written as

$$X = \begin{cases} X_0 + H_{\mathrm{for}}\bar{\xi} + h_1\left(\bar{\xi}\right)^n & \dot{\xi} \geq 0 \\ X_0 + H_{\mathrm{rev}}\bar{\xi} + h_1\left(\bar{\xi}\right)^n & \dot{\xi} < 0 \end{cases} \tag{16}$$

where X_0 is the initial transformation resistance, and H_{for} and H_{rev} are the linear hardening moduli during the forward and reverse transformations, respectively. In order to describe the increased hardening modulus during the cyclic deformation, H_{for} and H_{rev} are assumed to be changeable. The term $h_1\left(\bar{\xi}\right)^n$ reflects the nonlinear transformation hardening occurred at the end of forward transformation and in the beginning of reverse transformation. h_1 and n ($n > 1$) are two material parameters. $\bar{\xi}$ is the effective martensite volume fraction, and is defined as

$$\bar{\xi} = \frac{\xi - \xi_{\mathrm{r}}}{1 - \xi_{d} - \xi_{\mathrm{r}}} \tag{17}$$

It should be noted that the range of volume fraction ξ ($\xi_{\mathrm{r}} < \xi < 1 - \xi_{\mathrm{d}}$) will be changed during the cyclic deformation, since the ξ_{d} and ξ_{r} are not constants. However, from Eq. (17), it is seen that the range of effective martensite volume fraction $\bar{\xi}$ keeps unchanged (i.e., from 0 to 1).

As discussed by Yu et al. (2014, 2015), the super-elasticity degeneration of NiTi SMA during the cyclic deformation is caused by the increased dislocation density in the austenite phase. Thus, the evolutions of B, ξ_d and ξ_r are dependent on the current dislocation density ρ, and are assumed to be governed by the following equations:

$$\dot{B}_n = d_1 \left(c_1\sqrt{\rho} - B_n\right) \left|\dot{\xi}\right| \tag{18}$$

$$\dot{\xi}_d = d_1 \left(c_2\sqrt{\rho} - \xi_d\right) \left|\dot{\xi}\right| \tag{19}$$

$$\dot{\xi}_r = d_1 \left(c_3\sqrt{\rho} - \xi_r\right) \left|\dot{\xi}\right| \tag{20}$$

where d_1, c_1, c_2, and c_3 are material parameters.

For the internal variable Y, it can be decomposed into two parts, i.e.,

$$Y = Y_f + Y_\rho \tag{21}$$

where Y_f is the value of Y under the dislocation-free condition and is a constant; while Y_ρ represents the effect of dislocation slipping on the stress–strain hysteresis loop and is a function of current dislocation density. The evolution equation of Y_ρ can be written as

$$\dot{Y}_\rho = \begin{cases} -d_1 \left(c_4\sqrt{\rho} + Y_\rho\right) \left|\dot{\xi}\right| & \text{if } Y_\rho > -Y_{\lim} \\ 0 & \text{if } Y_\rho \leq -Y_{\lim} \end{cases} \tag{22}$$

where c_4 is a material parameter, $Y_{\lim}$ is the limited value of Y_ρ and is used to ensure the nonnegativity of the variable Y.

For the internal variables H_{for} and H_{rev}, they can be, respectively, decomposed into two parts, i.e.,

$$H_{for} = H_{for}^f + H_{for}^\rho \tag{23}$$

$$H_{rev} = H_{rev}^f + H_{rev}^\rho \tag{24}$$

where H_{for}^f and H_{rev}^f are the values of H_{for} and H_{rev} under the dislocation-free condition, respectively, and are constants; while H_{for}^ρ and H_{rev}^ρ represent the effects of dislocation slipping on the forward and reverse transformation hardening moduli, and are functions of current dislocation density. The evolution equations of H_{for}^ρ and H_{rev}^ρ are given as follows:

$$\dot{H}_{for}^\rho = d_1 \left(c_5\sqrt{\rho} - H_{for}^\rho\right) \left|\dot{\xi}\right| \tag{25}$$

$$\dot{H}_{rev}^\rho = d_1 \left(c_6\sqrt{\rho} - H_{rev}^\rho\right) \left|\dot{\xi}\right| \tag{26}$$

where c_5 and c_6 are two material parameters.

The dislocation slipping law of super-elastic NiTi SMA is given as follows:

$$\dot{\gamma} = (1-\xi)\gamma_0 \left(\frac{\boldsymbol{\sigma} : \boldsymbol{N}_{\mathrm{p}}}{\mu}\right)^2 \exp\left(\frac{-\Delta G_{\mathrm{slip}}}{k_{\mathrm{b}}\vartheta}\left(1-\left(\frac{\pi_{\mathrm{p}}}{\tau_0}\right)^p\right)^q\right)\left|\dot{\xi}\right| \quad \text{if} \quad \pi_{\mathrm{p}} > 0 \tag{27}$$

$$\dot{\gamma} = 0 \qquad \text{if} \quad \pi_{\mathrm{p}} \leq 0 \tag{28}$$

where γ_0, p and q are material parameters with the constraints of $0 < p \leq 1$ and $1 < q \leq 2$; μ is the shear modulus and $\mu = \frac{E}{2(1+\nu)}$, ν is the Poisson's ratio, ΔG_{slip} is the activation energy of dislocation slipping in a stress-free configuration, k_{b} is the Boltzmann's constant, and τ_0 is the resolved shear stress required to overcome the Peierls obstacles at $\vartheta = 0\,\mathrm{K}$. The term $\left|\dot{\xi}\right|$ in Eq. (27) reflects that the plasticity is induced by the martensite transformation and its reverse.

The evolution equation of dislocation density is described by the equation proposed by Mecking and Kocks (1981), i.e.,

$$\dot{\rho} = \left(k_1\sqrt{\rho} - k_2\rho\right)\dot{\gamma} \tag{29}$$

where k_1 and k_2 are two material parameters. The slipping resistance τ_{c} progressively increases during the cyclic deformation which is caused by the increasing density of forest dislocation, and can be written as (Franciosi 1985)

$$\tau_{\mathrm{c}} = \frac{1}{2}\mu b\sqrt{\rho} \tag{30}$$

where b is the magnitude of Burgers vector.

In order to reflect the internal heart production (caused by the transformation latent heat and mechanical dissipation), the equilibrium equation of heat in the form of temperature is derived from the first and second laws of thermodynamics and the constructed Helmholtz's energy (Yu et al. 2015):

$$c\dot{\vartheta} - k\nabla^2\vartheta = \pi_{\mathrm{tr}}\dot{\xi} + \pi_{\mathrm{p}}\dot{\gamma} - \Delta\eta^0\vartheta\dot{\xi} - \dot{\boldsymbol{\sigma}} : \boldsymbol{\alpha}\vartheta \tag{31}$$

where c and k are the specific heat at constant volume and the heat conductivity coefficient, respectively. From Eq. (31), it is seen that the first and second terms in the right side are the mechanical dissipations caused by the martensite transformation and dislocation slipping, respectively; the third one is the transformation latent heat; and the fourth one is caused by the thermal expansion.

3 Theoretical Analysis for NiTi SMA Wire-Type Components

In this section, two types of NiTi SMA wire-type structural components are considered, i.e., the single-wire and multi-wire ones with the same length. For the multi-wire one, the geometric dimension of each wire is assumed to be identical, as shown in Fig. 1. Then, the geometry of the component can be simply characterized by the number of wires, i.e., N. When $N = 1$, it represents the single-wire structure; when $N \geq 2$, it represents the multi-wire one.

Since only the NiTi SMA wire is used here, the constraint condition is written as follows:

$$\pi r_{\mathrm{s}}^2 = N\pi r_{\mathrm{m}}^2 \tag{32}$$

where r_{s} and r_{m} are the radii of the wires in the single-wire and multiple-wire components, respectively, and $r_{\mathrm{m}} = \frac{r_{\mathrm{s}}}{\sqrt{N}}$.

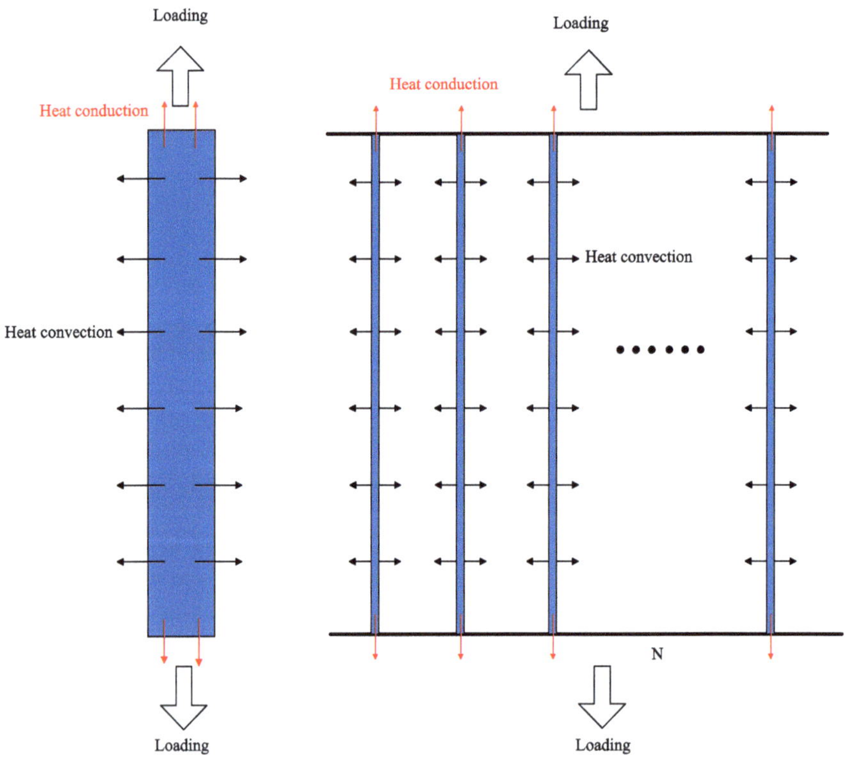

Fig. 1 Thermo-mechanical boundary conditions of single-wire and multiple-wire structures

The initial and boundary conditions of heat analysis in the wire-type structures are written as

$$\begin{cases} \boldsymbol{q} \cdot \boldsymbol{n}_{\text{col}} = h\left(\vartheta - \vartheta_{\text{r}}\right) & \text{in } S_{\text{col}} \\ \boldsymbol{q} \cdot \boldsymbol{n}_{\text{cro}} = \frac{\beta k}{L}\left(\vartheta - \vartheta_{\text{r}}\right) & \text{in } S_{\text{cro}} \\ \vartheta\left(t = 0\right) = \vartheta_{\text{r}} & \text{in } \Omega \end{cases} \tag{33}$$

where Ω is the domain occupied by the NiTi SMA wires. n_{col} and n_{cro} are the surface normal vectors of the columnar surface S_{col} and cross-sectional surface S_{cro}, respectively. ϑ_{r} is set as the room temperature. The first term in Eq. (33) reflects the heat convection through the wire's columnar surface, and h is the heat exchange coefficient of ambient media. The second term reflects the heat conduction through the wire's cross-sectional surface by referring to Yin et al. (2014), and β is a constant.

Integrating Eq. (31) in the cross section of the ith NiTi SMA wire in the structure, it yields

$$\begin{aligned} &\oiint_{A_{\text{cro}}^{i}} \left[c\dot{\vartheta}\left(x, y, z, t\right) - k\left(\frac{\partial^2 \vartheta\left(x, y, z, t\right)}{\partial x^2} + \frac{\partial^2 \vartheta\left(x, y, z, t\right)}{\partial y^2} + \frac{\partial^2 \vartheta\left(x, y, z, t\right)}{\partial z^2} \right) \right] \mathrm{d}y\, \mathrm{d}z \\ &= \oiint_{A_{\text{cro}}^{i}} \left[D_{\text{eff}}\left(x, y, z, t\right) \right] \mathrm{d}y \mathrm{d}z \end{aligned} \tag{34}$$

where A_{cro}^{i} is the cross section of ith NiTi SMA wire. $D_{\text{eff}} = \pi_{\text{tr}}\dot{\xi} + \pi_{\text{p}}\dot{\gamma} - \Delta\eta^0\vartheta\dot{\xi} - \dot{\boldsymbol{\sigma}} : \boldsymbol{\alpha}\vartheta$. The second term in the left side of Eq. (34) can be written as

$$\oiint_{A_{\text{cro}}^{i}} k \frac{\partial^2 \vartheta\left(x, y, z, t\right)}{\partial x^2} \mathrm{d}y \mathrm{d}z = kA \frac{\partial^2 \vartheta_{\text{av}}\left(x, t\right)}{\partial x^2} \tag{35}$$

where $\vartheta_{\text{av}}\left(x\right)$ is the average temperature in the cross section x and is a function of the x-coordinate. A is the area of cross section A_{cro}^{i} (noted that the geometric dimension of each wire is assumed to be identical), i.e.,

$$A = \pi r_m^2 = \frac{\pi r_s^2}{N} \tag{36}$$

$$\vartheta_{\text{av}}\left(x\right) = \frac{1}{A} \oiint_{A_{\text{cro}}^{i}} \vartheta\left(x, y, z, t\right) \mathrm{d}y \mathrm{d}z \tag{37}$$

The third and fourth terms in the left side of Eq. (34) can be written as follows using Gauss theory:

$$\begin{aligned} &\oiint_{A_{\text{cro}}^{i}} k \left[\frac{\partial^2 \vartheta\left(x, y, z, t\right)}{\partial y^2} + \frac{\partial^2 \vartheta\left(x, y, z, t\right)}{\partial z^2} \right] \mathrm{d}y \mathrm{d}z \\ &= \oint_{l_i} k \left[n_{\text{col}}^{y} \frac{\partial \vartheta\left(x, y, z, t\right)}{\partial y} + n_{\text{col}}^{z} \frac{\partial \vartheta\left(x, y, z, t\right)}{\partial z} \right] \mathrm{d}l \end{aligned} \tag{38}$$

where l_i is the boundary curve of cross section. n_{col}^y and n_{col}^z represent the components of n_{col} in the directions of y- and z-axes.

Considering the boundary condition (Eq. (33)), Eq. (38) can be rewritten as

$$\begin{aligned}
&\oint_{l_i} k\left[n_{\text{col}}^y \frac{\partial\vartheta(x,y,z,t)}{\partial y} + n_{\text{col}}^z \frac{\partial\vartheta(x,y,z,t)}{\partial z}\right] \mathrm{d}l \\
&= -\oint_{l_i} h\left[\vartheta(x,y,z,t) - \vartheta_{\text{r}}\right] \mathrm{d}l \\
&= -\frac{2\pi h r_s}{\sqrt{N}}\left[\bar{\vartheta}_{\text{av}}(x,t) - \vartheta_{\text{r}}\right]
\end{aligned} \tag{39}$$

where $\bar{\vartheta}_{\text{av}}(x,t)$ is the average temperature in the boundary curve of the cross section, and is defined as

$$\bar{\vartheta}_{\text{av}}(x,t) = \frac{\sqrt{N}}{2\pi r_s} \oint_{l_i} \vartheta(x,y,z,t)\,\mathrm{d}l \tag{40}$$

By Eqs. (35) and (39), the heat equilibrium equation (Eq. (34)) can be rewritten as

$$\begin{aligned}
&\oiint_{A_{\text{cro}}^i} \left[c\dot{\vartheta}(x,y,z,t)\right] \mathrm{d}y\mathrm{d}z - kA\frac{\partial^2\vartheta_{\text{av}}(x,t)}{\partial x^2} \\
&= \frac{2\pi h r_s}{\sqrt{N}}\left[\vartheta_{\text{r}} - \bar{\vartheta}_{\text{av}}(x,t)\right] + \oiint_{A_{\text{cro}}^i} \left[D_{\text{eff}}(x,y,z,t)\right] \mathrm{d}y\mathrm{d}z
\end{aligned} \tag{41}$$

In general, the temperature field in the wire-type structure is three-dimensional, i.e., $\vartheta = \vartheta(x,y,z,t)$. However, since the size of the wire-type component in the direction of length is much larger than that in the direction of cross section, the temperature field in the NiTi SMA wires can be regarded as one-dimensional for simplicity. By this assumption, the temperature of each material point in the cross section of the wire and its boundary curve can be considered to be the same as the average temperature in the cross section, i.e.,

$$\vartheta(x,y,z,t) = \vartheta(x,t) = \vartheta_{\text{av}}(x,t) = \bar{\vartheta}_{\text{av}}(x,t) \tag{42}$$

It should be noted that the stress field is also uniform in the cross section since the wires can be only subjected to a uniaxial loading, and the temperature field is assumed to be uniform in such cross section. Thus, all the internal variables can be regarded as uniform in the cross section. Finally, it yields

$$D_{\text{eff}}(x,y,z,t) = D_{\text{eff}}(x,t) \tag{43}$$

Using these simplifications, the heat equilibrium equation, initial and boundary conditions (Eqs. (31) and (33)) can be expressed in an one-dimensional (1D) form, i.e.,

$$c\dot{\vartheta} - k\frac{\partial^2 \vartheta}{\partial x^2} = \frac{2\sqrt{N}h}{r_s}(\vartheta_r - \vartheta) + \pi_{tr}\dot{\xi} + \pi_p\dot{\gamma} - \Delta\eta^0\vartheta\dot{\xi} - \dot{\boldsymbol{\sigma}} : \boldsymbol{\alpha}\vartheta \quad 0 < x < L \tag{44}$$

$$k\frac{d\vartheta}{dx} = \frac{\beta k}{L}(\vartheta - \vartheta_r) \quad x = 0 \tag{45}$$

$$-k\frac{d\vartheta}{dx} = \frac{\beta k}{L}(\vartheta - \vartheta_r) \quad x = L \tag{46}$$

$$\vartheta(x, 0) = \vartheta_r \tag{47}$$

From Eq. (44), it is seen that the evolution of temperature field in the NiTi SMA wire depends on the geometric parameter *N*. In this work, these 1D thermo-mechanically coupled equations are solved using the finite difference method, and 31 equally spaced nodes are used in the calculation.

4 Results and Discussions

The length of the wire-type structures is set to be 100 mm, and the radius of the single-wire is set to be 5 mm. The thermo-mechanically coupled constitutive model proposed by Yu et al. (2015) had been verified by simulating the super-elasticity degeneration of NiTi SMA at various loading rates. The material parameters used in the model (Yu et al. 2015) can be obtained from the cyclic stress–strain curves at specific loading rates. In this work, the parameters are set to be the same as those in Yu et al. (2015), and are listed in Table 1.

Table 1 Material parameters

Thermo-elastic constants: $E = 35\,\text{GPa}$, $v = 0.3$, $\alpha = 22 \times 10^{-6}/\text{K}$
Parameters related to martensite transformation: $g^{tr} = 0.0748$, $\Delta\eta^0 = -0.45\,\text{MPa/K}$, $c_1 = 115.58\,\text{N/m}$, $c_2 = 1.04 \times 10^{-7}\,\text{m}$, $c_3 = 9.04 \times 10^{-8}\,\text{m}$, $c_4 = 4.46\,\text{N/m}$, $c_5 = 5.52\,\text{N/m}$, $c_6 = 2.30\,\text{N/m}$; $Y_0 = 10.26\,\text{MPa}$, $Y_{lim} = 8\,\text{MPa}$, $X_0 = 29.9\,\text{MPa}$, $H^f_{for} = -1.41\,\text{MPa}$, $H^f_{rev} = 3.68\,\text{MPa}$, $h_1 = 7\,\text{MPa}$, $n = 10$, $m_{tr} = 100$, $d_1 = 0.3$
Parameters related to dislocation slipping: $\Delta G_{slip} = 2.5 \times 10^{-19}\text{J}$, $b = 3.6 \times 10^{-10}\text{m}$; $k_1 = 20 \times 10^7\,\text{m}^{-1}$, $k_2 = 5$, $\gamma_0 = 1000$, $p = 0.1$, $q = 1$, $\tau_0 = 1000\,\text{MPa}$, $\rho_0 = 1 \times 10^{10}\,\text{m}^{-2}$
Parameters related to heat transfer: $c_A = 2.86\,\text{MJ/(m}^3\text{K)}$, $c_M = 2.86\,\text{MJ/(m}^3\text{K)}$, $k = 18\,\text{Wm}^{-1}\text{K}^{-1}$, $h = 5\,\text{Wm}^{-2}\text{K}^{-1}$, $\beta = 3$

All the calculations are performed under the cyclic strain-controlled tension but stress-controlled unloading conditions and at room temperature (299 K, which is higher than the austenite finish temperature Af of 285.4 K). In the tensile part, nine strain rates, i.e., 1×10^{-6}/s, 5×10^{-6}/s, 1×10^{-5}/s, 5×10^{-5}/s, 7.5×10^{-5}/s, 1×10^{-4}/s, 1.5×10^{-4}/s, 5×10^{-4}/s, and 1×10^{-3}/s, are prescribed, respectively; in the unloading part, the stress rate is kept as a constant in each cycle, but varied cycle by cycle according to the requirement that the total unloading time should be equal to the total tensile time. So, the variation of loading rates in different loading cases can be represented only by the strain rates used in the corresponding tensile parts. At each loading rate, the calculations for the single-wire and multi-wire structures are performed with different number of wires (i.e., $N = 1, 10, 50, 100, 500$, and 1000). The maximum tensile strain and number of cycles are set as 9 % and 50 for each loading case, respectively.

In order to measure the dissipative property of the structures, the accumulated dissipation energy D is introduced here, i.e.,

$$D(N) = \sum_{n=1}^{\text{ncycle}} \sum_{i=1}^{N} \oint_n F_i \, \mathrm{d}l_i \tag{48}$$

where ncycle = 50 is the number of loading cycles; n and i represent the nth loading cycle and ith NiTi SMA wire in the structures, respectively; and F_i and l_i are the end force and end displacement of the ith NiTi SMA wire. It is seen that the accumulated dissipation energy is a function of the geometric parameter N.

Figure 2a–c shows the predicted cyclic stress–strain curves at a strain rate of 1×10^{-5}/s with the geometric parameter $N = 1$, 100, and 1000, respectively. It is seen that super-elasticity degeneration occurs, i.e., the residual strain accumulates, the transformation hardening modulus increases, but the start stress of martensite transformation and the width of hysteresis loop decreases with the increasing number of cycles. After certain cycles, a saturated state is reached too. Also, it is seen that the differences between the cyclic stress–strain curves are not obvious at the low strain rate. Figure 2d shows the evolution curves of the average temperature for the whole wire versus the number of cycles. The average temperature is obtained by averaging the temperatures at all finite difference nodes. It is seen that a temperature oscillation occurs during the cyclic deformation, which is caused by the release/absorption of transformation latent heat. The oscillation amplitude decreases with the increasing geometric dimension parameter N.

Figures 3 and 4 show the corresponding predicted results of the NiTi SMA structures at the strain rates of 1×10^{-4}/s and 1×10^{-3}/s, respectively. It is seen that the super-elasticity degeneration aggravates with the increasing strain rate, i.e., the saturated residual strain, the decrement in the start stress of martensite transformation, and the width of stress–strain hysteresis loop, and the increment in the transformation hardening modulus increases with the increasing strain rate. Meanwhile, the differences between the cyclic stress–strain curves with various geometric dimension parameters N become more and more apparent.

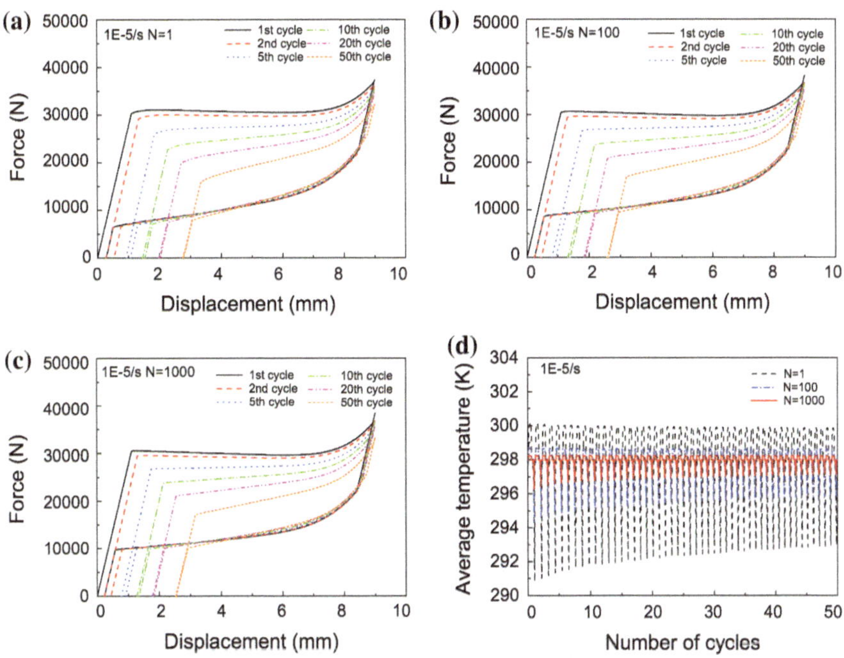

Fig. 2 Cyclic deformation of the super-elastic NiTi SMA structures at the strain rate of 1×10^{-5}/s: **a** the end force–displacement curves of the single-wire structure ($N = 1$); **b** the end force–displacement curves of the multiple-wire structure ($N = 100$); **c** the end force–displacement curves of the multiple-wire structure ($N = 1000$); **d** the evolution curves of the average temperature of structures

Figure 5a–c shows the evolution curves of the accumulated dissipation energy versus the values of geometric dimension parameter N at low (e.g., 1×10^{-6}/s, 5×10^{-6}/s, and 1×10^{-5}/s), moderate (e.g., 5×10^{-5}/s, 7.5×10^{-5}/s, 1×10^{-4}/s, and 1.5×10^{-4}/s), and high strain rates (e.g., 5×10^{-4}/s, and 1×10^{-3}/s), respectively. It is concluded that when the strain rate is low, the accumulated dissipation energy decreases with the increasing N, which means the dissipative property of the single-wire structure is much more advantageous than that of the multi-wire structure; at the moderate strain rates, the accumulated dissipation energy changes non-monotonically with the increasing N, the peak values occur at $N = 10, 50, 100,$ and 500 at the strain rates of 5×10^{-5}/s, 7.5×10^{-5}/s, 1×10^{-4}/s, and 1.5×10^{-4}/s, respectively; when the strain rate is high (e.g., 5×10^{-4}/s, and 1×10^{-3}/s), the accumulated dissipation energy increases with the increasing N; in the other word, the dissipative property of the multi-wire structure is much more advantageous than that of the single-wire one.

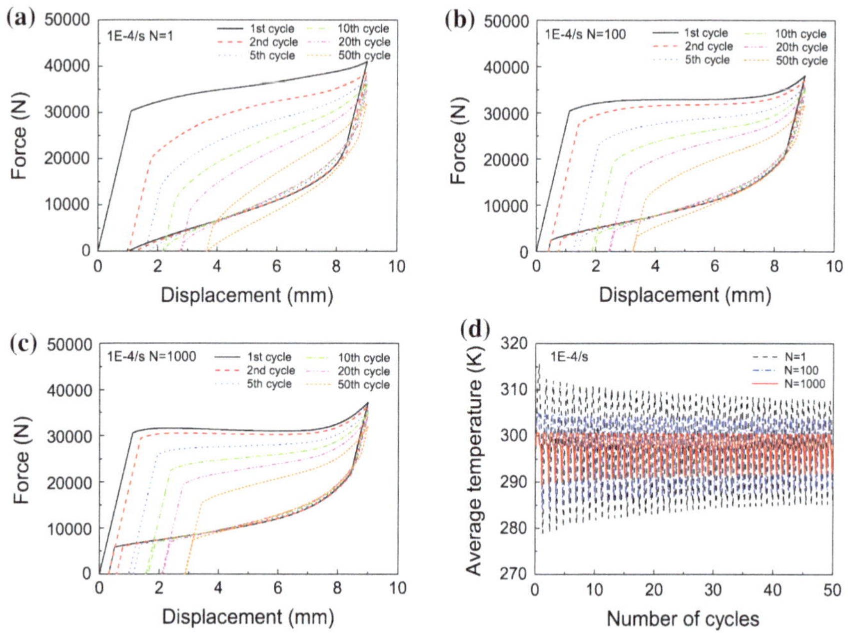

Fig. 3 Cyclic deformation of the super-elastic NiTi SMA structures at the strain rate of $1 \times 10Z^{-4}$/s: **a** the end force–displacement curves of the single-wire structure ($N = 1$); **b** the end force–displacement curves of the multiple-wire structure ($N = 100$); **c** the end force–displacement curves of the multiple-wire structure ($N = 1000$); **d** the evolution curves of the average temperature of structures

To sum up, it is concluded that in the practical design process of the NiTi SMA wire-type devices, if the range of applied loading rate is known in advance, the optimized dissipative property of the devices can be readily obtained by suitably choosing the number of wires matched with the prescribed loading rate.

5 Conclusions

1. The effect of geometric dimension on the dissipative property of NiTi SMA wire-type structures is predicted by a physical mechanism-based thermo-mechanically coupled constitutive model proposed by Yu et al. (2014, 2015). A three-dimensional equilibrium equation of heat is simplified to a one-dimensional form by the assumption of uniform temperature field in the cross section of the wires, and it is seen that the temperature field in the wire depends on its geometric dimension.

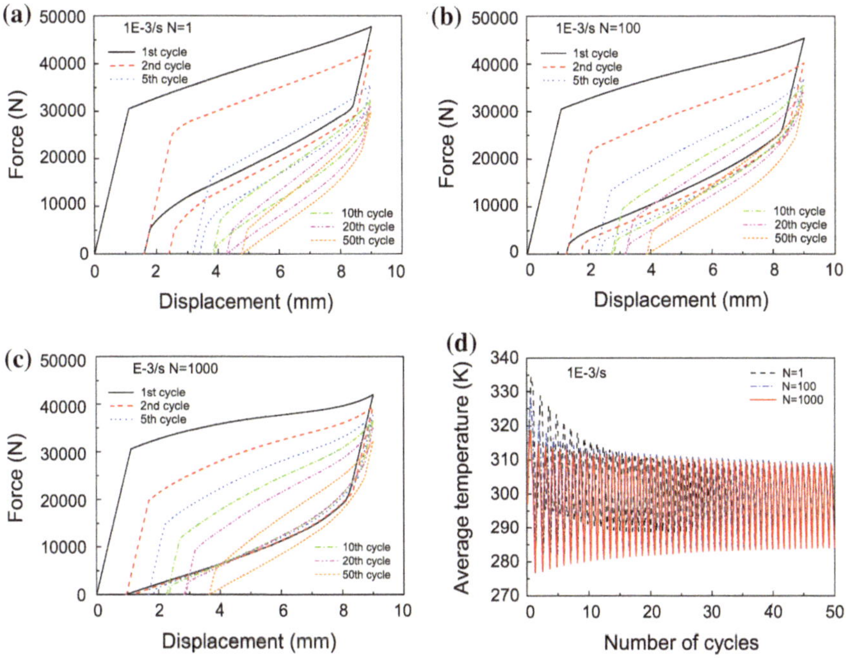

Fig. 4 Cyclic deformation of the super-elastic NiTi SMA structures at the strain rate of 1×10^{-3}/s: **a** the end force–displacement curves of the single-wire structure ($N = 1$); **b** the end force–displacement curves of the multiple-wire structure ($N = 100$); **c** the end force–displacement curves of the multiple-wire structure ($N = 1000$); **d** the evolution curves of the average temperature of structures

2. Two types of NiTi SMA structural components are considered, i.e., the single-wire and multi-wire ones. The dissipative property of the structure is measured by its accumulated dissipation energy during cyclic deformation.
3. At low (1×10^{-6}/s, 5×10^{-6}/s, 1×10^{-5}/s), moderate (5×10^{-5}/s, 7.5×10^{-5}/s, 1×10^{-4}/s, 1.5×10^{-4}/s), and high strain rates (5×10^{-4}/s, 1×10^{-3}/s), the accumulated dissipation energy decreases, changes non-monotonically, and increases with the increasing number of wires, respectively.
4. In practice, if the range of loading rate is known in advance, the optimized dissipative property of NiTi SMA wire-type structures can be obtained by choosing the number of wires suitably.

It should be noted that only the predicted results are provided in this work, but the results are still useful to design the NiTi SMA devices addressing the dissipation property. More solid analysis is needed further with the help of relative experimental observations in future work.

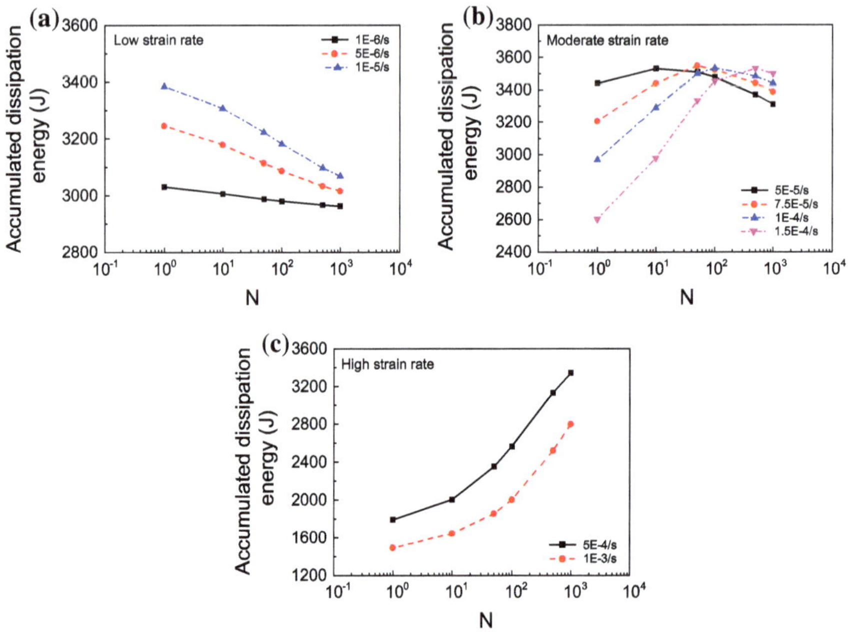

Fig. 5 Accumulated dissipation energy-geometric parameter (N) curves of super-elastic NiTi SMA wire-type structures: **a** at low strain rates; **b** at moderate strain rates; **c** at high strain rates

Acknowledgments Financial supports by the National Natural Science Foundation of China (11532010) and the project for Sichuan Provincial Youth Science and Technology Innovation Team, China (2013TD0004) are appreciated.

References

Auricchio, F., Marfia, S., Sacco, E.: Modelling of SMA materials: training and two way memory effects. Comput. Struct. **81**(24–25), 2301–2317 (2003). doi:10.1016/S0045-7949(03)00319-5

Auricchio, F., Reali, A., Stefanelli, U.: A three-dimensional model describing stress-induced solid phase transformation with permanent inelasticity. Int. J. Plast. **23**(2), 207–226 (2007). doi:10.1016/j.ijplas.2006.02.012

Delville, R., Malard, B., Pilch, J., Sittner, P., Schryvers, D.: Microstructure changes during non-conventional heat treatment of thin Ni-Ti wires by pulsed electric current studied by transmission electron microscopy. Acta Mater. **58**(13), 4503–4515 (2010). doi:10.1016/j.actamat.2010.04.046

Delville, R., Malard, B., Pilch, J., Sittner, P., Schryvers, D.: Transmission electron microscopy investigation of dislocation slip during superelastic cycling of Ni-Ti wires. Int. J. Plast. **27**(2), 282–297 (2011). doi:10.1016/j.ijplas.2010.05.005

Franciosi, P.: The concepts of latent hardening and strain hardening in metallic single crystals. Acta Metall. **33**(9), 1601–1612 (1985). doi:10.1016/0001-6160(85)90154-3

Gall, K., Maier, H.: Cyclic deformation mechanisms in precipitated niti shape memory alloys. Acta Mater. **50**(18), 4643–4657 (2002). doi:10.1016/S1359-6454(02)00315-4

Grabe, C., Bruhns, O.: On the viscous and strain rate dependent behavior of polycrystalline NiTi. Int. J. Solids Struct. **45**(7–8), 1876–1895 (2008). doi:10.1016/j.ijsolstr.2007.10.029

Hartl, D.J., Chatzigeorgiou, G., Lagoudas, D.C.: Three-dimensional modeling and numerical analysis of rate-dependent irrecoverable deformation in shape memory alloys. Int. J. Plast. **26**(10), 1485–1507 (2010). doi:10.1016/j.ijplas.2010.01.002

He, Y., Sun, Q.: Frequency-dependent temperature evolution in NiTi shape memory alloy under cyclic loading. Smart Mater. Struct. **19**(11), 115, 014 (2010a)

He, Y., Sun, Q.: Rate-dependent domain spacing in a stretched niti strip. Int. J. Solids Struct. **47**(20), 2775–2783 (2010b). doi:10.1016/j.ijsolstr.2010.06.006

He, Y., Sun, Q.: On non-monotonic rate dependence of stress hysteresis of superelastic shape memory alloy bars. Int. J. Solids Struct. **48**(11–12), 1688–1695 (2011). doi:10.1016/j.ijsolstr.2011.02.017

Jani, J.M., Leary, M., Subic, A., Gibson, M.A.: A review of shape memory alloy research, applications and opportunities. Mater. Des. **56**, 1078–1113 (2014). doi:10.1016/j.matdes.2013.11.084

Kan, Q., Kang, G.: Constitutive model for uniaxial transformation ratchetting of super-elastic niti shape memory alloy at room temperature. Int. J. Plast. **26**(3), 441–465 (2010). doi:10.1016/j.ijplas.2009.08.005

Kan, Q., Yu, C., Kang, G., Li, J., Yan, W.: Experimental observations on rate-dependent cyclic deformation of super-elastic niti shape memory alloy. Mech. Mater. Submitted for publication (2015)

Kang, G., Kan, Q., Qian, L., Liu, Y.: Ratchetting deformation of super-elastic and shape-memory niti alloys. Mech. Mater. **41**(2), 139–153 (2009). doi:10.1016/j.mechmat.2008.09.001

Lagoudas, D.C., Entchev, P.B.: Modeling of transformation-induced plasticity and its effect on the behavior of porous shape memory alloys. part i: constitutive model for fully dense SMAs. Mech. Mater. **36**(9), 865–892 (2004). doi:10.1016/j.mechmat.2003.08.006

Mecking, H., Kocks, U.: Kinetics of flow and strain-hardening. Acta Metall. **29**(11), 1865–1875 (1981). doi:10.1016/0001-6160(81)90112-7

Miyazaki, S., Imai, T., Igo, Y., Otsuka, K.: Effect of cyclic deformation on the pseudoelasticity characteristics of ti-ni alloys. Metall. Trans. A **17**(1), 115–120 (1986). doi:10.1007/BF02644447

Morin, C., Moumni, Z., Zaki, W.: A constitutive model for shape memory alloys accounting for thermomechanical coupling. Int. J. Plast. **27**(5), 748–767 (2011a). doi:10.1016/j.ijplas.2010.09.005

Morin, C., Moumni, Z., Zaki, W.: Thermomechanical coupling in shape memory alloys under cyclic loadings: Experimental analysis and constitutive modeling. Int. J. Plast. **27**(12), 1959–1980 (2011b). doi:10.1016/j.ijplas.2011.05.005. (special Issue In Honor of Nobutada Ohno)

Norfleet, D., Sarosi, P., Manchiraju, S., Wagner, M.X., Uchic, M., Anderson, P., Mills, M.: Transformation-induced plasticity during pseudoelastic deformation in ni-ti microcrystals. Acta Mater. **57**(12), 3549–3561 (2009). doi:10.1016/j.actamat.2009.04.009

Saint-Sulpice, L., Chirani, S.A., Calloch, S.: A 3d super-elastic model for shape memory alloys taking into account progressive strain under cyclic loadings. Mech. Mater. **41**(1), 12–26 (2009). doi:10.1016/j.mechmat.2008.07.004

Saleeb, A., Padula II, S.P., Kumar, A.: A multi-axial, multimechanism based constitutive model for the comprehensive representation of the evolutionary response of SMAs under general thermomechanical loading conditions. Int. J. Plast. **27**(5), 655–687 (2011). doi:10.1016/j.ijplas.2010.08.012

Shaw, J.A., Kyriakides, S.: Thermomechanical aspects of NiTi. J. Mech. Phys. Solids **43**(8), 1243–1281 (1995). doi:10.1016/0022-5096(95)00024-D

Simon, T., Kröger, A., Somsen, C., Dlouhy, A., Eggeler, G.: On the multiplication of dislocations during martensitic transformations in niti shape memory alloys. Acta Mater. **58**(5), 1850–1860 (2010). doi:10.1016/j.actamat.2009.11.028

Song, D., Kang, G., Kan, Q., Yu, C., Zhang, C.: The effect of martensite plasticity on the cyclic deformation of super-elastic NiTi shape memory alloy. Smart Mater. Struct. **23**(1), 015, 008 (2014a)

Song, D., Kang, G., Kan, Q., Yu, C., Zhang, C.: Non-proportional multiaxial transformation ratchetting of super-elastic niti shape memory alloy: experimental observations. Mech. Mater. **70**, 94–105 (2014b). doi:10.1016/j.mechmat.2013.12.003

Strnadel, B., Ohashi, S., Ohtsuka, H., Ishihara, T., Miyazaki, S.: Cyclic stress-strain characteristics of ti-ni and ti-ni-cu shape memory alloys. Mater. Sci. Eng.: A **202**(1–2), 148–156 (1995a). doi:10.1016/0921-5093(95)09801-1

Strnadel, B., Ohashi, S., Ohtsuka, H., Miyazaki, S., Ishihara, T.: Effect of mechanical cycling on the pseudoelasticity characteristics of ti-ni and ti-ni-cu alloys. Mater. Sci. Eng.: A **203**(1–2), 187–196 (1995b). doi:10.1016/0921-5093(95)09881-X

Wang, X., Xu, B., Yue, Z.: Micromechanical modelling of the effect of plastic deformation on the mechanical behaviour in pseudoelastic shape memory alloys. Int. J. Plast. **24**(8), 1307–1332 (2008). doi:10.1016/j.ijplas.2007.09.006

Yin, H., He, Y., Sun, Q.: Effect of deformation frequency on temperature and stress oscillations in cyclic phase transition of NiTi shape memory alloy. J. Mech. Phys. Solids **67**, 100–128 (2014). doi:10.1016/j.jmps.2014.01.013

Yu, C., Kang, G., Kan, Q.: A physical mechanism based constitutive model for temperature-dependent transformation ratchetting of niti shape memory alloy: one-dimensional model. Mech. Mater. **78**, 1–10 (2014). doi:10.1016/j.mechmat.2014.07.011

Yu, C., Kang, G., Kan, Q., Zhu, Y.: Rate-dependent cyclic deformation of super-elastic niti shape memory alloy: thermo-mechanical coupled and physical mechanism-based constitutive model. Int. J. Plast. **72**, 60–90 (2015). doi:10.1016/j.ijplas.2015.05.011

Zaki, W., Moumni, Z.: A 3d model of the cyclic thermomechanical behavior of shape memory alloys. J. Mech. Phys. Solids **55**(11), 2427–2454 (2007). doi:10.1016/j.jmps.2007.03.011

Part IV
Continuum Damage and Phase Fields

A Discussion on Gradient Damage and Phase-Field Models for Brittle Fracture

René de Borst and Clemens V. Verhoosel

Abstract Gradient-enhanced damage models find their roots in damage mechanics, which is a smeared approach from the onset, and gradients were added to restore well-posedness beyond a critical strain level. The phase-field approach to brittle fracture departs from a discontinuous description of failure, where the distribution function is regularised, which also leads to the inclusion of spatial gradients. Herein, we will consider both approaches, and discuss their similarities and differences.

1 Introduction

The numerical modelling of fracture can be approached from different points of view. Discrete models for fracture, where the geometrical discontinuity is modelled, i.e. by modifying the geometry of the original, intact structure, are perhaps intuitively the most appealing, and have been pursued since the late 1960s (Ngo and Scordelis 1967). Remeshing (Ingraffea and Saouma 1985) and the eXtended Finite Element Method, e.g. Belytschko and Black (1999), Moës et al. (1999), Remmers et al. (2003), Réthoré et al. (2007) have provided ways to decouple the crack path from the initial discretisation. Also, isogeometric finite element analysis beholds promise to flexibly model propagating cracks (Verhoosel et al. 2011).

Yet, issues remain such as the proper modelling of curved crack fronts in three dimensions, while the robust implementation of discrete cracks in a three-dimensional setting is a non-trivial task. Hence, smeared, or distributed, crack approaches have been put forward, where the discontinuity is distributed over a

R. de Borst (✉)
Department of Civil and Structural Engineering, University of Sheffield,
Sheffield, UK
e-mail: r.deborst@sheffield.ac.uk

C.V. Verhoosel
Department of Mechanical Engineering, Eindhoven University of Technology,
Eindhoven, The Netherlands
e-mail: c.v.verhoosel@tue.nl

K. Naumenko and M. Aßmus (eds.), *Advanced Methods of Continuum Mechanics for Materials and Structures*, Advanced Structured Materials 60,
DOI 10.1007/978-981-10-0959-4_14

In this formulation, all the other relations remain local: the local stress–strain relation, Eq. (1), the loading–unloading conditions, Eq. (3), and the dependence of the damage variable d on the history parameter, Eq. (4). As an alternative to Eq. (6), the local history parameter κ may be replaced in the damage-loading function f by a spatially averaged quantity $\bar{\kappa}$, such that

$$f(\epsilon, \bar{\kappa}) = \epsilon - \bar{\kappa}, \quad \bar{\kappa}(\mathbf{x}) = \frac{1}{\Psi(\mathbf{x})} \int_{\Omega} \psi(\mathbf{y}, \mathbf{x}) \kappa(\mathbf{y}) \mathrm{d}\Omega \tag{8}$$

2.3 Gradient-Enhanced Damage Models

Nonlocal constitutive relations can be considered as a point of departure for constructing gradient models, although we wish to emphasise that the latter class of models can also be defined directly by supplying higher order gradients in the damage-loading function. Yet, we will follow the first-mentioned route to underline the connection between integral- and differential-type nonlocal models. This is done either by expanding the kernel $\tilde{\epsilon}$ of the integral in Eq. (6) in a Taylor series, or by expanding of the history parameter κ in Eq. (8) as a Taylor series. If we truncate after the second-order terms and carry out the integration implied in Eq. (6) under the assumption of isotropy, the following relation ensues:

$$\bar{\epsilon} = \tilde{\epsilon} + g \nabla^2 \tilde{\epsilon} \tag{9}$$

where g is a gradient parameter of the dimension length squared.

Equation (9) bears the disadvantage that it requires the computation of second-order gradients of the local equivalent strain $\tilde{\epsilon}$. Since this quantity is a function of the strain tensor, and since the strain tensor involves first-order derivatives of the displacements, third-order derivatives of the displacements have to be computed, which would necessitate C^2-continuity of the shape functions. To obviate this problem, Eq. (9) is differentiated twice and the result is substituted again into Eq. (9). Again neglecting fourth-order terms leads to

$$\bar{\epsilon} - g \nabla^2 \bar{\epsilon} = \tilde{\epsilon} \tag{10}$$

In Peerlings et al. (2001) it has been shown that the implicit gradient formulation, Eq. (10), becomes formally identical to a fully nonlocal formulation when, in three dimensions, the weighting function is chosen as:

$$\psi(\mathbf{y}, \mathbf{x}) = \frac{1}{4\pi g r} \exp\left(-\frac{r}{\sqrt{g}}\right) \tag{11}$$

Indeed, the 'implicit gradient' formulation has a truly nonlocal character, which is different from the 'explicit gradient' formulation, Eq. (9).

In a fashion similar to the derivation of the gradient damage models based on the averaging of the equivalent strain $\tilde{\epsilon}$, we can elaborate a gradient approximation of Eq. (8), i.e. by developing $\bar{\kappa}$ in a Taylor series. For an isotropic, infinite medium, and truncating after the second term, we have (de Borst et al. 1996)

$$\bar{\kappa} = \kappa + g\nabla^2\kappa \tag{12}$$

Similar to the second-order gradient model where the nonlocal strain has been expanded, an 'implicit' version can be developed by differentiating Eq. (12) twice and substituting the result back into Eq. (12). Neglecting fourth-order terms then yields

$$\bar{\kappa} - g\nabla^2\bar{\kappa} = \kappa \tag{13}$$

A spatially averaged damage field $\bar{d}$ is now defined as a function of $\bar{\kappa}$: $\bar{d} = \bar{d}(\bar{\kappa})$. Taking the special case of a linear relation between $\bar{d}$ and $\bar{\kappa}$ (de Borst et al. 1996; Comi 1999):

$$\bar{d} = \bar{\kappa}_0 + a\bar{\kappa} \tag{14}$$

Equation (13) becomes

$$\bar{d} - \frac{g}{a}\nabla^2\bar{d} = \bar{\kappa}_0 + a\kappa \tag{15}$$

or, setting $d = \bar{\kappa}_0 + a\kappa$ and $g' = g/a$,

$$\bar{d} - g'\nabla^2\bar{d} = d \tag{16}$$

In the remainder, the prime will be omitted for notational convenience—noting that the gradient parameter g will have a different meaning depending on the exact gradient formulation—and we will write for the averaging equation of the nonlocal damage field:

$$\bar{d} - g\nabla^2\bar{d} = d \tag{17}$$

2.4 *Broadening of the Damage Zone*

We now consider the three-point bending beam of Fig. 1 (Askes et al. 2000). The beam has the dimensions $2000 \times 300\,\text{mm}^2$ and a thickness of 50 mm. It is supported by hinges on the left and right bottom corners, and is loaded by a distributed load $\bar{t}$ over the central section (100 mm) of the specimen.

A linear isotropic material is considered with modulus of elasticity $E = 20\,\text{GPa}$ in the undamaged state and Poisson's ratio $\nu = 0.2$. Plane–strain conditions have

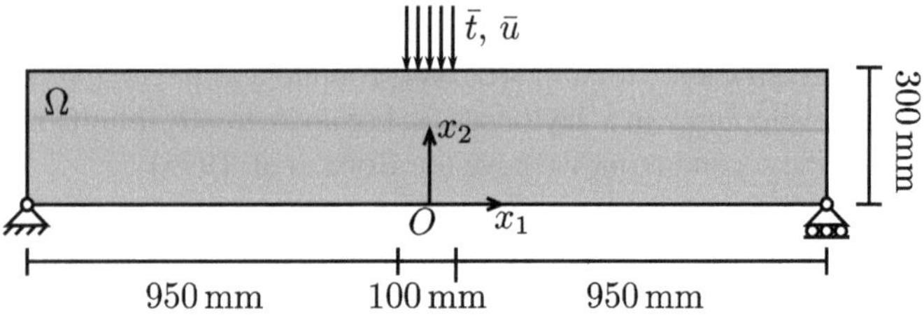

Fig. 1 Three-point bending specimen. The thickness of the specimen is 50 mm

been assumed, and the local equivalent strain is given by

$$\eta(\epsilon) = \sqrt{\langle \epsilon_i \rangle^2} = \sqrt{\langle \epsilon_1 \rangle^2 + \langle \epsilon_2 \rangle^2} \tag{18}$$

where ϵ_1 and ϵ_2 are the principal strains, and the Macaulay brackets distinguish between tension and compression. The damage law proposed by Geers et al. (1998) has been used

$$d(\kappa) = \begin{cases} 0 & \kappa \leq \kappa_0 \\ 1 - \frac{\kappa_0}{\kappa} \{(1-\alpha) + \alpha \exp[\beta(\kappa_0 - \kappa)]\} & \kappa > \kappa_0 \end{cases} \tag{19}$$

with $\kappa_0 = 10^{-4}$, $\alpha = 0.99$ and $\beta = 500$. The gradient parameter is taken as $g = 200$ mm.

The computations have been done on quadrilateral meshes with 2,388 elements and 9,552 elements, respectively. In the central region of the beam, where the damage zone evolves, hierarchical refinement (Vuong et al. 2011) has been used to improve the resolution. The total number of degrees of freedom with bilinear basis functions for both the displacement field and damage field is 7,246 for the coarse mesh and 28,816 for the fine mesh.

A force F is defined as the distributed load $\bar{t}$ times the area to which it has been applied. The displacement $\bar{u}$ has been taken as the average downward displacement of the loading region, see Fig. 1. This displacement has been used as the constraint in the path-following method used to trace the equilibrium path. The results obtained for the second-order gradient formulation are shown in Fig. 2. As can be seen, the results obtained using the coarse mesh are in good agreement with that of the fine mesh.

Upon propagation, the damage zone not only extends, but also broadens. The broadening effect is clearly visible in Fig. 3 and is insensitive to the mesh size. This artefact was first observed by Geers et al. (1998) and makes standard gradient damage formulations less suitable to mimick a sharp crack. As indicated in Geers et al. (1998) the broadening effect is a consequence of the continued increase of the local and nonlocal equivalent strain after damage has fully developed. The problem can be remedied by making the internal length scale a function of the local equivalent strain.

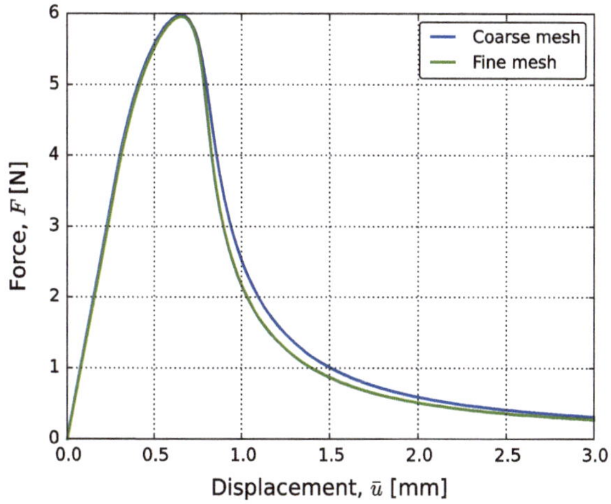

Fig. 2 Mesh convergence study for the three-point bending beam using the second-order gradient formulation

3 The Phase-Field Approach to Brittle Fracture

Below we shall give a succinct summary of phase-field models as applied to brittle fracture, including an example.

3.1 *Formulation*

The basic idea of phase-field models is to approximate a discontinuity Γ by a smeared surface Γ_ℓ. In a one-dimensional setting, the exponential function

$$\bar{d}(x) = e^{-\frac{|x|}{2\ell}} \tag{20}$$

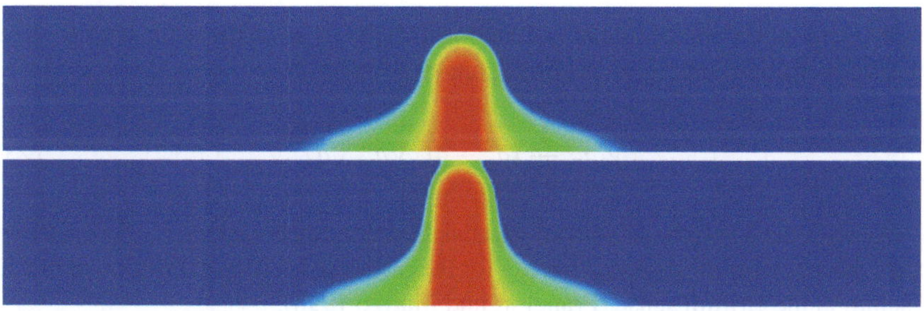

Fig. 3 Damage profiles for **a** $\bar{u} = 0.875$ mm, and **b** $\bar{u} = 2.00$ mm obtained using the second-order gradient formulation. Undamaged material is indicated in *blue*, fully damaged material in *red*

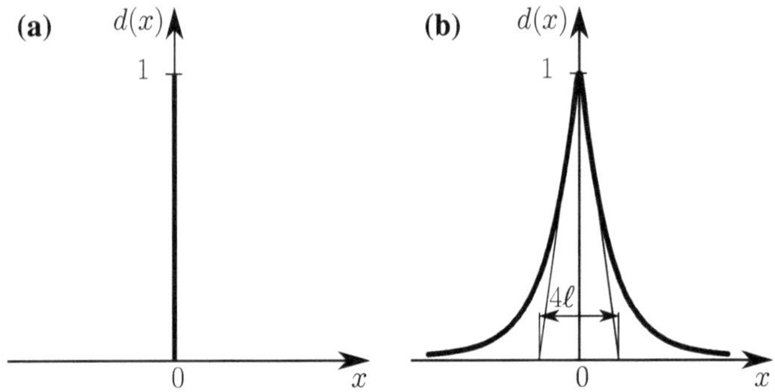

Fig. 4 **a** A sharp crack, and **b** smeared crack modelled with the length scale parameter ℓ

is used to approximate the discontinuous function of Fig. 4a, with ℓ the internal length scale parameter. The phase-field variable $\bar{d} \in [0, 1]$ describes the phase field. Following the earlier discussion on gradient-enhanced damage models, the bar indicates that a regularised (or spatially averaged) quantity is considered, and $\bar{d}$ is defined such that $\bar{d} = 0$ characterises the intact state of the material, while $\bar{d} = 1$ represents the fully broken material, similar to the definition commonly adopted in damage mechanics. In one dimension, Eq. (20) is the solution to:

$$\bar{d} - 4\ell^2 \bar{d}_{,xx} = 0 \tag{21}$$

where a comma denotes differentiation, subject to the boundary conditions: $\bar{d}(0) = 1$ and $\bar{d}(\pm\infty) = 0$. This can be demonstrated simply by applying the Ansatz function $\bar{d} = e^{-|\lambda||x|}$ to Eq. (21), solving for λ and subsequently using the boundary conditions to determine the constant parameter. Using Eq. (21) the discontinuity Γ can be approximated by the functional Γ_ℓ

$$\Gamma_\ell = \int_\Omega \underbrace{\frac{1}{4\ell}\left(\bar{d}^2 + 4\ell^2 \bar{d}_{,x}^2\right)}_{\gamma_\ell} \, \mathrm{dV} \tag{22}$$

with γ_ℓ the crack surface density function (Miehe et al. 2010b) for details. In a multi-dimensional setting γ_ℓ can be expanded as follows:

$$\gamma_\ell = \frac{1}{4\ell}\left(\bar{d}^2 + 4\ell^2 \nabla \bar{d} \cdot \nabla \bar{d}\right) \tag{23}$$

We consider a volume Ω with an internal discontinuity boundary Γ_d. As a starting point, we consider the potential energy for the case of a discrete description of brittle fracture in the Griffith sense (Francfort and Marigo 1998):

$$\Psi_{\text{pot}} = \int_{\Omega} \psi^e(\boldsymbol{\epsilon})\,\mathrm{d}V + \int_{\Gamma_d} \mathcal{G}_c\,\mathrm{d}A \tag{24}$$

with the elastic energy density ψ^e a function of the infinitesimal strain tensor $\boldsymbol{\epsilon}$: $\psi^e = \psi^e(\boldsymbol{\epsilon})$. The elastic energy density is expressed by Hooke's law for an isotropic linear elastic material as $\psi^e(\boldsymbol{\epsilon}) = \frac{1}{2}\lambda\epsilon_{ii}\epsilon_{jj} + \mu\epsilon_{ij}\epsilon_{ij}$, with λ and μ the Lamé constants, and the summation convention applies. In Eq. (24) the fracture energy, i.e. the amount of energy dissipated upon the creation of a unit of fracture surface, is denoted by $\mathcal{G}_c$. In the spirit of a regularised crack topology, the work required to create a unit crack area is expressed as a volume integral which depends on the phase field variable $\bar{d}$ and the fracture energy $\mathcal{G}_c$:

$$\int_{\Gamma_d} \mathcal{G}_c \mathrm{d}A = \int_{\Omega} \mathcal{G}_c \gamma_\ell(\bar{d}, \nabla\bar{d})\,\mathrm{d}V. \tag{25}$$

The next step is inspired by damage mechanics concepts and relies on the assumption that the evolution of the phase field is directly related to crack growth. As such, it can be thought of as a way to model the loss of stiffness of the bulk of the solid. For this purpose, a degradation function $h = h(\bar{d})$ is introduced, which must satisfy the following conditions:

$$\begin{cases} h : [0,1] \rightarrow [0,1] \\ h(0) = 1 \quad , \quad h(1) = 0 \\ h'(\bar{d}) < 0 \quad \bar{d} \in [0, 1[\\ h'(1) = 0 \end{cases} \tag{26}$$

These properties ensure damage propagation and provide an upper bound to the phase field $\bar{d}$ variable of one (Miehe et al. 2010a). A quadratic polynomial is widely used:

$$h(\bar{d}) = (1 - \bar{d})^2 \tag{27}$$

Bourdin et al. (2000) have multiplied the degradation function h by the elastic energy density of the undamaged state, ψ_0, such that the elastic energy density of the damaged state reads

$$\psi^e(\boldsymbol{\epsilon}, \bar{d}) = h(\bar{d})\psi_0(\boldsymbol{\epsilon}) \tag{28}$$

This formulation was refined to account for the fact that damage evolution occurs under different straining modes (Amor et al. 2009; Miehe et al. 2010b), and it was assumed that the elastic energy of the undamaged state can be additively decomposed into a damaged and an intact part, $\psi_0 = \psi_0^d + \psi_0^i$, such that the degradation function h only acts on the damaged part:

$$\psi^e(\boldsymbol{\epsilon}, \bar{d}) = h(\bar{d})\psi_0^d(\boldsymbol{\epsilon}) + \psi_0^i(\boldsymbol{\epsilon}) \tag{29}$$

Substitution of Eqs. (25) and (29) into Eq. (24) yields the total potential energy of the smeared formulation for brittle fracture:

$$\Psi_{\text{pot}} = \int_{\Omega} \left(h(\bar{d})\psi_0^d(\boldsymbol{\epsilon}) + \psi_0^i(\boldsymbol{\epsilon}) + \mathcal{G}_c\gamma_\ell(\bar{d}, \nabla\bar{d}) \right) \mathrm{d}V \tag{30}$$

Minimisation of Ψ_{pot} and introduction of a history field κ to enforce irreversibility (Miehe et al. 2010a) lead to the equilibrium equation:

$$\operatorname{div}\boldsymbol{\sigma} = \mathbf{0} \tag{31}$$

and

$$h'(\bar{d})\kappa + \frac{\mathcal{G}_c}{2\ell}\left(\bar{d} - 4\ell^2\nabla^2\bar{d}\right) = 0 \tag{32}$$

subject to the boundary conditions $\mathbf{n}\cdot\boldsymbol{\sigma} = \bar{\mathbf{t}}$, $\mathbf{u} = \bar{\mathbf{u}}$, $\mathbf{n}\cdot\nabla\bar{d} = 0$, with $\bar{\mathbf{t}}$ and $\bar{\mathbf{u}}$ the prescribed boundary tractions and displacements, respectively. We note that in the phase-field literature often the symbol $\mathcal{H}$ is used instead of κ to denote the history variable. The present notation is preferred, however, as it emphasises the similarity with gradient-enhanced damage models. The Cauchy stress $\boldsymbol{\sigma}$ and the driving force $\mathcal{F}$ are derived according to standard thermodynamic arguments:

$$\boldsymbol{\sigma} = \frac{\partial\psi^e}{\partial\boldsymbol{\epsilon}} = h(\bar{d})\frac{\partial\psi_0^d}{\partial\boldsymbol{\epsilon}} + \frac{\partial\psi_0^i}{\partial\boldsymbol{\epsilon}} \tag{33}$$

and

$$\mathcal{F} = -\frac{\partial\psi^e}{\partial\bar{d}} = -h'(\bar{d})\kappa \tag{34}$$

with ψ^e as defined in Eq. (29). The history field reads

$$\kappa = \max\psi_0^d(\boldsymbol{\epsilon}) \tag{35}$$

3.2 Numerical Example

We consider the single-edge notched specimen of Fig. 5 with pure shear boundary conditions. A prescribed horizontal displacement, $\bar{u}_s$, is applied to the top boundary. Plane–strain conditions and linear elasticity have been assumed with the Lamé parameters $\lambda = 121.15\,\text{kN/mm}^2$ and $\mu = 80.77\,\text{kN/mm}^2$. The critical energy release rate $\mathcal{G}_c = 2.7\times10^{-3}\,\text{kN/mm}$. The fracture length scale is equal to $\ell = 0.015\,\text{mm}$. To accurately capture the evolution of the phase field, the mesh is refined along the anticipated crack path, Fig. 5. In phase-field models for brittle fracture, the length

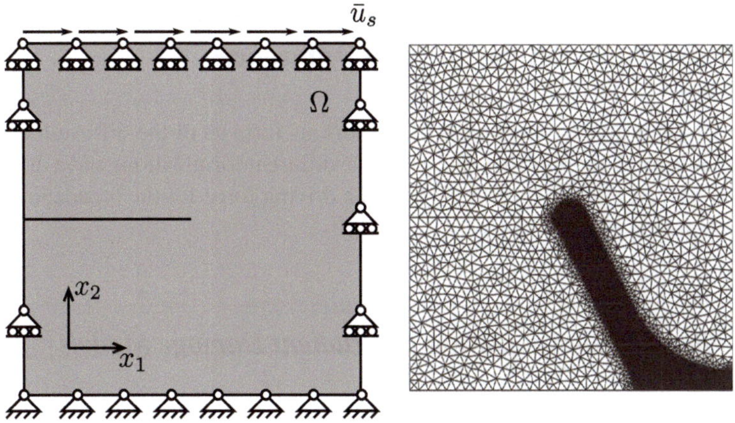

Fig. 5 Problem set-up and finite element mesh for the single-edge notched pure shear test

scale must be taken as small as possible to obtain an accurate approximation to the underlying linear elastic fracture mechanics problem. Evidently, very fine grids are then required. In this case, the characteristic element size in this refinement region $h_{\text{elem}} = \ell/4 = 0.00375\,\text{mm}$, which results in 26,472 elements.

Figure 6 gives the solution when the shear crack has developed almost completely. The force–displacement diagrams are shown for various step sizes. The overestimation of the amount of dissipated energy for large step sizes is a direct consequence of the employed staggered solution procedure (Vignollet et al. 2014). Most noteworthy, however, is that, different from the gradient damage model, the crack does not broaden in the wake of the crack tip upon propagation

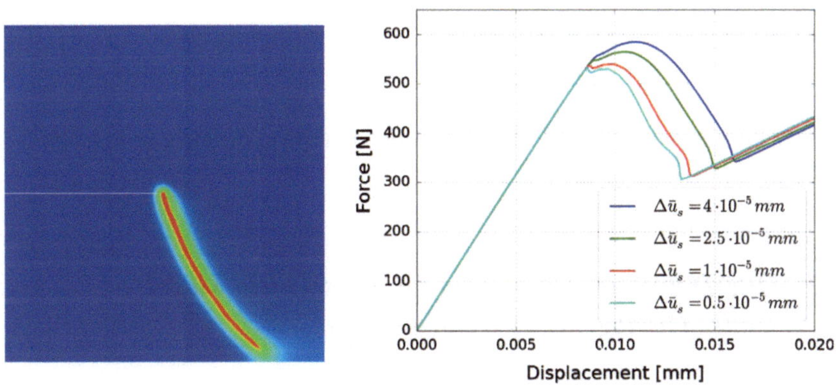

Fig. 6 Phase-field solution and step size study ($\Delta\bar{u}_s$) for the staggered algorithm (edge notched pure shear test)

4 Gradient Damage Models and Phase-Field Models for Fracture: Similarities and Differences

Below we shall give a comparison of the various formats of the diffusion equation for the damage/phase field that ensues for the different formulations and a discussion of the importance of the specific form of the driving force for the broadening of the damage zone.

4.1 *The Diffusion Equation in Gradient Damage Models*

The Euler–Lagrange equations that govern the phase-field evolution are, cf. Eq. (32):

$$\bar{d} - 4\ell^2 \nabla^2 \bar{d} = -\frac{2\ell h'(\bar{d})\kappa}{\mathcal{G}_c} \tag{36}$$

where implicit use has been made of the natural boundary condition $\boldsymbol{n}_\Gamma \cdot \nabla \bar{d} = 0$. In the context of phase-field models for fracture, this can be interpreted as a condition requiring cracks to be perpendicular to the external boundaries of the domain. The history parameter κ obeys the Kuhn–Tucker loading–unloading conditions with the loading function defined as

$$f(\boldsymbol{\epsilon}, \kappa) = \psi_0^d(\boldsymbol{\epsilon}) - \kappa \tag{37}$$

The partial differential equation that governs the evolution of the phase field, Eq. (36), is closely related to gradient damage models. In fact, when defining the local damage field as

$$d = -\frac{2\ell h'(\bar{d})\kappa}{\mathcal{G}_c} \tag{38}$$

and the gradient parameter $g = (2\ell)^2$, the implicit form of the damage equation (17) is recovered: $\bar{d} - g\nabla^2 \bar{d} = d$.

4.2 *One-Dimensional Solution for Phase-Field Models Under Homogeneous Deformations*

The behaviour of the phase-field model can further be illustrated from the homogeneous solution to the one-dimensional problem. Taking $\psi_0^d = \frac{1}{2}E\epsilon^2$, $\sigma = E\epsilon$ and $h(\bar{d}) = (1 - \bar{d})^2$, the homogeneous solution to Eq. (36) shown in Fig. 7 is obtained. Hence, after a sufficient amount of damage has accumulated, i.e. for $\bar{d}_c = \frac{1}{4}$, a critical

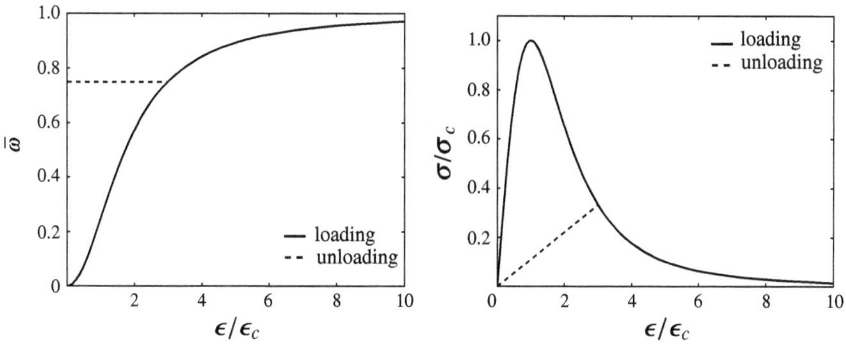

Fig. 7 Homogeneous solution to the one-dimensional phase-field formulation in brittle fracture for the degradation function $h(\bar{d}) = (1 - \bar{d})^2$

load level is reached, with

$$\sigma_c = \frac{9}{16}\sqrt{\frac{\mathcal{G}_c E}{6\ell}} \qquad \epsilon_c = \sqrt{\frac{\mathcal{G}_c}{6\ell E}} \tag{39}$$

after which softening occurs. In the phase-field formulation, unloading occurs using a secant stiffness, similar to damage models. From the expression for the critical stress it is clear that when the internal length scale ℓ goes to zero, the critical stress goes to infinity. This behaviour of the phase-field model correctly replicates linear elastic fracture mechanics, which is also not capable of nucleating fractures in the absence of singularities.

4.3 *The Driving Force*

As discussed, a marked difference between gradient-enhanced damage models and the phase-field approach to brittle fracture is the observation that the latter class of models does not lead to a broadening of the damage zone in the wake of the tip of the fracture process zone. This can be directly explained when comparing the expressions that result from the driving force, defined in Eq. (34).

For damage models, the degradation function typically has the format

$$h(\bar{d}) = 1 - \bar{d} \tag{40}$$

Substitution into Eq. (34) then results in

$$\mathcal{F} = \kappa \tag{41}$$

Hence, $\mathcal{F}$ does not vanish at complete loss of integrity, i.e. when $\bar{d} = 1$, which explains the continued broadening of the damage zone. This is different for phase-field models. Indeed, the conditions that are imposed on the degradation function, in particular Eq. $(26)_3$, make that the driving force vanishes when $\bar{d} = 1$. For instance, the quadratic degradation function of Eq. (27), $h(\bar{d}) = (1 - \bar{d})^2$, results in:

$$\mathcal{F} = (1 - \bar{d})\kappa \tag{42}$$

which becomes zero when $\bar{d} = 1$, and ensures a constant band width in the wake of the crack tip.

5 Concluding Remarks

The gradient damage and the phase-field approach to brittle fracture are almost identical in terms of their mathematical structure, and therefore the difference between gradient damage models and phase-field models is mainly in their interpretation. Whereas in gradient damage models, the left-hand side of Eq. (17) can be interpreted as a spatial averaging operator, the left-hand side in phase-field models, Eq. (36), follows from the regularised energy variation due to fracture evolution. The right-hand-side in Eq. (17) can therefore be interpreted as a local damage field, while that in Eq. (36) mimicks the thermodynamic driving force for the smeared fracture. It is in this right-hand side that the most relevant differences enter between gradient damage models and phase-field models for brittle fracture. From Eq. (38) we observe that the phase-field formulation naturally introduces a length scale into the driving force for the damage field. In combination with the vanishing derivative of the degradation function at complete loss of integrity this ensures that, once a phase-field fracture has fully developed, it does not broaden. Equation (38) suggests that the vanishing derivative of the degradation function is key to driving the internal length scale to zero, which closely resembles a strategy proposed in gradient damage modelling to avoid damage zone broadening (Geers et al. 1998).

References

Amor, H., Marigo, J.J., Maurini, J.: Regularized formulation of the variational brittle fracture with unilateral contact: numerical experiments. J. Mech. Phys. Solids **57**, 1209–1229 (2009)

Askes, H., Pamin, J., de Borst, R.: Dispersion analysis and element-free Galerkin solutions of second and fourth-order gradient-enhanced damage models. Int. J. Numer. Meth. Eng. **49**, 811–832 (2000)

Belytschko, T., Black, T.: Elastic crack growth in finite elements with minimal remeshing. Int. J. Numer. Meth. Eng. **45**, 601–620 (1999)

Bourdin, B., Francfort, G.A., Marigo, J.J.: Numerical experiments in revisited brittle fracture. J. Mech. Phys. Solids **48**, 797–826 (2000)

Bourdin, B., Francfort, G.A., Marigo, J.J.: The variational approach to fracture. J. Elast. **91**, 5–148 (2008)
Comi, C.: Computational modelling of gradient-enhanced damage in quasi-brittle materials. Mech. Cohesive-frict. Mater. **4**, 17–36 (1999)
de Borst, R., Benallal, A., Heeres, O.M.: A gradient-enhanced damage approach to fracture. J. Phys. IV **C6**, 491–502 (1996)
Francfort, G.A., Marigo, J.J.: Revisiting brittle fracture as an energy minimization problem. J. Mech. Phys. Solids **46**, 1319–1342 (1998)
Geers, M.G.D., de Borst, R., Brekelmans, W.A.M., Peerlings, R.H.J.: Strain-based transient-gradient damage model for failure analyses. Comput. Methods Appl. Mech. Eng. **160**, 133–153 (1998)
Ingraffea, A.R., Saouma, V.: Numerical modelling of discrete crack propagation in reinforced and plain concrete. Fracture Mechanics of Concrete, pp. 171–225. Martinus Nijhoff Publishers, Dordrecht (1985)
Miehe, C., Hofacker, M., Welschinger, F.: A phase field model for rate-independent crack propagation: Robust algorithmic implementation based on operator splits. Comput. Methods Appl. Mech. Eng. **199**, 2765–2778 (2010a)
Miehe, C., Welschinger, F., Hofacker, M.: Thermodynamically consistent phase-field models of fracture: variational principles and multi-field FE implementations. Int. J. Numer. Meth. Eng. **83**, 1273–1311 (2010b)
Moës, N., Dolbow, J., Belytschko, T.: A finite element method for crack growth without remeshing. Int. J. Numer. Meth. Eng. **46**, 131–150 (1999)
Ngo, D., Scordelis, A.C.: Finite element analysis of reinforced concrete beams. J. Am. Concr. Inst. **64**, 152–163 (1967)
Peerlings, R.H.J., de Borst, R., Brekelmans, W.A.M., de Vree, H.P.J.: Gradient-enhanced damage for quasi-brittle materials. Int. J. Numer. Meth. Eng. **39**, 3391–3403 (1996)
Peerlings, R.H.J., Geers, M.G.D., de Borst, R., Brekelmans, W.A.M.: A critical comparison of nonlocal and gradient-enhanced softening continua. Int. J. Solids Struct. **38**, 7723–7746 (2001)
Pijaudier-Cabot, G., Bažant, Z.P.: Nonlocal damage theory. ASCE J. Eng. Mech. **113**, 1512–1533 (1987)
Remmers, J.J.C., de Borst, R., Needleman, A.: A cohesive segments method for the simulation of crack growth. Comput. Mech. **31**, 69–77 (2003)
Réthoré, J., de Borst, R., Abellan, M.A.: A two-scale approach for fluid flow in fractured porous media. Int. J. Numer. Meth. Eng. **75**, 780–800 (2007)
Verhoosel, C.V., Scott, M.A., Hughes, T.J.R., de Borst, R.: An isogeometric analysis approach to gradient damage models. Int. J. Numer. Meth. Eng. **86**, 115–134 (2011)
Vignollet, J., May, S., de Borst, R., Verhoosel, C.V.: Phase-field models for brittle and cohesive fracture. Meccanica **49**, 2587–2601 (2014)
Vuong, A.V., Giannelli, G., Jüttler, B., Simeon, B.: A hierarchical approach to adaptive local refinement in isogeometric analysis. Comput. Methods Appl. Mech. Eng. **200**, 3554–3567 (2011)

Experiments on Damage and Failure Mechanisms in Ductile Metals at Different Loading Conditions

Michael Brünig, Steffen Gerke and Marco Schmidt

Abstract This paper deals with a phenomenological damage and failure model for ductile metals. The anisotropic continuum approach takes into account the effect of stress state on damage condition and damage rule corresponding to different mechanisms acting on the micro-scale. Different branches of the criteria are formulated depending on stress intensity, stress triaxiality, and the Lode parameter. A new experimental program will be discussed in detail to validate the proposed continuum framework. Experiments with aluminum alloys are performed using a biaxial testing machine allowing individual loading of flat specimens in two directions. Loads are recorded during loading of the specimens and digital image correlation technique has been used to analyze the strain states in critical regions of the specimens. The biaxial experiments cover a wide range of stress states in shear-tension and shear-compression regimes. They will extend understanding of stress-state-dependent damage and failure mechanisms in ductile metals.

1 Introduction

Damage and failure prediction is an important issue in analysis and design of metal forming processes. For example, it has been observed in many experiments with ductile metals that during loading or forming of material samples large, often localized inelastic deformations occur which may be accompanied by damage and failure processes acting on the micro-scale. In this context, theoretical models also taking into account negative stress triaxialities are of special interest since many industrial

M. Brünig (✉) · S. Gerke · M. Schmidt
Institut für Mechanik und Statik, Universität der Bundeswehr München,
85577 Neubiberg, Germany
e-mail: michael.bruenig@unibw.de

S. Gerke
e-mail: steffen.gerke@unibw.de

M. Schmidt
e-mail: m.schmidt@unibw.de

K. Naumenko and M. Aßmus (eds.), *Advanced Methods of Continuum Mechanics for Materials and Structures*, Advanced Structured Materials 60,
DOI 10.1007/978-981-10-0959-4_15

processes like rolling and forging involve compressive hydrostatic stress states. In these industrial processes, formation of damage and failure on the micro-level may then lead to macro-cracks and, thus, to final failure of structural elements. Many theoretical approaches are able to simulate in a phenomenological way growth and coalescence of micro-defects as well as corresponding macroscopic material softening under general loading conditions. They are based on internal scalar or tensorial damage variables whose increase is governed by evolution laws. Critical values of these damage variables may be seen as major parameters characterizing onset of fracture.

Recent investigations have elucidated that damage and failure mechanisms depend on the stress state acting in a material point. For example, during tension loading conditions with high positive stress triaxialities damage is mainly caused by nucleation, growth and coalescence of micro-voids, whereas during shear and compression loadings with small positive or negative stress triaxialities the prevailing damage mechanisms are formation and growth of micro-shear-cracks. In addition, combination of these basic microscopic mechanisms has been observed for moderate positive stress triaxialities, whereas no damage occurs in ductile metals for high negative stress states (see, for example, Bao and Wierzbicki 2004; Brünig et al. 2008, 2011). Thus, development of accurate and realistic phenomenological models requires detailed experimental analyses covering a wide range of tension, compression, and shear loading conditions. However, it is difficult to carry out experiments with negative stress triaxialities testing thin metal sheets and further research activities are also necessary in this field.

A number of experiments with carefully designed metal specimens has been proposed and discussed in the literature. For example, uniaxial tension tests with unnotched and differently notched specimens and corresponding numerical simulations have been performed to investigate dependence of damage and fracture on stress triaxiality (Becker et al. 1988; Bao and Wierzbicki 2004; Bonora et al. 2005; Bai and Wierzbicki 2008; Brünig et al. 2008, 2011; Gao et al. 2010; Dunand and Mohr 2011). However, these experiments on ductile metals with flat specimens showed stress triaxialities only in a small positive region. Thus, new geometries of uniaxially loaded specimens have been developed to analyze nearly zero stress triaxialities (Bao and Wierzbicki 2004; Brünig et al. 2008; Gao et al. 2010; Driemeier et al. 2010) where shear mechanisms occur in the critical regions. Furthermore, Driemeier et al. (2015) proposed a bi-failure specimen tested in a uniaxial tension machine generating first local failure at high stress triaxiality followed by final failure at low stress triaxiality.

Moreover, for other regions of stress triaxialities butterfly specimens with complex geometry have been developed (Mohr and Henn 2007; Bai and Wierzbicki 2008; Dunand and Mohr 2011) which can be loaded in uniaxial tension tests in different directions using special experimental equipment. Alternatively, two-dimensional experiments and different geometries of cruciform specimens have been investigated to analyze anisotropic plastic behavior of metal sheets for different loading paths (Demmerle and Boehler 1993; Müller and Pöhland 1996; Kuwabara 2007; Kulawinski et al. 2011). Further aspects for design of biaxially loaded specimens have been taken into account to develop stress-state-dependent damage criteria. In

Brünig et al. (2015a, b, c) biaxial experiments with new shear-tension specimens have been proposed to study in detail the stress-state-dependent damage and failure processes.

Since strain-based forming limit diagrams are strongly dependent on strain paths the application of stress-based fracture criteria is required to solve practical problems. Therefore, based on data taken from various experiments different continuum damage models have been proposed in the literature, see Chow and Wang (1987), Chaboche (1988), Lemaitre (1996), Voyiadjis and Kattan (1999), Brünig (2003), Naumenko et al. (2011), Altenbach et al. (2013), and Brünig et al. (2015b) among many others. It has been shown that anisotropic approaches based on tensorial damage variables are more suitable to model stress-state-dependent behavior but their practical applicability is often limited by large number of parameters and difficulties in their identification. Therefore, in the present paper a phenomenological continuum approach will be discussed taking into account stress-state-dependent damage functions. To validate the proposed framework a new experimental program is proposed where tests with biaxially loaded specimens are performed covering a wide range of two-dimensional loading conditions. Load–displacement curves are recorded and digital image correlation technique is used to analyze current strain states in critical regions of the specimens.

2 Continuum Damage Model

An extended version of the anisotropic continuum damage model proposed by Brünig (2003) is discussed predicting the inelastic deformation behavior as well as the evolution of ductile damage and failure in ductile metals. Phenomenological modeling of macroscopic damage and failure behavior corresponding to different processes acting on the micro-level caused by various loading conditions is based on combination of microscopic and macroscopic investigations. They take into account results of numerical simulations on the micro-level (Brünig et al. 2013, 2014) as well as results of various experiments with carefully designed specimens under different loading conditions (Brünig et al. 2008, 2015b, 2016; Driemeier et al. 2010). This phenomenological approach is briefly summarized in the present paper.

In the proposed continuum model the damage process is governed by the evolution of damage strains on the macro-level caused by various stress-state-dependent mechanisms on the micro-level. The continuum framework is based additive decomposition of the strain rate tensor into elastic, $\dot{\boldsymbol{H}}^{el}$, effective plastic, $\dot{\bar{\boldsymbol{H}}}^{pl}$, and damage parts, $\dot{\boldsymbol{H}}^{da}$. In addition, free energy functions with respect to undamaged and damaged configurations are introduced leading to an elastic law taking into account the deteriorating effect of damage on elastic material properties. Plastic behavior of ductile metals is governed by a yield condition and a flow rule and, in a similar way, damage behavior is characterized by a damage condition and a damage rule which will be discussed in more detail in the present paper.

Determination of onset and continuation of damage is based on the concept of damage surface formulated in stress space (Chow and Wang 1987; Brünig 2003). Thus, the damage condition

$$f^{da} = \alpha I_1 + \beta\sqrt{J_2} - \sigma = 0 \tag{1}$$

is expressed in terms of the stress invariants I_1 and J_2 of the stress tensor and the damage threshold σ represents the material toughness to micro-defect propagation, whereas the variables α and β denote damage mode parameters corresponding to the different damage mechanisms acting on the micro-level: shear modes for negative stress triaxialities, void-growth-dominated modes for large positive triaxialities and mixed modes (simultaneous growth of voids and evolution of micro-shear-cracks) for lower positive stress triaxialities. Therefore, the damage mode parameters α and β in Eq. (1) depend on the stress intensity $\sigma_{eq} = \sqrt{3J_2}$, the stress triaxiality

$$\eta = \frac{\sigma_m}{\sigma_{eq}} = \frac{I_1}{3\sqrt{3J_2}} \tag{2}$$

defined as the ratio of the mean stress σ_m and the von Mises equivalent stress σ_{eq} as well as on the Lode parameter

$$\omega = \frac{2\tilde{T}_2 - \tilde{T}_1 - \tilde{T}_3}{\tilde{T}_1 - \tilde{T}_3} \quad \text{with } \tilde{T}_1 \geq \tilde{T}_2 \geq \tilde{T}_3 \tag{3}$$

expressed in terms of the principal stress components $\tilde{T}_1$, $\tilde{T}_2$ and $\tilde{T}_3$.

Based on micro-mechanical calculations with aluminum alloys considering deformation behavior of micro-defects in three-dimensionally loaded void-containing unit cells (Brünig et al. 2013) as well as on results of experiments with biaxially loaded specimens (Brünig et al. 2015b, 2016) the parameter α is taken to be

$$\alpha(\eta) = \begin{cases} -0.15 & \text{for } \eta \leq 0 \\ 0.33 & \text{for } \eta > 0 \end{cases} \tag{4}$$

and the parameter β is given by the nonnegative function

$$\beta(\eta, \omega) = \beta_0(\eta, \omega = 0) + \beta_\omega(\omega) \geq 0, \tag{5}$$

with

$$\beta_0(\eta) = -1.28\,\eta + 0.85 \tag{6}$$

and

$$\beta_\omega(\omega) = -0.017\,\omega^3 - 0.065\,\omega^2 - 0.078\,\omega\,. \tag{7}$$

Furthermore, the damage strain rate tensor is given by

$$\dot{\boldsymbol{H}}^{da} = \dot{\mu}\left(\bar{\alpha}\,\frac{1}{\sqrt{3}}\,\mathbf{1} + \bar{\beta}\,\boldsymbol{N}\right) \tag{8}$$

where $\dot{\mu}$ is a nonnegative scalar-valued factor and $\boldsymbol{N} = \frac{1}{2\sqrt{J_2}}\,\text{dev}\tilde{\boldsymbol{T}}$ represents the normalized stress related deviatoric tensor. The parameters $\bar{\alpha}$ and $\bar{\beta}$ are kinematic variables describing the portion of volumetric and isochoric damage-based deformations. They are given by

$$\bar{\alpha}(\eta) = \begin{cases} 0 \ \text{ for } \ \eta \leq 0 \\ 0.5714\,\eta \ \text{ for } \ 0 < \eta \leq 1.75 \\ 1 \ \text{ for } \ \eta > 1.75 \end{cases} \tag{9}$$

and

$$\bar{\beta}(\eta,\,\omega) = \bar{\beta}_0(\eta) + f_\beta(\eta)\,\bar{\beta}_\omega(\omega) \tag{10}$$

with

$$\bar{\beta}_0(\eta) = \begin{cases} 0.87 \ \text{ for } \ \eta \leq \frac{1}{3} \\ 0.97875 - 0.32625\,\eta \ \text{ for } \ \frac{1}{3} < \eta \leq 3\,, \\ 0 \ \text{ for } \ \eta > 3 \end{cases} \tag{11}$$

$$f_\beta(\eta) = -0.0252 + 0.0378\,\eta \tag{12}$$

and

$$\bar{\beta}_\omega(\omega) = \begin{cases} 1 - \omega^2 \ \text{ for } \ \eta \leq \frac{2}{3} \\ 0 \ \text{ for } \ \eta > \frac{2}{3} \end{cases} \quad . \tag{13}$$

It can be seen that the macroscopic damage rule (8) takes into account volumetric parts (first term in Eq. (8)) corresponding to isotropic growth of voids on the micro-scale as well as deviatoric parts (second term in Eq. (8)) corresponding to anisotropic evolution of micro-shear-cracks, respectively. Therefore, both basic damage mechanisms discussed above (growth of isotropic voids and evolution of micro-shear-cracks) acting on the micro-level are involved in the macroscopic damage rule (8) of the phenomenological continuum model. In addition, it should be noted that the parameters α and β as well as $\bar{\alpha}$, $\bar{\beta}$ and $\bar{\delta}$ correspond to damage processes on the micro-level which are nearly identical for many ductile metals and, thus, can be used for this class of materials.

continuously increased by the velocity of 0.04 mm/min while the same displacement is applied on the cylinder 2.2 on opposite side of the same axis, see Fig. 3a. The generated force $F_{2.1}$ is now taken, multiplied by the factor α, and applied on the cylinder 1.1 as $F_{1.1}$ causing the machine displacement $u^M_{1.1}$. In continuation the same displacement is applied as $u^M_{1.2}$ on the cylinder 1.2 on the opposite side of the same axis. By applying this mainly displacement driven technique, a very stable procedure is achieved allowing to maintain an almost constant load ratio $F_1 : F_2$ throughout the entire experiment.

Within the experimental program discussed above digital image correlation (DIC) technique has been applied to analyze deformation behavior in the critical zones of the specimens. The full three-dimensional displacement field of the specimen surface was measured with a Q-400 digital 3D image correlation system provided by Dantec/Limess. The stereo setting consists of two 2.0Mpx cameras equipped with Ricoh FL-CC7528-2M 75 mm F/2.8 lenses and extension tubes of 10 mm. The evaluation was realized with the corresponding Istra 4D software provided with the system. One side of the specimen was first sprayed with a white acrylic lacquer and then the speckle pattern shown in Fig. 3c was sprayed on with a black acrylic lacquer using an air brush system to achieve a sufficiently fine pattern. Specimen preparation was realized shortly before the experiment was performed to avoid excessive curing and, thus, no exfoliation of the coating was observed during the experiment. Furthermore, the biaxial test machine equipment was augmented by an interface allowing the export of eight signals to the DIC system. Consequently, the displacements as well as the forces in axes directions were transferred and stored with the data sets of the image correlation system at a frequency of 0.25 Hz while the velocity of the machine was 0.04 mm/min. The respective loads are recorded during the tests. To analyze the behavior of the inelastically strained part of the specimen for different loading conditions the displacements are taken from movements of the red points in Fig. 3b, c: the horizontal displacement is $u_1 = u_{1.1} + u_{1.2}$ and the vertical one is

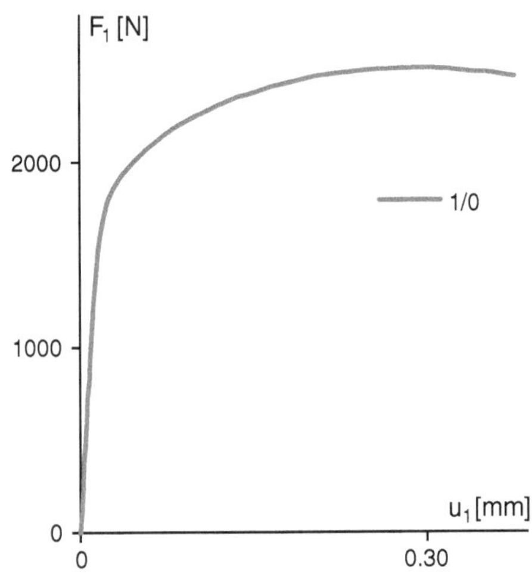

Fig. 4 Load–displacement curve $F_1 - u_1$ for tension loading

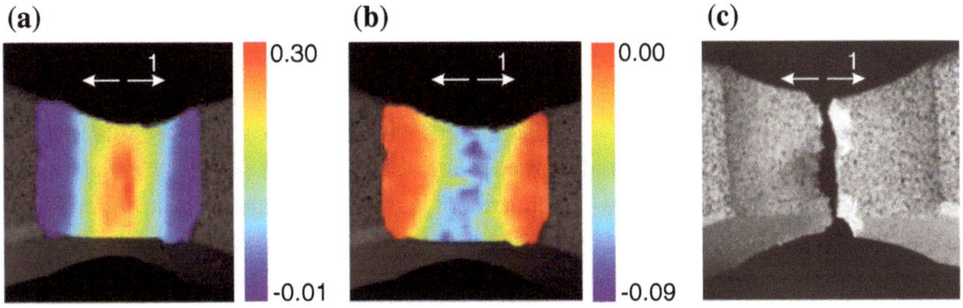

Fig. 5 Strains **a** ε_{xx}, **b** ε_{yy} before fracture and **c** fracture pattern under tension loading

$u_2 = u_{2.1} + u_{2.2}$. Furthermore, the logarithmic strains given by the Istra 4D software with respect to the coordinate system indicated in Fig. 3a are studied shortly before fracture occurrence.

First experiment is the uniaxial tension test with the load ratio $F_1 : F_2 = 1 : 0$. The corresponding load–displacement curve $F_1 - u_1$ is shown in Fig. 4. After elastic loading smaller increase in load is observed caused by inelastic behavior due to plastic and damage processes in the material. The load maximum of $F_1 = 2560$ N is reached at the displacement $u_1 = 0.30$ mm and after this point small decrease in load corresponds to fast increase in damage leading to fracture of the specimen at $u_1 = 0.38$ mm. In addition, for this tension test distributions of different strains shortly before fracture occurs analyzed by DIC are shown in Fig. 5. In particular, localization of the normal strains ε_{xx} in loading direction in the center of the specimen is observed with maximum strains of about 30 % (Fig. 5a). In this zone negative strains ε_{yy} perpendicular to the loading direction can be seen indicating necking of the specimen in this part (Fig. 5b). In this zone of localized normal strains fracture occurs. Figure 5c shows the fracture line nearly normal to the loading direction which is characteristic for failure of notched tensile specimens made of ductile metals.

For combined shear-tension and shear-compression tests load–displacement curves $F_1 - u_1$ are shown in Fig. 6. In particular, for the shear-tension test with $F_1 : F_2 = 2 : 1$ load maximum of $F_1 = 2100$N is reached shortly before fracture

Fig. 6 Load–displacement curves $F_1 - u_1$ for different loading conditions

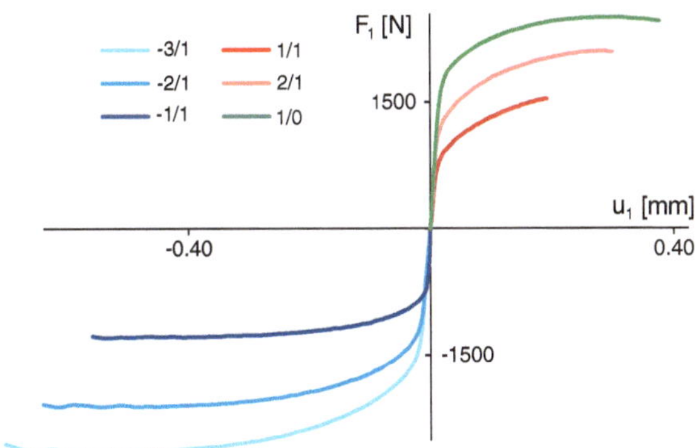

occurs at the displacement $u_1 = 0.30\,\text{mm}$ whereas for $F_1 : F_2 = 1 : 1$ with larger amount of shear loading the specimen failed at $F_1 = 1500\,\text{N}$ and $u_1 = 0.20\,\text{mm}$ but—surprisingly—no decrease in load is observed. The load–displacement curve for the pure tension test discussed above is also included in Fig. 6. It can be seen that increasing effect of shear leads to smaller maximum loads F_1 and smaller displacements u_1 at failure. In addition, load–displacement curves for experiments under shear-compression loading conditions are also shown in Fig. 6. In particular, for the shear-compression test with $F_1 : F_2 = -3 : 1$ load minimum of $F_1 = -2600$, N is reached at the displacement $u_1 = -0.52\,\text{mm}$. Then, a small increase in load is observed up to fracture at the displacement $u_1 = -0.66\,\text{mm}$. For $F_1 : F_2 = -2 : 1$ the load minimum is $F_1 = 2100\,\text{N}$ at $u_1 = -0.44\,\text{mm}$ and fracture occurs at $u_1 = -0.64\,\text{mm}$ whereas for $F_1 : F_2 = -1 : 1$ the minimum is $F_1 = 1250\,\text{N}$ at $u_1 = -0.40\,\text{mm}$ with final fracture at $u_1 = -0.56\,\text{mm}$. Thus, higher superimposed shear leads to smaller load minima and smaller displacements at fracture. In comparison with the shear-tension experiments the behavior of shear-compression loaded specimens is more ductile.

In addition, different load–displacement curves $F_2 - u_2$ for the biaxially loaded specimens are shown in Fig. 7. For the simple shear test ($F_1 : F_2 = 0 : 1$) the $F_2 - u_2$ curve even in the inelastic regime shows an increase in load and final fracture occurs at the load maximum of $F_2 = 1800$ N and at the fracture displacement $u_2 = 1.90$ mm. Decrease in load maxima is observed when the shear test is superimposed by tension loading. For example, for the load ratio $F_1 : F_2 = 1 : 1$ the maximum load is $F_2 = 1550$ N with the fracture displacement $u_2 = 1.06$ mm and for $F_1 : F_2 = 2 : 1$ it is only $F_2 = 1050$ N with the fracture displacement $u_2 = 0.76$ mm. This, again, indicates that with increasing superimposed tension loading the behavior of the biaxially loaded specimen becomes more brittle. On the other hand, decrease in load maxima is also observed when the shear test is superimposed by compression loading. In particular, for the load ratio $F_1 : F_2 = -1 : 1$ the maximum load is $F_2 = 1300$ N with the fracture displacement $u_2 = 1.36$ mm, for the load ratio $F_1 : F_2 = -2 : 1$ the maximum load is $F_2 = 1050$ N with the fracture displacement $u_2 = 1.16$ mm and for the load ratio $F_1 : F_2 = -3 : 1$ the maximum load is only $F_2 = 900$ N with

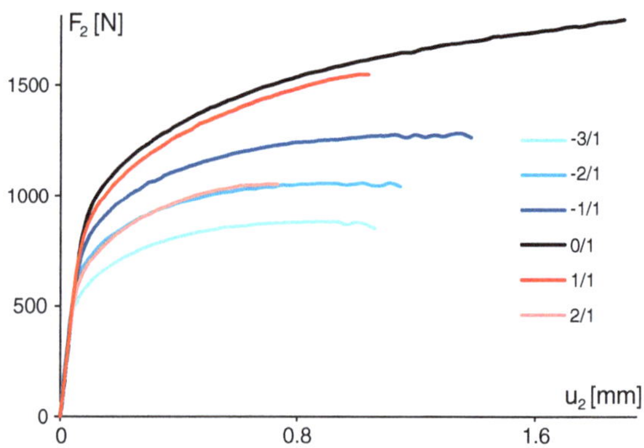

Fig. 7 Load–displacement curves $F_2 - u_2$ for different loading conditions

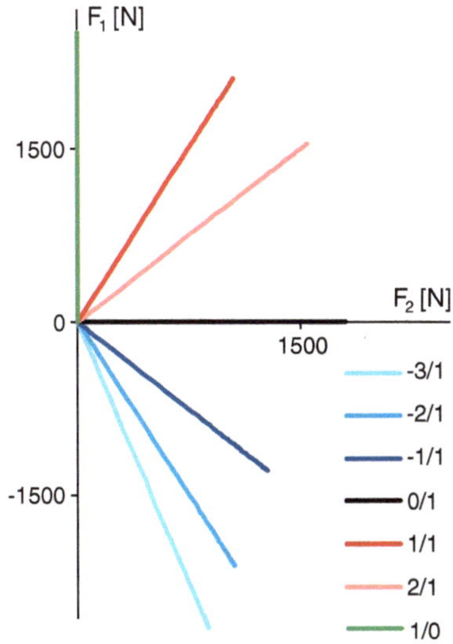

Fig. 8 $F_1 - F_2$ curves for different loading conditions

the fracture displacement $u_2 = 1.06$ mm. In these shear-compression experiments the $F_2 - u_2$ curves show pronounced maximum plateaus and shortly before fracture a slight decrease in load can be observed. Scattering in the curves may be caused by resistances between the grains during the shear processes on the micro-level.

Caused by the biaxial loading process discussed above the ratio of forces $F_1 : F_2$ remains constant during the entire respective experiment, see Fig. 8. Consequently, due to nonlinear material behavior the ratio of displacements $u_1 : u_2$ shown in Fig. 9 is a nonlinear function. In detailed analyses of the experimental results constant ratio of forces is required because this will lead to nearly constant stress states in the material sample. Then, the test data can be used to study the stress-state-dependent damage and failure behavior of ductile metals in an efficient way.

Distribution of different strain components in the notched central part of the specimens are shown in Figs. 10, 11 and 12 for the investigated loading conditions. For

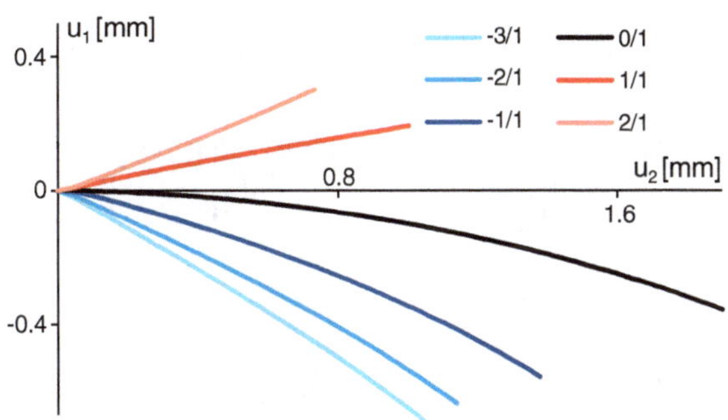

Fig. 9 $u_1 - u_2$ curves for different loading conditions

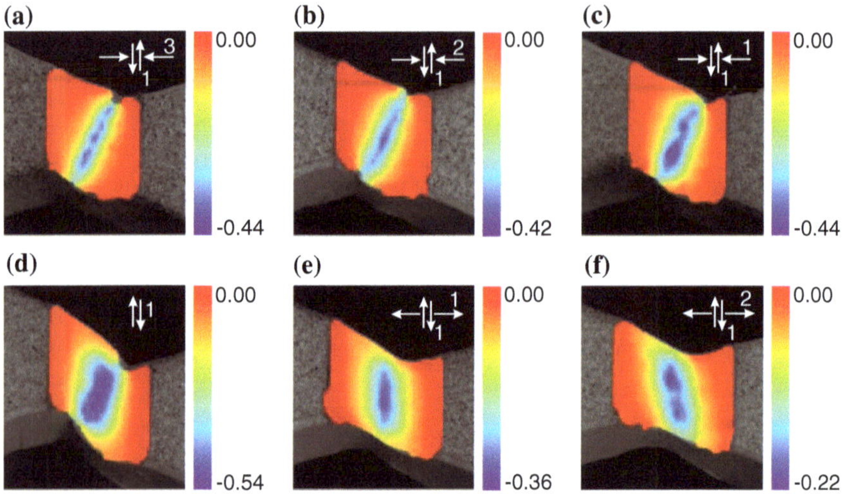

Fig. 10 Shear strains ε_{xy} for different loading conditions before fracture

example, localized bands of the shear strains ε_{xy} occur for each loading condition with values up to 54 % for $F_1 : F_2 = 0 : 1$, slightly smaller values in the shear-compression range and smaller values for shear-tension loading showing earlier failure as discussed above (Fig. 10). In these bands with larger shear strains the normal strains ε_{xx} also reach their extreme values. In the shear-compression range localized strain components up to $-50\,\%$ can be seen for $F_1 : F_2 = -3 : 1$ whereas under shear-tension loading normal strains up to 34 % occur in the experiments (Fig. 11).

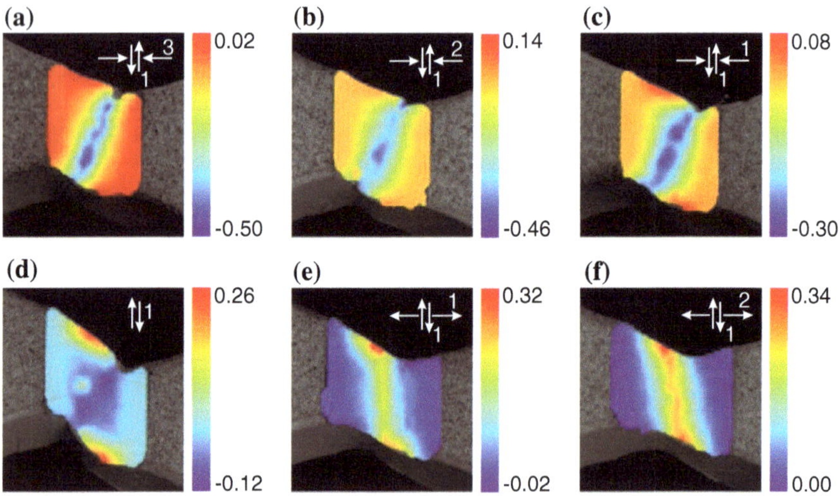

Fig. 11 Strains ε_{xx} for different loading conditions before fracture

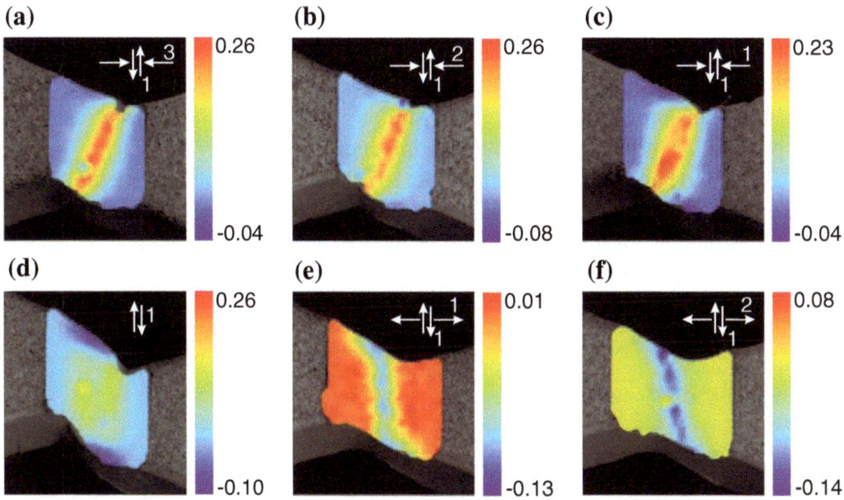

Fig. 12 Strains ε_{yy} for different loading conditions before fracture

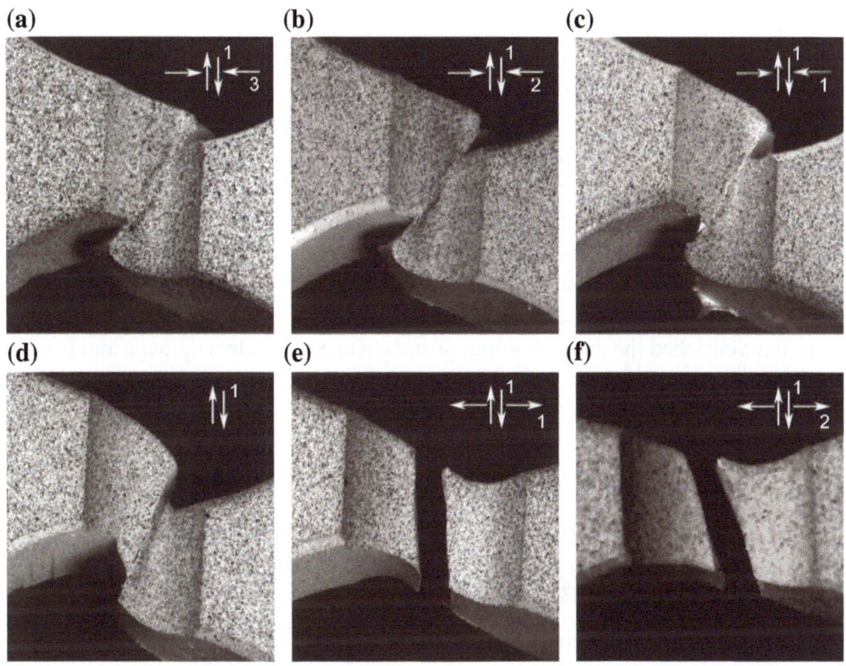

Fig. 13 Fracture patterns for different loading conditions

Similar behavior can be seen in Fig. 12 showing the distributions of the strain component ε_{yy} for the investigated loading conditions. For shear-compression loading strains up to 26 % can be observed in the experiments in the localized bands whereas under shear-tension conditions negative strains up to −14 % have been measured corresponding to the necking behavior in this region.

The localized bands of shear strains (Fig. 10) nicely correspond the fracture modes shown in Fig. 13. All tests are dominated by shear loading leading to shear fracture mechanisms which are superimposed by compression and tension modes. This leads to different angles of the fracture lines.

4 Conclusions

The paper has discussed a phenomenological continuum model taking into account the effect of stress state on damage and fracture mechanisms acting on different scales. Damage mode functions have been proposed which are able to simulate all relevant effects observed in different experiments.

A new experimental program has been proposed and series of tests with biaxially loaded specimens have been performed. Tests with shear-tension and shear-compression conditions cover a large range of positive and negative stress states. Digital image correlation technique has been used to visualize the strain fields in critical regions of the tested specimens. Detailed information on load-deformation behavior as well as on strain states have been delivered especially in the critical regions of the specimens where large inelastic deformations and ductile failure occurred. The results of the presented experimental program can be used to validate the proposed stress-state-dependent constitutive equations of the proposed continuum model. In addition, the experimental results and observations based on biaxially loaded specimens have enriched the understanding of stress-state-dependent damage and fracture processes in ductile metals.

References

Altenbach, H., Kozhar, S., Naumenko, K.: Modeling creep damage of an aluminum-silicon eutectic alloy. Int. J. Damage Mech. **22**, 683–698 (2013)

Bai, Y., Wierzbicki, T.: A new model of metal plasticity and fracture with pressure and Lode dependence. Int. J. Plast. **24**, 1071–1096 (2008)

Bao, Y., Wierzbicki, T.: On the fracture locus in the equivalent strain and stress triaxiality space. Int. J. Mech. Sci. **46**, 81–98 (2004)

Becker, R., Needleman, A., Richmond, O., Tvergaard, V.: Void growth and failure in notched bars. J. Mech. Phys. Solids **36**, 317–351 (1988)

Bonora, N., Gentile, D., Pirondi, A., Newaz, G.: Ductile damage evolution under triaxial state of stress: theory and experiments. Int. J. Plast. **21**, 981–1007 (2005)

Brünig, M.: An anisotropic ductile damage model based on irreversible thermodynamics. Int. J. Plast. **19**, 1679–1713 (2003)

Brünig, M., Chyra, O., Albrecht, D., Driemeier, L., Alves, M.: A ductile damage criterion at various stress triaxialities. Int. J. Plast. **24**, 1731–1755 (2008)

Brünig, M., Albrecht, D., Gerke, S.: Numerical analyses of stress-triaxiality-dependent inelastic deformation behavior of aluminum alloys. Int. J. Damage Mech. **20**, 299–317 (2011)

Brünig, M., Gerke, S., Hagenbrock, V.: Micro-mechanical studies on the effect of the stress triaxiality and the Lode parameter on ductile damage. Int. J. Plast. **50**, 49–65 (2013)

Brünig, M., Gerke, S., Hagenbrock, V.: Stress-state-dependence of damage strain rate tensors caused by growth and coalescence of micro-defects. Int. J. Plast. **63**, 49–63 (2014)

Brünig, M., Brenner, D., Gerke, S.: Modeling of stress-state-dependent damage and failure of ductile metals. Appl. Mech. Mater. **784**, 35–42 (2015a)

Brünig, M., Brenner, D., Gerke, S.: Stress state dependence of ductile damage and fracture behavior: experiments and numerical simulations. Eng. Fract. Mech. **141**, 152–169 (2015b)

Brünig, M., Gerke, S., Brenner, D.: Experiments and numerical simulations on stress-state-dependence of ductile damage criteria. In: Altenbach, H., Brünig, M. (eds.) Inelastic Behavior of Materials and Structures Under Monotonic and Cyclic Loading. Advanced Structured Materials, pp. 17–34. Springer, Berlin (2015c)

Brünig, M., Gerke, S., Schmidt, M. (2016) Biaxial experiments and phenomenological modeling of stress-state-dependent ductile damage and fracture. Int. J. Fract. (submitted for publication)

Chaboche, J.: Continuum damage mechanics. Part I: general concepts. J. Appl. Mech. **55**, 59–64 (1988)

Chow, C., Wang, J.: An anisotropic theory of continuum damage mechanics for ductile fracture. Eng. Fract. Mech. **27**, 547–558 (1987)

Demmerle, S., Boehler, J.: Optimal design of biaxial tensile cruciform specimens. J. Mech. Phys. Solids **41**, 143–181 (1993)

Driemeier, L., Brünig, M., Micheli, G., Alves, M.: Experiments on stress-triaxiality dependence of material behavior of aluminum alloys. Mech. Mater. **42**, 207–217 (2010)

Driemeier, L., Moura, R., Machado, I., Alves, M.: A bifailure specimen for accessing failure criteria performance. Int. J. Plast. **71**, 62–86 (2015)

Dunand, M., Mohr, D.: On the predictive capabilities of the shear modified Gurson and the modified Mohr-Coulomb fracture models over a wide range of stress triaxialities and Lode angles. J. Mech. Phys. Solids **59**, 1374–1394 (2011)

Gao, X., Zhang, G., Roe, C.: A study on the effect of the stress state on ductile fracture. Int. J. Damage Mech. **19**, 75–94 (2010)

Kulawinski, D., Nagel, K., Henkel, S., Hübner, P., Fischer, H., Kuna, M., Biermann, H.: Characterization of stress-strain behavior of a cast trip steel under different biaxial planar load ratios. Eng. Fract. Mech. **78**, 1684–1695 (2011)

Kuwabara, T.: Advances in experiments on metal sheet and tubes in support of constitutive modeling and forming simulations. Int. J. Plast. **23**, 385–419 (2007)

Lemaitre, J.: A Course on Damage Mechanics. Springer, Berlin (1996)

Mohr, D., Henn, S.: Calibration of stress-triaxiality dependent crack formation criteria: a new hybrid experimental-numerical method. Exp. Mech. **47**, 805–820 (2007)

Müller, W., Pöhland, K.: New experiments for determining yield loci of sheet metal. Mater. Process. Technol. **60**, 643–648 (1996)

Naumenko, K., Altenbach, H., Kutschke, A.: A combined model for hardening, softening, and damage processes in advanced heat resistence steels at elevated temperature. Int. J. Damage Mech. **20**, 578–597 (2011)

Voyiadjis, G., Kattan, P.: Advances in Damage Mechanics: Metals and Metal Matrix Composites. Elsevier, Amsterdam (1999)

Modeling of Damage Deactivation in Concrete Under Cyclic Compression

Artur Ganczarski and Marcin Cegielski

Abstract This paper was directly inspired by the presentation of Prof. Rodrigo Desmorat during The Second International Conference on Damage Mechanics, Troyes 2015 and the fruitful discussions between the first author and Profs. Holm Altenbach and Andre Dragon. The main goal of the work is not modeling the initial loading curve for concrete, but presenting the model of continuous damage deactivation able to capture subsequent unloading/reloading loops accompanying cyclic compression. A combination of conventional formulation of the universal curve of initial loading combined with Chaboche's type damage evolution law and the continuous damage deactivation of frictional character allows for a proper quantitative and qualitative mapping of the experimental data by Sinha et al. (1964).

1 Introduction

Concrete is a composite material consisting of the coarse aggregate and matrix of mortar which itself comprises a mixture of cement and sand. The physical behavior of concrete is very complex and generally depends on its structure determined by the ratio of water to cement, ratio of cement to aggregate, shape and size of aggregate, and the kind of cement. From the point of view continuum mechanics, the microstructure of concrete is ignored and is treated as the homogeneous and initially isotropic material.

Concrete belongs to the class of brittle materials, for which strain–stress curve strongly depends on the microcracks concentration (in sense of the continuous damage mechanics) and macrocracks (in sense of the fracture mechanics) existing both

A. Ganczarski (✉)
Institute of Applied Mechanics, Cracow University of Technology,
Al. Jana Pawła II 37, 31-864 Kraków, Poland
e-mail: artur.ganczarski@pk.edu.pl

M. Cegielski
Institute of Applied Informatics, Cracow University of Technology,
Al. Jana Pawła II 37, 31-864 Kraków, Poland
e-mail: mc@mech.pk.edu.pl

K. Naumenko and M. Aßmus (eds.), *Advanced Methods of Continuum Mechanics for Materials and Structures*, Advanced Structured Materials 60,
DOI 10.1007/978-981-10-0959-4_16

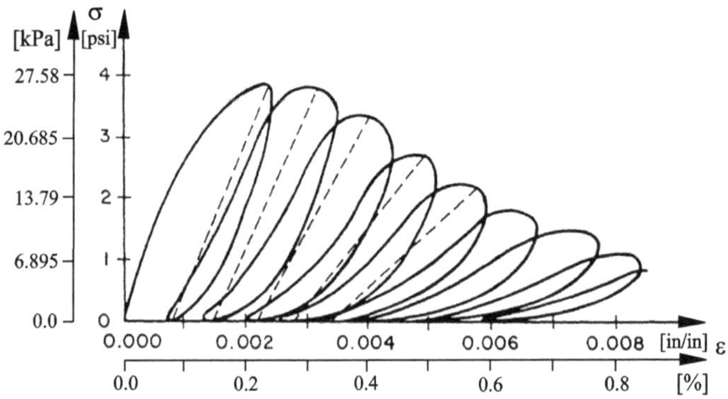

Fig. 1 Uniaxial cyclic compression of concrete, after Sinha et al. (1964)

inside and at the surface of the material. In particular, concrete containing initially lot of microcracks at the interfaces between coarse aggregates and the mortar exhibits strong segregation, shrinkage, and thermal expansion of a cement paste. Under applied loading, further microcracks appear at the aggregate and mortar interface, resulting in the macroscopic nonlinear behavior, which finally leads to macrocracks.

2 Test of the Uniaxial Cyclic Compression of Concrete, by Sinha et al. (1964)

The uniaxial compressive curve $\varepsilon - \sigma$ of concrete obtained by Sinha et al. (1964) is shown in Fig. 1. According to a convention used in the theory of brittle and granular materials, stress and strain corresponding to compression are considered as positive, hence the whole scheme was moved from the third quarter to the first quarter of the coordinate system. Curves referring to subsequent unloading/reloading phases are not straight line segments but loops that change size and have decreasing average slopes. If one assumes that the average slope is equal to a slope of straight line passing turning points and that the material behavior under unloading and reloading is linear elastic (dotted line in Fig. 1), then the average elastic modulus decreases with increasing strain. The above stiffness degradation behavior is strictly related to the damage growth, especially strongly manifesting in the post peak range.

3 Constitutive Model of a Material with Strain Softening

Many engineering materials such as concretes, rocks, and solids exhibit essential strain softening effect beyond the peak of maximum stress. Exemplary $\varepsilon - \sigma$ curves, obtained from the uniaxial compressive strain-controlled tests by Wischers (1978),

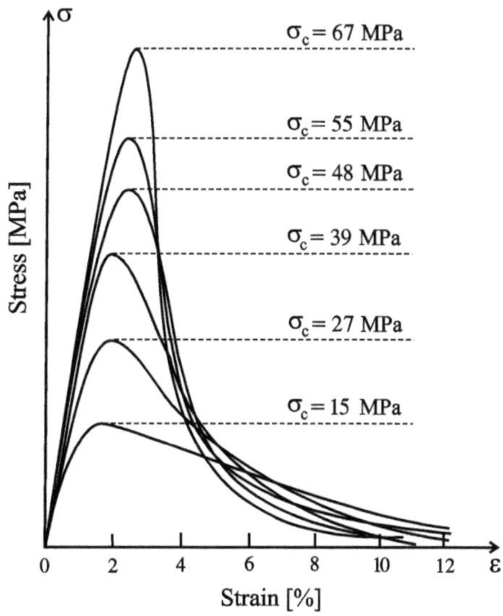

Fig. 2 Curves $\varepsilon - \sigma$ for uniaxial compression of concrete specimens, after Wischers (1978)

are presented in Fig. 2. The higher magnitude of maximum stress σ_c is the curves are characterized by the sharper decreasing of maximum stress at its right-hand branch. This effect strictly related to strain localization was mentioned by van Mier (1984), who concluded that the branch referring to material softening does not reflect true material behavior, but rather response of the specimen as a whole.

Although above remarks are true, in conventional formulation of the finite element method dedicated to concrete the effect of strain localization is not described by a separate theory, but considered in the equation of the universal curve of initial loading. One can distinguish two strain ranges corresponding to the beginning of crushing $\varepsilon \geqslant \varepsilon_c$ and the ultimate compressive strain $\varepsilon \geqslant \varepsilon_u$, see Fig. 3, such that the following relation can be used to describe the above curve (Bathe 1992)

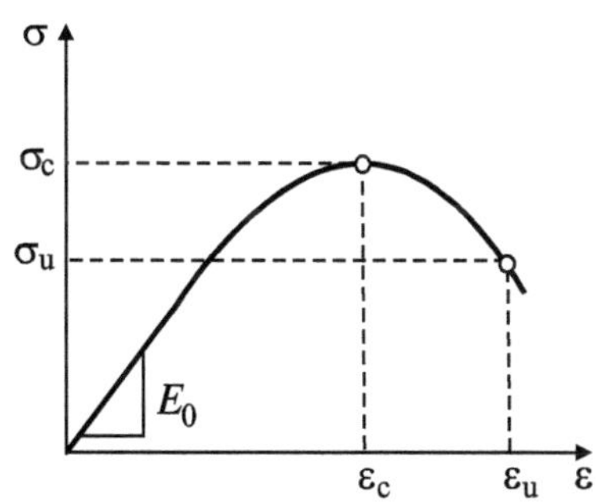

Fig. 3 Schematic curve of initial loading (envelope) for concrete

Table 1 Magnitudes of material constants describing curve of initial loading of concrete

E_0 [MPa]	ε_c [%]	ε_u [%]	σ_c [kPa]	σ_u [kPa]
24.8	−0.25	−0.85	−27.2	−6.3

$$\begin{aligned}
\frac{d\sigma}{d\varepsilon} &= E_0 \frac{1 - B\left(\frac{\varepsilon}{\varepsilon_c}\right)^2 - 2C\left(\frac{\varepsilon}{\varepsilon_c}\right)^3}{\left[1 + A\left(\frac{\varepsilon}{\varepsilon_c}\right) + B\left(\frac{\varepsilon}{\varepsilon_c}\right)^2 + C\left(\frac{\varepsilon}{\varepsilon_c}\right)^3\right]^2} \\
A &= \frac{\frac{E_0}{E_u} + \left(p^3 - 2p^2\right)\frac{E_0}{E_s} - \left(2p^3 - 3p^2 + 1\right)}{p\left(p^2 - 2p + 1\right)} \\
B &= 2\frac{E_0}{E_s} - 3 - 2A, \quad C = 2 - \frac{E_0}{E_s} + A \\
E_s &= \frac{\sigma_c}{\varepsilon_c}, \quad E_u = \frac{\sigma_u}{\varepsilon_u}, \quad p = \frac{\varepsilon_u}{\varepsilon_c}
\end{aligned} \tag{1}$$

Taking the magnitudes of material constants presented in Table 1, the following fitting of the curve of initial loading, given by Eq. (1) to the experimental data by Sinha et al. (1964), is seen in Fig. 4.

4 Equation of Damage Evolution

Concrete belongs to the class of materials characterized by plastic softening since it exhibits plastic behavior accompanied by the stiffness degradation. In order to consider both these effects, Bažant and Kim (1979) formulated the theory of a progressive plastic fracturing, in which plastic deformation is defined in a traditional plastic flow theory fashion, but stiffness degradation is modeled like in the Dougill (1975, 1976) theory, see Fig. 5a. However, the application of this theory is rather difficult since the

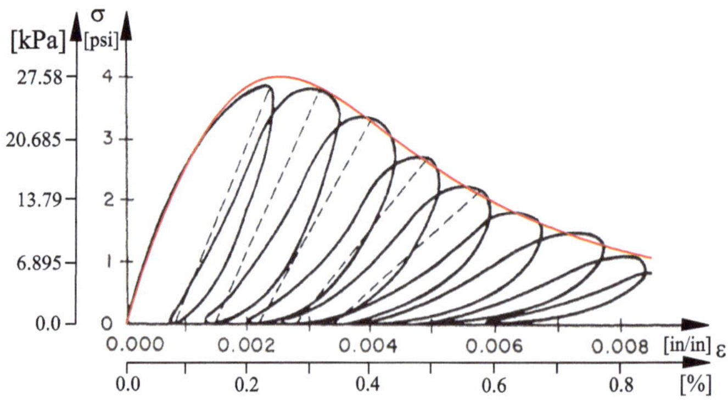

Fig. 4 Fitting of initial loading curve (*red line*) to results by Sinha et al. (1964)

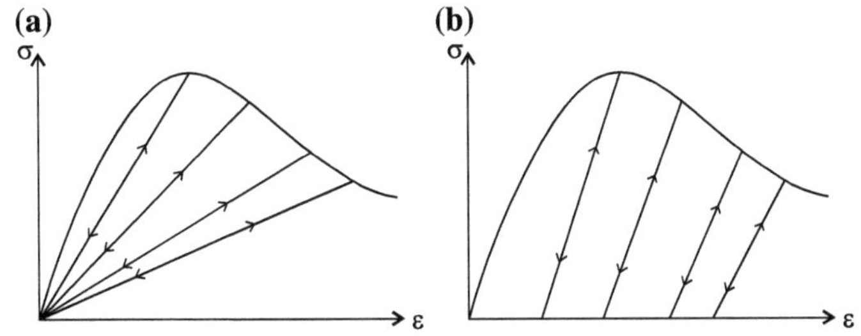

Fig. 5 Typical models of cyclic compression for concrete: **a** progressive plastic fracture after Bažant and Kim (1979), **b** elastic-plastic with stifness degradation accompanying to both hardening and strain softening, after Chen and Han (1988)

yield criterion is defined in the stress space, whereas the fracture criterion is defined in the strain space. In order to avoid this problem, Chen and Han (1988) suggested a consistent approach to elastic–plastic material with stiffness degradation accompanying both strain hardening and softening called plastic-fracturing behavior, in which both criteria are defined in the strain space, see Fig. 5b. The corresponding equation of damage evolution, analogous like in theory by Lemaitre–Chaboche, can be derived from Gibbs' potential (Murakami and Kamiya 1997; Al-Gadhib et al. 2000; Skrzypek and Kuna-Ciskał 2003; Voyiadjis et al. 2008).

An original model of concrete reinforced by the high-strength steel armature subjected to cyclic tension was proposed by Sun and Wagoner (2011). It was based on Chaboche's concept of the mixed isotropic-kinematic hardening generalized to the case of two limit surfaces defined in the stress space. Plastic potential was associated with Helmholtz's free energy, but not with Gibbs' complementary energy like in the traditional formulation. Results of numerical simulation compared to experimental data of cyclic tests are shown in Fig. 6.

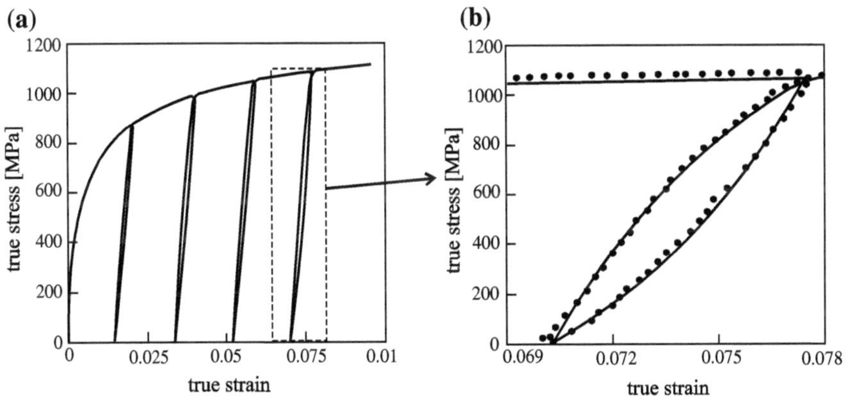

Fig. 6 Results of cyclic tension for reinforced concrete, after Sun and Wagoner (2011)

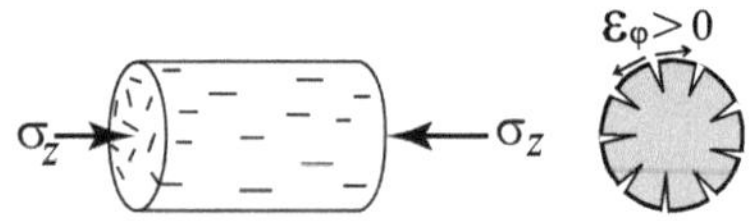

Fig. 7 Microcrack pattern of concrete specimen under compression

Previously presented general formulations are simultaneously too complicated, when compared to the uniaxial cyclic compression problem under consideration. Hence, authors of this work suggest simplified formulation based on the experimental tests done on concrete specimens that exhibit microcrack system in the direction perpendicular to compression σ_z caused by a positive hoop strain ε_φ (see Fig. 7).

The appropriate equation of damage evolution proposed by Chaboche (1985, 1988) takes the following format

$$\frac{\mathrm{d}D}{\mathrm{d}\varepsilon} = \begin{cases} \left(\dfrac{\varepsilon}{\varepsilon_0}\right)^s & \text{if } \varepsilon \leqslant \varepsilon_\mathrm{c} \wedge \mathrm{d}\varepsilon < 0 \\ 0 & \text{if } \varepsilon > \varepsilon_\mathrm{c} \vee \mathrm{d}\varepsilon \geqslant 0 \end{cases} \tag{2}$$

which is valid for positive strain ε_φ. The application of Eq. (2) to modeling of microcrack system presented in Fig. 7, which depends on strain ε_φ, leads to

$$\frac{\mathrm{d}D}{\mathrm{d}\varepsilon} = \begin{cases} \left(\dfrac{-\nu\varepsilon}{\varepsilon_0}\right)^s & \text{if } \varepsilon \leqslant \varepsilon_\mathrm{c} \wedge \mathrm{d}\varepsilon < 0 \\ 0 & \text{if } \varepsilon > \varepsilon_\mathrm{c} \vee \mathrm{d}\varepsilon \geqslant 0 \end{cases} \tag{3}$$

where the following assumptions: $\varepsilon = \varepsilon_z$ and $\varepsilon_\varphi = -\nu\varepsilon_z$ were done, and the symbol ν stands for Poisson's ratio.

5 Effect of Continuous Damage Deactivation in Concrete

A discussion on the classical, non-continuous damage deactivation effect should be initiated by invoking the concept by Lemaitre (1992). In the case of uniaxial tensile stress and scalar damage, the effective stress is defined as follows:

$$\widetilde{\sigma} = \frac{\sigma}{1 - D} \tag{4}$$

whereas the corresponding effective modulus of elasticity takes a format

$$\widetilde{E} = E\,(1 - D)\,. \tag{5}$$

Aforementioned relations are also valid in the case when microcracks stay opened under uniaxial compressive stress. However, in the case of a broad class of materials subjected to compressive stress microcracks close. This effect is characteristic for brittle materials and has nothing to do with the crushing effect in concrete. In the case when microcracks close completely, two sets of conditions must be defined, one for tensile stress and other for compressive stress

$$\widetilde{\sigma} = \begin{cases} \dfrac{\sigma}{1-D} \\ \sigma \end{cases} \qquad \widetilde{E} = \begin{cases} E\,(1-D) & \text{if } \sigma > 0 \\ E & \text{if } \sigma < 0\,. \end{cases} \tag{6}$$

Microcracks appearing in a real material are usually of sophisticated shape such that complete closure is not possible. In order to take into account this effect, a crack closure parameter h $(0 \leqslant h \leqslant 1)$ is introduced and it modifies the conditions for compressive stress

$$\widetilde{\sigma} = \begin{cases} \dfrac{\sigma}{1-D} \\ \dfrac{\sigma}{1-Dh} \end{cases} \qquad \widetilde{E} = \begin{cases} E\,(1-D) & \text{if } \sigma > 0 \\ E\,(1-Dh) & \text{if } \sigma < 0\,. \end{cases} \tag{7}$$

The application of such model to the description of unloading path leads to a linear relation between drop of stress and corresponding drop of strain, which is characterized by the modulus $\widetilde{E}^{+}$. An entering to the compressive range is accompanied by the switch to another branch of unloading path characterized by modulus $\widetilde{E}^{-}$, see Fig. 8.

A real material does not exhibit such bilinear path. The concept of a continuous damage deactivation by Hansen and Schreyer (1995), eliminating switch between $\widetilde{E}^{+}$ and $\widetilde{E}^{-}$, includes in a change of parameter h by function $h(\varepsilon)$ according to the rule

$$h\,(\varepsilon) = \begin{cases} 0.0 & \text{if } \varepsilon \le \varepsilon_l \\ 0.5\left[1-\cos\left(\pi\dfrac{\varepsilon-\varepsilon_l}{\varepsilon_h-\varepsilon_l}\right)\right] & \text{if } \varepsilon_l < \varepsilon < \varepsilon_h \\ 1.0 & \text{if } \varepsilon_h \le \varepsilon, \end{cases} \tag{8}$$

what graphically corresponds to a change of Heaviside's function by a smooth function presented in Fig. 9.

The effect of continuous damage deactivation in concrete has more complex nature when compared to conventional metallic materials. This is mainly due to the fact that microcracks are caused by positive strain along direction perpendicular to the axis

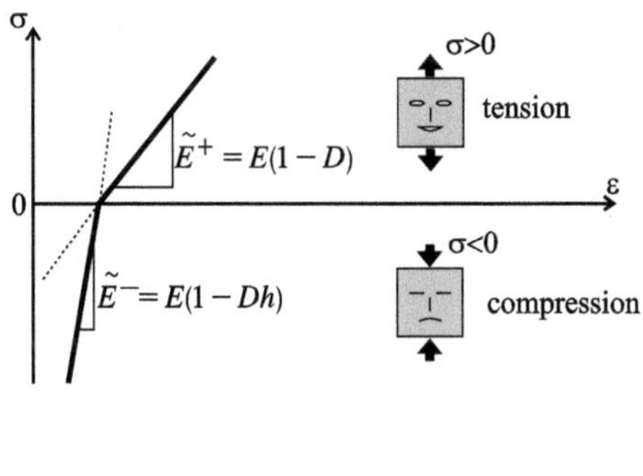

Fig. 8 Bilinear unloading path

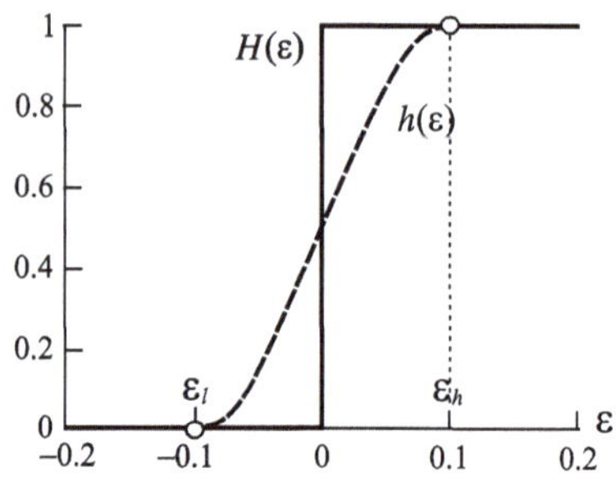

Fig. 9 Concept of continuous damage deactivation by Hansen and Schreyer (1995)

of a compressed specimen and their edges are very rough so the effect of continuous damage deactivation has frictional nature, which is directly associated with both positive strain ε and its increment sign $\mathrm{d}\varepsilon$. The graphical interpretation of the proposed effect of a continuous damage deactivation in a concrete is shown in Fig. 10, whereas the corresponding mathematical relations are given below

$$\begin{aligned} &\widetilde{E} = E_0\,(1 - D) \qquad \text{for initial loading curve given by Eq. (1)} \\ &\frac{\mathrm{d}\sigma}{\mathrm{d}\varepsilon} = E_0\,(1 - Dh) \text{ where } h(\varepsilon) = \begin{cases} 1 - \dfrac{\varepsilon}{\varepsilon_\mathrm{b}} & \text{unloading path } \widehat{\mathrm{AB}} \text{ in Fig. 10} \\ \dfrac{\varepsilon}{\varepsilon_\mathrm{b}} & \text{reloading path } \widehat{\mathrm{BA}} \text{ in Fig. 10} \end{cases} \end{aligned} \tag{9}$$

Let us follow in detail the sequence of phenomena modeled by subsequent segments of the $\varepsilon - \sigma$ curve. In the virgin state, a material is free from damage $D = 0$, what

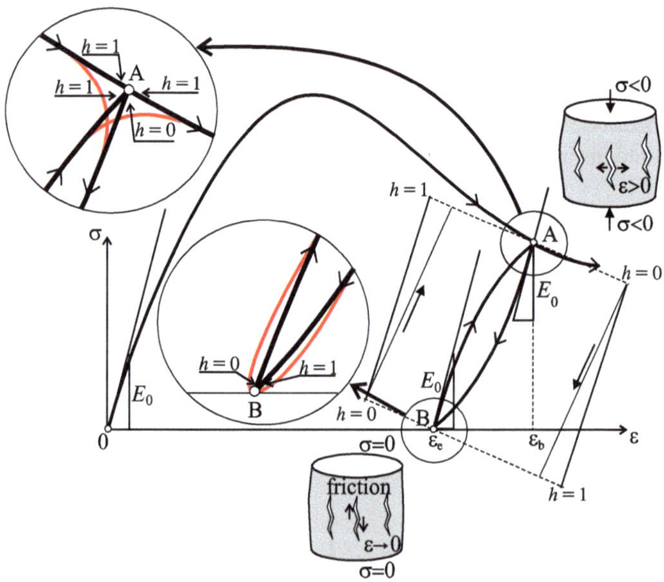

Fig. 10 Graphical interpretation of continuous damage deactivation effect for concrete

manifests by the slope of the tangent line at the origin 0 equal to E_0. Next, following the gradual increase of a load, concrete characteristics becomes step by step more nonlinear, what is obviously the consequence of damage growth $D > 0$. After that the peak stress curve of initial loading significantly decreases hence in this range both strain $\varepsilon > 0$ and its increment $\mathrm{d}\varepsilon > 0$ are positive, what corresponds to damage activation $h = 1$. In case of the unloading process, a real material exhibits certain effect of "inertia," which results in the fact that strain increment is still positive $\mathrm{d}\varepsilon > 0$; however, its magnitude is smaller than the increment following the curve of initial loading. Consequently, the initial range of unloading process follows an arc colored in red, see details magnified in the upper window of Fig. 10. However, from the continuous damage deactivation point of view, this material behavior is modeled in a discrete way, such that the unloading process begins at point A, after passing of which the strain increment instantaneously changes sign $\mathrm{d}\varepsilon < 0$, what corresponds to damage deactivation $h = 0$ and simultaneous initial stiffness recovery E_0 by the material. Further stage of unloading follows the arc $\widehat{\mathrm{AB}}$ and is accompanied by a linear increase of the damage deactivation parameter, which achieves the value $h = 1$ at point B. At this point the reloading process begins, which is accompanied by the change of sign by strain increment $\mathrm{d}\varepsilon > 0$ and instantaneous damage deactivation $h = 0$, see details in the lower window of Fig. 10. As a consequence, for point B, they obey all conditions that have been defined for point A. Further stage of reloading process follows the arc $\widehat{\mathrm{BA}}$ and is accompanied by a linear increase of the damage deactivation parameter, which attains value $h = 1$ at point A. This point belongs to the curve of initial loading; hence, the damage deactivation parameter saves its magnitude h and the strain increment does not change sign $\mathrm{d}\varepsilon > 0$.

Multiple application of the above model leads to the subsequent unloading/reloading loops that are characterized by decreasing average slope, which depends on the current damage

$$E_0 < \widetilde{E}_1 = E_0(1 - D_1) < \widetilde{E}_2 = E_0(1 - D_2) < \cdots < \widetilde{E}_4 = E_0(1 - D_4) \quad (10)$$

The progressive magnitude of dissipated energy per unit volume on ith loop is equal to

$$\Phi_i = \oint_{\widehat{\mathrm{A}_i\mathrm{B}_i}}^{\widehat{\mathrm{B}_i\mathrm{A}_i}} \sigma \mathrm{d}\varepsilon = \int_{\varepsilon_{\mathrm{b}i}}^{\varepsilon_{\mathrm{e}i}} E_0\left(1 - D_i \frac{\varepsilon - \varepsilon_{\mathrm{b}i}}{\varepsilon_{\mathrm{b}i} - \varepsilon_{\mathrm{e}i}}\right) \varepsilon \mathrm{d}\varepsilon + \int_{\varepsilon_{\mathrm{e}i}}^{\varepsilon_{\mathrm{b}i}} E_0\left(1 - D_i \frac{\varepsilon - \varepsilon_{\mathrm{e}i}}{\varepsilon_{\mathrm{e}i} - \varepsilon_{\mathrm{b}i}}\right) \varepsilon \mathrm{d}\varepsilon = E_0 D_i \frac{\varepsilon_{\mathrm{b}i}^2 - \varepsilon_{\mathrm{e}i}^2}{2} \quad (11)$$

Introduction results shown in Table 2

$$\Phi_1 = 21.68 \text{ Pa} < \Phi_2 = 40.10 \text{ Pa} < \Phi_3 = 73.86 \text{ Pa} < \Phi_4 = 134.44 \text{ Pa} \quad (12)$$

leads to the scheme presented in Fig. 11.

Table 2 Magnitudes of strain at initial ε_{bi} and final ε_{ei} points of ith unloading/reloading loop and corresponding damage parameter D_i

Loop number i	ε_{bi} [%]	ε_{ei} [%]	D_i
1	0.25×10^{-2}	0.12×10^{-2}	0.26352
2	0.35×10^{-2}	0.22×10^{-2}	0.43653
3	0.45×10^{-2}	0.33×10^{-2}	0.63640
4	0.55×10^{-2}	0.42×10^{-2}	0.85991

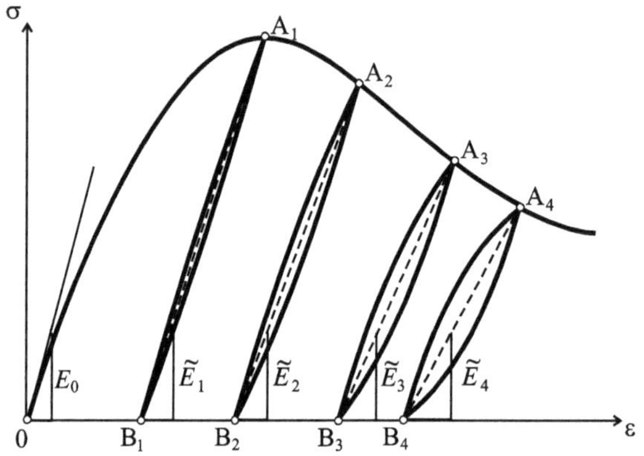

Fig. 11 Evolution of average slope and dissipative energy in subsequent loops of unloading/reloading

6 Results

Both numerical integration of Eqs. (1) and (2) by the use of routine odeint.for (Press et al. 1993) and implementation to FEM code bar.for (Owen and Hinton 1980; Ganczarski and Skrzypek 2009), based on the initial stiffness matrix method (partly cited by Bathe 1992), presented in Table 3 allows to model $\varepsilon - \sigma$ curve for an uniaxial cyclic compression of concrete specimen as shown in Fig. 12. All characteristic details of initial loading curve $\varepsilon - \sigma$, being an envelope of the diagram presented in Fig. 1 as well as the subsequent unloading/reloading loops, which exhibit both a progressive increase of the area of a surface referring to dissipative energy per cycle and the simultaneous drop of the average slope of the line connecting turning points, are well mapped.

Table 3 Material constants for damage model of concrete

ν	ε_0 [%]	s
0.25	2.5×10^{-6}	0.5

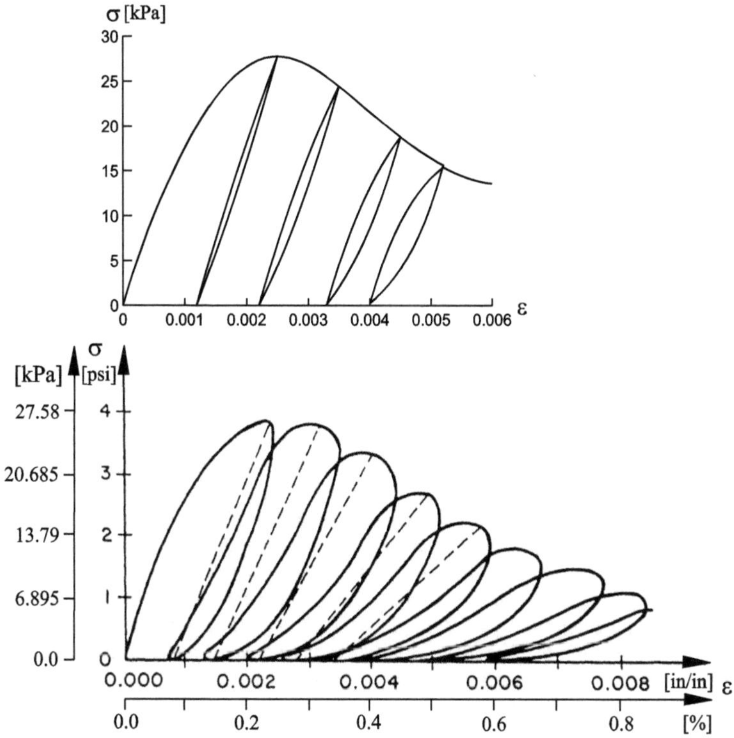

Fig. 12 Curve $\varepsilon - \sigma$ of uniaxial cyclic compression for concrete with continuous damage deactivation effect included (3), after Ganczarski and Cegielski (2008)

7 Conclusions

The following conclusions can be formulated:

- the proposed model including effects of plastic softening and Chaboche's type damage evolution can successfully be applied to model a cyclic compression of concrete,
- the original concept of strain-controlled continuous damage deactivation is efficient in modeling the subsequent unloading/reloading paths,
- the precise identification of material parameters allows to capture, not only qualitatively but also quantitatively, good agreement with experimental evidences comprising distinction between loading and reloading paths, which form loops of the progressive increase of areas of a surface referring to dissipative energy per cycle and decreasing average slopes.

References

Al-Gadhib, A., Baluch, M., Shaalan, A., Khan, A.: Damage model for monotonic and fatigue response of high strength concrete. Int. J. Damage Mech. **9**, 57–74 (2000)

Bathe, K.: Adina, theory and modeling guide. Report ARD pp. 92–8 (1992)

Bažant, Z., Kim, S.: Plastic-fracturing theory for concrete. J. Eng. Mech. Div. ASCE **105**(EM3), 407–428 (1979)

Chaboche, J.L.: Méchanique des matériaux solides. Bordas, Paris (1985)

Chaboche, J.L.: Continuum damage mechanics: Part i: General concepts, part ii: Damage growth, crack initiation, and crack growth. J. Appl. Mech. **55**(3), 59–71 (1988)

Chen, X., Han, D.: Plasticity for Structural Engineering. Springer, Berlin (1988)

Dougill, J.: Some remarks on path independence in the small in plasticity. Q. Appl. Math. **32**, 233–243 (1975)

Dougill, J.: On stable progressively fracturing solids. Zeitschrift für Angewandte Mathematik und Physik **27**, 423–437 (1976)

Ganczarski, A., Cegielski, M.: Continuous damage deactivation in modelling of low cycle fatigue of metallic and concrete materials. In: Proceedings of WCCM8 ECCOMAS 2008 on CD ROM (2008)

Ganczarski, A., Skrzypek, J.: Plasticity of Engineering materials. Wydawnictwo PK, Kraków (2009)

Hansen, N., Schreyer, H.: Damage deactivation. Trans. ASME **62**, 450–458 (1995)

Lemaitre, J.: A Course on Damage Mechanics. Springer, Berlin (1992)

Murakami, S., Kamiya, K.: Constitutive and damage evolution equations of elastic-brittle materials based on irreversible thermodynamics. Int. J. Solids Struct. **39**(4), 473–486 (1997)

Owen, D., Hinton, E.: Finite Elements in Plasticity: Theory and Practice. Pineridge Press Ltd., Swansea (1980)

Press, W., Teukolsky, S., Vetterling, W., Flannery, B.: Numerical Recipes in Fortran. Cambridge Press, Cambridge (1993)

Sinha, B., Gerstle, K., Tulin, L.: Stress-strain relations for concrete under cyclic loading. ACI J. **61**(2), 195–211 (1964)

Skrzypek, J., Kuna-Ciskał, H.: Anisotropic elastic-brittle-damage and fracture models based on irreversible thermodynamic. In: Skrzypek, J.J., Ganczarski, A. (eds.) Anisotropic behaviour of damaged materials. Springer, Berlin (2003)

Sun, L., Wagoner, R.: Complex unloading behavior: nature of the deformation and its consistent constitutive representation. Int. J. Plast. **27**, 1126–1144 (2011)

van Mier, J.: Complete stress-strain behavior and damaging status of concrete under multiaxial condititons. RILEM-CEB-CRNS, International Conference On Concrete Under Multiaxial Conditions, Presses de l'Université Paul Sabatier, Toulouse, France **1**, 75–85 (1984)

Voyiadjis, G., Taqieddin, Z., Kattan, P.: Anisotropic damage-plasticity model for concrete. Int. J. Plast. **24**, 1946–1965 (2008)

Wischers, G.: Application of effects of compressive loads on concrete. Betontech **2**, 3 (1978)

Phase-Field Modelling of Damage and Fracture—Convergence and Local Mesh Refinement

Markus Kästner, Paul Hennig, Thomas Linse and Volker Ulbricht

Abstract In this contribution, we outline the combination of a phase-field model of brittle fracture with adaptive spline-based approximations. The phase-field method provides a convenient way to model crack propagation without topological updates of the used discretisation as the crack is represented implicitly in terms of an order parameter field that can be interpreted as damage variable. For the accurate approximation of the order parameter field that may exhibit steep gradients, we utilise locally refined hierarchical B-splines in conjunction with Bézier extraction. The latter allows for the implementation of the approach in any standard finite element code. Moreover, standard procedures of adaptive finite element analysis for error estimation and marking of elements are directly applicable due to the strict use of an element viewpoint. Two different demonstration problems are considered. At first we examine the convergence properties of the phase-field approach and explain the influence of the domain size and the discretisation for the one-dimensional problem of a bar. Afterwards, results of the adaptive local refinement are compared with uniformly refined Lagrangian and spline-based discretisations. In the second example, the developed algorithms are applied to simulate crack propagation in a two-dimensional single-edge notched, shear loaded plate.

M. Kästner (✉) · P. Hennig · T. Linse · V. Ulbricht
Institute of Solid Mechanics, Technische Universität Dresden, 01062 Dresden, Germany
e-mail: markus.kaestner@tu-dresden.de

P. Hennig
e-mail: paul.hennig@tu-dresden.de

T. Linse
e-mail: thomas.linse@tu-dresden.de

V. Ulbricht
e-mail: volker.ulbricht@tu-dresden.de

K. Naumenko and M. Aßmus (eds.), *Advanced Methods of Continuum Mechanics for Materials and Structures*, Advanced Structured Materials 60,
DOI 10.1007/978-981-10-0959-4_17

1 Introduction

Phase-field models have become a powerful tool for the simulation of phase transformations and morphological changes in different fields of physics as well as materials and engineering science. Compared to sharp interface models, their advantage is that topological changes are avoided. Instead, interfaces are treated in a diffuse manner in terms of an order parameter that varies continously in the complete domain. Typical applications include the modelling and simulation of solidification processes, spinodal decomposition, coarsening of precipitate phases, shape memory effects, recrystallisation, and dislocation dynamics as summarised in the reviews of, e.g. Chen (2002); Emmerich (2008); Moelans et al. (2008); Steinbach (2009). More recently, the phase-field approach has also been used to model crack propagation which is in the focus of this contribution as reviewed by, e.g. Ambati et al. (2015). Different from standard sharp crack representations, the discrete crack is smeared in terms of a phase-field order parameter which can be interpreted as damage variable.

In all these examples, the phase-field approach allows to fully capture the physics of the individual interfaces without the need to explicitly track them. However, the variability of the approach comes at the cost of highly refined meshes that are required to properly resolve the transition zone in a numerical finite element model. In this contribution, we therefore utilise spline-based approximations to discretise a phase-field model of fracture. Compared to Lagrangian polynomials, the spline basis provides a more efficient resolution of steep gradients (Kästner et al. 2016a). Moreover, the higher order continuity of the basis allows for a direct discretisation of weak forms containing higher than first-order derivatives that may result form surface energy contributions. In this way, mixed formulations which introduce additional degrees of freedom and are prone to stability issues can be avoided.

The efficiency of the numerical solution procedure is further increased by a local refinement of the mesh in the vicinity of a crack based on hierarchical splines. An implementation developed in Hennig et al. (2016), which is based on Bézier extraction (Borden et al. 2011), allows for a strict element viewpoint and hence the application of standard adaptive finite element procedures.

The paper is organised as follows: in Sect. 2, we briefly review the phase-field approach to fracture and explain the hierarchical spline approximation used to resolve the crack phase field. The numerical properties of the phase-field representation and the adaptive discretisation are studied in Sect. 3. In particular, we present a numerical Γ-convergence study and explain certain peculiarities observed by May et al. (2015) using a one-dimensional example. Then, the efficiency of the hierarchical approximation is analysed and the developed algorithms for local refinement are applied to a two-dimensional demonstration problem. The paper is completed by conclusions in Sect. 4.

2 Adaptive Phase-Field Modelling of Brittle Fracture

The modelling of crack propagation, using standard finite elements in conjunction with a sharp crack representation, requires an update of the mesh topology if the crack evolves. This restriction can be partly resolved with the extended finite element method (XFEM) (Moës et al. 1999; Belytschko et al. 2001; Wells and Sluys 2001; Fries and Belytschko 2010; Kästner et al. 2016b), that uses a nodal enrichment of the approximation, or element-based enrichments (Ortiz et al. 1987; Simo et al. 1993; Jirásek 2000; Linder et al. 2011) to model the displacement discontinuity across a crack. Although these methods are used in principal non-conforming regular meshes, topological updates are still required, e.g. for integration purposes or to update the position of the crack tip. Consequently, the simulation of complex crack patterns that involve branching and coalescence is problematic, in particular for three-dimensional problems.

Different from these approaches, the diffuse phase-field approximation used in the following does not involve the crack topology in an explicit way. Instead, the sharp crack representation, Fig. 1a, is regularised in terms of a phase-field order parameter d which can be interpreted as damage variable. The transition zone between the undamaged ($d = 0$) and the fully broken material ($d = 1$) is characterised by an internal length scale ℓ_0, Fig. 1b, which has to be considered as a material parameter. By coupling the scalar phase-field to a mechanical boundary value problem, the initiation and propagation of the crack are described by the evolution of the crack phase-field and the cumbersome numerical tracking of the crack path is avoided, Fig. 1c.

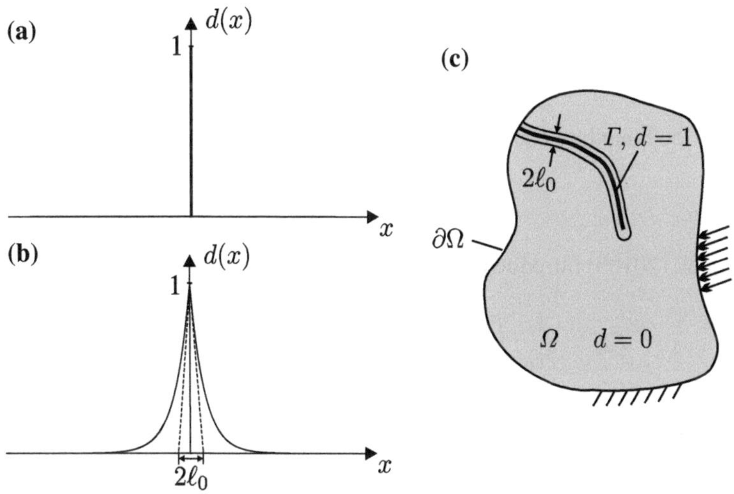

Fig. 1 Phase-field approach to brittle fracture: **a** one-dimensional sharp crack, **b** diffuse crack representation in terms of the order parameter $d \in [0, 1]$, and **c** phase-field representation of a crack in the two-dimensional domain Ω

2.1 Phase-Field Modelling of Brittle Fracture

The phase-field modelling of fracture in brittle materials can be perceived as a regularised version of the variational formulation of the Griffith theory presented by Francfort and Marigo (1998)

$$\Pi = \int_{\Omega} \psi \, \mathrm{d}V + \int_{\Gamma} \mathcal{G}_\mathrm{c} \mathrm{d}A \tag{1}$$

where ψ is the bulk free energy, $\mathcal{G}_\mathrm{c}$ is the fracture energy required to create a new crack surface of unit area, Ω is the domain of the body under investigation and Γ is the newly created fracture surface, see Fig. 1c. Bourdin et al. (2000) suggested to regularise the sharp crack representation

$$\int_{\Gamma} \mathcal{G}_\mathrm{c} \mathrm{d}A \approx \int_{\Omega} \mathcal{G}_\mathrm{c} \gamma_{\ell_0}(d) \, \mathrm{d}V \tag{2}$$

where $\gamma_{\ell_0}(d)$ represents a crack surface density. This function depends on the regularisation of the sharp crack in terms of the phase-field variable $d \in [0, 1]$, i.e. a purely geometrical feature. In a one-dimensional setting, the exponential approximation

$$d(x) = \begin{cases} \mathrm{e}^{\frac{x}{\ell_0}} & x \leq 0 \\ \mathrm{e}^{-\frac{x}{\ell_0}} & 0 < x \end{cases} \tag{3}$$

according to Fig. 1b is used which is the solution to the differential equation

$$d - \ell_0^2 d'' = 0 \quad \text{in } \Omega = [-\infty; +\infty] \tag{4}$$

subject to the boundary conditions

$$d(0) = 1 \text{ and } d(\pm\infty) = 0 \,. \tag{5}$$

Miehe et al. (2010b) introduced a functional

$$\Gamma_{\ell_0} = \frac{1}{2\ell_0} \int_{\Omega} \left(d^2 + \ell_0^2 d'^2\right) \mathrm{dV} = \frac{1}{2\ell_0} \int_{\Omega} \gamma_{\ell_0}(d) \, \mathrm{dV} \tag{6}$$

which is interpreted as the crack surface. Hence, the crack surface density $\gamma_{\ell_0}(d) = d^2 + \ell_0^2 d'^2$ can be related to the phase-field order parameter d. It is noted that Eq. (6) is exact for one-dimensional problems, i.e.

$$\Gamma_{\ell_0} = \frac{1}{2\ell_0} \int_{-\infty}^{+\infty} \left(d^2 + \ell_0^2 d'^2\right) A\mathrm{d}x = A = \Gamma, \tag{7}$$

independent of ℓ_0 for a function $d(x)$ that satisfies the boundary conditions according to Eq. (5). Note that the fracture surface of one-dimensional problem (e.g. of a broken bar) is identical to the cross section area, therefore the identity $A = \Gamma$ has been used in (7). However, the finite size of the domain Ω and the discretisation of the numerical model will introduce errors that are examined numerically in Sect. 3.1. A generalisation of the crack surface density to a multi-dimensional setting reads

$$\gamma_{\ell_0}(d) = d^2 + \ell_0^2 |\nabla d|^2. \tag{8}$$

As a consequence of the regularisation of the discrete crack in terms of the order parameter d, the crack is smeared over the complete domain Ω. The effect of d on the bulk material is accounted in terms of a transition rule (Miehe et al. 2015)

$$\boldsymbol{F}(d) = (1-d)^m \boldsymbol{F}^{\mathrm{s}} + d^m \boldsymbol{F}^{\mathrm{c}} \tag{9}$$

between the constitutive tensors $\boldsymbol{F}^{\mathrm{s}}$ of the intact solid (s) and the cracked (c) material $\boldsymbol{F}^{\mathrm{c}}$. Contributions $\boldsymbol{F}^{\mathrm{c}}$ from the cracked domain may exist for non-mechanical problems, e.g. for heat conduction. In the purely mechanical case, $\boldsymbol{F}^{\mathrm{c}} = \boldsymbol{0}$ holds and the similarity of Eq. (9) with a degradation law of damage mechanics becomes obvious. Most frequently, the exponent $m = 2$ is used in phase-field approaches to fracture but Borden et al. (2012) have found that a cubic degradation function might be more suitable for brittle fracture as it yields a more abrupt failure.

Eventually, the regularised energy functional is given by

$$\Pi_{\ell_0}(\boldsymbol{u}, d, \nabla d) = \int_{\Omega} (1-d)^2 \psi(\boldsymbol{u})\, \mathrm{d}V + \frac{\mathcal{G}_{\mathrm{c}}}{\ell_0} \int_{\Omega} \left[d^2 + \ell_0^2 |\nabla d|^2\right] \mathrm{d}V. \tag{10}$$

It contains contributions from the elastic energy in the damaged bulk material and the fracture energy which is smeared across the transition zone of length ℓ_0.

In order to complete the phase-field model, an evolution equation for crack propagation has to be defined. Miehe et al. (2015) suggest the general form

$$\frac{\mathrm{d}}{\mathrm{d}t}\Gamma_{\ell_0} = \frac{1}{\ell_0} \int_{\Omega} [(1-d)\mathcal{H} - \mathcal{R}]\, \dot{d}\, \mathrm{d}V \tag{11}$$

with $\mathcal{H}$ a local crack driving force, and $\mathcal{R}$ a local viscous crack resistance. For the crack surface density defined in Eq. (8) and the constitutive assumption of $\mathcal{R} = \eta\dot{d}$, $\eta \geq 0$, the evolution of the crack phase-field is given by

$$\eta\dot{d} = (1-d)\mathcal{H} - \left(d - \ell_0^2 \Delta d\right) \tag{12}$$

and the homogeneous boundary condition $\nabla d \cdot \boldsymbol{n} = 0$ on $\partial\Omega$. Miehe et al. (2015) present different choices for the crack driving force $\mathcal{H}$, that can depend on the history of the energetic or stress state and may include threshold values. Here, the energetic driving force

$$H(\boldsymbol{x}, t) = \max_{\tau \in [0,t]} \frac{2\psi^+(\boldsymbol{x}, t)}{\mathcal{G}_c/\ell_0} \tag{13}$$

has been used where ψ^+ is the tensile part of the free energy (Miehe et al. 2010b). This choice prevents crack healing and the tension/compression split avoids unphysical crack branching. The approach leads to a coupled system of two differential equations for the displacement field $\boldsymbol{u}$ and the phase-field order parameter d. The quasi static solution can be found by using a staggered (Miehe et al. 2010a) or monolithic (Borden et al. 2012) solution scheme. A recent review of different formulations can be found in Ambati et al. (2015).

2.2 *Adaptive Spline-Based Discretisations*

In this section, we give an overview on the hierarchical spline-based discretisation that provides an efficient, locally refined approximation of the steep gradients of the phase-field order parameter. A multi-dimensional spline basis is defined in terms of the tensor product of one-dimensional basis functions which will prohibit a truly local refinement. Various approaches have been developed to overcome the restrictive tensor product structure (Hennig et al. 2016). Among them T-splines (Sederberg et al. 2003, 2004), which result from the insertion of extra vertices into the tensor product mesh, and hierarchical (H)B-splines (Forsey and Bartels 1988) that use basis functions defined by different levels of nested meshes are currently most prevalent. Local mesh refinement using T-splines and hierarchical B-splines is illustrated in Fig. 2 for the benchmark case of a sharp internal layer. It is noted that the

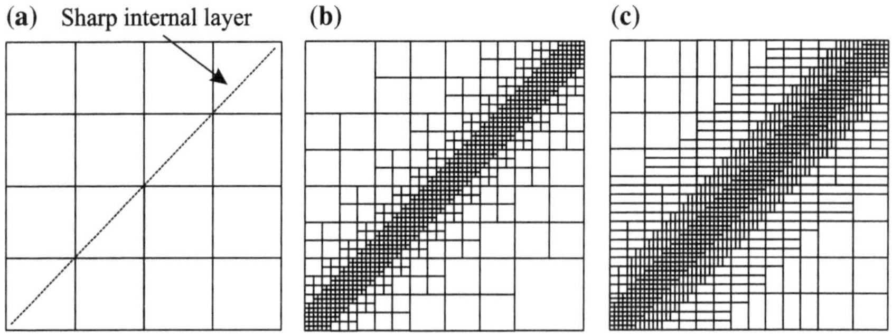

Fig. 2 Two approaches to local mesh refinement: **a** initial mesh with a cubic B-spline basis, degree of freedom DOF = 76 and an internal discontinuity. **b** Local refinement (*four levels*) using hierarchical B-splines (DOF = 1444), and **c** T-splines (DOF = 2058)

T-spline-based refinement is not as local as with hierarchical B-splines (Dörfel et al. 2010; Evans et al. 2015). The refinement propagates along the parametric directions as a consequence of the restrictions imposed by the required analysis suitability, i.e. the linear independence of the basis. Therefore a hierarchical B-spline basis is applied in this contribution. In particular, truncated hierarchical splines introduced by Giannelli et al. (2012) that reduce the interaction between basis functions of different refinement levels and improve the condition of the coefficient matrix are utilised.

2.2.1 Multi-Level B-Splines and Bézier Extraction

A univariate B-spline basis is defined in terms of a knot vector $\Xi = \{\xi_I\}_{I=1}^{s}$ of size s as outlined in the textbook by Piegl and Tiller (1996), Fig. 3a. Consider a hierarchy of M knot vectors $\Xi^m, m = 0 \ldots M-1$ created by successive uniform h-refinement starting from the knot vector Ξ^0 of the coarsest mesh (level $m = 0$). The resulting knot vectors indicated by the symbol $\times$ in Fig. 3a are nested, i.e. $\Xi^m \subset \Xi^{m+1}$. Each knot vector Ξ^m defines a set of k^m B-spline basis functions $\mathsf{n}^m = \{N_I^m\}_{I=1}^{k^m}$ of order p. These sets of B-spline basis functions span the approximation spaces $\mathcal{N}^m$ of each corresponding level and are nested, $\mathcal{N}^m \subset \mathcal{N}^{m+1}$. They are referred to as multi-level basis. An exemplary univariate three-level basis is illustrated in Fig. 3a for the case of second-order B-splines ($p = 2$).

The introduced one-dimensional B-spline basis functions are defined globally over a knot vector. The central idea of Bézier extraction introduced by Borden et al. (2011) is to represent the globally defined B-spline basis functions

$$\mathsf{n}(\xi) = \mathsf{C}\mathsf{b}(\xi) \tag{14}$$

by Bernstein element shape functions $\mathsf{b}(\xi)$. The linear operator C is called Bézier extraction operator and maps the piecewise C^0-continuous Bernstein polynomials onto the B-spline basis of arbitrary continuity. The elements that support the Bernstein polynomial shape functions are defined by non-zero sub-domains $[\xi_I \; \xi_{I+1}]$ These $e = 1, \ldots, n_{\mathrm{e}}$ sub-domains, denoted by Ω_{p}^e, are referred to as Bézier elements. The Bernstein shape functions $\mathsf{b}(\xi)$ are the same for each element. As a result, the spline approximation can be implemented in every existing finite element code using Bézier extraction. It can be applied to B-splines, NURBS and T-splines and therefore represents a canonical approach to the implementation of isogeometric analysis (Hughes et al. 2005). In this contribution we will employ Bézier extraction to implement the truncated hierarchical basis.

2.2.2 Hierarchical B-Splines and Global System of Equations

In the hierarchical approach to local refinement, basis functions from the approximation spaces of different levels $\mathcal{N}^m$ are combined to span the hierarchical basis space $\mathcal{A}$. However, building the hierarchical basis cannot be achieved by simply

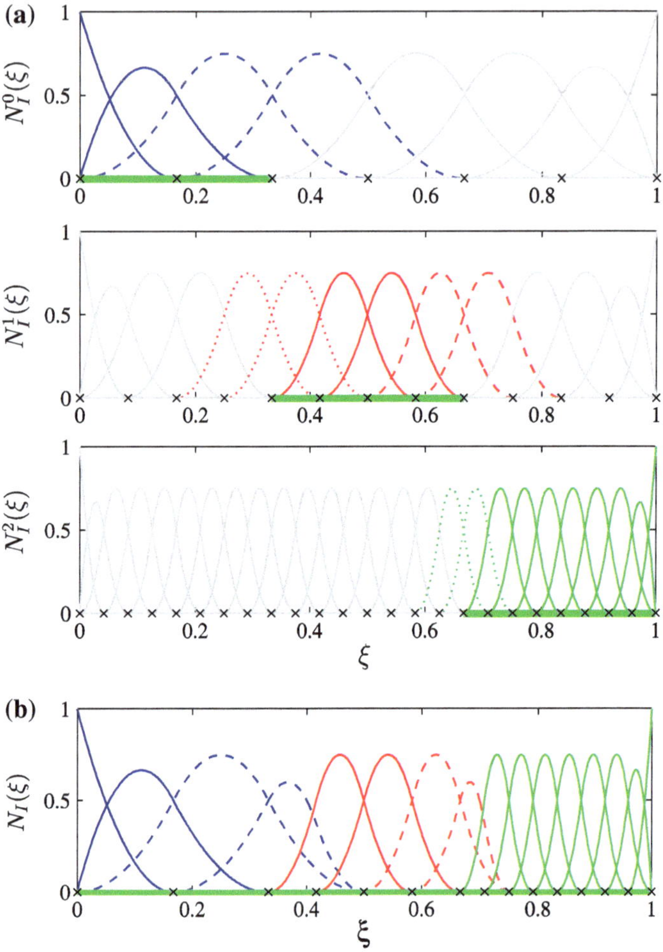

Fig. 3 Local refinement based on hierarchical B-splines: **a** Three-level basis $\mathcal{N}^m(\xi)$ and associated nested knot vectors Ξ^m. Active elements are indicated in *green* and basis functions $\mathcal{A}^m$ belonging to active elements are plotted as coloured lines. While basis functions in solid colour contribute directly to the hierarchical approximation, *dashed* and *dotted* lines represent linearly dependent basis functions. This issue is resolved automatically during the assembly of the global system of equation and results in a truncated hierarchical basis **b**. This basis is indentical the proposal of Giannelli et al. (2012)

adding basis functions from finer hierarchy levels to the coarse scale basis. This would result in the loss of linear independence, because coarse level basis functions are linear combinations of the fine-scale functions.

The selection of basis functions contributing to the hierarchical approximation is essentially element-based which is exemplarily illustrated in Fig. 3. By some criterion elements $\mathcal{E}_e^m := \Omega_{\mathrm{P}}^{e,m}$ of different hierarchy levels have to be chosen to discretise the analysis domain. These elements $\mathcal{E}_e^m$ define sets $\mathcal{M}^m = \bigcup_e \mathcal{E}_e^m$. They are called *active elements* and cover the domain $\Omega_P = \bigcup_{m=0}^{L-1} \mathcal{M}^m$ without any overlap. This can be seen in Fig. 3a where active elements are indicated in green. Every active element is associated to a number of $p + 1$ basis functions in the multi-level basis. The union of these basis functions on each level m forms a set

$$\mathcal{A}^m = \{N_I^m \in \mathcal{N}^m : \text{supp} N_I^m \cap \mathcal{M}^m \neq \emptyset\} \tag{15}$$

and is plotted in colour in Fig. 3a. However, to ensure a linearly independent basis not all of these functions can contribute to the hierarchical approximation, i.e. attention has to be paid to basis functions whose support overlaps with the domains of active elements on finer or coarser hierarchy levels. Their contributions are correctly accounted during the assembly of the hierarchical system of equations using Bézier extraction as explained in more detail in Hennig et al. (2016).

Based on Bézier extraction, the global system of equations defined by the hierarchical approximation $\mathcal{A}$ can be computed implicitly. This procedure consists of three steps:

1. At first the element matrices of all active elements are computed without considering information on whether the basis function contributes to the hierarchical basis or not. This ensures the applicability of standard Bézier extraction.
2. Once element matrices for all active elements of one level m have been obtained, they are assembled to form sub-systems

$$\mathsf{K}^m \mathbf{u}^m = \mathbf{f}^m \tag{16}$$

   for each hierarchy level m with $\mathbf{u}^m$ and $\mathbf{f}^m$ column vectors containing the control values of generalised degrees of freedom and the associated generalised forces.
3. In the multi-level system of Eq. (16) there is no communication between individual levels. This interconnection is introduced in terms of the hierarchical refinement operator M_h explained in Hennig et al. (2016). Its entries result from the refinement or subdivision operator M (Scott et al. 2012; Schillinger et al. 2012) that relates the basis functions on different levels of the nested meshes

$$\mathsf{n}^m = \mathsf{M}^{m,m+1} \mathsf{n}^{m+1}. \tag{17}$$

   It acts as a transformation matrix on the multi-level system, i.e. a simple matrix multiplication

$$\mathsf{K}_\mathrm{h} = \mathsf{M}_\mathrm{h} \mathsf{K} \mathsf{M}_\mathrm{h}^\mathrm{T} \tag{18}$$

$$\mathsf{f}_\mathrm{h} = \mathsf{M}_\mathrm{h} \mathsf{f} \tag{19}$$

produces the hierarchical system of equations

$$\mathsf{K}_{\mathrm{h}} \mathbf{u}_{\mathrm{h}} = \mathsf{f}_{\mathrm{h}}. \tag{20}$$

It ensures that only basis functions in $\mathcal{A}$ contribute to the approximation.

3 Analysis

In this section, we will examine the convergence properties of the phase-field approach. It is expected, that the functional Π_{ℓ_0} of the regularised crack approaches the functional Π of the discrete crack for the limiting case $\ell_0 \rightarrow 0$. Moreover, $\Gamma_{\ell_0} = \Gamma = A$, i.e. the diffuse representation of the crack surface is exactly the discrete surface, should hold in the one-dimensional case. These properties have been investigated numerically by May et al. (2015). We will start from their results and give explanations for the observed peculiarities. Afterwards, results of the adaptive local refinement algorithm are compared with results obtained from uniformly refined Lagrangian and spline-based discretisations. In the third example, the phase-field approach to brittle fracture is applied to simulate crack propagation in a two-dimensional single-edge notched shear test.

3.1 *Convergence of the Phase Field Approximation*

Miehe et al. (2010b) analysed the phase-field representation of a crack with the surface Γ in a two-dimensional unit square plate regarding the accuracy of Γ / Γ_{ℓ_0}. In their elementary study, they found that a reasonable approximation of $\Gamma_{\ell_0} \approx \Gamma$ requires a minimum size of finite elements, e.g. an element size of

$$h < \frac{\ell_0}{2} \tag{21}$$

is needed for a two-dimensional problem using quadrilateral four node elements.

A numerical convergence study of Γ_{ℓ_0} regarding ℓ_0 was carried out by May et al. (2015) for a one-dimensional structure. They applied a tensile load to an elastic bar with length $L = 1\,\mathrm{mm}$, cross section $A = 1\,\mathrm{mm}^2$ and a reduced cross section $A/2$ in the centre part. A driving force $\mathcal{H}$ related to the elastic energy density according to Eq. (13) has been used in the evolution Eq. (12) to simulate quasistatic crack growth ($\eta = 0$). Consequently, the governing system of equations of the analysed problem can be compared to gradient damage models of brittle fracture, e.g. Peerlings et al. (1996), when d is interpreted as a damage variable.

May et al. (2015) found a minimum of the error $\Gamma_E = \left|\Gamma_{\ell_0} - \Gamma\right| / \Gamma$ at a length scale $\ell_0 = 0.05\,\mathrm{mm}$. The same minimum was calculated for different fine discretizations satisfying (21), i.e. the minimum occurred at $\ell_0 = 0.05\,\text{mm}$ independently of the element size h. A further decrease of the ratio ℓ_0/h resulted in an increased error Γ_E.

Thus, May et al. (2015) concluded that Γ-convergence is not necessarily attained, because their numerically obtained final crack surface Γ_{ℓ_0} evaluated at $\max(d) = 0.99$ fails to converge towards $\Gamma = A/2$, i.e.

$$\Gamma_{\ell_0}\Big|_{\ell_0 \to 0} \neq \Gamma. \tag{22}$$

Since May et al. (2015) presented no explanation on why a minimum of the error Γ_E occurs at a certain length scale ℓ_0, we focus on that point in the following. For a one-dimensional problem with a finite length L, the boundary condition (5) has to be reformulated to

$$d(0) = 1 \text{ and } d(\pm L/2) = 0 \tag{23}$$

to account for the finite domain $\Omega = [-L/2; +L/2]$. The analytic solution of the differential Eq. (4) subject to the boundary conditions (23) now reads

$$d(x) = \frac{1}{1 - \mathrm{e}^{-\frac{L}{\ell_0}}} \begin{cases} \mathrm{e}^{\frac{x}{\ell_0}} - \mathrm{e}^{-\frac{(x+L)}{\ell_0}} & x \leq 0 \\ \mathrm{e}^{-\frac{x}{\ell_0}} - \mathrm{e}^{-\frac{(L-x)}{\ell_0}} & 0 < x. \end{cases} \tag{24}$$

As a consequence of the finite domain and the analytic solution (24), the functional (7) now evaluates as

$$\Gamma_{\ell_0} = \frac{1}{2\ell_0} \int_{-\frac{L}{2}}^{+\frac{L}{2}} \left(d^2 + \ell_0^2 d'^2\right) A\,\mathrm{dx} = A \frac{1 - \mathrm{e}^{-\frac{2L}{\ell_0}}}{\left(1 - \mathrm{e}^{-\frac{L}{\ell_0}}\right)^2}, \tag{25}$$

i.e. the approximated crack surface depends on the size of the domain L.

Below, we solve Eq. (4) for the boundary conditions (23) using standard finite elements with C^0-continuous Lagrangian polynomial shape functions of order p for the one-dimensional bar problem with $L = 1\,\text{mm}$ and $A = 1\,\text{mm}^2$. The numerical results are compared to the analytic solution (24) by evaluating Eqs. (6) and (25). Figure 4 shows that the absolute error $\Gamma_{\ell_0} - A$ also shows a minimum as found by May et al. (2015): for a fixed discretisation, e.g. a constant element size h, the error $\Gamma_{\ell_0} - A$ follows the analytic solution (25) for decreasing ℓ_0/L (and thus decreasing ℓ_0/h) down to a certain point until $\Gamma_{\ell_0} - A$ increases again. However, notice that for a fixed characteristic length ℓ_0 the error $\Gamma_{\ell_0} - A$ converges to the analytic solution (25) for decreasing element sizes h (see black symbols in Fig. 4). Thus, the minima seen in Fig. 4 result from boundary effects for larger values ℓ_0/h and too coarse discretisations for lower values of ℓ_0/L. In the following section we will show that a local refinement in the vicinity of the crack will significantly reduce this discretisation error.

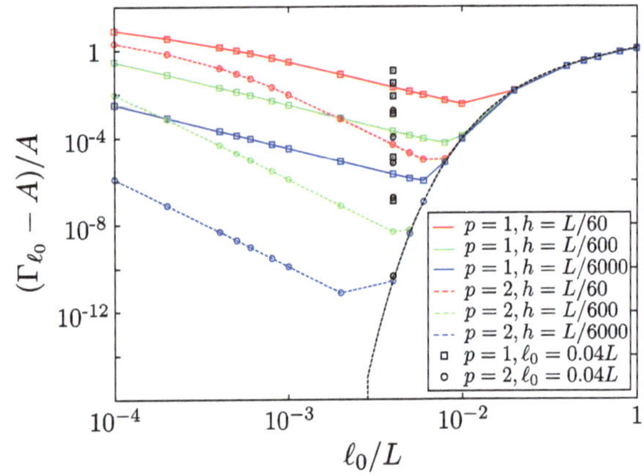

Fig. 4 Comparison of numerical (polynomial shape functions of order $p = 1$ and $p = 2$) and analytic solution (*black dotted line*) of the functional Γ_{ℓ_0}. Black symbols: numerical solution for a constant characteristic length $\ell_0 = 0.04L$ and varying element size h

3.2 *Convergence of the Adaptive Refinement*

It is noted from the previous section that small elements are needed in the vicinity of $d = 1$ for a reasonable approximation of the phase-field variable d (24). This motivates an adaptive local refinement in this region. We will therefore compare the approximation properties of uniformly refined C^0-continuous Lagrangian polynomials and C^{p-1}-continuous B-splines with locally refined hierarchical B-splines according to Sect. 2.2 in the following.

Again, the differential Eq. (4) will be solved on the domain $\Omega = [-L/2, L/2]$ with boundary conditions (23) and the parameters $L = 1\,\text{mm}$, $A = 1\,\text{mm}^2$ and $\ell_0 = 0.04L$. The error between the numerical d_h and the analytic solution d according to Eq. (24) is measured in the H^1-norm

$$|d - d_h|_{H^1} = \sqrt{\int_\Omega (d - d_h)^2 dx + \int_\Omega (d' - d_h')^2 dx}. \tag{26}$$

For the computations, the domain is discretized by 25 up to 2500 evenly spaced finite and 32 up to 2048 Bézier elements. To apply the boundary condition $d(0) = 1$, the B-spline basis is chosen to be C^0-continuous, i.e. interpolatory, at the control point at $x = 0$. In Fig. 5a the error in the H^1-norm is illustrated for linear, quadratic and cubic basis functions with respect to the total number of degrees of freedom. Both approximations, based on Lagrangian polynomials and B-splines, converge with the expected convergence rate of $-p$. However, the approximation with B-splines is more efficient than Lagrangian polynomial, as less degrees of freedom are required to reach a certain error level for the same order of approximation. Also note that the error regarding the characteristic element size h would be the same for IGA and FEM. This conclusion has been also found for the numerical solution of the Cahn-Hilliard equation in Kästner et al. (2016a).

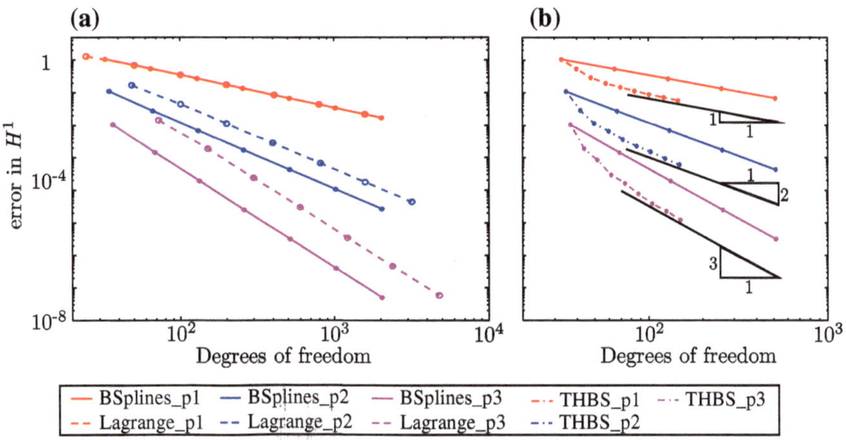

Fig. 5 Comparison of the convergence of the discretisation error in the H^1-norm for uniformly refined B-splines with **a** uniformly refined Lagrangian polynomials, and **b** locally refined hierarchical B-splines for linear, quadratic and cubic basis functions

For the adaptive local refinement a multi-level basis with $M = 9$ hierarchical levels is used, cf. Sect. 2.2. Starting from the coarsest mesh, the error in the H^1-norm (26) η_e is computed for each single element e. To mark the elements for refinement, quantile marking

$$\Theta = \alpha - \text{quantile}(\eta_e) \tag{27}$$

is used where $\alpha = 0.8$ is chosen in this example. Hence, the 20 % of elements with the largest element errors η_e are refined in each refinement step. The results for a linear, quadratic and cubic basis are shown in Fig. 5b. Due to strong local gradients, an improvement in the convergence rate is observed in the pre-asymptotic range until the theoretical rate of $-p$ is recovered in the asymptotic domain. In this way, adaptive local refinement can cut off the necessary degrees of freedom by up to 70 % without loss of precision. Note that the pre-asymptotic range increases with the locality of the gradients in the computational domain as demonstrated in Hennig et al. (2016) for the example of a circular plate with hole.

3.3 *Demonstration*

The final two-dimensional demonstration problem which considers a square specimen that is notched on the left edge has previously been considered by Miehe et al. (2010a). The plate is clamped at the bottom edge while a given horizontal displacement is applied on the top edge. The initial notch is prescribed as a discrete crack in terms of a C^{-1}-continuity line, Fig. 6. The Lamé parameters for the material are $\lambda = 121, 15\,\text{kN}\,\text{mm}^{-2}$ and $\mu = 80, 77\,\text{kN}\,\text{mm}^{-2}$ and the phase-field parameters are given by the fracture toughness $\mathcal{G}_c = 2.7 \times 10^{-3}\,\text{kN}\,\text{mm}^{-1}$ and the characteristic

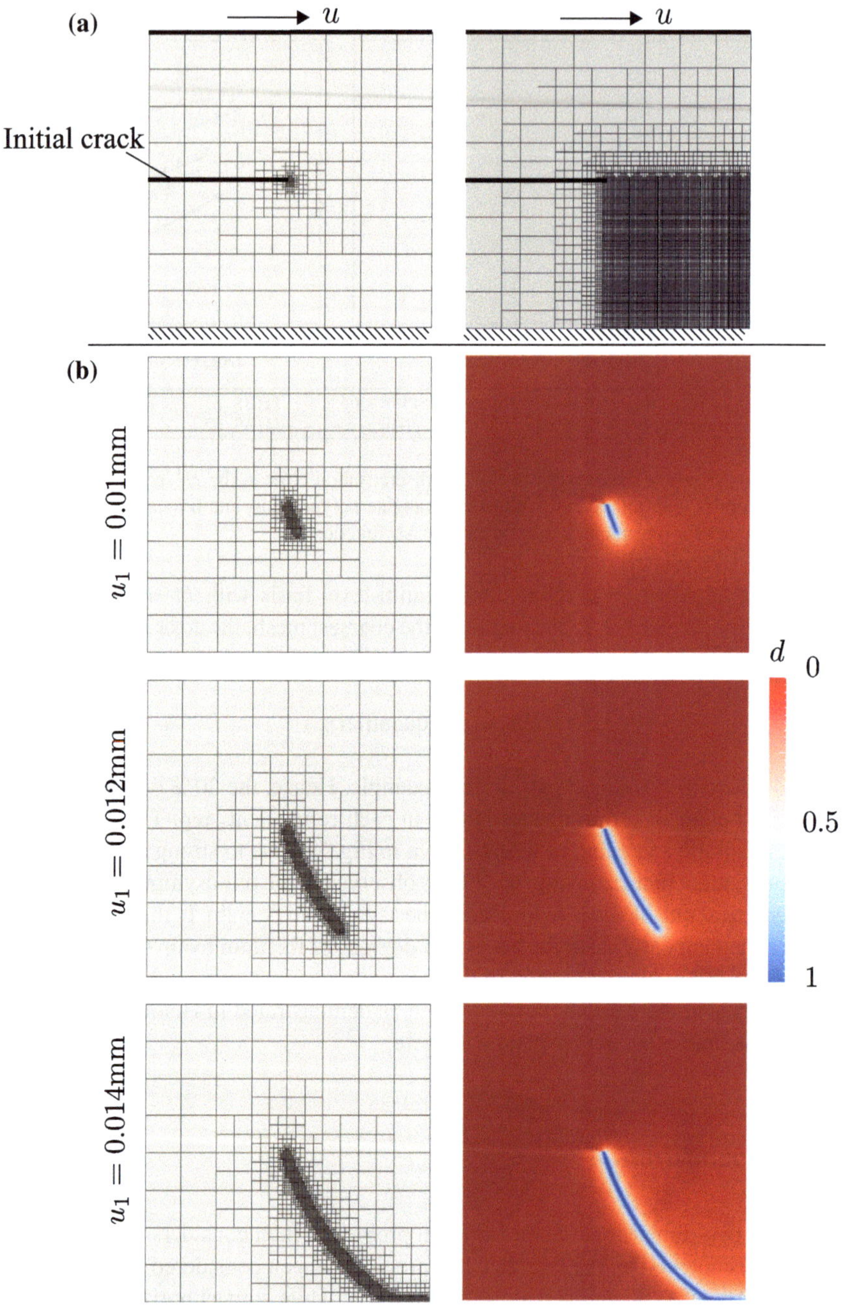

Fig. 6 Meshes and solution of the single edge notched shear test: The initial mesh in **a** on the left is adaptively refined by the evolution of the crack (**b**). The phase-field variable on the uniformly refined mesh is given in **b** on the right

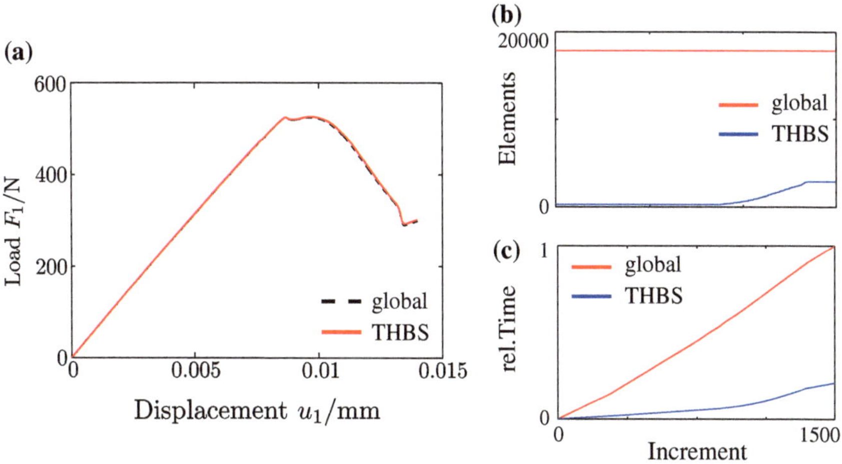

Fig. 7 Computational results of the single edge notched shear test: **a** The adaptive computation is consistent with the uniformly refined comparative computation **b**, reduces the number of elements **c** and hence the overall computational time **c**

length $\ell_0 = 0.75 \times 10^{-2}$ mm. These parameters are identical to the work of Miehe et al. (2010a) which allows for a quantitative comparison.

A staggered solution scheme is used with a displacement increment of $\Delta u = 1 \times 10^{-5}$ mm per time step. The initial mesh is pre-refined in the vicinity of the crack tip, Fig. 6a (left). The threshold for the adaptive mesh refinement is set to $d = 0.5$, i.e. elements are marked for refinement if the phase-field parameter at any quadrature point of the element becomes larger than 0.5.

A comparison between two solutions, one obtained on an adaptively refined mesh and one computed on a uniformly pre-refined mesh is presented in Fig. 6. Both solutions produce the same crack pattern and it can be seen that the adaptive algorithm resolves the crack path. Moreover, the shape of the crack is identical to the numerical solution in Miehe et al. (2010a).

In addition to the qualitative agreement of the crack paths, the force-displacement curves obtained with both solutions is identical, Fig. 7a. The adaptive approach significantly reduces the number of elements which improves the efficiency compared to uniform refinement, Fig. 7b. It is observed from Fig. 7c that the computational time is cut by 76 %. Note also that knowledge on the expected crack path was used to generate the uniformly pre-refined mesh which is impossible for problems of arbitrary complexity.

4 Conclusions

We have presented the combination of a phase-field model of brittle fracture with adaptive spline-based approximations. A numerical convergence study was performed for a one-dimensional problem of a bar for which $\Gamma_{\ell_0} = \Gamma = A$ should hold. We demonstrated how boundary effects and the discretisation error prevent numerical solutions from being exact. Moreover, the characteristic minimum in the error $\Gamma_\mathrm{E} = |\Gamma_{\ell_0} - \Gamma|/\Gamma$ found in May et al. (2015) was explained, i.e. with increasing ratios ℓ_0/L, boundary effects due to the finite domain size deteriorate the phase-field approximation. On the other hand the error increases for small characteristic length scales ℓ_0 because of discretisation errors.

In order to limit this discretisation error, we have utilised hierarchical B-splines that can be refined in the vicinity of the crack. The use of an element viewpoint and the application of Bézier extraction to a multi-level mesh facilitate the implementation into any existing finite element code. Moreover, standard procedures of adaptive finite element analysis are directly applicable. While Bézier extraction provided a multi-level system of equation with independent levels, it was shown that a simple matrix multiplication produces a hierarchical system of equations that is identical to the use of the truncated hierarchical basis. It is noted that the explicit computation of the basis is avoided.

In a second numerical convergence study, we have demonstrated the superiority of the spline discretisation over Lagrangian polynomials already for uniform mesh refinement. The local refinement was found to increase the convergence rate in the pre-asymptotic domain while optimal convergence rates are recovered in the asymptotic range. In this way the total number of degrees of freedom required to achieve a certain error level can be significantly reduced compared to uniform meshes. The developed algorithms were eventually applied to simulate crack propagation in a two-dimensional single-edge notched, shear loaded plate. Here, the adaptive local refinement produced exactly the same crack path and load-displacement curve as a simulation based on a highly refined uniform mesh. As the locally refined mesh involved considerably fewer elements, a significantly shorter computation time, that considers also the numerical effort for marking and refining elements, was found in relation to the uniform solution which illustrates the numerical efficiency of the approach.

Acknowledgments The authors gratefully acknowledge support by the Deutsche Forschungsgemeinschaft (DFG) in the Priority Programme 1748 "Reliable simulation techniques in solid mechanics. Development of non-standard discretization methods, mechanical and mathematical analysis" under the project KA3309/3-1.

References

Ambati, M., Gerasimov, T., De Lorenzis, L.: A review on phase-field models of brittle fracture and a new fast hybrid formulation. Comput. Mech. **55**(2), 383–405 (2015)

Belytschko, T., Moës, N., Usui, S., Parimi, C.: Arbitrary discontinuities in finite elements. Int. J. Numer. Meth. Eng. **50**, 993–1013 (2001)

Borden, M.J., Scott, M.A., Evans, J.A., Hughes, T.J.R.: Isogeometric finite element data structures based on Bézier extraction of NURBS. Int. J. Numer. Meth. Eng. **87**(1–5), 15–47 (2011)

Borden, M.J., Verhoosel, C.V., Scott, M.A., Hughes, T.J., Landis, C.M.: A phase-field description of dynamic brittle fracture. Comput. Methods Appl. Mech. Eng. **217–220**, 77–95 (2012)

Bourdin, B., Francfort, G., Marigo, J.J.: Numerical experiments in revisited brittle fracture. J. Mech. Phys. Solid. **48**, 797–826 (2000)

Chen, L.Q.: Phase-field models for microstructure evolution. Annu. Rev. Mater. Res. **32**(1), 113–140 (2002)

Dörfel, M.R., Jüttler, B., Simeon, B.: Adaptive isogeometric analysis by local h-refinement with T-splines. Comput. Method. Appl. Mech. Eng. **199**(5–8), 264–275 (2010)

Emmerich, H.: Advances of and by phase-field modelling in condensed-matter physics. Adv. Phys. **57**(1), 1–87 (2008)

Evans, E.J., Scott, M.A., Li, X., Thomas, D.C.: Hierarchical T-splines: Analysis-suitability, Bézier extraction, and application as an adaptive basis for isogeometric analysis. Comput. Method. Appl. Mech. Eng. **284**, 1–20 (2015)

Forsey, D.R., Bartels, R.H.: Hierarchical B-spline refinement. SIGGRAPH Comput. Graph. **22**(4), 205–212 (1988)

Francfort, G., Marigo, J.J.: Revisiting brittle fracture as an energy minimization problem. J. Mech. Phys. Solid. **46**, 1319–1342 (1998)

Fries, T.P., Belytschko, T.: The extended/generalized finite element method: an overview of the method and its applications. Int. J. Numer. Meth. Engng. **84**, 253–304 (2010)

Giannelli, C., Jüttler, B., Speleers, H.: THB-splines: the truncated basis for hierarchical splines. Comput. Aided Geom. D **29**(7), 485–498 (2012)

Hennig, P., Müller. S., Kästner, M.: Bézier extraction and adaptive refinement of truncated hierarchical NURBS. Comput. Method. Appl. Mech. Eng. (2016). doi:10.1016/j.cma.2016.03.009

Hughes, T.J.R., Cottrell, J.A., Bazilevs, Y.: Isogeometric analysis: CAD, finite elements, NURBS, exact geometry and mesh refinement. Comput. Method. Appl. Mech. Eng. **194**(39–41), 4135–4195 (2005)

Jirásek, M.: Comparative study on finite elements with embedded discontinuities. Comput. Method. Appl. Mech. Eng. **188**(1–3), 307–330 (2000)

Kästner, M., Müller, S., Hirsch, F., Pap, J.S., Jansen, I., Ulbricht, V.: XFEM modeling of interface failure in adhesively bonded fiber-reinforced polymers. Adv. Eng. Mater. **18**, 417–426 (2016b)

Kästner, M., Metsch, P., de Borst, R.: Isogeometric analysis of the cahn-hilliard equation - a convergence study. J. Comput. Phys. **305**, 360–371 (2016a)

Linder, C., Rosato, D., Miehe, C.: New finite elements with embedded strong discontinuities for the modeling of failure in electromechanical coupled solids. Comput. Method. Appl. Mech. Eng. **200**, 141–161 (2011)

May, S., Vignollet, J., de Borst, R.: A numerical assessment of phase-field models for brittle and cohesive fracture: Γ-convergence and stress oscillations. Eur. J. Mech. A-Solid. **52**, 72–84 (2015)

Miehe, C., Hofacker, M., Welschinger, F.: A phase field model for rate-independent crack propagation: robust algorithmic implementation based on operator splits. Comput. Method. Appl. Mech. Eng. **199**(45–48), 2765–2778 (2010a)

Miehe, C., Welschinger, F., Hofacker, M.: Thermodynamically consistent phase-field models of fracture: variational principles and multi-field FE implementations. Int. J. Numer. Meth. Eng. **83**(10), 1273–1311 (2010b)

Miehe, C., Schänzel, L.M., Ulmer, H.: Phase field modeling of fracture in multi-physics problems. Part I. Balance of crack surface and failure criteria for brittle crack propagation in thermo-elastic solids. Comput. Method. Appl. Mech. Eng. **294**, 449–485 (2015)
Moelans, N., Blanpain, B., Wollants, P.: An introduction to phase-field modeling of microstructure evolution. Calphad **32**(2), 268–294 (2008)
Moës, N., Dolbow, J., Belytschko, T.: A finite element method for crack growth without remeshing. Int. J. Numer. Meth. Eng. **46**, 131–150 (1999)
Ortiz, M., Leroy, Y., Needleman, A.: A finite element method for localized failure analysis. Comput. Method. Appl. Mech. Eng. **61**(2), 189–214 (1987)
Peerlings, R., de Borst, R., Brekelmans, W., de Vree, J.: Gradient enhanced damage for quasi-brittle materials. Int. J. Numer. Method. Eng. **39**(19), 3391–3403 (1996)
Piegl, L., Tiller, W.: The NURBS Book. Springer, Berlin (1996)
Schillinger, D., Dede, L., Scott, M.A., Evans, J.A., Borden, M.J., Rank, E., Hughes, T.J.R.: An isogeometric design-through-analysis methodology based on adaptive hierarchical refinement of NURBS, immersed boundary methods, and T-spline CAD surfaces. Comput. Method. Appl. Mech. Eng. **249**, 116–150 (2012)
Scott, M.A., Li, X., Sederberg, T.W., Hughes, T.J.R.: Local refinement of analysis-suitable T-splines. Comput. Method. Appl. Mech. Eng. **213–216**, 206–222 (2012)
Sederberg, T.W., Zheng, J., Bakenov, A., Nasri, A.: T-splines and T-NURCCs. ACM Trans. Graph **22**(3), 477–484 (2003)
Sederberg, T.W., Cardon, D.L., Finnigan, G.T., North, N.S., Zheng, J., Lyche, T.: T-spline simplification and local refinement. ACM Trans. Graph **23**(3), 276–283 (2004)
Simo, J.C., Oliver, J., Armero, F.: An analysis of strong discontinuities induced by strain-softening in rate-independent inelastic solids. Comput. Mech. **12**(5), 277–296 (1993)
Steinbach, I.: Phase-field models in materials science. Model. Simul. Mater. Sci. **17**(7), 073001 (2009)
Wells, G.N., Sluys, L.J.: A new method for modelling cohesive cracks using finite elements. Int. J. Numer. Meth. Eng. **50**(12), 2667–2682 (2001)

Part V
Fracture Mechanics and Failure Criteria

Prestressed Orthotropic Material Containing an Elliptical Hole

Eduard-Marius Craciun

Abstract Based on the representation of the incremental stress fields by complex potentials and conformal mapping technique, the fundamental solutions for an unbounded, homogeneous, orthotropic elastic body containing an elliptical hole subjected to uniform remote loads are determined. The orthotropic body is under by uniform remote tensile, tangential, and antiplane shear loads—cases corresponding to Mode I, Mode II, and Mode III of fracture. The solutions are obtained in a compact and elementary form.

1 Introduction

The problem of an isotropic body with an elliptical hole was studied by many authors using Kolosov–Muskhelishvili's complex potentials Muskhelishvili (1953)–Bertoldi et al. (2007) or the integral transform method Singh et al. (2012). In what follows our results Craciun and Soós (2006), Craciun and Barbu (2015) for a prestressed elastic composite material under by uniform distributed remote loads are presented and extended. To get the complex potentials describing the incremental stress and displacement fields, $\Psi_1 = \Psi_1(z_1)$ and $\Psi_2 = \Psi_2(z_2)$ for the plane problem, and $\Psi_3 = \Psi_3(z_3)$ for the antiplane problem, a technique based on the conformal mapping of the exterior of the elliptical hole in the planes on the exterior of the unit circle is used. The unknown potentials are represented by two Laurent series in the complex planes and their coefficients are determined from the boundary conditions. The compact closed-form analytical solutions, i.e., the complex potentials, of the considered boundary value problems for an unbounded, homogeneous, prestressed orthotropic elastic composite with an elliptic hole are obtained.

E.-M. Craciun (✉)
Ovidius University of Constanta, Bldv. Mamaia, No. 124, 900527 Constanta, Romania
e-mail: mcraciun@univ-ovidius.ro

K. Naumenko and M. Aßmus (eds.), *Advanced Methods of Continuum Mechanics for Materials and Structures*, Advanced Structured Materials 60,
DOI 10.1007/978-981-10-0959-4_18

2 Representation of the Incremental Stress Fields

The representation of elastic fields by complex potentials in the classical case of anisotropic elastic bodies was given by Lekhnitski (1963). This representation was used, for instance, by Sih and Leibowitz (1968) to analyze problems concerning the existence of a crack in an anisotropic elastic solid. The results obtained by Lekhnitski (1963) were generalized for the case of a prestressed material by Guz (1983), who also has analyzed the influence of the initial applied stresses on the behavior of a solid body containing cracks. Guz's representation of the incremental stress fields by complex potentials Guz (1983), Cristescu et al. (2004) is presented. In what follows it is considered an elastic composite with an elliptical hole prestressed with the initial applied stress $\overset{\circ}{\sigma}_{11}$, in the direction of Ox_1 axis, i.e., along the major semi-axis of the ellipse. The initial deformed equilibrium configuration of the body is assumed to be homogeneous and locally stable. The paper starts with representation of the incremental stress fields corresponding to the antiplane state, by a single complex potential $\Psi_3 = \Psi_3(z_3)$ depending on the complex variable $z_3 = x_1 + \mu_3 x_2$. The complex parameter μ_3 is the root of the characteristic equation of the differential equilibrium equation and has the following form, see Guz (1983)–Cristescu et al. (2004):

$$\mu_3 = \frac{1}{\omega_{2332}}\left[-\omega_{1332} + i\sqrt{\omega_{1331}\omega_{2332} - \omega_{1332}\omega_{2331}}\right], \tag{1}$$

where ω_{klmn} $(k, l, m, n = 1, 2, 3)$ are the instantaneous elasticities of the material in its free reference configuration and can be expressed through engineering constants of the composite and initial applied stress $\overset{\circ}{\sigma}_{11}$ and i denotes the imaginary unit, see Cristescu et al. (2004).

Taking into account the antiplane state relative to the plane x_1x_2 the instantaneous elasticities of the material have to satisfy the following restrictions:

$$\sqrt{\omega_{1331}\omega_{2332} - \omega_{1332}\omega_{2331}} > 0, \omega_{2332} > 0. \tag{2}$$

The corresponding components θ_{13} and θ_{23} of the nominal stress are then given by

$$\theta_{13} = 2Re\left\{q\Psi_3(z_3)\right\}, \theta_{23} = 2Re\left\{\Psi_3(z_3)\right\}, q = \frac{\rho_1}{\rho_2},$$

$$\rho_1 = \omega_{1331} + \mu_3\omega_{1323}, \rho_2 = \omega_{2313} + \mu_3\omega_{2323}. \tag{3}$$

It is assumed that the initial deformed composite material is in *plane state* relative to the x_1x_2 plane.

The representation of the incremental stress fields by two arbitrary analytical complex potentials $\Psi_j = \Psi_j(z_j)$, $j = 1, 2$ has the following form:

$$\theta_{22} = 2Re\left\{\Psi_1(z_1) + \Psi_2(z_2)\right\}, \tag{4}$$

$$\theta_{21} = -2Re\left\{a_1\mu_1\Psi_1(z_1) + a_2\mu_2\Psi_2(z_2)\right\}, \tag{5}$$

$$a_j = \frac{\omega_{2112}\omega_{1122}\mu_j^2 - \omega_{1111}\omega_{1212}}{B_j\mu_j^2}, \tag{6}$$

$$\theta_{12} = -2Re\left\{\mu_1\Psi_1(z_1) + \mu_2\Psi_2(z_2)\right\}, \tag{7}$$

$$\theta_{11} = 2Re\left\{a_1\mu_1^2\Psi_1(z_1) + a_2\mu_2^2\Psi_2(z_2)\right\}, \tag{8}$$

$$B_j = \omega_{2222}\omega_{2112}\mu_j^2 + \omega_{1111}\omega_{2222} - \omega_{1122}(\omega_{1122} + \omega_{1212}), \tag{9}$$

where μ_1 and μ_2 are the roots of characteristic equation of equilibrium, see Cristescu et al. (2004). The instantaneous elasticities of the material ω_{klmn} $(k, l, m, n = 1, 2)$ can be expressed through engineering constants of the composite and initial applied stress $\overset{\circ}{\sigma}_{11}$, by the following relations, see Cristescu et al. (2004):

$$\omega_{1111} = \frac{1 - \nu_{23}\nu_{32}}{E_2E_3H} + \overset{\circ}{\sigma}_{11}, \quad \omega_{2222} = \frac{1 - \nu_{13}\nu_{31}}{E_1E_3H},$$

$$\omega_{1122} = \frac{\nu_{12} + \nu_{32}\nu_{13}}{E_1E_3H},$$

$$\omega_{1212} = \omega_{1221} = \omega_{2112} = G_{12},$$

with

$$H = \frac{1 - \nu_{12}\nu_{21} - \nu_{23}\nu_{32} - \nu_{31}\nu_{13} - \nu_{21}\nu_{32}\nu_{13} - \nu_{12}\nu_{23}\nu_{31}}{E_1E_2E_3}.$$

In these relations E_1, E_2, E_3 are Young's moduli in the corresponding symmetry directions of the material, G_{12} is the shear modulus in the symmetry plane Ox_1x_2 and ν_{12}, ..., ν_{32} are the Poisson's ratios.

Also, for an orthotropic material the roots μ_1 and μ_2 usually are not equal. In what follows the case of non-equal roots is considered

$$\mu_1 \neq \mu_2.$$

3 Antiplane State

In this section the plane problem of antiplane shear loads, i.e., the case corresponding to the third mode of fracture, is studied. Let us consider an unbounded, homogeneous, prestressed orthotropic elastic composite containing an elliptical hole which is acted by an antiplane constant shear load $\tau > 0$ in the direction of the x_3 axis at large distances. The boundary of the elliptical hole is free from stresses.

Let us write the boundary conditions corresponding to the mechanical problem:

$$\lim_{|z_3|\to\infty} \theta_{13}(z_3) = 0, \qquad \lim_{|z_3|\to\infty} \theta_{23}(z_3) = \tau > 0,$$

$$n_1\theta_{13}(z_3) + n_2\theta_{23}(z_3) = 0, \tag{10}$$

on the hole boundary, where n_1 and n_2 are the components of the unit exterior normal to the boundary.

In order to find the complex potential $\Psi_3 = \Psi_3(z_3)$ is considered the *conformal mapping* of the exterior of the elliptical hole onto the exterior of the unit circle, having the form

$$z = x_1 + ix_2 = \frac{a+b}{2}\zeta + \frac{a-b}{2}\frac{1}{\zeta}$$

$$z_3 = x_1 + \mu_3 x_2 = \frac{a - i\mu_3 b}{2}\zeta_3 + \frac{b + i\mu_3 b}{2}\frac{1}{\zeta_3}. \tag{11}$$

The inverse mapping is given by

$$\zeta = \frac{z + \sqrt{z^2 - a^2 + b^2}}{a+b}, \qquad \zeta_3 = \frac{z_3 + \sqrt{z_3^2 - a^2 - \mu_3 b^2}}{a - i\mu_3 b}. \tag{12}$$

Let $a, b, b \leq a$ be the two semi-axis of the elliptical hole and if $b \to 0$ the considered hole obviously becomes the mathematical model of an usual, classical Griffith-Irwin crack given by a segment of length $2a$.

Let us introduce now the complex potential $\Psi_3(\zeta_3)$ through the relation

$$\Psi_3(\zeta_3) = \Psi_3(z_3(\zeta_3)), \tag{13}$$

where for simplicity it is used the same notation Ψ_3 for the complex potential depending on z_3 or on ζ_3. The boundary conditions (10) by means of the potential $\Psi_3(\zeta_3)$ and the mapping formula (11) become

$$\lim_{|z_3|\to\infty} \theta_{13}(\zeta_3) = 0, \qquad \lim_{|z_3|\to\infty} \theta_{23}(\zeta_3) = \tau \tag{14}$$

and

$$(\cos\theta)\theta_{13}(\zeta_3) + (\sin\theta)\theta_{23}(\zeta_3) = 0, \quad \zeta_3 = e^{i\theta}, 0 \leq \theta \leq 2\pi, \tag{15}$$

where $(\cos\theta, \ \sin\theta)$ is the unit exterior to the unit circle in the complex plane ζ_3, and θ is obviously the angle between this normal and the x_1 axis.

The complex potential $\Psi_3 = \Psi_3(\zeta_3)$ is an analytic function on the exterior of the unit circle, and thus we may write it as

$$\Psi_3(\zeta_3) = A_0 + \sum_{m=1}^{\infty} A_m \zeta_3^{-m}, \tag{16}$$

where $A_0,\ A_1,\ A_2, \ldots$ are unknown complex constants to be determined from the boundary conditions. The second boundary condition from (10) at large distance from the hole leads to the following restrictions on A_0:

$$qA_0 + \bar{q}\bar{A}_0 = 0,\ A_0 + \bar{A}_0 = \tau \tag{17}$$

therefore,

$$A_0 = -\tau\frac{\bar{q}}{q-\bar{q}} = \frac{\tau\bar{q}i}{2r_2},\ q = r_1 + ir_2. \tag{18}$$

The third boundary condition from (10) imposes an additional restriction on the coefficients of the potential $\Psi_3(z_3)$

$$\begin{aligned}(q+i)A_1 &+ (\bar{q}-i)\bar{A}_1 + (q+i)A_2e^{-i\theta} + (\bar{q}-i)\ \bar{A}_2\ e^{i\theta} \\ &+ \sum_{m=2}^{\infty}[(q+i)A_{m+1} + (q-i)A_{m-1}]e^{-im\theta} + \sum_{m=2}^{\infty}[(\bar{q}-i)\bar{A}_{m+1} + (\bar{q}+i)\bar{A}_{m-1}]e^{im\theta} \\ &= -A_0[(q+i)e^{i\theta} + (q-i)e^{-i\theta}] - \bar{A}_0[(\bar{q}+i)e^{i\theta} + (\bar{q}-i)e^{-i\theta}],\end{aligned} \tag{19}$$

for $0 \le \theta \le 2\pi$. Condition (19) is fulfilled if and only if the following relations are satisfied:

$$(q+i)A_1 + (\bar{q}-i)\bar{A}_1 = 0 \tag{20}$$

$$(q+i)A_2 = -[A_0(q-i) + \bar{A}_0(\bar{q}-i)],$$

$$(\bar{q}-i)\bar{A}_2 = -[A_0(q+i) + \bar{A}_0(\bar{q}+i)] \tag{21}$$

$$A_{1+2m} = s^m A_1,\ A_{2+2m} = s^m A_2,\ m = 1, 2, 3... \tag{22}$$

with

$$s = -\frac{q-i}{q+i}. \tag{23}$$

Equation (21) becomes

$$A_2 = -\frac{A_0\,(q-i) + \overline{A}_0\,(\overline{q}-i)}{q+i} \tag{24}$$

and taking into account (24) the expression of the constant A_2 has the following form:

$$A_2 = i\frac{\tau}{q+i}. \tag{25}$$

Let γ and δ denote the real and the imaginary part of A_1, respectively, i.e.,

$$A_1 = \gamma + i\delta. \tag{26}$$

Now, from (20) we get the following value for A_1:

$$A_1 = \gamma\left(1 + i\frac{r_1}{r_2 + 1}\right) \tag{27}$$

with

$$q = r_1 + ir_2. \tag{28}$$

Let us remark that the real number γ remains undetermined in the expression (27) of A_1. This is not an unexpected result since we have a boundary value problem in stress where such indetermination generally occurs. After some laborious manipulations, using (22) into (14) the *final* form of the complex potential $\Psi_3(\zeta_3)$ has the following form:

$$\Psi_3(\zeta_3) = A_0 + \frac{A_1\zeta_3 + A_2}{\zeta_3^2 - s}. \tag{29}$$

The basic complex potential $\Psi_3(z_3)$ may then be obtained by introducing the expression of ζ_3 given by (12) into the right-hand side of (29) and the problem is completely solved.

4 Plane State

In this section the plane problem of a uniform distributed remote tensile load, i.e., the case corresponding to the first opening mode of fracture, is studied. Let us consider an unbounded, homogeneous, prestressed orthotropic elastic composite containing an elliptical hole which is acted by a uniform constant normal tensile load $p > 0$ in the direction of the x_2 axis at large distances. The boundary of the elliptical hole is free from stresses.

Let us write the boundary conditions corresponding to our mechanical problem:

$$\lim_{|z|\to\infty} \theta_{11}(z) = \lim_{|z|\to\infty} \theta_{12}(z) = \lim_{|z|\to\infty} \theta_{21}(z) = 0, \quad \lim_{|z|\to\infty} \theta_{22}(z) = p > 0, \tag{30}$$

$$n_1\theta_{11}(z) + n_2\theta_{21}(z) = 0, \;\; n_1\theta_{12}(z) + n_2\theta_{22}(z) = 0,$$

on the hole boundary, where n_1 and n_2 are the components of the unit exterior normal to the boundary.

The complex potentials $\Psi_j = \Psi_j(z_j)$, $j = 1, 2$, must be determined not in the region of the infinite prestressed orthotropic plate with an elliptical hole, denoted by

S, but in the regions S_j, $j = 1, 2$ obtained from S by the affine transformations:

$$x_1^j = x_1 + \alpha_j x_2,\ x_2^j = \beta_j x_2,\ j = 1, 2. \tag{31}$$

The regions S_j are also planes with elliptical holes whose contours are given by the equations

$$x_1^j = a\cos\theta + \alpha_j b\sin\theta,\ x_2^j = \beta_j \sin\theta,\ 0 \le \theta \le 2\pi,\ j = 1, 2. \tag{32}$$

The following conformal mapping of the regions S, S_1, and S_2 onto the exterior of the unit circle is used:

$$z = x_1 + ix_2 = \frac{a+b}{2}\zeta + \frac{a-b}{2}\frac{1}{\zeta},$$

$$z_j = x_1 + \mu_j x_2 = \frac{a - i\mu_j b}{2}\zeta_j + \frac{a + i\mu_j b}{2}\frac{1}{\zeta_j},\ j = 1, 2. \tag{33}$$

The inverse mapping is given by

$$\zeta = \frac{z + \sqrt{z^2 - a^2 + b^2}}{a+b},\ \zeta_j = \frac{z_j + \sqrt{z_j^2 - a^2 - \mu_j b^2}}{a - i\mu_j b},\ j = 1, 2. \tag{34}$$

When the x_1 and x_2 are running along the contour of the ellipse taking the values $\frac{x_1^2}{a^2} + \frac{x_2^2}{b^2} = 1$, the functions defined by the (34) take the values $\zeta = \zeta_1 = \zeta_2 = e^{i\theta}$.

Let us introduce now the complex potentials $\Psi_j(\zeta_j)$ through the relations $\Psi_j(\zeta_j) = \Psi_j(z_j(\zeta_j))$, $j = 1, 2$, where for simplicity we use the same notation Ψ_j for the complex potentials depending on z_j or on ζ_j. The boundary conditions (30) by means of the potential $\Psi_j(\zeta_j)$ and the mapping formulae (33) become

$$\lim_{|\zeta|\to\infty} \theta_{11}(\zeta) = \lim_{|\zeta|\to\infty} \theta_{12}(\zeta) = \lim_{|\zeta|\to\infty} \theta_{21}(\zeta) = 0,\quad \lim_{|\zeta|\to\infty} \theta_{22}(\zeta) = p > 0, \tag{35}$$

$$(\cos\theta)\theta_{11}(\zeta) + (\sin\theta)\theta_{21}(\zeta) = 0,\ (\cos\theta)\theta_{12}(\zeta) + (\sin\theta)\theta_{22}(\zeta) = 0, \tag{36}$$

on the unit circle $\zeta = e^{i\theta}, 0 \le \theta < 2\pi$, where $(\cos\theta,\ \sin\theta)$ is the unit exterior normal to the unit circle in the complex plane ζ, and θ is obviously the angle between this normal and the x_1 axis.

The complex potentials $\Psi_j = \Psi_j(\zeta_j)$ $j = 1, 2$ are analytical functions on the exterior of the unit circle, i.e.,

$$\Psi_1(\zeta_1) = A_0 + \sum_{m=1}^{\infty} A_m \zeta_1^{-m},$$

$$\Psi_2(\zeta_2) = B_0 + \sum_{m=1}^{\infty} B_m \zeta_2^{-m}, \tag{37}$$

where A_k and B_k, $k = 0, 1, \ldots$ are unknown complex constants to be determined from the boundary conditions.

Using Eqs. (35)–(37) the following restrictions on A_0, $\overline{A}_0$, B_0 and $\overline{B}_0$ are imposed:

$$A_0 + \overline{A}_0 + B_0 + \overline{B}_0 = p, \tag{38}$$

$$\begin{aligned} \mu_1 A_0 + \overline{\mu}_1 \overline{A}_0 + \mu_2 B_0 + \overline{\mu}_2 \overline{B}_0 &= 0, \\ a_1 \mu_1 A_0 + \overline{a}_1 \overline{\mu}_1 \overline{A}_0 + a_2 \mu_2 B_0 + \overline{a}_2 \overline{\mu}_2 \overline{B}_0 &= 0, \\ a_1 \mu_1^2 A_0 + \overline{a}_1 \overline{\mu}_1^2 \overline{A}_0 + a_2 \mu_2^2 B_0 + \overline{a}_2 \overline{\mu}_2^2 \overline{B}_0 &= 0. \end{aligned} \tag{39}$$

Using the representation formulae, the expressions of the complex potentials, and boundary conditions (36), the following expressions are obtained:

$$A_{2m+1} = \xi_1^m A_1, \;\; A_{2m+2} = \xi_1^m A_2, \; B_{2m+1} = \xi_2^m B_1, \;\; B_{2m+2} = \xi_2^m B_2, \tag{40}$$

with

$$\xi_k = \frac{i - \mu_k}{\mu_k + i}, \; k = 1, 2,$$

$$B_1 = 2\frac{\overline{a}_2 \overline{\mu}_2 Re((\mu_1 + i)A_1) - Re(a_1 \mu_1 (\mu_1 + i)A_1)}{(\mu_2 + i)(a_2 \mu_2 - \overline{a}_2 \overline{\mu}_2)}. \tag{41}$$

Finally, from system (38), the coefficients A_2 and B_2 can be determined.

The complex coefficient A_1 remains undetermined in the expressions (37) of complex potentials of $\Psi_j = \Psi_j\left(z_j\right)$, $j = 1, 2$. This is not an unexpected result since it is considered a boundary value problem in stress where such indetermination generally occurs.

The expression of the complex potentials $\Psi_j = \Psi_j\left(z_j\right)$, $j = 1, 2$ may now be written using (40) into (37):

$$\begin{aligned} \Psi_1(\zeta_1) &= A_0 + \frac{A_1 \zeta_1 + A_2}{\zeta_1^2 - \xi_1}, \\ \Psi_2(\zeta_2) &= B_0 + \frac{B_1 \zeta_2 + B_2}{\zeta_2^2 - \xi_2}. \end{aligned} \tag{42}$$

In the last part of the paper, the plane problem of uniform remote tangential shear loads, i.e., the case corresponding to the second mode of fracture, is studied.

Let us consider an unbounded, homogeneous, anisotropic elastic body containing an elliptical hole under by a uniform remote constant tangential shear load $h > 0$ in the direction of the x_1 axis. The boundary of the elliptical hole is free from stress. The boundary condition (36) remains unchanged and the far-field conditions (35) become

$$\lim_{|z|\to\infty} \theta_{11}(z) = \lim_{|z|\to\infty} \theta_{21}(z) = \lim_{|z|\to\infty} \theta_{22}(z) = 0, \quad \lim_{|z|\to\infty} \theta_{12}(z) = h > 0. \tag{43}$$

Using the same formalism as in the previous case, the same expressions of the complex potentials are obtained. The coefficient B_1 has the same form as before and the coefficient A_1 rests undetermined. To find the coefficients A_0, B_0, A_2, and B_2, it will use the same procedure, as for the plane problem of uniform remote tensile load. From the far-field conditions the following restrictions are obtained:

$$\begin{gathered} A_0 + \overline{A}_0 + B_0 + \overline{B}_0 = 0, \quad \mu_1 A_0 + \overline{\mu}_1 \overline{A}_0 + \mu_2 B_0 + \overline{\mu}_2 \overline{B}_0 = -h, \\ a_1 \mu_1 A_0 + \overline{a}_1 \overline{\mu}_1 \overline{A}_0 + a_2 \mu_2 B_0 + \overline{a}_2 \overline{\mu}_2 \overline{B}_0 = 0, \\ a_1 \mu_1^2 A_0 + \overline{a}_1 \overline{\mu}_1^2 \overline{A}_0 + a_2 \mu_2^2 B_0 + \overline{a}_2 \overline{\mu}_2^2 \overline{B}_0 = 0. \end{gathered} \tag{44}$$

Let us observe that the above system could be a determinate system, an indeterminate system, and in the case of resonance due to the initial applied stress $\overset{\circ}{\sigma}_{11}$ an incompatible system. Finally, the values of the complex coefficients A_2 and B_2 are obtained.

The *final* forms of the complex potentials $\Psi_j(\zeta_j)$, $j = 1, 2$ are thus determined by elementary calculus in both situations of uniform remote tensile and tangential shear loads, respectively. The basic complex potentials $\Psi_j(z_j)$ $j = 1, 2$ may be then obtained by introducing the expression of ζ_j given by (34) into the right-hand side of (42) and the problem is completely solved.

5 Final Remarks

Compact closed-form analytical solutions of the considered boundary value problems for an unbounded, homogeneous, prestressed orthotropic elastic composite containing an elliptical hole, subjected to uniform remote tensile, tangential, and antiplane shear loads (Mode I, Mode II, and Mode III of Fracture), are obtained.

The general results are practically relevant, e.g., for the study of incremental stress, strain, and displacement fields in the vicinity of the elliptical hole, can be applied to study a variety of composite mechanics problems, and can be extended for prestressed thermoelastic, ferromagnetic, or piezoelectric materials with elliptical holes.

1 Introduction

Experimental methods provide a limited number of failure points under different loadings. This number is often not sufficient to fit the modern strength hypotheses, which are necessary for the dimensioning and optimizing of structural components. Some loading cases, e.g., the hydrostatic tensile loading, cannot be realized and there are no clear statements regarding the material behaviour at this stress state. The adjustment of the parameters of the chosen criterion is precarious in the case of limited data sets.

Numerical methods based on the real microstructure of a foam, on the cell size and cell irregularities, and on the properties of the bulk polymer allow calculating failure points under any loading. These points can be validated by available measured data, cf. Fahlbusch (2015).

In the current approach, a finite element (FE) model based on a tetracaidecahedron is generated. The model is consisting of a single unit cell. Periodic boundary conditions are implemented and a strain-energy homogenization concept is applied (Hohe and Becker 2005). The model is adapted to geometrical parameters of the real microstructure, which result from an analysis of high-resolution images. The distribution of the geometrical parameters on the microstructure is taken into account by the numerical evaluation of several so-called test volume elements (Monte Carlo method, Landau and Binder 2009).

For an engineering application the numerically calculated sets of points should be approximated with a limit surface. The most useful criteria for hard foams are the Ehlers criterion (Droste 2004; Ehlers 1993, 1995), Bigoni–Piccolroaz criterion (Bigoni and Piccolroaz 2004; Penasa et al. 2014; Piccolroaz and Bigoni 2009), and the cosine ansatz of Altenbach–Bolchoun–Kolupaev (Altenbach et al. 2014). These criteria, however, have different limitations in the geometry of the shape in the π-plane and of the meridian and cannot be applied to arbitrary isotropic materials.

For a versatile use a criterion is desirable which can describe limit surfaces with a large number of different shapes in the π-plane. This surface is composed based on the convex combination of two extremal limit surfaces, of Capurso (Capurso 1967; Sayir 1970) and of Haythornthwaite (Candland 1975; Haythornthwaite 1961). The Capurso–Haythornthwaite generalization proposed by Altenbach–Bolchoun–Kolupaev (Altenbach et al. 2014; Bolchoun et al. 2011) includes several known criteria like Tresca (Reuss 1933; Tresca 1868), von Mises (Huber 1904; von Mises 1913), Schmidt–Ishlinsky (Burzyński 1928; de Araújo 1962), Mariotte (Benvenuto 1991), and Ko (Ko 1963; Sayir 1970). This criterion contains also the bicubic criterion of hexagonal symmetry (Altenbach 2012), which is similar to the unified yield criterion of Yu (Yu 2002, 2004). The criteria of Ehlers, Bigoni–Piccolroatz, and the cosine ansatz of Altenbach–Bolchoun–Kolupaev can be approximated with this generalization.

The dependence of the failure behaviour of hard foams on the hydrostatic stress should be taken in account. The simple substitution based on the position of the hydrostatic nodes (hydrostatic poles of the surface on the hydrostatic axis) is used to define closed surfaces in the principal stress space.

The comparison of the numerical points for the cuts perpendicular to the hydrostatic axis shows the dependence of the geometry of the cross sections on the hydrostatic stress. So the parameters of the shape in the π-plane are formulated as a function of the axiator of the stress tensor I_1. This function can be numerically identified. A surface with a varying shape in the π-plane was the result. The isotropic assumption remains unaltered.

In the current approach, a method is introduced which allows deducing the parameters of the failure criterion based on the numerical points. Characteristics of the behaviour like failure under hydrostatic stress and varying cross sections in the π-plane in dependence on the first invariant of the stress tensor are considered. The determination of the parameters follows with a general routine. The method is shown for data sets of hard foams but is not limited to this material. The FE model is based on the material data of polymethacrylimide (PMI) hard foam ROHACELL® 71IG, Evonik Industries AG, Darmstadt (Rohacell 2010). The results of the approximations are visualized and compared in the Burzyński-plane (invariant plane $(I_1, \sqrt{3\, I_2'})$), π-plane (cross sections with $I_1 = \text{const.}$), and in the σ_{I}-σ_{II} diagram (Altenbach et al. 2014; Chen and Zhang 1991).

2 Generalized Strength Hypothesis

2.1 Invariants of Stress Tensor

Failure criteria for isotropic material behaviour should be invariant with respect to an arbitrary rotation of the coordinate system (Życzkowski 1981). Hence, the criteria are built up using the invariants of the symmetric second-rank stress tensor. From the solution of the eigenvalue problem, the principal values (principal stresses or principal invariants) are obtained and denoted by σ_{I}, σ_{II}, and σ_{III} (Altenbach et al. 1995; Altenbach and Kolupaev 2014). The following order is assumed

$$\sigma_{\text{I}} \geq \sigma_{\text{II}} \geq \sigma_{\text{III}}. \tag{1}$$

Three stress invariants, the trace I_1 of the stress tensor (axiator) and the invariants I_2', I_3' of the stress deviator (Altenbach et al. 1995; Altenbach and Kolupaev 2014; Życzkowski 1981), as a function of the principal stresses

$$I_1 = \sigma_{\text{I}} + \sigma_{\text{II}} + \sigma_{\text{III}}, \tag{2}$$

$$I_2' = \frac{1}{6}\left[(\sigma_{\text{I}} - \sigma_{\text{II}})^2 + (\sigma_{\text{II}} - \sigma_{\text{III}})^2 + (\sigma_{\text{III}} - \sigma_{\text{I}})^2\right], \tag{3}$$

and

$$I_3' = \left(\sigma_{\mathrm{I}} - \frac{1}{3} I_1\right)\left(\sigma_{\mathrm{II}} - \frac{1}{3} I_1\right)\left(\sigma_{\mathrm{III}} - \frac{1}{3} I_1\right) \tag{4}$$

are often used in the modelling. The criteria applied in the current approach are formulated in the axiatoric-deviatoric invariants (2)–(4)

$$\Phi(I_1,\, I_2',\, I_3',\, \sigma_{\mathrm{eq}}) = 0. \tag{5}$$

For criteria independent of the hydrostatic stress it follows

$$\Phi(I_2',\, I_3',\, \sigma_{\mathrm{eq}}) = 0. \tag{6}$$

Other formulations are discussed in Altenbach et al. (1995, 2014), Altenbach and Kolupaev (2014), Yu (2004), Życzkowski (1981) for instance.

With the invariant relations a geometrical meaning of the loading cases in the stress space can be defined, e.g. Altenbach et al. (2014), Altenbach and Kolupaev (2014):

- the scaled axiator I_1 of the stress tensor and the scaled root of the second invariant of the stress deviator

$$\chi = I_1/\sqrt{3} \qquad \text{and} \qquad \rho = \sqrt{2\, I_2'}, \tag{7}$$

  which are used as axes in the Burzyński-plane (Życzkowski 1981),
- the stress angle θ in the π-plane (Fig. 1) (Chen and Zhang 1991; Novozhilov 1951; Ottosen and Ristinmaa 2005; Życzkowski 1981)

$$\cos 3\theta = \frac{3\sqrt{3}}{2} \frac{I_3'}{(I_2')^{3/2}}, \qquad \theta \in \left[0,\, \frac{\pi}{3}\right], \tag{8}$$

  and

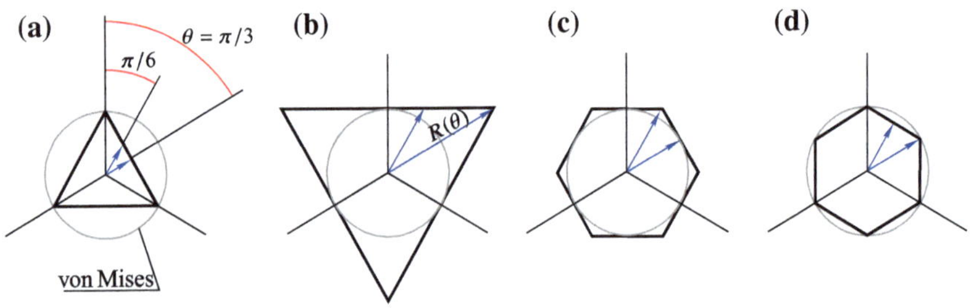

Fig. 1 Particular cross sections in the π-plane with the cut $I_1 = \text{const.}$ relating to the von Mises hypothesis with $R(0) = R(\pi/3) = R(\pi/6)$. **a** Ko, **b** Mariotte, **c** Schmidt–Ishlinsky, **d** Tresca

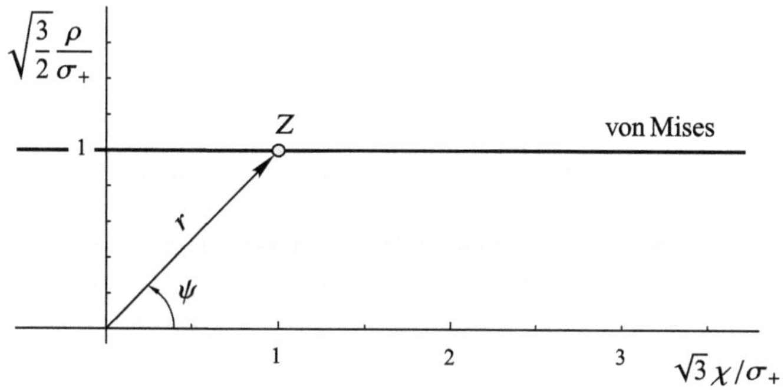

Fig. 2 Von Mises hypothesis in the Burzyński-plane. The axes are scaled by the uniaxial tensile strength σ_+ in order to obtain $I_1/\sigma_+ = \sqrt{3\,I_2'}/\sigma_+ = 1$ at the uniaxial tension (point Z). The radius r together with the angles θ and ψ defines a spherical coordinate system

- the elevation ψ in the Burzyński-plane (Fig. 2) (Altenbach and Kolupaev 2014; Fahlbusch 2015; Mendera 1966; Życzkowski 1981)

$$\tan\psi = \frac{\sqrt{3\,I_2'}}{I_1}, \qquad \psi \in [\,0,\ \pi\,]. \tag{9}$$

The function $\tan\psi$ is often denoted as stress triaxility factor (Davis and Connelly 1959; Du Bois et al. 2006; Finnie and Heller 1959; Odqvist and Hult 1962). Further functions of the invariants are given in Altenbach et al. (1995), Altenbach and Kolupaev (2014), Matsuoka and Nakai (1974), Życzkowski (1981), among others.

2.2 Stress Relations

For analytical comparison of different material properties some dimensionless values are introduced. These values provide a simple way to characterize limit surfaces.

Nine tests are chosen for the analysis and comparison of the limit surfaces Φ:

- two loadings corresponding to one-dimensional stress states (tension Z and compression D),
- five loadings corresponding to the plane stress states (torsion T, two balanced (2D uniform) plane states at tension B_Z and compression B_D, two tests on thin-walled tube specimens with closed ends under inner I_Z and outer U_D pressure), and
- two loadings corresponding to hydrostatic (3D balanced) tension and compression.

For more details see Altenbach et al. (2014), Altenbach and Kolupaev (2014), Kolupaev et al. (2013). The relevant stress components are listed in Table 1. All these loading cases can be assumed as basic tests. They are sufficient for the comparison of the most important features of the surfaces. Several restrictions are introduced on the basis of these relations (Altenbach et al. 2014; Altenbach and Kolupaev 2014).

Table 1 Basic stress states, stress relations, normalized coordinates in the principal stress space, normalized axiatoric-deviatoric invariants, the stress angle θ, and the elevation ψ (8) (Altenbach et al. 2014)

Loading	Stress	Label	Stress relation	$\left(\frac{\sigma_{\rm I}}{\sigma_+}, \frac{\sigma_{\rm II}}{\sigma_+}, \frac{\sigma_{\rm III}}{\sigma_+}\right)$	$\frac{I_1}{\sigma_+}$	$\frac{\sqrt{3I_2'}}{\sigma_+}$	$\frac{3\sqrt[3]{I_3'/2}}{\sigma_+}$	θ	ψ
Tension	σ_+	Z	1	(1, 0, 0)	1	1	1	0	$\frac{\pi}{4}$
Compression	σ_-	D	d	$(-d, 0, 0)$	$-d$	d	d	$\frac{\pi}{3}$	$\frac{3\pi}{4}$
Torsion	τ_*	T	k	$\left(\frac{k}{\sqrt{3}}, -\frac{k}{\sqrt{3}}, 0\right)$	0	k	0	$\frac{\pi}{6}$	$\frac{\pi}{2}$
Biaxial tension[a]	$\sigma_{\rm BZ}$	$B_{\rm Z}$	$b_{\rm Z}$	$(b_{\rm Z}, b_{\rm Z}, 0)$	$2\,b_{\rm Z}$	$b_{\rm Z}$	$-b_{\rm Z}$	$\frac{\pi}{3}$	$0.15\,\pi$
Biaxial compression[a]	$\sigma_{\rm BD}$	$B_{\rm D}$	$b_{\rm D}$	$(-b_{\rm D}, -b_{\rm D}, 0)$	$-2\,b_{\rm D}$	$b_{\rm D}$	$b_{\rm D}$	0	$0.85\,\pi$
Inner pressure[b]	$\sigma_{\rm IZ}$	$I_{\rm Z}$	$i_{\rm Z}$	$\left(\frac{2}{\sqrt{3}} i_{\rm Z}, \frac{1}{\sqrt{3}} i_{\rm Z}, 0\right)$	$\sqrt{3}\, i_{\rm Z}$	$i_{\rm Z}$	0	$\frac{\pi}{6}$	$\frac{\pi}{6}$
Outer pressure[b]	$\sigma_{\rm UD}$	$U_{\rm D}$	$u_{\rm D}$	$\left(-\frac{2}{\sqrt{3}} u_{\rm D}, -\frac{1}{\sqrt{3}} u_{\rm D}, 0\right)$	$-\sqrt{3}\, u_{\rm D}$	$u_{\rm D}$	0	$\frac{\pi}{6}$	$\frac{2\pi}{3}$
Hydrostatic tension	$\sigma_{\rm AZ}$	$A_{\rm Z}$	$a_+^{\rm hyd}$	$\left(a_+^{\rm hyd}, a_+^{\rm hyd}, a_+^{\rm hyd}\right)$	$3\, a_+^{\rm hyd}$	0	0	–	0
Hydrostatic compression	$\sigma_{\rm AD}$	$A_{\rm D}$	$a_-^{\rm hyd}$	$\left(-a_-^{\rm hyd}, -a_-^{\rm hyd}, -a_-^{\rm hyd}\right)$	$-3\, a_-^{\rm hyd}$	0	0	–	0

[a] Balanced plane loading
[b] Loadings in a thin-walled tube specimen with closed ends under pressure

Their selection is, however, not unambiguous and can be expanded according to the available equipment and expected phenomena, cf. Altenbach et al. (1995), Altenbach and Kolupaev (2014), Blumenauer (1996).

The values (Table 1) relating the respective stresses to σ_+ are introduced in order to obtain

$$k = d = i_Z = u_D = b_Z = b_D = 1 \quad \text{and} \quad a_-^{\text{hyd}},\, a_+^{\text{hyd}} \to \infty \tag{10}$$

for the von Mises type material (von Mises 1913, 1928)

$$\sigma_{\text{eq}}^2 = 3\, I_2' = \frac{1}{2}\left[(\sigma_{\text{I}} - \sigma_{\text{II}})^2 + (\sigma_{\text{II}} - \sigma_{\text{III}})^2 + (\sigma_{\text{III}} - \sigma_{\text{I}})^2\right]. \tag{11}$$

The loading points

- Z and B_D belong to the meridian with a stress angle $\theta = 0$ (0-meridian),
- D and B_Z belong to the $\pi/3$-meridian, and
- T, I_Z, and U_Z belong to the $\pi/6$-meridian.

For the points A_Z and A_D the stress angle θ (8) is indefinite.

For criteria independent of the hydrostatic stress (6) the values on the meridians defined by the stress angles $\theta = 0, \pi/6$, and $\pi/3$ are computed to Backhaus (1983), Kolupaev (2006), Życzkowski (1981)

$$b_D = 1, \qquad k = i_Z = u_D, \qquad \text{and} \qquad d = b_Z. \tag{12}$$

For the classical hypotheses (normal stress hypothesis,Tresca hypothesis, von Mises hypothesis, and Schmidt–Ishlinsky hypotheses) it follows (Altenbach et al. 2014)

$$b_Z = 1, \qquad b_D = d. \tag{13}$$

The well-known criteria are compared in Table 2 on the basis of these values.

For further discussions the values d_{inc} and k_{inc} are introduced. The geometrical meaning follows with (Fig. 1)

$$d_{\text{inc}} = \frac{R(\pi/3)}{R(0)}, \qquad k_{\text{inc}} = \frac{R(\pi/6)}{R(0)} \qquad \text{for} \qquad I_1 = \text{const}. \tag{14}$$

The values d_{inc} and k_{inc} describe the position of the meridians with $\theta = \pi/3$ and $\pi/6$ in relation to the 0-meridian in the Burzyński-plane with $I_1 = \text{const}$. In case of material behaviour which has not a varying shape in the π-plane, the parameters are constant ($d_{\text{inc}}, k_{\text{inc}} = \text{const}$). For material behaviour which is also insensitive to hydrostatic stress ($A_D, A_Z \to \infty$) the parameters are $d_{\text{inc}} = d$ and $k_{\text{inc}} = k$.

The cross sections of the criteria can be compared in the d_{inc}-k_{inc} diagram (Fig. 3) (Altenbach et al. 2014; Bolchoun et al. 2011; Kolupaev and Altenbach 2010). In this diagram the criteria of Haythornthwaite and Capurso limit the convex shapes in the π-plane.

Table 2 Cross sections in the π-plane (Fig. 1) and the values k_{inc} and d_{inc} for the criteria of material behaviour, which are insensitive to hydrostatic stress (Fig. 3)

Cross section	k_{inc} (–)	d_{inc} (–)	Criterion	Source
a	$1/\sqrt{3}$	$1/2$	Ko	Ko (1963), Kolupaev et al. (2009), Sayir (1970), Ziegler (1969)
b	$2/\sqrt{3}$	2	Mariotte	Altenbach et al. (1995), Benvenuto (1991), Birger et al. (1993), Finnie and Heller (1959), Ivlev (1959), Mariotte (1700), Shanley (1957), Sauter and Winterger (1990), Sayir (1970), Sayir and Ziegler (1969), Sähn et al. (1993), Życzkowski (1981)
c	$2/\sqrt{3}$	1	Schmidt–Ishlinsky	Annin (1999), Burzyński (1928), de Araújo (1962), Haythornthwaite (1961), Hill (1950), Ishlinsky (1940), Schmidt (1932), Yu (1961, 1983a, b, 2004)
d	$\sqrt{3}/2$	1	Tresca	Coulomb (1776), Reuss (1933), Tresca (1868), Prager and Hodge (1954)
m	1	1	von Mises	Föppl and Föppl (1920), Hencky (1924), Huber (1904), von Mises (1913), Yagn (1933), Skrzypek (1993), Życzkowski (1981)

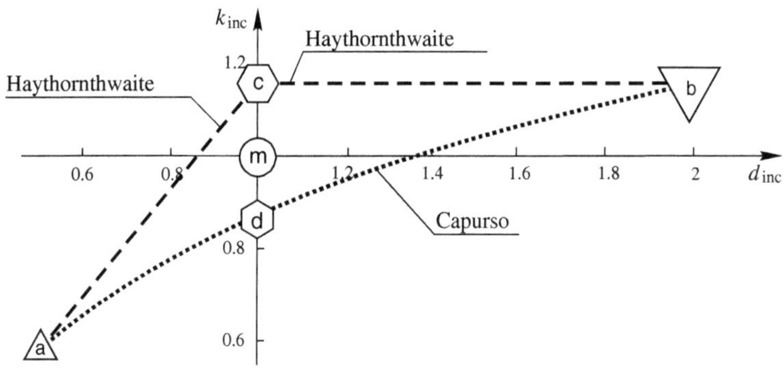

Fig. 3 Diagram d_{inc}-k_{inc} for convex criteria of material behaviour for cuts with $I_1 = \text{const.}$ compared with the hypothesis of von Mises. Certain cross sections in the π-plane are visualized in order to achieve a better understanding (cf. Fig. 1 and Table 2)

2.3 Capurso–Haythornthwaite Generalization

The Haythornthwaite criterion (Altenbach et al. 2014; Haythornthwaite 1961)

$$\begin{aligned}\Phi_H = {} & \frac{3^6}{2^6}\,\frac{1}{d_{\text{inc}}^3}\,(I_3')^2 - \frac{3^5}{2^5}\,\frac{d_{\text{inc}}-1}{d_{\text{inc}}^3}\,I_2'\,I_3'\,\sigma_{\text{eq}} - \frac{3^4}{2^4}\,\frac{1}{d_{\text{inc}}^2}\,(I_2')^2\,\sigma_{\text{eq}}^2 \\ & - \frac{3^3}{2^3}\,\frac{1-d_{\text{inc}}^3}{d_{\text{inc}}^3}\,I_3'\,\sigma_{\text{eq}}^3 + \frac{3^2}{2^2}\,\frac{1+d_{\text{inc}}^2}{d_{\text{inc}}^2}\,I_2'\,\sigma_{\text{eq}}^4 - \sigma_{\text{eq}}^6 \end{aligned} \tag{15}$$

and the Capurso criterion (Capurso 1967; Sayir 1970)

$$\Phi_C = \alpha_4 \sigma_{eq}^4 I_2' + \alpha_3 \sigma_{eq}^3 I_3' + \alpha_2 \sigma_{eq}^2 \left(I_2'\right)^2 + \alpha_1 \sigma_{eq} I_2' I_3' + \beta_2 \left(I_2'\right)^3 + \beta_1 \left(I_3'\right)^2 - \sigma_{eq}^6 \tag{16}$$

with the parameters

$$\alpha_4 = 6 \frac{(d_{inc} - 1)^2 + d_{inc}}{d_{inc}^2}, \qquad \alpha_3 = 3^3 \frac{d_{inc} - 1}{d_{inc}^2},$$

$$\alpha_2 = -3^2 \frac{\left((d_{inc} - 1)^2 + d_{inc}\right)^2}{d_{inc}^4}, \qquad \alpha_1 = -3^4 \frac{(d_{inc} - 1) \left((d_{inc} - 1)^2 + d_{inc}\right)}{d_{inc}^4},$$

$$\beta_2 = \frac{(1 - 2\, d_{inc})^2\, (d_{inc} - 2)^2\, (1 + d_{inc})^2}{d_{inc}^6}, \quad \beta_1 = -3^3 \frac{\left((d_{inc} - 1)^2 + d_{inc}\right)^3}{d_{inc}^6} \tag{17}$$

are functions of the value $d_{inc} \in [1/2,\ 2]$ (Altenbach et al. 2014; Capurso 1967). With the linear combination (Altenbach et al. 2014; Bolchoun et al. 2011; Kolupaev et al. 2013)

$$\Phi_{CH} = (1 - \xi)\, \Phi_C + \xi\, \Phi_H, \qquad \xi \in [0,\ 1] \tag{18}$$

a criterion follows which describes all points in the d_{inc}-k_{inc} diagram (Fig. 3) with a convex shape of the surface in the π-plane by the use of two parameters d_{inc} and ξ. The parameter ξ is a function of the stress relations d_{inc} and k_{inc}. The analytical solution of the equation for $\xi(d_{inc}, k_{inc})$ is too large and hence omitted.

The criterion contains (Table 2)

- the bicubic criterion of hexagonal symmetry in the π-plane with $d_{inc} = 1$ (Fig. 3) and the parameter[1]

$$\xi = 1 - \frac{2}{k_{inc}^2} + \frac{270 - 115\, k_{inc}^2}{216 - 189\, k_{inc}^2 + 64\, k_{inc}^4}, \qquad k_{inc} \in \left[\frac{\sqrt{3}}{2},\ \frac{2}{\sqrt{3}}\right], \tag{21}$$

  including the Tresca hypothesis with $k_{inc} = \sqrt{3}/2$ and the Schmidt–Ishlinsky hypothesis with $k_{inc} = 2/\sqrt{3}$,

---

[1]Equation (18) is linear with respect to ξ. If the stresses for the torsion loading (Table 1)

$$\sigma_I = -\sigma_{II} = k_{inc} \sigma_+ / \sqrt{3}, \qquad \sigma_{III} = 0 \tag{19}$$

are inserted in Eq. (18), it follows for $d = 1$

$$-\frac{1}{2}\, k_{inc}^2\, (\xi - 2^2) + k_{inc}^4 \left(\frac{7\,\xi}{2^4} - 1\right) - \frac{2^2}{3^3}\, k_{inc}^6\, (\xi - 1) = 1. \tag{20}$$

Solving Eq. (20) with the respect to ξ yields the monotonic increasing function (21) in the region $k_{inc} \in [\sqrt{3}/2,\ 2/\sqrt{3}]$.

- the von Mises hypothesis resulting with $d_{\text{inc}} = k_{\text{inc}} = 1$ to $\xi = 2^6/(7 \cdot 13) \approx 0.7033$,
- the Ko criterion with $d_{\text{inc}} = 1/2$ and any given ξ, and
- the Mariotte criterion with $d_{\text{inc}} = 2$ and any given ξ.

This criterion is suitable for multi-purpose use and can be applied in different generalizations.

2.4 *Generalization of Yield Criteria*

A pressure sensible generalization of the criteria is obtained by the substitution (Altenbach et al. 2014; Altenbach and Kolupaev 2014; Kolupaev et al. 2013)

$$\sigma_{\text{eq}} \rightarrow \sqrt[6]{\left|\frac{\sigma_{\text{eq}} - \gamma_1 I_1}{1 - \gamma_1}\right|^{6-l-m} \left|\frac{\sigma_{\text{eq}} - \gamma_2 I_1}{1 - \gamma_2}\right|^{l} \sigma_{\text{eq}}^{m}}. \tag{22}$$

The parameters γ_1 and γ_2 determine the position of the nodes A_Z and A_D on the hydrostatic axis with the restrictions $\gamma_1 \in [0, 1[$ and $\gamma_2 < 0$ for closed surfaces. The powers l and m are chosen to be positive ($l > 0, m > 0, l + m < 6$). They define the curvature of the meridian through appropriate weighting.

Significantly improved fitting results can be obtained with the combined substitution

$$\sigma_{\text{eq}} \rightarrow \begin{cases} \sqrt[6]{\left|\frac{\sigma_{\text{eq}} - \gamma_1 I_1}{1 - \gamma_1}\right|^{6-l-m} \left|\frac{\sigma_{\text{eq}} - \gamma_2 I_1}{1 - \gamma_2}\right|^{l} \sigma_{\text{eq}}^{m}}, & I_1 > I_T, \\ \sqrt[6]{\left|\frac{\tilde{\sigma}_+ - \tilde{\gamma}_1 I_1}{1 - \tilde{\gamma}_1}\right|^{6-\tilde{l}-\tilde{m}} \left|\frac{\tilde{\sigma}_+ - \tilde{\gamma}_2 I_1}{1 - \tilde{\gamma}_2}\right|^{\tilde{l}} \tilde{\sigma}_+^{\tilde{m}}}, & I_1 \leq I_T \end{cases} \tag{23}$$

and the continuously differentiable transition at the cross section $I_T \in [-d\,\sigma_+, \sigma_+]$. This substitution has more adjustment possibilities.

3 Fitting Procedure

A fitting procedure for the approximation of given data sets is shown in Fig. 4. It is assumed that the data include several failure points under multi-axial loadings. In the first step the failure points on the 0-meridian are chosen and described by the substitution (22) in (11)

$$(3I_2')^3 = \left|\frac{\sigma_{\text{eq}} - \gamma_1 I_1}{1 - \gamma_1}\right|^{6-l-m} \left|\frac{\sigma_{\text{eq}} - \gamma_2 I_1}{1 - \gamma_2}\right|^{l} \sigma_{\text{eq}}^{m}. \tag{24}$$

For the current data sets the advanced formulation of the functions in two areas is used (Eq. (F1)). The transition at $I_1 = 0$ in the Burzyński-plane proves to be favourable using the classical partition in a tension and compression area. The fitting parameters $l, \tilde{l}, m, \tilde{m}, \gamma_2, \tilde{\gamma}_1$, and $\tilde{\sigma}_+$ are determined by a nonlinear optimization approach. The parameters γ_1 and $\tilde{\gamma}_2$ are set by the hydrostatic nodes A_Z and A_D, which are known from numerical investigation ($\gamma_1 = 1/3a_+^{hyd}$ and $\tilde{\gamma}_2 = -\tilde{\sigma}_+/(3a_-^{hyd}\sigma_+)$). The best fit is obtained through the least square method.

The parameter of the shape ξ (21) can be assumed in a good approximation as a constant value independent of I_1. For $d_{inc}(I_1) = 1$ the parameter ξ is obtained using Eq. (21) or (F2). In the case $d_{inc}(I_1^*) = 1$ the parameter ξ is calculated for the coordinate I_1^* which results from a cut of the meridians $\theta = 0$ and $\theta = \pi/3$ in the Burzyński-plane.

The shape of the limit surfaces in the π-plane is considered by d_{inc} as a function of I_1 (F4). The division in two areas with a continuously differentiable transition at $I_1 = 0$ is adopted, cf. Eq. (23) with $I_T = 0$. The function profile is determined with a cubic spline approximated to the numerical points and the evaluation of the least square method.

With these three steps all parameters of the Capurso–Haythornthwaite generalization combined with the substitution to describe the shape of the 0-meridian and the formulation of the geometrical parameter ξ and the function $d_{inc}(I_1)$ are determined. The method provides limit surfaces considering failure under hydrostatic tensile and compression stress and varying cross sections in the π-plane in dependence on the first invariant of the stress tensor.

A restriction of the method results from the number of parameters in the criterion. If only a limited number of failure points is given, additional assumptions have to be made considering the specific material behaviour. In the following possible assumptions are discussed for the example of a polymer foam and a further optimization step over the whole procedure is introduced. The proposed fitting procedure is not limited to the Capurso–Haythornthwaite generalization. It can be used also for other criteria.

4 Hard Foams

The general approach of the proposed limit surface and the determination of its parameters is shown for a polymer hard foam. The data points are calculated numerically. The main attributes of the implemented FE model are summarized in the following. More details can be found in Fahlbusch (2015).

4.1 Experimental Data of the Microstructure

The microstructure of a foam can be characterized by high-resolution pictures. Figure 5 shows a μCT image of the hard foam ROHACELL® 71 IG. The diameter

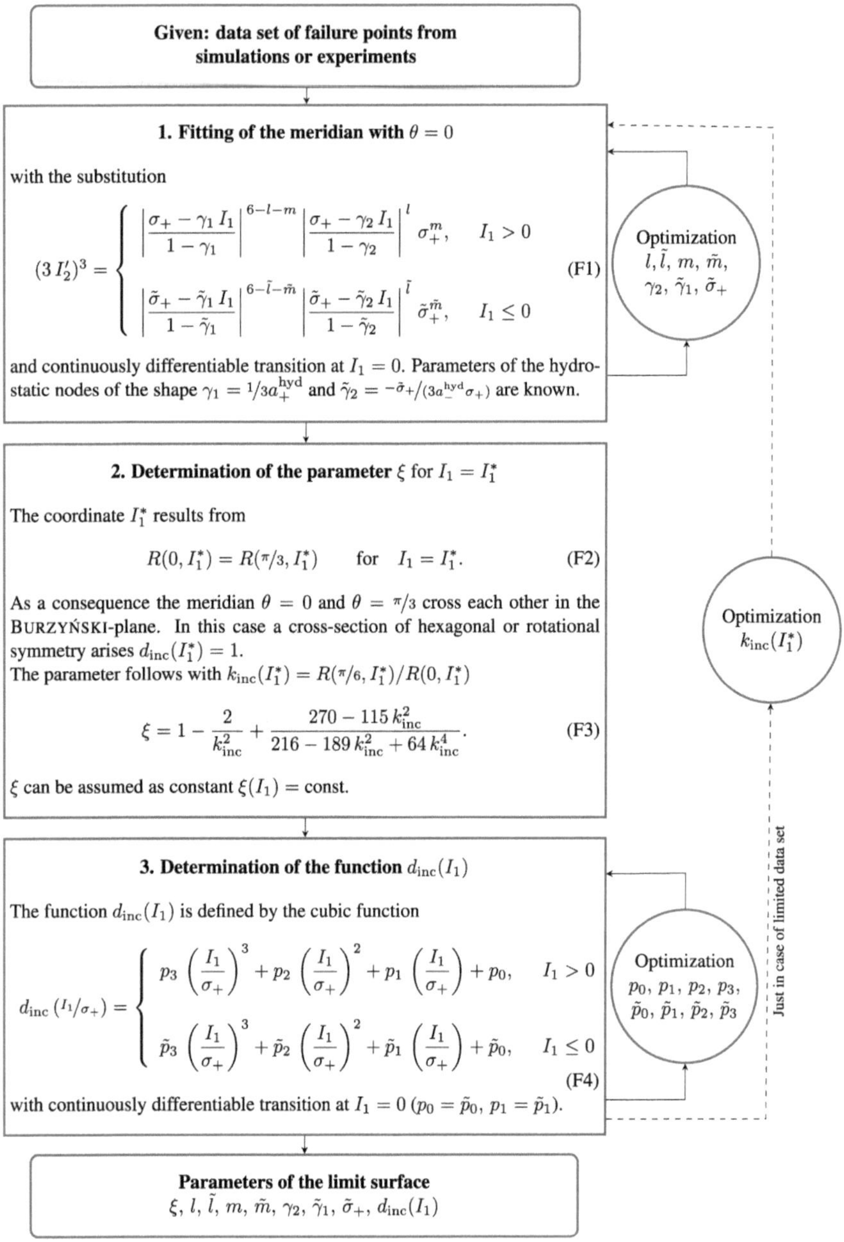

Fig. 4 Fitting procedure of the parameters of the Capurso–Haythornthwaite generalization

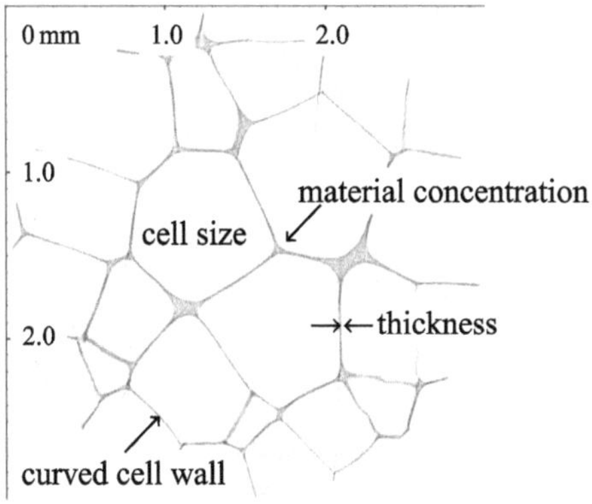

Fig. 5 μCT image of ROHACELL® 71 IG with the geometrical characteristics of the microstructure (Fahlbusch 2015; Fahlbusch et al. 2015)

of the cells is about 0.5 mm. The microstructure has also some further specific geometrical parameters, which are called imperfections. This microstructure includes variations in the cell wall thickness, material concentrations towards the edges and curvatures of the cell walls, and edges. Each geometrical parameter is identified and the expected value and the standard deviation are calculated. These values are the important parameters to specify the microstructure of the investigated foam.

Beside the classification of the geometry the material in the cell walls should be experimentally tested. Due to the cell size it is hard to investigate the material out of the cell walls. Material data of the foamed bulk material are not known. The elastic properties of the unfoamed material are identified by a compression test: the elastic modulus $E = 5985\,\text{MPa}$, the Poisson's ratio $\nu = 0.3$, and the density $\rho = 1249\,\text{kg/m}^3$ (Fahlbusch 2015). The strength of a uniaxial tensile test amounts $\sigma_+ = 147.5\,\text{MPa}$ (Schlimper 2014) and of a compression test $\sigma_- = 213.0\,\text{MPa}$ (Fahlbusch 2015). Both strength values equal the maximum stresses achieved in the tests. These data are used as input data for the FE model, although it can not be excluded that the foaming process leads to a change in the material properties.

4.2 *FE Model for Polymer Foams*

FE models on the basis of the CT data are suitable to present the characteristic microstructure and allow a reliable calculation of the effective elastic properties. Thereby numerical models have a high level of adaptability to the image data, whereas for example analytical models mostly simplify the microstructure, cf. Grenestedt (1999).

Mono-dispers models are based on one initial geometry which can be periodically repeated, cf. Grenestedt (1999). As geometry the tetracaidecahedron is mainly used because it is space filling and approximately isotropic, cf. Fig. 6 (Daxner et al. 2006; Simone and Gibson 1998). Following the many examples in literature an FE model is implemented based on the tetracaidecahedron, cf. Grenestedt and Bassinet (2000),

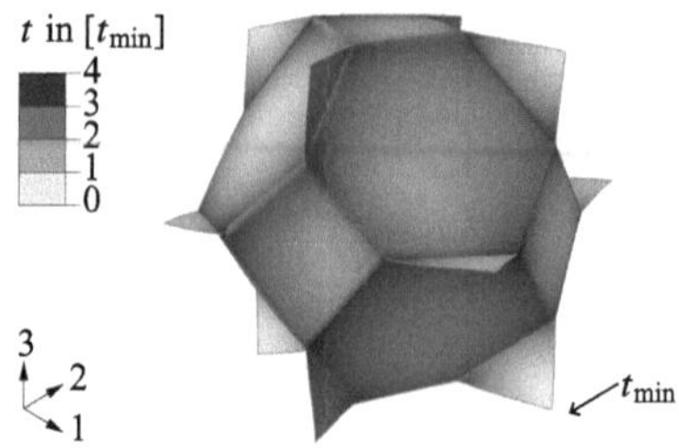

Fig. 6 Cell wall thickness distribution with the minimal thickness t_{min} of an FE model with imperfections (Fahlbusch 2015)

Mills (2010), Simone and Gibson (1998). In contrast to previous studies the proposed model is extensively adapted to the data of the image analysis.

Initially, an FE model without any geometrical irregularities is generated. To guarantee periodic repeatability in the deformed state, periodic boundary conditions are applied. The effective properties on the macroscale are calculated by a strain-energy based homogenization concept (Hohe and Becker 2005). The geometrical data of the microstructure are taken into account by attributing different shell thicknesses to the single-cell faces. The shell elements are getting thicker towards the cell edges. An initial curvature of the cell walls and edges is introduced. By the displacement of single-corner nodes the edge length is varied. An example of an FE model is given in Fig. 6.

To consider the variation in the geometrical data several FE models are generated and the average value is calculated (Monte Carlo method Landau and Binder 2009). With the macroscopical value $\bar{\Psi}_i = \bar{\Psi}(\boldsymbol{X}_i)$ of the N different models this means that the specified value results in

$$\bar{\Psi} = \frac{1}{N}\sum_{i=1}^{N}\bar{\Psi}_i. \tag{25}$$

For the calculation of the strength for several load situations different macroscopical deformation states are applied to the cell. The strength equals the macroscopical stress which is reached when the von Mises hypothesis or the Pisarenko–Lebedev criterion is fulfilled on the micro-level. In this case the von Mises hypothesis is given by Eq. (11) and the Pisarenko–Lebedev criterion (Altenbach et al. 1995, 2014; Pisarenko and Lebedev 1976) is

$$\sigma_{\mathrm{eq}} = \frac{1}{d}\sqrt{\frac{3}{2}s_{ij}s_{ij}} + \left(1 - \frac{1}{d}\right)\max\left(\sigma_{\mathrm{I}}, \sigma_{\mathrm{II}}, \sigma_{\mathrm{III}}\right) = \sigma_{\mathrm{PL}} \tag{26}$$

with the value

$$d = \frac{|\sigma_-|}{\sigma_+} \geq 1. \tag{27}$$

This criterion describes the intermediate range between the absolute brittle (normal stress hypothesis with $d \to \infty$) and ideal-ductile (the von Mises hypothesis with $d = 1$) material behaviour. The different criteria on the micro-level provide limit surfaces which are symmetric ($a_+^{\text{hyd}} = a_-^{\text{hyd}}$) as well as asymmetric ($a_+^{\text{hyd}} \neq a_-^{\text{hyd}}$) to the axis $I_1 = 0$ in the Burzyński-plane. The data sets consist of 13–5000 points and are thus a good example to illustrate the proposed method.

The following three data sets are investigated:

- data set 1: $d = 1$ and $a_+^{\text{hyd}} = a_-^{\text{hyd}}$
  failure points calculated with the ideal FE model without imperfections, geometrical linear simulations, and the von Mises hypothesis on the micro-level,
- data set 2: $d > 1$ and $a_+^{\text{hyd}} < a_-^{\text{hyd}}$
  failure points calculated with the ideal FE model without imperfections, geometrical linear simulations, and the Pisarenko–Lebedev criterion on the micro-level, and
- data set 3: $d < 1$ and $a_+^{\text{hyd}} > a_-^{\text{hyd}}$
  failure points calculated with an FE model adapted to the geometrical and material data of ROHACELL® 71IG, geometrical nonlinear simulations, and the Pisarenko–Lebedev criterion on the micro-level.

5 Results

The **data set 1** includes 5000 failure points. These points are shown in Fig. 7 in the principal stress space. The surface is obviously closed. For the determination of the failure criterion points on the meridians $\theta = 0$, $\pi/6$, and $\pi/3$ are sufficient. With the three approximation steps (Fig. 4) the function parameters of the criterion are calculated. Table 3 summarizes the parameters as well as the stress relations.

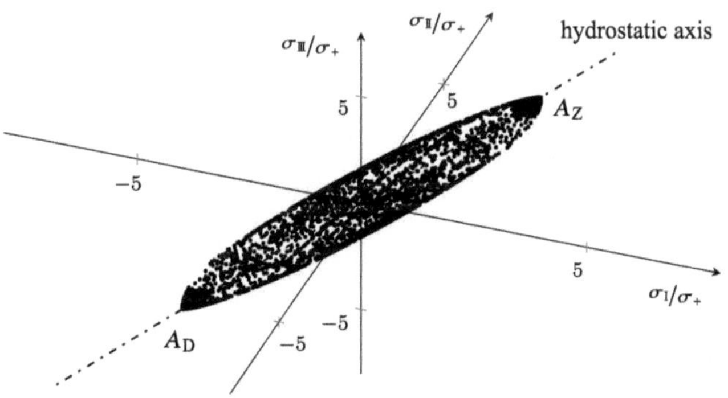

Fig. 7 Limit surface in the principal stress space, given data set 1 with $\sigma_+ = 2.66\,\text{MPa}$

Table 3 Parameters of the Capurso–Haythornthwaite generalization (18) with the substitution (22) and varying cross section in the π-plane (F4) as well as the stress relations

Data set	1		2		3	
Tensile strength σ_+ (MPa)	2.6623		2.0335		1.7981	
Parameters of the failure criterion						
ξ	0.9230		0.8946		0.9098	
$\tilde{\sigma}_+$ and σ_+	$0.9766\,\sigma_+$	σ_+	$1.0081\,\sigma_+$	σ_+	$1.5127\,\sigma_+$	σ_+
range I_1/σ_+	[−8.76, 0]	[0, 8.76]	[−11.4864, 0]	[0, 7.7009]	[−2.7, 0]	[0, 4.34]
$\tilde{l}$ and l	3.4219	2.0194	3.5215	8.0847×10^{-2}	3.4020	1.5265×10^{-1}
$\tilde{m}$ and m	2.0675	1.7938	0	3.9418	1.3799	7.2685×10^{-2}
$\tilde{\gamma}_1$ and γ_1	5.9812×10^{-1}	1/8.76	1.5964×10^{-1}	1/7.7009	8.2633×10^{-2}	1/4.34
$\tilde{\gamma}_2$ and γ_2	−0.9766/8.76	−0.1622	−1.0081/11.4864	−2.1131	−1.5127/2.70	−16.5352
$\tilde{p}_0$ and p_0	1	1	1.0368	1.0368	1	1
$\tilde{p}_1$ and p_1	-3.1628×10^{-2}	-3.1628×10^{-2}	-4.7321×10^{-2}	-4.7321×10^{-2}	5.4×10^{-2}	5.4×10^{-2}
$\tilde{p}_2$ and p_2	3.1529×10^{-3}	-2.6189×10^{-3}	-6.1879×10^{-3}	-2.0202×10^{-3}	-8.3759×10^{-2}	-1.8071×10^{-2}
$\tilde{p}_3$ and p_3	-7.1553×10^{-4}	-3.2680×10^{-5}	-8.1565×10^{-4}	-1.1504×10^{-4}	-1.3027×10^{-2}	1.3530×10^{-2}
Stress relation in (–)						
d	1		1.13		0.87	
k	1.07		1.11		1.28	
b_Z	0.93		0.91		0.88	
b_D	0.93		1.03		0.76	
i_Z	1.03		1.00		0.93	
u_D	1.03		1.14		0.76	
a_+^{hyd}	8.76/3		7.70/3		4.34/3	
a_-^{hyd}	8.76/3		11.48/3		2.70/3	

In the visualizations Figs. 7, 8, 9, 10, 11, 12, 13, 14, 15, and 16 the axes are normalized by the tensile strength σ_+ for better comparison. The stress relations correspond to the coordinates of the loadings A_D, B_D, U_D, T, I_Z, B_Z, and A_Z

The data set describes a symmetric limit surface to the cut $I_1 = 0$, which becomes evident with the failure points in the Burzyński-plane (Fig. 8). In the whole range $I_1/\sigma_+ \in [-8.76, 8.76]$ the points are well mapped by the criterion. Just small differences occur for the $\pi/6$-meridian and $I_1/\sigma_+ \rightarrow -3a_-^{\text{hyd}}$ (the point with $I_1/\sigma_+ = -3a_-^{\text{hyd}}$ is exactly approximated). Figure 9 completes the visualization with the plane stress states in the σ_{I}-σ_{II} plane ($\sigma_{\text{III}} = 0$).[2]

With the introduction of d_{inc} as function of the hydrostatic stress the changing cross section can be described with the criterion. Figure 10 visualizes the function $d_{\text{inc}}(I_1)$. The value of d_{inc} relates to the shape of the cross section of the limit surface. Close to the hydrostatic nodes the surface describes triangles in the cross section and for the cut $I_1 = 0$ hexagonal symmetry. With the triangles a and b in Fig. 1 the extremal values of the function d_{inc} are defined $d_{\text{inc}} \in [1/2, 2]$.

The method is not limited to symmetric surfaces to the $\sqrt{3\, I_2'}$-axis in the Burzyński-plane. The failure points of **data set 2** as well as the approximation are shown in the Burzyński-plane (Fig. 11) and in the σ_{I}-σ_{II} plane ($\sigma_{\text{III}} = 0$) (Fig. 12). The function d_{inc} is visualized in Fig. 13. As before it is set $d_{\text{inc}} \in [1/2, 2]$ with the same cross sections in the π-plane at the hydrostatic nodes. For $I_1 = 0$ it follows from the cross of the meridians $\theta = 0$ and $\theta = \pi/3$ the parameter $d_{\text{inc}}(I_1 = 0) = 1.04$. Table 3 includes all parameters of this surface. Concluding, it can be stated that a high approximation quality is achieved with the proposed method.

The next analysis is an example how to apply the method with a limited data set. In this case the numerical consideration of imperfections by the Monte Carlo method and of geometrical nonlinearities during the simulations is computing-intensive. Also, the number of experiments is often restricted and leads to a limited number of failure points. The **data set 3** consists of the strength values for the stress states Z, D, T, B_{Z}, B_{D}, A_{Z}, A_{D} (cf. Table 4), and the values of the plane stress states with the step of the stress angle $\Delta\theta = \pi/24$, so that $\theta = \pi/24, \pi/12, \pi/8, 5\pi/24, \pi/4$, and $7\pi/24$, cf. Eq. (8) (Fahlbusch et al. 2015). Due to the limited number of the points, assumptions for $d_{\text{inc}}(I_1^*)$, $d_{\text{inc}}(I_1/\sigma_+ = 3a_+^{\text{hyd}})$ and $d_{\text{inc}}(I_1/\sigma_+ = -3a_-^{\text{hyd}})$ have to be made. The parameter $d_{\text{inc}}(I_1^* = 0) = 1$ and thereby

$$R(0, I_1^*) = R(\pi/3, I_1^*) \tag{28}$$

(Eq. 14) is set pragmatically. The meridians $\theta = 0$ and $\theta = \pi/3$ cross each other at $I_1 = 0$ in the Burzyński-plane. Additionally, the assumption of a triangle in the cross section at the hydrostatic nodes as before is made. It results $d_{\text{inc}}(I_1/\sigma_+ = 3a_+^{\text{hyd}}) = 2$ and $d_{\text{inc}}(I_1/\sigma_+ = -3a_-^{\text{hyd}}) = 1/2$, which is opposite to the approximations of the sets 1 and 2.

[2]Visualization tools in CAS Wolfram-Mathematica are free available.

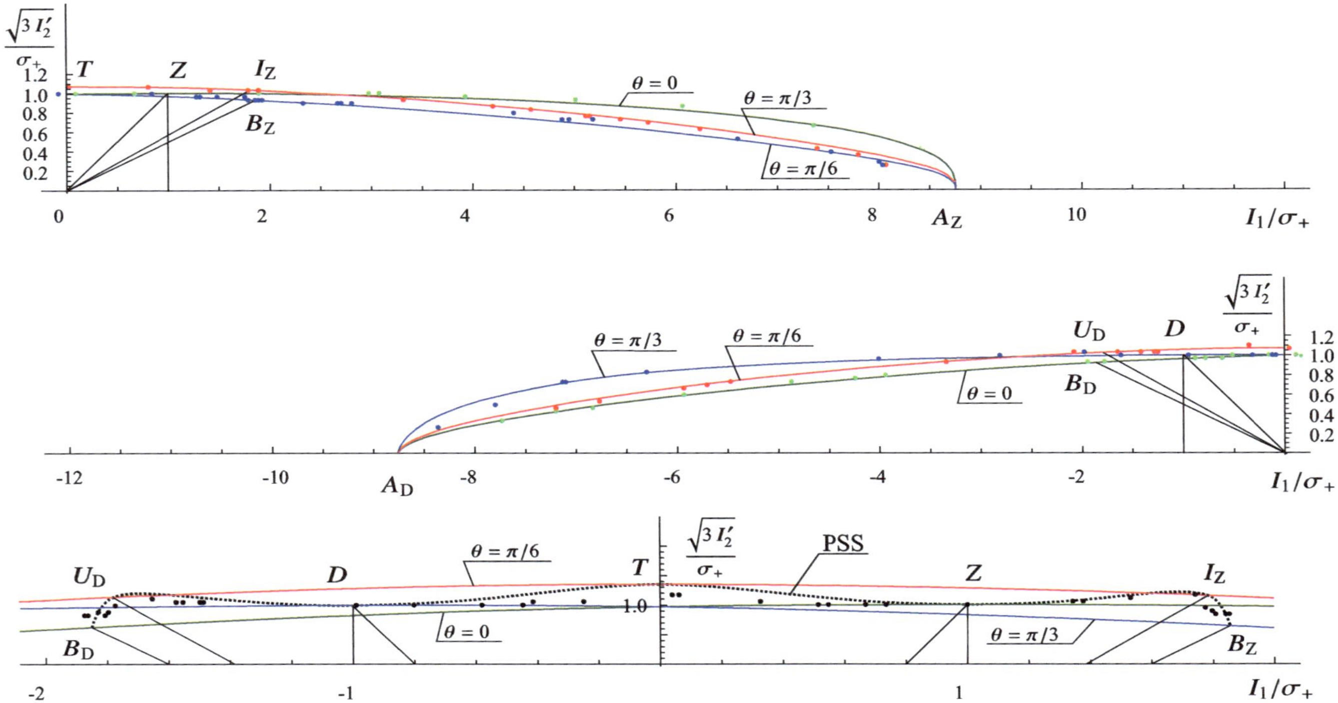

Fig. 8 Approximation of the data set 1 with $\sigma_+ = 2.66\,\text{MPa}$ and the numerical points. Burzyński-plane with the meridians $\theta = 0, \pi/6, \pi/3$ and the curve of the plane stress states (PSS). The diagram is divided into two regions for clarity. The PSS region is enlarged

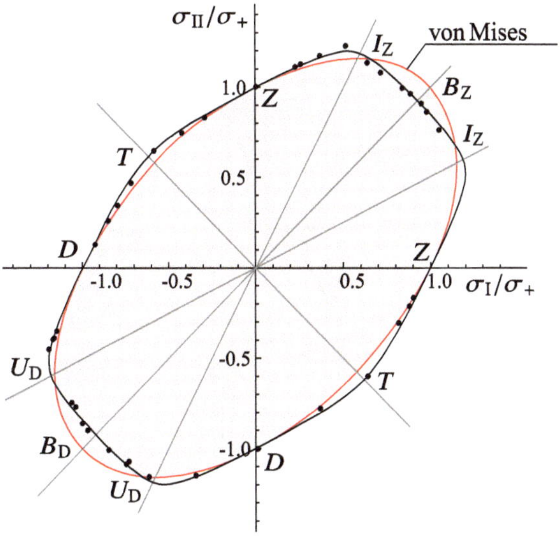

Fig. 9 Approximation of the data set 1 with $\sigma_+ = 2.66$ MPa and the numerical points. σ_I-σ_{II} plane ($\sigma_{III} = 0$). For comparison the von Mises hypothesis is shown

The further determination of the parameters follows the known steps (Fig. 4). The calculation of the 0-meridian is based on the stress points A_D, B_D, Z, and A_Z. Two more conditions result from the requirement of a continuously differentiable transition of the two subsections at $I_1 = 0$. Only one additional condition for the determination of the parameters is missing. Therefore, the point for $R(0, I_1^*)$ is introduced which describes the point of intersection of the 0-meridian with the $\sqrt{3I_2^D}/\sigma_+$ axis. The point is not known yet.

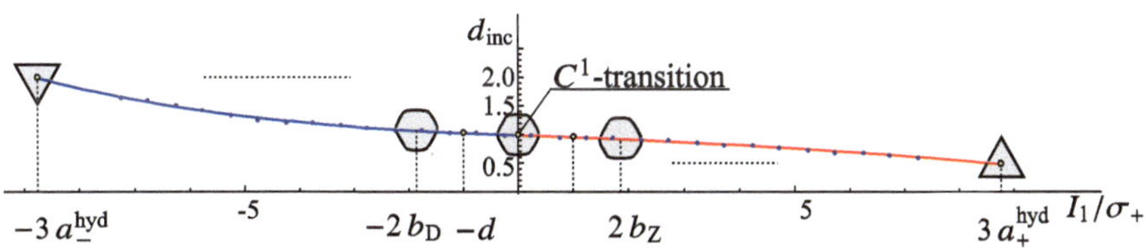

Fig. 10 d_{inc}-I_1 diagram for the data set 1. The approximation of $d_{inc}(I_1)$ is visualized with the cross sections in the π-plane, $\sigma_+ = 2.66$ MPa

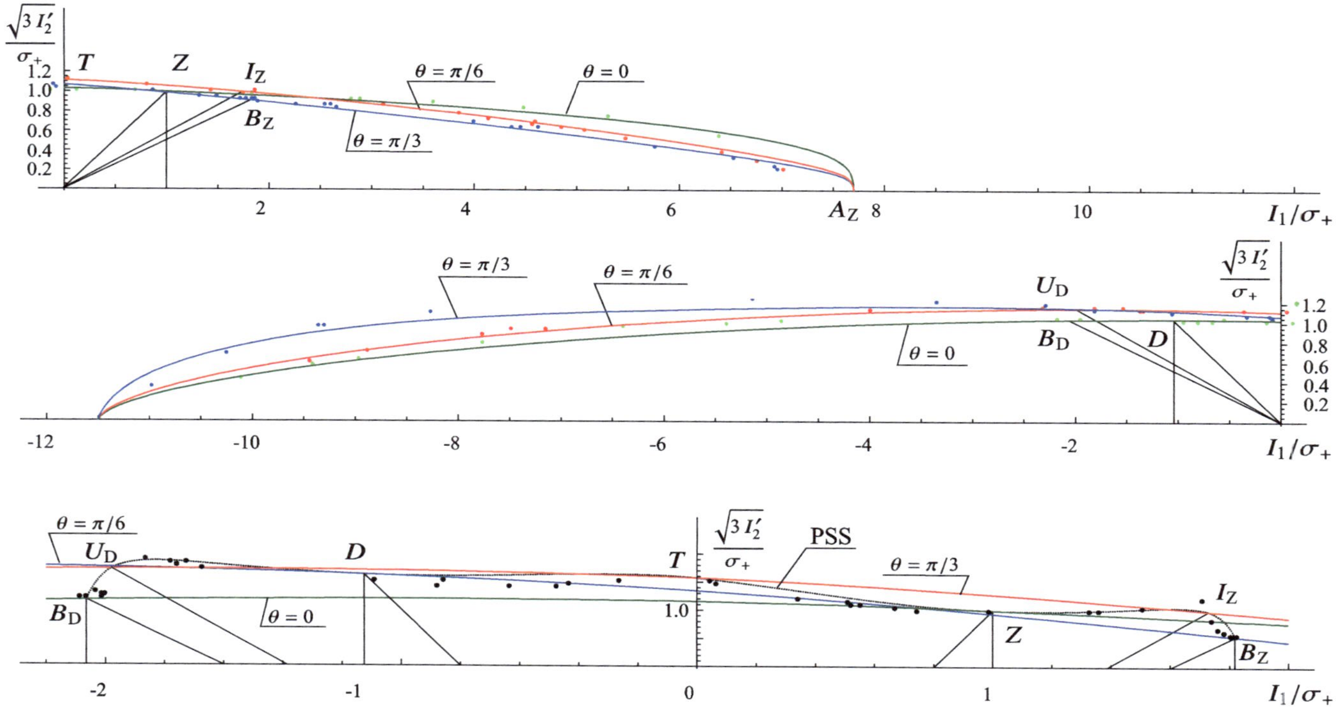

Fig. 11 Approximation of the data set 2 with $\sigma_+ = 2.03$ MPa and the numerical points. Burzyński-plane with the meridians $\theta = 0, \pi/6, \pi/3$ and the curve of the plane stress states (PSS). The diagram is divided into two regions for clarity. The PSS region is enlarged

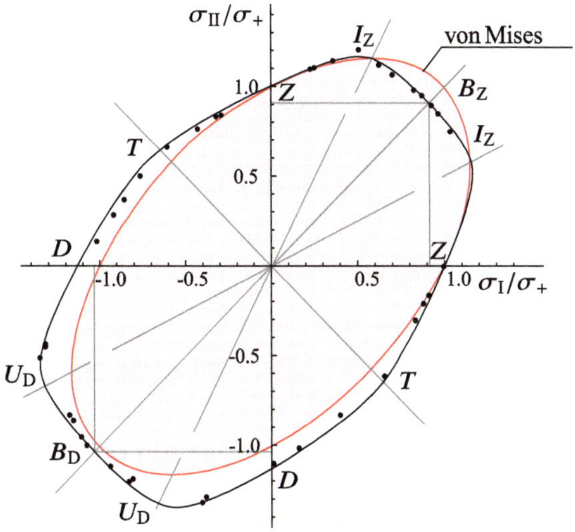

Fig. 12 Approximation of the data set 2 with $\sigma_+ = 2.03$ MPa and the numerical points. σ_I-σ_{II} plane ($\sigma_{III} = 0$). For comparison the von Mises hypothesis is shown

For $I_1 = I_1^* = 0$, the ordinate of the $\pi/6$-meridian $R(\pi/6, I_1^*) = 1.28$ results from the point T. The value $k_{inc}(I_1^*)$ is defined by (cf. Eq. (14))

$$k_{inc}(I_1^*) = \frac{R(\pi/6, I_1^*)}{R(0, I_1^*)}, \tag{29}$$

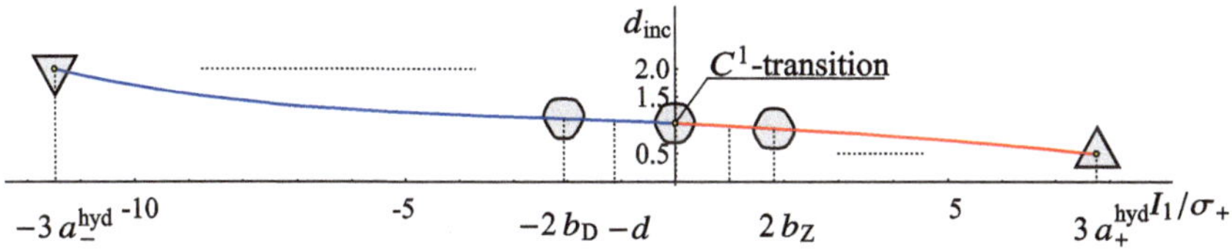

Fig. 13 d_{inc}-I_1 diagram for the data set 2. The approximation of $d_{inc}(I_1)$ is visualized with the cross sections in the π-plane, $\sigma_+ = 2.03$ MPa

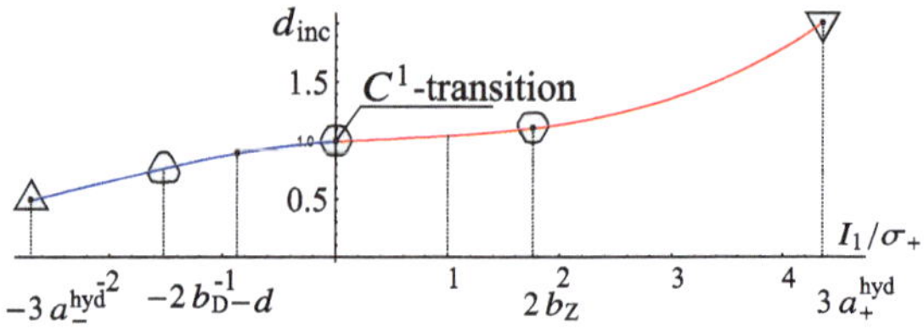

Fig. 14 d_{inc}-I_1 diagram for the data set 3. The approximation of $d_{inc}(I_1)$ is visualized with the cross sections in the π-plane ($\sigma_+ = 1.8\,\mathrm{MPa}$)

thereby generally the restriction

$$k_{inc}(I_1^*) \in \left[\frac{\sqrt{3}}{2}, \frac{2}{\sqrt{3}}\right] \qquad \text{for} \qquad d_{inc} = 1 \tag{30}$$

follows (Eq. 21). The limits are derived from the convexity condition (Fig. 3). With the parameter $k_{inc}(I_1^*)$ the point $R(0, I_1^*)$ is fixed which is necessary for the calculation of the parameters of the 0-meridian in Eq. (F1).

The parameter $k_{inc}(I_1^*)$ is identified through an optimization about the complete three steps (Fig. 4). The best fit is obtained by the least square method between the approximation of the limit criterion and the numerically calculated values. Thereby all parameters in Eq. (F1) are determined.

Table 3 contains the parameters of the failure criterion for the material ROHACELL® 71IG using the Capurso–Haythornthwaite generalization and the related values (stress relations). For the data set 3 in case of the biaxial tension it follows $b_Z \approx 0.88$ and in case of the hydrostatic tension $a_+^{hyd} \approx 1.45$, cf. Table 4. These stress states are often estimated with the normal stress hypothesis with $b_Z = a_+^{hyd} = 1$ (Eq. 13) due to missing material data. In case of biaxial tension the current approach leads to a more conservative interpretation than a dimensioning with the normal stress hypothesis. A validation with experimental tests would be desirable here knowing of the difficult realization of a 2D tension test (Kolupaev et al. 2015).

Figure 14 shows the approximation of the function $d_{inc}(I_1)$ with the resulting cross sections in the π-plane. The curve runs from $d_{inc} = 1/2$ for $I_1/\sigma_+ = -3a_-^{hyd}$ to $d_{inc} = 2$ for $I_1/\sigma_+ = 3a_+^{hyd}$ or from the triangle a to the triangle b (Fig. 1). These cross sections correspond to the known description of the cellular materials at the

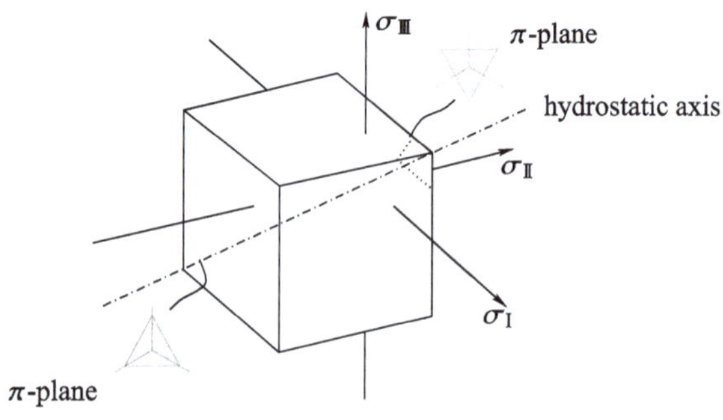

Fig. 15 Schematic representation of the normal stress hypothesis in the principal stress space with two cross sections perpendicular to the hydrostatic axis (π-plane)

Table 4 Numerically calculated stresses for ROHACELL®71 IG (data set 3) with ZT for superimposed tension and torsion, DT for compression and torsion loadings

Label	σ_{I} (MPa)	σ_{II} (MPa)	σ_{III} (MPa)	I_1 (MPa)	I'_2 (MPa^2)	$\sqrt{3\,I'_2}$ (MPa)	I'_3 (MPa^3)	$\cos 3\theta$ (–)	θ (degree)	$\tan\psi$ (–)	ψ (degree)
Z	1.80	0	0	1.80	1.08	1.80	0.43	1	0	1	45
D	0	0	−1.56	−1.56	0.81	1.56	−0.28	−1	60	−1	135
T	1.33	0	−1.33	0	1.78	2.31	0	0	30	∞	90
B_{Z}	1.59	1.59	0	3.18	0.84	1.59	−0.30	−1	60	0.5	26.6
B_{D}	0	−1.37	−1.37	−2.73	0.62	1.37	0.19	1	0	−0.5	153.4
A_{Z}	2.60	2.60	2.60	7.80	0	0	0	–	–	0	0
A_{D}	−1.62	−1.62	−1.62	−4.85	0	0	0	–	–	0	180
ZT	1.82	0	−0.29	1.53	1.31	1.98	0.54	0.93	7.3	1.3	52.3
ZT	1.78	0	−0.72	1.06	1.66	2.23	0.54	0.66	16.2	2.1	64.5
ZT	1.64	0	−1.02	0.62	1.79	2.32	0.36	0.39	22.3	3.7	75.1
DT	1.06	0	−1.66	−0.60	1.88	2.38	−0.37	−0.37	37.3	−3.9	104.2
DT	0.60	0	−1.63	−1.03	1.32	1.99	−0.42	−0.71	45.0	−1.9	117.4
DT	0.26	0	−1.71	−1.45	1.14	1.85	−0.44	−0.94	53.1	−1.3	128.1

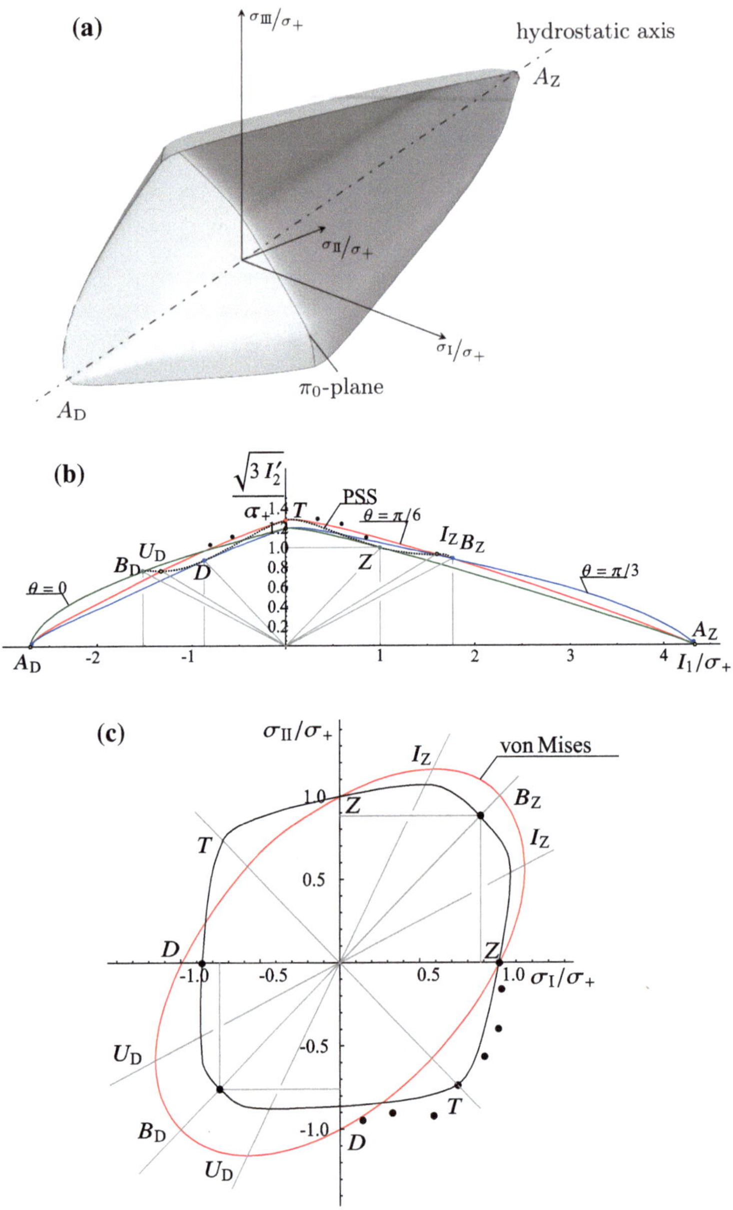

Fig. 16 Approximation of the data set 3 ($\sigma_+ = 1.8\,\text{MPa}$). **a** Surface in the principal stress space (π_0-plane: cut with $I_1 = 0$). **b** Burzyński-plane with the meridians $\theta = 0, \pi/6, \pi/3$ and the curve of the plane stress states (PSS). **c** σ_I-σ_{II} plane ($\sigma_{III} = 0$). For comparison the von Mises hypothesis is shown

hydrostatic nodes with the normal stress hypothesis (Fig. 15), cf. Shaw and Sata (1966). Figure 16 illustrates the limit surface in the principal stress space, in the Burzyński- and σ_{I}-σ_{II} plane ($\sigma_{III} = 0$).

The phenomenological limit criterion provides a good approximation of the data set. The comparison with the numerical points shows the conservative prediction of the failure points. The limit surface has changing cross sections along the hydrostatic axis with $I_1/\sigma_+ \in [-2.70,\ 4.34]$.

With the proposed formulation of the limit surface and a continuously differentiable transition in two sections the adjustment options are sufficient. The extension in four sections with continuously differentiable transitions is possible. The assumption of convex meridians and convex shapes in the π-plane underlies the proposed method. Cuntze allows also non-convex shapes and achieves a high approximation quality of data sets for ROHACELL® 71IG using his failure mode concept (Cuntze 2015). However, the requirement of convexity due to Drucker's postulate is often applied for limit surfaces (e.g. Drucker 1957; Irgens 2008). For most applications convex surfaces are preferred (Yu 2004). Non-convex surfaces are discussed in Bigoni (2012), Bolchoun et al. (2011) and are possible to be applied with the discussed method.

6 Conclusion

In the current approach, a generalization of the Capurso–Haythornthwaite criterion to approximate limit surfaces to given failure points is proposed. For the determination of the parameters of the criterion a fitting procedure is developed. The failure points can result from experiments or simulations. The method is validated for data sets of a closed cell foam, which are calculated numerically. The limit surface of the material shows some special characteristics like the sensitivity to hydrostatic stress, the strength-differential effect ($\sigma_+ \neq \sigma_-$) and cross sections changing along the hydrostatic axis. With the generalized criterion the material characteristics are considered and the given failure points are well approximated. The method provides a rather accurate description.

The developed formulation of the criterion is based on a generalization of the known Capurso–Haythornthwaite criterion, which has been modified and extended. This approach provides a high degree of generalization. It is not limited to a special kind of material and can be applied to materials like aerated concrete, cellular ceramics, metal ceramics and to data sets with a different number of failure points. In case of a limited number of points further assumptions have to be made which should consider the material peculiarities. Possible assumptions in the case of the polymer foam are discussed. An extension of the formulation into more than two areas is addressed in the paper. The criterion can be implemented in a computer software for the dimensioning of structures out of the polymer foam.

Acknowledgments The work reported here was supported by Deutsche Forschungsgemeinschaft (DFG), Germany under Grant No. BE 1090/27-1, KO 3382/6-1 which is gratefully acknowledged.

References

Altenbach, H.: Kontinuumsmechanik: Einführung in die materialunabhängigen und materialabhängigen Gleichungen, 2nd edn. Springer, Vieweg, Berlin, Heidelberg (2012)
Altenbach, H., Kolupaev, V.A.: Classical and non-classical failure criteria. In: Altenbach, H., Sadowski, T. (eds.) Failure and Damage Analysis of Advanced Materials. CISM, pp. 1–66. Springer, Heidelberg (2014)
Altenbach, H., Altenbach, J., Zolochevsky, A.: Erweiterte Deformationsmodelle und Versagenskriterien der Werkstoffmechanik. Deutscher Verlag für Grundstoffindustrie, Stuttgart (1995)
Altenbach, H., Bolchoun, A., Kolupaev, V.A.: Phenomenological yield and failure criteria. In: Altenbach, H., Öchsner, A. (eds.) Plasticity of Pressure-Sensitive Materials. ASM, pp. 49–152. Springer, Heidelberg (2014)
Annin, B.D.: Theory of ideal plasticity with a singular yield surface. J. Appl. Mech. Tech. Phys. **40**(2), 347–353 (1999)
Backhaus, G.: Deformationsgesetze. Akademie-Verlag, Berlin (1983)
Benvenuto, E.: An Introduction to the History of Structural Mechanics. Springer, New York (1991)
Bigoni, D.: Nonlinear Solid Mechanics: Bifurcation Theory and Material Instability. Cambridge University Press, Cambridge (2012)
Bigoni, D., Piccolroaz, A.: Yield criteria for quasibrittle and frictional materials. Int. J. Solids Struct. **41**(11), 2855–2878 (2004)
Birger, I.A., Shopp, B.F., Iosilevich, G.B.: Strength Computations for Machine Components (in Russ.: Raschet na prochnost' detalej mashin, Spravochnik). Mashinostroenie, Moscow (1993)
Blumenauer, H.: Werkstoffprüfung. Dt. Verl. für Grundstoffindustrie, Leipzig (1996)
Bolchoun, A., Kolupaev, V.A., Altenbach, H.: Convex and non-convex flow surfaces (in German: Konvexe und nichtkonvexe Fließflächen). Forschung im Ingenieurwesen **75**(2), 73–92 (2011)
Burzyński, W.: Studjum nad hipotezami wytężenia. Akademia Nauk Technicznych, Lwów (1928)
Candland, C.T.: Implications of macroscopic failure criteria which are independent of hydrostatic stress. Int. J. Fract. **11**(3), 540–543 (1975)
Capurso, M.: Yield conditions for incompressible isotropic and orthotropic materials with different yield stress in tension and compression. Meccanica **2**(2), 118–125 (1967)
Chen, W.F., Zhang, H.: Structural Plasticity - Theory, Problems, and CAE Software. Springer, New York (1991)
Coulomb, C.A.: Essai sur une application des règles des maximis et minimis à quelques problèmes de statique relatifs, à l'architecture. Mem Acad Roy Div Sav **7**, 343–387 (1776)
Cuntze, R.G.: Fracture failure surface of the foam Rohacell. Veranstaltung Carbon Composites, AGs Zerstörungsfreie Werkstoff- und Bauteilprüfung, Klebetechnik & Engineering, Stuttgart, 17. März (2015)
Davis, E.A., Connelly, F.M.: Stress distribution and plastic deformation in rotating cylinders of strainhardening material. J. Appl. Mech. - Trans. ASME **26**, 25–30 (1959)
Daxner, T., Bitsche, R.D., Böhm, H.J.: Space-filling polyhedra as mechanical models for solidified dry foams. Mater. Trans. **47**(9), 2213–2218 (2006)
de Araújo, F.C.: Elasticidade e plasticidade. Imprensa Portuguesa, Porto (1962)
Droste, A.: Beschreibung und Anwendung eines elastisch-plastischen Materialmodells mit Schädigung für hochporöse Metallschäume. Bericht Nr. II-9, Inst. für Mechanik (Bauwesen), Stuttgart (2004)
Drucker, D.C.: A definition of stable inelastic material. Technical report 2, Division of Engineering, Brown University, Providence, Rhode Island (1957)

Du Bois, P., Feucht, M., Haufe, A., Kolling, S.: A generalized damage and failure formulation for SAMP. In: Fifth German LS-DYNA Forum, Ulm, pp. A–II: 77–110 (2006)
Ehlers, W.: Constitutive equations for granular materials in geomechanical context. In: Hutter, K. (ed.) Continuum Mechanics in Environmental Science and Geophysics. CISM, vol. 337, pp. 313–402. Springer, Wien (1993)
Ehlers, W.: A single-surface yield function for geomaterials. Arch. Appl. Mech. **65**(4), 246–259 (1995)
Fahlbusch, N.-C.: Entwicklung und Analyse mikromechanischer Modelle zur Beschreibung des Effektivverhaltens von geschlossenzelligen Polymerschäumen. Ph.D. thesis, Studienbereich Mechanik, Technischen Universität Darmstadt (2015)
Fahlbusch, N.-C., Becker, W., Kolupaev, V.A., Geertz, G.: Non-linear material behaviour and failure of closed-cell polymer foams. Acta Mech. 1–9 (2015) doi:10.1007/s00707-015-1533-x
Finnie, I., Heller, W.R.: Creep of Engineering Materials. McGraw-Hill, New York (1959)
Föppl, A., Föppl, L.: Drang und Zwang: Eine höhrere Festigkeitslehre für Ingenieure. R. Oldenbourg, München (1920)
Grenestedt, J.L.: Effective elastic behavior of some models for perfect cellular solids. Int. J. Solids Struct. **36**(10), 1471–1501 (1999)
Grenestedt, J.L., Bassinet, F.: Influence of cell wall thickness variations on elastic stiffness of closed-cell cellular solids. Int. J. Mech. Sci. **42**(7), 1327–1338 (2000)
Haythornthwaite, R.M.: Range of yield condition in ideal plasticity. Proc. ASCE J. Eng. Mech. Div. **87**(EM6), 117–133 (1961)
Hencky, H.: Zur Theorie plastischer Deformationen und der hierdurch im Material hervorgerufenen Nachspannungen. ZAMM **4**(4), 323–334 (1924)
Hill, R.: On the inhomogeneous deformation of a plastic lamina in a compression test. Philos. Mag. Ser. 7 **41**(319), 733–744 (1950)
Hohe, J., Becker, W.: A probabilistic approach to the numerical homogenization of irregular solid foams in the finite strain regime. Int. J. Solids Struct. **42**(11), 3549–3569 (2005)
Huber, M.T.: Specific strain work as a measurment of material effort (in Polish: Właściwa praca odkształcenia jako miara wytężenia materyału). Czasopismo Techniczne 22:34–40, 49–50, 61–62, 80–81 (1904)
Irgens, F.: Continuum Mechanics. Springer, New York (2008)
Ishlinsky, A.Y.: Hypothesis of strength of shape change (in Russ.: Gipoteza prochnosti formoizmenenija). Uchebnye Zapiski Moskovskogo Universiteta, Mekhanika **46**, 104–114 (1940)
Ivlev, D.D.: The theory of fracture of solids. J. Appl. Math. Mech. **23**(3), 884–895 (1959)
Ko, W.L.: Application of the finite elastic theory to the behavior of rubber-like materials. Ph.D. thesis, California Institute of Technology, Pasadena (1963)
Kolupaev, V.A.: 3D-Creep Behaviour of Parts Made of Non-Reinforced Thermoplastics (in German: Dreidimensionales Kriechverhalten von Bauteilen aus unverstärkten Thermoplasten). Ph.D. thesis, Martin-Luther-Universität Halle-Wittenberg, Halle (2006)
Kolupaev, V.A., Altenbach, H.: Considerations on the Unified Strength Theory due to Mao-Hong Yu (in German: Einige Überlegungen zur Unified Strength Theory von Mao-Hong Yu). Forschung im Ingenieurwesen **74**(3), 135–166 (2010)
Kolupaev, V.A., Bolchoun, A., Altenbach, H.: New trends in application of strength hypotheses (in German: Aktuelle Trends beim Einsatz von Festigkeitshypothesen). Konstruktion **61**(5), 59–66 (2009)
Kolupaev, V.A., Yu, M.H., Altenbach, H.: Visualisation of the unified strength theory. Arch. Appl. Mech. **83**(7), 1061–1085 (2013)
Kolupaev, V.A., Becker, W., Massow, H., Kiegelmann, E.M.: Reliable designs in foam (in German: Mit Schaumstoffen zuverlässig konstruieren). Magazine for Plastics, Kunststoffe international **105**(1–2), 32–35 (2015)
Landau, D.P., Binder, K.: A Guide to Monte Carlo Simulations in Statistical Physics, 3rd edn. Cambridge University Press, Cambridge (2009)
Mariotte, M.: Traité du mouvement des eaux et des autres corps fluides. J. Jambert, Paris (1700)

Matsuoka, H., Nakai, T.: Stress-deformation and strength characteristics of soil under three different principal stresses. Proc. Jpn. Soc. Civ. Eng. JSCE **232**, 59–70 (1974)

Mendera, Z.: Wytężenie spoiny czołowej w interpretacji powierzchni granicznych. Przegląd Spawalnictwa SIMP **XVIII**(1), 6–13 (1966)

Mills, N.J.: Deformation mechanisms and the yield surface of low-density, closed-cell polymer foams. J. Mater. Sci. **45**, 5831–5843 (2010)

Novozhilov, V.V.: On the principles of the statical analysis of the experimental results for isotropic materials (in Russ.: O prinzipakh obrabotki rezultatov staticheskikh ispytanij izotropnykh materialov). Prikladnaja Matematika i Mechanika **XV**(6), 709–722 (1951)

Odqvist, F.G., Hult, J.: Kriechfestigkeit metallischer Werkstoffe. Springer, Berlin (1962)

Ottosen, N.S., Ristinmaa, M.: The Mechanics of Constitutive Modeling. Elsevier Science, London (2005)

Penasa, M., Piccolroaz, A., Argani, L., Bigoni, D.: Integration algorithms of elastoplasticity for ceramic powder compaction. J. Eur. Ceram. Soc. **34**, 2775–2788 (2014)

Piccolroaz, A., Bigoni, D.: Yield criteria for quasibrittle and frictional materials: a generalization to surfaces with corners. Int. J. Solids Struct. **46**(20), 3587–3596 (2009)

Pisarenko, G.S., Lebedev, A.A.: Deformation and Strength of Materials under Complex Stress State (in Russ.: Deformirovanie i prochnost' materialov pri slozhnom naprjazhennom sostojanii). Naukowa Dumka, Kiev (1976)

Prager, W., Hodge, P.: Theorie ideal plastischer Körper. Springer, Wien (1954)

Reuss, A.: Vereinfachte Beschreibung der plastischen Formänderungsgeschwindigkeiten bei Voraussetzung der Schubspannungsfließbedingung. ZAMM **13**(5), 356–360 (1933)

Rohacell (2010) Product information ROHACELL® IG/IG-F. Evonik Industries, Evonik Röhm GmbH, Performance Polymers Business Unit. Darmstadt. http://www.rohacell.com

Sähn, S., Göldner, H., Nickel, J., Fischer, K.: Bruch- und Beurteilungskriterien in der Festigkeitslehre. Fachbuchverlag, Leipzig (1993)

Sauter, J., Winterger, N.: Neue und alte statische Festigkeitshypothesen. VDI, Reihe 1: Konstruktionstechnik/Maschinenelente Nr. 191, Düsseldorf (1990)

Sayir, M.: Zur Fließbedingung der Plastizitätstheorie. Ingenieur-Archiv **39**, 414–432 (1970)

Sayir, M., Ziegler, H.: Der Verträglichkeitssatz der Plastizitätstheorie und seine Anwendung auf räumlich unstetige Felder. Zeitschrift für angewandte Mathematik und Physik ZAMP **20**(1), 78–93 (1969)

Schlimper, R.: Verformungs- und Versagensverhalten geschlossenzelliger Polymerhartschaumstoffe. Shaker, Aachen (2014)

Schmidt, R.: Über den Zusammenhang von Spannungen und Formänderungen im Verfestigungsgebiet. Ingenieur-Archiv **3**(3), 215–235 (1932)

Shanley, F.R.: Strength of Materials. McGraw-Hill, New York (1957)

Shaw, M.C., Sata, T.: The plastic behavior of cellular materials. Int. J. Mech. Sci. **8**, 469–478 (1966)

Simone, A.E., Gibson, L.J.: Effects of solid distribution on the stiffness and strength of metallic foams. Acta Mater. **46**(6), 2139–2150 (1998)

Skrzypek, J.J.: Plasticity and Creep: Theory, Examples and Problems. CRC Press, Boca Raton (1993)

Tresca, H.: Mémoire sur l'ecoulement des corps solides. Mémoires Pres par Div Savants **18**, 733–799 (1868)

von Mises, R.: Mechanik des festen Körpers im plastischen deformablen Zustand. Nachrichten der Königlichen Gesellschaft der Wissenschaften Göttingen, Mathematisch-physikalische Klasse, pp. 589–592 (1913)

von Mises, R.: Mechanik der plastischen Formänderung von Kristallen. ZAMM **8**, 161–185 (1928)

Yagn, Y.I.: Strength of Materials: Theory and Problems (in Russ.: Soprotivlenie materialov: teorja i zadachnik). Kubuch, Leningrad (1933)

Yu, M.H.: General behaviour of isotropic yield function (in Chinese). Scientific and Technological Research Paper of Xi'an Jiaotong University, pp. 1–11 (1961)

Yu, M.H.: Twin shear stress yield criterion. Int. J. Mech. Sci. **25**(1), 71–74 (1983a)

Yu, M.H.: Twin shear stress yield criterion. Reply to Prof. Hill's comments. Int. J. Mech. Sci. **25**(11), 845–846 (1983b)
Yu, M.H.: Advances in strength theories for materials under complex stress state in the 20th century. Appl. Mech. Rev. **55**(5), 169–218 (2002)
Yu, M.H.: Unified Strength Theory and Its Applications. Springer, Berlin (2004)
Ziegler, H.: Zum plastischen Potential der Bodenmechanik. Zeitschrift für angewandte Mathematik und Physik ZAMP **20**, 659–675 (1969)
Życzkowski, M.: Combined Loadings in the Theory of Plasticity. PWN-Polish Scientific Publishers, Warszawa (1981)

On the Problem of Cracking in 2-Phase Ceramic Matrix Composite Materials

Tomasz Sadowski, Liviu Marsavina and Eduard-Marius Craciun

Abstract The aim of the paper is to present a theoretical analysis of phenomena occurring in the 2-phase ceramic composite with the gradual degradation of the material properties under the uniaxial tension process. Ceramic composite materials have a nonlinear and complex overall response to applied loads. It is caused by the following factors: existence of an inital porosity, development of limited plasticity, different phases and internal microdefects. These microdefects cause stress concentrations and locally change the state of stress, which results in the development of mesocracks leading to macrocracks. In this contribution, a multiscale approach was applied in modelling of such material response to depict phenomena at micro- meso- and macro-scales. In experiments it was shown that defects developed mainly intergranularly what resulted in inhomogeneity and induced anisotropy of the material.

1 Motivation for Research

Permanent innovations of different branches of engineering technologies require application of new multiphase engineering materials with specially designed composition to satisfy limitation of extremal exploitation conditions of critical structural parts in airplanes, cars and buildings etc. The novel materials are different types of composites obtained as mixtures of various phases and further subjected to specific

---

T. Sadowski (✉)
Department of Solid Mechanics, Lublin University of Technology,
Nadbystrzycka 40 Str., Lublin, Poland
e-mail: t.sadowski@pollub.pl

L. Marsavina
Department Strength of Materials, Politehnica University of Timisoara,
Blvd. M. Viteazu, Nr.1, 300222 Timisoara, Romania
e-mail: msvina@mec.upt.ro

E.-M. Craciun
Faculty of Mechanical, Industrial and Martime Engineering, Ovidius University
of Constanta, Bd. Mamaia 124, 900527 Constanta, Romania
e-mail: mcraciun@univ-ovidius.ro

K. Naumenko and M. Aßmus (eds.), *Advanced Methods of Continuum Mechanics for Materials and Structures*, Advanced Structured Materials 60,
DOI 10.1007/978-981-10-0959-4_20

technological process in the manufacturing, e.g. Gömze and Gömze (2009, 2013). The materials engineering allows for designing of almost arbitrary internal structure of the composites, particularly important for industrial demands. A very important example is functionally graded materials, possessing gradation of the material properties, e.g. Birman and Byrd (2007), Sadowski and Neubrand (2004), Sadowski et al. (2007a, 2009), Nakonieczny and Sadowski (2008, 2009), Bîrsan et al. (2012), Ivanov et al. (2013), Petrova et al. (2012), Petrova and Sadowski (2012), Bîrsan et al. (2013a, b), Burlayenko et al. (2015a, b), Sadowski et al. (2015), Petrova and Sadowski (2014), Taczała et al. (2015, 2016), Sadowski and Pankowski (2016). Other examples are brittle composite materials built on the basis of cement or clay matrix, e.g. Golewski and Sadowski (2014), Golewski et al. (2012), Lenci et al. (2011, 2012), Sadowski et al. (2005).

Advanced composites can be built up as a sequence of layers or other types of gradation of material properties. Introduction of layers between material components (e.g. Fic and Szeląg 2015; Sadowski et al. 2006, 2007b; Sadowski and Nowicki 2008; Dębski and Sadowski 2014; Felten et al. 2008; Sadowski and Golewski 2013, 2015; Postek and Sadowski 2011; Bieniaś et al. 2012; Gajewski and Sadowski 2014) and as TBC covering of turbine blades (e.g. Sadowski and Samborski 2008; Sadowski and Golewski 2011, 2012a, b, c; Kalatur et al. 2014; Savchenko et al. 2014), etc. are other examples of improvements in composites or joining technologies. Nanocomposites (e.g. Lau and Hui 2002) or adhesives with different nanoparticles or carbon nanotubes Wu et al. (2009), Zhao et al. (2007) are good applications of particle reinforced composites.

Two-phase ceramics composites (CMC) possess many practical applications. For example—the Al_2O_3/ZrO_2 composites are frequently used as a thermal barrier coating in different structures. In this paper, it is carried out modelling of the crack propagation in the chosen CMC, subjected to uniaxial stress. Considered CMC material consists of matrix Al_2O_3—and the second phase ZrO_2 (with volume content up to 20 %). As the two components have different thermal expansion coefficients, it is very difficult to produce perfectly continuous CMC. Usually microcracks and microporosity appear in initial internal structure due to the thermal mismatch. These microdefects cause stress concentrations which results in the development of mesocracks leading to macrocracks.

As different phenomena develop at micro-, meso- and macro-scales, the most suitable method for modelling of such CMC material is multiscale approach with averaging procedure over the Representative Volume Element (RVE).

2 Micro-, Meso-, Macro-Correlation to Formulate Constitutive Equations

Formulation of constitutive equations for Al_2O_3/ZrO_2 (as an example of two-phase ceramic composites) includes the elastic deformations of initially micro-porous CMC

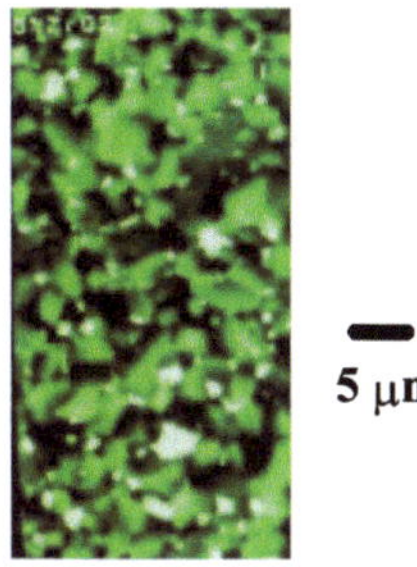

$Al_2O_3 + (5\%)ZrO_2$

Fig. 1 Scanning electron microscopy photographs of the considered composites

and the deformations created by growing cracks in CMC, Fig. 1. Evolution of the internal structure of RES due to mechanical loading should be described at microscopic, mesoscopic and macroscopic levels:

1. The microscopic level concerns the degradation phenomena occuring at the length scale on the level of the single grain. Cracks are initiated by microcracks and micropores inside the grain or at the grain boundaries. Furthermore, mismatch of the thermal expansion coefficient of different phases creates additional stress concentration at the grain boundaries, which decrease toughness of material in this area. Microcracks nucleated inside grains move to grain boundaries, at which can be arrested or can spread along the nearest grain boundary. It results from the significantly less fracture toughness of this part of composite. Alternatively, the microcracks can pass through a grain boundary to the adjacent grain. At this microscopic level, separation of a single grain from the polycrystal can be described.
2. The mesoscopic level obeys the RVE, which is created by a set of grains. Mesocracks are the basic elements of the defect structure inside polycrystal. Their diameters correspond to the single straight facet of the grain boundaries structure. Kinked and wing (zig–zag) cracks result from the development of deformation inside the composite.
3. The macroscopic level refers to the dimensions of the tested sample. The composite material is regarded as a continuum with properties of the polycrystal. Analytical micromechanical model or finite element analysis are used for calculations done as averaged values over of RVE.

Methodology used for monolithic ceramics was presented in Sadowski (1994a, b, 1995). In case of two-phase materials, the following constitutive rule can be proposed (using Voigt's notation) in order to describe general features of the quasi-static deformation process of the material

$$\bar{\varepsilon}_i = \bar{S}^{e}_{ik}(\sigma_m, p_f, p_o, N_c^{(s)})\sigma_k \quad \text{for} \quad i, k = 1, \ldots, 6 \quad \text{and} \quad j = 1, 2, 3, \tag{1}$$

where $\bar{\varepsilon}_i$ is the strain vector, $\bar{S}^{e}_{ik}$ is the elastic compliance matrix, σ_k is the stress vector, p_f is the volume content of the second phase (ZrO_2), p_o is the porosity parameter

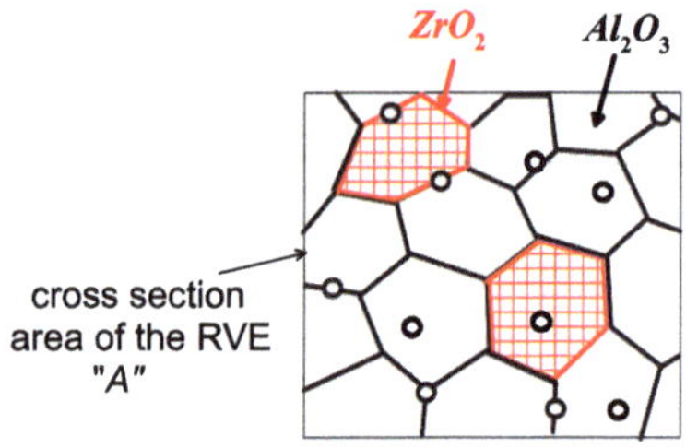

Fig. 2 Initial structure RVE of CMC including both phases and porosity

and $N_c^{(s)}$ are sets of parameters defining the presence of different kinds of defects "s" developing inside the material. In the advanced state of deformation the averaged elastic compliance tensor can be split into the following parts:

$$\bar{S}_{ik}^{e}(x_j, p_f, p_o, N_c^{(s)}) = (1 - p_f)S_{ik}^{C1} + p_f S_{ik}^{C2}(x_j, p_f) + S_{ik}^{p_o}(x_j, \sigma_r, p_o) + S_{ik}^{cr}(x_j, \sigma_r, p_f, p_o, N_c^{(s)}), \quad (2)$$

where S_{ik}^{C1} is a compliance matrix of the first phase (C1)—matrix material Al_2O_3, $S_{ik}^{C2}(x_j, p_f, p_o)$ describes phase two (C2) – ZrO_2, $S_{ik}^{p_o}(x_j, \sigma_r, p_o)$ is the part reflecting the porosity existence (e.g. Kachanov 1993a, b). The $S_{ik}^{cr}(x_j, \sigma_r, p_f, p_o, N_c^{(s)})$ is additional part of the compliance matrix due to set of different types of cracks growing in the CMC.

Figures 2, 3, 4 and 5 present scenario of gradual degradation of the considered polycrystalline CMC. All phases will be described by constitutive equations.

3 Initial Elastic Properties and Porosity of the CMC, Fig. 1

The initial elastic properties of the two-phase CMCs without porosity can be estimated according to rule of mixture, e.g. Kachanov (1993a, b), Sadowski and de Borst (2008):

$$\bar{S}_{ik}^{CMCe}(x_j, p_f) = (1 - p_f)S_{ik}^{C1} + p_f S_{ik}^{C2}(x_j, p_f). \quad (3)$$

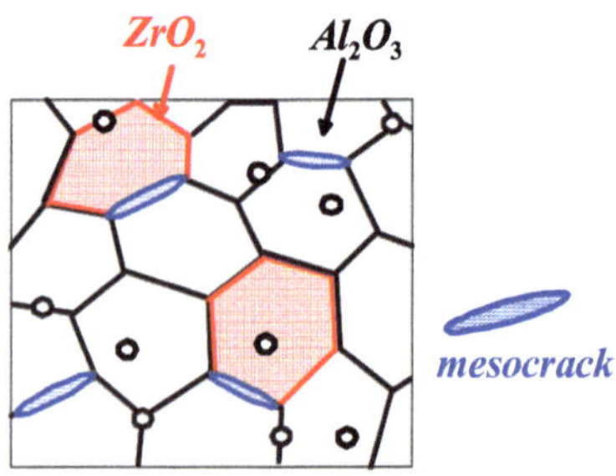

Fig. 3 Early stage deformation process with set of mesocracks

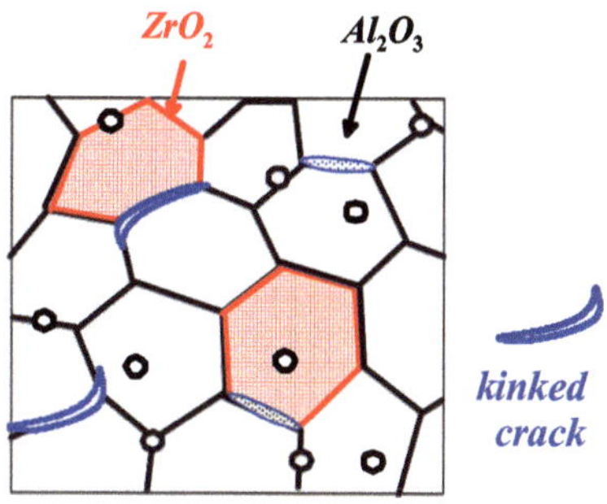

Fig. 4 Advanced stage of deformation process of the CMC including set of kinked cracks

According to Kachanov (1993a, b) the material porosity can be described by the following compliance matrix $S_{ik}^{\mathrm{p_o}}(x_\mathrm{j}, \sigma_\mathrm{rt}, p_\mathrm{o})$:

$$S_{ik}^{\mathrm{p_o}} = \frac{4(1-\nu_0^2)}{E}\frac{p_o}{1-p_\mathrm{o}}\begin{bmatrix} \frac{3}{4} & -\frac{1}{4} & -\frac{1}{4} & 0 & 0 & 0 \\ -\frac{1}{4} & \frac{3}{4} & -\frac{1}{4} & 0 & 0 & 0 \\ -\frac{1}{4} & -\frac{1}{4} & \frac{3}{4} & 0 & 0 & 0 \\ 0 & 0 & 0 & 1 & 0 & 0 \\ 0 & 0 & 0 & 0 & 1 & 0 \\ 0 & 0 & 0 & 0 & 0 & 1 \end{bmatrix}, \tag{4}$$

where porosity parameter is calculated from $p_\mathrm{o} = \frac{1}{A}\sum_{s=1}^{N^{(\mathrm{p_o})}} \pi r_{(\mathrm{s})}^2$ and $N^{(\mathrm{p_o})}$ is the number of pores inside the RVE. The pores diameter is denoted by $r_{(\mathrm{s})}$, whereas A is the surface area cross-section of the considered RVE (Fig. 2).

4 Nucleation of Mesocracks in the CMC, Figs. 3 and 6

Let us consider that mesocracks start from the stress concentrations at the pores edge. At the grain boundaries crack nucleus initiate and spread along interfaces of grains, Fig. 6. It results from significantly less fracture toughness in comparison to

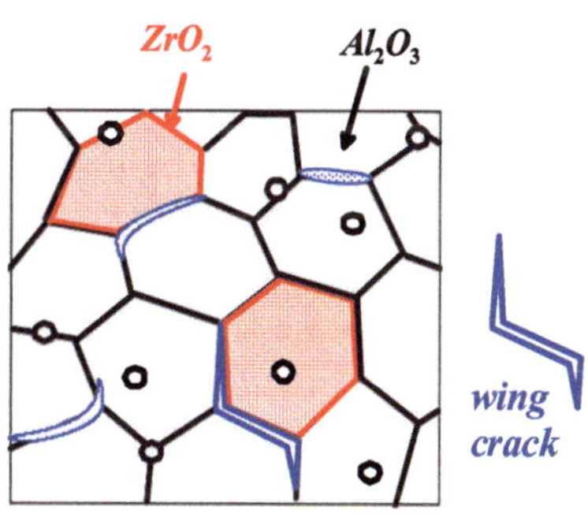

Fig. 5 Final stage of deformation process of the CMC before failure including set of wing cracks

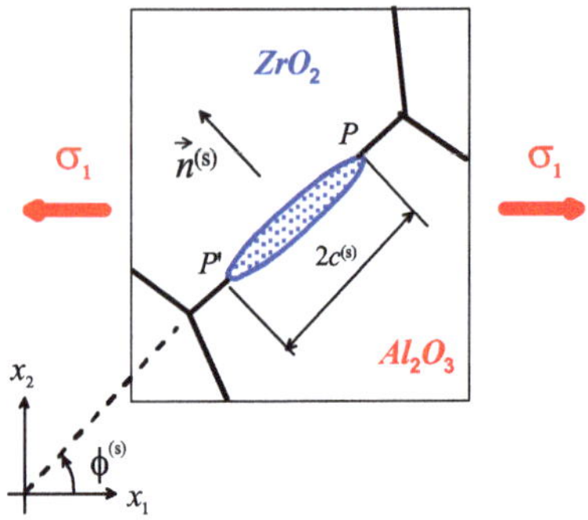

Fig. 6 Mesocrack at grain boundary

pure grain. For a pore situated inside grain, cracks start and grow along the trajectory perpendicular to local maximum tensile stress and move to grain boundary (Marsavina and Sadowski 2007a, b, 2009). Consequently, the crack can in general have one deflection (Fig. 4) or two deflections or penetrate the nearest grain. The fracture surface energy of the grain boundary γ_{gb} in the polycrystalline CMC is significantly less in comparison to fracture surface energy of the grain γ_g. It means that cracks grow mainly inter-granularly. In the advanced stage of deformation one can distinguish three different types of cracks developed in RVE, Fig. 5, i.e. mesocracks, kinked cracks and wing cracks.

The mesocracks are the most important in case of uniaxial loading. If we denote by $N_c^{(s)\text{meso}}$ the number of mescoracks, their presence in RVE can be described by the following components of the compliance matrix for 2D:

$$S_{ij}^{\text{cr(meso)}}(x_k, \sigma_r, p_f, p_o, N_c^{(s)\text{meso}}) = \sum_{s=1}^{N_c^{(s)\text{meso}}} \frac{2\pi (c^{(s)})^2}{A} \begin{bmatrix} (n_1^{(s)})^2 \bar{S}_{11}^{\text{CMCe}} & 0 & 0 \\ 0 & (n_2^{(s)})^2 \bar{S}_{22}^{\text{CMCe}} & 0 \\ \frac{1}{2}(n_1^{(s)} n_2^{(s)}) \bar{S}_{11}^{\text{CMCe}} & \frac{1}{2}(n_1^{(s)} n_2^{(s)}) \bar{S}_{22}^{\text{CMCe}} & 0 \end{bmatrix}, \tag{5}$$

where $i, j = 1, 2, 3$, $n_1^{(s)}$ and $n_2^{(s)}$ are components of the vector $n^{(s)}$ specifying space orientation of the mesocrack "s", Fig. 6.

Considering cracks with kink (Fig. 7), if the nondimensional opening displacement components $\bar{B}_1^{(k)}$ and $\bar{B}_2^{(k)}$ of the kink are known, the additional normal strains ($\bar{\varepsilon}_1$ and $\bar{\varepsilon}_2$) due to kink creation in RVE can be calculated according to

$$\bar{\varepsilon}_1^{(k)} = \sum_{k=1}^{N_c^{(k)}} \frac{\bar{L}(c^{(k)})^2}{2A} \left\{ \left[(n_1^{(k)})^2 s_2^{(k)} + n_1^{(k)} n_2^{(k)} s_1^{(k)} \right] \bar{B}_1^{(k)} + \left[(n_1^{(k)})^2 s_1^{(k)} - n_1^{(k)} n_2^{(k)} s_2^{(k)} \right] \bar{B}_2^{(k)} \right\} \tag{6}$$

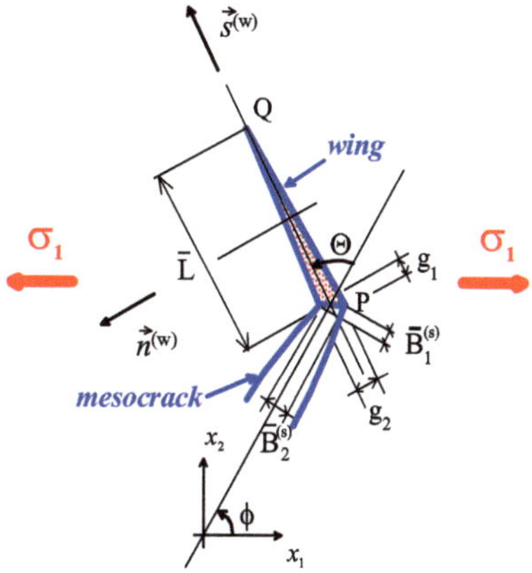

Fig. 7 Kink of the mesocrack (PQ) at the interface of CMC grains

$$\bar{\varepsilon}_2^{(k)} = \sum_{k=1}^{N_c^{(k)}} \frac{\bar{L}(c^{(k)})^2}{2A} \left\{ \left[(n_2^{(k)})^2 s_2^{(k)} - n_1^{(k)} n_2^{(k)} s_1^{(k)} \right] \bar{B}_1^{(s)} + \left[(n_2^{(k)})^2 s_1^{(w)} + n_1^{(w)} n_2^{(w)} s_2^{(w)} \right] \bar{B}_2^{(k)} \right\} \tag{7}$$

The shear strain $\bar{\varepsilon}_3$ are equal to:

$$\bar{\varepsilon}_3^{(k)} = \sum_{k=1}^{N_c^{(k)}} \frac{\bar{L}(c^{(k)})^2}{4A} \left\{ \left[2n_2^{(k)} n_2^{(k)} s_2^{(k)} + \left\{ (n_2^{(k)})^2 - (n_1^{(k)})^2 \right\} s_1^{(k)} \right] \bar{B}_1^{(k)} + \left[2n_2^{(k)} n_2^{(k)} s_1^{(k)} - \left\{ (n_2^{(k)})^2 - (n_1^{(k)})^2 \right\} s_2^{(k)} \right] \bar{B}_2^{(k)} \right\}. \tag{8}$$

In Eqs. (6), (7) and (8) $N_c^{(k)}$ is the number of kinks in RVE, whereas $c^{(k)}$ is the half length of the mesocrack, which creates the kink.

5 Growth of Mesocracks

The nucleated mesocracks or kinks develop along grain facets changing their direction. The crack propagation strongly depends on the grain boundary porosity spread along grain boundaries as well as thermal residual stresses. The mesocrack at grain boundaries (Fig. 3) can grow if the energy release rate G^{cr} satisfies the following condition:

$$G^{cr}(\sigma_r, c^{(s)}, \phi^{(s)}) \geq \gamma_{gb}, \tag{9}$$

where $\gamma_{gb} = (0.5 - 0.75)\gamma_g$ is the critical value of the grain boundary fracture surface energy and is considerably less in comparison to grain surface energy γ_g. Kinks propagate when, Fig. 4:

$$G^{cr}(\sigma_r, c^{(k)}, \phi^{(k)}, \bar{L}^{(k)}, \theta^{(k)}) \geq \gamma_{gb}. \tag{10}$$

Cracks propagate under mixed mode.

6 Deformation Process of Polycrystalline Al_2O_3 + 20 %vc ZrO_2 Containing Porosity Under Uniaxial Tensile Loading

Taking into account the above formulated procedure the mechanical response of the considered material was described. The numerical example describing the deformation process of porous Al_2O_3 + 20 %vc ZrO_2 was performed in order to show the capability of the material modelling by micromechanical approach. The initial mechanics features are collected in Table 1. Numerical calculations were performed according to the theoretical formulation for the following data: $1/\bar{S}_{11}^{CMCe} = 287$ GPa, $\bar{\nu}^{CMCe} = 0.226$, $K_{Ic}^{CMC} = 8,5\ \mathrm{MPa\,m}^{1/2}$. It was assumed that RVE contains hexagonal grains of the mean diameter $2\bar{D} = 5\,\mu$m. Figure 8 presents results of the numerical model simulation. It includes pure Al_2O_3 and ZrO_2 samples responses without

Table 1 Material properties of the CMC components

Material	E_0 (GPa)	ν_0 (–)	K_{Ic} (MPam$^{1/2}$)
Al_2O_3	315	0.22	4.5
ZrO_2	175	0.25	12.0

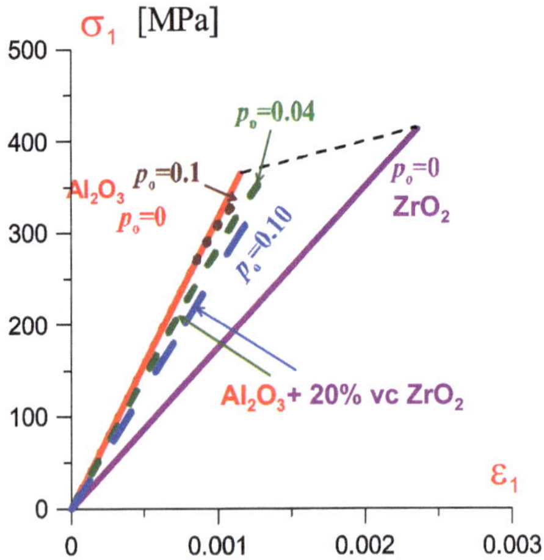

Fig. 8 Tensile stress–strain curve for response in porous composite Al_2O_3 + 20 %vc ZrO_2

porosity ($p_o = 0$), treated as a reference composite materials. Moreover, we analysed polycrystalline Al_2O_3 with small amount porosity equal to $p_o = 0.1$. The basic material was $Al_2O_3 + 20\,\%vc\ ZrO_2$ with the porosity up to 10 %.

7 Conclusions

The multiscale model of complex CMC, i.e. porous $Al_2O_3 + 20\,\%vc\ ZrO_2$ was proposed in the paper. The modelling starts from the information about internal CMC structure at micro-level. The obtained results lead to the following conclusions:

1. Volume contents of both phases constituting the CMC and initial porosity strongly influenced the material behaviour under mechanical loading,
2. Pores existing in the polycrystalline CMC act as the sources of micro-cracks initiations,
3. Micro-cracks develop when criterion of cracks propagation (9) is satisfied to create meso-cracks,
4. The dominant mode of meso-cracks propagation is inter-granular by kinking and creation of the wing cracks,
5. Alternatively, trans-granular processes of cracks propagation can be incorporated in the CMC model,
6. Numerical extension of the CMC model can be formulated with application of the cohesive zone model (e.g. Sadowski and Golewski 2011, 2012b, c; Ivanov et al. 2016; Burlayenko and Sadowski 2009, 2010; Vignollet et al. 2014; Burlayenko and Sadowski 2014a, b; Allix et al. 2014; Burlayenko and Sadowski 2012b; Sadowski et al. (2016); Burlayenko and Sadowski 2012a) at interfaces.

Acknowledgments This work was financially supported by Ministry of Science and Higher Education within the statutory research number S/20/2016.

References

Allix, O., Feld, N., Baranger, E., Guimard, J., Ha-Minh, C.: The compressive behaviour of composites including fiber kinking: modelling across the scales. Meccanica **49**(11), 2571–2586 (2014)
Bieniaś, J., Dębski, H., Surowska, B., Sadowski, T.: Analysis of microstructure damage in carbon/epoxy composites using FEM. Comput. Mater. Sci. **64**, 168–172 (2012)
Birman, V., Byrd, L.: Modeling and analysis of functionally graded materials and structures. Appl. Mech. Rev. **60**(5), 195–216 (2007)
Bîrsan, M., Altenbach, H., Sadowski, T., Eremeyev, V., Pietras, D.: Deformation analysis of functionally graded beams by the direct approach. Compos. Part B: Eng. **43**(3), 1315–1328 (2012)
Bîrsan, M., Sadowski, T., Marsavina, L., Linul, E., Pietras, E.: Mechanical behavior of sandwich composite beams made of foams and functionally graded materials. Int. J. Solids Struct. **50**(3), 519–530 (2013a)
Bîrsan, M., Sadowski, T., Pietras, D.: Thermoelastic deformations of cylindrical multi-layered shells using a direct approach. J. Therm. Stresses **36**(8), 749–789 (2013b)

Burlayenko, V., Sadowski, T.: Analysis of structural performance of sandwich plates with foam-filled aluminum hexagonal honeycomb core. Comput. Mater. Sci. **45**(3), 658–662 (2009)
Burlayenko, V., Sadowski, T.: Influence of skin/core debonding on free vibration behavior of foam and honeycomb cored sandwich plates. Int. J. Non-Linear Mech. **45**(10), 959–968 (2010)
Burlayenko, V., Sadowski, T.: Finite element nonlinear dynamic analysis of sandwich plates with partially detached facesheet and core. Finite Elem. Anal. Des. **62**, 49–64 (2012a)
Burlayenko, V., Sadowski, T.: A numerical study of the dynamic response of sandwich plates initially damaged by low-velocity impact. Comput. Mater. Sci. **52**(1), 212–216 (2012b)
Burlayenko, V., Sadowski, T.: Nonlinear dynamic analysis of harmonically excited debonded sandwich plates using finite element modelling. Compos. Struct. **108**, 354–366 (2014a)
Burlayenko, V., Sadowski, T.: Transient dynamic response of debonded sandwich plates predicted with finite element analysis. Meccanica **49**(11), 2617–2633 (2014b)
Burlayenko, V., Altenbach, H., Sadowski, T.: An evaluation of displacement-based finite element models used for free vibration analysis of homogeneous and composite plates. J. Sound Vib. **358**, 152–175 (2015a)
Burlayenko, V., Altenbach, H., Sadowski, T., Dimitrova, S.: Computational simulations of thermal shock cracking by the virtual crack closure technique in a functionally graded plate. Comput. Mater. Sci. **116**, 11–21 (2015b)
Dębski, H., Sadowski, T.: Modelling of microcracks initiation and evolution along interfaces of the WC/Co composite by the finite element method. Comput. Mater. Sci. **83**, 403–411 (2014)
Felten, F., Schneider, G., Sadowski, T.: Estimation of R-curve in WC/Co cermet by CT test. Int. J. Refract. Met. Hard Mater. **26**(1), 55–60 (2008)
Fic, S., Szeląg, M.: Analysis of the development of cluster cracks caused by elevated temperatures in cement paste. Constr. Build. Mater. **83**, 223–229 (2015)
Gajewski, J., Sadowski, T.: Sensitivity analysis of crack propagation in pavement bituminous layered structures using a hybrid system integrating artificial neural networks and finite element method. Comput. Mater. Sci. **82**, 114–117 (2014)
Golewski, G., Sadowski, T.: An analysis of shear fracture toughness K IIc and microstructure in concretes containing fly-ash. Constr. Build. Mater. **51**, 207–214 (2014)
Golewski, G., Golewski, P., Sadowski, T.: Numerical modelling crack propagation under mode II fracture in plain concretes containing siliceous fly-ash additive using XFEM method. Comput. Mater. Sci. **62**, 75–78 (2012)
Gömze, L., Gömze, L.: Alumina-based hetero-modulus ceramic composites with extreme dynamic strength. Zeitschrift für Angewandte Mathematik und Mechanik **61**, 38–42 (2009)
Gömze, L., Gömze, L.: Ceramic based lightweight composites with extreme dynamic strength. IOP Conf. Ser.: Mater. Sci. Eng. **47**(1), 12,033–12,038 (2013)
Ivanov, I., Sadowski, T., Pietras, D.: Crack propagation in functionally graded strip under thermal shock. Eur. Phys. J. Spec. Top. **222**(7), 1587–1595 (2013)
Ivanov, I., Velchev, D., Georgiev, N., Ivanov, I., Sadowski, T.: A plate finite element for modelling of triplex laminated glass and comparison with other computational models. Meccanica **51**(2), 341–358 (2016)
Kachanov, M.: Elastic solids with many cracks and related problems. Adv. Appl. Mech. **30**, 259–445 (1993a)
Kachanov, M.: On the effective moduli of solids with cavities and cracks. Int. J. Fract. **59**(1), R17–R21 (1993b)
Kalatur, E., Buyakova, S., Kulkov, S., Gotman, I., Kocserha, I.: Porosity and mechanical properties of zirconium ceramics. Építöanyag **66**(2), 31–34 (2014)
Lau, A., Hui, D.: The revolutionary creation of new advanced materials - carbon nanotube composites. Compos. Part B: Eng. **33**(4), 263–277 (2002)
Lenci, S., Piattoni, Q., Clementi, F., Sadowski, T.: An experimental study on damage evolution of unfired dry earth under compression. Int. J. Fract. **172**(2), 193–200 (2011)
Lenci, S., Clementi, F., Sadowski, T.: Experimental determination of the fracture properties of unfired dry earth. Eng. Fract. Mech. **87**, 62–72 (2012)

Marsavina, L., Sadowski, T.: Effect of biaxial load on crack deflection/penetration at bi-material ceramic interface. Int. J. Fract. **148**(1), 79–84 (2007a)
Marsavina, L., Sadowski, T.: Stress intensity factors for an interface kinked crack in a bi-material plate loaded normal to the interface. Int. J. Fract. **145**(3), 237–243 (2007b)
Marsavina, L., Sadowski, T.: Kinked crack at a bi-material ceramic interface - numerical determination of fracture parameters. Comput. Mater. Sci. **44**(3), 941–950 (2009)
Nakonieczny, K., Sadowski, T.: Thermal shock response of FGM cylindrical plates with various grading patterns. Comput. Mater. Sci. **43**(1), 171–178 (2008)
Nakonieczny, K., Sadowski, T.: Modelling of thermal shocks in composite materials using a mesh-free FEM. Comput. Mater. Sci. **44**(4), 1307–1311 (2009)
Petrova, V., Sadowski, T.: Theoretical analysis of mode II cracks in a compact shear specimen. Comput. Mater. Sci. **64**, 248–252 (2012)
Petrova, V., Sadowski, T.: Theoretical modeling and analysis of thermal fracture of semi-infinite functionally graded materials with edge cracks. Meccanica **49**(11), 2603–2615 (2014)
Petrova, V., Marsavina, L., Sadowski, T.: Revisit of compact mode II crack specimen: analysis and fracture interpretation. Theor. Appl. Fract. Mech. **59**(1), 41–48 (2012)
Postek, E., Sadowski, T.: Assessing the influence of porosity in the deformation of metal-ceramic composites. Compos. Interfaces **18**(1), 57–76 (2011)
Sadowski, T.: Mechanical response of semi-brittle ceramics subjected to tension-compression state. Part i: theoretical modeling. Int. J. Damage Mech. **3**(2), 212–233 (1994a)
Sadowski, T.: Modelling of semi-brittle MgO ceramic behaviour under compression. Mech. Mater. **18**(1), 1–16 (1994b)
Sadowski, T.: Mechanical response of semi-brittle ceramics subjected to tension-compression state. Part ii: description of deformation process. Int. J. Damage Mech. **4**(4), 293–318 (1995)
Sadowski, T., de Borst, R.: Lecture Notes on Composite Materials: Current Topics and Achievements. Springer Science & Business Media, New York (2008)
Sadowski, T., Bîrsan, M., Pietras, D.: Numerical analysis of multilayered and FGM structural elements under mechanical and thermal loads. Comparison of the finite elements and analytical models. Arch. Civ. Mech. Eng. **15**, 1180–1192 (2015)
Sadowski, T., Golewski, P.: Multidisciplinary analysis of the operational temperature increase of turbine blades in combustion engines by application of the ceramic thermal barrier coatings (TBC). Comput. Mater. Sci. **50**(4), 1326–1335 (2011)
Sadowski, T., Golewski, P.: The analysis of heat transfer and thermal stresses in thermal barrier coatings under exploitation. Defect Diffus. Forum **326**, 530–535 (2012a)
Sadowski, T., Golewski, P.: Detection and numerical analysis of the most efforted places in turbine blades under real working conditions. Comput. Mater. Sci. **64**, 285–288 (2012b)
Sadowski, T., Golewski, P.: The influence of quantity and distribution of cooling channels of turbine elements on level of stresses in the protective layer TBC and the efficiency of cooling. Comput. Mater. Sci. **52**(1), 293–297 (2012c)
Sadowski, T., Golewski, P.: Heat transfer and stress concentrations in a two-phase polycrystalline composite structure. Part i: theoretical modelling of heat transfer. Materialwissenschaft und Werkstofftechnik **44**(5), 497–505 (2013)
Sadowski, T., Golewski, P.: Description of non-stationary heat transfer in two-phase polycrystalline metal-ceramic composites. Acta Phys. Pol. A **128**(4), 624–628 (2015)
Sadowski, T., Neubrand, A.: Estimation of the crack length after thermal shock in FGM strip. Int. J. Fract. **127**(2), L135–L140 (2004)
Sadowski, T., Nowicki, T.: Numerical investigation of local mechanical properties of WC/Co composite. Comput. Mater. Sci. **43**(1), 235–241 (2008)
Sadowski, T., Samborski, S.: Development of damage state in porous ceramics under compression. Computational Materials Science **43**(1), 75–81 (2008)
Sadowski, T., Hardy, S., Postek, E.: Prediction of the mechanical response of polycrystalline ceramics containing metallic intergranular layers under uniaxial tension. Comput. Mater. Sci. **34**(1), 46–63 (2005)

Sadowski, T., Hardy, S., Postek, E.: A new model for the time-dependent behaviour of polycrystalline ceramic materials with metallic inter-granular layers under tension. Mater. Sci. Eng.: A **424**(1), 230–238 (2006)
Sadowski, T., Boniecki, M., Librant, Z., Nakonieczny, K.: Theoretical prediction and experimental verification of temperature distribution in FGM cylindrical plates subjected to thermal shock. Int. J. Heat Mass Transf. **50**(21), 4461–4467 (2007a)
Sadowski, T., Postek, E., Denis, C.: Stress distribution due to discontinuities in polycrystalline ceramics containing metallic inter-granular layers. Comput. Mater. Sci. **39**(1), 230–236 (2007b)
Sadowski, T., Ataya, S., Nakonieczny, K.: Thermal analysis of layered FGM cylindrical plates subjected to sudden cooling process at one side – comparison of two applied methods for problem solution. Comput. Mater. Sci. **45**(3), 624–632 (2009)
Sadowski, T., Craciun, E., Răbâea, A., Marsavina, L.: Mathematical modeling of three equal collinear cracks in an orthotropic solid. Meccanica **51**(2), 329–339 (2016)
Sadowski, T., Pankowski, B.: Numerical modelling of two-phase ceramic composite response under uniaxial loading. Compos. Struct. **134**, 338–394 (2016)
Savchenko, N., Sevostyanova, I., Sablina, T., Molnar, L., Geber, R., Gömze, L., Kulkov, S., Gömze, L.: The influence of porosity on the elasticity and strength of alumina ceramics. Építöanyag **66**(2), 44–47 (2014)
Taczała, M., Buczkowski, R., Kleiber, M.: Postbuckling analysis of functionally graded plates on an elastic foundation. Compos. Struct. **132**, 842–847 (2015)
Taczała, M., Buczkowski, R., Kleiber, M.: Nonlinear free vibration of pre-and post-buckled FGM plates on two-parameter foundation in the thermal environment. Compos. Struct. **137**, 85–92 (2016)
Vignollet, J., May, S., de Borst, R., Verhoosel, C.: Phase-field models for brittle and cohesive fracture. Meccanica **49**(11), 2587–2601 (2014)
Wu, Z., Li, J., Timmer, D., Lozano, K., Bose, S.: Study of processing variables on the electrical resistivity of conductive adhesives. Int. J. Adhes. Adhes. **29**(5), 488–494 (2009)
Zhao, H., Liang, T., Liu, B.: Synthesis and properties of copper conductive adhesives modified by SiO2 nanoparticles. Int. J. Adhes. Adhes. **27**(6), 429–433 (2007)

On Failure Theories for Composite Materials

Ramesh Talreja

Abstract Limitations inherent in failure theories formulated on homogenized description of composite materials are discussed. Failure mechanisms in composite materials, as understood today, are reviewed. Based on this knowledge, arguments are put forth to abandon the classical approach to formulation of failure theories for composite materials, and to instead use a computation-based failure assessment methodology. Such a methodology is proposed. In conjunction with this, the idea of virtual testing to supplement experimental determination of material response characteristics is discussed.

1 Introduction

As applications of composite materials have expanded from the aerospace to non-aerospace fields, such as automotive, wind turbines, and subsea structures, design for safe performance has become critical. Current application of carbon-epoxy composites in civilian aircraft is essentially a replacement of aluminum (coined as "black aluminum") with limited use of the potential of these materials. The main roadblock to a higher level of utilization is lack of reliable failure analysis capabilities in the industry. The current approach in the aerospace applications is to conduct extensive tests to demonstrate safe performance, which has forced the industry to keep the load levels excessively low. Although the low weight of polymer composites still gives

R. Talreja (✉)
Department of Aerospace Engineering, Texas A&M University,
College Station, TX 77843, USA
e-mail: talreja@tamu.edu

R. Talreja
Department of Materials Science and Engineering, Texas A&M University,
College Station, TX 77843, USA

R. Talreja
Department of Engineering Science and Mathematics, Luleå University of Technology,
SE 971 87 Luleå, Sweden

K. Naumenko and M. Aßmus (eds.), *Advanced Methods of Continuum Mechanics for Materials and Structures*, Advanced Structured Materials 60,
DOI 10.1007/978-981-10-0959-4_21

significant fuel savings in aircraft, other applications cannot afford such conservative designs.

A compelling case was made in a recently completed survey of composite failure theories (Soden et al. 1998; Hinton et al. 2002, 2004; Kaddour et al. 2004) that their predictive capabilities judged against test data are not reliable. This is a sad commentary on the state of affairs in view of the long history of the development of the theories. Perhaps this long history is the reason for the problems underlying the lack of success of the theories. As will be argued below, the path taken in developing the failure theories has been flawed. The nature of the flaws is such that efforts to "improve" the theories are frustrated by lack of agreement with some or the other test data. An alternative path must be taken, and it may lead to the painful realization that the current theories are best left behind.

The following sections will discuss first the fundamentals of the common theories of failure in unidirectional composites, drawing upon a previous treatment (Talreja 2014). The physical nature of the failure in these composites will be summarized next. The models that account for the known mechanisms will be reviewed and the challenges remaining in this direction will be described. Finally, a scheme to connect the models into a failure assessment methodology will be presented.

2 The Classical Failure Theories for Unidirectional Composites

The first significant theory of failure in unidirectional composites appeared in 1965 (Azzi and Tsai 1965), known as the Tsai–Hill theory. In fact, associating Hill's name with this theory is questionable since Hill (1948) did not in any way suggest that his theory, developed for yielding of anisotropic metals (metals with texture), may be applied to fiber-reinforced polymers. Hill's yield criterion for anisotropic solids was a simple mathematical generalization of the von Mises criterion for isotropic yielding when expressed in terms of the deviatoric stress components. In this generalization, the important connection with the distortional energy density was lost. This fact is significant because having energy concept to capture the physics of metal plasticity strengthens the basis of the yield criterion. In any case, the weak form of the criterion in Hill's proposal still only applies to yielding that is driven by shear at the microscopic level. The underlying mechanism of yielding for anisotropic (orthotropic) metals in Hill's formulation manifests itself in six yield constants—three normal stress thresholds and three shear stress thresholds—as a generalization of a single yield stress in the isotropic case.

Adopting the Hill criterion to another material (a unidirectional composite) can only be justified if that material yields (or in another way becomes critical) as a result of a single shear-driven mechanism. Although a polymer may satisfy this requirement, it is doubtful that it will continue to do so when reinforced with fibers. In fact even an unreinforced polymer shows pressure sensitivity in its inelastic behavior, thereby violating the shear-driven yield requirement. The presence of fibers significantly alters "failure" in the form of deviation from elasticity, as will be discussed later.

After reducing the Hill criterion to two-dimensional form and assuming isotropy in the cross-sectional plane of unidirectional composites, the criterion in (Azzi and Tsai 1965) takes the following form

$$\left(\frac{\sigma_1}{X}\right)^2 - \left(\frac{\sigma_1\sigma_2}{X}\right)^2 + \left(\frac{\sigma_2}{Y}\right)^2 + \left(\frac{\sigma_{12}}{S}\right)^2 = 1 \tag{1}$$

where σ_1 and σ_2 are the normal stresses in the fiber and transverse directions, respectively, and σ_{12} is the in-plane shear stress. X, Y, and S are the "yield" stresses corresponding to σ_1, σ_2, and σ_{12}, respectively.

Equation (1) was shown in Azzi and Tsai (1965) to agree well with test data for a unidirectional composite loaded in tension at an angle inclined to the fiber direction. However, other data generated later did not show good agreement, suggesting a need for another theory.

Observing that the stress components in Eq. (1) were "interacting" in a quadratic manner, a more general quadratic polynomial than what appears here could improve the lacking agreement with test data. This idea seemed to have led to the search for a more general formulation, which was found in the tensor polynomial for strength proposed in Goldenblatt and Kopnov (1965). A simplification of the polynomial to plane stress state resulted in the quadratic expression proposed in Tsai and Wu (1971) as

$$F_1\sigma_1 + F_2\sigma_2 + F_{11}\sigma_1^2 + F_{22}\sigma_2^2 + F_{12}\sigma_1\sigma_2 + F_{66}\sigma_{12}^2 = 1 \tag{2}$$

after using the fact that the sign of the shear stress has no effect on the shear strength. The coefficients of the terms in Eq. (2) are inverses of strength values and these can be found by testing with one imposed stress component at a time, except the constant F_{12}, which requires a biaxial stress test. Since a biaxial test can be done with an arbitrary combination of the two normal stresses, determination of this constant becomes non-unique. This flaw in the theory cannot be removed, but can only be "fixed" by adopting a biaxial test as a convention. The consequence of the lack of a physical reasoning for choosing the biaxial stress for determining F_{12} was discussed among others by Hashin (1980), who also pointed out that a single surface in the σ_1-σ_2-σ_{12}-space, represented by Eq. (2), was inadequate in describing failure in different combinations of stresses. He suggested instead to formulate the failure criteria in a piecewise smooth form with each branch of the failure surface dealing with a certain failure mode. He suggested to separate fiber failure mode from the matrix failure mode because of different governing mechanisms and to formulate quadratic interaction equations in each case.

Hashin's work (Hashin 1980) provided a direction away from the totally curve-fitting schemes toward utilizing some understanding of the failure process. In particular, his suggestion to consider a failure plane in the matrix of a composite to formulate the matrix mode failure criteria was taken up by Puck and coworkers (Puck and Schürmann 1998, 2002). These authors devised elaborate procedures for determining additional material constants needed to characterize the failure envelopes in different stress combinations.

Many other works have also proposed formulations of failure criteria using various assumptions regarding how criticality of failure occurs. As mentioned above, assessments of a large number of such failure theories reported in Soden et al. (1998), Hinton et al. (2002, 2004), and Kaddour et al. (2004) concluded that no single theory was able to agree with all test data used for comparison.

One can ask why the proposed theories are not satisfactory. Perhaps the answer lies in the failure mechanisms, which are rich in detail and varied in how they initiate, progress, and become critical. The next section will discuss the failure mechanisms.

3 Failure Mechanisms in Unidirectional Composites

3.1 *Fiber Failure in Tension*

Fiber failure in a unidirectional composite under imposed overall axial tension has been studied extensively. A recent work using high resolution X-ray microtomography (Aroush et al. 2006) provides clear evidence of the statistical nature of this process (Fig. 1). As reported in that work, fiber breaks appear initially at low loads as single failures at discrete locations because of failures at weak points. On increasing the applied load, more fibers fail, mostly near the previously broken fibers, and the so-called doublets form. This process continues until one or more of the broken fiber clusters grow unstably to failure of the composite.

3.2 *Fiber Failure in Compression*

Although the strength of a unidirectional composite under axial compression is characterized as a fiber failure mode, as by Hashin (1980) and subsequently by many, the failure mechanisms involved in this case depend significantly on the matrix behavior. As described by Jelf and Fleck (1992), the fiber failure in compression may be categorized as elastic microbuckling or plastic microbuckling, depending on whether the matrix stress–strain behavior is linear or nonlinear. For the latter case, the compressive strength is predicted well by Budiansky's kink band model (Budiansky 1983), which has the fiber misalignment and matrix shear yield stress as parameters. Kyriakides et al. (1995) further verified the dependence of the compressive strength on fiber misalignment angle and generalized it to fiber waviness as the microstructure (defect) scale. The microstructural features of the kink bands and mechanisms of their formation were clarified by these authors. Their analysis found that in the presence

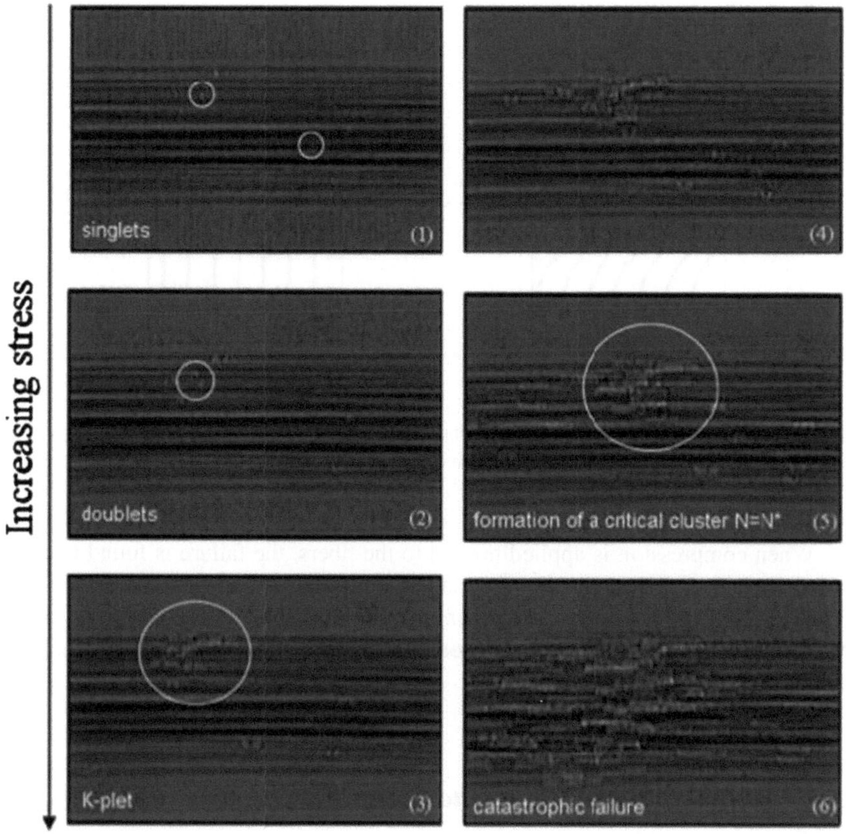

Fig. 1 Sequence of fiber failures in a unidirectional composite under axial tension (from Aroush et al. 2006)

of fiber waviness, localization of shear deformation occurs in the matrix. This forms bands and the flow of the matrix in the bands results in bending of fibers and eventual breakage, which results in the formation of the observed kink bands. Figure 2 illustrates schematically the early stage of microbuckling leading to formation of a kink band (Berbinau et al. 1999).

3.3 Matrix Failure in Transverse Tension or Compression

On loading a unidirectional composite under tension normal to fibers, failure occurs suddenly and at low stress levels. The mechanisms leading to the catastrophic failure are conveniently studied by observing the initiation and progression of cracks within the plies of a laminate. The appearance of these cracks is typified by the images shown in Fig. 3 (Gamstedt and Sjögren 1999).

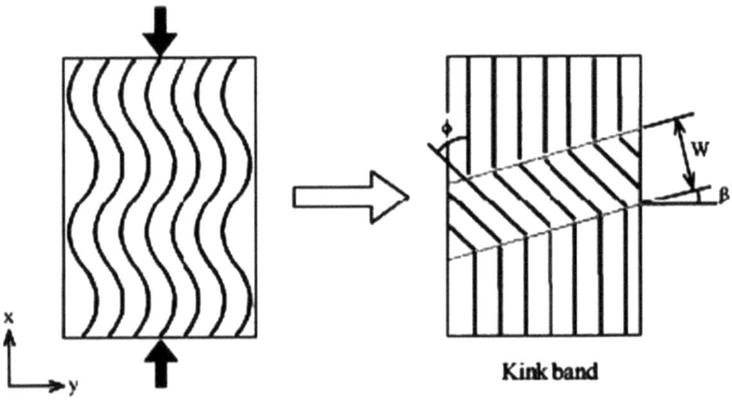

Fig. 2 Schematic illustration of fiber microbuckling and kink band formation in a unidirectional composite subjected to axial compression (Berbinau et al. 1999)

When compression is applied normal to the fibers, the failure is found to occur along a plane that is inclined to the loading direction, as reported in González and LLorca (2007) (Fig. 4). On closer examination, it is found that microscopic cracks formed due to shear and the coalescence of these cracks led to failure along the inclined plane.

3.4 Matrix Failure in In-Plane Shear

Under an in-plane shear stress, a unidirectional composite displays nonlinear behavior. A part of this is due to a shear-induced flow of the matrix and the other part is caused by cracks formed in the matrix. Such cracks are illustrated in the image shown in Fig. 5 (Redon 2000). These cracks, as shown in the figure, form multiple

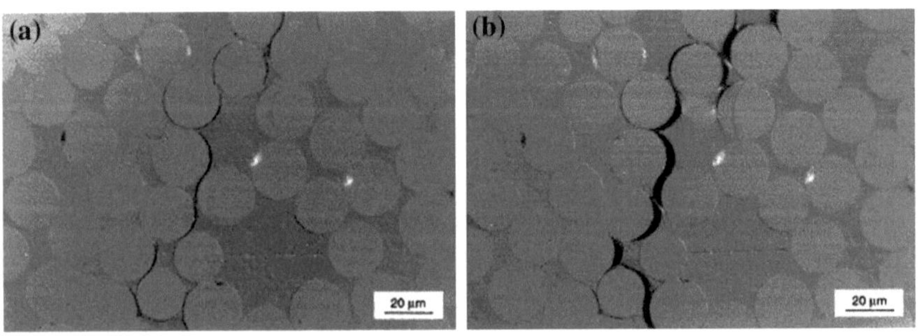

Fig. 3 Images of cracks formed under a tensile load normal to fibers. Image (a) is at a low load, while image (b) is taken at the same location at a higher load. From Gamstedt and Sjögren (1999)

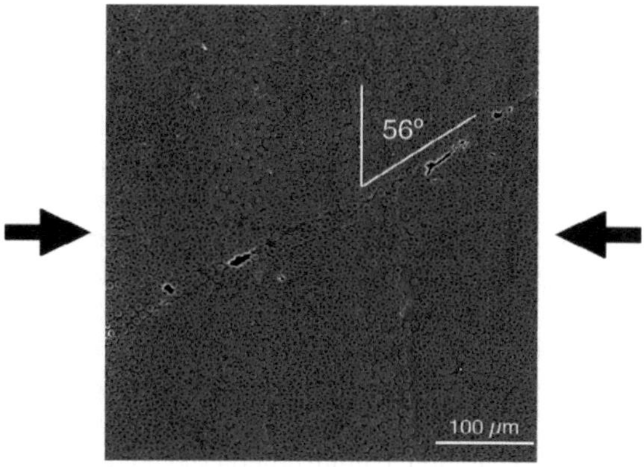

Fig. 4 Matrix failure on an inclined plane under compression normal to fibers (González and LLorca 2007)

cracks with their planes inclined to the fiber axis. On increased loading, the cracks turn along the fibers and merge together, forming a failure plane. Such a failure plane is also formed under compression normal to fibers (Fig. 4), indicating the role of the shear stress on the plane.

4 Formulation of Failure Criteria for Unidirectional Composites

In view of the brief overview of the failure mechanisms in unidirectional composites described above in Sect. 3, one can scrutinize the failure theories discussed in Sect. 2. Following points can be made.

(a) There is no basis for a single failure criterion represented by a single smooth surface in the stress space of the three stress components in the principal coordinates of the composite. Thus, the two criteria given by Eqs. (1) and (2) are each not physically based. There would be a physical basis, e.g., for an orthotropic solid where only one failure mechanism operates, e.g., yielding. Thus, the Hill criterion for yielding of a metal with texture (produced, for example, by rolling in one direction), from which Eq. (1) was derived, is a physically based criterion, while Eq. (1) for unidirectional composite failure is not.

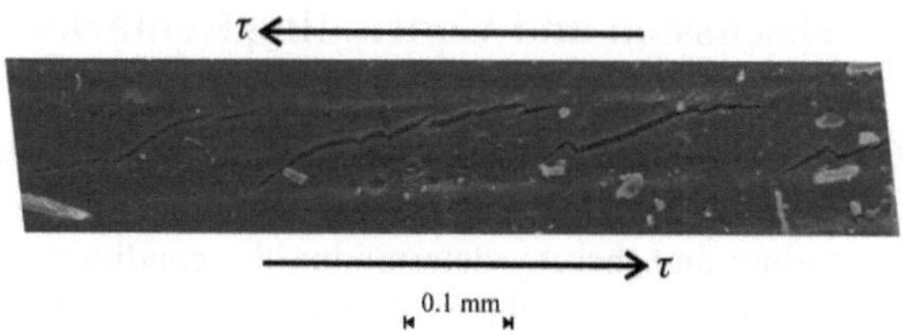

Fig. 5 Matrix cracks inclined to the fiber direction (*horizontal*) formed under in-plane shear in a unidirectional composite (Redon 2000)

(b) Under combined loading, i.e., when two or three stress components in the principal coordinates of a unidirectional composite are simultaneously applied, the "interaction" cannot a priori be assumed as quadratic, as in Eq. (2). Hashin (1980) had argued that not knowing the nature of this interaction, assuming it to be linear would be too restrictive. Therefore, quadratic interaction should be assumed, also for the reason that it would allow more flexibility in curve-fitting with experimental data. To be sure, Hashin did caution against taking the quadratic form of the criterion to be physically based, pointing out that justification for this existed only for the metal yielding.

(c) Separating the failure of a unidirectional composite into fiber failure mode and matrix failure mode, as suggested by Hashin (1980), is a good first approximation, but the mechanisms underlying the two failure modes are not amenable to such separation. Failure criteria should result from analysis of the failure mechanisms, and the critical condition for each mechanism should define failure.

The observations (a)–(c) above suggest that the current (classical) approach of formulating failure criteria based on a homogenized description of unidirectional composites is simply not capable of accounting for the physical nature of the failure mechanisms. A common feature of all failure mechanisms is that they are governed by "local" conditions, i.e., by the stress states at the microscopic level. For instance, in the tensile fiber failure, the formation of the multiple fiber failures and their critical cluster size, illustrated in Fig. 1, depend on the local (triaxial) stress state, which is responsible for the fiber failure progression (Zhuang et al. 2016). In the compression fiber failure mode, the formation of the kink band, illustrated in Fig. 2, also depends on the local (triaxial) stress state developed due to microbuckling of fibers. In this mechanism, the role of fiber imperfections has been found to be significant in triggering the formation of a kink band (Kyriakides et al. 1995).

For the matrix failure mode, Hashin (1980) suggested failure to occur on a failure plane, which is inclined to the fibers but does not intersect the fibers. Such a plane is shown in Fig. 4. The failure of this plane was proposed by Hashin to take place under the combined action of the traction components acting on the plane. A closer examination of the failure mechanisms, described above, however suggest that this may be inaccurate. The effect of shear on a plane depends on the local (triaxial) stress state resulting in inclined cracks shown in Fig. 5. Thus, although macroscopically the failure seems to occur on a plane, the initiation, progression, and criticality of the failure is governed by microscopic scale conditions.

5 Discussion and Concluding Remarks

Composite material failure is fundamentally a failure process at the scale of the fibers and matrix. Homogenizing the two constituents removes the possibility of analyzing the failure and thereby determining the conditions of criticality. Efforts since the first proposed failure theory (Azzi and Tsai 1965) in 1965 have led to increasing complexity of formulations resulting from assumptions to capture the features of the essentially microscopic scale process at the macroscopic (homogenized) scale. Not only have such efforts not succeeded (Soden et al. 1998; Hinton et al. 2002, 2004; Kaddour et al. 2004), they have introduced a great deal of uncertainty for the designers of composite structures. Each new theory coming out in the literature promises to fit the data well until another theory finds discrepancy and proposes to improve the "predictions" by yet other assumptions.

The way forward must be a multi-scale approach that is guided by the observations of the failure process at the microscopic scale. In such an approach, one must avoid making assumptions like the cohesive zones that are not evidenced by the observations. Instead, the local stress states should be calculated by the finite element type of models and failure criteria for fiber and matrix failure should be applied based on known mechanisms of failure. The techniques for observing failure at small scales are quite sharp today, e.g., based on micro-focus computed X-ray radiography (Scott et al. 2014). Constructing a scheme to analyze failure should be guided by the details revealed by such high resolution observations.

Reliance on testing for validation of the failure theories is part of the classical approach. This principle has guided the development of the failure theories in the past. The new developments in computational simulation provide another way to validate (and develop) failure theories. This is known as "virtual" testing (LLorca et al. 2011). Essentially, the idea is to conduct tests by simulation on the computer. This allows examining failure under combined loading that would be difficult, if not impossible, to do in physical tests. The risk, of course, is to do "virtual" tests that would not correspond to reality. This can happen if either the stress analysis is incorrect or if the failure criteria at the microscopic scale are not physically justified, or both. A good approach would be to validate the virtual testing approach by simpler physical tests before expanding the approach to more general cases. Still, this cannot guarantee the validity of the virtual testing.

References

Aroush, D.R.B., Maire, E., Gauthier, C., Youssef, S., Cloetens, P., Wagner, H.: A study of fracture of unidirectional composites using in situ high-resolution synchrotron X-ray microtomography. Compos. Sci. Technol. **66**(10), 1348–1353 (2006). doi:10.1016/j.compscitech.2005.09.010

Azzi, V.D., Tsai, S.W.: Anisotropic strength of composites. Exp. Mech. **5**(9), 283–288 (1965). doi:10.1007/BF02326292

Berbinau, P., Soutis, C., Guz, I.: Compressive failure of 0° unidirectional carbon-fibre-reinforced plastic (CFRP) laminates by fibre microbuckling. Compos. Sci. Technol. **59**(9), 1451–1455 (1999). doi:10.1016/S0266-3538(98)00181-X

Budiansky, B.: Micromechanics. Comput. Struct. **16**(1–4), 3–12 (1983). doi:10.1016/0045-7949(83)90141-4

Gamstedt, E., Sjögren, B.: Micromechanisms in tension-compression fatigue of composite laminates containing transverse plies. Compos. Sci. Technol. **59**(2), 167–178 (1999). doi:10.1016/S0266-3538(98)00061-X

Goldenblatt, I.I., Kopnov, V.A.: Strength criteria for anisotropic materials. Izvestia Academy Nauk USSR Mechanika **6**, 77–83 (1965)

González, C., LLorca, J.: Mechanical behavior of unidirectional fiber-reinforced polymers under transverse compression: microscopic mechanisms and modeling. Compos. Sci. Technol. **67**(13), 2795–2806 (2007). doi:10.1016/j.compscitech.2007.02.001

Hashin, Z.: Failure criteria for unidirectional fiber composites. J. Appl. Mech. **47**(2), 329–334 (1980). doi:10.1115/1.3153664

Hill, R.: A theory of the yielding and plastic flow of anisotropic metals. Proc. Royal Soc. Lond. A: Math., Phys. Eng. Sci. **193**(1033), 281–297 (1948). doi:10.1098/rspa.1948.0045

Hinton, M., Kaddour, A., Soden, P.: A comparison of the predictive capabilities of current failure theories for composite laminates, judged against experimental evidence. Compos. Sci. Technol. **62**(12–13), 1725–1797 (2002). doi:10.1016/S0266-3538(02)00125-2

Hinton, M., Kaddour, A., Soden, P.: A further assessment of the predictive capabilities of current failure theories for composite laminates: comparison with experimental evidence. Compos. Sci. Technol. **64**(3–4), 549–588 (2004). doi:10.1016/S0266-3538(03)00227-6. Failure criteria in fibre reinforced polymer composites Part C: additional theories conclusions and recommendations

Jelf, P., Fleck, N.A.: Compression failure mechanisms in unidirectional composites. J. Compos. Mater. **26**(18), 2706–2726 (1992). doi:10.1177/002199839202601804

Kaddour, A., Hinton, M., Soden, P.: A comparison of the predictive capabilities of current failure theories for composite laminates: additional contributions. Compos. Sci. Technol. **64**(3–4), 449–476 (2004). doi:10.1016/S0266-3538(03)00226-4. Failure criteria in fibre reinforced polymer composites Part C: additional theories conclusions and recommendations

Kyriakides, S., Arseculeratne, R., Perry, E., Liechti, K.: On the compressive failure of fiber reinforced composites. Int. J. Solids Struct. **32**(6–7), 689–738 (1995). doi:10.1016/0020-7683(94)00157-R. Time dependent problems in mechanics

LLorca, J., González, C., Molina-Aldareguía, J.M., Segurado, J., Seltzer, R., Sket, F., Rodríguez, M., Sádaba, S., Muñoz, R., Canal, L.P.: Multiscale modeling of composite materials: a roadmap towards virtual testing. Adv. Mater. **23**(44), 5130–5147 (2011). doi:10.1002/adma.201101683

Puck, A., Schürmann, H.: Failure analysis of FRP laminates by means of physically based phenomenological models. Compos. Sci. Technol. **58**(7), 1045–1067 (1998). doi:10.1016/S0266-3538(96)00140-6

Puck, A., Schürmann, H.: Failure analysis of FRP laminates by means of physically based phenomenological models. Compos. Sci. Technol. **62**(12–13), 1633–1662 (2002). doi:10.1016/S0266-3538(01)00208-1

Redon, O.: Fatigue damage development and failure in unidirectional and angle-ply glass fibre/carbon fibre hybrid laminates. Master's thesis, Materials research department, Risø National Laboratory, Roskilde, Denmark (2000)

Scott, A., Sinclair, I., Spearing, S., Mavrogordato, M., Hepples, W.: Influence of voids on damage mechanisms in carbon/epoxy composites determined via high resolution computed tomography. Compos. Sci. Technol. **90**, 147–153 (2014). doi:10.1016/j.compscitech.2013.11.004

Soden, P., Hinton, M., Kaddour, A.: A comparison of the predictive capabilities of current failure theories for composite laminates. Compos. Sci. Technol. **58**(7), 1225–1254 (1998). doi:10.1016/S0266-3538(98)00077-3

Talreja, R.: Assessment of the fundamentals of failure theories for composite materials. Compos. Sci. Technol. **105**, 190–201 (2014). doi:10.1016/j.compscitech.2014.10.014

Tsai, S.W., Wu, E.M.: A general theory of strength for anisotropic materials. J. Compos. Mater. **5**(1), 58–80 (1971)

Zhuang, L., Talreja, R., Varna, J.: Tensile failure of a unidirectional composite from a local fracture plane, to be submitted (2016)

Part VI
Rods, Plates and Shells

On the Dislocation Density Tensor in the Cosserat Theory of Elastic Shells

Mircea Bîrsan and Patrizio Neff

Abstract We consider the Cosserat continuum in its finite strain setting and discuss the dislocation density tensor as a possible alternative curvature strain measure in three-dimensional Cosserat models and in Cosserat shell models. We establish a close relationship (one-to-one correspondence) between the new shell dislocation density tensor and the bending-curvature tensor of 6-parameter shells.

1 Indroduction

The Cosserat-type theories have recently seen a tremendous renewed interest for their prospective applicability to model physical effects beyond the classical ones. These comprise notably the so-called size-effects ("smaller is stiffer").

In a finite strain Cosserat-type framework, the group of proper rotations SO(3) has a dominant place. The original idea of the Cosserat brothers (Cosserat and Cosserat 1909) to consider independent rotational degrees of freedom in addition to the macroscopic displacement was heavily motivated by their treatment of plate and shell theory. Indeed, in shell theory it is natural to attach a preferred orthogonal frame (triad) at any point of the surface, one vector of which is the normal to the midsurface, the other two vectors lying in the tangent plane. This is the notion of the "trièdre caché".

M. Bîrsan (✉)
Lehrstuhl für Nichtlineare Analysis und Modellierung, Fakultät für Mathematik,
Universität Duisburg-Essen, Thea-Leymann Str. 9, 45127 Essen, Germany
e-mail: mircea.birsan@uni-due.de

M. Bîrsan
Department of Mathematics, Alexandru Ioan Cuza University of Iaşi,
Blvd. Carol I, no. 11, 700506 Iaşi, Romania

P. Neff
Head of Lehrstuhl für Nichtlineare Analysis und Modellierung,
Fakultät für Mathematik, Universität Duisburg-Essen, Thea-Leymann Str. 9,
45127 Essen, Germany
e-mail: patrizio.neff@uni-due.de

K. Naumenko and M. Aßmus (eds.), *Advanced Methods of Continuum Mechanics for Materials and Structures*, Advanced Structured Materials 60,
DOI 10.1007/978-981-10-0959-4_22

The idea to consider then an orthogonal frame which is not strictly linked to the surface, but constitutively coupled, leads to the notion of the "trièdre mobile". And this then is already giving rise to a prototype Cosserat shell (6-parameter) theory. For an insightful review of various Cosserat-type shell models, we refer to Altenbach et al. (2010).

However, the Cosserat brothers have never proposed any more specific constitutive framework, apart from postulating euclidean invariance (frame-indifference) and hyperelasticity. For specific problems it is necessary to choose a constitutive framework and to determine certain strain and curvature measures. This task is still not conclusively done, see e.g. Pietraszkiewicz and Eremeyev (2009).

Among the existing models for Cosserat-type shells, we mention the theory of simple elastic shells (Altenbach and Zhilin 2004), which has been developed by Zhilin (1976, 2006) and Altenbach and Zhilin (1982, 1988). Later, this theory has been successfully applied to describe the mechanical behaviour of laminated, functionally graded, viscoelastic or porous plates in Altenbach (2000), Altenbach and Eremeyev (2008, 2009, 2010) and of multi-layered, orthotropic, thermoelastic shells in Bîrsan and Altenbach (2010, 2011), Bîrsan et al. (2013), Sadowski et al. (2015). Another remarkable approach is the general 6-parameter theory of elastic shells presented in Libai and Simmonds (1998), Chróścielewski et al. (2004), Eremeyev and Pietraszkiewicz (2004). Although the starting point is different, one can see that the kinematical structure of the nonlinear 6-parameter shell theory is identical to that of a Cosserat shell model, see also Bîrsan and Neff (2014a, b).

In this paper, we would like to draw attention to alternative curvature measures, motivated by dislocation theory, which can also profitably be used in the three-dimensional Cosserat model and the Cosserat shell model. The object of interest is Nye's dislocation density tensor $\operatorname{Curl}\boldsymbol{P}$. Within the restriction to proper rotations it turns out that Nye's tensor provides a complete control of all spatial derivatives of rotations (Neff and Münch 2008) and we rederive this property for micropolar continua using general curvilinear coordinates. Then, we focus on shell-curvature measures and define a new shell dislocation density tensor using the surface Curl operator. Then, we prove that a relation analogous to Nye's formula holds also for Cosserat (6-parameter) shells.

The paper is structured as follows. In Sect. 2 we present the kinematics of a three-dimensional Cosserat continuum, as well as the appropriate strain measures and curvature strain measures, written in curvilinear coordinates. Here, we show the close relationship between the wryness tensor and the dislocation density tensor, including the corresponding Nye's formula. In Sect. 3, we define the Curl operator on surfaces and present several representations using surface curvilinear coordinates. These relations are then used in Sect. 4 to introduce the new shell dislocation density tensor and to investigate its relationship to the elastic shell bending-curvature tensor of 6-parameter shells.

2 Strain Measures of a Three-Dimensional Cosserat Model in Curvilinear Coordinates

Let $\mathcal{B}$ be a Cosserat elastic body which occupies in its reference (initial) configuration the domain $\Omega_\xi \subset \mathbb{R}^3$. A generic point of Ω_ξ will be denoted by (ξ_1, ξ_2, ξ_3). The deformation of the Cosserat body is described by a vectorial map φ_ξ and a microrotation tensor $\boldsymbol{R}_\xi$,

$$\varphi_\xi : \Omega_\xi \to \Omega_c, \quad \boldsymbol{R}_\xi : \Omega_\xi \to \mathrm{SO}(3),$$

where Ω_c is the deformed (current) configuration. Let (x_1, x_2, x_3) be some general curvilinear coordinates system on Ω_ξ. Thus, we have a parametric representation $\boldsymbol{\Theta}$ of the domain Ω_ξ

$$\boldsymbol{\Theta} : \Omega \to \Omega_\xi, \quad \boldsymbol{\Theta}(x_1, x_2, x_3) = (\xi_1, \xi_2, \xi_3),$$

where $\Omega \subset \mathbb{R}^3$ is a bounded domain with Lipschitz boundary $\partial\Omega$. The covariant base vectors with respect to these curvilinear coordinates are denoted by $\boldsymbol{g}_i$ and the contravariant base vectors by $\boldsymbol{g}^j$ $(i, j = 1, 2, 3)$, i.e.

$$\boldsymbol{g}_i = \frac{\partial \boldsymbol{\Theta}}{\partial x_i} = \boldsymbol{\Theta}_{,i}, \quad \boldsymbol{g}^j \cdot \boldsymbol{g}_i = \delta_i^j,$$

where δ_i^j is the Kronecker symbol. We employ the usual conventions for indices: the Latin indices $i, j, k, \ldots$ range over the set $\{1, 2, 3\}$, while the Greek indices $\alpha, \beta, \gamma, \ldots$ are confined to the range $\{1, 2\}$; the comma preceding an index i denotes partial derivatives with respect to x_i; the Einstein summation convention over repeated indices is also used.

Introducing the deformation function φ by the composition

$$\varphi := \varphi_\xi \circ \boldsymbol{\Theta} : \Omega \to \Omega_c, \quad \varphi(x_1, x_2, x_3) := \varphi_\xi\big(\boldsymbol{\Theta}(x_1, x_2, x_3)\big),$$

we can express the (elastic) deformation gradient $\boldsymbol{F}$ as follows:

$$\boldsymbol{F} := \nabla_\xi \, \varphi_\xi(\xi_1, \xi_2, \xi_3) = \nabla_x \, \varphi(x_1, x_2, x_3) \cdot \big[\nabla_x \boldsymbol{\Theta}(x_1, x_2, x_3)\big]^{-1}.$$

Using the direct tensor notation, we can write

$$\nabla_x \varphi = \varphi_{,i} \otimes \boldsymbol{e}_i, \quad \nabla_x \boldsymbol{\Theta} = \boldsymbol{g}_i \otimes \boldsymbol{e}_i, \quad \big[\nabla_x \boldsymbol{\Theta}\big]^{-1} = \boldsymbol{e}_j \otimes \boldsymbol{g}^j,$$

where $\boldsymbol{e}_i$ are the unit vectors along the coordinate axes Ox_i in the parameter domain Ω. Then, the deformation gradient can be expressed by

$$\boldsymbol{F} = \varphi_{,i} \otimes \boldsymbol{g}^i.$$

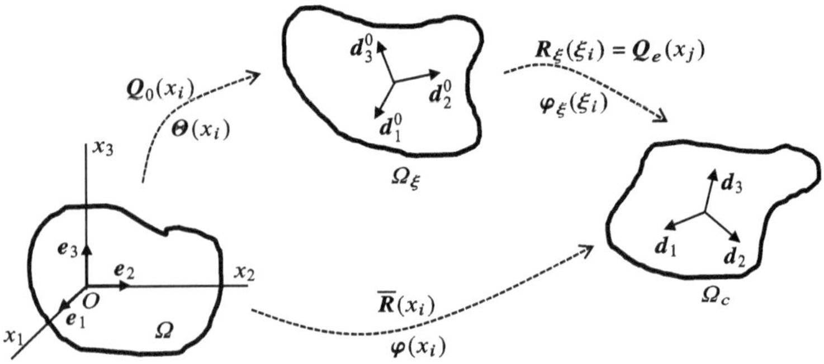

Fig. 1 The reference (initial) configuration Ω_ξ of the Cosserat continuum, the deformed (current) configuration Ω_c and the parameter domain Ω of the curvilinear coordinates (x_1, x_2, x_3). The triads of directors $\{\boldsymbol{d}_i\}$ and $\{\boldsymbol{d}_i^0\}$ satisfy the relations $\boldsymbol{d}_i = \boldsymbol{Q}_e\boldsymbol{d}_i^0 = \overline{\boldsymbol{R}}\boldsymbol{e}_i$ and $\boldsymbol{d}_i^0 = \boldsymbol{Q}_0\boldsymbol{e}_i$, where $\boldsymbol{Q}_e$ is the elastic microrotation field, $\boldsymbol{Q}_0$ the initial microrotation, and $\overline{\boldsymbol{R}}$ the total microrotation field

The orientation and rotation of points in Cosserat (micropolar) media can also be described by means of triads of orthonormal vectors (called *directors*) attached to every point. We denote by $\{\boldsymbol{d}_i^0\}$ the triad of directors ($i = 1, 2, 3$) in the reference configuration Ω_ξ and by $\{\boldsymbol{d}_i\}$ the directors in the deformed configuration Ω_c, see Fig. 1. We introduce the *elastic microrotation* $\boldsymbol{Q}_e$ as the composition

$$\boldsymbol{Q}_e := \boldsymbol{R}_\xi \circ \boldsymbol{\Theta} : \Omega \to \mathrm{SO}(3), \quad \boldsymbol{Q}_e(x_1, x_2, x_3) := \boldsymbol{R}_\xi\big(\boldsymbol{\Theta}(x_1, x_2, x_3)\big),$$

which can be characterized with the help of the directors by the relations

$$\boldsymbol{Q}_e\boldsymbol{d}_i^0 = \boldsymbol{d}_i\,, \quad \text{i.e.,} \quad \boldsymbol{Q}_e = \boldsymbol{d}_i \otimes \boldsymbol{d}_i^0.$$

Let $\boldsymbol{Q}_0$ be the *initial microrotation* (describing the position of the directors in the reference configuration Ω_ξ)

$$\boldsymbol{Q}_0\boldsymbol{e}_i = \boldsymbol{d}_i^0\,, \quad \text{i.e.,} \quad \boldsymbol{Q}_0 = \boldsymbol{d}_i^0 \otimes \boldsymbol{e}_i.$$

Then, the *total microrotation* $\overline{\boldsymbol{R}}$ is given by

$$\overline{\boldsymbol{R}} : \Omega \to \mathrm{SO}(3), \quad \overline{\boldsymbol{R}}(x_i) := \boldsymbol{Q}_e(x_i)\,\boldsymbol{Q}_0(x_i) = \boldsymbol{d}_j(x_i) \otimes \boldsymbol{e}_j.$$

The non-symmetric Biot-type stretch tensor (the elastic *first Cosserat deformation tensor*, see Cosserat and Cosserat (1909), p. 123, Eq. (43)) is now

$$\overline{\boldsymbol{U}}_e := \boldsymbol{Q}_e^T \boldsymbol{F} = \big(\boldsymbol{d}_i^0 \otimes \boldsymbol{d}_i\big)\,\big(\boldsymbol{\varphi}_{,j} \otimes \boldsymbol{g}^j\big) = \big(\boldsymbol{\varphi}_{,j} \cdot \boldsymbol{d}_i\big)\,\boldsymbol{d}_i^0 \otimes \boldsymbol{g}^j.$$

and the non-symmetric *strain tensor* for nonlinear micropolar materials is defined by

$$\overline{\boldsymbol{E}}_e := \overline{\boldsymbol{U}}_e - \mathbf{1}_3 = \left(\boldsymbol{\varphi}_{,j} \cdot \boldsymbol{d}_i - \boldsymbol{g}_j \cdot \boldsymbol{d}_i^0\right) \boldsymbol{d}_i^0 \otimes \boldsymbol{g}^j,$$

where $\mathbf{1}_3 = \boldsymbol{g}_i \otimes \boldsymbol{g}^i = \boldsymbol{d}_i^0 \otimes \boldsymbol{d}_i^0$ is the unit three-dimensional tensor. As a strain measure for curvature (orientation change) one can employ the so-called *wryness tensor* $\boldsymbol{\Gamma}$ given by:

$$\boldsymbol{\Gamma} := \operatorname{axl}\left(\boldsymbol{Q}_e^T \boldsymbol{Q}_{e,i}\right) \otimes \boldsymbol{g}^i = \boldsymbol{Q}_0\left[\operatorname{axl}\left(\overline{\boldsymbol{R}}^T \overline{\boldsymbol{R}}_{,i}\right) - \operatorname{axl}\left(\boldsymbol{Q}_0^T \boldsymbol{Q}_{0,i}\right)\right] \otimes \boldsymbol{g}^i, \tag{1}$$

where $\operatorname{axl}(\boldsymbol{A})$ denotes the axial vector of any skew-symmetric tensor $\boldsymbol{A}$. For a detailed discussion on various strain measures of nonlinear micropolar continua we refer to the paper Pietraszkiewicz and Eremeyev (2009).

As an alternative to the wryness tensor $\boldsymbol{\Gamma}$ one can make use of the Curl operator to define the so-called *dislocation density tensor* $\overline{\boldsymbol{D}}_e$ by (Neff and Münch 2008)

$$\overline{\boldsymbol{D}}_e := \boldsymbol{Q}_e^T \operatorname{Curl} \boldsymbol{Q}_e\,, \tag{2}$$

which is another curvature measure for micropolar continua. Note that the Curl operator has various definitions in the literature, but we will make its significance clear in the next Sect. 2.1, where we present the Curl operator in curvilinear coordinates. The use of the dislocation density tensor $\overline{\boldsymbol{D}}_e$ instead of the wryness tensor in conjuction with micropolar and micromorphic media has several advantages, as it was shown in Ghiba et al. (2015), Neff et al. (2014), Madeo et al. (2015). The relationship between the wryness tensor $\boldsymbol{\Gamma}$ and the dislocation density tensor $\overline{\boldsymbol{D}}_e$ is discussed in Sect. 2.2 in details.

Using the strain and curvature tensors $(\overline{\boldsymbol{E}}_e\,, \overline{\boldsymbol{D}}_e)$ the elastically stored energy density W for the isotropic nonlinear Cosserat model can be expressed as (Neff et al. 2015; Lankeit et al. 2016)

$$\begin{aligned} W(\overline{\boldsymbol{E}}_e, \overline{\boldsymbol{D}}_e) &= W_{\mathrm{mp}}(\overline{\boldsymbol{E}}_e) + W_{\mathrm{curv}}(\overline{\boldsymbol{D}}_e), \qquad \text{where} \\ W_{\mathrm{mp}}(\overline{\boldsymbol{E}}_e) &= \mu \,\|\operatorname{dev}_3 \operatorname{sym} \overline{\boldsymbol{E}}_e\,\|^2 + \mu_c \,\|\operatorname{skew} \overline{\boldsymbol{E}}_e\,\|^2 + \frac{\kappa}{2}\left(\operatorname{tr} \overline{\boldsymbol{E}}_e\right)^2, \\ W_{\mathrm{curv}}(\overline{\boldsymbol{D}}_e) &= \mu\, L_c^p \left(a_1 \|\operatorname{dev}_3 \operatorname{sym} \overline{\boldsymbol{D}}_e\|^2 + a_2 \|\operatorname{skew} \overline{\boldsymbol{D}}_e\|^2 + a_3\left(\operatorname{tr} \overline{\boldsymbol{D}}_e\right)^2\right)^{p/2}, \end{aligned} \tag{3}$$

where μ is the shear modulus, κ is the bulk modulus of classical isotropic elasticity, and μ_c is called the *Cosserat couple modulus*, which are assumed to satisfy

$$\mu > 0, \qquad \kappa > 0, \qquad \text{and} \qquad \mu_c > 0\,.$$

The parameter L_c introduces an internal length which is characteristic for the material, $a_i > 0$ are dimensionless constitutive coefficients and $p \geq 2$ is a constant exponent. Here, $\operatorname{dev}_3 \boldsymbol{X} := \boldsymbol{X} - \frac{1}{3}\,(\operatorname{tr}\boldsymbol{X})\,\mathbf{1}_3$ is the deviatoric part of any second order tensor $\boldsymbol{X}$.

Under these assumptions on the constitutive coefficients, the existence of minimizers to the corresponding minimization problem of the total energy functional has been shown, e.g. in Neff et al. (2015), Lankeit et al. (2016).

2.1 The Curl Operator

For a vector field $\boldsymbol{v}$, the (coordinate-free) definition of the vector curl $\boldsymbol{v}$ is

$$(\operatorname{curl}\boldsymbol{v})\cdot\boldsymbol{c} = \operatorname{div}(\boldsymbol{v}\times\boldsymbol{c}) \qquad \text{for all constant vectors } \boldsymbol{c}, \tag{4}$$

where $\cdot$ denotes the scalar product and $\times$ the vector product. The Curl of a tensor field $\boldsymbol{T}$ is the tensor field defined by

$$\big(\operatorname{Curl}\boldsymbol{T}\big)^T\boldsymbol{c} = \operatorname{curl}\big(\boldsymbol{T}^T\boldsymbol{c}\big) \qquad \text{for all constant vectors } \boldsymbol{c}. \tag{5}$$

Remark 22.1 The operator $\operatorname{Curl}\boldsymbol{T}$ given by (5) coincides with the Curl operator defined in Svendsen (2002), Mielke and Müller (2006). However, for other authors the Curl of $\boldsymbol{T}$ is the transpose of $\operatorname{Curl}\boldsymbol{T}$ defined by (5), see e.g. Gurtin (1981), Eremeyev et al. (2013).

Then, from (4) and (5) we obtain the following formulas

$$\operatorname{curl}\boldsymbol{v} = -\boldsymbol{v}_{,i}\times\boldsymbol{g}^i, \qquad \operatorname{Curl}\boldsymbol{T} = -\boldsymbol{T}_{,i}\times\boldsymbol{g}^i. \tag{6}$$

Indeed, the Definition (4) yields

$$(\operatorname{curl}\boldsymbol{v})\cdot\boldsymbol{c} = \operatorname{div}(\boldsymbol{v}\times\boldsymbol{c}) = (\boldsymbol{v}\times\boldsymbol{c})_{,i}\cdot\boldsymbol{g}^i = (\boldsymbol{v}_{,i}\times\boldsymbol{c})\cdot\boldsymbol{g}^i = (\boldsymbol{g}^i\times\boldsymbol{v}_{,i})\cdot\boldsymbol{c},$$

and the Eq. $(6)_1$ holds. Further, from (5) we get

$$\big(\operatorname{Curl}\boldsymbol{T}\big)^T\boldsymbol{c} = \operatorname{curl}\big(\boldsymbol{T}^T\boldsymbol{c}\big) = \boldsymbol{g}^i\times\big(\boldsymbol{T}^T\boldsymbol{c}\big)_{,i} = \boldsymbol{g}^i\times\big(\boldsymbol{T}^T_{,i}\boldsymbol{c}\big) = \big(\boldsymbol{g}^i\times\boldsymbol{T}^T_{,i}\big)\boldsymbol{c},$$

so it follows $\operatorname{Curl}\boldsymbol{T} = \big(\boldsymbol{g}^i\times\boldsymbol{T}^T_{,i}\big)^T = -\boldsymbol{T}_{,i}\times\boldsymbol{g}^i$ and the relations (6) are proved.

In order to write the components of curl $\boldsymbol{v}$ and Curl $\boldsymbol{T}$ in curvilinear coordinates, we introduce the following notations

$$g_{ij} = \boldsymbol{g}_i\cdot\boldsymbol{g}_j\,, \quad g = \det\big(g_{ij}\big)_{3\times 3} > 0.$$

The alternating (Ricci) third-order tensor is

$$\boldsymbol{\epsilon} = -\mathbf{1}_3 \times \mathbf{1}_3 = \epsilon_{ijk}\, \boldsymbol{g}^i \otimes \boldsymbol{g}^j \otimes \boldsymbol{g}^k = \epsilon^{ijk} \boldsymbol{g}_i \otimes \boldsymbol{g}_j \otimes \boldsymbol{g}_k\,, \quad \text{where}$$

$$\epsilon_{ijk} = \sqrt{g}\, e_{ijk}\,, \quad \epsilon^{ijk} = \frac{1}{\sqrt{g}}\, e_{ijk}\,, \quad e_{ijk} = \begin{cases} 1, & (i,j,k) \text{ is even permutation} \\ -1, & (i,j,k) \text{ is odd permutation} \\ 0, & (i,j,k) \text{ is no permutation} \end{cases}.$$

The covariant, contravariant, and mixed components of any vector field $\boldsymbol{v}$ and any tensor field $\boldsymbol{T}$ are introduced by

$$\boldsymbol{v} = v_k\, \boldsymbol{g}^k = v^k \boldsymbol{g}_k\,, \quad \boldsymbol{T} = T_{jk}\, \boldsymbol{g}^j \otimes \boldsymbol{g}^k = T^{jk} \boldsymbol{g}_j \otimes \boldsymbol{g}_k = T^{j}_{\cdot k}\, \boldsymbol{g}_j \otimes \boldsymbol{g}^k.$$

For the partial derivatives with respect to x_i we have the well-known expressions

$$\boldsymbol{v}_{,i} = v_{k|i}\, \boldsymbol{g}^k\,, \qquad \boldsymbol{T}_{,i} = T_{jk|i}\, \boldsymbol{g}^j \otimes \boldsymbol{g}^k = T^{j}_{\cdot k|i}\, \boldsymbol{g}_j \otimes \boldsymbol{g}^k, \tag{7}$$

where a subscript bar preceding the index i denotes covariant derivative w.r.t. x_i.

Using the relations (7) in (6), we can write the components of curl $\boldsymbol{v}$ and Curl $\boldsymbol{T}$ as follows

$$\operatorname{curl} \boldsymbol{v} = \epsilon^{ijk} v_{j|i}\, \boldsymbol{g}_k\,, \qquad \operatorname{Curl} \boldsymbol{T} = \epsilon^{ijk} T_{sj|i}\, \boldsymbol{g}^s \otimes \boldsymbol{g}_k = \epsilon^{ijk} T^{s}_{\cdot j|i}\, \boldsymbol{g}_s \otimes \boldsymbol{g}_k. \tag{8}$$

Indeed, from $(6)_1$ and $(7)_1$ we find

$$\operatorname{curl} \boldsymbol{v} = -\big(v_{k|i}\, \boldsymbol{g}^k\big) \times \boldsymbol{g}^i = -v_{k|i}\big(\boldsymbol{g}^k \times \boldsymbol{g}^i\big) = -v_{k|i}\big(\epsilon^{kij} \boldsymbol{g}_j\big) = \epsilon^{ijk} v_{j|i}\, \boldsymbol{g}_k.$$

Analogously, from $(6)_2$ and $(7)_2$ we get

$$\operatorname{Curl} \boldsymbol{T} = -\big(T_{sk|i}\, \boldsymbol{g}^s \otimes \boldsymbol{g}^k\big) \times \boldsymbol{g}^i = -T_{sk|i}\, \boldsymbol{g}^s \otimes \big(\boldsymbol{g}^k \times \boldsymbol{g}^i\big) = \epsilon^{ijk} T_{sj|i}\, \boldsymbol{g}^s \otimes \boldsymbol{g}_k.$$

Thus, Eq. (8) is proved.

Remark 22.2 In the special case of Cartesian coordinates, the relations (6) and (8) admit the simple form

$$\operatorname{curl} \boldsymbol{v} = -\boldsymbol{v}_{,i} \times \boldsymbol{e}_i = e_{ijk} v_{j,i}\, \boldsymbol{e}_k, \quad \operatorname{Curl} \boldsymbol{T} = -\boldsymbol{T}_{,i} \times \boldsymbol{e}_i = e_{ijk}\, T_{sj,i}\, \boldsymbol{e}_s \otimes \boldsymbol{e}_k,$$

where $\boldsymbol{v} = v_i \boldsymbol{e}_i$ and $\boldsymbol{T} = T_{ij} \boldsymbol{e}_i \otimes \boldsymbol{e}_j$ are the corresponding coordinates. Moreover, in this case one can write

$$\operatorname{Curl} \boldsymbol{T} = \boldsymbol{e}_i \otimes \operatorname{curl}\big(\boldsymbol{T}_i\big) \qquad \text{for} \qquad \boldsymbol{T} = \boldsymbol{e}_i \otimes \boldsymbol{T}_i, \tag{9}$$

where $\boldsymbol{T}_i = T_{ij}\, \boldsymbol{e}_j$ are the three rows of the 3×3 matrix $\big(T_{ij}\big)_{3\times 3}$. The relation (9) shows that Curl is defined row-wise (Neff and Münch 2008): the rows of the 3×3 matrix Curl $\boldsymbol{T}$ are, respectively, the three vectors $\operatorname{curl}\big(\boldsymbol{T}_i\big)$, $i = 1, 2, 3$.

Remark 22.3 In order to write the corresponding formula in curvilinear coordinates which is analogous to (9), we introduce the vectors $\boldsymbol{T}_i := T_{ij}\,\boldsymbol{g}^j$ and $\boldsymbol{T}^i := T^{ij}\,\boldsymbol{g}_j = T^i_{\cdot j}\,\boldsymbol{g}^j$ such that it holds

$$\boldsymbol{T} = \boldsymbol{g}^i \otimes \boldsymbol{T}_i \qquad \text{and} \qquad \boldsymbol{T} = \boldsymbol{g}_i \otimes \boldsymbol{T}^i. \tag{10}$$

If we differentiate $(10)_1$ with respect to x_j we get

$$\boldsymbol{T}_{,j} = \boldsymbol{g}^r_{\cdot,j} \otimes \boldsymbol{T}_r + \boldsymbol{g}^i \otimes \boldsymbol{T}_{i,j} = -\Gamma^r_{ji}\boldsymbol{g}^i \otimes \boldsymbol{T}_r + \boldsymbol{g}^i \otimes \boldsymbol{T}_{i,j} = \boldsymbol{g}^i \otimes \big(\boldsymbol{T}_{i,j} - \Gamma^r_{ji}\,\boldsymbol{T}_r\big),$$

where Γ^r_{ij} are the Christoffel symbols of the second kind. Hence, it follows

$$\boldsymbol{T}_{,j} = \boldsymbol{g}^i \otimes \boldsymbol{T}_{i|j} \qquad \text{with} \qquad \boldsymbol{T}_{i|j} := \boldsymbol{T}_{i,j} - \Gamma^r_{ji}\,\boldsymbol{T}_r = T_{ik|j}\,\boldsymbol{g}^k. \tag{11}$$

Taking the vector product of $(11)_1$ with $\boldsymbol{g}^j$ we obtain

$$\operatorname{Curl}\boldsymbol{T} = -\boldsymbol{T}_{,j} \times \boldsymbol{g}^j = -\big(\boldsymbol{g}^i \otimes \boldsymbol{T}_{i|j}\big) \times \boldsymbol{g}^j, \quad \text{i.e.}$$

$$\operatorname{Curl}\boldsymbol{T} = \boldsymbol{g}^i \otimes \operatorname{curl}_{\text{cov}}\big(\boldsymbol{T}_i\big) \quad \text{where} \quad \operatorname{curl}_{\text{cov}}\big(\boldsymbol{T}_i\big) := -\boldsymbol{T}_{i|j} \times \boldsymbol{g}^j. \tag{12}$$

The relation (12) is the analogue of (9) for curvilinear coordinates. Similarly, by differentiating $(10)_2$ with respect to x_j one can obtain the relation

$$\operatorname{Curl}\boldsymbol{T} = \boldsymbol{g}_i \otimes \operatorname{curl}_{\text{cov}}\big(\boldsymbol{T}^i\big) \quad \text{where we denote} \tag{13}$$

$$\operatorname{curl}_{\text{cov}}\big(\boldsymbol{T}^i\big) := -\boldsymbol{T}^i_{\cdot|j} \times \boldsymbol{g}^j \quad \text{and} \quad \boldsymbol{T}^i_{\cdot|j} := \boldsymbol{T}^i_{\cdot,j} + \Gamma^i_{rj}\,\boldsymbol{T}^r = T^i_{\cdot k|j}\,\boldsymbol{g}^k.$$

2.2 Relation Between the Wryness Tensor and the Dislocation Density Tensor

Let $\boldsymbol{A} = A_{ij}\boldsymbol{g}^i \otimes \boldsymbol{g}^j$ be an arbitrary skew-symmetric tensor and $\operatorname{axl}(\boldsymbol{A}) = a_k\boldsymbol{g}^k$ its axial vector. Then, the following relations hold

$$\begin{aligned}
\boldsymbol{A} \quad &= \operatorname{axl}(\boldsymbol{A}) \times \boldsymbol{1}_3 = \boldsymbol{1}_3 \times \operatorname{axl}(\boldsymbol{A}),\\
\operatorname{axl}(\boldsymbol{A}) &= -\tfrac{1}{2}\,\boldsymbol{\epsilon} : \boldsymbol{A} = -\tfrac{1}{2}\,\epsilon^{ijk}A_{ij}\,\boldsymbol{g}_k\,,\\
\boldsymbol{A} \quad &= -\boldsymbol{\epsilon}\,\operatorname{axl}(\boldsymbol{A}) = -\epsilon^{ijk}a_k\,\boldsymbol{g}_i \otimes \boldsymbol{g}_j\,,
\end{aligned} \tag{14}$$

where the double dot product ":" of two tensors $\boldsymbol{B} = B^{ijk}\,\boldsymbol{g}_i \otimes \boldsymbol{g}_j \otimes \boldsymbol{g}_k$ and $\boldsymbol{T} = T_{ij}\,\boldsymbol{g}^i \otimes \boldsymbol{g}^j$ is defined as $\boldsymbol{B} : \boldsymbol{T} = B^{ijk}T_{jk}\,\boldsymbol{g}_i$.

Using these relations, we can derive the close relationship between the wryness tensor and the dislocation density tensor: it holds

$$\overline{\boldsymbol{D}}_e = -\boldsymbol{\Gamma}^T + (\operatorname{tr}\boldsymbol{\Gamma})\,\mathbf{1}_3, \quad \text{or equivalently,} \tag{15}$$

$$\boldsymbol{\Gamma} = -\overline{\boldsymbol{D}}_e^T + \frac{1}{2}\,(\operatorname{tr}\overline{\boldsymbol{D}}_e)\,\mathbf{1}_3. \tag{16}$$

Indeed, in view of the Eq. $(14)_3$ and the Definition (1) we have

$$\begin{aligned}\boldsymbol{Q}_e^T\boldsymbol{Q}_{e,k}\otimes\boldsymbol{g}^k &= -\boldsymbol{\epsilon}\,\operatorname{axl}(\boldsymbol{Q}_e^T\boldsymbol{Q}_{e,k})\otimes\boldsymbol{g}^k = -\boldsymbol{\epsilon}\,\boldsymbol{\Gamma}\\ &= -\big(\epsilon_{ijr}\,\boldsymbol{g}^i\otimes\boldsymbol{g}^j\otimes\boldsymbol{g}^r\big)\big(\Gamma^s_{.k}\boldsymbol{g}_s\otimes\boldsymbol{g}^k\big) = -\epsilon_{ijs}\,\Gamma^s_{.k}\boldsymbol{g}^i\otimes\boldsymbol{g}^j\otimes\boldsymbol{g}^k.\end{aligned}$$

Hence, we deduce

$$\boldsymbol{Q}_e^T\boldsymbol{Q}_{e,k} = -\epsilon_{ijs}\,\Gamma^s_{.k}\boldsymbol{g}^i\otimes\boldsymbol{g}^j. \tag{17}$$

In view of $(6)_2$, the Definition (2) can be written in the form

$$\overline{\boldsymbol{D}}_e = \boldsymbol{Q}_e^T\big(-\boldsymbol{Q}_{e,k}\times\boldsymbol{g}^k\big) = -(\boldsymbol{Q}_e^T\boldsymbol{Q}_{e,k})\times\boldsymbol{g}^k. \tag{18}$$

Inserting (17) in (18), we obtain

$$\begin{aligned}\overline{\boldsymbol{D}}_e &= \epsilon_{ijs}\,\Gamma^s_{.k}\,(\boldsymbol{g}^i\otimes\boldsymbol{g}^j)\times\boldsymbol{g}^k = \epsilon_{ijs}\,\Gamma^s_{.k}\boldsymbol{g}^i\otimes\big(\epsilon^{jkr}\boldsymbol{g}_r\big) = \big(\epsilon_{jsi}\,\epsilon^{jkr}\big)\Gamma^s_{.k}\boldsymbol{g}^i\otimes\boldsymbol{g}_r\\ &= \big(\delta^k_s\,\delta^r_i - \delta^r_s\,\delta^k_i\big)\Gamma^s_{.k}\boldsymbol{g}^i\otimes\boldsymbol{g}_r = \Gamma^s_{.s}\boldsymbol{g}^i\otimes\boldsymbol{g}_i - \Gamma^s_{.i}\boldsymbol{g}^i\otimes\boldsymbol{g}_s = (\operatorname{tr}\boldsymbol{\Gamma})\,\mathbf{1}_3 - \boldsymbol{\Gamma}^T.\end{aligned}$$

Thus, the relation (15) is proved. If we apply the trace operator and the transpose in (15) we obtain also the relation (16). For infinitesimal strains this formula is well-known under the name *Nye's formula*, and $(-\boldsymbol{\Gamma})$ is also called *Nye's curvature tensor* (Nye 1953). This relation has been first established in Neff and Münch (2008).

Let us find the components of the wryness tensor and the dislocation density tensor in curvilinear coordinates. To this aim, we write first the skew-symmetric tensor

$$\begin{aligned}\boldsymbol{Q}_e^T\boldsymbol{Q}_{e,i} &= \big(\boldsymbol{d}_j^0\otimes\boldsymbol{d}_j\big)\big(\boldsymbol{d}_{k,i}\otimes\boldsymbol{d}_k^0+\boldsymbol{d}_k\otimes\boldsymbol{d}^0_{k,i}\big) = \big(\boldsymbol{d}_j\cdot\boldsymbol{d}_{k,i}\big)\,\boldsymbol{d}_j^0\otimes\boldsymbol{d}_k^0+\boldsymbol{d}_j^0\otimes\boldsymbol{d}^0_{j,i}\\ &= \big(\boldsymbol{d}_j\cdot\boldsymbol{d}_{k,i}-\boldsymbol{d}_j^0\cdot\boldsymbol{d}^0_{k,i}\big)\,\boldsymbol{d}_j^0\otimes\boldsymbol{d}_k^0.\end{aligned} \tag{19}$$

Then, we obtain for the axial vector the equation

$$\operatorname{axl}\big(\boldsymbol{Q}_e^T\boldsymbol{Q}_{e,i}\big) = -\frac{1}{2}\,e_{jks}\,\big(\boldsymbol{d}_j\cdot\boldsymbol{d}_{k,i}-\boldsymbol{d}_j^0\cdot\boldsymbol{d}^0_{k,i}\big)\,\boldsymbol{d}_s^0\,. \tag{20}$$

Indeed, according to $(14)_2$ and (19) we can write

$$\begin{aligned}\operatorname{axl}(\boldsymbol{Q}_e^T\boldsymbol{Q}_{e,i}) &= -\frac{1}{2}\,\epsilon : (\boldsymbol{Q}_e^T\boldsymbol{Q}_{e,i})\\ &= -\frac{1}{2}\left(e_{sjk}\,\boldsymbol{d}_s^0\otimes\boldsymbol{d}_j^0\otimes\boldsymbol{d}_k^0\right) : \left[\left(\boldsymbol{d}_l\cdot\boldsymbol{d}_{r,i}-\boldsymbol{d}_l^0\cdot\boldsymbol{d}_{r,i}^0\right)\boldsymbol{d}_l^0\otimes\boldsymbol{d}_r^0\right]\\ &= -\frac{1}{2}\,e_{jks}\left(\boldsymbol{d}_j\cdot\boldsymbol{d}_{k,i}-\boldsymbol{d}_j^0\cdot\boldsymbol{d}_{k,i}^0\right)\boldsymbol{d}_s^0\end{aligned}$$

and the relation (20) is proved. Using (20) in the Definition (1) we find the following formula for the wryness tensor

$$\boldsymbol{\Gamma} = \frac{1}{2}\,e_{jks}\left(\boldsymbol{d}_{j,i}\cdot\boldsymbol{d}_k-\boldsymbol{d}_{j,i}^0\cdot\boldsymbol{d}_k^0\right)\boldsymbol{d}_s^0\otimes\boldsymbol{g}^i. \tag{21}$$

To obtain an expression for the components of $\overline{\boldsymbol{D}}_e$ we insert (19) in (18) and we get

$$\begin{aligned}\overline{\boldsymbol{D}}_e &= -\left(\boldsymbol{d}_j\cdot\boldsymbol{d}_{k,i}-\boldsymbol{d}_j^0\cdot\boldsymbol{d}_{k,i}^0\right)\left(\boldsymbol{d}_j^0\otimes\boldsymbol{d}_k^0\right)\times\boldsymbol{g}^i\\ &= \left(\boldsymbol{d}_{j,i}\cdot\boldsymbol{d}_k-\boldsymbol{d}_{j,i}^0\cdot\boldsymbol{d}_k^0\right)\boldsymbol{d}_j^0\otimes\left(\boldsymbol{d}_k^0\times\boldsymbol{g}^i\right).\end{aligned} \tag{22}$$

We rewrite the last vector product as

$$\boldsymbol{d}_k^0\times\boldsymbol{g}^i=\boldsymbol{d}_k^0\times\left[\left(\boldsymbol{g}^i\cdot\boldsymbol{d}_r^0\right)\boldsymbol{d}_r^0\right]=\left(\boldsymbol{g}^i\cdot\boldsymbol{d}_r^0\right)\boldsymbol{d}_k^0\times\boldsymbol{d}_r^0=e_{krs}\left(\boldsymbol{g}^i\cdot\boldsymbol{d}_r^0\right)\boldsymbol{d}_s^0$$

and we insert it in (22) to find the following expression for the dislocation density tensor

$$\overline{\boldsymbol{D}}_e = e_{krs}\left(\boldsymbol{d}_{j,i}\cdot\boldsymbol{d}_k-\boldsymbol{d}_{j,i}^0\cdot\boldsymbol{d}_k^0\right)\left(\boldsymbol{g}^i\cdot\boldsymbol{d}_r^0\right)\boldsymbol{d}_j^0\otimes\boldsymbol{d}_s^0. \tag{23}$$

Remark 22.4 In the special case of Cartesian coordinates one can identify $\boldsymbol{d}_i^0 = \boldsymbol{e}_i$, $\boldsymbol{g}^i=\boldsymbol{g}_i=\boldsymbol{e}_i$, and the relations (21) and (22) simplify to the forms

$$\begin{aligned}\boldsymbol{\Gamma} &= \frac{1}{2}\,e_{iks}\left(\boldsymbol{d}_{k,j}\cdot\boldsymbol{d}_s\right)\boldsymbol{e}_i\otimes\boldsymbol{e}_j,\\ \overline{\boldsymbol{D}}_e &= e_{ijk}\left(\boldsymbol{d}_{j,i}\cdot\boldsymbol{d}_s\right)\boldsymbol{e}_s\otimes\boldsymbol{e}_k.\end{aligned}$$

Remark 22.5 One can find various definitions of the wryness tensor in the literature, see e.g. Tambača and Velčić (2010), where $\boldsymbol{\Gamma}$ is called the *curvature strain tensor*. Thus, one can alternatively define the wryness tensor by

$$\boldsymbol{\Gamma} = \boldsymbol{Q}_e^T\,\boldsymbol{\omega}, \tag{24}$$

where $\boldsymbol{\omega}$ is the second order tensor given by

$$\boldsymbol{\omega} = \boldsymbol{\omega}_i\otimes\boldsymbol{g}^i \qquad \text{with} \qquad \boldsymbol{Q}_{e,i} = \boldsymbol{\omega}_i\times\boldsymbol{Q}_e. \tag{25}$$

If we compare the Definition (1) with (24), (25), we see that indeed $\boldsymbol{Q}_e^T\,\boldsymbol{\omega}_i = \mathrm{axl}\big(\boldsymbol{Q}_e^T\boldsymbol{Q}_{e,i}\big)$, i.e.

$$\boldsymbol{\omega}_i \;=\; \boldsymbol{Q}_e\,\mathrm{axl}\big(\boldsymbol{Q}_e^T\boldsymbol{Q}_{e,i}\big) = \mathrm{axl}\big(\boldsymbol{Q}_{e,i}\,\boldsymbol{Q}_e^T\big). \tag{26}$$

By a straightforward but lengthy calculation, one can prove that the vectors $\boldsymbol{\omega}_i$ are expressed in terms of the directors by

$$\boldsymbol{\omega}_i \;=\; \frac{1}{2}\big[\boldsymbol{d}_j\times\boldsymbol{d}_{j,i} - \boldsymbol{Q}_e\big(\boldsymbol{d}_j^0\times\boldsymbol{d}_{j,i}^0\big)\big]. \tag{27}$$

Inserting (27) in $(25)_1$ and (24), we obtain the expression of the wryness tensor written with the help of the directors $\boldsymbol{d}_i$

$$\boldsymbol{\Gamma} \;=\; \frac{1}{2}\big[\boldsymbol{Q}_e^T\,(\boldsymbol{d}_j\times\boldsymbol{d}_{j,i}) - \boldsymbol{d}_j^0\times\boldsymbol{d}_{j,i}^0\big]\otimes\boldsymbol{g}^i. \tag{28}$$

3 The Curl Operator on Surfaces

Let S be a smooth surface embedded in the Euclidean space $\mathbb{R}^3$ and let $\boldsymbol{y}_0(x_1,x_2)$, $\boldsymbol{y}_0:\omega\rightarrow\mathbb{R}^3$, be a parametrization of this surface. We denote the covariant base vectors in the tangent plane by $\boldsymbol{a}_1,\boldsymbol{a}_2$ and the contravariant base vectors by $\boldsymbol{a}^1,\boldsymbol{a}^2$:

$$\boldsymbol{a}_\alpha = \frac{\partial\boldsymbol{y}_0}{\partial x_\alpha} = \boldsymbol{y}_{0,\alpha},\quad \boldsymbol{a}_\alpha\cdot\boldsymbol{a}^\beta = \delta_\alpha^\beta$$

and let

$$\boldsymbol{a}_3 = \boldsymbol{a}^3 = \boldsymbol{n}_0 = \frac{\boldsymbol{a}_1\times\boldsymbol{a}_2}{|\boldsymbol{a}_1\times\boldsymbol{a}_2|},$$

where $\boldsymbol{n}_0$ is the unit normal to the surface. Further, we designate by

$$a_{\alpha\beta} = \boldsymbol{a}_\alpha\cdot\boldsymbol{a}_\beta,\qquad a^{\alpha\beta} = \boldsymbol{a}^\alpha\cdot\boldsymbol{a}^\beta,\qquad a = \sqrt{\det\,(a_{\alpha\beta})_{2\times2}} = |\boldsymbol{a}_1\times\boldsymbol{a}_2|\,>0$$

and we have

$$\boldsymbol{a}^\alpha\times\boldsymbol{a}^\beta = \epsilon^{\alpha\beta}\boldsymbol{a}_3,\;\; \boldsymbol{a}^3\times\boldsymbol{a}^\alpha = \epsilon^{\alpha\beta}\boldsymbol{a}_\beta,\;\; \boldsymbol{a}_\alpha\times\boldsymbol{a}_\beta = \epsilon_{\alpha\beta}\boldsymbol{a}^3,\;\; \boldsymbol{a}_3\times\boldsymbol{a}_\alpha = \epsilon_{\alpha\beta}\boldsymbol{a}^\beta, \tag{29}$$

where $\epsilon^{\alpha\beta} = \dfrac{1}{a}\,e_{\alpha\beta}$, $\epsilon_{\alpha\beta} = a\,e_{\alpha\beta}$ and $e_{\alpha\beta}$ is the two-dimensional alternator given by $e_{12} = -e_{21} = 1,\; e_{11} = e_{22} = 0$.

Then, $\boldsymbol{a} = a_{\alpha\beta}\boldsymbol{a}^\alpha \otimes \boldsymbol{a}^\beta = a^{\alpha\beta}\boldsymbol{a}_\alpha \otimes \boldsymbol{a}_\beta = \boldsymbol{a}_\alpha \otimes \boldsymbol{a}^\alpha$ represents the first fundamental tensor of the surface S, while the second fundamental tensor $\boldsymbol{b}$ is defined by

$$\boldsymbol{b} = -\text{Grad}_s\, \boldsymbol{n}_0 = -\boldsymbol{n}_{0,\alpha} \otimes \boldsymbol{a}^\alpha = b_{\alpha\beta}\, \boldsymbol{a}^\alpha \otimes \boldsymbol{a}^\beta = b^\alpha_\beta\, \boldsymbol{a}_\alpha \otimes \boldsymbol{a}^\beta, \quad \text{with}$$
$$b_{\alpha\beta} = -\boldsymbol{n}_{0,\beta} \cdot \boldsymbol{a}_\alpha = b_{\beta\alpha}, \qquad b^\alpha_\beta = -\boldsymbol{n}_{0,\beta} \cdot \boldsymbol{a}^\alpha.$$

The surface gradient Grad_s and surface divergence Div_s operators are defined for a vector field $\boldsymbol{v}$ by

$$\text{Grad}_s\, \boldsymbol{v} = \frac{\partial \boldsymbol{v}}{\partial x_\alpha} \otimes \boldsymbol{a}^\alpha = \boldsymbol{v}_{,\alpha} \otimes \boldsymbol{a}^\alpha, \qquad \text{Div}_s\, \boldsymbol{v} = \text{tr}\big[\text{Grad}_s\, \boldsymbol{v}\big] = \boldsymbol{v}_{,\alpha} \cdot \boldsymbol{a}^\alpha. \tag{30}$$

We also introduce the so-called *alternator tensor* $\boldsymbol{c}$ of the surface (Zhilin 2006)

$$\boldsymbol{c} = -\boldsymbol{n}_0 \times \boldsymbol{a} = -\boldsymbol{a} \times \boldsymbol{n}_0 = \epsilon^{\alpha\beta}\, \boldsymbol{a}_\alpha \otimes \boldsymbol{a}_\beta = \epsilon_{\alpha\beta}\, \boldsymbol{a}^\alpha \otimes \boldsymbol{a}^\beta. \tag{31}$$

The tensors $\boldsymbol{a}$ and $\boldsymbol{b}$ are symmetric, while $\boldsymbol{c}$ is skew-symmetric and satisfies $\boldsymbol{cc} = -\boldsymbol{a}$. Note that the tensors $\boldsymbol{a}$, $\boldsymbol{b}$, and $\boldsymbol{c}$ defined above are *planar*, i.e. they are tensors in the tangent plane of the surface. Moreover, $\boldsymbol{a}$ is the identity tensor in the tangent plane.

We define **the surface Curl operator** curl_s for vector fields $\boldsymbol{v}$ and, respectively, Curl_s for tensor fields $\boldsymbol{T}$ by

$$\big(\text{curl}_s\, \boldsymbol{v}\big) \cdot \boldsymbol{k} = \text{Div}_s\big(\boldsymbol{v} \times \boldsymbol{k}\big) \qquad \text{for all constant vectors } \boldsymbol{k}, \tag{32}$$
$$\big(\text{Curl}_s\, \boldsymbol{T}\big)^T \boldsymbol{k} = \text{curl}_s\big(\boldsymbol{T}^T \boldsymbol{k}\big) \qquad \text{for all constant vectors } \boldsymbol{k}. \tag{33}$$

Thus, $\text{curl}_s\, \boldsymbol{v}$ is a vector field, while $\text{Curl}_s\, \boldsymbol{T}$ is a tensor field.

Remark 22.6 These definitions are analogous to the corresponding Definitions (4), (5) in the three-dimensional case. Notice that the curl operator on surfaces has a different significance for other authors, see e.g. Backus et al. (1996).

From the Definitions (32) and (33) it follows

$$\text{curl}_s\, \boldsymbol{v} = -\boldsymbol{v}_{,\alpha} \times \boldsymbol{a}^\alpha, \quad \text{Curl}_s\, \boldsymbol{T} = -\boldsymbol{T}_{,\alpha} \times \boldsymbol{a}^\alpha. \tag{34}$$

Indeed, in view of (30) and (32) we have

$$\big(\text{curl}_s\, \boldsymbol{v}\big) \cdot \boldsymbol{k} = \text{Div}_s\big(\boldsymbol{v} \times \boldsymbol{k}\big) = \big(\boldsymbol{v} \times \boldsymbol{k}\big)_{,\alpha} \cdot \boldsymbol{a}^\alpha = \big(\boldsymbol{v}_{,\alpha} \times \boldsymbol{k}\big) \cdot \boldsymbol{a}^\alpha$$
$$= \big(\boldsymbol{a}^\alpha \times \boldsymbol{v}_{,\alpha}\big) \cdot \boldsymbol{k} = \big(-\boldsymbol{v}_{,\alpha} \times \boldsymbol{a}^\alpha\big) \cdot \boldsymbol{k} \quad \text{for all constant vectors } \boldsymbol{k}$$

and also

$$\big(\mathrm{Curl}_s\,\boldsymbol{T}\big)^T\boldsymbol{k} = \mathrm{curl}_s\big(\boldsymbol{T}^T\boldsymbol{k}\big) = \boldsymbol{a}^\alpha \times \big(\boldsymbol{T}^T\boldsymbol{k}\big)_{,\alpha} = \boldsymbol{a}^\alpha \times \big(\boldsymbol{T}^T_{,\alpha}\boldsymbol{k}\big) = \big(\boldsymbol{a}^\alpha \times \boldsymbol{T}^T_{,\alpha}\big)\boldsymbol{k},$$

which implies $\mathrm{Curl}_s\,\boldsymbol{T} = \big(\boldsymbol{a}^\alpha \times \boldsymbol{T}^T_{,\alpha}\big)^T = -\boldsymbol{T}_{,\alpha} \times \boldsymbol{a}^\alpha$, so the relations (34) hold true.

To write the components of $\mathrm{curl}_s\,\boldsymbol{v}$ and $\mathrm{Curl}_s\,\boldsymbol{T}$ we employ the covariant derivatives on the surface. Let $\boldsymbol{v} = v_i\,\boldsymbol{a}^i$ be a vector field on $\mathcal{S}$. Then, we have

$$\begin{aligned} &\boldsymbol{a}^\alpha_{\cdot,\beta} = -\Gamma^\alpha_{\beta\gamma}\,\boldsymbol{a}^\gamma + b^\alpha_\beta\,\boldsymbol{a}^3, \qquad \boldsymbol{a}_{3,\beta} = -b^\alpha_\beta\,\boldsymbol{a}_\alpha = -b_{\alpha\beta}\,\boldsymbol{a}^\alpha,\\ &\boldsymbol{v}_{,\alpha} = (v_{\beta|\alpha} - b_{\alpha\beta}\,v_3)\boldsymbol{a}^\beta + (v_{3,\alpha} + b^\beta_\alpha\,v_\beta)\boldsymbol{a}^3, \end{aligned} \tag{35}$$

where $v_{\beta|\alpha} = v_{\beta,\alpha} - \Gamma^\gamma_{\alpha\beta}\,v_\gamma$ is the covariant derivative with respect to x_α. Inserting this relation in $(34)_1$ and using $(29)_{1,2}$ we obtain

$$\mathrm{curl}_s\,\boldsymbol{v} \;=\; \epsilon^{\alpha\beta}\big[(v_{3,\beta} + b^\gamma_\beta\,v_\gamma)\boldsymbol{a}_\alpha + v_{\beta|\alpha}\,\boldsymbol{a}_3\big]. \tag{36}$$

For a tensor field $\boldsymbol{T} = T_{ij}\,\boldsymbol{a}^i \otimes \boldsymbol{a}^j = T^{ij}\,\boldsymbol{a}_i \otimes \boldsymbol{a}_j = T^{i}_{\cdot j}\,\boldsymbol{a}_i \otimes \boldsymbol{a}^j$ on the surface, the derivative $\boldsymbol{T}_{,\gamma}$ can be expressed as

$$\begin{aligned} \boldsymbol{T}_{,\gamma} = &\big(T_{\alpha\beta|\gamma} - b_{\alpha\gamma}\,T_{3\beta} - b_{\beta\gamma}\,T_{\alpha3}\big)\boldsymbol{a}^\alpha \otimes \boldsymbol{a}^\beta + \big(T_{\alpha3|\gamma} + b^\beta_\gamma\,T_{\alpha\beta} - b_{\alpha\gamma}\,T_{33}\big)\boldsymbol{a}^\alpha \otimes \boldsymbol{a}^3\\ &+\big(T_{3\alpha|\gamma} + b^\beta_\gamma\,T_{\beta\alpha} - b_{\alpha\gamma}\,T_{33}\big)\boldsymbol{a}^3 \otimes \boldsymbol{a}^\alpha + \big(T_{33,\gamma} + b^\alpha_\gamma\,T_{\alpha3} + b^\alpha_\gamma\,T_{3\alpha}\big)\boldsymbol{a}^3 \otimes \boldsymbol{a}^3, \end{aligned} \tag{37}$$

where the covariant derivatives are

$$\begin{aligned} &T_{\alpha\beta|\gamma} = T_{\alpha\beta,\gamma} - \Gamma^\delta_{\beta\gamma}\,T_{\alpha\delta} - \Gamma^\delta_{\alpha\gamma}\,T_{\delta\beta},\\ &T_{\alpha3|\gamma} = T_{\alpha3,\gamma} - \Gamma^\beta_{\alpha\gamma}\,T_{\beta3}, \qquad T_{3\alpha|\gamma} = T_{3\alpha,\gamma} - \Gamma^\beta_{\alpha\gamma}\,T_{3\beta}. \end{aligned}$$

Using (37) in $(34)_2$ we obtain with the help of $(29)_{1,2}$ the decomposition

$$\begin{aligned} \mathrm{Curl}_s\,\boldsymbol{T} = &\,\epsilon^{\beta\gamma}\big(T_{\alpha3|\gamma} + b^\sigma_\gamma\,T_{\alpha\sigma} - b_{\alpha\gamma}\,T_{33}\big)\boldsymbol{a}^\alpha \otimes \boldsymbol{a}_\beta + \epsilon^{\gamma\beta}\big(T_{\alpha\beta|\gamma} - b_{\alpha\gamma}\,T_{3\beta}\big)\boldsymbol{a}^\alpha \otimes \boldsymbol{a}_3\\ &+ \epsilon^{\beta\gamma}\big(T_{33,\gamma} + b^\alpha_\gamma\,T_{\alpha3} + b^\alpha_\gamma\,T_{3\alpha}\big)\boldsymbol{a}^3 \otimes \boldsymbol{a}_\beta + \epsilon^{\gamma\beta}\big(T_{3\beta|\gamma} + b^\alpha_\gamma\,T_{\alpha\beta}\big)\boldsymbol{a}^3 \otimes \boldsymbol{a}_3. \end{aligned} \tag{38}$$

Alternatively, one can use the mixed components $T^{i}_{\cdot j}$ and write $\mathrm{Curl}_s\,\boldsymbol{T}$ in the tensor basis $\{\boldsymbol{a}_i \otimes \boldsymbol{a}_j\}$

$$\begin{aligned} \mathrm{Curl}_s\,\boldsymbol{T} = &\,\epsilon^{\beta\gamma}\big(T^\alpha_{\cdot3|\gamma} + b^\sigma_\gamma\,T^\alpha_{\cdot\sigma} - b^\alpha_\gamma\,T^3_{\cdot3}\big)\boldsymbol{a}_\alpha \otimes \boldsymbol{a}_\beta + \epsilon^{\gamma\beta}\big(T^\alpha_{\cdot\beta|\gamma} - b^\alpha_\gamma\,T^3_{\cdot\beta}\big)\boldsymbol{a}_\alpha \otimes \boldsymbol{a}_3\\ &+ \epsilon^{\beta\gamma}\big(T^3_{\cdot3,\gamma} + b_{\alpha\gamma}\,T^\alpha_{\cdot3} + b^\alpha_\gamma\,T^3_{\cdot\alpha}\big)\boldsymbol{a}_3 \otimes \boldsymbol{a}_\beta + \epsilon^{\gamma\beta}\big(T^3_{\cdot\beta|\gamma} + b_{\alpha\gamma}\,T^\alpha_{\cdot\beta}\big)\boldsymbol{a}_3 \otimes \boldsymbol{a}_3. \end{aligned} \tag{39}$$

where

$$T^{\alpha}_{\cdot\beta|\gamma} = T^{\alpha}_{\cdot\beta,\gamma} + \Gamma^{\alpha}_{\gamma\sigma}\, T^{\sigma}_{\cdot\beta} - \Gamma^{\sigma}_{\beta\gamma}\, T^{\alpha}_{\cdot\sigma},$$
$$T^{\alpha}_{\cdot 3|\gamma} = T^{\alpha}_{\cdot 3,\gamma} + \Gamma^{\alpha}_{\gamma\sigma}\, T^{\sigma}_{\cdot 3}, \qquad T^{3}_{\cdot\beta|\gamma} = T^{3}_{\cdot\beta,\gamma} - \Gamma^{\sigma}_{\beta\gamma}\, T^{3}_{\cdot\sigma}.$$

Remark 22.7 In order to obtain a formula analogous to (9) and (12), (13) for Curl_s we write $\boldsymbol{T}$ in the form

$$\boldsymbol{T} = \boldsymbol{a}^i \otimes \boldsymbol{T}_i = \boldsymbol{a}_i \otimes \boldsymbol{T}^i \qquad \text{with} \qquad \boldsymbol{T}_i = T_{ij}\, \boldsymbol{a}^j, \quad \boldsymbol{T}^i = T^{i}_{\cdot j}\, \boldsymbol{a}^j.$$

By differentiating the first equation with respect to x_γ we get

$$\boldsymbol{T}_{,\gamma} = \boldsymbol{a}^i_{\ ,\gamma} \otimes \boldsymbol{T}_i + \boldsymbol{a}^i \otimes \boldsymbol{T}_{i,\gamma} = \big(-\Gamma^{\alpha}_{\beta\gamma}\, \boldsymbol{a}^\beta + b^{\alpha}_{\gamma}\, \boldsymbol{a}^3\big) \otimes \boldsymbol{T}_\alpha - b_{\alpha\gamma}\, \boldsymbol{a}^\alpha \otimes \boldsymbol{T}_3 + \boldsymbol{a}^i \otimes \boldsymbol{T}_{i,\gamma}$$
$$= \boldsymbol{a}^\alpha \otimes \big(\boldsymbol{T}_{\alpha,\gamma} - \Gamma^{\beta}_{\alpha\gamma}\, \boldsymbol{T}_\beta - b_{\alpha\gamma}\, \boldsymbol{T}_3\big) + \boldsymbol{a}^3 \otimes \big(\boldsymbol{T}_{3,\gamma} + b^{\alpha}_{\gamma}\, \boldsymbol{T}_\alpha\big).$$

Taking the vector product with $\boldsymbol{a}^\gamma$ and using $(34)_2$ we find

$$\mathrm{Curl}_s\, \boldsymbol{T} = -\big[\boldsymbol{a}^\alpha \otimes \big(\boldsymbol{T}_{\alpha|\gamma} - b_{\alpha\gamma}\, \boldsymbol{T}_3\big) + \boldsymbol{a}^3 \otimes \big(\boldsymbol{T}_{3,\gamma} + b^{\alpha}_{\gamma}\, \boldsymbol{T}_\alpha\big)\big] \times \boldsymbol{a}^\gamma, \tag{40}$$

with $\boldsymbol{T}_{\alpha|\gamma} := \boldsymbol{T}_{\alpha,\gamma} - \Gamma^{\beta}_{\alpha\gamma}\, \boldsymbol{T}_\beta$. Similarly, we obtain

$$\mathrm{Curl}_s\, \boldsymbol{T} = -\big[\boldsymbol{a}_\alpha \otimes \big(\boldsymbol{T}^{\alpha}_{\cdot|\gamma} - b^{\alpha}_{\gamma}\, \boldsymbol{T}^3\big) + \boldsymbol{a}_3 \otimes \big(\boldsymbol{T}^{3}_{\cdot,\gamma} + b_{\alpha\gamma}\, \boldsymbol{T}^\alpha\big)\big] \times \boldsymbol{a}^\gamma, \tag{41}$$

with $\boldsymbol{T}^{\alpha}_{\cdot|\gamma} := \boldsymbol{T}^{\alpha}_{\cdot,\gamma} + \Gamma^{\alpha}_{\beta\gamma}\, \boldsymbol{T}^\beta$. The Eqs. (40) and (41) are the counterpart of the relations (12) and, respectively, (13) in the three-dimensional theory.

4 The Shell Dislocation Density Tensor

Let us present first the kinematics of Cosserat-type shells, which coincides with the kinematics of the 6-parameter shell model, see Chróścielewski et al. (2004), Eremeyev and Pietraszkiewicz (2006), Bîrsan and Neff (2014b).

We consider a deformable surface $\omega_\xi \subset \mathbb{R}^3$ which is identified with the midsurface of the shell in its reference configuration and denote with (ξ_1, ξ_2, ξ_3) a generic point of the surface. Each material point is assumed to have 6 degrees of freedom (3 for translations and 3 for rotations). Thus, the deformation of the Cosserat-type shell is determined by a vectorial map $\boldsymbol{m}_\xi$ and the microrotation tensor $\boldsymbol{R}_\xi$

$$\boldsymbol{m}_\xi : \omega_\xi \to \omega_c, \quad \boldsymbol{R}_\xi : \omega_\xi \to \mathrm{SO}(3),$$

where ω_c denotes the deformed (current) configuration of the surface. We consider a parametric representation $\boldsymbol{y}_0$ of the reference configuration ω_ξ

$$\boldsymbol{y}_0 : \omega \to \omega_\xi\,, \quad \boldsymbol{y}_0(x_1, x_2) = (\xi_1, \xi_2, \xi_3),$$

where $\omega \subset \mathbb{R}^2$ is the bounded domain of variation (with Lipschitz boundary $\partial\omega$) of the parameters (x_1, x_2). Using the same notations as in Sect. 3, we introduce the base vectors $\boldsymbol{a}_i$, $\boldsymbol{a}^j$ and the fundamental tensors $\boldsymbol{a}$, $\boldsymbol{b}$ for the reference surface ω_ξ.

The deformation function $\boldsymbol{m}$ is then defined by the composition

$$\boldsymbol{m} = \boldsymbol{m}_\xi \circ \boldsymbol{y}_0 : \omega \to \omega_c, \quad \boldsymbol{m}(x_1, x_2) := \boldsymbol{m}_\xi\big(\boldsymbol{y}_0(x_1, x_2)\big).$$

According to (30), the surface gradient of the deformation has the expression

$$\mathrm{Grad}_s\, \boldsymbol{m} \;=\; \boldsymbol{m}_{,\alpha} \otimes \boldsymbol{a}^\alpha. \tag{42}$$

As in the three-dimensional case (see Sect. 2) we define the *elastic microrotation* $\boldsymbol{Q}_e$ by the composition

$$\boldsymbol{Q}_e = \boldsymbol{R}_\xi \circ \boldsymbol{y}_0 : \omega \to \mathrm{SO}(3), \quad \boldsymbol{Q}_e(x_1, x_2) := \boldsymbol{R}_\xi\big(\boldsymbol{y}_0(x_1, x_2)\big),$$

the *total microrotation* $\overline{\boldsymbol{R}}$ by

$$\overline{\boldsymbol{R}} : \omega \to \mathrm{SO}(3), \quad \overline{\boldsymbol{R}}(x_1, x_2) = \boldsymbol{Q}_e(x_1, x_2)\, \boldsymbol{Q}_0(x_1, x_2),$$

where $\boldsymbol{Q}_0 : \omega \to \mathrm{SO}(3)$ is the *initial microrotation*, which describes the orientation of points in the reference configuration.

To characterize the orientation and rotation of points in Cosserat-type shells, one employs (as in the three-dimensional case) a triad of orthonormal directors attached to each point. We denote by $\boldsymbol{d}_i^0(x_1, x_2)$ the directors in the reference configuration ω_ξ and by $\boldsymbol{d}_i(x_1, x_2)$ the directors in the deformed configuration ω_c $(i = 1, 2, 3)$. The domain ω is referred to an orthogonal Cartesian frame $Ox_1x_2x_3$ such that $\omega \subset Ox_1x_2$ and let $\boldsymbol{e}_i$ be the unit vectors along the coordinate axes Ox_i. Then, the microrotation tensors can be expressed as follows

$$\boldsymbol{Q}_e = \boldsymbol{d}_i \otimes \boldsymbol{d}_i^0, \qquad \overline{\boldsymbol{R}} = \boldsymbol{Q}_e\, \boldsymbol{Q}_0 = \boldsymbol{d}_i \otimes \boldsymbol{e}_i, \qquad \boldsymbol{Q}_0 = \boldsymbol{d}_i^0 \otimes \boldsymbol{e}_i. \tag{43}$$

Remark 22.8 The initial directors $\boldsymbol{d}_i^0$ are usually chosen such that

$$\boldsymbol{d}_3^0 = \boldsymbol{n}_0, \quad \boldsymbol{d}_\alpha^0 \cdot \boldsymbol{n}_0 = 0, \tag{44}$$

i.e. $\boldsymbol{d}_3^0$ is orthogonal to ω_ξ and $\boldsymbol{d}_\alpha^0$ belong to the tangent plane. This assumption is not necessary in general, but it will be adopted here since it simplifies many of the subsequent expressions. In the deformed configuration, the director $\boldsymbol{d}_3$ is no longer orthogonal to the surface ω_c (the Kirchhof–Love condition is not imposed). One convenient choice of the initial microrotation tensor $\boldsymbol{Q}_0 = \boldsymbol{d}_i^0 \otimes \boldsymbol{e}_i$ such that the conditions (44) be satisfied is $\boldsymbol{Q}_0 = \text{polar}\big(\boldsymbol{a}_i \otimes \boldsymbol{e}_i\big)$, as it was shown in Remark 10 of (Bîrsan and Neff 2014a).

Let us present next the shell strain and curvature measures. In the 6-parameter shell theory the *elastic shell strain tensor* $\boldsymbol{E}_e$ is defined by (Chróścielewski et al. 2004, Eremeyev and Pietraszkiewicz 2006)

$$\boldsymbol{E}_e \;=\; \boldsymbol{Q}_e^T\, \text{Grad}_s \boldsymbol{m} - \boldsymbol{a}. \tag{45}$$

To write the components of $\boldsymbol{E}_e$ we insert (42) and $(43)_1$ into (45)

$$\boldsymbol{E}_e \;=\; \big(\boldsymbol{d}_i^0 \otimes \boldsymbol{d}_i\big)\big(\boldsymbol{m}_{,\alpha} \otimes \boldsymbol{a}^\alpha\big) - \boldsymbol{a}_\alpha \otimes \boldsymbol{a}^\alpha = \big(\boldsymbol{m}_{,\alpha} \cdot \boldsymbol{d}_i - \boldsymbol{a}_\alpha \cdot \boldsymbol{d}_i^0\big)\, \boldsymbol{d}_i^0 \otimes \boldsymbol{a}^\alpha .$$

As a measure of orientation (curvature) change, the *elastic shell bending-curvature tensor* $\boldsymbol{K}_e$ is defined by (Chróścielewski et al. 2004, Eremeyev and Pietraszkiewicz 2006, Bîrsan and Neff 2014b)

$$\boldsymbol{K}_e = \text{axl}\big(\boldsymbol{Q}_e^T \boldsymbol{Q}_{e,\alpha}\big) \otimes \boldsymbol{a}^\alpha = \boldsymbol{Q}_0\big[\text{axl}\big(\overline{\boldsymbol{R}}^T \overline{\boldsymbol{R}}_{,\alpha}\big) - \text{axl}\big(\boldsymbol{Q}_0^T \boldsymbol{Q}_{0,\alpha}\big)\big]. \tag{46}$$

We remark the analogy to the Definition (1) of the wryness tensor $\boldsymbol{\Gamma}$ in the three-dimensional theory. Following the analogy to (2), we employ next the surface curl operator Curl_s defined in Sect. 3 to introduce the new *shell dislocation density tensor* $\boldsymbol{D}_e$ by

$$\boldsymbol{D}_e = \boldsymbol{Q}_e^T\, \text{Curl}_s\, \boldsymbol{Q}_e. \tag{47}$$

In view of relation $(34)_2$, we can write this definition in the form

$$\boldsymbol{D}_e = \boldsymbol{Q}_e^T\big(-\boldsymbol{Q}_{e,\alpha} \times \boldsymbol{a}^\alpha\big) = -\big(\boldsymbol{Q}_e^T \boldsymbol{Q}_{e,\alpha}\big) \times \boldsymbol{a}^\alpha. \tag{48}$$

The tensor $\boldsymbol{D}_e$ given by (47) represents an alternative strain measure for orientation (curvature) change in Cosserat-type shells.

In what follows, we want to establish the relationship between the shell bending-curvature tensor $\boldsymbol{K}_e$ and the shell dislocation density tensor $\boldsymbol{D}_e$. We observe that this relationship is analogous to the corresponding relations (19), (20) in the three-dimensional theory. More precisely, in the shell theory it holds

$$\boldsymbol{D}_e = -\boldsymbol{K}_e^T + \big(\text{tr}\, \boldsymbol{K}_e\big)\mathbf{1}_3 \quad \text{or equivalently,} \quad \boldsymbol{K}_e = -\boldsymbol{D}_e^T + \frac{1}{2}\big(\text{tr}\, \boldsymbol{D}_e\big)\mathbf{1}_3. \tag{49}$$

To prove (49), we designate the components of the shell bending-curvature tensor by $\boldsymbol{K}_e = K_{i\alpha}\, \boldsymbol{d}_i^0 \otimes \boldsymbol{a}^\alpha$ and use $(16)_3$ to write

$$(\boldsymbol{Q}_e^T\boldsymbol{Q}_{e,\alpha})\otimes\boldsymbol{a}^\alpha = -\epsilon\,\text{axl}(\boldsymbol{Q}_e^T\boldsymbol{Q}_{e,\alpha})\otimes\boldsymbol{a}^\alpha = -\epsilon\,\boldsymbol{K}_e$$
$$= -\big(e_{ijk}\,\boldsymbol{d}_i^0\otimes\boldsymbol{d}_j^0\otimes\boldsymbol{d}_k^0\big)\big(K_{s\alpha}\,\boldsymbol{d}_s^0\otimes\boldsymbol{a}^\alpha\big) = -e_{ijs}\,K_{s\alpha}\,\boldsymbol{d}_i^0\otimes\boldsymbol{d}_j^0\otimes\boldsymbol{a}^\alpha,$$

which implies

$$\boldsymbol{Q}_e^T\boldsymbol{Q}_{e,\alpha} = -e_{ijs}\,K_{s\alpha}\,\boldsymbol{d}_i^0\otimes\boldsymbol{d}_j^0$$

We substitute the last relation into (48) and derive

$$\begin{aligned}
\boldsymbol{D}_e &= \big(e_{ijs}\,K_{s\alpha}\,\boldsymbol{d}_i^0\otimes\boldsymbol{d}_j^0\big)\times\boldsymbol{a}^\alpha = \big(e_{ijs}\,K_{s\alpha}\,\boldsymbol{d}_i^0\otimes\boldsymbol{d}_j^0\big)\times\big[(\boldsymbol{a}^\alpha\cdot\boldsymbol{d}_\beta^0)\,\boldsymbol{d}_\beta^0\big]\\
&= (\boldsymbol{a}^\alpha\cdot\boldsymbol{d}_\beta^0)\big[e_{ijs}\,K_{s\alpha}\,\boldsymbol{d}_i^0\otimes(\boldsymbol{d}_j^0\times\boldsymbol{d}_\beta^0)\big] = (\boldsymbol{a}^\alpha\cdot\boldsymbol{d}_\beta^0)\big[e_{ijs}\,e_{j\beta m}\,K_{s\alpha}\,\boldsymbol{d}_i^0\otimes\boldsymbol{d}_m^0\big]\\
&= (\boldsymbol{a}^\alpha\cdot\boldsymbol{d}_\beta^0)\big[(\delta_{im}\,\delta_{s\beta}-\delta_{i\beta}\,\delta_{sm})K_{s\alpha}\,\boldsymbol{d}_i^0\otimes\boldsymbol{d}_m^0\big]\\
&= (\boldsymbol{a}^\alpha\cdot\boldsymbol{d}_\beta^0)\big[-K_{s\alpha}\,\boldsymbol{d}_\beta^0\otimes\boldsymbol{d}_s^0+K_{\beta\alpha}\,\boldsymbol{d}_i^0\otimes\boldsymbol{d}_i^0\big]\\
&= -K_{i\alpha}\big[(\boldsymbol{a}^\alpha\cdot\boldsymbol{d}_\beta^0)\,\boldsymbol{d}_\beta^0\big]\otimes\boldsymbol{d}_i^0+K_{\beta\alpha}\big(\boldsymbol{d}_\beta^0\cdot\boldsymbol{a}^\alpha\big)\,\boldsymbol{1}_3\big]\\
&= -\big(K_{i\alpha}\,\boldsymbol{d}_i^0\otimes\boldsymbol{a}^\alpha\big)^T+\text{tr}\big(K_{i\alpha}\,\boldsymbol{d}_i^0\otimes\boldsymbol{a}^\alpha\big)\,\boldsymbol{1}_3 = -\boldsymbol{K}_e^T+\big(\text{tr}\,\boldsymbol{K}_e\big)\boldsymbol{1}_3,
\end{aligned}$$

which shows that $(49)_1$ holds true. Applying the trace operator to Eq. $(49)_1$ we get $\text{tr}\,\boldsymbol{K}_e = \frac{1}{2}\,\text{tr}\,\boldsymbol{D}_e$. Inserting this into $(49)_1$ we obtain $(49)_2$. The proof is complete.

Remark 22.9 As a consequence of relations (49) we deduce the relations between the norms, traces, symmetric and skew-symmetric parts of the two tensors in the forms

$$\begin{aligned}
&\|\boldsymbol{D}_e\|^2 = \|\boldsymbol{K}_e\|^2+\big(\text{tr}\,\boldsymbol{K}_e\big)^2, \qquad \|\boldsymbol{K}_e\|^2 = \|\boldsymbol{D}_e\|^2-\frac{1}{4}\big(\text{tr}\,\boldsymbol{D}_e\big)^2,\\
&\text{tr}\,\boldsymbol{D}_e = 2\,\text{tr}\,\boldsymbol{K}_e,\quad \text{skew}\,\boldsymbol{D}_e = \text{skew}\,\boldsymbol{K}_e,\quad \text{dev}_3\text{sym}\,\boldsymbol{D}_e = -\text{dev}_3\text{sym}\,\boldsymbol{K}_e.
\end{aligned} \tag{50}$$

Indeed the relations (50) can be easily proved if we apply the operators tr, $\|\cdot\|$, skew, dev_3, and sym to the Eq. $(49)_1$. In view of $(50)_1$ and $\big(\text{tr}\,\boldsymbol{K}_e\big)^2\leq 3\,\|\boldsymbol{K}_e\|^2$, we obtain the estimate

$$\|\boldsymbol{K}_e\|\leq\|\boldsymbol{D}_e\|\leq 2\,\|\boldsymbol{K}_e\|. \tag{51}$$

In what follows, we write the components of the tensors $\boldsymbol{K}_e$ and $\boldsymbol{D}_e$. To this aim, we use the relations

$$\begin{aligned}
\boldsymbol{Q}_e^T\boldsymbol{Q}_{e,\alpha} &= \big(\boldsymbol{d}_i^0\otimes\boldsymbol{d}_i\big)\big(\boldsymbol{d}_{k,\alpha}\otimes\boldsymbol{d}_k^0+\boldsymbol{d}_k\otimes\boldsymbol{d}_{k,\alpha}^0\big)\\
&= (\boldsymbol{d}_i\cdot\boldsymbol{d}_{k,\alpha})\boldsymbol{d}_i^0\otimes\boldsymbol{d}_k^0+\boldsymbol{d}_i^0\otimes\boldsymbol{d}_{i,\alpha}^0 = (\boldsymbol{d}_i\cdot\boldsymbol{d}_{k,\alpha}-\boldsymbol{d}_i^0\cdot\boldsymbol{d}_{k,\alpha}^0)\boldsymbol{d}_i^0\otimes\boldsymbol{d}_k^0,
\end{aligned} \tag{52}$$

which can be proved in the same way as Eq. (19). We compute the axial vector of the skew-symmetric tensor (52) and find (similar to (20))

$$\operatorname{axl}\big(\boldsymbol{Q}_e^T\boldsymbol{Q}_{e,\alpha}\big) = -\frac{1}{2}\, e_{ijk}\,\big(\boldsymbol{d}_j\cdot\boldsymbol{d}_{k,\alpha} - \boldsymbol{d}_j^0\cdot\boldsymbol{d}_{k,\alpha}^0\big)\,\boldsymbol{d}_i^0. \tag{53}$$

By virtue of (53) the Definition (46) yields

$$\begin{aligned}\boldsymbol{K}_e &= \frac{1}{2}\, e_{ijk}\,\big(\boldsymbol{d}_{j,\alpha}\cdot\boldsymbol{d}_k - \boldsymbol{d}_{j,\alpha}^0\cdot\boldsymbol{d}_k^0\big)\,\boldsymbol{d}_i^0\otimes\boldsymbol{a}^\alpha\\ &= \big(\boldsymbol{d}_{2,\alpha}\cdot\boldsymbol{d}_3 - \boldsymbol{d}_{2,\alpha}^0\cdot\boldsymbol{d}_3^0\big)\,\boldsymbol{d}_1^0\otimes\boldsymbol{a}^\alpha + \big(\boldsymbol{d}_{3,\alpha}\cdot\boldsymbol{d}_1 - \boldsymbol{d}_{3,\alpha}^0\cdot\boldsymbol{d}_1^0\big)\,\boldsymbol{d}_2^0\otimes\boldsymbol{a}^\alpha\\ &\quad + \big(\boldsymbol{d}_{1,\alpha}\cdot\boldsymbol{d}_2 - \boldsymbol{d}_{1,\alpha}^0\cdot\boldsymbol{d}_2^0\big)\,\boldsymbol{d}_3^0\otimes\boldsymbol{a}^\alpha,\end{aligned} \tag{54}$$

which gives the components $K_{i\alpha}$ of the shell bending-curvature tensor $\boldsymbol{K}_e$ in the tensor basis $\{\boldsymbol{d}_i^0\otimes\boldsymbol{a}^\alpha\}$.

For the components of $\boldsymbol{D}_e$, we insert the relation (52) in the Eq. (48)

$$\boldsymbol{D}_e = -\big(\boldsymbol{d}_i\cdot\boldsymbol{d}_{k,\alpha} - \boldsymbol{d}_i^0\cdot\boldsymbol{d}_{k,\alpha}^0\big)\big(\boldsymbol{d}_i^0\otimes\boldsymbol{d}_k^0\big)\times\boldsymbol{a}^\alpha.$$

Using that $\boldsymbol{d}_k^0\times\boldsymbol{a}^\alpha = \boldsymbol{d}_k^0\times\big[\big(\boldsymbol{a}^\alpha\cdot\boldsymbol{d}_\beta^0\big)\,\boldsymbol{d}_\beta^0\big] = \big(\boldsymbol{a}^\alpha\cdot\boldsymbol{d}_\beta^0\big)\,e_{k\beta j}\,\boldsymbol{d}_j^0$, we obtain

$$\boldsymbol{D}_e = e_{jk\beta}\big(\boldsymbol{d}_{i,\alpha}\cdot\boldsymbol{d}_k - \boldsymbol{d}_{i,\alpha}^0\cdot\boldsymbol{d}_k^0\big)\big(\boldsymbol{a}^\alpha\cdot\boldsymbol{d}_\beta^0\big)\,\boldsymbol{d}_i^0\otimes\boldsymbol{d}_j^0\,, \tag{55}$$

which shows the components of the shell dislocation density tensor in the tensor basis $\{\boldsymbol{d}_i^0\otimes\boldsymbol{d}_j^0\}$.

5 Remarks and Discussion

Herein we present some other ways to express the shell dislocation density tensor, the shell bending-curvature tensor and discuss their close relationship.

Remark 22.10 It is sometimes useful to express the components of the shell dislocation density tensor $\boldsymbol{D}_e$ in the tensor basis $\{\boldsymbol{a}^i\otimes\boldsymbol{a}_j\}$. If we multiply the relation $(49)_2$ with $\boldsymbol{n}_0$ and take into account that $\boldsymbol{K}_e\boldsymbol{n}_0 = \boldsymbol{0}$, then we find $\boldsymbol{0} = -\boldsymbol{D}_e^T\boldsymbol{n}_0 + \frac{1}{2}\,\big(\operatorname{tr}\boldsymbol{D}_e\big)\boldsymbol{n}_0\,$, which means

$$\boldsymbol{n}_0\,\boldsymbol{D}_e \;=\; \frac{1}{2}\,\big(\operatorname{tr}\boldsymbol{D}_e\big)\boldsymbol{n}_0.$$

It follows that the components of $\boldsymbol{D}_e$ in the directions $\boldsymbol{n}_0\otimes\boldsymbol{a}_\alpha$ are zero, i.e. $\boldsymbol{D}_e$ has the structure

$$\boldsymbol{D}_e = \boldsymbol{D}_\parallel + D_{\alpha 3}\,\boldsymbol{a}^\alpha\otimes\boldsymbol{n}_0 + \frac{1}{2}\,\big(\operatorname{tr}\boldsymbol{D}_e\big)\boldsymbol{n}_0\otimes\boldsymbol{n}_0, \tag{56}$$

where $\boldsymbol{D}_\parallel = \boldsymbol{D}_e \boldsymbol{a} = D^{\alpha\beta}\boldsymbol{a}_\alpha \otimes \boldsymbol{a}_\beta = D_{\alpha\cdot}^{\ \beta}\, \boldsymbol{a}^\alpha \otimes \boldsymbol{a}_\beta$ is the planar part of $\boldsymbol{D}_e$ (the part in the tangent plane). If we insert (56) into $(49)_1$ and use $\frac{1}{2}\operatorname{tr}\boldsymbol{D}_e = \operatorname{tr}\boldsymbol{K}_e$, we get

$$\boldsymbol{D}_\parallel + D_{\alpha 3}\,\boldsymbol{a}^\alpha \otimes \boldsymbol{n}_0 + \big(\operatorname{tr}\boldsymbol{K}_e\big)\boldsymbol{n}_0 \otimes \boldsymbol{n}_0 = -K_{i\alpha}\,\boldsymbol{a}^\alpha \otimes \boldsymbol{d}_i^0 + \big(\operatorname{tr}\boldsymbol{K}_e\big)\big(\boldsymbol{a} + \boldsymbol{n}_0 \otimes \boldsymbol{n}_0\big),$$

which implies (in view of (54)) that

$$D_{\alpha 3} = -K_{3\alpha} = \boldsymbol{d}_1 \cdot \boldsymbol{d}_{2,\alpha} - \boldsymbol{d}_1^0 \cdot \boldsymbol{d}_{2,\alpha}^0 \quad \text{and} \quad \boldsymbol{D}_\parallel = -\big(\boldsymbol{K}_\parallel\big)^T + \big(\operatorname{tr}\boldsymbol{K}_e\big)\boldsymbol{a}, \tag{57}$$

where $\boldsymbol{K}_\parallel = \boldsymbol{a}\,\boldsymbol{K}_e = K_{\beta\alpha}\boldsymbol{d}_\beta^0 \otimes \boldsymbol{a}^\alpha$ is the planar part of $\boldsymbol{K}_e$.

Remark 22.11 We observe that between the planar part $\boldsymbol{D}_\parallel$ of $\boldsymbol{D}_e$ and the planar part $\boldsymbol{K}_\parallel$ of $\boldsymbol{K}_e$ there exists a special relationship. The tensor $\boldsymbol{D}_\parallel$ is the cofactor of the tensor $\boldsymbol{K}_\parallel$. Let us explain this in more details: for any planar tensor $\boldsymbol{S} = S^{\alpha}_{\cdot\beta}\,\boldsymbol{a}_\alpha \otimes \boldsymbol{a}^\beta$ we introduce the transformation

$$T(\boldsymbol{S}) = -\boldsymbol{S}^T + \big(\operatorname{tr}\boldsymbol{S}\big)\,\boldsymbol{a}. \tag{58}$$

One can prove that this transformation has the properties

$$T\big(T(\boldsymbol{S})\big) = \boldsymbol{S} \qquad \text{and} \qquad T(\boldsymbol{S}) = -\boldsymbol{c}\,\boldsymbol{S}\,\boldsymbol{c}, \tag{59}$$

where the alternator $\boldsymbol{c}$ is defined in (31). Moreover, in view of $(59)_2$ and (31) we can write $T(\boldsymbol{S})$ in the tensor basis $\{\boldsymbol{a}^\alpha \otimes \boldsymbol{a}_\beta\}$ as follows

$$T(\boldsymbol{S}) = S^{2}_{\cdot 2}\,\boldsymbol{a}^1 \otimes \boldsymbol{a}_1 - S^{2}_{\cdot 1}\,\boldsymbol{a}^1 \otimes \boldsymbol{a}_2 - S^{1}_{\cdot 2}\,\boldsymbol{a}^2 \otimes \boldsymbol{a}_1 + S^{1}_{\cdot 1}\,\boldsymbol{a}^2 \otimes \boldsymbol{a}_2,$$

which shows that the 2×2 matrix of the components of $T(\boldsymbol{S})$ in the basis $\{\boldsymbol{a}^\alpha \otimes \boldsymbol{a}_\beta\}$ is the cofactor of the matrix of components of $\boldsymbol{S}$ in the basis $\{\boldsymbol{a}_\alpha \otimes \boldsymbol{a}^\beta\}$, since

$$\begin{pmatrix} S^{2}_{\cdot 2} & -S^{2}_{\cdot 1} \\ -S^{1}_{\cdot 2} & S^{1}_{\cdot 1} \end{pmatrix} = \operatorname{Cof}\begin{pmatrix} S^{1}_{\cdot 1} & S^{1}_{\cdot 2} \\ S^{2}_{\cdot 1} & S^{2}_{\cdot 2} \end{pmatrix}.$$

If the tensor $\boldsymbol{S}$ is invertible, then from the Cayley–Hamilton relation $\big(\boldsymbol{S}^T\big)^2 - \big(\operatorname{tr}\boldsymbol{S}\big)\boldsymbol{S}^T + \det\boldsymbol{S} = \boldsymbol{0}$ and (58) we deduce

$$T(\boldsymbol{S}) = -\boldsymbol{S}^T + \big(\operatorname{tr}\boldsymbol{S}\big)\,\boldsymbol{a} = \big(\det\boldsymbol{S}\big)\,\boldsymbol{S}^{-T} =: \operatorname{Cof}\big(\boldsymbol{S}\big). \tag{60}$$

In our case, for the shell bending-curvature tensor $\boldsymbol{K}_e$ we have $\operatorname{tr}\boldsymbol{K}_e = \operatorname{tr}\big(\boldsymbol{a}\boldsymbol{K}_e\big) = \operatorname{tr}\big(\boldsymbol{K}_\parallel\big)$, in view of (54). Then, the relation $(57)_2$ yields

$$\boldsymbol{D}_\parallel = -\big(\boldsymbol{K}_\parallel\big)^T + \big(\operatorname{tr}\boldsymbol{K}_\parallel\big)\,\boldsymbol{a}.$$

Using the relations (58)–(60) we see that $\boldsymbol{D}_{\parallel}$ is the image of $\boldsymbol{K}_{\parallel}$ under the transformation T, so that it holds

$$\begin{aligned} \boldsymbol{D}_{\parallel} &= T\big(\boldsymbol{K}_{\parallel}\big) = -\boldsymbol{c}\,\big(\boldsymbol{K}_{\parallel}\big)\,\boldsymbol{c} = \mathrm{Cof}\big(\boldsymbol{K}_{\parallel}\big), \\ \boldsymbol{K}_{\parallel} &= T\big(\boldsymbol{D}_{\parallel}\big) = -\boldsymbol{c}\,\big(\boldsymbol{D}_{\parallel}\big)\,\boldsymbol{c} = \mathrm{Cof}\big(\boldsymbol{D}_{\parallel}\big). \end{aligned} \tag{61}$$

From (56), (57) we can write

$$\boldsymbol{D}_e = \mathrm{Cof}\big(\boldsymbol{K}_{\parallel}\big) - K_{3\alpha}\,\boldsymbol{a}^{\alpha} \otimes \boldsymbol{n}_0 + \big(\mathrm{tr}\,\boldsymbol{K}_{\parallel}\big)\boldsymbol{n}_0 \otimes \boldsymbol{n}_0, \tag{62}$$

which expresses once again the close relationship between the shell dislocation density tensor $\boldsymbol{D}_e$ and the shell bending-curvature tensor $\boldsymbol{K}_e$.

Remark 22.12 The shell bending-curvature tensor $\boldsymbol{K}_e$ can also be expressed in terms of the directors $\boldsymbol{d}_i$. In this respect, an analogous relation to the formula (28) for the wryness tensor (see Remark 22.5) holds

$$\boldsymbol{K}_e \;=\; \frac{1}{2}\big[\boldsymbol{Q}_e^T\big(\boldsymbol{d}_i \times \boldsymbol{d}_{i,\alpha}\big) - \boldsymbol{d}_i^0 \times \boldsymbol{d}_{i,\alpha}^0\big] \otimes \boldsymbol{a}^{\alpha}. \tag{63}$$

To prove (63), we write the two terms in the brackets in the following form

$$\begin{aligned} \boldsymbol{Q}_e^T\big(\boldsymbol{d}_i \times \boldsymbol{d}_{i,\alpha}\big) &= \big(\boldsymbol{d}_k^0 \otimes \boldsymbol{d}_k\big)\big(\boldsymbol{d}_i \times \boldsymbol{d}_{i,\alpha}\big) = \big[\boldsymbol{d}_k \cdot \big(\boldsymbol{d}_i \times \boldsymbol{d}_{i,\alpha}\big)\big]\,\boldsymbol{d}_k^0 \\ &= \big[\boldsymbol{d}_{i,\alpha} \cdot \big(\boldsymbol{d}_k \times \boldsymbol{d}_i\big)\big]\,\boldsymbol{d}_k^0 = e_{kij}\,\big(\boldsymbol{d}_{i,\alpha} \cdot \boldsymbol{d}_j\big)\,\boldsymbol{d}_k^0 \end{aligned}$$

and similarly

$$\boldsymbol{d}_i^0 \times \boldsymbol{d}_{i,\alpha}^0 = \big[\boldsymbol{d}_k^0 \cdot \big(\boldsymbol{d}_i^0 \times \boldsymbol{d}_{i,\alpha}^0\big)\big]\,\boldsymbol{d}_k^0 = \big[\boldsymbol{d}_{i,\alpha}^0 \cdot \big(\boldsymbol{d}_k^0 \times \boldsymbol{d}_i^0\big)\big]\,\boldsymbol{d}_k^0 = e_{kij}\,\big(\boldsymbol{d}_{i,\alpha}^0 \cdot \boldsymbol{d}_j^0\big)\,\boldsymbol{d}_k^0.$$

Inserting the last two relations into (63) we obtain

$$\boldsymbol{K}_e \;=\; \frac{1}{2}\,e_{ijk}\,\big[\big(\boldsymbol{d}_{j,\alpha} \cdot \boldsymbol{d}_k\big)\,\boldsymbol{d}_i^0 - \big(\boldsymbol{d}_{j,\alpha}^0 \cdot \boldsymbol{d}_k^0\big)\,\boldsymbol{d}_i^0\big] \otimes \boldsymbol{a}^{\alpha},$$

which holds true, by virtue of (54). Thus, (63) is proved.

We can put the relation (63) in the form

$$\boldsymbol{K}_e \;=\; \boldsymbol{Q}_e^T\,\boldsymbol{\omega} \qquad \text{where we define} \tag{64}$$

$$\boldsymbol{\omega} = \boldsymbol{\omega}_{\alpha} \otimes \boldsymbol{a}^{\alpha} \qquad \text{with} \qquad \boldsymbol{\omega}_{\alpha} = \frac{1}{2}\big[\boldsymbol{d}_i \times \boldsymbol{d}_{i,\alpha} - \boldsymbol{Q}_e\big(\boldsymbol{d}_i^0 \times \boldsymbol{d}_{i,\alpha}^0\big)\big]. \tag{65}$$

If we compare the relations (64) and the Definition (46), we derive

$$\boldsymbol{\omega}_{\alpha} = \boldsymbol{Q}_e\,\mathrm{axl}\big(\boldsymbol{Q}_e^T\boldsymbol{Q}_{e,\alpha}\big) = \mathrm{axl}\big(\boldsymbol{Q}_{e,\alpha}\,\boldsymbol{Q}_e^T\big).$$

Then, from (16) we deduce $\boldsymbol{Q}_{e,\alpha}\boldsymbol{Q}_e^T = \boldsymbol{\omega}_\alpha \times \mathbf{1}_3$ and by multiplication with $\boldsymbol{Q}_e$ we find

$$\boldsymbol{Q}_{e,\alpha} = \boldsymbol{\omega}_\alpha \times \boldsymbol{Q}_e, \qquad \alpha = 1, 2. \tag{66}$$

Thus, the Eqs. (64), (65) can be employed for an alternative definition of the shell bending-curvature tensor, namely

$$\boldsymbol{K}_e = \boldsymbol{Q}_e^T \boldsymbol{\omega}\,, \quad \text{where} \quad \boldsymbol{\omega} = \boldsymbol{\omega}_\alpha \otimes \boldsymbol{a}^\alpha \quad \text{and} \quad \boldsymbol{Q}_{e,\alpha} = \boldsymbol{\omega}_\alpha \times \boldsymbol{Q}_e. \tag{67}$$

This is the counterpart of the relations (24), (25) for the wryness tensor in the three-dimensional theory of Cosserat continua. The relations (67) were used to define the corresponding shell bending-curvature tensor, e.g. in Altenbach and Zhilin (2004), Zhilin (2006).

Remark 22.13 As shown by relations (3) for the three-dimensional case, one can introduce the elastically stored shell energy density W as a function of the shell strain tensor and the shell dislocation density tensor

$$W = W\big(\boldsymbol{E}_e\,,\,\boldsymbol{D}_e\big). \tag{68}$$

If (68) is assumed to be a quadratic convex and coercive function, then the existence of solutions to the minimization problem of the total energy functional for Cosserat shells can be proved in a similar manner as in Theorem 6 of Bîrsan and Neff (2014a). In the proof, one should employ decisively the estimate (51) and the expressions of the shell dislocation density tensor $\boldsymbol{D}_e$ established in the previous sections.

References

Altenbach, H.: An alternative determination of transverse shear stiffnesses for sandwich and laminated plates. Int. J. Solids Struct. **37**, 3503–3520 (2000)

Altenbach, H., Eremeyev, V.: Direct approach-based analysis of plates composed of functionally graded materials. Arch. Appl. Mech. **78**, 775–794 (2008)

Altenbach, H., Eremeyev, V.: On the bending of viscoelastic plates made of polymer foams. Acta Mech. **204**, 137–154 (2009)

Altenbach, H., Eremeyev, V.: On the effective stiffness of plates made of hyperelastic materials with initial stresses. Int. J. Non-Linear Mech. **45**, 976–981 (2010)

Altenbach, H., Zhilin, P.: Eine nichtlineare Theorie dünner Dreischichtschalen und ihre Anwendung auf die Stabilitätsuntersuchung eines dreischichtigen Streifens. Technische Mechanik **3**, 23–30 (1982)

Altenbach, H., Zhilin, P.: A general theory of elastic simple shells (in Russian). Uspekhi Mekhaniki **11**, 107–148 (1988)

Altenbach, H., Zhilin, P.: The theory of simple elastic shells. In: Kienzler, R., Altenbach, H., Ott, I. (eds.) Theories of Plates and Shells. Critical Review and New Applications, Euromech Colloquium 444, pp. 1–12. Springer, Heidelberg (2004)

Altenbach, J., Altenbach, H., Eremeyev, V.: On generalized Cosserat-type theories of plates and shells: a short review and bibliography. Arch. Appl. Mech. **80**, 73–92 (2010)
Backus, G., Parker, R., Constable, C.: Foundations of Geomagnetism. Cambridge University Press, Cambridge (1996)
Bîrsan, M., Altenbach, H.: A mathematical study of the linear theory for orthotropic elastic simple shells. Math. Methods Appl. Sci. **33**, 1399–1413 (2010)
Bîrsan, M., Altenbach, H.: Analysis of the deformation of multi-layered orthotropic cylindrical elastic shells using the direct approach. In: Altenbach, H., Eremeyev, V. (eds.) Shell-like Structures: Non-classical Theories and Applications, pp. 29–52. Springer, Berlin (2011)
Bîrsan, M., Neff, P.: Existence of minimizers in the geometrically non-linear 6-parameter resultant shell theory with drilling rotations. Math. Mech. Solids **19**(4), 376–397 (2014a)
Bîrsan, M., Neff, P.: Shells without drilling rotations: a representation theorem in the framework of the geometrically nonlinear 6-parameter resultant shell theory. Int. J. Eng. Sci. **80**, 32–42 (2014b)
Bîrsan, M., Sadowski, T., Pietras, D.: Thermoelastic deformations of cylindrical multi-layered shells using a direct approach. J. Therm. Stress. **36**, 749–789 (2013)
Chróścielewski, J., Makowski, J., Pietraszkiewicz, W.: Statics and Dynamics of Multifold Shells: Nonlinear Theory and Finite Element Method (in Polish). Wydawnictwo IPPT PAN, Warsaw (2004)
Cosserat, E., Cosserat, F.: Théorie des corps déformables. Hermann et Fils, Paris (1909) (Reprint 2009)
Eremeyev, V., Pietraszkiewicz, W.: The nonlinear theory of elastic shells with phase transitions. J. Elast. **74**, 67–86 (2004)
Eremeyev, V., Pietraszkiewicz, W.: Local symmetry group in the general theory of elastic shells. J Elast. **85**, 125–152 (2006)
Eremeyev, V., Lebedev, L., Altenbach, H.: Foundations of Micropolar Mechanics. Springer, Heidelberg (2013)
Ghiba, I., Neff, P., Madeo, A., Placidi, L., Rosi, G.: The relaxed linear micromorphic continuum: existence, uniqueness and continuous dependence in dynamics. Math. Mech. Solids **20**, 1171–1197 (2015)
Gurtin, M.: An Introduction to Continuum Mechanics., Mathematics in Science and Engineering, vol. 158, 1st edn. Academic, London (1981)
Lankeit, J., Neff, P., Osterbrink, F.: Integrability conditions between the first and second cosserat deformation tensor in geometrically nonlinear micropolar models and existence of minimizers (2016). arXiv:150408003
Libai, A., Simmonds, J.: The Nonlinear Theory of Elastic Shells, 2nd edn. Cambridge University Press, Cambridge (1998)
Madeo, A., Neff, P., Ghiba, I., Placidi, L., Rosi, G.: Wave propagation in relaxed linear micromorphic continua: modelling metamaterials with frequency band gaps. Contin. Mech. Thermodyn. **27**, 551–570 (2015)
Mielke, A., Müller, S.: Lower semi-continuity and existence of minimizers in incremental finite-strain elastoplasticity. Z. Angew. Math. Mech. **86**, 233–250 (2006)
Neff, P., Münch, I.: Curl bounds Grad on SO(3). ESAIM: Control, Optim. Calc. Var. **14**, 148–159 (2008)
Neff, P., Ghiba, I., Madeo, A., Placidi, L., Rosi, G.: A unifying perspective: the relaxed linear micromorphic continua. Contin. Mech. Thermodyn. **26**, 639–681 (2014)
Neff, P., Bîrsan, M., Osterbrink, F.: Existence theorem for geometrically nonlinear Cosserat micropolar model under uniform convexity requirements. J. Elast. **121**, 119–141 (2015)
Nye, J.: Some geometrical relations in dislocated crystals. Acta Metall. **1**, 153–162 (1953)
Pietraszkiewicz, W., Eremeyev, V.: On natural strain measures of the non-linear micropolar continuum. Int. J. Solids Struct. **46**, 774–787 (2009)
Sadowski, T., Bîrsan, M., Pietras, D.: Multilayered and FGM structural elements under mechanical and thermal loads. Part I: comparison of finite elements and analytical models. Arch. Civ. Mech. Eng. **15**, 1180–1192 (2015)

Svendsen, B.: Continuum thermodynamic models for crystal plasticity including the effects of geometrically necessary dislocations. J. Mech. Phys. Solids **50**(25), 1297–1329 (2002)
Tambača, J., Velčić, I.: Existence theorem for nonlinear micropolar elasticity. ESAIM: Control, Optim. Calc. Var. **16**, 92–110 (2010)
Zhilin, P.: Mechanics of deformable directed surfaces. Int. J. Solids Struct. **12**, 635–648 (1976)
Zhilin, P.: Applied Mechanics—Foundations of Shell Theory (in Russian). State Polytechnical University Publisher, Sankt Petersburg (2006)

Direct Approach Versus Consistent Approximation

Reinhold Kienzler and Patrick Schneider

Abstract Relations between plate theories resulting from the direct approach and the consistent approximation are established and the resulting equations are compared. By introducing a scalar measure for the thickness strain, both theories can be reconciled within a consistent second-order approximation.

1 Introduction

Plates are important structural elements and are defined in the following manner:

- *Geometry*: Plates belong to the family of thin plane structures. Let us consider a three-dimensional plate continuum with constant thickness h as depicted in Fig. 1. The mid-surface of the plate is plane and embedded in the (x_1, x_2)-plane of a three-dimensional coordinate system (x_1, x_2, x_3). The characteristic in-plane dimension, say a, is much larger than the characteristic out-of-plane dimension h, especially the plate parameter c^2 is much smaller than unity

$$c^2 = \frac{h^2}{12a^2} \ll 1. \tag{1}$$

  Plates with variable thickness are also admissible but are not considered in the following.
- *Material*: The material behavior must be symmetric with respect to the midplane, i.e., the most general plate material is a monotropic material with the plane of symmetry coinciding with the (x_1, x_2)-plane (13 independent material parameters).

R. Kienzler (✉) · P. Schneider
Bremen Institute for Mechanical Engineering, University of Bremen, Bremen, Germany
e-mail: rkienzler@uni-bremen.de

P. Schneider
e-mail: pasch@uni-bremen.de

K. Naumenko and M. Aßmus (eds.), *Advanced Methods of Continuum Mechanics for Materials and Structures*, Advanced Structured Materials 60,
DOI 10.1007/978-981-10-0959-4_23

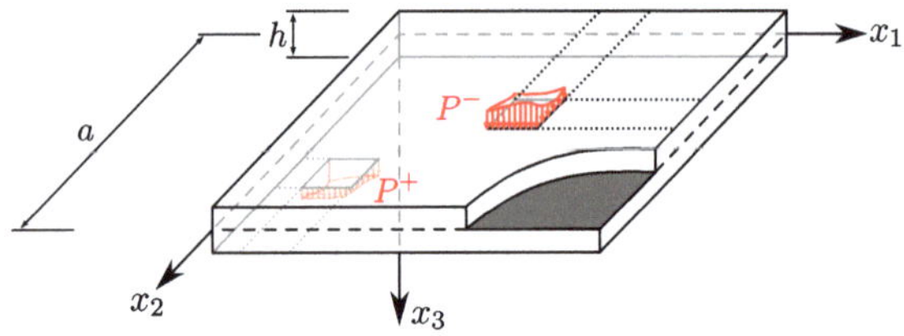

Fig. 1 Plate continuum

For simplicity, we treat isotropic material behavior in the following (2 independent material parameters, e.g., Young's modulus E and Poisson's ratio ν).

- *Load*: The loading condition has to obey certain symmetry and anti-symmetry conditions, for details see Schneider and Kienzler (2015b). We restrict ourselves to transverse loading $P_3^{+/-} =: P^{+/-}$ through the plate faces at $x_3 = \pm h/2$, respectively. Arbitrary transverse loads may be decomposed in an symmetric and anti-symmetric part as depicted in Fig. 2. The symmetric part causes "quenching" and belongs to the class of disc problems (plane stress) and is not subject of the matter treated here. We combine the face loading to

$$P = 2P^a. \tag{2}$$

Moment loading applied by antisymmetric shear forces ($x_3 = \pm h/2 : \ P_\alpha^{+/-}, \ \alpha = 1, 2$) could easily be incorporated.

Plate theories attempt to describe the state of stress and deformation within a three-dimensional plate continuum by quantities defined on the mid-surface. The process is therefore a dimension-reduction problem and as such inherently approximative. In general, three methods are pursued to derive plate theories:

- *Classical or engineering approach*: Point of departure is a set of a priori assumptions for the displacement and stress distributions in thickness direction. These assumptions are based either on experimental observations or on the ingenuity of its proponent. For, e.g., transverse shear stresses are neglected completely or its influence is considered by shear-correction factors. Due to the brilliance of the scholars involved, the accuracy of derived theories are remarkable. A historical account of classical plate theories may be found, e.g., in Szabó (1987). Refined theories are dealt with in Schneider and Kienzler (2015a).

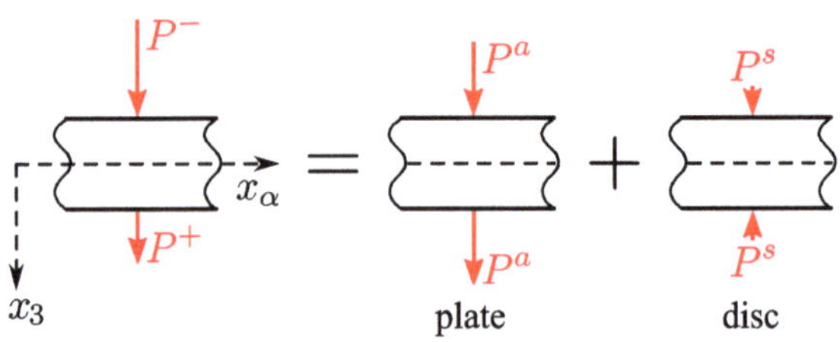

Fig. 2 Arbitrary transverse loading decomposed in an antisymmetric part (plate loading) and a symmetric part (disc loading), ($\alpha = 1, 2$)

- *Direct approach*: The plate mid-plane is considered as a Cosserat surface endowed with a set of deformable directors attached to each point of the plane. The stress resultants and the strain quantities "live" on the plane. Under the assumption that the strain-energy density is a homogeneous quadratic form (linear elasticity), constitutive relations are established involving effective stiffness parameters. Despite this mathematical elegance, the main drawback of the theory is that the identification of the material parameters is not straight forward. Instead of physical experiments often recourse is made to mathematical reflections, which render it necessary to return from the two-dimensional space to the three-dimensional theory. Suitable test problems of the classical theory of elasticity with favorably closed-form solutions are used. The question, however, which test problems should be used, is by far not resolved. The method thus contains a certain degree of arbitrariness. Regularly, the shear-correction factor is the first, nontrivial eigenvalue of a Sturm–Liouville problem (Altenbach 2000), which can be solved analytically only for special cases. An excellent overview over the theories relying on the direct approach is given in Altenbach et al. (2009) with an extended bibliography.
- *Consistent approach*: Displacement, stresses, and strains are developed in direction of thickness into Fourier series with respect to a suitable basis. After introducing proper nondimensional quantities and integrating over the plate thickness, the dimensionless, small plate parameter c^2 (1) evolves quite naturally. The elastic potential appears as power series in this plate parameter. It can be shown (Schneider et al. 2014) that the infinite set of corresponding two-dimensional Euler-Lagrange equations is indeed equivalent to the problem of the three-dimensional theory of elasticity. This infinite series can be truncated at different orders giving rise to hierarchical plate theories. In a forth-coming paper, an a priori error estimate for truncated theories will be given. In combination with the pseudo-reduction method (for details see Schneider and Kienzler 2011), plate-differential equations and equations for the stress resultants are derived without any recourse to either a priori assumptions or shear-correction factors.

In several papers, plate theories from the engineering approach are compared with those of the consistent approach (e.g., Schneider and Kienzler 2015a; Kienzler 2002; Kienzler and Schneider 2012), whereas a comparison between theories of the direct approach and the consistent approach is still missing.

The honoree of this volume has contributed much to plate theories. Many of these papers, which will be referenced and discussed at different places within this volume, are devoted to theories relying on the direct approach, whereas the presenting authors of this paper are firm proponents of the consistent approach.Therefore, it is striking to contribute to a kind of unification of both theories.

In this paper, we adopt the theory of W.A. Palmow and H. Altenbach (1982). There is no reason that we just choose this paper of all, except that the Sturm–Liouville problem is explained in some detail and all equations needed for the following are given. In order to keep the equations as simple as possible, we restrict ourselves to the special case:

- No moment loads, the load is merely applied transversely through the plate faces $P = 2P^a$ (cf. Fig. 2).
- The plate material is homogeneous and isotropic.

Since the notation used in Palmow and Altenbach (1982) deviates to some extent to that of the authors, we provide a list of correspondence in Sect. 2, where the theory is recapitulated. For simplicity, when identical symbols are used for different items, the upper-right capital letter A is used for Altenbach and K for Kienzler (in all modesty). Since coordinates and displacements are nondimensionalized by the characteristic length a (in K), the letter a appears in this list frequently.

We compare the results of Palmow/Altenbach (A) with those of Schneider/Kienzler (K) and Marguerre/Henckey (M) in Sect. 3 by just reporting (without any derivations) the respective equations. In Sect. 4, we supply evidence for the introduction of a measure of a thickness strain with which all equations coincide. The extended theory is thus equivalent within a consistent second-order approximation.

2 Palmow/Altenbach Theory

We more or less copy Fig. 1 of Palmow and Altenbach (1982) as Fig. 3a and juxtapose it with definitions usually found in textbooks on plates and shells, cf., e.g., Eschenauer et al. (1997), Maguerre and Woernle (1995).

2.1 Statics

First, we consider the list of correspondence for coordinates, differential operators and static quantities in Table 1, where also an index notation is used. Greek indices range from 1 to 2, the summation convention is applied for repeated indices, and a comma with a following index stands for partial differentiation with respect to the

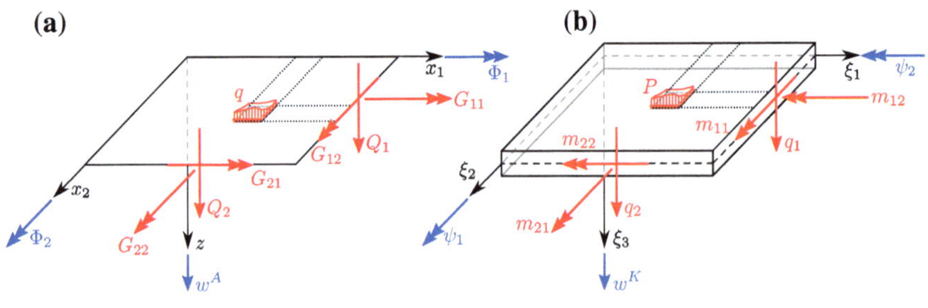

Fig. 3 Displacements and stress resultants defined in Palmow/Altenbach (**a**) and Schneider/Kienzler (**b**)

Table 1 List of correspondence for static quantities

	Palmow/Altenbach	Schneider/Kienzler
Coordinates	x_1 x_2 z	$\xi_1 = {}^{x_1}/_a$ $\xi_2 = {}^{x_2}/_a$ $\xi_3 = {}^{z}/_a$
Transverse shear forces	$\underline{Q} = (Q_\alpha) = \begin{pmatrix} Q_1 \\ Q_2 \end{pmatrix}$ $\underline{Q}_x = (Q_{x\alpha}) = \begin{pmatrix} -Q_2 \\ Q_1 \end{pmatrix}$	$\underline{q} = (q_\alpha) = \underline{Q}$ $Q_{x\alpha} = \varepsilon_{3\beta\alpha} q_\beta$
Moments	$\underline{\underline{G}} = (G_{\alpha\beta}) = \begin{bmatrix} G_{11} & G_{12} \\ G_{21} & G_{22} \end{bmatrix}$	$\underline{\underline{m}} = (m_{\alpha\beta}) = \begin{bmatrix} m_{11} & m_{12} \\ m_{21} & m_{22} \end{bmatrix}$ $(G_{\alpha\beta}) = (\varepsilon_{3\gamma\beta}\, m_{\alpha\gamma}) = \begin{bmatrix} -m_{12} & m_{11} \\ -m_{22} & m_{21} \end{bmatrix}$
Loads	q	$P = 2P^a = q$
Differential-Operators	$\vec{\nabla}() = \begin{pmatrix} \frac{\partial()}{\partial x_1} \\ \frac{\partial()}{\partial x_2} \end{pmatrix}$ $\vec{\nabla}_x() = \begin{pmatrix} -\frac{\partial()}{\partial x_2} \\ \frac{\partial()}{\partial x_1} \end{pmatrix}$ $\vec{\nabla} \bullet \underline{c} = \frac{\partial c_1}{\partial x_1} + \frac{\partial c_2}{\partial x_2}$ $\vec{\nabla}\underline{c} = \begin{bmatrix} \frac{\partial c_1}{\partial x_1} & \frac{\partial c_2}{\partial x_1} \\ \frac{\partial c_1}{\partial x_2} & \frac{\partial c_2}{\partial x_2} \end{bmatrix}$	$((),_\alpha) = \begin{pmatrix} \frac{\partial()}{\partial \xi_1} \\ \frac{\partial()}{\partial \xi_2} \end{pmatrix} = a \begin{pmatrix} \frac{\partial()}{\partial x_1} \\ \frac{\partial()}{\partial x_2} \end{pmatrix} = a\vec{\nabla}()$ $(\varepsilon_{3\beta\alpha}(),_\beta) = a\vec{\nabla}_x()$ $(c_{\alpha,\alpha}) = a\vec{\nabla} \bullet \underline{c}$ $(c_{\alpha,\beta})^T = \begin{bmatrix} c_{1,1} & c_{2,1} \\ c_{1,2} & c_{2,2} \end{bmatrix} = a\vec{\nabla}\underline{c}$

coordinate ξ_α indicated. The tensor of unity is given either by $\underline{\underline{I}}$ (A) or by Kronecker's delta $\delta_{\alpha\beta}(K)$, and $\varepsilon_{3\alpha\beta}$ is the two-dimensional permutation tensor

$$\delta_{\alpha\beta} = \begin{bmatrix} 1 & 0 \\ 0 & 1 \end{bmatrix}, \qquad \varepsilon_{3\alpha\beta} = \begin{bmatrix} 0 & 1 \\ -1 & 0 \end{bmatrix}. \tag{3}$$

The equilibrium equations (1.4) and (1.6) of Palmow and Altenbach (1982) thus transform to

$$\begin{aligned} \vec{\nabla} \bullet \underline{Q} + q &= 0 \;\triangleq\; \tfrac{1}{a} q_{\alpha,\alpha} + P = 0, \\ \vec{\nabla} \bullet \underline{\underline{G}} - \underline{Q}_x &= 0 \;\triangleq\; \tfrac{1}{a} m_{\alpha\beta,\alpha} - q_\beta = 0, \end{aligned} \tag{4}$$

and read rather classical.

2.2 Kinematics

Now we consider the list of correspondence for the kinematic quantities in Table 2. Together with the definitions of Table 1, several relations are revealed and included in the Table. In anticipation of the following, the quantity ψ^K, which is a key quantity in the Schneider/Kienzler theory is introduced. This quantity

$$\psi^K = \varepsilon_{3\alpha\beta}\psi_{\beta,\alpha} \tag{5}$$

serves as a measure for the shear deformation. In the classical theory, i.e., in the first-order approximation, ψ^K vanishes by invoking Kirchhoff's celebrated normal hypothesis

$$\psi_\beta = -w_{,\beta}, \tag{6}$$

whereas in the second-order approximation, ψ^K is a fast decaying function and describes edge effects.

2.3 Constitutive Equations

Within the linear theory of elasticity, it is postulated that the strain-energy density per unit of plate area is a homogeneous quadratic form of the components of $\underline{\Lambda}$ and $\underline{\underline{M}}$ (cf. Table 2). For an isotropic plate, the strain-energy density depends merely on the modul of $\underline{\Lambda}$, i.e., $\underline{\Lambda} \bullet \underline{\Lambda}$ and on the invariants of $\underline{\underline{M}}$, i.e., $(\underline{\underline{I}} \bullet \bullet \underline{\underline{M}}^{Sym})^2$, $\underline{\underline{M}}^{Sym} \bullet \bullet \underline{\underline{M}}^{Sym}$ and $\underline{\underline{M}}^{Ske} \bullet \bullet \underline{\underline{M}}^{Ske}$. Thus, we have four independent parameters denoted by A, B, Γ, and F. Following the algebra in Palmow and Altenbach (1982) and introducing our denotation, we end up with the following constitutive relations:

$$\begin{aligned} m_{\alpha\beta} &= \tfrac{1}{a}\left\{2\Gamma\psi_{\beta,\alpha} + (F-\Gamma)\,\psi_{\gamma,\gamma}\delta_{\alpha\beta} + B\varepsilon_{3\alpha\beta}\varepsilon_{3\gamma\delta}\psi_{\delta,\gamma}\right\}, \\ q_\alpha &= A\left(w_{,\alpha} + \psi_\alpha\right). \end{aligned} \tag{7}$$

(Note that in (2.6), w has to be replaced by $\vec{\nabla} w$ and $-F$ by $+F$ in (2.9).)

2.4 Plate-Differential Equation

Before insertion of the constitutive equations (7) into the equilibrium equations (4), the authors represent the vector of rotations with a Helmholtz decomposition by two scalar functions ϕ^A and ψ^A according to

Table 2 List of correspondence for kinematic quantities

	Palmow/Altenbach	Schneider/Kienzler
Displacements	w^A	$w^K = w^A/a$
Angles of rotation	$\underline{\Phi} = (\Phi_\alpha) = \begin{pmatrix} \Phi_1 \\ \Phi_2 \end{pmatrix}$ $\underline{\Phi}_x = (\Phi_{x\alpha}) = \begin{pmatrix} -\Phi_2 \\ \Phi_1 \end{pmatrix}$	$\underline{\psi} = (\psi_\alpha) = \begin{pmatrix} \psi_1 \\ \psi_2 \end{pmatrix}$ $(\Phi_\alpha) = (\varepsilon_{3\beta\alpha}\psi_\beta) = \begin{pmatrix} -\psi_2 \\ \psi_1 \end{pmatrix}$ $(\Phi_{x\alpha}) = (-\psi_\alpha) = \begin{pmatrix} -\psi_1 \\ -\psi_2 \end{pmatrix}$
Distortion tensor	$\underline{\underline{M}} = (M_{\alpha\beta}) = \vec{\nabla}\,\underline{\Phi} = \begin{bmatrix} \frac{\partial \Phi_1}{\partial x_1} & \frac{\partial \Phi_2}{\partial x_1} \\ \frac{\partial \Phi_1}{\partial x_2} & \frac{\partial \Phi_2}{\partial x_2} \end{bmatrix}$	$(M_{\alpha\beta}) = \left(\frac{1}{a}\Phi_{\beta,\alpha}\right) = \left(\frac{1}{a}\varepsilon_{3\gamma\beta}\psi_{\gamma,\alpha}\right) = \frac{1}{a}\begin{bmatrix} -\psi_{2,1} & \psi_{1,1} \\ -\psi_{2,2} & \psi_{1,2} \end{bmatrix}$
Symmetric part	$\underline{\underline{M}}^{Sym} = \frac{1}{2}\left(\frac{\partial \Phi_\alpha}{\partial x_\beta} + \frac{\partial \Phi_\beta}{\partial x_\alpha}\right)$ $= \begin{bmatrix} \frac{\partial \Phi_1}{\partial x_1} & \frac{1}{2}\left(\frac{\partial \Phi_2}{\partial x_1} + \frac{\partial \Phi_1}{\partial x_2}\right) \\ \frac{1}{2}\left(\frac{\partial \Phi_1}{\partial x_2} + \frac{\partial \Phi_2}{\partial x_1}\right) & \frac{\partial \Phi_2}{\partial x_2} \end{bmatrix}$	$M^{Sym}_{\alpha\beta} = \frac{1}{2a}(\varepsilon_{3\gamma\beta}\psi_{\gamma,\alpha} + \varepsilon_{3\gamma\alpha}\psi_{\gamma,\beta})$ $= \frac{1}{2a}(\varepsilon_{3\alpha\beta}\psi_{\gamma,\gamma} - 2\varepsilon_{3\alpha\gamma}\psi_{\gamma,\beta})$ $= \frac{1}{a}\begin{bmatrix} -\psi_{2,1} & \frac{1}{2}(\psi_{1,1} - \psi_{2,2}) \\ \frac{1}{2}(\psi_{1,1} - \psi_{2,2}) & \psi_{1,2} \end{bmatrix}$
Trace	$\underline{\underline{I}} \bullet\bullet \underline{\underline{M}} = \underline{\underline{I}} \bullet\bullet \underline{\underline{M}}^{Sym} = \vec{\nabla} \bullet \underline{\Phi}$	$\psi^K = \varepsilon_{3\alpha\beta}\psi_{\beta,\alpha} = \psi_{2,1} - \psi_{1,2}$ $= -aM^{Sym}_{\alpha\alpha} = -aM_{\alpha\alpha} = -\Phi_{\alpha,\alpha} = -a\vec{\nabla} \bullet \underline{\Phi}$
Skewsymmetric part	$\underline{\underline{M}}^{Ske} = \frac{1}{2}\left(\frac{\partial \Phi_\alpha}{\partial x_\beta} - \frac{\partial \Phi_\beta}{\partial x_\alpha}\right)$ $= \begin{bmatrix} 0 & \frac{1}{2}\left(\frac{\partial \Phi_2}{\partial x_1} - \frac{\partial \Phi_1}{\partial x_2}\right) \\ -\frac{1}{2}\left(\frac{\partial \Phi_2}{\partial x_1} + \frac{\partial \Phi_1}{\partial x_2}\right) & 0 \end{bmatrix}$	$M^{Ske}_{\alpha} = \frac{1}{2a}(\varepsilon_{3\gamma\beta}\psi_{\gamma,\alpha} - \varepsilon_{3\gamma\alpha}\psi_{\gamma,\beta})$ $= \frac{1}{2a}\varepsilon_{3\alpha\beta}\psi_{\gamma,\gamma}$ $= \frac{1}{a}\begin{bmatrix} 0 & \frac{1}{2}(\psi_{1,1} + \psi_{2,2}) \\ -\frac{1}{2}(\psi_{1,1} + \psi_{2,2}) & 0 \end{bmatrix}$
Shear deformation	$\underline{\Lambda} = \vec{\nabla}(w^A) - \underline{\Phi}_x$ $= \begin{pmatrix} \frac{\partial w^A}{\partial x_1} + \Phi_2 \\ \frac{\partial w^A}{\partial x_2} - \Phi_1 \end{pmatrix}$	$(\Lambda_\alpha) = (w^K_{,\alpha} + \psi_\alpha) = \begin{pmatrix} w^K_{,1} + \psi_1 \\ w^K_{,2} + \psi_2 \end{pmatrix}$

$$\underline{\Phi} = \vec{\nabla}\phi^A + \vec{\nabla}_x\psi^A = \begin{pmatrix}\Phi_1\\ \Phi_2\end{pmatrix} = \begin{pmatrix}\frac{\partial\phi^A}{\partial x_1} - \frac{\partial\psi^A}{\partial x_2}\\ \frac{\partial\phi^A}{\partial x_2} + \frac{\partial\psi^A}{\partial x_1}\end{pmatrix}. \tag{8}$$

Transforming to our notation, we find

$$\psi_\alpha = \psi^A_{,\alpha} + \varepsilon_{3\alpha\beta}\phi^A_{,\beta}, \tag{9}$$

and by simple algebra, we assemble some helpful relations for the following:

$$\begin{aligned}\psi_{\alpha,\beta} &= \psi^A_{,\alpha\beta} + \varepsilon_{3\alpha\gamma}\phi^A_{,\gamma\beta},\\ \psi_{\alpha,\alpha} &= \Delta\psi^A,\\ \varepsilon_{3\alpha\beta}\,\psi_{\beta,\alpha} &= \psi^K = -\Delta\phi^A.\end{aligned} \tag{10}$$

Finally, three partial differential equations are derived for the calculation of the three unknowns w, ψ^A, ϕ^A (see (3.7) and (3.5))

$$\begin{aligned}(F+\Gamma)\,\Delta\psi^A - a^2A\left(w+\psi^A\right) &= 0,\\ (2\Gamma+B)\,\Delta\phi^A - a^2A\,\phi^A &= 0,\\ A\Delta\left(w+\psi^A\right) + aP &= 0.\end{aligned} \tag{11}$$

2.5 *Identification of the Elastic Moduli*

Two test cases are considered: the rectangular plate under pure bending and the circular plate under pure twist. For a homogeneous isotropic material, it turns out that the test cases are verified, if the following values are chosen for the material moduli, cf. (4.8), (4.24), and (4.26),

$$\begin{aligned}F &= \frac{1+\nu}{2}K,\\ \Gamma &= \frac{1-\nu}{2}K,\\ B &= -\frac{1-\nu}{2}K,\\ A &= \mu^2\Gamma = \frac{\pi^2}{h^2}\frac{1-\nu}{2}K = \frac{\pi^2}{24}\frac{1-\nu}{a^2c^2}K.\end{aligned} \tag{12}$$

K is the classical plate stiffness

$$K = \frac{Eh^3}{12(1-\nu^2)} \tag{13}$$

and μ^2

$$\mu^2 = \frac{\pi^2}{h^2} \tag{14}$$

follows as smallest, nontrivial eigenvalue of the Sturm–Liouville problem

$$\begin{aligned} &\frac{d^2V}{dz^2} + \mu^2 V = 0, \\ z = \pm\frac{h}{2}: \quad &\frac{dV}{dz} = 0, \end{aligned} \tag{15}$$

which has this simple form, if the material is homogeneous and isotropic. The factor μ^2 is strongly related to the shear correction factor κ

$$\kappa^{Mi} = \frac{\mu h^2}{12} = \frac{\pi^2}{12} \cong 0.822,$$

which appears first in Mindlin's work (Mindlin 1951), where he investigated vibrations of plates. The value itself does not differ much from Reisser's shear-correction factor

$$\kappa^{Re} = \frac{5}{6} \simeq 0.833,$$

which is based on an energetic mean of the shear-stress distribution over the plate thickness. Replacing A, B, F, and Γ in (7) and (11), the constitutive equations and the equilibrium equations, respectively, yield

$$\begin{aligned} m_{\alpha\beta} &= \frac{K}{a}\left((1-\nu)\psi_{\beta,\alpha} + \nu\, \psi_{\gamma,\gamma}\delta_{\alpha\beta} - \frac{1-\nu}{2}\varepsilon_{3\alpha\beta}\, \psi^K\right), \\ q_\alpha &= \frac{1-\nu}{2}\frac{K\pi^2}{h^2}\left(w_{,\alpha} + \psi_\alpha\right) = \frac{1-\nu}{2}\frac{K}{a^2c^2}\frac{\pi^2}{12}\left(w_{,\alpha} + \psi_\alpha\right); \end{aligned} \tag{16}$$

$$\begin{aligned} w + \psi^A &= c^2 \tfrac{12}{\pi^2}\tfrac{2}{1-\nu}\Delta\psi^A, \\ \phi^A &= c^2 \tfrac{12}{\pi^2}\Delta\phi^A, \\ \tfrac{\pi^2}{12}\tfrac{1}{c^2}\tfrac{1-\nu}{2}K\Delta\left(w + \psi^A\right) &= a^3 P. \end{aligned} \tag{17}$$

Applying the Laplacian Δ to $(17)_1$ and replacing $\Delta(w + \psi^A)$ in $(17)_3$ leads to a single equation for ψ^A as

$$K\Delta\Delta\psi^A = -a^3 P. \tag{18}$$

After integrating the decoupled differential equation system $(17)_2$ and (18) for ϕ^A and ψ^A, the transverse displacement w follows from $(17)_1$ without further integration.

Before proceeding further, we want to estimate the order of magnitude of the terms involved. We introduce the tension/compression stiffness D

$$D = \frac{Eh}{1 - \nu^2}, \tag{19}$$

and obtain

$$\begin{aligned} K, B, F, \Gamma &\sim a^2 D \cdot O(c^2), \\ A &\sim \quad D \cdot O(c^0). \end{aligned} \tag{20}$$

From Schneider and Kienzler (2015a), Schneider et al. (2014), Kienzler (2002), Kienzler and Schneider (2012) we further know

$$\begin{aligned} P &= && O(c^2), \\ \psi^K &= -\Delta\phi^A = && O(c^2), \\ c^2 \Delta\psi^K &= && O(c^2). \end{aligned} \tag{21}$$

With it, the equations read

$$\begin{aligned} m^{\alpha\beta} &= \tfrac{K}{a}\left((1-\nu)\psi_{\beta,\alpha} + \nu\psi_{\gamma,\gamma}\delta_{\alpha\beta}\right) - \tfrac{1-\nu}{2}\varepsilon_{3\alpha\beta}\psi^K + D \cdot O(c^4), \\ q_\alpha &= \tfrac{K}{a^2}\tfrac{1-\nu}{2}\tfrac{\pi^2}{12c^2}(w_{,\alpha} + \psi_\alpha) + D \cdot O(c^2); \end{aligned} \tag{22}$$

$$\begin{aligned} K\Delta\Delta\psi^A &= -a^3 P + O(c^4), \\ \phi^A &= \tfrac{12c^2}{\pi^2}\Delta\phi^A + O(c^6), \\ w + \psi^A &= c^2 \tfrac{24}{\pi^2(1-\nu)}\Delta\psi^A + O(c^4). \end{aligned} \tag{23}$$

It turns out that the equations approximate the three-dimensional theory on quite different scales and are far from the idea of a uniform-approximation technique. We come back to this later. Now, we derive a different representation of the equations in order to compare them with other theories.

We multiply $(23)_3$ with c^2 and apply the biharmonic operator $\Delta\Delta$

$$c^2\Delta\Delta(w + \psi^A) = c^4 \frac{24}{\pi^2(1-\nu)}\Delta\Delta\Delta\psi^A + O(c^6). \tag{24}$$

Terms of order c^6 are neglected in a consistent second-order approximation. Therefore, it follows

$$c^4\Delta\Delta(w + \psi^A) = 0 + O(c^6) \tag{25}$$

and the term $c^4\Delta\Delta\Delta\psi^A$ on the right-hand side of (24) can be replaced by $-c^4\Delta\Delta\Delta w$, and, combined with Eq. $(23)_1$ multiplied by c^2, we have

$$K\Delta\Delta\psi^A = -K\Delta\Delta w - c^2 \frac{24}{\pi^2(1-\nu)} a^3 \Delta P = -a^3 P, \tag{26}$$

leading finally with

$$K\Delta\Delta w = a^3\left(P - c^2\frac{24}{\pi^2(1-\nu)}\Delta P\right) + O(c^6) \tag{27}$$

to one partial differential equation in w. In $(23)_2$, we apply the operator Δ and replace $\Delta\phi^A$ by $-\psi^K$, with the result

$$c^2\left(\psi^K - \frac{12c^2}{\pi^2}\Delta\psi^K\right) = 0 + O(c^6)\ . \tag{28}$$

Using a similar procedure, we replace $\psi_{\beta,\alpha}$ by $(10)_1$ in the bending tensor $(16)_1$ and use $(10)_3$ multiplied by c^2 giving rise to

$$\begin{aligned} m^A_{\alpha\beta} = &-\frac{K}{a}\left\{\left(1 + \frac{24}{\pi^2}\frac{c^2}{1-\nu}\Delta\right)\left[(1-\nu)w_{,\alpha\beta} + \nu w_{,\gamma\gamma}\delta_{\alpha\beta}\right]\right. \\ &\left. + c^2\frac{6}{\pi^2}(1-\nu)\left(\varepsilon_{3\alpha\beta}\psi^K_{,\beta} + \varepsilon_{3\beta\gamma}\psi^K_{,\alpha}\right)_{,\gamma}\right\} \end{aligned} \tag{29}$$

For the transverse shear forces $(16)_2$ we use (9), $(23)_3$, $(23)_2$, (5), and $(23)_1$

$$q^A_\alpha = -\frac{K}{a^2}\left(\Delta w_{,\alpha} + \frac{1-\nu}{2}\varepsilon_{3\alpha\beta}\psi_{,\beta}\right) - c^2 a\frac{24}{\pi^2(1-\nu)}P_{,\alpha}\ . \tag{30}$$

3 Comparison of the Palmow/Altenbach Theory with the Theories of Schneider/Kienzler and Marguerre/Hencky

In textbook (Maguerre and Woernle 1995), the theory can be found that we call the theory of Marguerre/Hencky, although this theory is most probably attributed to other scientists. We beg the pardon of any reader who knows better and are grateful for any hints. The derivation is straight forward. The hypothesis is used that straight fibers vertical to the plate mid-plane remain straight and are not stretched, i.e., the thickness strains are neglected ($w = w(\xi_1, \xi_2)$) and that the shear strains are constant over the plate thickness, what is not tolerable, since they have to vanish at the plate faces (Eschenauer et al. 1997). This deficiency is counter balanced using Reissner's (1944) shear-correction factor $\kappa^{Re} = 5/6$. We compile the resulting equations without detailed derivation and add the corresponding equation of the consistent second-order approximation, cf. Kienzler and Schneider (2016) ϑ.

$$\begin{aligned} K\Delta\Delta w^A &= a^3\left(1 - c^2\frac{12}{\pi^2}\frac{2}{1-\nu}\Delta\right)P, \\ K\Delta\Delta w^M &= a^3\left(1 - c^2\frac{6}{5}\frac{2}{1-\nu}\Delta\right)P, \\ K\Delta\Delta w^K &= a^3\left(1 - c^2\frac{6}{5}\frac{2-\nu}{1-\nu}\Delta\right)P; \end{aligned} \tag{31}$$

$$\begin{aligned} (A):&\quad \psi^K - \tfrac{12}{\pi^2}\Delta\psi^K = 0, \\ (M):&\quad \psi^K - \tfrac{6}{5}\Delta\psi^K = 0, \\ (K):&\quad \psi^K - \tfrac{6}{5}\Delta\psi^K = 0; \end{aligned} \tag{32}$$

$$\begin{aligned} q_\alpha^A &= -\frac{K}{a^2}\left(\Delta w_{,\alpha} + \frac{1-\nu}{2}\varepsilon_{3\alpha\beta}\psi^K_{,\beta}\right) - c^2 a\frac{12}{\pi^2}\frac{2}{1-\nu}P_{,\alpha}, \\ q_\alpha^M &= -\frac{K}{a^2}\left(\Delta w_{,\alpha} + \frac{1-\nu}{2}\varepsilon_{3\alpha\beta}\psi^K_{,\beta}\right) - c^2 a\frac{6}{5}\frac{2}{1-\nu}P_{,\alpha}, \\ q_\alpha^K &= -\frac{K}{a^2}\left(\Delta w_{,\alpha} + \frac{1-\nu}{2}\varepsilon_{3\alpha\beta}\psi^K_{,\beta}\right) - c^2 a\frac{6}{5}\frac{2-\nu}{1-\nu}P_{,\alpha}; \end{aligned} \tag{33}$$

$$\begin{aligned} m_{\alpha\beta}^A = -\tfrac{K}{a}\Big\{&\left(1 + \tfrac{24}{\pi^2(1-\nu)}c^2\Delta\right)\left[(1-\nu)w_{,\alpha\beta} + \nu w_{,\gamma\gamma}\delta_{\alpha\beta}\right] \\ &+ c^2\tfrac{6}{\pi^2}(1-\nu)\left(\varepsilon_{3\alpha\gamma}\psi^K_{,\beta} + \varepsilon_{3\beta\gamma}\psi^K_{,\alpha}\right)_{,\gamma}\Big\}, \\ m_{\alpha\beta}^M = -\tfrac{K}{a}\Big\{&\left(1 + \tfrac{12}{5(1-\nu)}c^2\Delta\right)\left[(1-\nu)w_{,\alpha\beta} + \nu w_{,\gamma\gamma}\delta_{\alpha\beta}\right] \\ &+ c^2\tfrac{3}{5}(1-\nu)\left(\varepsilon_{3\alpha\gamma}\psi^K_{,\beta} + \varepsilon_{3\beta\gamma}\psi^K_{,\alpha}\right)_{,\gamma}\Big\}, \\ m_{\alpha\beta}^K = -\tfrac{K}{a}\Big\{&\left(1 + \tfrac{12}{5(1-\nu)}c^2\Delta\right)\left[(1-\nu)w_{,\alpha\beta} + \nu w_{,\gamma\gamma}\delta_{\alpha\beta}\right] \\ &+ c^2\tfrac{3}{5}(1-\nu)\left(\varepsilon_{3\alpha\gamma}\psi^K_{,\beta} + \varepsilon_{3\beta\gamma}\psi^K_{,\alpha}\right)_{,\gamma}\Big\} \\ &+ c^2\tfrac{6}{5}\tfrac{\nu}{1-\nu}a^2P\delta_{\alpha\beta}. \end{aligned} \tag{34}$$

The factors $ac^2P_{,\alpha}$ and a^2c^2P can be replaced by $c^2\frac{K}{a^2}\Delta\Delta w_{,\alpha}$ and $c^2\frac{K}{a}\Delta\Delta w$, respectively, without loss of approximation accuracy, cf. Schneider et al. (2014).

If $\frac{1}{\kappa^{Mi}} = 12/\pi^2 \cong 1.22$ is replaced by $\frac{1}{\kappa^{Re}} = 6/5 = 1.20$ (the values differ by about 1 %) in the Palmow/Altenbach equations, they coincide with the Marguerre/Hencky equations and vice versa. The factor $5/6$ comes from an energetic reasoning, whereas $\pi^2/12$ is the result from a test example involving a circular plate. If another test example, e.g., a rectangular plate, would have been considered, the shear-correction factor would probably not involve the factor π^2. Therefore, it is immediately obvious that the Palmow/Altenbach and the Marguerre/Hencky theories differ only by the shear-correction factor. Next, let us exemplarily compare the Maguerre/Hencky theory with the consistent approximation. It is seen that the differential equations (32) for the shear-strain measure ψ^K are identical, whereas the differential equations for the transverse displacement w (31), the stress resultants Q_α (33) and $M_{\alpha\beta}$ (34) differ all by the amount

$$c^2\frac{6}{5}\frac{\nu}{1-\nu}\hat{P} = c^2\frac{1}{a^3}K\frac{6}{5}\frac{\nu}{1-\nu}\Delta\Delta\hat{w}, \tag{35}$$

either in the form $\hat{P} = \Delta P$ in (31) in the form $\hat{P} = aP_{,\alpha}$ in (33) or $\hat{P} = a^2P$ in (34). The same reasoning applies for the Palmow/Altenbach theory. In the following, we will interpret this term and show, how to extend the Palmow/Altenbach theory in order to be consistent within a second-order approximation.

4 Extension of the Palmow/Altenbach Theory

In difference to the consistent approach, the Palmow/Altenbach theory does not take strains in thickness direction into account,say ε_{33}. The transverse displacement w is considered an energetically motivated average over the plate thickness but assumed to be constant, i.e., $\varepsilon_{33} = 0$. (Within the Cosserat theory, this could be included by a suitable director.) Since plane-stress conditions are assumed simultaneously (which is a contradiction in itself), a first-order correction in the thickness strain, which is proportional to $c^2\Delta w$, is thus included, cf., e.g., Kienzler (2002). The deviation from the plane-stress state is, therefore, of the order $O(c^4)$ and should be proportional to $\Delta\Delta w$. Let us illustrate the problem by considering a bended plate element (strongly bent for reasons of illustration) in Fig. 4. The longitudinal fibers at the bottom of the plate $\xi_3 = +h/2a$ will be stretched and the fibers on the upper plane $\xi_3 = -h/2a$ are compressed. Indeed for the antisymmetric loading (see Fig. 2), it is an immediate consequence of the three-dimensional problem of elasticity (cf. Schneider and Kienzler 2015b), that u_1 and also u_2 are generally odd functions in

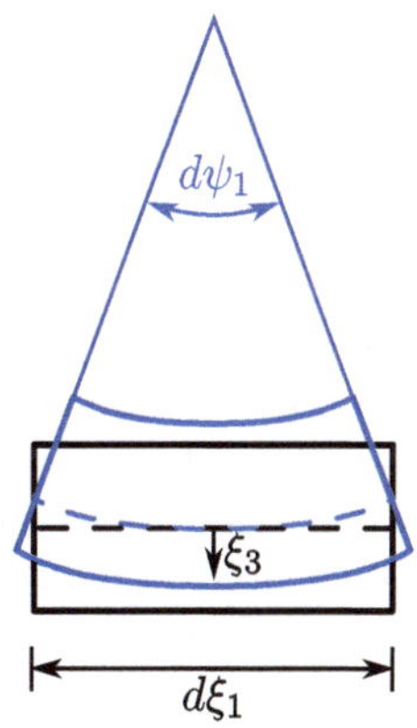

Fig. 4 Bent plate element

ξ_3, whereas u_3 is an even function in ξ_3. Therefore, the resulting in-plane strains $\varepsilon_{\alpha\beta}$ are odd functions, too, i.e.,

$$\varepsilon_{\alpha\beta}(+\xi_3) = -\varepsilon_{\alpha\beta}(-\xi_3). \tag{36}$$

Due to transverse contraction (Poisson's ratio ν), $\varepsilon_{11}$and ε_{22} are accompanied by transverse strains ε_{33}, which are also odd functions in ξ_3, leading therefore to no overall thickness change

$$\int_{-\frac{h}{2a}}^{+\frac{h}{2a}} \varepsilon_{33} d\xi_3 = 0.$$

However, transverse strains are not absent locally except in the neutral plane $\xi_3 = 0$. Since isotropic material is assumed throughout, Hooke's law, cf., e.g., Altenbach et al. (2004)

$$\sigma_{ij} = E_{ijkl}\, \varepsilon_{kl} \tag{37}$$

does not couple shear strains $\varepsilon_{kl}(k \neq l)$ with longitudinal strains $\varepsilon_{kl}(k = l)$ and a measure for the thickness strain will be proportional to

$$\frac{E_{1133}}{E_{1111}} = \frac{E_{2233}}{E_{1111}} = \frac{\nu}{1-\nu}. \tag{38}$$

The in-plane strains in plate theory are proportional to $\psi_{\alpha,\beta}$ (see $(16)_1$)

$$\varepsilon_{\alpha\beta} \sim \psi_{\alpha,\beta}. \tag{39}$$

The one and only possibility to include a scalar (ε_{33}) thickness strain measure in the kinematics is, therefore, to modify the distorsion tensor in an isotropic manner. After these preliminary reflections, we try the following educated guess

$$\begin{aligned} (\psi^*_{\alpha,\beta}) &= \left(\psi_{\alpha,\beta} + c^2\alpha\frac{\nu}{1-\nu}\Delta\Delta w\delta_{\alpha\beta}\right) \\ &= \begin{bmatrix} \psi_{1,1} + c^2\alpha\frac{\nu}{1-\nu}\Delta\Delta w & \psi_{1,2} \\ \psi_{2,1} & \psi_{2,2} + c^2\alpha\frac{\nu}{1-\nu}\Delta\Delta w \end{bmatrix}. \end{aligned} \tag{40}$$

The shortcoming of this ansatz is that material parameters (here: ν) are involved in the kinematic equations, but some a priori assumptions have to be introduced to cover the thickness strain properly. An alternative would be to introduce the thickness strain measure as independent kinematical variable, introduce an energetically complementary stress resultant, e.g., ${}^1m_{33}$ (cf. Kienzler 2002; Schneider and Kienzler 2015a; Kienzler and Schneider 2012), and to develop additional constitutive equations from the extended strain-energy density U. However, we follow the much less pretentious ansatz (40) with a constant α, which will be accommodated in the following.

By inspection of the strain measures in Table 2, it turns out that merely the skew-symmetric part $\underline{\underline{M}}^{Ske}$ is changed, whereas all other strain measures are not affected

$$\underline{\underline{M}}^{Ske*} = \begin{bmatrix} 0 & \frac{1}{2}(\psi_{1,1} + \psi_{2,2}) + c^2\alpha\frac{\nu}{1-\nu}\Delta\Delta w \\ -\frac{1}{2}(\psi_{1,1} + \psi_{2,2}) - c^2\alpha\frac{\nu}{1-\nu}\Delta\Delta w & 0 \end{bmatrix},$$

$$M^{Ske*}_{\alpha\beta} = \varepsilon_{3\alpha\beta}\left(\frac{1}{2}\psi_{\gamma,\gamma} + c^2\alpha\frac{\nu}{1-\nu}\Delta\Delta w\right). \tag{41}$$

In the strain energy, only the term involving the material parameter F is changed and this change affects merely the tensor of bending moments. Introducing (40) into (7) reads

$$m^*_{\alpha\beta} = \frac{1}{a}\left\{2\Gamma\psi_{\beta,\alpha} + (F-\Gamma)\psi_{\gamma,\gamma}\delta_{\alpha\beta} + 2Fc^2\alpha\frac{\nu}{1-\nu}\Delta\Delta w\delta_{\alpha\beta} + B\varepsilon_{3\alpha\beta}\psi^K\right\},$$
$$q^*_\alpha = q_\alpha. \tag{42}$$

The equilibrium equations (4) in the form $\frac{1}{a}m^*_{\beta\alpha,\beta} - q^*_\alpha = 0$ deliver with ψ^A and ϕ^A as introduced in (19)

$$\begin{aligned} &(F+\Gamma)\Delta\psi^A - a^2A(w+\psi^A) + 2Fc^2\alpha\tfrac{\nu}{1-\nu}\Delta\Delta w = 0,\\ &(2\Gamma + B)\Delta\phi^A - a^2A\,\phi^A = 0,\\ &A\Delta(w+\psi^A) = -aP. \end{aligned} \tag{43}$$

It may be mentioned that the order-of-magnitude estimation (20) and (21), stays valid, although the material parameters are not identified yet. Note that in comparison with (11), only $(11)_1$ has been extended by a term of the order $D \cdot O(c^4)$.

$$w + \psi^A = \frac{F+\Gamma}{a^2A}\Delta\psi^A + \frac{2F}{a^2A}c^2\alpha\frac{\nu}{1-\nu}\Delta\Delta w + O(c^6), \tag{44}$$

and comparing with (23) shows that the term $O(c^4)$ is present and the equation is now accurate within a consistent second-order approach. Multiplying by c^2 and c^4 leads to

$$c^2(w+\psi^A) = c^2\frac{F+\Gamma}{a^2A}\Delta\psi^A + O(c^6), \tag{45}$$

$$c^4\psi^A = -c^4 w + O(c^6), \tag{46}$$

respectively.

We use (9), $(10)_1$, (45), (46), $(43)_2$, and $(10)_3$ to rearrange $m^*_{\alpha\beta}$ (42)

$$\begin{aligned}
m^*_{\alpha\beta} &= \frac{1}{a}\{2\Gamma\ (\psi^A_{,\alpha\beta} + \varepsilon_{3\beta\gamma}\phi^A_{,\gamma\alpha}) \\
&+ (F-\Gamma)\psi^A_{,\gamma\gamma}\delta_{\alpha\beta} + B\varepsilon_{3\alpha\beta}\psi^K \\
&+ 2Fc^2\alpha\frac{\nu}{1-\nu}\ \Delta\Delta w\delta_{\alpha\beta}\} + O(c^6) \\
&= \frac{1}{a}\left\{2\Gamma\left(-w_{,\alpha\beta} - c^2\frac{F+\Gamma}{a^2A}\Delta w_{,\alpha\beta}\right)\right. \\
&+ (F-\Gamma)\left(-w_{,\gamma\gamma} - c^2\frac{F+\Gamma}{a^2A}\Delta w_{,\gamma\gamma}\right)\delta_{\alpha\beta} \\
&- 2\Gamma\frac{2\Gamma+B}{a^2A}\varepsilon_{3\beta\gamma}\psi^K_{,\gamma\alpha} + B\frac{2\Gamma+B}{a^2A}\varepsilon_{3\alpha\beta}\psi^K_{,\gamma\gamma} \\
&+ 2Fc^2\alpha\frac{\nu}{1-\nu}\ \Delta\Delta w\delta_{\alpha\beta}\} \\
&= -\frac{1}{a}\left\{\left(1+\frac{F+\Gamma}{a^2A}\Delta\right)(2\Gamma w_{,\alpha\beta} + (F-\Gamma)w_{,\gamma\gamma}\delta_{\alpha\beta})\right. \\
&+ \frac{2\Gamma+B}{a^2A}\left(2\Gamma\varepsilon_{3\beta\gamma}\psi^K_{,\gamma\alpha} - B\varepsilon_{3\alpha\beta}\psi^K_{,\gamma\gamma}\right) \\
&\left.- 2Fc^2\alpha\ \frac{\nu}{1-\nu}\Delta\Delta w\delta_{\alpha\beta}\right\} + D\cdot O(c^6).
\end{aligned} \tag{47}$$

Now we use the consistent second-order plate theory as "test example" to identify the material parameters. Comparing (47) with $(34)_3$, we find with the identity

$$\varepsilon_{3\alpha\beta}\psi^K_{,\gamma\gamma} = \varepsilon_{3\alpha\gamma}\psi^K_{,\gamma\beta} - \varepsilon_{3\beta\gamma}\psi^K_{,\gamma\alpha} \tag{48}$$

the following relations

$$\begin{aligned}
2\Gamma &= K(1-\nu), \\
F-\Gamma &= \nu K, \\
\tfrac{F+\Gamma}{a^2A} &= \tfrac{12c^2}{5(1-\nu)}, \\
\tfrac{2\Gamma+B}{a^2A}(2\Gamma+B) &= Kc^2\tfrac{3}{5}(1-\nu), \\
\tfrac{2\Gamma+B}{a^2A}B &= -Kc^2\tfrac{3}{5}(1-\nu), \\
2Fc^2\alpha\tfrac{\nu}{1-\nu}\Delta\Delta w &= Kc^2\tfrac{6}{5}\tfrac{\nu}{1-\nu}\Delta\Delta w.
\end{aligned} \tag{49}$$

This though overdetermined system of algebraic equations has a unique solution

$$\Gamma = \frac{1-\nu}{2}K,$$

$$
\begin{aligned}
F &= \frac{1+\nu}{2}K, \\
B &= -\frac{1-\nu}{2}K, \\
A &= \frac{5}{12}\frac{1-\nu}{a^2c^2}K, \\
\alpha &= \frac{6}{5}\frac{1}{1+\nu}.
\end{aligned} \tag{50}
$$

In comparison to (12), it turns out that Γ, F, and B are not changed, whereas in A, the term $\frac{\pi^2}{24} \cong 0.411$ is replaced by $\frac{5}{12} \cong 0.417$, i.e., κ^{Mi} by κ^{Re}, and the thickness strain measure, which has to be added in the distorsion tensor (45), is given by

$$
c^2\frac{6}{5}\frac{\nu}{1-\nu^2}\Delta\Delta w.
$$

It may be mentioned that a less educated though also less intriguing guess, i.e.,

$$
c^2\beta\vartheta(\xi_1\xi_2)
$$

would lead to the same result, since ϑ would be identified by $\Delta\Delta w$. After identifying all constants, let us explore the change in q^*_α. Starting with $(7)_2$, we have with (9)

$$
q^*_\alpha = A(w_{,\alpha} + \psi^A_{,\alpha} + \varepsilon_{3\alpha\beta}\, \phi^A_{,\beta}).
$$

Preceding further, the extended Eq. (44) has to be invoked rather than $(11)_1$ leading to

$$
q^*_\alpha = \frac{1}{a^2}\left\{(F+\Gamma)\Delta\psi^A_{,\alpha} + 2Fc^2\alpha\frac{\nu}{1-\nu}\Delta\Delta w_{,\alpha} + a^2A\, \varepsilon_{3\alpha\beta}\phi^A_{,\beta}\right\}. \tag{51}
$$

Using finally (45), (43), and $(10)_3$, we arrive at

$$
\begin{aligned}
q^*_\alpha = -\frac{1}{a^2}\Bigg\{&(F+\Gamma)\left(1+\frac{F+\Gamma}{a^2A}\Delta\right)\Delta w_{,\alpha} - 2Fc^2\alpha\frac{\nu}{1-\nu}\Delta\Delta w_{,\alpha} \\
&+ (B+2\Gamma)\varepsilon_{3\alpha\beta}\psi^K_{,\beta}\Bigg\}
\end{aligned} \tag{52}
$$

Introducing the already identified constants (50), we recover

$$
\begin{aligned}
q^*_\alpha &= \frac{K}{a^2}\left\{\Delta w_{,\alpha} + c^2\frac{6(2-\nu)}{5(1-\nu)}\Delta\Delta w_{,\alpha} + \frac{1-\nu}{2}\varepsilon_{3\alpha\beta}\psi^K_{,\beta}\right\} \\
&\overset{!}{=} q^K_\alpha + O(c^6).
\end{aligned} \tag{53}
$$

Finally, we apply the pseudo-reduction technique to the differential equation system (43). First, we replace the moduli by (50) and obtain

$$K\Delta\psi^A - \frac{5(1-\nu)}{12c^2}K(w+\psi^A) + c^2K\frac{6}{5}\frac{\nu}{1-\nu}\Delta\Delta w = 0$$
$$\frac{1-\nu}{2}\Delta\phi^A - \frac{5(1-\nu)}{12c^2}K\phi^A = 0 \tag{54}$$
$$\frac{5(1-\nu)}{12a^2c^2}K\Delta(w+\psi^A) = -aP.$$

We apply the Laplacian to $(54)_1$ and eliminate $\Delta(w+\psi^A)$ by $(54)_3$

$$K\Delta\Delta\psi^A + a^3P + c^2K\frac{6}{5}\frac{\nu}{1-\nu}\Delta\Delta\Delta w + O(c^6) = 0. \tag{55}$$

Application of the operator $c^2\Delta$ to $(54)_3$ delivers

$$K\Delta\Delta\psi^A = -K\Delta\Delta w - a^3\Delta P\frac{12}{5}c^2\frac{1}{1-\nu} + O(c^6). \tag{56}$$

Finally, $c^2K\Delta\Delta\Delta w$ may be replaced by $c^2a^3\Delta P$ leading to

$$K\Delta\Delta w = a^3\left(1 - c^2\frac{6}{5}\frac{2-\nu}{1-\nu}\Delta\right)P + O(c^6), \tag{57}$$

coinciding with the differential equation of the consistent second-order plate theory.

5 Conclusion

A comparison between a plate theory based on the direct approach and on the consistent second-order plate theory is given. When adopting the Palmow/Altenbach theory, the identification of the elastic moduli leads to a shear-correction factor, which is determined to $\pi^2/12$. Thickness strains are not considered. In the paper, it is shown how a small addition in the distorsion tensor by a scalar-valued thickness measure extends the Palmow/Altenbach theory to a consistent second-order plate theory. The same reasoning applies for the Marguerre/Hencky-plate theory.

References

Altenbach, H.: On the determination of transverse shear stiffnesses of orthotropic plates. ZAMP—Zeitschrift für angewandte Mathematik und Physik **51**(4), 629–649 (2000). doi:10.1007/s000330050021

Altenbach, H., Altenbach, J., Kissing, W.: Mechanics of Composite Structural Elements. Springer, Berlin (2004). doi:10.1007/978-3-662-08589-9

Altenbach, J., Altenbach, H., Eremeyev, V.A.: On generalized Cosserat-type theories of plates and shells: a short review and bibliography. Arch. Appl. Mech. **80**(1), 73–92 (2009). doi:10.1007/s00419-009-0365-3

Eschenauer, H., Olhoff, N., Schnell, W.: Applied Structural Mechanics: Fundamentals of Elasticity, Load-Bearing Structures, Structural Optimization. Springer, Berlin (1997). doi:10.1007/978-3-642-59205-8

Kienzler, R.: On consistent plate theories. Arch. Appl. Mech. **72**(4), 229–247 (2002). doi:10.1007/s00419-002-0220-2

Kienzler, R., Schneider, P.: Consistent theories of isotropic and anisotropic plates. J. Theor. Appl. Mech. **50**(3), 755–768 (2012)

Kienzler, R., Schneider, P.: Consistent plate theories—a matter still not settled? Advances in Mechanics: Theoretical, Computational and Interdisciplinary Issues, pp. 377–390. CRC Press, London (2016). doi:10.1007/978-3-642-21855-2_25

Maguerre, K., Woernle, H.T.: Elastische Platten. Technical report, Bibliographisches Institut, Mannheim (1995)

Mindlin, R.D.: Influence of rotary inertia and shear in flexural motion of isotropic, elastic plates. J. Appl. Mech. (ASME) **18**, 31–38 (1951)

Palmow, W.A., Altenbach, H.: Über eine Cosseratsche Theorie für elastische Platten. Technische Mechanik **3**(3), 5–9 (1982)

Reissner, E.: On the theory of bending of elastic plates. J. Math. Phys. **23**, 184–191 (1944)

Schneider, P., Kienzler, R.: An algorithm for the automatisation of pseudo reductions of PDE systems arising from the uniform-approximation technique. Shell-Like Structures: Non-classical Theories and Applications, pp. 377–390. Springer, Berlin (2011). doi:10.1007/978-3-642-21855-2_25

Schneider, P., Kienzler, R.: Comparison of various linear plate theories in the light of a consistent second-order approximation. Math. Mech. Solids **20**(7), 871–882 (2015a). doi:10.1177/1081286514554352

Schneider, P., Kienzler, R.: On exact rod/beam/shaft-theories and the coupling among them due to arbitrary material anisotropies. Int. J. Solids Struct. **56–57**, 265–279 (2015b). doi:10.1016/j.ijsolstr.2014.10.022

Schneider, P., Kienzler, R., Böhm, M.: Modeling of consistent second-order plate theories for anisotropic materials. ZAMM—J. Appl. Math. Mech./Zeitschrift für Angewandte Mathematik und Mechanik **94**(1–2), 21–42 (2014). doi:10.1002/zamm.201100033

Szabó, I.: Geschichte der mechanischen Prinzipien: und ihrer wichtigsten Anwendungen. Birkhäuser, Basel (1987). doi:10.1007/978-3-0348-5998-1

Modelling of a Rotating Active Thin-Walled Composite Beam System Subjected to High Electric Fields

Jarosław Latalski

Abstract An electromechanical coupled theory is used to develop the equations of motion of a rotating thin-walled composite beam with surface bonded/embedded piezoelectric transducers. The higher order constitutive relations for the piezoceramic material are used to take into account the impact of a high electric field. In the mathematical model of the hybrid structure, the non-classical effects like material anisotropy, rotary inertia and transverse shear deformation as well as an arbitrary beam pitch angle are incorporated. Moreover, the model considers the hub mass moment of inertia and a non-constant rotating speed case. This approach results in an additional equation of motion for the hub sub-system and enhances the generality of the formulation. It is shown that final equations of motion of the hub–beam system are mutually coupled and form a nonlinear system of partial differential equations. Comparing to the purely mechanical model, the proposed electromechanical one introduces additional stiffness-type couplings between individual degrees of freedom of the system.

1 Introduction

The development of smart materials offers a great potential for advanced modern structural systems. The engineering applications of electromechanical or magneto-mechanical materials are met in aerospace, automotive, civil, mechanical, biomedical and communication engineering disciplines.

One of the most popular materials exhibiting electromechanical properties are piezocrystals and piezoceramics. In smart systems, these materials are embedded in the host structure and used as actuators and sensors by taking advantage of direct and converse piezoelectric effects. Therefore, the smart material is not only expected to bear mechanical loadings but also to feature the capability of strain actuating, structural health monitoring, nondestructive sensing, etc.

J. Latalski (✉)
Department of Applied Mechanics, Lublin University of Technology,
ul. Nadbystrzycka 36, 20-618 Lublin, Poland
e-mail: j.latalski@pollub.pl

K. Naumenko and M. Aßmus (eds.), *Advanced Methods of Continuum Mechanics for Materials and Structures*, Advanced Structured Materials 60,
DOI 10.1007/978-981-10-0959-4_24

Further benefits in structural static and dynamic characteristics may be achieved by a combination of smart materials and composite material technology. Results of scientific research from the past years confirm synergistic interactions of anisotropic composite material tailoring and adaptive material properties. Excellent examples are improvement in helicopter rotor vibration that can be achieved through the implementation of active control technology using smart materials combined with both light-weight and directional properties of multilayered laminates Brockmann (2009).

To fully benefit from the unique adaptive capabilities of piezoelectric composite structures, much research is focused on developing advanced mathematical models of these systems. The initial studies of hybrid systems did not consider the electric and mechanic field couplings, e.g. Detwiler et al. (1995). Also, Song et al. (2001), Librescu and Song (2006) studied thin-walled composites with embedded piezoceramic layers. In the proposed mathematical model of the smart system, authors separately considered mechanical and electrical domains. The mechanical properties of the hybrid material resulted solely from the combination of stiffnesses of two materials. On the other hand, the adaptive nature of the structure was captured by introducing a dynamic moment acting at the tip of the beam. In control algorithms, this moment was related, via a prescribed functional relationship, with one of the various kinematical quantities characterizing the response of the structure. This method, referred to as the boundary moment control methodology, does not consider electromechanical properties of piezoceramics and neglects the electrical and mechanical domains coupling. Moreover, this approach is not capable of taking into consideration any electrical circuitry connected to the piezoelectric device.

In general, the errors that result from using uncoupled models, as opposed to coupled ones, are relatively moderate. However, there are some cases in which very large differences exist when using the two approaches. One such case is for high-frequency vibrations or thick piezoelectric material layers (Thornburgh et al. 2004).

The coupled electromechanical models of smart composite materials were studied by e.g. Mitchell and Reddy (1995). Authors presented a method of enhancing current plate theories to include the charge equations of electrostatics. In this way, a physically correct model of the piezoelectric effect was included. Further studies on coupled domain models of piezoelectric composite plates were done by e.g. Li et al. (1998) and Zhou et al. (2000). Chattopadhyay et al. (1999) elaborated a refined higher order displacement field model of a plate with eccentricity to investigate the behaviour of smart helicopter rotor blades. The proposed theory was based on a three-dimensional model; so no typical one-dimensional beam equations were formulated. The procedure was implemented using finite element method.

Further improvement in modelling hybrid systems may be achieved by taking into account the possible non-linear effects. This is especially important in case of high power electromechanical systems often operated under extreme conditions. The common linear models provide a reasonable approximation of the characteristics of piezoelectric materials at low levels of applied electric field and stress. Unfortunately, these relations become increasingly inaccurate as the electric field and stress levels increase. This is manifested in numerous laboratory experiments on the piezoelectric materials and piezoelectric devices—see e.g. Arafa and Baz (2004), Wagner and

Hagedorn (2003), Yao et al. (2004). Concerning the electric field, the discrepancies between linear and non-linear models are apparent at relatively high electric field levels ($\approx 500\,{}^{\mathrm{V}}/_{\mathrm{mm}}$) (Birman 2005). Nonetheless, as reported by Wagner and Hagedorn (2003), non-linear effects can be observed even if the electric field remains small but the piezoceramic actuators are excited in resonance.

A comprehensive review of non-linear effects in piezoelectric ceramics was given by Hall (2001). Author discussed the most important dielectric and piezoelectric non-linearities in piezoelectric ceramics as well as less known elastic ones. Studies were concluded with the observation that in most ferroelectric ceramics the dielectric, elastic and piezoelectric coefficients increased considerably with electric field and stress amplitude. It was in many cases that the relevant coefficients increased in an almost linear way with field/stress amplitude. In author's opinion, this provides the fairly straightforward means of modelling the non-linear effects in piezoceramic components.

The similar approach to model the non-linear effects in active elements by presuming the coupling coefficients to depend on the electric field has been proposed by other authors too. Priya et al. (2001) found that the elastic and electromechanical properties of the material were dependent on the square of the strain magnitude. Williams (2004) formulated explicit expressions for piezoelectric coefficients as functions of peak-to-peak voltage amplitude squared. Non-linear formula for the piezoelectric coefficients with respect to actuator strain has been used by Silva et al. (2015) to study the non-linear behaviour of piezoelectric energy harvesters.

An alternative approach to consider non-linearities in piezoceramics is to consider a non-linear constitutive equations system based on higher order electromechanical coupling effect. This method was incorporated by e.g. Tiersten is his study of plates (Tiersten 1993). Author proposed using a linear theory in terms of structural displacements but cubic in electric field. Yang et al. (2007) presented an approximate non-linear analysis of a simple plate thickness-shear mode transformer near resonance. Non-linear constitutive equations of electroelasticity were implemented. Birman (2005) considered a physically nonlinear piezoelectric material behaviour and discussed its possible application for a piezoelectric rod subjected to periodic electric field. In finite element studies of composite smart systems, the non-linear constitutive equations were adopted by e.g. Samal et al. (2005).

In the present contribution, the equations of motion of a rotating thin-walled beam with surface mounted/embedded piezoelectric transducer are developed by the Hamilton's method. The higher order constitutive relations for the piezoceramic material are used to take into account the impact of high electric field. In the mathematical model of the hybrid structure, the non-classical effects like material anisotropy, rotary inertia and transverse shear deformation as well as an arbitrary beam pitch angle are incorporated.

To derive governing equations of the system, we take the similar approach as it was taken for conventional composite beam in the previous author's papers (Georgiades et al. 2014; Latalski et al. 2016). That is, the constitutive model at the lamina level in the fibre coordinate system is set, and then transformed to the global coordinate system. Next, the non-linear constitutive model for the piezoceramic material is used to find the stress formula at the level of smart material layer. These relations are finally combined for all layers to establish the constitutive model of the hybrid laminate with the embedded piezoceramic layer.

The expressions for 2-D stress resultants and stress couples are established. At the next step, these are integrated around the profile cross-section mid-line circumference to formulate the final governing equations. The resulting equations are similar to the regular beam ones obtained previously, however, additional terms arise in this case due to the presence of electromechanical coupling.

2 Problem Formulation

Let us consider a slender, straight and elastic composite thin-walled beam clamped at the rigid hub of radius R_0 and inertia J_h rotating about a fixed frame axis CZ_0 as shown in Fig. 1. The hub's current position is described by an angle $\psi(t)$ with respect to an inertial reference frame (X_0, Y_0, Z_0) and the rotational speed of the system $\dot{\psi}(t)$ is assumed to be arbitrary, i.e. not necessarily constant. The system is driven by an external torque T_{ext} applied to the hub.

The beam has a cylindrical or prismatic cross section, spanwise uniform and without initial twist in its natural state. Presetting angle of the beam with respect to a rotation plane (X_0, Y_0) is denoted by θ—see Fig. 1b. The laminae material of the beam is linearly elastic (Hookean) and its principal axes are skewed with respect to the wall local coordinate system by the arbitrary angle α as measured from the cross-section circumferential direction axis s (see Fig. 3 in the following section).

Apart from the fibre unidirectional layers, there are additional piezoceramic layers of thickness h_p. Piezoelectric material is oriented with its polarization axis normal to the plane of the beam and has fully electroded major surfaces. These elements are either symmetrically embedded into the laminate or bonded to the outer surfaces. This is the usual geometry for transversely operating piezoelectric actuators and sensors. Moreover, it is supposed that these piezoelectric layers span the total length

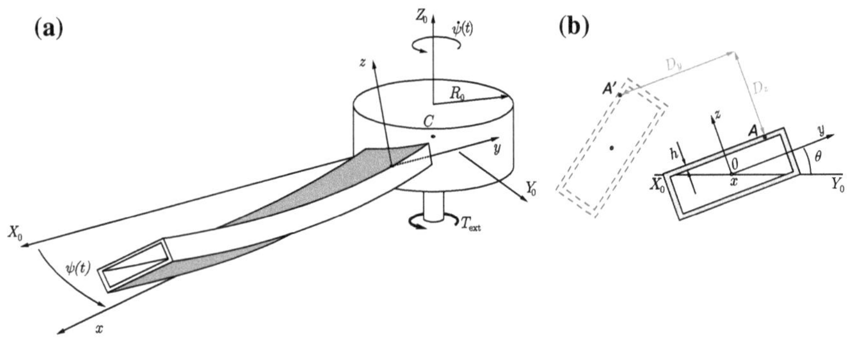

Fig. 1 **a** Rotating thin-walled composite beam with electroded surfaces; the coordinate systems (X_0, Y_0, Z_0) and (x, y, z) are inertial and local one related to the beam, respectively; **b** displacements of a representative point A due to elastic deformation of the specimen

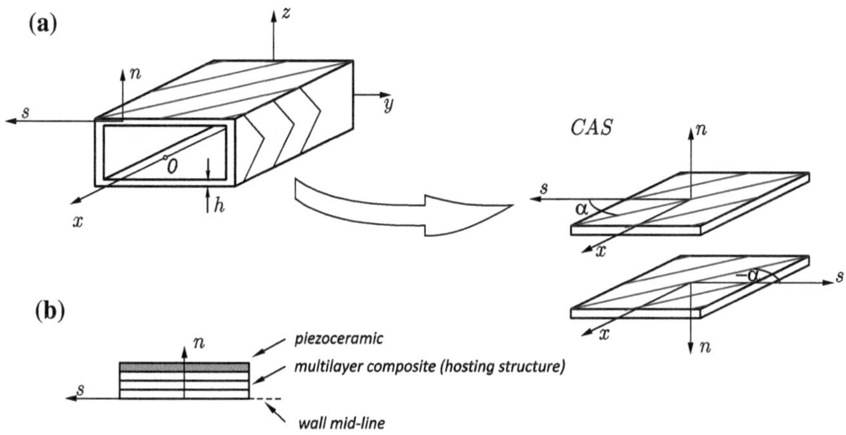

Fig. 2 Rectangular cross-section thin-walled beam with circumferentially asymmetric stiffness (CAS) lamination scheme and surface bonded piezoactuators

of the hosting beam so that the mechanical properties of the structure are spanwise uniform.

For the discussed configuration of the piezoelement, the electric displacement becomes zero along the two in-plane directions and only the out-of-plane component of the electric displacement vector is different from 0.

3 Mathematical Model of the System

For the derivation of the mathematical model of the system, the following kinematic and static assumptions are postulated:

1. the original shape of the cross-section is maintained in its plane, but is allowed to warp out of the plane,
2. the concept of the non-uniform torsional deformation is applied. Therefore, the rate of beam twist $\varphi' = \frac{\mathrm{d}\varphi}{\mathrm{d}x}$ depends in general on the spanwise coordinate x,
3. in addition to the primary warping effects (related to the cross-section shape) a secondary warping related to the wall thickness is also considered,
4. the transverse beam shear deformations γ_{xy} and γ_{xz} are taken into account. These are assumed to be uniform over the beam cross-section,
5. the ratio of wall thickness to the radius of curvature at any point of the beam wall is negligibly small while compared to unity. In a special case of the prismatic beams made of planar segments, this ratio is exactly 0,
6. the stress in transverse normal (σ_{nn}) direction and the hoop stress resultant (N_{ss}) are very small and can be neglected,
7. the piezoceramic layers are perfectly bonded to the hosting structure.

3.1 Kinematic Relations

The mechanical equations of motion of the rotating beam and charge equation of electrostatics are derived according to the extended Hamilton's principle of the least action

$$\delta J = \int_{t_1}^{t_2} \left(\delta T - \delta U + \delta W_{\mathrm{ext}}\right)\mathrm{d}t = 0, \tag{1}$$

where J is the action, T is the kinetic energy, U is the potential energy including mechanical (U_{m}) and electrical components (U_{e}), and the work done by the external loads is given by the W_{ext} term.

To write down the formulas for both energies let us consider an arbitrary point A—see Fig. 1b—located on the beam profile that is specified by its position vector $\mathbf{r} = \{x, Y, Z\}^{\mathsf{T}}$ in the local beam coordinate system $(0, x, y, z)$. As the whole structure rotates by $\psi(t)$ about the CZ_0 axis and simultaneously the beam experiences an elastic deformation the discussed point A occupies a new position in space A′ that is defined in the inertial frame of reference (CX_0, Y_0, Z_0) as

$$\mathbf{R} = \left\{ \begin{array}{c} (D_x + x + R_0)\cos\psi - (D_y + Y)\cos\theta\sin\psi + (D_z + Z)\sin\theta\sin\psi \\ (D_x + x + R_0)\sin\psi + (D_y + Y)\cos\theta\cos\psi - (D_z + Z)\sin\theta\cos\psi \\ (D_y + Y)\sin\theta + (D_z + Z)\cos\theta \end{array} \right\}. \tag{2}$$

D_x, D_y and D_z are displacements of the given point A expressed in the local coordinates frame $(0, x, y, z)$ as

$$\begin{aligned} D_x &= u_0(x,t) + \vartheta_y(x,t)\left(z - n\tfrac{\mathrm{d}y}{\mathrm{d}s}\right) + \vartheta_z(x,t)\left(y + n\tfrac{\mathrm{d}z}{\mathrm{d}s}\right) - G(n,s)\varphi'(x,t) \\ D_y &= v_0(x,t) - \left(y + n\tfrac{\mathrm{d}z}{\mathrm{d}s}\right)\left(1 - \cos\varphi(x,t)\right) - \left(z - n\tfrac{\mathrm{d}y}{\mathrm{d}s}\right)\sin\varphi(x,t) \\ &\approx v_0(x,t) - \tfrac{1}{2}\left(y + n\tfrac{\mathrm{d}z}{\mathrm{d}s}\right)\varphi^2(x,t) - \left(z - n\tfrac{\mathrm{d}y}{\mathrm{d}s}\right)\varphi(x,t) \\ D_z &= w_0(x,t) + \left(y + n\tfrac{\mathrm{d}z}{\mathrm{d}s}\right)\sin\varphi(x,t) - \left(z - n\tfrac{\mathrm{d}y}{\mathrm{d}s}\right)\left(1 - \cos\varphi(x,t)\right) \\ &\approx w_0(x,t) + \left(y + n\tfrac{\mathrm{d}z}{\mathrm{d}s}\right)\varphi(x,t) - \tfrac{1}{2}\left(z - n\tfrac{\mathrm{d}y}{\mathrm{d}s}\right)\varphi^2(x,t) \end{aligned} \tag{3}$$

where off-mid-line coordinates Y, Z have been replaced by their profile mid-line counterparts y, z as $Y = y + n\frac{\mathrm{d}z}{\mathrm{d}s}$ and $Z = z - n\frac{\mathrm{d}y}{\mathrm{d}s}$, respectively (Georgiades et al. 2014; Librescu and Song 2006).

In formulas (3) variables u_0, v_0, w_0 are displacements of the point 0 located on the beam axis and belonging to the same cross-section as the discussed point A, $\varphi(x,t)$ denotes rotation of the cross section (profile twist). Angles $\vartheta_y(x,t) = \gamma_{xz} - w_0'$ and $\vartheta_z(x,t) = \gamma_{xy} - v_0'$ represent cross-section rotations about respective local axes y and z considering the shear effect. These six variables constitute a set of basic mechanical unknowns of the problem. Besides, the first equation in the above set includes warping effect due to the warping function term $G(s,n)$, where n stands for the distance from the cross-section wall mid-line to the location of the considered

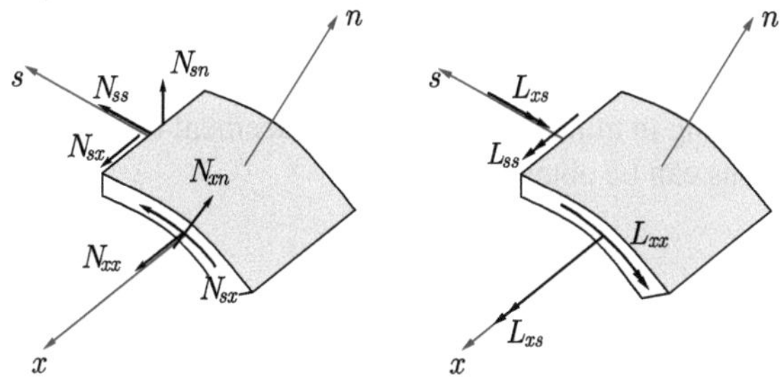

Fig. 3 In-plane and transversal stress resultants and stress couples acting on a beam wall representative element

point A, and s is a circumferential coordinate (measured counterclockwise—see also Fig. 3 in the next section).

Moreover, in the second and the third equation, the approximation $\cos\varphi \approx 1 - {}^1\!/\!_2\,\varphi^2$ is adopted. The final linearization of resulting state equations is performed after accepting this approximation. This approach allows to account for all terms present within a linear mathematical model of the structure.

3.2 Velocity Vector

The velocity vector of the discussed point A of the elastic body in the inertial reference frame can be obtained by differentiating the position vector (2) with respect to time as

$$\begin{aligned} \dot{R}_x &= \left[-(D_x + x + R_0)\sin\psi - (D_y + Y)\cos\theta\cos\psi + (D_z + Z)\sin\theta\cos\psi(t)\right]\dot{\psi} \\ &\quad + \dot{D}_x\cos\psi - \dot{D}_y\cos\theta\sin\psi + \dot{D}_z\sin\theta\sin\psi, \\ \dot{R}_y &= \left[(D_x + x + R_0)\cos\psi - (D_y + Y)\cos\theta\sin\psi + (D_z + Z)\sin\theta\sin\psi\right]\dot{\psi} \\ &\quad + \dot{D}_x\sin\psi + \dot{D}_y\cos\theta\cos\psi - \dot{D}_z\sin\theta\cos\psi, \\ \dot{R}_z &= \dot{D}_y\sin\theta + \dot{D}_z\cos\theta, \end{aligned} \tag{4}$$

where overdot means time derivative, so $\dot{D}_x$, $\dot{D}_y$ and $\dot{D}_z$ terms correspond to velocities of deformation and $\dot{\psi} = \dot{\psi}(t)$ is the angular velocity of the rigid hub.

Given above expressions allow to express the total kinetic energy of the system—i.e. the energy of the beam and the hub

$$T = \frac{1}{2}J_\mathrm{h}\dot{\psi}^2(t) + \frac{1}{2}\int_V \rho\,\dot{\mathbf{R}}^\mathsf{T}\dot{\mathbf{R}}\,\mathrm{d}V, \tag{5}$$

where designation ρ refers to average beam material density and V refers to representative volume element and infinitesimal element is $\mathrm{d}V = \mathrm{d}n\mathrm{d}s\mathrm{d}x$; J_h is the mass moment of hub inertia.

3.3 Strains and Stresses

Bearing in mind above given displacement relations (3), the following strain expressions can be obtained

$$\begin{aligned}
\varepsilon_{xx} &= \varepsilon_{xx}^{(0)} + n\varepsilon_{xx}^{(1)} = u_0' + z\vartheta_y' + y\vartheta_z' - G^{(0)}(s)\varphi'' + \underline{\tfrac{1}{2}\left(v_0' - z\varphi'\right)^2} \\
&\quad + \underline{\tfrac{1}{2}\left(w_0' + y\varphi'\right)^2} + n\left(\tfrac{\mathrm{d}z}{\mathrm{d}s}\vartheta_z' - \tfrac{\mathrm{d}y}{\mathrm{d}s}\vartheta_y' - G^{(1)}(s)\varphi''\right) \\
\gamma_{xs} &= \gamma_{xs}^{(0)} + n\gamma_{xs}^{(1)} = (\vartheta_y + w_0')\tfrac{\mathrm{d}z}{\mathrm{d}s} + (\vartheta_z + v_0')\tfrac{\mathrm{d}y}{\mathrm{d}s} + g^{(0)}(s)\varphi' + ng^{(1)}(s)\varphi' \\
\gamma_{xn} &= \gamma_{xn}^{(0)} \quad = -(\vartheta_y + w_0')\tfrac{\mathrm{d}y}{\mathrm{d}s} + (\vartheta_z + v_0')\tfrac{\mathrm{d}z}{\mathrm{d}s}
\end{aligned} \tag{6}$$

where the prime symbol corresponds to differentiation with respect to space variable x, and superscripts $()^{(0)}$ and $()^{(1)}$ denote mid-line and off-mid-line components, respectively.

Although the mathematical model of the beam is limited to the linear case, higher order terms associated with the lateral and transversal displacements in axial strain ε_{xx} are taken into account—these are emphasized by underline in (6). These terms play a crucial role in proper modelling of the blade stiffening effect arising from the system rotation $\dot{\psi}(t)$. This approach is one of several methodologies that are used in linear models of rotating systems to capture this phenomena. For further reading on this and other centrifugal effect treatments please consult appropriate papers, e.g. Mayo et al. (2004).

One can easily check that according to the appropriate linear definitions, the three remaining strains ε_{yy}, ε_{zz} and γ_{yz} are identically zero. This observation conforms the cross-section non-deformability assumption 1, posed at the very beginning.

To write down the expressions for stresses in piezoceramic and in laminate layers, individual constitutive relations have to be used. Polarized ceramics like PZT-5, exhibit transverse isotropic properties. Based on the fundamental work by Maugin (1985) and by Joshi (1992), a set of full non-linear constitutive relations for this type of material can be written. For the specific case of ceramics subjected to electric field in polling direction x_3 (thickness-wise) and limiting the analysis to linear strains and cubic non-linearities in the electric field, variable constitutive equations for converse effect are simplified to the following form

$$\begin{Bmatrix} \sigma_{ss} \\ \sigma_{xx} \\ \sigma_{nn} \\ \sigma_{xn} \\ \sigma_{sn} \\ \sigma_{xs} \end{Bmatrix}_{(\mathrm{p})} = \begin{bmatrix} \tilde{C}_{11} & \tilde{C}_{12} & \tilde{C}_{13} & 0 & 0 & 0 \\ \tilde{C}_{12} & \tilde{C}_{11} & \tilde{C}_{13} & 0 & 0 & 0 \\ \tilde{C}_{13} & \tilde{C}_{13} & \tilde{C}_{33} & 0 & 0 & 0 \\ 0 & 0 & 0 & \tilde{C}_{44} & 0 & 0 \\ 0 & 0 & 0 & 0 & \tilde{C}_{55} & 0 \\ 0 & 0 & 0 & 0 & 0 & \tilde{C}_{66} \end{bmatrix} \begin{Bmatrix} \varepsilon_{ss} \\ \varepsilon_{xx} \\ \varepsilon_{nn} \\ \gamma_{xn} \\ \gamma_{sn} \\ \gamma_{xs} \end{Bmatrix} - \begin{bmatrix} 0 & 0 & e_{31} \\ 0 & 0 & e_{31} \\ 0 & 0 & e_{33} \\ 0 & e_{15} & 0 \\ e_{15} & 0 & 0 \\ 0 & 0 & 0 \end{bmatrix} \begin{Bmatrix} 0 \\ 0 \\ E_3 \end{Bmatrix}$$

$$
- \begin{bmatrix} \hat{b}_{11} & \hat{b}_{12} & \hat{b}_{31} & 0 & 0 & 0 \\ \hat{b}_{12} & \hat{b}_{11} & \hat{b}_{31} & 0 & 0 & 0 \\ \hat{b}_{13} & \hat{b}_{13} & \hat{b}_{33} & 0 & 0 & 0 \\ 0 & 0 & 0 & \hat{b}_{44} & 0 & 0 \\ 0 & 0 & 0 & 0 & \hat{b}_{44} & 0 \\ 0 & 0 & 0 & 0 & 0 & \hat{b}_{66} \end{bmatrix} \begin{Bmatrix} 0 \\ 0 \\ E_3^2 \\ 0 \\ 0 \\ 0 \end{Bmatrix} - \begin{bmatrix} f_{11} & f_{12} & f_{31} & 0 & 0 & 0 \\ f_{12} & f_{11} & f_{31} & 0 & 0 & 0 \\ f_{13} & f_{13} & f_{33} & 0 & 0 & 0 \\ 0 & 0 & 0 & f_{44} & 0 & 0 \\ 0 & 0 & 0 & 0 & f_{44} & 0 \\ 0 & 0 & 0 & 0 & 0 & f_{66} \end{bmatrix} \begin{Bmatrix} 0 \\ 0 \\ E_3^3 \\ 0 \\ 0 \\ 0 \end{Bmatrix} \tag{7}
$$

where $\tilde{C}_{ij}$ stands for the second order piezoceramic elasticity tensor at constant electric field, e_{ij} is the tensor of piezoelectric coefficients, $\hat{b}_{ij}$ is effective electrostrictive constants tensor and f_{ij} is fourth order piezoelectric tensor. Subscript (p) is used to denote stresses in the actuator and to make a clear distinction between the piezoceramic and the laminate layers discussed later.

For the direct piezoelectric effect, the relation is given as follows

$$
\begin{aligned}
\begin{Bmatrix} D_1 \\ D_2 \\ D_3 \end{Bmatrix} &= \begin{bmatrix} 0 & 0 & 0 & 0 & e_{15} & 0 \\ 0 & 0 & 0 & e_{15} & 0 & 0 \\ e_{31} & e_{31} & e_{33} & 0 & 0 & 0 \end{bmatrix} \begin{Bmatrix} \varepsilon_{ss} \\ \varepsilon_{xx} \\ \varepsilon_{nn} \\ \gamma_{xn} \\ \gamma_{sn} \\ \gamma_{xs} \end{Bmatrix} + \begin{bmatrix} \xi_{11} & 0 & 0 \\ 0 & \xi_{11} & 0 \\ 0 & 0 & \xi_{33} \end{bmatrix} \begin{Bmatrix} 0 \\ 0 \\ E_3 \end{Bmatrix} \\
&+ \begin{bmatrix} 0 & 0 & 0 & 0 & \chi_{15} & 0 \\ 0 & 0 & 0 & \chi_{15} & 0 & 0 \\ \chi_{15} & \chi_{15} & \chi_{33} & 0 & 0 & 0 \end{bmatrix} \begin{Bmatrix} 0 \\ 0 \\ E_3^2 \\ 0 \\ 0 \\ 0 \end{Bmatrix} \\
&+ E_3^3 \left(\hat{\chi}_1 + \hat{\chi}_2 \right) \begin{Bmatrix} 0 \\ 0 \\ 1 \end{Bmatrix} + E_3^2 \left(\hat{\chi}_2 + \hat{\chi}_3 \right) \begin{Bmatrix} 0 \\ 0 \\ E_3 \end{Bmatrix}
\end{aligned} \tag{8}
$$

where ξ_{ij} is second order permittivity tensor, χ and $\hat{\chi}$ are third and fourth order electric susceptibility tensors, respectively. Considered equations, if limited to the second order nonlinearities, get simplified to the form given by Tiersten (1993). Full form of transverse isotropic ceramic constitutive equations up to third order electric non-linearities and arbitrary electric field vector has been given by Yang (1999).

Following the assumption 6 on pp. XXX one arrives at the set of reduced constitutive equations for the piezoceramics

$$
\begin{Bmatrix} \sigma_{ss} \\ \sigma_{xx} \\ \sigma_{xs} \end{Bmatrix}_{(p)} = \begin{bmatrix} \tilde{Q}_{11} & \tilde{Q}_{12} & 0 \\ \tilde{Q}_{12} & \tilde{Q}_{11} & 0 \\ 0 & 0 & \tilde{Q}_{66} \end{bmatrix} \begin{Bmatrix} \varepsilon_{ss} \\ \varepsilon_{xx} \\ \gamma_{ss} \end{Bmatrix} - \begin{Bmatrix} \tilde{e}_{31} E_3 \\ \tilde{e}_{31} E_3 \\ 0 \end{Bmatrix} - \begin{Bmatrix} \tilde{b}_{31} E_3^2 \\ \tilde{b}_{31} E_3^2 \\ 0 \end{Bmatrix} - \begin{Bmatrix} \tilde{f}_{31} E_3^3 \\ \tilde{f}_{31} E_3^3 \\ 0 \end{Bmatrix} \tag{9}
$$

$$
\sigma_{xn_{(p)}} = \tilde{C}_{44} \gamma_{xn} \tag{10}
$$

where

$$\tilde{Q}_{11} = \tilde{Q}_{22} = \tilde{C}_{11} - \frac{\tilde{C}_{13}^2}{\tilde{C}_{33}} \qquad \tilde{e}_{31} = e_{31} - \frac{\tilde{C}_{13}e_{33}}{\tilde{C}_{33}}$$

$$\tilde{Q}_{12} = \tilde{C}_{12} - \frac{\tilde{C}_{13}^2}{\tilde{C}_{33}} \qquad \tilde{b}_{31} = \hat{b}_{31} - \frac{\tilde{C}_{13}\hat{b}_{33}}{\tilde{C}_{33}}$$

$$\tilde{Q}_{66} = \tilde{C}_{66} \qquad \tilde{f}_{31} = f_{31} - \frac{\tilde{C}_{13}f_{33}}{\tilde{C}_{33}}$$

Similar reduced constitutive equations may be formulated for individual composite laminae modelled as an orthotropic layer with material principal axis out of the structure reference system

$$\begin{Bmatrix} \sigma_{ss} \\ \sigma_{xx} \\ \sigma_{xs} \end{Bmatrix}_{(\mathrm{c})} = \begin{bmatrix} \overline{Q}_{11} & \overline{Q}_{12} & \overline{Q}_{16} \\ \overline{Q}_{12} & \overline{Q}_{22} & \overline{Q}_{26} \\ \overline{Q}_{16} & \overline{Q}_{26} & \overline{Q}_{66} \end{bmatrix} \begin{Bmatrix} \varepsilon_{ss} \\ \varepsilon_{xx} \\ \gamma_{ss} \end{Bmatrix} \tag{11}$$

$$\sigma_{xn_{(\mathrm{c})}} = \overline{C}_{44}\gamma_{xn} \qquad \sigma_{sn_{(\mathrm{c})}} = \overline{C}_{45}\gamma_{xn} \tag{12}$$

where

$$\overline{Q}_{ij} = \overline{C}_{ij} - \frac{\overline{C}_{i3}\overline{C}_{j3}}{\overline{C}_{33}} \quad \text{for } i, j = 1, 2, 6;$$

and $\overline{C}_{ij}$ denotes members of laminae elasticity tensor.

The 2-D stress resultants and stress couples—see Fig. 3—in the combined hybrid laminate-piezoceramic system can be obtained via the integration of individual stress components through the total thickness of the beam wall. Thus, these are given as follows

- membrane

$$\begin{Bmatrix} N_{ss} \\ N_{xx} \\ N_{xs} \end{Bmatrix} = \sum_{k=1}^{N} \int_{n_{(k-1)}}^{n_k} \begin{Bmatrix} \sigma_{ss}^{(k)} \\ \sigma_{xx}^{(k)} \\ \sigma_{xs}^{(k)} \end{Bmatrix}_{(\mathrm{c})} \mathrm{d}n + \int_{h_\mathrm{p}} \begin{Bmatrix} \sigma_{ss} \\ \sigma_{xx} \\ \sigma_{xs} \end{Bmatrix}_{(\mathrm{p})} \mathrm{d}n \tag{13}$$

- transverse

$$\begin{Bmatrix} N_{xn} \\ N_{sn} \end{Bmatrix} = \sum_{k=1}^{N} \int_{n_{(k-1)}}^{n_k} \begin{Bmatrix} \sigma_{xn}^{(k)} \\ \sigma_{sn}^{(k)} \end{Bmatrix}_{(\mathrm{c})} \mathrm{d}n + \int_{h_\mathrm{p}} \begin{Bmatrix} \sigma_{xn} \\ 0 \end{Bmatrix}_{(\mathrm{p})} \mathrm{d}n \tag{14}$$

- stress couples

$$\begin{Bmatrix} L_{xx} \\ L_{xs} \end{Bmatrix} = \sum_{k=1}^{N} \int_{n_{(k-1)}}^{n_k} \begin{Bmatrix} \sigma_{xx}^{(k)} \\ \sigma_{xs}^{(k)} \end{Bmatrix}_{(\mathrm{c})} n\mathrm{d}n + \int_{h_\mathrm{p}} \begin{Bmatrix} \sigma_{xx} \\ \sigma_{xs} \end{Bmatrix}_{(\mathrm{p})} n\mathrm{d}n \tag{15}$$

In the above formulas, the used (k) superscript refers to the subsequent laminate layer numbers and h_p denotes the piezoceramic layer thickness. Regarding the not specified term L_{ss}, one observes that it tends to zero due to the prismatic (or at least shallow shell) cross-section shape assumption—see point 5 on pp. XXX.

Inserting constitutive equations for the piezoceramic (9), (10) and for the laminae (11), (12) to the above given definitions (13)–(15) and next decomposing strains into their mid-line $()^{(0)}$ and off-mid-line components $()^{(1)}$ as proposed in (6) one gets

$$\begin{Bmatrix} N_{xx} \\ N_{xs} \\ N_{xn} \\ L_{xx} \\ L_{xs} \end{Bmatrix} = \begin{bmatrix} K_{11} & K_{12} & K_{13} & K_{14} & 0 \\ K_{21} & K_{22} & K_{23} & K_{24} & 0 \\ 0 & 0 & 0 & 0 & A_{44} \\ K_{41} & K_{42} & K_{43} & K_{44} & 0 \\ K_{51} & K_{52} & K_{53} & K_{54} & 0 \end{bmatrix} \begin{Bmatrix} \varepsilon_{xx}^{(0)} \\ \bar{\gamma}_{xs}^{(0)} \\ \varphi' \\ \varepsilon_{xx}^{(1)} \\ \gamma_{xn}^{(0)} \end{Bmatrix} + \begin{bmatrix} K_{1e} & K_{1b} & K_{1f} \\ K_{2e} & K_{2b} & K_{2f} \\ 0 & 0 & 0 \\ K_{4e} & K_{4b} & K_{4f} \\ K_{5e} & K_{5b} & K_{5f} \end{bmatrix} \begin{Bmatrix} E_3 \\ E_3^2 \\ E_3^3 \end{Bmatrix} \tag{16}$$

where the detailed definitions of stiffnesses K_{ij} $(i = 1, 2, 4, 5;\ j = 1, \ldots, 4)$ are given in Georgiades et al. (2014). The new stiffness terms resulting from the electro-mechanical properties of the piezo-layers are as follows:

$$K_{1\lambda} = \left(\frac{A_{12}}{A_{11}} - 1\right)\lambda \int_{h_p} \mathrm{d}n, \qquad K_{2\lambda} = \frac{A_{16}}{A_{11}}\lambda \int_{h_p} \mathrm{d}n \tag{17}$$

$$K_{4\lambda} = \left(\frac{B_{12}}{A_{11}} \int_{h_p} \mathrm{d}n - \int_{h_p} n\mathrm{d}n\right)\lambda, \quad K_{5\lambda} = \frac{B_{16}}{A_{11}}\lambda \int_{h_p} \mathrm{d}n \qquad \text{for } \lambda \in \left(\tilde{e}_{31}, \tilde{b}_{31}, \tilde{f}_{31}\right).$$

Integrating obtained stress resultants and stress couples (16) along the mid-line contour and incorporating coordinate transformations yields the following 1-D generalized loadings

$$T_x = \int_c N_{xx}\mathrm{d}s \qquad \text{the axial force}$$

$$Q_y = \int_c \left(N_{xs}\frac{\mathrm{d}y}{\mathrm{d}s} + N_{xn}\frac{\mathrm{d}z}{\mathrm{d}s}\right)\mathrm{d}s \qquad \text{the shear force in } y\text{-direction}$$

$$Q_z = \int_c \left(N_{xs}\frac{\mathrm{d}z}{\mathrm{d}s} - N_{xn}\frac{\mathrm{d}y}{\mathrm{d}s}\right)\mathrm{d}s \qquad \text{the shear force in } z\text{-direction}$$

$$M_x = \int_c \left(g^{(0)}(s)N_{xs} + g^{(1)}(s)L_{xs}\right)\mathrm{d}s \qquad \text{the twisting moment about } x\text{-axis} \tag{18}$$

$$M_y = \int_c \left(zN_{xx} - L_{xx}\frac{\mathrm{d}y}{\mathrm{d}s}\right)\mathrm{d}s \qquad \text{the bending moment in } y\text{-axis}$$

$$M_z = \int_c \left(yN_{xx} + L_{xx}\frac{\mathrm{d}z}{\mathrm{d}s}\right)\mathrm{d}s \qquad \text{the bending moment in } z\text{-axis}$$

$$B_w = \int_c \left(G^{(0)}(s)N_{xx} + G^{(1)}(s)L_{xx}\right)\mathrm{d}s \quad \text{the warping torque (bimoment)}$$

$$T_r = \int_c N_{xx}(y^2 + z^2)\mathrm{d}s$$

In the above notation, script c in the integral limit denotes profile mid-line circumference.

The mechanical potential energy of the elastic system under consideration is given by

$$\begin{aligned} U_\mathrm{m} &= \tfrac{1}{2}\int_0^l\int_c\int_{h+h_\mathrm{p}} \left(\sigma_{xx}\varepsilon_{xx} + \sigma_{xn}\gamma_{xn} + \sigma_{xs}\gamma_{xs}\right)\mathrm{d}n\mathrm{d}s\mathrm{d}x \\ &= \tfrac{1}{2}\int_0^l\int_c \left[N_{xx}\varepsilon_{xx}^{(0)} + L_{xx}\varepsilon_{xx}^{(1)} + N_{xn}\gamma_{xn}^{(0)} + N_{xs}\gamma_{xs}^{(0)} + L_{xs}\gamma_{xs}^{(1)}\right]\mathrm{d}s\mathrm{d}x. \end{aligned} \tag{19}$$

Based on quoted definitions of generalized loadings (18) and strain definitions (6), after performing some mathematical manipulations, one arrives at the expression for the potential energy U_m of the flexible system

$$\begin{aligned} U_\mathrm{m} = \tfrac{1}{2}\int_0^l \Big\{ & T_x\left[u_0' + \tfrac{1}{2}{v_0'}^2 + \tfrac{1}{2}{w_0'}^2\right] + Q_y\left(v_0' + \vartheta_z\right) + Q_z\left(w_0' + \vartheta_y\right) \\ & + M_x\varphi' + M_y\left(\vartheta_y' - \varphi' v_0'\right) + M_z\left(\vartheta_z' + \varphi' w_0'\right) \\ & + \tfrac{1}{2}T_r\varphi'^2 - B_\omega\varphi''\Big\}\mathrm{d}x. \end{aligned} \tag{20}$$

On the other hand, quantities T_x, Q_y, Q_z, T_r and M_x, M_y, M_z, B_ω may be also expressed in terms of fundamental problem unknowns. After substituting for the 2-D stress and stress couple resultants (13)–(15) and inserting the strain formulas (6) and final integration along the cross-section mid-line one arrives at

$$\begin{aligned} T_x &= a_{11}u_0' + a_{15}\vartheta_y + a_{13}\vartheta_y' + a_{14}\vartheta_z + a_{12}\vartheta_z' + a_{14}v_0' + a_{15}w_0' + a_{17}\varphi' \\ &\quad - a_{16}\varphi'' + a_{1e}E_3 + a_{1b}E_3^2 + a_{1f}E_3^3 \\ Q_y &= a_{14}u_0' + a_{45}\vartheta_y + a_{34}\vartheta_y' + a_{44}\vartheta_z + a_{24}\vartheta_z' + a_{44}v_0' + a_{45}w_0' + a_{47}\varphi' \\ &\quad - a_{46}\varphi'' + a_{4e}E_3 + a_{4b}E_3^2 + a_{4f}E_3^3 \\ Q_z &= a_{15}u_0' + a_{55}\vartheta_y + a_{35}\vartheta_y' + a_{45}\vartheta_z + a_{25}\vartheta_z' + a_{45}v_0' + a_{55}w_0' + a_{57}\varphi' \\ &\quad - a_{56}\varphi'' + a_{5e}E_3 + a_{5b}E_3^2 + a_{5f}E_3^3 \\ M_x &= a_{17}u_0' + a_{57}\vartheta_y + a_{37}\vartheta_y' + a_{47}\vartheta_z + a_{27}\vartheta_z' + a_{47}v_0' + a_{57}w_0' + a_{77}\varphi' \end{aligned}$$

$$
\begin{aligned}
&- a_{67}\varphi'' + a_{7e}E_3 + a_{7b}E_3^2 + a_{7f}E_3^3 \\
M_y &= a_{13}u_0' + a_{35}\vartheta_y + a_{33}\vartheta_y' + a_{34}\vartheta_z + a_{23}\vartheta_z' + a_{34}v_0' + a_{35}w_0' + a_{37}\varphi' \\
&- a_{36}\varphi'' + a_{3e}E_3 + a_{3b}E_3^2 + a_{3f}E_3^3 \\
M_z &= a_{12}u_0' + a_{25}\vartheta_y + a_{23}\vartheta_y' + a_{24}\vartheta_z + a_{22}\vartheta_z' + a_{24}v_0' + a_{25}w_0' + a_{27}\varphi' \\
&- a_{26}\varphi'' + a_{2e}E_3 + a_{2b}E_3^2 + a_{2f}E_3^3 \\
B_w &= a_{16}u_0' + a_{56}\vartheta_y + a_{36}\vartheta_y' + a_{46}\vartheta_z + a_{26}\vartheta_z' + a_{46}v_0' + a_{56}w_0' + a_{67}\varphi' \\
&- a_{66}\varphi'' + a_{6e}E_3 + a_{6b}E_3^2 + a_{6f}E_3^3
\end{aligned}
\tag{21}
$$

where the cross-sectional stiffness terms a_{ij} $(i, j = 1, \ldots, 7)$ are defined in previous research Georgiades et al. (2014). Stiffness coefficients resulting from the electrical properties of the specimen are defined

$$
\begin{aligned}
a_{1\lambda} &= \int_c K_{1\lambda}\mathrm{d}s, & a_{5\lambda} &= \int_c \frac{\mathrm{d}z}{\mathrm{d}s}K_{2\lambda}\mathrm{d}s \\
a_{2\lambda} &= \int_c \left(K_{1\lambda}y + \frac{\mathrm{d}z}{\mathrm{d}s}K_{4\lambda}\right)\mathrm{d}s, & a_{6\lambda} &= \int_c \left(K_{1\lambda}G^{(0)} + K_{4\lambda}G^{(1)}\right)\mathrm{d}s \\
a_{3\lambda} &= \int_c \left(K_{1\lambda}z - \frac{\mathrm{d}y}{\mathrm{d}s}K_{4\lambda}\right)\mathrm{d}s, & a_{7\lambda} &= \int_c \left(K_{2\lambda}g^{(0)} + K_{5\lambda}g^{(1)}\right)\mathrm{d}s \\
a_{4\lambda} &= \int_c \frac{\mathrm{d}y}{\mathrm{d}s}K_{2\lambda}\mathrm{d}s & & \text{for } \lambda \in (e, b, f).
\end{aligned}
\tag{22}
$$

3.4 Virtual Work

The term of virtual work present in the Hamilton's principle (1) is the work of all external forces as given by

$$
\begin{aligned}
\delta W_{\text{ext}} = \int_0^l &\left[P_{\text{ext},x}\delta u_0 + m_{\text{ext},y}\delta\vartheta_y + m_{\text{ext},z}\delta\vartheta_z + P_{\text{ext},y}\delta v_0\right. \\
&\left. + P_{\text{ext},z}\delta w_0 + \left(m_{\text{ext},x} + m'_{\text{ext},w}\right)\delta\varphi\right]\mathrm{d}x + T_{\text{ext}}\delta\psi(t) - \left[m_{\text{ext},w}\delta\varphi\right]\Big|_{x=0}^{x=l}
\end{aligned}
\tag{23}
$$

where $P_{\text{ext},y}$, $P_{\text{extt},z}$ are applied external shear forces per unit span, $P_{\text{ext},x}$ is unit axial force, $m_{\text{ext},x}$, $m_{\text{ext},y}$, $m_{\text{ext},z}$ are moments about x, y, z axes, respectively, $m_{\text{ext},w}$ is bimoment and $T_{\text{ext},z}$ is driving torque applied at hub as introduced earlier. Definitions and detailed derivation of appropriate terms in the above expression are given in Georgiades et al. (2014).

3.5 Electric Field

Contribution of the electric field to the total potential energy of the considered electro-mechanical system is given by

$$U_e = -\frac{1}{2}\int_V D_3 E_3 \mathrm{d}V \tag{24}$$

In the above formula the assumption on the piezoceramic layer to be electroded in the thickness direction has been adopted.

The electric displacement D_3 can be expressed by problem unknowns based on the direct piezoelectric effect formula (8) and strain definitions (6). Next, adopting the $\sigma_{nn} = 0$ condition (see assumption 6 on pp. XXX) one arrives at the following formula

$$\begin{aligned} D_3 = {} & a_{E1}u_0' + a_{E2}\vartheta_z' + a_{E3}\vartheta_y' + a_{E4}(\vartheta_z + v_0') + a_{E5}(\vartheta_y + w_0') \\ & + a_{E7}\varphi' - a_{E6}\varphi'' + a_{Ee}E_3 + a_{Eb}E_3^2 + a_{Ef}E_3^3 \end{aligned} \tag{25}$$

where the appropriate proportionality coefficients are

$$\begin{aligned} a_{E1} &= \int_c\int_{h_p}\left(1-\frac{A_{12}}{A_{11}}\right)\tilde{e}_{31}\,\mathrm{d}s\,\mathrm{d}n \\ a_{E2} &= \int_c\int_{h_p}\left[\frac{\mathrm{d}z}{\mathrm{d}s}\left(-\frac{B_{12}}{A_{11}}+n\right)+\left(1-\frac{A_{12}}{A_{11}}\right)y\right]\tilde{e}_{31}\,\mathrm{d}s\,\mathrm{d}n \\ a_{E3} &= \int_c\int_{h_p}\left[\frac{\mathrm{d}y}{\mathrm{d}s}\left(\frac{B_{12}}{A_{11}}-n\right)+\left(1-\frac{A_{12}}{A_{11}}\right)z\right]\tilde{e}_{31}\,\mathrm{d}s\,\mathrm{d}n \\ a_{E4} &= \int_c\int_{h_p}\left(-\frac{A_{16}}{A_{11}}\right)\frac{\mathrm{d}y}{\mathrm{d}s}\tilde{e}_{31}\,\mathrm{d}s\,\mathrm{d}n \\ a_{E5} &= \int_c\int_{h_p}\left(-\frac{A_{16}}{A_{11}}\right)\frac{\mathrm{d}z}{\mathrm{d}s}\tilde{e}_{31}\,\mathrm{d}s\,\mathrm{d}n \\ a_{E6} &= \int_c\int_{h_p}\left[G^{(1)}(s)\left(\frac{B_{12}}{A_{11}}-n\right)+G^{(0)}(s)\left(\frac{A_{12}}{A_{11}}-1\right)\right]\tilde{e}_{31}\,\mathrm{d}s\,\mathrm{d}n \\ a_{E7} &= \int_c\int_{h_p}\left(-\frac{A_{16}}{A_{11}}g^{(0)}-\frac{B_{16}}{A_{11}}g^{(1)}\right)\tilde{e}_{31}\,\mathrm{d}s\,\mathrm{d}n \\ a_{Ee} &= \int_c\int_{h_p}\left(\xi_{33}+\frac{e_{33}^2}{\tilde{C}_{33}}+\frac{h_p}{A_{11}}\tilde{e}_{31}^2\right)\mathrm{d}s\,\mathrm{d}n \\ a_{Eb} &= \int_c\int_{h_p}\left(\chi_{33}+\frac{e_{33}\widehat{b}_{33}}{\tilde{C}_{33}}+\frac{h_p}{A_{11}}\tilde{b}_{31}\tilde{e}_{31}\right)\mathrm{d}s\,\mathrm{d}n \\ a_{Ef} &= \int_c\int_{h_p}\left(\widehat{\chi}_1+2\widehat{\chi}_2+\widehat{\chi}_3+\frac{e_{33}f_{33}}{\tilde{C}_{33}}+\frac{h_p}{A_{11}}\tilde{f}_{31}\tilde{e}_{31}\right)\mathrm{d}s\,\mathrm{d}n \end{aligned} \tag{26}$$

These are the combined form of relations (22) and (17).

4 Governing Equations

Using the extended Hamilton's least action principle (1), the minimization of the total energy given by its components (5), (20) and (24) and the minimization of the virtual work of external loads, (23) yields a system of eight electromechanically coupled differential equations. Six of these equations are equations of motion of the flexible smart composite beam structure, one describes the motion of the rigid hub and the last equation is an electrostatic equation. These are specified as resulting from variation of individual independent problem variables.

- $\delta\psi(t)$

$$
\begin{aligned}
&-J_{\rm h}\ddot{\psi}(t) - B_{22}\ddot{\psi}(t)\\
&-B_{14}l\ddot{\psi}(t)\cos^2\theta - B_{13}l\ddot{\psi}(t)\sin^2\theta + 2B_{15}l\ddot{\psi}(t)\cos\theta\sin\theta\\
&-\int_0^l \Big[2B_1(R_0+x)u_0\ddot{\psi}(t) + 2\big(B_{12}\cos^2\theta - B_{11}\sin\theta\cos\theta\big)v_0\ddot{\psi}(t)\\
&+2\big(B_{11}\sin^2\theta - B_{12}\sin\theta\cos\theta\big)w_0\ddot{\psi}(t) + 2B_{11}(R_0+x)\vartheta_y\ddot{\psi}(t)\\
&+2B_{12}(R_0+x)\vartheta_z\ddot{\psi}(t) + 2\big(B_{17}\sin^2\theta - B_{18}\cos^2\theta\big)\varphi\ddot{\psi}(t)\ddot{\psi}(t)\\
&-2B_7(R_0+x)\varphi' + 2\big(B_{16} - B_{19}\big)\varphi\ddot{\psi}(t)\sin\theta\cos\theta\\
&+\big(B_{11}\sin\theta - B_{12}\cos\theta\big)\ddot{u}_0 + B_1(R_0+x)\ddot{v}_0\cos\theta - B_1(R_0+x)\ddot{w}_0\sin\theta\\
&+\big(B_{13}\sin\theta - B_{15}\cos\theta\big)\ddot{\vartheta}_y + \big(B_{15}\sin\theta - B_{14}\cos\theta\big)\ddot{\vartheta}_z\\
&+\big(B_9\cos\theta - B_8\sin\theta\big)\ddot{\varphi}' - \big(B_3\sin\theta + B_2\cos\theta\big)(R_0+x)\ddot{\varphi}\\
&+2B_1(R_0+x)\dot{u}_0\dot{\psi}(t) - 2B_{11}\dot{v}_0\dot{\psi}(t)\sin\theta\cos\theta + 2B_{12}\dot{v}_0\dot{\psi}(t)\cos^2\theta\\
&-2B_{12}\dot{w}_0\dot{\psi}(t)\sin\theta\cos\theta + 2B_{11}\dot{w}_0\dot{\psi}(t)\sin^2\theta + 2B_{11}(R_0+x)\dot{\vartheta}_y\dot{\psi}(t)\\
&+2B_{12}(R_0+x)\dot{\vartheta}_z\dot{\psi}(t) - 2B_7(R_0+x)\dot{\varphi}'\dot{\psi}(t) + 2B_{17}\dot{\varphi}\dot{\psi}(t)\sin^2\theta\\
&-2B_{18}\dot{\varphi}\dot{\psi}(t)\cos^2\theta + 2(B_{16} - B_{19})\dot{\varphi}\dot{\psi}(t)\sin\theta\cos\theta\Big]\mathrm{d}x + T_{\mathrm{ext},z} = 0
\end{aligned}
\tag{27}
$$

- δu_0,

$$
\begin{aligned}
&-B_1\ddot{u}_0 - B_{12}\ddot{\vartheta}_z - B_{11}\ddot{\vartheta}_y + B_7\ddot{\varphi}' + 2B_1\dot{v}_0\dot{\psi}(t)\cos\theta - 2B_2\dot{\varphi}\dot{\psi}(t)\cos\theta\\
&-2B_1\dot{w}_0\dot{\psi}(t)\sin\theta - 2B_3\dot{\varphi}\dot{\psi}(t)\sin\theta + B_1(R_0+x+u_0)\dot{\psi}^2(t)\\
&+B_{12}\vartheta_z\dot{\psi}^2(t) + B_{11}\vartheta_y\dot{\psi}^2(t) - B_7\varphi'\dot{\psi}^2(t) - B_1w_0\ddot{\psi}(t)\sin\theta\\
&-B_3\varphi\ddot{\psi}(t)\sin\theta - B_{11}\ddot{\psi}(t)\sin\theta + B_1v_0\ddot{\psi}(t)\cos\theta\\
&-B_2\varphi\ddot{\psi}(t)\cos\theta + B_{12}\ddot{\psi}(t)\cos\theta + T_x' + P_{\mathrm{ext},x} = 0
\end{aligned}
\tag{28}
$$

with boundary conditions,

$$
u_0\big|_{x=0} = 0, \qquad T_x\big|_{x=l} = 0 \tag{29}
$$

• δv_0

$$\begin{aligned}&-B_1\ddot{v}_0+B_2\ddot{\varphi}-2B_1\dot{u}_0\dot{\psi}(t)\cos\theta-2B_{12}\dot{\vartheta}_z\dot{\psi}(t)\cos\theta\\&-2B_{11}\dot{\vartheta}_y\dot{\psi}(t)\cos\theta+2B_7\dot{\varphi}'\dot{\psi}(t)\cos\theta-B_1w_0\dot{\psi}^2(t)\sin\theta\cos\theta\\&-B_3\varphi\dot{\psi}^2(t)\sin\theta\cos\theta-B_{11}\dot{\psi}^2(t)\sin\theta\cos\theta+B_1v_0\dot{\psi}^2(t)\cos^2\theta\\&-B_2\varphi\dot{\psi}^2(t)\cos^2\theta+B_{12}\dot{\psi}^2(t)\cos^2\theta-B_1(R_0+x+u_0)\ddot{\psi}(t)\cos\theta\\&-B_{12}\vartheta_z\ddot{\psi}(t)\cos\theta-B_{11}\vartheta_y\ddot{\psi}(t)\cos\theta+B_7\varphi'\ddot{\psi}(t)\cos\theta\\&+Q_y'+(T_xv_0')'+P_{\mathrm{ext},y}=0\end{aligned}\tag{30}$$

with boundary conditions,

$$v_0\big|_{x=0}=0,\qquad \left[Q_y+T_xv_0'\right]\big|_{x=l}=0\tag{31}$$

• δw_0

$$\begin{aligned}&-B_1\ddot{w}_0-B_3\ddot{\varphi}+2B_1\dot{u}_0\,\dot{\psi}(t)\sin\theta+2B_{12}\dot{\vartheta}_z\dot{\psi}(t)\sin\theta\\&+2B_{11}\dot{\vartheta}_y\dot{\psi}(t)\sin\theta-2B_7\dot{\varphi}'\dot{\psi}(t)\sin\theta+B_1w_0\dot{\psi}^2(t)\sin^2\theta\\&+B_3\varphi\dot{\psi}^2(t)\sin^2\theta+B_{11}\dot{\psi}^2(t)\sin^2\theta-B_1v_0\dot{\psi}^2(t)\sin\theta\cos\theta\\&+B_2\varphi\dot{\psi}^2(t)\sin\theta\cos\theta-B_{12}\dot{\psi}^2(t)\sin\theta\cos\theta+B_1(R_0+x+u_0)\ddot{\psi}(t)\sin\theta\\&+B_{12}\vartheta_z\ddot{\psi}(t)\sin\theta+B_{11}\vartheta_y\ddot{\psi}(t)\sin\theta-B_7\varphi'\ddot{\psi}(t)\sin\theta\\&+Q_z'+(T_xw_0')'+P_{\mathrm{ext},z}=0\end{aligned}\tag{32}$$

with boundary conditions,

$$w_0\big|_{x=0}=0,\qquad \left[Q_z+T_xw_0'\right]\big|_{x=l}=0\tag{33}$$

• $\delta\vartheta_y$

$$\begin{aligned}&-B_{11}\ddot{u}_0-B_{13}\ddot{\vartheta}_y-B_{15}\ddot{\vartheta}_z+B_8\ddot{\varphi}'+2B_{11}\,\dot{v}_0\,\dot{\psi}(t)\cos\theta\\&-2B_{16}\dot{\varphi}\dot{\psi}(t)\cos\theta-2B_{11}\dot{w}_0\dot{\psi}(t)\sin\theta-2B_{17}\dot{\varphi}\dot{\psi}(t)\sin\theta\\&+B_{11}(R_0+x+u_0)\dot{\psi}^2(t)+B_{15}\vartheta_z\dot{\psi}^2(t)+B_{13}\vartheta_y\dot{\psi}^2(t)-B_8\varphi'\dot{\psi}^2(t)\\&-B_{11}w_0\ddot{\psi}(t)\sin\theta-B_{17}\varphi\ddot{\psi}(t)\sin\theta+B_{11}v_0\ddot{\psi}(t)\cos\theta\\&-B_{16}\varphi\ddot{\psi}(t)\cos\theta-B_{13}\ddot{\psi}(t)\sin\theta+B_{15}\ddot{\psi}(t)\cos\theta-Q_z+M_y'+m_{\mathrm{ext},y}=0\end{aligned}\tag{34}$$

with boundary conditions,

$$\vartheta_y\big|_{x=0} = 0, \qquad M_y\big|_{x=l} = 0 \tag{35}$$

- $\delta\vartheta_z$

$$\begin{aligned}
&-B_{12}\ddot{u}_0 - B_{15}\ddot{\vartheta}_y - B_{14}\ddot{\vartheta}_z + B_9\ddot{\varphi}' + 2B_{12}\dot{v}_0\,\dot{\psi}(t)\cos\theta \\
&-2B_{18}\dot{\varphi}\dot{\psi}(t)\cos\theta - 2B_{12}\dot{w}_0\dot{\psi}(t)\sin\theta - 2B_{19}\dot{\varphi}\dot{\psi}(t)\sin\theta \\
&+B_{12}(R_0 + x + u_0)\dot{\psi}^2(t) + B_{14}\vartheta_z\dot{\psi}^2(t) + B_{15}\vartheta_y\dot{\psi}^2(t) - B_9\varphi'\dot{\psi}^2(t) \\
&-B_{12}w_0\ddot{\psi}(t)\sin\theta - B_{19}\varphi\ddot{\psi}(t)\sin\theta + B_{12}v_0\ddot{\psi}(t)\cos\theta - B_{18}\varphi\ddot{\psi}(t)\cos\theta \\
&-B_{15}\ddot{\psi}(t)\sin\theta + B_{14}\ddot{\psi}(t)\cos\theta - Q_y + M_z' + m_{\text{ext},z} = 0
\end{aligned} \tag{36}$$

with boundary conditions,

$$\vartheta_z\big|_{x=0} = 0, \qquad M_z\big|_{x=l} = 0 \tag{37}$$

- $\delta\varphi$

$$\begin{aligned}
&B_2\ddot{v}_0 - B_3\ddot{w}_0 - B_4\ddot{\varphi} - B_5\ddot{\varphi} + (R_0 + x)\varphi\ddot{\psi}(t)(B_3\cos\theta - B_2\sin\theta) \\
&+(R_0 + x + u_0)(B_2\cos\theta + B_3\sin\theta)\ddot{\psi}(t) \\
&-(B_{20}\cos\theta + B_{21}\sin\theta)\varphi'\ddot{\psi}(t) + (B_{16}\cos\theta + B_{17}\sin\theta)\vartheta_y\ddot{\psi}(t) \\
&+(B_{18}\cos\theta + B_{19}\sin\theta)\vartheta_z\ddot{\psi}(t) + (B_{17}\sin^2\theta - B_{18}\cos^2\theta)\dot{\psi}^2(t) \\
&+\left(B_{16} - B_{19}\right)\dot{\psi}^2(t)\cos\theta\sin\theta - (B_2\cos^2\theta + B_3\sin\theta\cos\theta)v_0\dot{\psi}^2(t) \\
&+(B_2\sin\theta\cos\theta + B_3\sin^2\theta)w_0\dot{\psi}^2(t) + (B_4 - B_{19})\varphi\dot{\psi}^2(t)\cos^2\theta \\
&+(B_5 - B_{16})\varphi\dot{\psi}^2(t)\sin^2\theta + \left[B_7(R_0 + x + u_0)\right]'\dot{\psi}^2(t) \\
&+(2B_6 + B_{17} + B_{18})\varphi\dot{\psi}^2(t)\sin\theta\cos\theta + \left(B_8\vartheta_y\right)'\dot{\psi}^2(t) + \left(B_9\vartheta_z\right)'\dot{\psi}^2(t) \\
&+2(B_2\cos\theta + B_3\sin\theta)\dot{u}_0\dot{\psi}(t) + 2(B_{16}\cos\theta + B_{17}\sin\theta)\dot{\vartheta}_y\dot{\psi}(t) \\
&+2(B_{18}\cos\theta + B_{19}\sin\theta)\dot{\vartheta}_z\dot{\psi}(t) - 2(B_{20}\cos\theta + B_{21}\sin\theta)\dot{\varphi}'\dot{\psi}(t) \\
&+\left(B_7v_0\right)'\ddot{\psi}(t)\cos\theta - \left(B_7w_0\right)'\ddot{\psi}(t)\sin\theta - \left(B_{20}\varphi\right)'\ddot{\psi}(t)\cos\theta \\
&-\left(B_{21}\varphi\right)'\ddot{\psi}(t)\sin\theta + 2\left(B_7\dot{v}_0\right)'\dot{\psi}(t)\cos\theta - 2\left(B_7\dot{w}_0\right)'\dot{\psi}(t)\sin\theta \\
&-2\left(B_{20}\dot{\varphi}\right)'\dot{\psi}(t)\cos\theta - 2\left(B_{21}\dot{\varphi}\right)'\dot{\psi}(t)\sin\theta - \left(B_8\ddot{\vartheta}_y\right)' \\
&-\left(B_9\ddot{\vartheta}_z\right)' - \left(B_7\ddot{u}_0\right)' + \left(B_{10}\ddot{\varphi}'\right)' - \left(B_{10}\varphi'\right)'\dot{\psi}^2(t) \\
&+M_x' + B_w'' + (T_r\varphi')' + m_{\text{ext},x} + m_{\text{ext},w}' = 0
\end{aligned} \tag{38}$$

with boundary conditions,

$$
\begin{aligned}
&\big[(B_9 \cos\theta - B_8 \sin\theta)\ddot{\psi}(t) + B_7 v_0 \ddot{\psi}(t) \cos\theta \\
&- B_7 w_0 \ddot{\psi}(t) \sin\theta - (B_{20} \cos\theta + B_{21} \sin\theta)\varphi\ddot{\psi}(t) \\
&+ (R_0 + x + u_0) B_7 \dot{\psi}^2(t) + B_8 \vartheta_y \dot{\psi}^2(t) + B_9 \vartheta_z \dot{\psi}^2(t) \\
&- B_{10}\varphi' \dot{\psi}^2(t) + 2B_7 \dot{v}_0 \dot{\psi}(t) \cos\theta - 2B_7 \dot{w}_0 \dot{\psi}(t) \sin\theta \\
&- 2(B_{20} \cos\theta + B_{21} \sin\theta)\dot{\varphi}\dot{\psi}(t) - B_7 \ddot{u}_0 - B_8 \ddot{\vartheta}_y - B_9 \ddot{\vartheta}_z + B_{10} \ddot{\varphi}' \\
&+ M_x + B_w' + T_r \varphi' - m_{ext,w}\big]\big|_{x=l} = 0, \\
&B_w\big|_{x=l} = 0, \qquad \varphi'\big|_{x=0} = 0, \qquad \varphi\big|_{x=0} = 0.
\end{aligned}
\tag{39}
$$

where B_i ($i = 1 \ldots 21$) terms are inertia coefficients as defined in Georgiades et al. (2014).

• δE_3

$$
\begin{aligned}
&a_{E1} u_0' + a_{E2} \vartheta_z' + a_{E3} \vartheta_y' + a_{E4}(\vartheta_z + v_0') + a_{E5}(\vartheta_y + w_0') \\
&+ a_{E7}\varphi' - a_{E6}\varphi'' + a_{Ee} E_3 + a_{Eb} E_3^2 + a_{Ef} E_3^3 = 0
\end{aligned}
\tag{40}
$$

Relations (27)–(40) form a nonlinear system of PDEs with all equations mutually coupled. Although governing equations look similar to the purely mechanical system, the principal difference stays in the definition of 1-D generalized loadings which include electrical components as expressed in (21). New addition to the governing system is the electrostatic equation (40).

4.1 Symmetric Composite Uniform Box-Beam

In this section, a specific case of a composite symmetric beam with rectangular closed cross-section as presented in Fig. 2 is examined. Piezoelectric patches are embedded on the cross-section flanges (full width) on the inner and outer surfaces, symmetrically with respect to wall mid-line—Fig. 2b.

The circumferentially asymmetric stiffness (CAS) composite configuration is assumed. This arrangement implies the ply-angle distribution $\alpha(z) = \alpha(z)$ in the top and bottom walls of the box-beam (flanges) and $\alpha(y) = \alpha(y)$ in the lateral walls (webs). As reported in the literature (see e.g. Librescu and Song 2006), the discussed fabric configuration decouples the full set of six beam equations of motion into two independent sub-systems: one exhibiting flapwise bending–shear–twisting coupling and the second one where axial stretching and chordwise bending–shear modes coexist. Thus, clamping the beam to the hub at $\theta = 90°$ angle, as considered in the following, makes the flexible body to exhibit lead–lag deformation to be coupled with twisting. Moreover, in this configuration piezoelectric transducers can excite the lead–lag bending. Considering simplifications resulting from the cross-section symmetry (see Georgiades et al. 2014 for inertia terms calculations), the equations of motion are given by

- for the rigid hub

$$J_\text{h}\ddot{\psi}(t) + (B_{22} + B_4)\ddot{\psi}(t) \\ + \int_0^l \Big\{ b_1(R_0 + x)\big[2u_0\ddot{\psi}(t) + 2\dot{u}_0\dot{\psi}(t) - \ddot{w}_0\big] + B_4\ddot{\vartheta}_y \Big\} \mathrm{d}x = T_{\text{ext},z}(t) \tag{41}$$

- displacement in lead-lag plane w_0

$$b_1\ddot{w}_0 - 2b_1\dot{u}_0\,\dot{\psi}(t) - b_1 w_0\dot{\psi}^2(t) - b_1(R_0+x+u_0)\ddot{\psi}(t) \\ -a_{55}\vartheta_y' - a_{55}w_0'' - \left(T_x w_0'\right)' = 0 \tag{42}$$

with boundary conditions $w_0\big|_{x=0} = 0, \quad \left(\vartheta_y + w_0'\right)\big|_{x=l} = 0$

- transverse shear ϑ_y

$$B_4\ddot{\vartheta}_y - B_4\vartheta_y\dot{\psi}^2(t) + B_4\ddot{\psi}(t) + a_{55}\left(\vartheta_y + w_0'\right) \\ -(a_{33}\vartheta_y' + a_{37}\varphi' + a_{3e}E_3 + a_{3b}E_3^2 + a_{3f}E_3^3)' = 0 \tag{43}$$

with boundary conditions $\vartheta_y\big|_{x=0} = 0,$

$$\left(a_{33}\vartheta_y' + a_{37}\varphi' + a_{3e}E_3 + a_{3b}E_3^2 + a_{3f}E_3^3\right)\big|_{x=l} = 0$$

- profile twist angle φ

$$(B_4 + B_5)\ddot{\varphi} + (B_4 - B_5)\varphi\dot{\psi}^2(t) - a_{37}\vartheta_y'' - a_{77}\varphi'' - \left(T_r\varphi'\right)' = 0 \tag{44}$$

with boundary conditions,

$$\varphi\big|_{x=0} = 0, \quad \left(a_{37}\vartheta_y' + a_{77}\varphi'\right)\big|_{x=l} = 0$$

- electrostatic equation E_3

$$a_{E3}\vartheta_y' + a_{Ee}E_3 + a_{Eb}E_3^2 + a_{Ef}E_3^3 = 0 \tag{45}$$

The dynamic equation of axial motion may be simplified if the axial inertia term is disregarded. This can be done, bearing in mind the axial vibration modes are usually much higher compared to the bending and torsional ones. Therefore, the axial force T_x can be found by direct integration of (28) in space domain and taking into account the second boundary condition (29). Therefore, the term $T_x(x)$ present in (42) is defined as

$$T_x(x) = b_1(l - x)\big[R_0 + \tfrac{1}{2}(l + x)\big]\dot{\psi}^2(t).$$

It corresponds to system stiffening resulting from rotational transportation motion and it results from taking into account axial strain higher order terms—see summands underlined in (6).

Derived equations of motion for the rigid hub—thin-walled smart composite beam constitute a system of partial differential equations. Individual equations of bending and profile twist are coupled through the third d.o.f.—i.e. shear deformation. There is also an electrostatic equation in the discussed governing system. The electric field as an independent degree of freedom is present in bending equation (43), since the assumed placement of active elements results in bending moment in lead–lag plane only.

5 Conclusions

The provided text presents the derivation of governing equations for the rotating smart composite thin-walled beam system. In the performed analysis, the general case of non-zero pitch angle and variable angular velocity are considered. Moreover, non-classical effects like material anisotropy, transverse shear deformation, cross-section warping and electromechanical couplings are taken into account. The derived model is linear with respect to mechanical variables, however electric properties of the embedded piezoceramic layers are described by third order relations. Due to this enhancement, the model is capable to take into account the strong electric field loadings. The derived formula allows to consider the general case regarding placement of piezoelectric actuators—i.e. symmetric or not regarding the beam wall mid-line.

Obtained equations of motion form a non-linear system of eight partial differential equations. Six of these equations describe the motion of the flexible smart beam, the seventh represents the motion of the rigid hub and the last one is the electrostatic equation. The detailed analysis of derived equations shows these are mutually coupled. The coupling arises from the non-zero pitch angle as well as from the non-constant rotating speed. Additional couplings in the system of equations of motion are observed, due to electric properties of the piezoelement layer.

Acknowledgments The work is financially supported by grant DEC-2012/07/B/ST8/03931 from the Polish National Science Centre. Author would also like to thank Professor Jerzy Warmiński for his comments and the valuable discussions while preparing the manuscript.

References

Arafa, M., Baz, A.: On the nonlinear behavior of piezoelectric actuators. J. Vib. Control **10**(3), 387–398 (2004)

Birman, V.: Physically Nonlinear Behavior of Piezoelectric Actuators Subject to High Electric Fields. Research Triangle Park (NC) (2005)

Brockmann, T.H.: Theory of Adaptive Fiber Composites: From Piezoelectric Material Behavior to Dynamics of Rotating Structures, Solid Mechanics and its Applications, vol. 161. Springer, Dordrecht (2009)

Chattopadhyay, A., Gu, H., Liu, Q.: Modeling of smart composite box beams with nonlinear induced strain. Compos. Part B: Eng. **30**(6), 603–612 (1999)

Detwiler, D.T., Shen, M.H., Venkayya, V.B.: Finite element analysis of laminated composite structures containing distributed piezoelectric actuators and sensors. Finite Elem. Anal. Des. **20**(2), 87–100 (1995)

Georgiades, F., Latalski, J., Warmiński, J.: Equations of motion of rotating composite beams with a nonconstant rotation speed and an arbitrary preset angle. Meccanica **49**(8), 1833–1858 (2014)

Hall, D.A.: Review: nonlinearity in piezoelectric ceramics. J. Mater. Sci. **36**(19), 4575–4601 (2001)

Joshi, S.P.: Non-linear constitutive relations for piezoceramic materials. Smart Mater. Struct. **1**(1), 80–83 (1992)

Latalski, J., Warmiński, J., Rega, G.: Bending-twisting vibrations of a rotating hub–thin-walled composite beam system. Math. Mech. Solids (2016). Accessed 3 Mar 2016. doi: 10.1177/1081286516629768

Li, X., Jiang, W., Shui, Y.: Coupled mode theory for nonlinear piezoelectric plate vibrations. IEEE Trans. Ultrason. Ferroelectr. Freq. Control **45**(3), 800–805 (1998)

Librescu, L., Song, O.: Thin-Walled Composite Beams: Theory and Application. Springer, Dordrecht and the Netherlands (2006)

Maugin, G.A.: Nonlinear Electromechanical Effects and Applications. Series in theoretical and applied mechanics. World Scientific Publishing, Singapore (1985)

Mayo, J., García-Vallejo, D., Domínguez, J.: Study of the geometric stiffening effect: comparison of different formulations. Multibody Syst. Dyn. **11**(4), 321–341 (2004)

Mitchell, J.A., Reddy, J.N.: A refined hybrid plate theory for composite laminates with piezoelectric laminae. Int. J. Solids Struct. **32**(16), 2345–2367 (1995)

Priya, S., Viehland, D., Carazo, A.V., Ryu, J., Uchino, K.: High-power resonant measurements of piezoelectric materials: Importance of elastic nonlinearities. J. Appl. Phys. **90**(3), 1469 (2001)

Samal, M.K., Seshu, P., Parashar, S., Uv, Wagner, Hagedorn, P., Dutta, B.K., Kushwaha, H.S.: A finite element model for nonlinear behaviour of piezoceramics under weak electric fields. Finite Elem. Anal. Des. **41**(15), 1464–1480 (2005)

Silva, L.L., Savi, M.A., Monteiro, Paulo C.C., Netto, T.A.: On the nonlinear behavior of the piezoelectric coupling on vibration-based energy harvesters. Shock Vib. **2015**, 1–15 (2015)

Song, O., Kim, J.B., Librescu, L.: Synergistic implications of tailoring and adaptive materials technology on vibration control of anisotropic thin-walled beams. Int. J. Eng. Sci. **39**(1), 79–94 (2001)

Thornburgh, R., Chattopadhyay, A., Ghoshal, A.: Transient vibration of smart structures using a coupled piezoelectric-mechanical theory. J. Sound Vib. **274**(1–2), 53–72 (2004)

Tiersten, H.F.: Electroelastic equations for electroded thin plates subject to large driving voltages. J. Appl. Phys. **74**(5), 3389–3393 (1993)

Wagner, U.v., Hagedorn, P.: Nonlinear effects of piezoceramics excited by weak electric fields. Nonlinear Dyn. **31**(2), 133–149 (2003)

Williams, R.B.: Nonlinear mechanical and actuation characterization of piezoceramic fiber composites: Ph.d. (2004)

Yang, J.s.: Equations for the extension and flexure ofelectroelastic plates under strong electric fields. Int. J. Solids Struct. **36**(21), 3171–3192 (1999)

Yang, J.s., Chen, Z., Hu, Y., Jiang, S., Guo, S.: Weakly nonlinear behavior of a plate thickness-mode piezoelectric transformer. IEEE Trans. Ultrason. Ferroelectr. Freq. Control **54**(4), 877–881 (2007)

Yao, L.Q., Zhang, J.G., Lu, L., Lai, M.O.: Nonlinear static characteristics of piezoelectric bending actuators under strong applied electric field. Sens. Actuators A: Phys. **115**(1), 168–175 (2004)

Zhou, X.U., Chattopadhyay, A., Thornburgh, R.: Analysis of piezoelectric smart composites using a coupled piezoelectric-mechanical model. J. Intell. Mater. Syst. Struct. **11**(3), 169–179 (2000)

On a Description of Deformable Junction in the Resultant Nonlinear Shell Theory

Wojciech Pietraszkiewicz

Abstract The virtual work principle for two regular shell elements joined together along a part of their boundaries is proposed within the general nonlinear resultant shell theory. It is assumed that translations across the junction curve are smooth, but no restrictions are enforced on the rotations. For stiff and hinge type junctions, the curvilinear integral along the junction curve vanishes identically. In the case of deformable junction, the 1D constitutive type relation is proposed, where the constitutive function should be established by experiments for each particular engineering construction of the junction.

1 Introduction

By junctions of shells we mean design elements used for assembling regular shell parts along some of their boundaries into the complex multi-shell structure.

It follows from the review by Pietraszkiewicz and Konopińska (2015) that different shell models available in the literature require special forms of jump conditions at the singular surface curves modelling the shell junctions. Jump conditions corresponding to different shell models may lead to different stress and strain distributions near the junction. But the review also indicates that almost in all descriptions of shell junctions available in the literature the stiff junction conditions were enforced. Deformability of the junction itself was explicitly indicated and used in only a few papers based on the simplest shell models. This is in sharp contrast to the analyses and design of one-dimensional steel framed structures, where various semi-rigid beam-to-column connections were discussed in a number of papers, summarised in several books e.g. (Chen et al. 1996; Faella et al. 1999) and were even introduced into Eurocode 3 (1993).

Within the resultant nonlinear six-field shell model, the mechanical theory of compound multi-shell structures was initiated by Makowski and Stumpf (1994) and

W. Pietraszkiewicz (✉)
Institute of Fluid-Flow Machinery, PASci, ul. Gen. J. Fiszera 14, 80-231 Gdańsk, Poland
e-mail: pietrasz@imp.gda.pl

K. Naumenko and M. Aßmus (eds.), *Advanced Methods of Continuum Mechanics for Materials and Structures*, Advanced Structured Materials 60,
DOI 10.1007/978-981-10-0959-4_25

developed in the book by Chróścielewski et al. (2004). In this approach several regular shell elements may be joined at the common junction, deformability of any of the shell branches at the junction may individually be defined, and the junction curve itself may be equipped with additional mechanical properties independent from the adjacent shell branches. Unfortunately, the BVP of such a general theory became extremely complex and virtually useless for engineering applications. Even relatively simple cases of branching and self-intersecting shells developed in Konopińska and Pietraszkiewicz (2007) and Pietraszkiewicz and Konopińska (2011) led to complex shell relations, which were still hardly readable for engineering community. This explains why in the review (Pietraszkiewicz and Konopińska 2015) only a few papers on compound shell structures with deformable junctions was noted.

In this paper, we formulate the variational principle of virtual work for the simple compound shell structure under the following assumptions:

1. The structure consists of only two regular shell elements joined along a part of their boundaries.
2. Translations of the whole base surface, including the junction curve, are smooth.

By further constraining the junction behaviour the stiff junction, the hinge junction, and the deformable junction are described and the corresponding reduced forms of the PVW are derived.

2 Notation and Basic Shell Equilibrium Conditions

The system of notations used here and basic shell relations are compatible with the ones used in the book by Chróścielewski et al. (2004) and papers by Konopińska and Pietraszkiewicz (2007), and Pietraszkiewicz and Konopińska (2011, 2014).

A shell is a 3D solid body identified in the undeformed placement with a region B of the physical space $\mathcal{E}$ having the translation vector space E. The position vectors $\mathbf{x}$ and $\mathbf{y} = \chi(\mathbf{x})$ relative to some origin $o \in \mathcal{E}$ of any material particle in the undeformed and deformed placement, respectively, are represented by

$$\mathbf{x} = \boldsymbol{x} + \xi \boldsymbol{t} , \qquad \mathbf{y} = \boldsymbol{y}(\boldsymbol{x}) + \boldsymbol{\zeta}(\boldsymbol{x}, \xi), \qquad \boldsymbol{\zeta}(\boldsymbol{x}, 0) = \mathbf{0}. \tag{1}$$

Here $\boldsymbol{x}$ and $\boldsymbol{y}$ are position vectors of some shell base surfaces M and $N = \chi(M)$, $\boldsymbol{\zeta}$ is a deviation vector from N, $\boldsymbol{n}$ is the unit vector normal to M and orienting it, $\boldsymbol{t}$ is the unit vector not necessary normal to M with $\boldsymbol{t} \cdot \boldsymbol{n} > 0$, and $\xi \in [-h^-, h^+]$ is the distance from M along $\boldsymbol{t}$ with $h = h^- + h^+$ the initial shell thickness measured along ξ.

The shell base surface M may be irregular one, in general, consisting of regular parts $M_1, M_2, \ldots, M_n$ joined together along some parts of their edges. The junction curves form together a net of singular surface curves Γ along which the junction jump (or continuity) conditions should be formulated.

The resultant 2D equilibrium equations in the referential description, which are satisfied for any part $\Pi \in M \backslash \Gamma$, are

$$\mathrm{Div}_s \boldsymbol{N} + \boldsymbol{f} = \boldsymbol{0}, \qquad \mathrm{Div}_s \boldsymbol{M} + \mathrm{ax}\left(\boldsymbol{N}\boldsymbol{F}^T - \boldsymbol{F}\boldsymbol{N}^T\right) + \boldsymbol{c} = \boldsymbol{0}, \tag{2}$$

where $(\boldsymbol{N},\ \boldsymbol{M}) \in E \otimes T_x M$ are the referential stress resultant and stress couple tensors, $\boldsymbol{f}$ and $\boldsymbol{c}$ are the external resultant surface force and couple vectors applied on N, but measured per unit area of M, $\boldsymbol{F} = \mathrm{Grad}_s \boldsymbol{y} \in E \otimes T_x M$ is the shell deformation gradient tensor, Div_s is the divergence surface operator on M, and $\mathrm{ax}(\cdot)$ is the axial vector associated with the skew tensor $(\cdot)$.

The resultant static boundary conditions satisfied along ∂M_f are

$$\boldsymbol{n}^* - \boldsymbol{N}\boldsymbol{\nu} = \boldsymbol{0}, \qquad \boldsymbol{m}^* - \boldsymbol{M}\boldsymbol{\nu} = \boldsymbol{0}, \tag{3}$$

where $\boldsymbol{n}^*$ and $\boldsymbol{m}^*$ are the external resultant boundary force and couple vectors applied along $\partial N_f = \chi(\partial M_f)$, but measured per unit length of ∂M_f having the outward unit normal $\boldsymbol{\nu}$, and $(\cdot)^*$ means the prescribed field.

There may be in general k shell elements with regular base surfaces $M_1, M_2, \ldots, M_n$ joined together by parts of their edges $\partial M_i,\ i = 1, 2, \ldots, k \leq n$, along the singular curve Γ. In such a general case the resultant static continuity conditions across the curve Γ are (Chróścielewski et al. 2004; Konopińska and Pietraszkiewicz 2007)

$$[\boldsymbol{N}\boldsymbol{\nu}] + \boldsymbol{f}_\Gamma = \boldsymbol{0}, \qquad [\boldsymbol{M}\boldsymbol{\nu}] + [\boldsymbol{y} \times \boldsymbol{N}\boldsymbol{\nu}] + \boldsymbol{c}_\Gamma = \boldsymbol{0}, \tag{4}$$

$$[\boldsymbol{N}\boldsymbol{\nu}] = \sum_{i=1}^{k\leq n} \boldsymbol{N}_i \boldsymbol{\nu}_i, \qquad [\boldsymbol{M}\boldsymbol{\nu}] = \sum_{i=1}^{k\leq n} \boldsymbol{M}_i \boldsymbol{\nu}_i, \tag{5}$$

where $[\boldsymbol{N}\boldsymbol{\nu}]$ and $[\boldsymbol{M}\boldsymbol{\nu}]$ are the jumps of $\boldsymbol{N}\boldsymbol{\nu}$ and $\boldsymbol{M}\boldsymbol{\nu}$ at each regular point of Γ, and $\boldsymbol{f}_\Gamma$,$\boldsymbol{c}_\Gamma$ are some 1D compensating force and couple vector fields applied along $\Gamma \cap \Pi$. Explicit definitions for $\boldsymbol{f}_\Gamma$, $\boldsymbol{c}_\Gamma$ in case of branching and self-intersecting shells are given in Konopińska and Pietraszkiewicz (2007).

3 Kinematic Relations at the Shell Junction

In order to keep the junction relations in focus, we discuss here only two shell elements with regular base surfaces M_1 and M_2 connected together along their common edges coinciding with Γ, see Fig. 1.

If Γ is oriented consistently with M_1, then at any point $x_\Gamma \in \Gamma$ we have $\boldsymbol{\tau}_\Gamma = \boldsymbol{\tau}_1$, $\boldsymbol{n}_\Gamma = \boldsymbol{n}_1$, and $\boldsymbol{\nu}_\Gamma = \boldsymbol{\nu}_1 = \boldsymbol{\tau}_\Gamma \times \boldsymbol{n}_\Gamma$. At the same point $x_\Gamma \in \Gamma$ the orthonormal triad $\boldsymbol{\nu}_2,\ \boldsymbol{\tau}_2,\ \boldsymbol{n}_2$ of the edge $\partial\Pi_2 \cap \Gamma \subset \partial M_2$ does not coincide with the triad $\boldsymbol{\nu}_1,\ \boldsymbol{\tau}_1,\ \boldsymbol{n}_1$. In fact, we have,

$$\boldsymbol{\tau}_2 = -\boldsymbol{\tau}_1, \qquad \boldsymbol{n}_2 = \boldsymbol{n}_1 \cos\alpha + \boldsymbol{\nu}_1 \sin\alpha\,, \qquad \boldsymbol{\nu}_2 = -\boldsymbol{\nu}_1 \cos\alpha + \boldsymbol{n}_1 \sin\alpha. \tag{6}$$

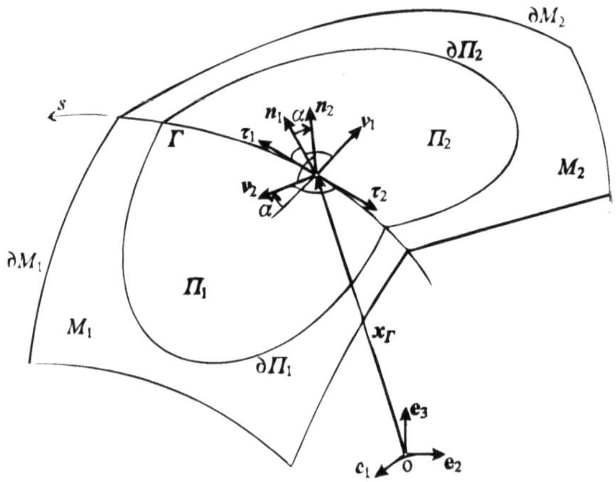

Fig. 1 The irregular surface $M = M_1 \cup M_2$ with the fold Γ, Pietraszkiewicz and Konopińska (2014)

Hence, in this case the static jumps in (5) across Γ are defined as follows:

$$[\boldsymbol{N}\boldsymbol{\nu}] = \boldsymbol{N}_1\boldsymbol{\nu}_1 + \boldsymbol{N}_2\boldsymbol{\nu}_2 = -\left(\boldsymbol{N}_2\cos\alpha - \boldsymbol{N}_1\right)\boldsymbol{\nu}_1,$$
$$[\boldsymbol{M}\boldsymbol{\nu}] = \boldsymbol{M}_1\boldsymbol{\nu}_1 + \boldsymbol{M}_2\boldsymbol{\nu}_2 = -\left(\boldsymbol{M}_2\cos\alpha - \boldsymbol{M}_1\right)\boldsymbol{\nu}_1. \qquad (7)$$

For only two surface elements connected along the fold Γ, the direct through-the-thickness integration of 3D equilibrium equations along the skew coordinate ξ indicated in (Konopińska and Pietraszkiewicz 2007) can be performed *exactly*. It is only sufficient to understand the transverse coordinate ξ as a rectilinear one, not necessarily orthogonal to the surfaces M_1 and M_2, but which coincides on the both sides of the junction region. In the case of only two shell elements there is also no necessity of introducing additional compensating force $\boldsymbol{f}_\Gamma$ and couple $\boldsymbol{c}_\Gamma$ vectors, as well as there is no additional concentrated loadings $\boldsymbol{n}_i$, $\boldsymbol{m}_i$ and $\boldsymbol{n}_e$, $\boldsymbol{m}_e$ at the initial and end points of Γ within M as well, which were necessary in the case of branching and self-intersecting shells, see Konopińska and Pietraszkiewicz (2007). As a result, in this case the static continuity conditions (4) are simplified to

$$[\boldsymbol{N}\boldsymbol{\nu}] = \boldsymbol{0}, \qquad [\boldsymbol{M}\boldsymbol{\nu}] + [\boldsymbol{y} \times \boldsymbol{N}\boldsymbol{\nu}] = \boldsymbol{0}, \qquad (8)$$

where the jumps are defined by (7).

In this paper, we additionally assume the deformed position vector field $\boldsymbol{y}$ to be always smooth, so that $[\boldsymbol{y}] = \boldsymbol{0}$ across Γ. By this requirement we prevent the shell to be decomposed along Γ during deformation. As a result, the static continuity conditions (8) are reduced to

$$[\boldsymbol{N}\boldsymbol{\nu}] = \boldsymbol{0}, \qquad [\boldsymbol{M}\boldsymbol{\nu}] = \boldsymbol{0}. \qquad (9)$$

Let $(\boldsymbol{v}, \boldsymbol{w}) \in E$ be two vector fields smooth at the regular points of $M\backslash\Gamma$, and $(\boldsymbol{v}_\Gamma, \boldsymbol{w}_\Gamma) \in E$ be two other vector fields smooth along Γ. Then, for any part $\Pi \in M$

containing the fold Γ we can set the integral identity

$$\iint_{\Pi\backslash\Gamma} \left\{(\mathrm{Div}_s \boldsymbol{N} + \boldsymbol{f}) \cdot \boldsymbol{v} + \left(\mathrm{Div}_s \boldsymbol{M} + \mathrm{ax}(\boldsymbol{N}\boldsymbol{F}^{\mathrm{T}} - \boldsymbol{F}\boldsymbol{N}^{\mathrm{T}}) + \boldsymbol{c}\right) \cdot \boldsymbol{w}\right\} \mathrm{d}a$$
$$+ \int_{\Pi\cap\partial M_f} \left\{\left(\boldsymbol{n}^* - \boldsymbol{N}\boldsymbol{\nu}\right) \cdot \boldsymbol{v} + \left(\boldsymbol{m}^* - \boldsymbol{M}\boldsymbol{\nu}\right) \cdot \boldsymbol{w}\right\} \mathrm{d}s$$
$$- \int_{\Pi\cap\Gamma} \{[\boldsymbol{N}\boldsymbol{\nu}] \cdot \boldsymbol{v}_\Gamma + [\boldsymbol{M}\boldsymbol{\nu}] \cdot \boldsymbol{w}_\Gamma\} \, \mathrm{d}s = 0. \tag{10}$$

By simple algebra we have

$$\begin{aligned} (\mathrm{Div}_s \boldsymbol{N}) \cdot \boldsymbol{v} &= \boldsymbol{N} \bullet \mathrm{Grad}_s \boldsymbol{v}, \\ (\mathrm{Div}_s \boldsymbol{M}) \cdot \boldsymbol{w} &= \boldsymbol{M} \bullet \mathrm{Grad}_s \boldsymbol{w}, \\ \mathrm{ax}(\boldsymbol{N}\boldsymbol{F}^T - \boldsymbol{F}\boldsymbol{N}^T) \cdot \boldsymbol{w} &= \boldsymbol{N} \bullet (\boldsymbol{W}\boldsymbol{F}), \end{aligned} \tag{11}$$

where $\bullet$ is the scalar product in the tensor space such that for any $\boldsymbol{A}, \boldsymbol{B} \in E \otimes T_x M$ we have $\boldsymbol{A} \bullet \boldsymbol{B} = \mathrm{tr}(\boldsymbol{A}^T \boldsymbol{B})$, and $\boldsymbol{W} = \boldsymbol{w} \times \mathbf{1}$ is the skew tensor, where $\mathbf{1}$ means the identity tensor of the space $E \otimes E$.

Since Π is an arbitrarily chosen part of M containing Γ, transforming (10) with the help of (11) and applying the surface divergence theorems (see Konopińska and Pietraszkiewicz (2007), f. (23)–(26)) we obtain

$$- \iint_{M\backslash\Gamma} \{\boldsymbol{N} \bullet (\mathrm{Grad}_s \boldsymbol{v} - \boldsymbol{W}\boldsymbol{F}) + \boldsymbol{M} \bullet \mathrm{Grad}_s \boldsymbol{w}\} \, \mathrm{d}a + \iint_{M\backslash\Gamma} (\boldsymbol{f} \cdot \boldsymbol{v} + \boldsymbol{c} \cdot \boldsymbol{w}) \, \mathrm{d}a$$
$$+ \int_{\partial M_f} \left(\boldsymbol{n}^* \cdot \boldsymbol{v} + \boldsymbol{m}^* \cdot \boldsymbol{w}\right) \mathrm{d}s + \int_{\partial M_d} (\boldsymbol{N}\boldsymbol{\nu} \cdot \boldsymbol{v} + \boldsymbol{M}\boldsymbol{\nu} \cdot \boldsymbol{w}) \, \mathrm{d}s$$
$$+ \int_{\Gamma} \{[\boldsymbol{N}\boldsymbol{\nu} \cdot \boldsymbol{v}] - [\boldsymbol{N}\boldsymbol{\nu}] \cdot \boldsymbol{v}_\Gamma + [\boldsymbol{M}\boldsymbol{\nu} \cdot \boldsymbol{w}] - [\boldsymbol{M}\boldsymbol{\nu}] \cdot \boldsymbol{w}_\Gamma\} \, \mathrm{d}s = 0. \tag{12}$$

The real shell deformation is described by the translation vector $\boldsymbol{u} = \boldsymbol{y} - \boldsymbol{x}$ and the rotation tensor $\boldsymbol{Q} = \boldsymbol{d}_i \otimes \boldsymbol{t}_i$, $i = 1, 2, 3$, of M, where $\boldsymbol{d}_i$, $\boldsymbol{t}_i$ are orthonormal triads of directors associated with M in the deformed and undeformed placements, respectively. Then the vectors $\boldsymbol{v}, \boldsymbol{w}$ in (12) may be interpreted as the kinematically admissible virtual translations and rotations,

$$\begin{aligned} \boldsymbol{v} &\equiv \delta \boldsymbol{u}, \\ \boldsymbol{W} &= (\delta \boldsymbol{Q})\boldsymbol{Q}^T \equiv \boldsymbol{\Omega} = \boldsymbol{\omega} \times \mathbf{1}, \\ \boldsymbol{w} &= \frac{1}{2}(\mathbf{1} \times \mathbf{1}) \bullet \boldsymbol{W} \equiv \boldsymbol{\omega} = \frac{1}{2}(\mathbf{1} \times \mathbf{1}) \bullet \boldsymbol{\Omega}, \end{aligned} \tag{13}$$

where δ is the symbol of virtual change (variation).

Since $\delta \boldsymbol{u} = \boldsymbol{\omega} = \mathbf{0}$ along $\partial M_d = \partial M \backslash \partial M_f$, the integral over ∂M_d in (12) vanishes identically. Moreover, it was found in Chróścielewski et al. (2004), Pietraszkiewicz et al. (2005) that

$$\mathrm{Grad}_s \delta \boldsymbol{u} - \boldsymbol{\Omega} \boldsymbol{F} = \delta^c \boldsymbol{E}, \qquad \mathrm{Grad}_s \boldsymbol{\omega} = \delta^c \boldsymbol{K}, \tag{14}$$

where $\delta^c(\cdot) = \boldsymbol{Q}\delta\{\boldsymbol{Q}^T(\cdot)\}$ is the co-rotational variation of $(\cdot)$, and the 2D shell stretch and bending tensors are defined by

$$\boldsymbol{E} = \boldsymbol{J}\mathbf{F} - \boldsymbol{Q}\boldsymbol{I}, \qquad \boldsymbol{K} = \boldsymbol{C}\mathbf{F} - \boldsymbol{Q}\boldsymbol{B}. \tag{15}$$

In (15), $\boldsymbol{I} = \mathrm{Grad}_s \boldsymbol{x} \in E \otimes T_x M$ and $\boldsymbol{J} = \mathrm{grad}_s \boldsymbol{y} \in E \otimes T_y N$ are the inclusion operators on $M \backslash \Gamma$ and $N \backslash \chi(\Gamma)$, $\boldsymbol{B}$ and $\boldsymbol{C}$ are the structure curvature tensors of the shell base surface in the undeformed and deformed placement, respectively, and $\mathbf{F} \in T_y N \otimes T_x M$ is the tangential surface deformation gradient such that $\mathrm{d}y = \mathbf{F}\mathrm{d}x$.

Introducing the virtual strain energy density in $M \backslash \Gamma$ defined by

$$\sigma = \boldsymbol{N} \bullet \delta^c \boldsymbol{E} + \boldsymbol{M} \bullet \delta^c \boldsymbol{K}, \tag{16}$$

the principle of virtual work following from (12) takes the form

$$\begin{aligned} \iint_{M \backslash \Gamma} \sigma \,\mathrm{d}a = & \iint_{M \backslash \Gamma} (\boldsymbol{f} \cdot \delta \boldsymbol{u} + \boldsymbol{c} \cdot \boldsymbol{\omega}) \,\mathrm{d}a + \int_{\partial M_f} \left(\boldsymbol{n}^* \cdot \delta \boldsymbol{u} + m^* \cdot \boldsymbol{\omega}\right) \mathrm{d}s \\ & + \int_{\Gamma} \{[\boldsymbol{N}\boldsymbol{\nu} \cdot \delta \boldsymbol{u}] - [\boldsymbol{N}\boldsymbol{\nu}] \cdot \boldsymbol{v}_\Gamma + [\boldsymbol{M}\boldsymbol{\nu} \cdot \boldsymbol{\omega}] - [\boldsymbol{M}\boldsymbol{\nu}] \cdot \boldsymbol{w}\} \,\mathrm{d}s. \end{aligned} \tag{17}$$

The curvilinear integral over Γ in (17) includes the jump terms which describe the shell–junction interaction between two joined shell elements with base surfaces M_1 and M_2. Explicit expressions of the jump terms depend on the type of junction modelled by this approach.

The large variety of types of 1D structural elements, which can be used as junctions in compound shell structures, together with complex kinematics required within the resultant six-field shell model, makes the general nonlinear BVP of such structures to be very complex and hardly readable in engineering applications.

The compound jump terms in (17) can be decomposed as follows:

$$\begin{aligned} [\boldsymbol{N}\boldsymbol{\nu} \cdot \delta \boldsymbol{u}] &= [\boldsymbol{N}\boldsymbol{\nu}] \cdot < \delta \boldsymbol{u} > + < \boldsymbol{N}\boldsymbol{\nu} > \cdot [\delta \boldsymbol{u}], \\ [\boldsymbol{M}\boldsymbol{\nu} \cdot \boldsymbol{\omega}] &= [\boldsymbol{M}\boldsymbol{\nu}] \cdot < \boldsymbol{\omega} > + < \boldsymbol{M}\boldsymbol{\nu} > \cdot [\boldsymbol{\omega}], \end{aligned} \tag{18}$$

where $< \boldsymbol{a} >$ is the average value of $\boldsymbol{a} \in E$ at Γ. In our special case of smooth translations everywhere discussed here, the translation at the junction curve Γ may be interpreted as the average translation of both edges $\partial M_1 \cap \Gamma$ and $\partial M_2 \cap \Gamma$, so that $< \delta \boldsymbol{u} > \equiv \delta \boldsymbol{u}$. But the rotation tensors $\boldsymbol{Q}_1 = \boldsymbol{Q}|_{\partial M_1 \cap \Gamma}$ and $\boldsymbol{Q}_2 = \boldsymbol{Q}|_{\partial M_2 \cap \Gamma}$ of the edges at the same $x_\Gamma \in \Gamma$ may be different, in general, $\boldsymbol{Q}_1 \neq \boldsymbol{Q}_2$.

With (18) the PVW (17) can be reduced to

$$\iint_{M\setminus\Gamma} \sigma \, \mathrm{d}a = \iint_{M\setminus\Gamma} (\boldsymbol{f} \cdot \delta\boldsymbol{u} + \boldsymbol{c} \cdot \boldsymbol{\omega}) \, \mathrm{d}a + \int_{\partial M_f} \left(\boldsymbol{n}^* \cdot \delta\boldsymbol{u} + \boldsymbol{m}^* \cdot \boldsymbol{\omega}\right) \mathrm{d}s$$
$$+ \int_{\Gamma} \{[\boldsymbol{M}\boldsymbol{\nu}] (< \boldsymbol{\omega} > - \boldsymbol{w}_{\Gamma}) + < \boldsymbol{M}\boldsymbol{\nu} > \cdot [\boldsymbol{\omega}]\} \, \mathrm{d}s. \tag{19}$$

Let us introduce explicitly the net rotation tensor $\boldsymbol{Q}_{\Gamma}$ of Γ such that $\boldsymbol{Q}_2 = \boldsymbol{Q}_{\Gamma}\boldsymbol{Q}_1$ at any $x_{\Gamma} \in \Gamma$ when x_{Γ} is approached from both sides of Γ, respectively. Since $\boldsymbol{Q}_2$, $\boldsymbol{Q}_{\Gamma}$, $\boldsymbol{Q}_1$ are all proper orthogonal tensors, we have

$$\boldsymbol{Q}_2\boldsymbol{Q}_2^T = \boldsymbol{1}, \qquad \boldsymbol{Q}_{\Gamma}\boldsymbol{Q}_{\Gamma}^T = \boldsymbol{1}, \qquad \boldsymbol{Q}_1\boldsymbol{Q}_1^T = \boldsymbol{1}. \tag{20}$$

Virtual changes of these orthogonality relations lead to

$$\begin{aligned} \delta\boldsymbol{Q}_2\boldsymbol{Q}_2^T &= -\boldsymbol{Q}_2\delta\boldsymbol{Q}_2^T = \boldsymbol{\omega}_2 \times \boldsymbol{1}, \\ \delta\boldsymbol{Q}_{\Gamma}\boldsymbol{Q}_{\Gamma}^T &= -\boldsymbol{Q}_{\Gamma}\delta\boldsymbol{Q}_{\Gamma}^T = \boldsymbol{\omega}_{\Gamma} \times \boldsymbol{1}, \\ \delta\boldsymbol{Q}_1\boldsymbol{Q}_1^T &= -\boldsymbol{Q}_1\delta\boldsymbol{Q}_1^T = \boldsymbol{\omega}_1 \times \boldsymbol{1}, \end{aligned} \tag{21}$$

$$\boldsymbol{\omega}_2 = \boldsymbol{\omega} + \boldsymbol{Q}_{\Gamma}\boldsymbol{\omega}_1. \tag{22}$$

The virtual rotations $\boldsymbol{\omega}_2$, $\boldsymbol{\omega}_{\Gamma}$ and $\boldsymbol{\omega}_1$ are all defined in the shell deformed placement.

Let the virtual rotation $\boldsymbol{w}_{\Gamma}$ at Γ be interpreted in terms of $\boldsymbol{\omega}$ as

$$< \boldsymbol{\omega} > = \frac{1}{2}\left\{\boldsymbol{\omega}_{\Gamma} + (\boldsymbol{Q}_{\Gamma} + \boldsymbol{1})\boldsymbol{\omega}_1\right\} \equiv \boldsymbol{w}_{\Gamma}. \tag{23}$$

Then the PVW (19) can be further reduced to

$$\iint_{M\setminus\Gamma} \sigma \, \mathrm{d}a = \iint_{M\setminus\Gamma} (\boldsymbol{f} \cdot \delta\boldsymbol{u} + \boldsymbol{c} \cdot \boldsymbol{\omega}) \, \mathrm{d}a + \int_{\partial M_f} \left(\boldsymbol{n}^* \cdot \delta\boldsymbol{u} + \boldsymbol{m}^* \cdot \boldsymbol{\omega}\right) \mathrm{d}s$$
$$+ \int_{\Gamma} < \boldsymbol{M}\boldsymbol{\nu} > \cdot [\boldsymbol{\omega}] \, \mathrm{d}s. \tag{24}$$

The variational statement (24) governs the simplified BVP of two regular shell elements with base surfaces M_1 and M_2 joined along the junction Γ. This PVW has been constructed under the assumption that the joint translations are smooth everywhere during deformations. As a result, kinematic description of the junction has been reduced to characterising how the rotations $\boldsymbol{Q}_1$ and $\boldsymbol{Q}_2$ of the neighbouring points of the junction are related to each other during deformation. This still allows one for a variety of possible characterisations of the junction. Some of the simplest particularly appealing junction characterisations are discussed below.

4 Description of the Junction

4.1 The Stiff Junction

The shell junction along Γ is called *stiff* if the shell deformation is continuous on the whole $M = M_1 \cup M_2$ including Γ. In this case

$$[\delta \boldsymbol{u}] = \boldsymbol{0}, \qquad [\boldsymbol{\omega}] = \boldsymbol{0}, \qquad \boldsymbol{u}_1 = \boldsymbol{u}_2, \qquad \boldsymbol{Q}_1 = \boldsymbol{Q}_2, \tag{25}$$

and the curvilinear integral along Γ in (24) vanishes. The correspondingly simplified PVW is reduced to

$$\iint_M \sigma \,\mathrm{d}a = \iint_M (\boldsymbol{f} \cdot \delta \boldsymbol{u} + \boldsymbol{c} \cdot \boldsymbol{\omega}) \,\mathrm{d}a + \int_{\partial M_f} \left(\boldsymbol{n}^* \cdot \delta \boldsymbol{u} + \boldsymbol{m}^* \cdot \boldsymbol{\omega}\right) \mathrm{d}s. \tag{26}$$

The physical meaning of (26) is that in this case the junction along Γ does not contribute to the virtual work of the compound shell structure. The mechanical behaviour of the junction itself is enforced by the behaviour of stiffly joined shell lateral boundary surfaces of regular shell parts with surface elements M_1 and M_2. This is exactly the case of almost all types of shell junctions reviewed in Pietraszkiewicz and Konopińska (2015). In particular, within the resultant nonlinear six-field shell theory several folded and multi-shell structures with stiff junctions were modelled and analysed with FEM by Chróścielewski et al. (1997). Nonlinear dynamic problems of such structures were discussed by Chróścielewski et al. (2002). A number of linear and nonlinear FE solutions of multi-shells with stiff junctions was summarised in the book of Chróścielewski et al. (2004).

4.2 The Hinge Junction

The *hinge* junction along Γ is understood when $\boldsymbol{u}$ is continuous across Γ, that is $[\delta \boldsymbol{u}] = \boldsymbol{0}$, $\boldsymbol{u}_1 = \boldsymbol{u}_2$, but $\boldsymbol{Q}_1$, $\boldsymbol{Q}_2$ are entirely unconstrained when approaching Γ along a path on corresponding M_1 , M_2. In this case $[\boldsymbol{\omega}] \neq \boldsymbol{0}$, in general. However, in order the entirely unconstrained rotations $\boldsymbol{Q}_1$, $\boldsymbol{Q}_2$ to happen, from equilibrium it follows that no moments at both sides of Γ should be allowed,

$$\boldsymbol{M}_1 \boldsymbol{\nu}_1 = \boldsymbol{0}, \qquad \boldsymbol{M}_2 \boldsymbol{\nu}_2 = \boldsymbol{0}, \tag{27}$$

so that $< \boldsymbol{M}\boldsymbol{\nu} >= \boldsymbol{0}$ and hence $< \boldsymbol{M}\boldsymbol{\nu} > \cdot [\boldsymbol{\omega}] = \boldsymbol{0}$ along Γ. As a result, in the corresponding PVW (24) the curvilinear integral along Γ vanishes as well reducing it again formally to (26). However, the important difference to the stiff junction is that in the case of hinge junction along Γ the additional static equilibrium conditions (27) have to be enforced in the process of solution of BVP.

Some special cases of such BVP for the linear elastic plate problems were proposed in the literature. For example, linear boundary value problems for anisotropic elastic Kirchoff plates with internal line hinges were discussed by Grossi (2011). Natural vibrations of the rectangular plate with a hinge line were analysed by Huang et al. (2009) and Grossi and Raffo (2013) within the linear Kirchhoff plate model, while Xiang and Reddy (2003) used the linear first order shear deformation plate model for this purpose.

4.3 *The Deformable Junction*

In the PVW (24) both ingredients $<\boldsymbol{M}\boldsymbol{\nu}>$ and $[\boldsymbol{\omega}]$ in the last integral may not together identically vanish, in general, that is $<\boldsymbol{M}\boldsymbol{\nu}>\neq \boldsymbol{0}$ and $[\boldsymbol{\omega}]\neq\boldsymbol{0}$. In this general case the shell junction along Γ may be called *deformable*.

From engineering point of view, the junctions can be classified according to

1. The type of medium used: bolted, welded, riveted, glued, adhesively bonded etc.;
2. The type of internal forces the junction is expected to transmit: membrane, shear, moment (stiff, deformable);
3. The type of elements the junction is joining: regular shell elements, transition stiffening ringbeams, special junction constructions.

This leads to a large variety of constructions of junctions in compound shell structures. Mechanical and/or deformability properties of each particular case of such junction should be known in advance before the analyses take place.

Let us differentiate the orthogonality relations (21) along Γ,

$$\begin{aligned}(\boldsymbol{Q}_2)'\boldsymbol{Q}_2^T &= -\boldsymbol{Q}_2(\boldsymbol{Q}_2^T)' = \boldsymbol{\kappa}_2\times\boldsymbol{1},\\ (\boldsymbol{Q}_\Gamma)'\boldsymbol{Q}_\Gamma^T &= -\boldsymbol{Q}_\Gamma(\boldsymbol{Q}_\Gamma^T)' = \boldsymbol{\kappa}_\Gamma\times\boldsymbol{1},\\ (\boldsymbol{Q}_1)'\boldsymbol{Q}_1^T &= -\boldsymbol{Q}_1(\boldsymbol{Q}_1^T)' = \boldsymbol{\kappa}_1\times\boldsymbol{1},\quad (.)' = \frac{d}{ds}(.),\end{aligned} \tag{28}$$

$$\boldsymbol{\kappa}_2 = \boldsymbol{\kappa}_\Gamma + \boldsymbol{Q}_\Gamma\boldsymbol{\kappa}_1. \tag{29}$$

The vector $\boldsymbol{\kappa}_\Gamma$ describes the bending properties of the junction curve Γ during shell deformation.

The mechanical behaviour of the deformable junction can be characterised by the relation

$$<\boldsymbol{M}\boldsymbol{\nu}> = f(\boldsymbol{\kappa}_\Gamma)\,, \tag{30}$$

where f is a smooth vector function of vectorial argument at any $x_\Gamma \in \Gamma$. The relation (30) is the kind of 1D constitutive equation modelling deformability properties of real engineering junctions. It is apparent that due to possible complexity of engineering junction constructions the function f should be established from appropriate experiments for each particular type of the junction.

With (30) and (22) the PVW (19) takes the modified form

$$\iint_{M\setminus\Gamma} \sigma \, \mathrm{d}a = \iint_{M\setminus\Gamma} (\boldsymbol{f}\cdot\delta\boldsymbol{u} + \boldsymbol{c}\cdot\boldsymbol{\omega}) \, \mathrm{d}a + \int_{\partial M_f} \left(\boldsymbol{n}^*\cdot\delta\boldsymbol{u} + \boldsymbol{m}^*\cdot\boldsymbol{\omega}\right) \mathrm{d}s + \int_\Gamma f(\boldsymbol{\kappa}_\Gamma)\cdot\left\{\boldsymbol{\omega}_\Gamma + \left(\boldsymbol{Q}_\Gamma - \boldsymbol{1}\right)\boldsymbol{\omega}_1\right\} \mathrm{d}s. \tag{31}$$

If there exists a scalar function $W(\boldsymbol{\kappa}_\Gamma)$ such that $f(\boldsymbol{\kappa}_\Gamma) = \partial W/\partial\boldsymbol{\kappa}_\Gamma$, the junction along Γ may be called *elastic*. The function W may be quite complex nonlinear function of $\boldsymbol{\kappa}_\Gamma$, so that such a junction is *nonlinearly elastic*, in general. But in some cases W may become a quadratic function such that

$$W = \frac{1}{2}\boldsymbol{L}\bullet(\boldsymbol{\kappa}_\Gamma\otimes\boldsymbol{\kappa}_\Gamma), \qquad f(\boldsymbol{\kappa}_\Gamma) = \boldsymbol{L}\boldsymbol{\kappa}_\Gamma, \tag{32}$$

where $\boldsymbol{L}$ is the second-order tensor of rotational material properties along Γ. In this case the shell junction can be called *linearly elastic*.

Special cases of elastic junction conditions within the linear Kirchhoff-type theory of elastic plates were discussed by Bernadou (1996), Titeux and Sanchez-Palencia (2000) and Nardinocchi (2002). Elastic junctions between two thin linearly elastic shells of Koiter type were asymptotically analysed by Akian (2005) and Merabet et al. (2010). Within the nonlinear theory of thin shells of Kirchhoff–Love type, description of several types of shell junctions were given by Makowski et al. (1999), and explicit numerical solutions of the shell of revolution with deformable elasto-plastic junctions were given by Chróścielewski et al. (2011a, b). Within the nonlinear six-field shell theory, the deformable junction of one branch at the shell branching was kinematically classified in Pietraszkiewicz and Konopińska (2011) as locally elastic, non-locally elastic and dissipative.

5 Conclusions

Within the general nonlinear resultant shell theory, we have formulated the virtual work principle for two regular shell elements joined together along their common boundaries. It has been assumed that translations across the junction curve are smooth, but no restrictions are enforced for the rotations. It has been shown that for stiff and hinge type junctions the curvilinear integral along the junction curve vanishes identically and does not bring an additional virtual work to the shell BVP. In the case of deformable junction, the 1D constitutive type relation has been proposed for the junction moments in terms of net rotations of the junction curve. The constitutive function should be established by experiments for each engineering construction of the deformable junction. As special cases, description of nonlinearly elastic and linearly elastic junctions have been noted.

The proposed description of shell junctions should allow development of appropriate numerical FEM programs for nonlinear analyses of multi-shells with various types of junctions.

Acknowledgments The research reported in this paper was supported by the National Science Centre of Poland with the grant DEC–2012/05/D/ST8/02298.

References

Akian, J.L.: Asymptotic analysis of bending-dominated shell junctions. Journal de Mathématiques Pures et Appliquées **84**(6), 667–716 (2005). doi:10.1016/j.matpur.2004.10.001

Bernadou, M.: Finite Element Methods for Thin Shell Problems. Wiley, Chichester (1996)

Chen, W., Goto, Y., Liew, J.: Stability Design of Semi-Rigid Frames. Wiley, New York (1996)

Chróścielewski, J., Makowski, J., Stumpf, H.: Finite element analysis of smooth, folded and multi-shell structures. Comput. Methods Appl. Mech. Eng. **141**(1–2), 1–46 (1997). doi:10.1016/S0045-7825(96)01046-8

Chróścielewski, J., Makowski, J., Pietraszkiewicz, W.: Non-linear dynamics of flexible shell structures. Comput. Assist. Mech. Eng. Sci. **9**, 341–357 (2002)

Chróścielewski, J., Makowski, J,, Pietraszkiewicz, W.: Statics and Dynamics of Multifold Shells: Nonlinear Theory and Finite Element Method. IFTR PASci Press, Warsaw, in Polish (2004)

Chróścielewski, J., Konopińska, V., Pietraszkiewicz, W.: On elasto-plastic analysis of thin shells with deformable junctions. In: Altenbach, H., Eremeyev, V.A. (eds.) Shell-Like Structures: Non-classical Theories and Applications, pp. 441–452. Springer, Berlin (2011a). doi: 10.1007/978-3-642-21855-2_28

Chróścielewski, J., Konopińska, V., Pietraszkiewicz, W.: On modelling and non-linear elasto-plastic analysis of thin shells with deformable junctions. ZAMM - J. Appl. Math. Mech. / Zeitschrift für Angewandte Mathematik und Mechanik **91**(6), 477–484 (2011b). doi:10.1002/zamm.201000225

Eurocode 3 EN 1993-1-8: Design of Steel Structures - part 1–8: Design of joints. CEN, Brussels (1993)

Faella, C., Piluso, V., Rizzano, G.: Structural Steel Semirigid Connections: Theory, Design, and Software. CRC Press, Boca Raton (1999). New Directions in Civil Engineering

Grossi, R.O.: Boundary value problems for anisotropic plates with internal line hinges. Acta Mech. **223**(1), 125–144 (2011). doi:10.1007/s00707-011-0552-5

Grossi, R.O., Raffo, J.: Natural vibrations of anisotropic plates with several internal line hinges. Acta Mech. **224**(11), 2677–2697 (2013). doi:10.1007/s00707-013-0892-4

Huang, M., Ma, X., Sakiyama, T., Matsuda, H., Morita, C.: Natural vibration study on rectangular plates with a line hinge and various boundary conditions. J. Sound Vib. **322**(1–2), 227–240 (2009). doi:10.1016/j.jsv.2008.11.006

Konopińska, V., Pietraszkiewicz, W.: Exact resultant equilibrium conditions in the non-linear theory of branching and self-intersecting shells. Int. J. Solids Struct. **44**(1), 352–369 (2007). doi:10.1016/j.ijsolstr.2006.04.030

Makowski, J., Stumpf, H.: Mechanics of irregular shell structures. Mitteilungen aus dem Institut für Mechanik Nr. 95, Ruhr-Universität Bochum, Bochum (1994)

Makowski, J., Pietraszkiewicz, W., Stumpf, H.: Jump conditions in the non-linear theory of thin irregular shells. J. Elast. **54**(1), 1–26 (1999). doi:10.1023/A:1007645414077

Merabet, I., Chacha, D., Nicaise, S.: Singular layers for transmission problems in thin shallow shell theory: elastic junction case. Comptes Rendus Mécanique **338**(5), 277–282 (2010). doi:10.1016/j.crme.2010.04.007

Nardinocchi, P.: Modelling junctions of thin plates. Eur. J. Mech. - A/Solids **21**(3), 523–534 (2002). doi:10.1016/S0997-7538(02)01222-6

Pietraszkiewicz, W., Konopińska, V.: On unique kinematics for the branching shells. Int. J. Solids Struct. **48**(14–15), 2238–2244 (2011). doi:10.1016/j.ijsolstr.2011.03.029

Pietraszkiewicz, W., Konopińska, V.: Singular curves in the resultant thermomechanics of shells. Int. J. Eng. Sci. **80**, 21–31 (2014). doi:10.1016/j.ijengsci.2014.02.019. Special issue on Nonlinear and Nonlocal Problems, in occasion of 70th birthday of Prof. Leonid Zubov

Pietraszkiewicz, W., Konopińska, V.: Junctions in shell structures: a review. Thin-Walled Struct. **95**, 310–334 (2015). doi:10.1016/j.tws.2015.07.010

Pietraszkiewicz W, Chróscielewski, J., Makowski, J.: On dynamically and kinematically exact theory of shells. In: Pietraszkiewicz, W., Szymczak, C. (eds.) Shell Structures: Theory and Applications, pp. 163–167. Taylor & Francis, London (2005)

Titeux, I., Sanchez-Palencia, E.: Junction of thin plates. Eur. J. Mech. - A/Solids **19**(3), 377–400 (2000). doi:10.1016/S0997-7538(00)00175-3

Xiang, Y., Reddy, J.: Natural vibration of rectangular plates with an internal line hinge using the first order shear deformation plate theory. J. Sound Vib. **263**(2), 285–297 (2003). doi:10.1016/S0022-460X(02)01124-0

Analytical and Numerical Modelling of a Sub- and Supersonic Moving Load Front Along a Rod's Skin

Wolfgang E. Weber, Yannick F. Fangye, Daniel Balzani and Bernd W. Zastrau

Abstract For both civil and mechanical engineering dynamic loads of structures are a major source of inner material damage. If (fibre) reinforced composite materials are exposed to such dynamic loads a pull-out of the reinforcing elements may occur. This dynamic pull-out of reinforcing elements is characterized by, amongst others, moving boundaries between regions of (partly) damaged and perfect bonding of reinforcement and surrounding matrix. To adequately describe these moving boundaries leads to enormous challenges. Within this contribution a simplified mechanical problem is investigated, which however provides some of the main phenomena of the dynamic pull-out. In detail, the stress and displacement fields within a rod of semi-infinite extent under a distributed load are evaluated. Herein, the front of the constant longitudinal load moves along the rod in longitudinal direction. The investigations are performed both analytically and numerically thus validating the model idealization included in the analytical solution.

W.E. Weber (✉) · Y.F. Fangye · D. Balzani · B.W. Zastrau
Institute of Mechanics and Shell Structures, Technische Universität Dresden,
George-Bähr-Straße 1, 01187 Dresden, Germany
e-mail: wolfgang.weber@tu-dresden.de

Y.F. Fangye
e-mail: yannick_francis.fangye@tu-dresden.de

D. Balzani
Dresden Center for Computational Materials Science,
Technische Universität Dresden, Dresden, Germany
e-mail: daniel.balzani@tu-dresden.de

B.W. Zastrau
e-mail: bernd.zastrau@tu-dresden.de

K. Naumenko and M. Aßmus (eds.), *Advanced Methods of Continuum Mechanics for Materials and Structures*, Advanced Structured Materials 60,
DOI 10.1007/978-981-10-0959-4_26

Fig. 1 Pedestrian bridge at Oschatz, Germany, made from TRC, by courtesy of SFB 528

1 Introduction

The adequate description of the wave propagation in rod-like or reinforced structures is a matter of enormous practical relevance. Practical applications involve, but are not restricted to, pile driving for offshore wind turbines and dynamic pull-out of reinforcing elements in composite materials, see e.g. Weber and Curbach (2013), Sridhar et al. (2003), Yang et al. (2006). Herein, the waves propagating through the material result from dynamic loads as, e.g. impact or repeated loading acting on the respective structure or structural member. Figure 1 shows a pedestrian bridge at Oschatz, Germany, which is made of textile reinforced concrete (TRC). TRC is a compound material developed recently and involving multifilament yarns of glass or carbon acting as reinforcement of a brittle fine-grained concrete matrix. For details see, amongst others, Curbach and Jesse (2009). If such a structure is subjected to vertical (dynamic) loads the resulting bending tensile stresses may lead to crack initiation in the concrete matrix. Subsequently, these cracks—which grow in the vertical direction—are bridged by the reinforcement. In this contribution, we assume no further growth of the crack(s). The situation is depicted in Fig. 2. As can readily be seen, this example of a dynamic pull-out gives significant challenges for both the analytical and the numerical modelling and so do similar problems of wave propagation in rod-like or reinforced structures. These challenges arise, e.g. from the

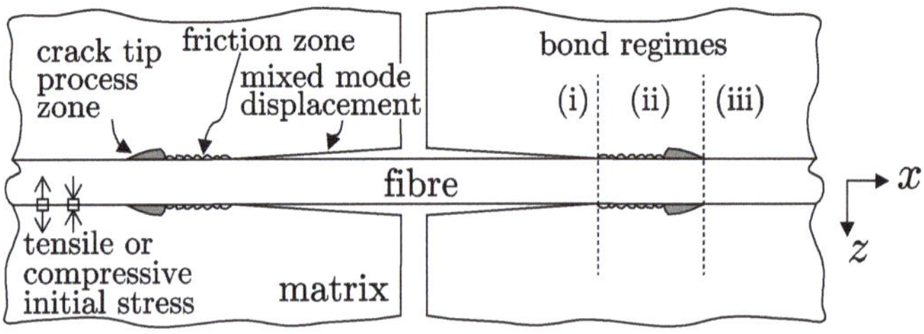

Fig. 2 Typical regions along a reinforcing element resulting from crack bridging under dynamic loads: (i) completely damaged bond, (ii) partly damaged bond, (iii) perfect bond, cf. Sridhar et al. (2003)

interaction of the travelling wave in the rod with the host in which the rod is embedded. This interaction depending on the load amplitude, e.g. leads to a successive failure of the bond between rod and surrounding host and thus a moving crack tip, see also Fig. 2. The rod then may be subdivided into three parts: (i) one with a completely damaged bond, (ii) one with a partly damaged bond, and (iii) another part with a still perfect bond. The boundaries between these parts will travel with certain speeds which are of interest in future investigations. These moving boundaries will change the geometrical dimensions of the respective constraints within the model with increasing time. Due to this type of nonlinearity advanced numerical methods such as the Spectral Finite Element Method (SFEM), the Boundary Element Method or the Scaled Boundary Finite Element Method currently are not able to play off their efficiency in dealing with wave propagation phenomena, see Krawczuk et al. (2006), Manolis et al. (2014), Gravenkamp et al. (2014), respectively. Thus, our solution procedure of choice is the standard FEM. To solve the given task, brick and surface elements are combined.

In order to deal with such and other phenomena of wave propagation in a first step a rather simple idealization is investigated, which however involves the problem of a moving boundary, too. It consists of a semi-infinite rod at whose surface a constantly distributed longitudinal load is travelling from the free end to infinity, see Fig. 3. For this purpose, the front of this load q propagates with c_q and thus continuously increases the surface area to which the load is applied. The longitudinal wave speed within the rod is c_L. With respect to Fig. 2 this model equals a situation within which

- the failure stress of the bond matrix—reinforcement has the same value q as the shear stress along the friction zone and
- no further bond degradation takes place.

Thus, the length of the bond regime (i) according to Fig. 2 is zero whereas the length of the bond regime (ii) at time t is $c_q t$. At the current stage no surrounding host is modelled. The shear stress in the bond regime (ii) resulting from the relative displacement between reinforcement and surrounding host consequently is applied directly by means of a constantly distributed longitudinal load whose front propagates with the speed c_q as sketched in Fig. 3. Herein, it is assumed that the speed the crack tip propagates within the longitudinal direction is correlated to the propagation speed of a longitudinal wave in the surrounding host—if it were existent. For simplicity, the speed the crack propagates with can be chosen to be identical to the longitudinal

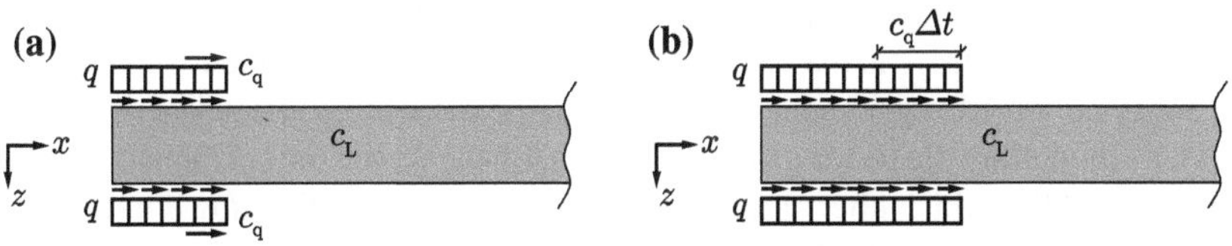

Fig. 3 Sketch of the pull-out idealization: with increasing time t the rod's surface area affected by the shear load increases, too. **a** $t = \bar{t}$. **b** $t = \bar{t} + \Delta t$

wave speed. Depending on the materials the reinforced composite consists of this speed may be smaller, equal or higher than the speed of the longitudinal wave in the reinforcement. Hence both the subsonic problem with $c_q \leq c_L$ and the supersonic problem with $c_q \geq c_L$ are investigated analytically. The material behaviour of the homogeneous rod is characterized as linearly elastic and isotropic. In order to obtain a reference solution for the numerical model a simplified analytical model is developed in Sect. 2. The numerical model consists of a cylindrical rod with a circular cross section with a radius r and a given density ϱ. The shear load q is applied at the surface of the rod and results in a normal stress distribution in the rod. The displacements are evaluated both in the longitudinal and in the radial direction. To ensure that the numerical approach gives reliable results, an analytical model is developed. This analytical model is confined to displacements along the centroidal axis. Consequently, it would be sufficient to model the system as a thin rod with the mass being allocated along the centroidal axis and a line load being equivalent to the shear load q. However, in order to provide results (e.g. normal stresses instead of a normal force) which may serve as a validation for the numerical model, a cross-sectional area is assigned. Owing to the Poisson effect, lateral expansions (and contractions) arise from the normal stress. For simplicity the lateral inertia effects are neglected in the analytical model (Graff 1991). This analytical model is capable of treating both the subsonic and the supersonic problems. Between both problems a smooth transition is revealed, which can hardly be detected by means of numerical methods. The results gained from the analytical model serve as an approximation of a reference solution for the numerical model. Numerical results are obtained in Sect. 3 and involve lateral inertia. The comparison, however, is restricted to the subsonic case. The results of both the analytical and the numerical model show a good agreement especially under consideration of the effects of lateral inertia. Deviations between the results of these two models are discussed. As a result, the numerical model can be used for more advanced future investigations. Finally, the conclusions and an outlook to future work are provided in Sect. 4.

2 Analytical Modelling—Method of Characteristics

2.1 Introduction

To provide an analytical model capable of validating the results of the numerical model given in Sect. 3.1 is the task of this section. The given problem according to Fig. 3 may be solved, e.g. by means of the theory of wave propagation in a rod (Graff 1991), by modelling the rod as a vibration chain, by the wave number transform solution (Doyle 1997) which afterwards has to be transformed into the time-domain (Pao and Mow 1971, Weber and Zastrau 2013) or the method of characteristics. Within this contribution the latter method is used. For a prescribed harmonic excitation the readers are referred to Azzam (2016).

The rod is assumed to be of semi-infinite extent taking into account that also within the numerical model no reflections other than from the free (left) end of the rod occur, see also Fig. 3. The material of the semi-infinite rod is characterized as isotropic and linearly elastic. The material parameters are the YOUNG's modulus E and the density ϱ. Lateral inertia is neglected and thus the longitudinal wave speed is $c_L = \sqrt{E/\varrho}$. As shown in Fig. 3, the load front of the longitudinal load of constant magnitude q travels along the rod's surface in positive x-direction with a constant speed c_q.

2.2 Subsonic Speed of the Load Front

The first case investigated is a load front which propagates with a subsonic speed with respect to the longitudinal wave speed of the rod. Thus $c_q < c_L = \sqrt{E/\varrho}$ holds. At every position x the load front arrives, it induces a longitudinal pulse. Due to the constant impedance along the rod, a half-portion of this load acts as a compressive load at the section of the rod in front of the load front whereas the other half-portion acts as a tensile load at the section of the rod behind that load front. Both waves travel with the speed c_L along the rod's axis. The latter wave is reflected at the rod's free end and then subsequently travels as compressive wave in positive x-direction, too. This situation is depicted in Fig. 4a, in which the distance a certain wave has travelled is plotted versus the time t. Exemplarily, the time t_L at which the longitudinal wave of the initial load front (that is, the load applied at $x = 0$ at time instance $t = 0$) passes an arbitrarily chosen reference point x is denoted within this diagram. Please note at this time instance the load front itself did not reach x, yet. This occurs at t_q, see Fig. 4a.

Clearly, a point x will be stress free as long as $t < t_L = x/c_L$. In what follows, a time instance in the time interval $t_L \leq t < t_q = x/c_q$ will be looked at. According to Fig. 4a at this time instance t two load portions arrived at the point x of the rod. The first portion is due to the load applied from $x_0 = 0$ to $x = x_1$ which directly propagates towards x. The second portion is due to the load applied from $x_0 = 0$ to $x = x_2$ which first propagated in the negative x-direction, then was reflected at the rod's free end and subsequently propagates towards x, too. Thus, for the normal

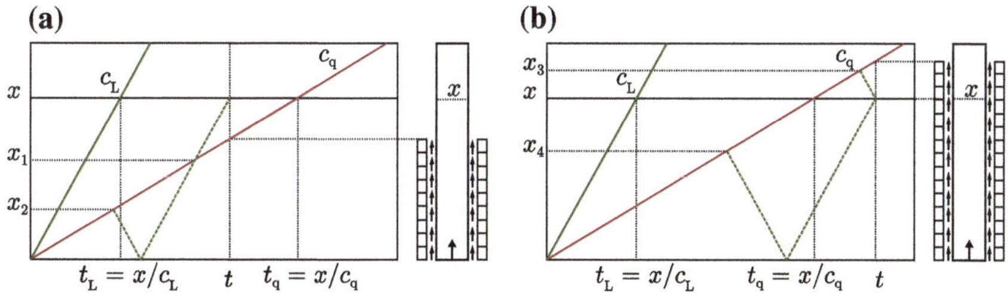

Fig. 4 Distance-time diagram, subsonic case

stress $\sigma_x =: \sigma$ we have

$$\sigma(x,t) = -\left\{ \underbrace{\frac{1}{2} q x_1}_{\text{direct}} + \underbrace{\frac{1}{2} q x_2}_{\text{reflected}} \right\}, \tag{1}$$

which we want to reformulate in terms of x and t. Due to the confinement to displacements in the longitudinal direction in this analytical model, an assignment of a cross-sectional area would not be necessary. However, in order to provide results which may serve as a validation for the numerical model, a cross-sectional area is assigned leading to a normal stress instead of a longitudinal force. The same holds for the shear load q according to Fig. 3. The distributed longitudinal load in the analytical model is related to an equivalent tangential load along the surface of the rod in the numerical model.

Concerning the first portion of the load in Eq. (1), it is induced by the load front moving with speed c_q from $x_0 = 0$ to $x = x_1$. The single wave portions then immediately start to propagate towards the reference point x with speed c_L. The total time t consequently splits off into two parts

$$t = \frac{x_1}{c_\mathrm{q}} + \frac{x - x_1}{c_\mathrm{L}} = x_1 \left(\frac{1}{c_\mathrm{q}} - \frac{1}{c_\mathrm{L}} \right) + \frac{x}{c_\mathrm{L}} \tag{2}$$

leading to

$$x_1 = \frac{c_\mathrm{q}}{c_\mathrm{L} - c_\mathrm{q}} \left(c_\mathrm{L} t - x \right). \tag{3}$$

The second portion of the load arriving at x at a time t was induced along the rod from $x_0 = 0$ to $x = x_2$, then propagated in negative x-direction before being reflected at the rod's free end and subsequently propagating towards x with speed c_L, see also Fig. 4a. The time history consequently consists of three sections

$$t = \frac{x_2}{c_\mathrm{q}} + \frac{x_2}{c_\mathrm{L}} + \frac{x}{c_\mathrm{L}} = x_2 \left(\frac{1}{c_\mathrm{q}} + \frac{1}{c_\mathrm{L}} \right) + \frac{x}{c_\mathrm{L}}. \tag{4}$$

This allows to calculate x_2 as

$$x_2 = \frac{c_\mathrm{q}}{c_\mathrm{L} + c_\mathrm{q}} \left(c_\mathrm{L} t - x \right). \tag{5}$$

Inserting Eqs. (3), (5) into Eq. (1) gives

$$\begin{aligned} \sigma(x,t) &= -\frac{1}{2} q c_\mathrm{q} \left(c_\mathrm{L} t - x \right) \left[\frac{1}{c_\mathrm{L} - c_\mathrm{q}} + \frac{1}{c_\mathrm{L} + c_\mathrm{q}} \right] \\ &= -q \frac{c_\mathrm{L} c_\mathrm{q}}{c_\mathrm{L}^2 - c_\mathrm{q}^2} \left(c_\mathrm{L} t - x \right) = -q \frac{c_\mathrm{L}}{c_\mathrm{L} + c_\mathrm{q}} \frac{c_\mathrm{q}}{c_\mathrm{L} - c_\mathrm{q}} \left(c_\mathrm{L} t - x \right). \end{aligned} \tag{6}$$

The third stress regime can be expected after the load front has passed the point x, that is for $t \geq t_q = x/c_q$. Now, there are three load portions arriving at point x. The first two are the same as in the precedent time interval. The new portion is due to the load induced at the rod in front of the position x which propagates as a tensile wave in negative x-direction and consequently passes the point x. According to Fig. 4b we then have

$$\sigma(x,t) = -\Bigg\{ \underbrace{\frac{1}{2}qx}_{\text{direct}} + \underbrace{\frac{1}{2}qx_4}_{\text{reflected}} - \underbrace{\frac{1}{2}q(x_3 - x)}_{\text{propagating backwards}} \Bigg\} = -\frac{1}{2}q(2x + x_4 - x_3), \quad (7)$$

which again will be reformulated in terms of x and t in what follows. Concerning x_4 we have

$$x_4 = \frac{c_q}{c_L + c_q}(c_L t - x), \quad (8)$$

see also Eq. (5). For the wave portion which stems from the waves propagating in the negative x-direction,

$$t = \frac{x_3}{c_q} + \frac{x_3 - x}{c_L} = x_3\left(\frac{1}{c_q} + \frac{1}{c_L}\right) - \frac{x}{c_L} \quad (9)$$

and thus

$$x_3 = \frac{c_q}{c_L + c_q}(c_L t + x) \quad (10)$$

holds. Inserting Eqs. (8), (10) into Eq. (7) gives the final formula for evaluating the stress at the position x within the third time interval $t \geq t_q = x/c_q$

$$\sigma(x,t) = -\frac{1}{2}q\left[2x + \frac{c_q}{c_L + c_q}(c_L t - x) - \frac{c_q}{c_L + c_q}(c_L t + x)\right] = -q\frac{c_L}{c_L + c_q}x. \quad (11)$$

Obviously, the stress within the third time interval is constant in time as the (tensile) waves propagating in negative x-direction are nullified by the (former tensile) waves which were reflected at the rod's free end.

In summary, at an arbitrary point x the stresses are

$$\sigma(x,t) = \begin{cases} 0 & \text{for } t < t_L \\ -q\dfrac{c_L}{c_L + c_q}\dfrac{c_q}{c_L - c_q}(c_L t - x) & \text{for } t_L \leq t < t_q \\ -q\dfrac{c_L}{c_L + c_q}x & \text{for } t_q \leq t \end{cases}. \quad (12)$$

Fig. 5 Normal stress along the rod for a given time t, subsonic case

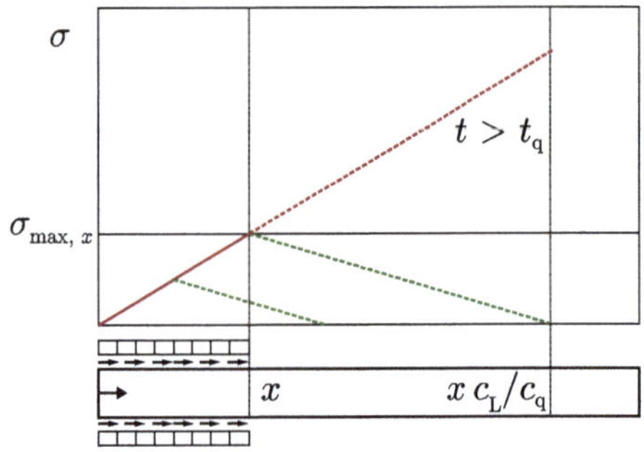

Consequently, at an arbitrary point x three different stress regimes can be distinguished: (i) zero stress until arrival of the first wave portion at $t_\mathrm{L} = x/c_\mathrm{L}$, (ii) increasing stress level due to the subsequent arrival of the other (partly reflected) wave packages, and (iii) constant stress level $\sigma_{\max,x}$ beginning from $t_\mathrm{q} = x/c_\mathrm{q}$. This maximum normal stress is

$$\sigma_{\max,x} = -q\frac{c_\mathrm{L}}{c_\mathrm{L} + c_\mathrm{q}}x. \tag{13}$$

This situation can also be deduced from Fig. 5, where the bi-linear stress distribution within the regimes (ii) and (iii) is clearly observable. Obviously, the maximum stress $\sigma_{\max,x}$ increases with increasing x.

The resulting displacements at the point x can be calculated by integrating Eq. (12) along the rod's axis

$$u(x,t) = \int_x^{c_\mathrm{L}t} \frac{1}{E}\sigma(\tilde{x},t)\,\mathrm{d}\tilde{x}, \tag{14}$$

which also gives three different regimes for the longitudinal displacements.

Within the time interval $0 < t = x/c_\mathrm{L} < t_\mathrm{L}$ no displacements occur at x. Within the second time interval the load front has passed the point x and reached a position $c_\mathrm{L}t$. Hence, between x and the load front the rod is compressed. The resulting displacement at the position x is evaluated by integrating Eq. (6)

$$\begin{aligned} u(x,t) &= \frac{1}{E}\int_x^{c_\mathrm{L}t} q\frac{c_\mathrm{L}c_\mathrm{q}}{c_\mathrm{L}^2 - c_\mathrm{q}^2}(c_\mathrm{L}t - \tilde{x})\,\mathrm{d}\tilde{x} \\ &= \frac{1}{E}q\frac{c_\mathrm{L}c_\mathrm{q}}{c_\mathrm{L}^2 - c_\mathrm{q}^2}\left[\frac{1}{2}(c_\mathrm{L}t)^2 - c_\mathrm{L}tx + \frac{1}{2}x^2\right] = \frac{1}{2E}q\frac{c_\mathrm{L}c_\mathrm{q}}{c_\mathrm{L}^2 - c_\mathrm{q}^2}(x - c_\mathrm{L}t)^2. \end{aligned} \tag{15}$$

The displacement of the point x at a time within the third interval $t > t_\mathrm{q} = x/c_\mathrm{q}$ consists of two parts. As it can be seen from Fig. 4b, in the third interval the load front passed point x, thus at time t the portion of the rod between x and $c_q t$ is under the stress regime according to Eq. (11). However, at the same time t the first wave package reached the position $c_L t$. Hence, also the portion of the rod between the positions $c_q t$ and $c_L t$ is under compression, in detail it is under the stress regime according to Eq. (6). The displacement of the point x at a time within the time interval $t > t_q = x/c_q$ thus reads as

$$\begin{aligned} u(x,t) &= \frac{1}{E}\int_{x}^{c_q t} q\frac{c_L}{c_L + c_q}\tilde{x}\,\mathrm{d}\tilde{x} + \frac{1}{E}\int_{c_q t}^{c_L t} q\frac{c_L c_q}{c_L^2 - c_q^2}\left(c_L t - \tilde{x}\right)\mathrm{d}\tilde{x} \\ &= \frac{1}{2E}q\frac{c_L}{c_L + c_q}\left[\left(c_q t\right)^2 - x^2\right] + \frac{1}{E}q\frac{c_L}{c_L^2 - c_q^2}c_q\left[\frac{1}{2}\left(c_L t\right)^2 - c_L c_q + \frac{1}{2}c_q^2\right] \\ &= \frac{1}{2E}q\frac{c_L}{c_L + c_q}\left[\left(c_q t\right)^2 - x^2\right] + \frac{1}{2E}q\frac{c_L}{c_L^2 - c_q^2}c_q t^2\left(c_L - c_q\right)^2 \\ &= \frac{1}{2E}q\frac{c_L}{c_L + c_q}\left(c_L c_q t^2 - x^2\right). \end{aligned} \tag{16}$$

In summary, the displacements of a point x at the three time intervals at hand are

$$u(x,t) = \begin{cases} 0 & \text{for } t < t_L \\ \dfrac{1}{2E}q\dfrac{c_L c_q}{c_L^2 - c_q^2}\left(c_L t - x\right)^2 & \text{for } t_L \le t < t_q \\ \dfrac{1}{2E}q\dfrac{c_L}{c_L + c_q}\left(c_L c_q t^2 - x^2\right) & \text{for } t_q \le t \end{cases}. \tag{17}$$

A 3D diagram showing the longitudinal displacement over the coordinate x along the rod's axis and over time t is displayed in Fig. 6. For t = const the bi-linear stress distribution results in both a decreasing and an increasing slope of the displacement function. Please note that in the present model lateral inertia is neglected. Additionally, the rod is modelled as semi-infinite, thus only reflections of waves at the left free end of the rod (at $x = 0$) are taken into account.

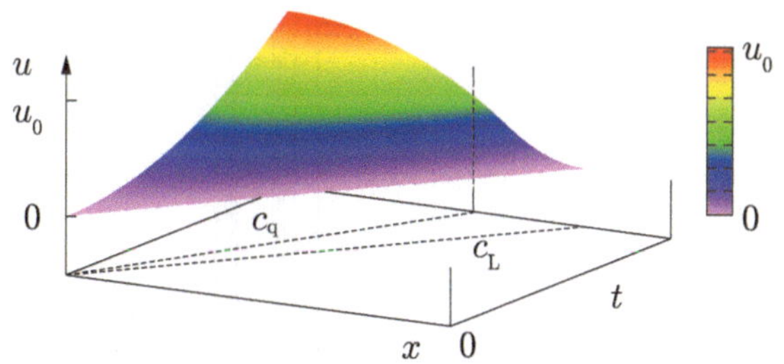

Fig. 6 Longitudinal displacement u over time t and coordinate x, subsonic case

2.3 Supersonic Speed of the Load Front

Within this second case the speed c_q the load front propagates with is higher than the speed c_L of a longitudinal wave. With respect to the sketch in Fig. 2, this situation is at hand if the (longitudinal) wave propagation speed within the matrix material is higher than the wave propagation speed within the directed inclusion. Such material combinations are known from some composite materials. For this supersonic case with $c_q \geq c_L$ the following stress regimes can be distinguished at an arbitrary reference point x: (i) zero normal stress until arrival of the load front at time $t = t_q = x/c_q$, (ii) increasing normal stress after the load front has passed x, and (iii) temporally constant normal stress beginning from $t = t_L = x/c_L$ when the information of the load applied at the rod's free end has reached x. The corresponding distance-time diagrams are given in Fig. 7.

Within the time interval $t_q \leq t \leq t_L$ the load front itself passed the point x, whereas the information that the load front also passed the points with coordinates less than x arrives successively. Additionally, as before the load front induces one (pressure) wave propagating in the positive x-direction and one (tensile) wave propagating in the negative x-direction. The latter wave which stems from points with coordinates greater than x also successively arrives at x, see also Fig. 7a. We thus have

$$\sigma(x,t) = -\left\{ \underbrace{\frac{1}{2}q\,(x - x_1)}_{\text{direct}} - \underbrace{\frac{1}{2}q\,(x_2 - x)}_{\text{propagating backwards}} \right\}. \tag{18}$$

With respect to the first term

$$t = \frac{x_1}{c_q} + \frac{x - x_1}{c_L} \tag{19}$$

and hence

$$x_1 = \frac{x - c_L t}{c_q - c_L} c_q \tag{20}$$

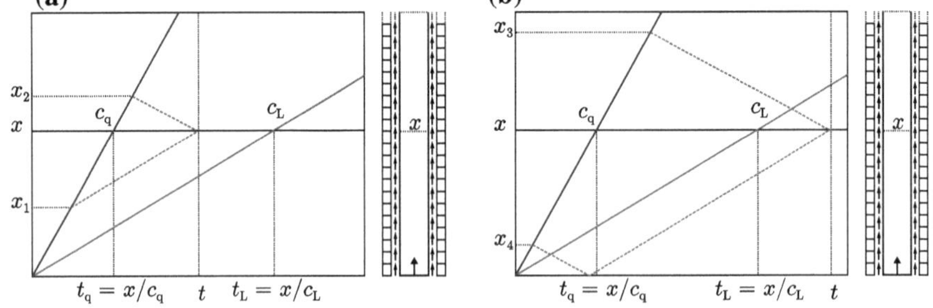

Fig. 7 Distance-time diagram, supersonic case

holds. The wave propagating backwards fulfils

$$t = \frac{x_2}{c_q} + \frac{x_2 - x}{c_L} \tag{21}$$

and thus

$$x_2 = \frac{x + c_L t}{c_q + c_L} c_q. \tag{22}$$

Inserting Eqs. (20), (22) into Eq. (1) gives

$$\sigma(x,t) = -q \frac{c_L^2}{c_q^2 - c_L^2} \left(c_q t - x\right) = -q \frac{c_L}{c_q + c_L} \frac{c_L}{c_q - c_L} \left(c_q t - x\right). \tag{23}$$

The third stress regime starts at time $t_L = x/c_L$, the arrival of the information of the initial wave package from $x_0 = 0$ at point x. Within this stress regime, again three load portions act at point x: the first two portions are the same as in the precedent time interval. The new portion is due to the load applied at the rod at coordinates less than x which propagates as a tensile wave towards the rod's free end, is reflected and consequently propagates towards x as compressive wave passing the coordinate x. According to Fig. 7b it is

$$\sigma(x,t) = -\left\{ \underbrace{\frac{1}{2} q x}_{\text{direct}} + \underbrace{\frac{1}{2} q x_4}_{\text{reflected}} - \underbrace{\frac{1}{2} q \left(x_3 - x\right)}_{\text{propagating backwards}} \right\} = -\frac{1}{2} q \left(2x + x_4 - x_3\right), \tag{24}$$

which again will be reformulated in terms of x and t in what follows. For x_4, the relation

$$t = \frac{x_4}{c_q} + \frac{x_4 + x}{c_L} \tag{25}$$

holds and thus

$$x_4 = \frac{c_q}{c_L + c_q} \left(c_L t - x\right) \tag{26}$$

comes out. For the wave portion which stems from the waves propagating in the negative x-direction,

$$t = \frac{x_3}{c_q} + \frac{x_3 - x}{c_L} = x_3 \left(\frac{1}{c_q} + \frac{1}{c_L}\right) - \frac{x}{c_L} \tag{27}$$

and thus

$$x_3 = \frac{c_q}{c_L + c_q}\,(c_L t + x) \tag{28}$$

holds. Inserting Eqs. (26), (28) into Eq. (24) leads to the final formula for evaluating the stress at the point x within the third interval $t \geq t_q = x/c_q$

$$\sigma\,(x,t) = -\frac{1}{2}q\left[2x + \frac{c_q}{c_L + c_q}\,(c_L t - x) - \frac{c_q}{c_L + c_q}\,(c_L t + x)\right] = -q\frac{c_L}{c_L + c_q}x. \tag{29}$$

As in the subsonic case, the stress within the third time interval is constant in time as the (tensile) waves propagating in negative x-direction are nullified by the (former tensile) waves which were reflected at the rod's free end.

The normal stress at an arbitrary point x at an arbitrary time t then is

$$\sigma\,(x,t) = \begin{cases} 0 & \text{for } t < t_q \\ -q\,\dfrac{c_L^2}{c_q^2 - c_L^2}\,(c_q t - x) & \text{for } t_q \leq t < t_L \\ -q\,\dfrac{c_L}{c_L + c_q}x & \text{for } t_L \leq t \end{cases}. \tag{30}$$

Integrating the strains resulting from these stresses, one obtains the normal displacements

$$u\,(x,t) = \begin{cases} 0 & \text{for } t < t_q \\ \dfrac{1}{2E}q\,\dfrac{c_L^2}{c_q^2 - c_L^2}\,(c_q t - x)^2 & \text{for } t_q \leq t < t_L \\ \dfrac{1}{2E}q\,\dfrac{c_L}{c_L + c_q}\,(c_L\,c_q t^2 - x^2) & \text{for } t_L \leq t \end{cases}. \tag{31}$$

As can be seen from Eqs. (17) and (31), the normal displacements of a spatial point x of the rod within the respective stress regimes are qualitatively similar for both the sub- and the supersonic case. However, the time instances at which there is a change in the stress regimes differs for the sub- and supersonic case. That is, the stress regime (ii) at point x starts at $t = t_L = x/c_L$ in the subsonic case, whereas in the supersonic case the stress regime (ii) is beginning from $t = t_q = x/c_q$. The stress regime (iii) at a point x for the subsonic case starts at $t = t_q = x/c_q$, whereas in the supersonic case it begins at $t = t_L = x/c_L$.

As could be expected, for $c_q = c_L =: c_{\text{transition}}$ the Eqs. (12) and (30) merge into one another. This aspect can hardly be obtained by a purely numerical approach.

The normal displacements for the supersonic case are plotted within a 3D diagram given with Fig. 8. As can readily be seen, the qualitative evolution of the displacements along the rod and for arbitrary time instances equals the one depicted in Fig. 6,

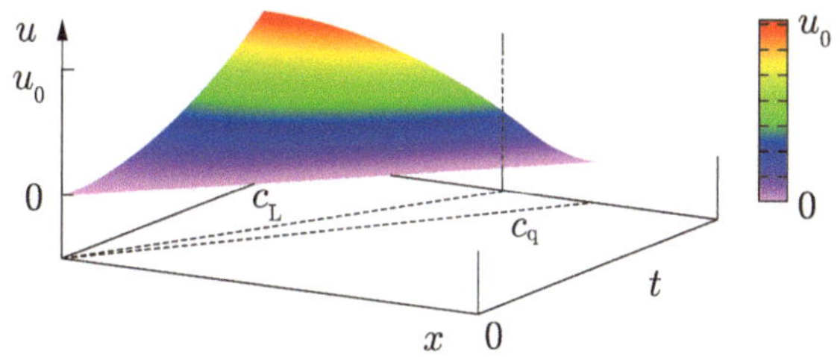

Fig. 8 Longitudinal displacement u over time t and coordinate x, supersonic case

whereas the respective cause is different. Due to this qualitative concordance and for the sake of brevity the numerical investigations taking place in the next section will be restricted to the subsonic case.

3 Numerical Modelling

3.1 Introduction and Convergence Study

Within this section the numerical model to investigate the given problem is introduced. The calculations are performed using the FE program FEAP, Taylor (2013). An implicit time integration using the NEWMARK-beta method with the parameters $\beta = 0.25$ and $\gamma = 0.5$ is applied. The employed solver is PARADISO. In order to further enhance an in-house code, no commercial FE program is employed. As sketched in Sect. 1 the present problem serves as a simplified model of the more complicated task giving the motivation for these investigations, that is the pull-out of reinforcement from a brittle surrounding matrix. Hence, as much as possible of the preliminary model could be used in future investigations. Consequently, the semi-infinite rod dealt with here is discretized as a finite rod. However, its length is chosen such that no reflections other than from the free (left) end of the rod according to Fig. 3 occur. This also avoids the implementation of, e.g. absorbing boundary conditions (Alonso-Mallo and Portillo 2015) or infinite elements.

For the beginning, a steel rod of length $l = 25$ cm and circular cross section with radius $r = 0.75$ cm is looked at. For the numerical investigation this rod is discretized using 20-node brick elements. The finite element discretization as well as the boundary conditions of the rod are shown in Fig. 9. Due to the symmetry of the rod only one quarter of the body is modelled. Although taking into account rotational symmetry with rotational FE would be more efficient for the present problem, a (nearly) full 3D discretization is used. This procedure allows for more advanced calculations taking into account, e.g. multifilament yarns of elliptic cross section as motivated in Weber and Zastrau (2009) and depicted in Fig. 10. Additionally, the inner structure of the reinforcement (Curosu et al. 2016) and eccentric set-ups (Weber and Zastrau 2010, 2011) can be modelled more realistically by means of a 3D-discretization, see also Fig. 10.

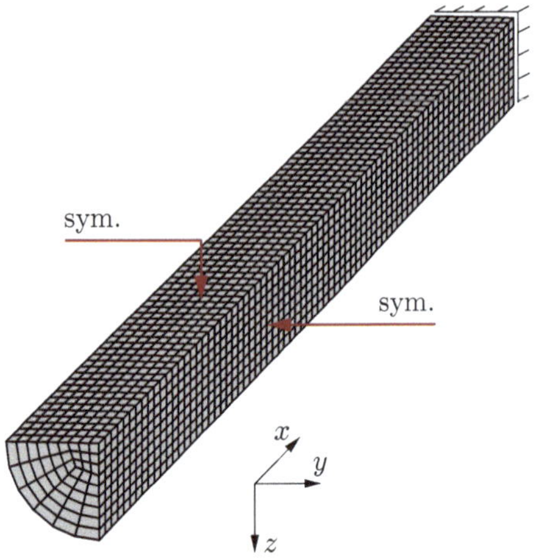

Fig. 9 Finite element discretization and boundary conditions of the rod

The material parameters of the rod are set as: Young's modulus $E = 2.1 \times 10^7\,\mathrm{N/cm^2}$, Poisson's ratio $\nu = 0.3$ and density $\varrho = 7.85\,\mathrm{g/cm^3}$. In order to apply the moving load a special 8-node quadrilateral surface element is implemented and given zero stiffness.

At time t, the current position of the load front is computed by means of the load speed c_q. Its position is compared with the coordinates of the Gauss points of the surface elements. These Gauss points which the load front has passed or just reached are given the full load. Contrary, for all Gauss points which have not been reached by the load front, yet, the load remains zero. Obviously, the approximation of the load distribution at each finite element is better the more Gauss points are used. Consequently, within this contribution for each of the 8-node quadrilateral surface elements 25 Gauss points are used.

For the present example the longitudinal wave speed of the material at hand is $c_\mathrm{L} = 5\,170$ m/s. Exemplary, the speed of the load propagation is set to $c_\mathrm{q} = 2\,586$ m/s $\approx 0.5 c_\mathrm{L}$. In order to prevent reflections of waves at the right end (in positive

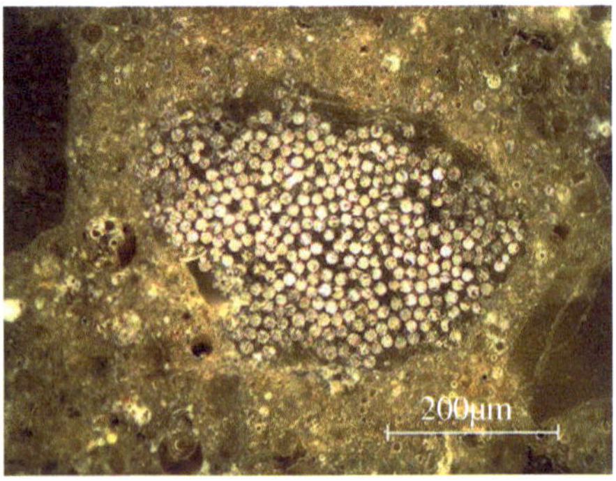

Fig. 10 Typical detail of a multifilament yarn including numerous filaments and interstitial zones, by courtesy of IfB, TU Dresden

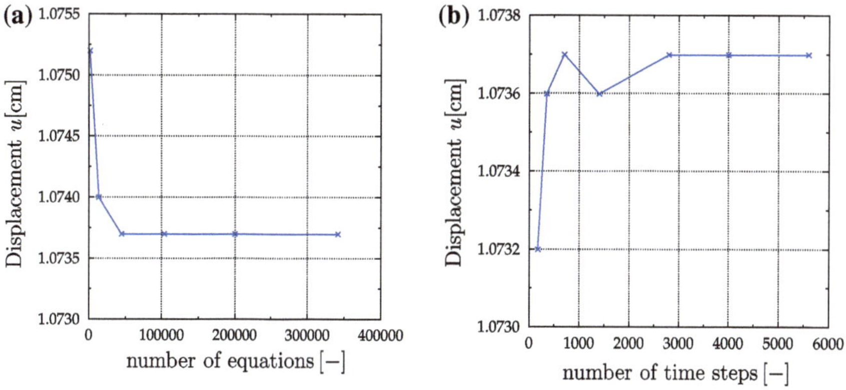

Fig. 11 Convergence studies (For the convergence studies an exemplarily load was chosen. All calculations are performed in the linear elastic regime using linearized strain measures, even though the computed deformation appears to be not small.). **a** Convergence study in space with 70 time steps. **b** Convergence study in time with 44 982 equations

x-direction as given in Fig. 9) of the discretized finite rod the load duration time was set to $t_{max} = 7 \times 10^{-5}$ s which is due to typical wave speeds for the material at hand.

As a first step, a convergence study in both space and time was applied, see also Fig. 11a, b. There, some results of the convergence study are plotted and show the convergence of the displacement u in positive x-direction of the node lying at the centre axis and the left free end of the rod. For a fixed time increment and starting from a rather coarse mesh the mesh was successively refined which resulted in an increasing number of equations. Additionally, for a fixed meshing the time increment was successively reduced thus increasing the number of time steps. As can be seen from Fig. 11a, b the displacement of the reference node converges quite fast where the convergence in space is more pronounced as the convergence in time.

The convergence study resulted in a discretization with 3 375 brick elements. A total of 450 surface elements is applied and overall the numerical model consists of 16 510 nodes with in total 44 982 degrees of freedom, see Fig. 11a. A time increment $\Delta t = 0.25 \times 10^{-7}$ s was obtained by means of the convergence study, see Fig. 11b. As can be seen from these figures with the chosen number of both finite elements and time steps robust and stable results for the displacement of the rod's free end are achieved.

3.2 Numerical Results

The main interest of this contribution is on the validation of the numerical approach by means of an analytical model. After the successful validation the numerical model

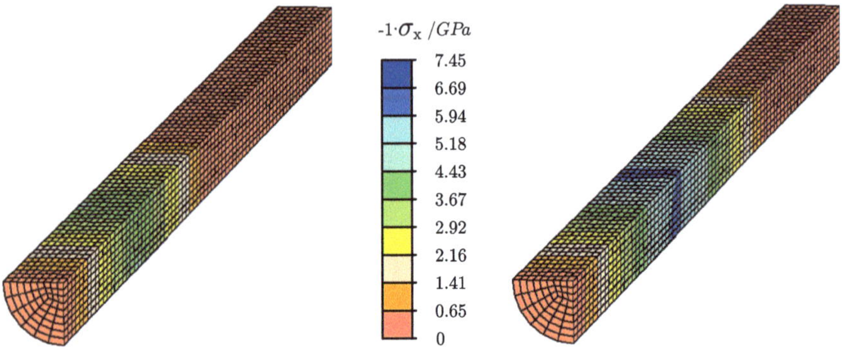

Fig. 12 Normal stress along the rod at $t = 2.5 \times 10^{-5}$ s (*left*) and $t = 3.5 \times 10^{-5}$ s (*right*)

allows for more complicated tasks, e.g. the numerical modelling of pull-out tasks in future investigations. In order to apply the shear load as indicated in Sect. 1, additional surface elements were implemented in the numerical model. In order to validate the numerical model, within this section the numerical results are compared with the results gained from the analytical model which was derived in the precedent Sect. 2. However, this analytical model is a simplified model with respect to the numerical one, as effects resulting from the lateral inertia are neglected.

First, the distribution of the normal stress along the rod's longitudinal axis is looked at. The respective results are given in Fig. 12 and show the normal stresses for two time instances chosen exemplarily. As can readily be seen both the maximum normal stress and its spatial occurrence increase with increasing time t. Additionally, the bi-linear normal stress distribution is observable. It starts with $\sigma_x\ (x_0 = 0,\ t)$ at the free end, increases linearly, and—after reaching its maximum value at the current position of the load front—decreases to zero and thus indicates the front of the initial longitudinal wave. From this position and the respective time t the effective wave speed c_L can be calculated. These results are in accordance with those obtained analytically. For a shortened discussion of these results the reader is referred to Weber et al. (2015).

Figure 12 indicates that there is a non-uniform distribution of the normal stress across the cross section of the rod. This is due to the consideration of lateral inertia, too. To make this effect more visible from the second time instance in Fig. 12 a portion of the rod beginning from the zone of maximum normal stress is enlarged and given in Fig. 13. Obviously, at a given coordinate x the normal stress at the centre axis is slightly lower than the corresponding stress at the rod's skin. This is due to the load being applied at the skin and thus the evolution of normal stress along the centre axis in the time domain is behind that along the skin.

The distribution of the longitudinal displacement for an arbitrarily chosen time instance is given in Fig. 14. The three regimes discussed for the analytical model in Sect. 2 are also observable from the numerical results: (i) until arrival of the front of the initial longitudinal wave there are no displacements, (ii) then the displacements

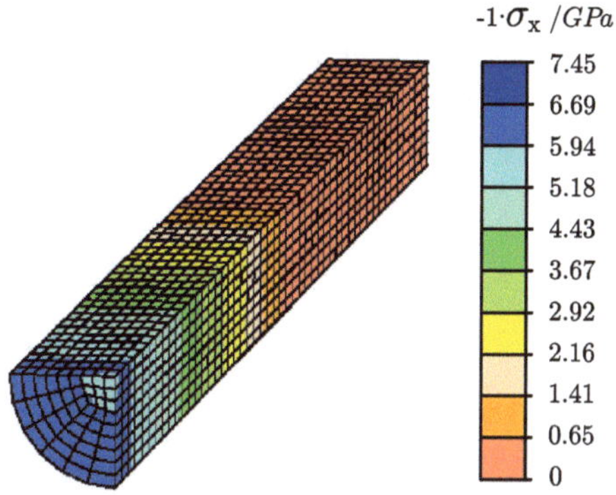

Fig. 13 Normal stress along the rod at time $t = 3.5 \times 10^{-5}$ s, starting from $x = 9$ cm

increase quite fast whereas in (iii) the gradient of the displacements reduces successively. This change in the gradient of the displacement field is also documented in the analytical model, see Figs. 6 and 8, respectively. However, the graphical representation of the longitudinal displacement distribution in Fig. 14 is not suitable to investigate this spatial change in the gradient in detail. Thus, Fig. 15 compares the longitudinal displacement distribution of the analytical and the numerical model. Within this figure the longitudinal displacement is plotted over the position along the rod and results are given for several time instances which coincide with the time instances presented within the precedent and forthcoming figures. The single values of the longitudinal displacement of the finite element model stem from the nodes. The numerical results are plotted for the nodes along the rod's central axis (within the figure denoted as "FEM-Middle") and along the rod's skin (denoted as "FEM-Border"). Only the longitudinal displacement distribution is presented as in the analytical model no lateral inertia is assumed. Please note that there is a very good qualitative and a good quantitative agreement between the results from the numerical and analytical model. Concerning the three longitudinal displacement regimes in the numerical results mentioned before it is now clearly observable that there

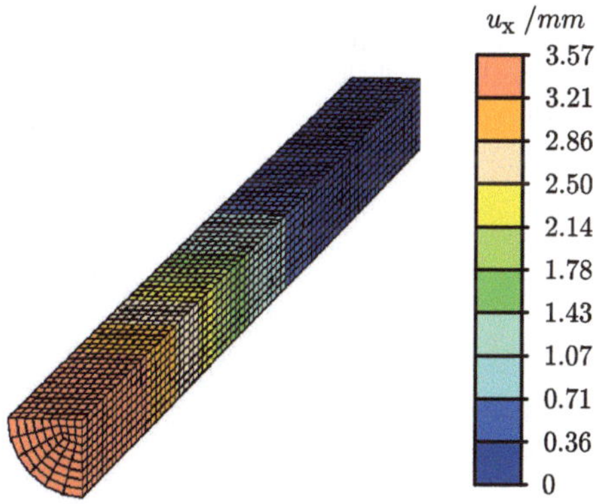

Fig. 14 Longitudinal displacement along the rod at $t = 4.0 \times 10^{-5}$ s

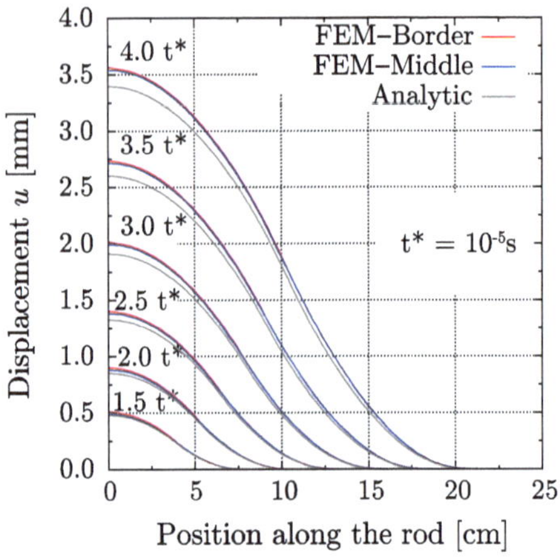

Fig. 15 Comparison of longitudinal displacements between numerical and analytical model

are increasing and decreasing slopes as predicted by the analytical model. However, the longitudinal displacements are slightly higher for the numerical results and the absolute difference increases with increasing time.

The relative deviation d between the finite element and the analytical results is given with Fig. 16 and is calculated by

$$d = \begin{cases} \left| \dfrac{u_{\text{analytical}} - u_{\text{numerical}}}{u_{\text{numerical}}} \right| & \text{for } u_{\text{numerical}} \neq 0 \\ 1 & \text{for } u_{\text{numerical}} = 0 \end{cases} . \tag{32}$$

As can readily be seen for all plotted times the relative deviation for the case of both the displacements along the central axis and along the rod's skin starts with $\approx 5 \ldots 7\,\%$, slightly increases in the direction of the load front and finally rapidly increases to $100\,\%$ at the current respective position of the expected load front.

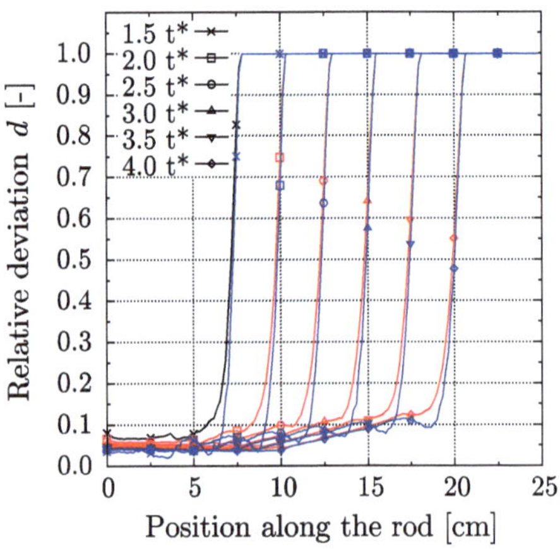

Fig. 16 Relative deviation of longitudinal displacements between numerical and analytical model

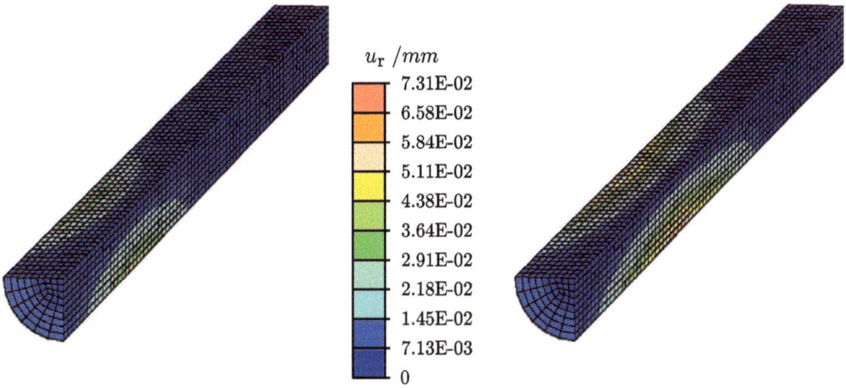

Fig. 17 Radial displacement along the rod at $t = 2.5 \times 10^{-5}$ s (*left*) and $t = 3.5 \times 10^{-5}$ s (*right*)

Obviously, for $u_{\text{numerical}} \neq 0$ at this position $u_{\text{analytical}} = 0$ must hold according to Eq. (32). If the longitudinal displacement in the analytical model is zero but in the numerical model it is $u_{\text{numerical}} \neq 0$ the wave speed in the finite element model must be higher than that in the analytical one. Note that in this case the deviations are overestimated already for small deviations $u_{\text{numerical}} \approx 0$. In the analytical model lateral inertia is neglected and consequently $c_{\text{L,analytical}} = \sqrt{E/\varrho}$ holds. If however lateral inertia is at hand, the material gets stiffer due to the restricted lateral strain and consequently the wave speed approaches $c_{\text{bulk}} = \sqrt{E\left(1-\nu^2\right)/\left[\varrho\left(1-\nu-2\nu^2\right)\right]}$ for an elastic half space, Graff (1991). For the material parameters introduced in Sect. 3.1 it is $c_{\text{bulk}} = 1.16\, c_{\text{L}}$. From Fig. 17 it is observable that lateral strain and thus lateral displacements u_r are restricted but not prevented and thus the wave speed in the numerical model is within the bounds $c_{\text{L,analytical}} < c_{\text{L,numerical}} < c_{\text{bulk}}$. As the wave speed in the finite element model is higher than in the analytical one at the same time instance t the rod segment under influence of a normal stress is longer than the respective length in the analytical model. Consequently, the displacement at the free end of the rod is bigger in the numerical case compared to the analytical one and this difference increases with increasing time t.

Figure 17 additionally shows that (i) the value and the spatial occurrence of the maximum lateral displacement and (ii) the length of the rod showing non-zero lateral displacements both increase with increasing time. This is plausible with respect to the normal stress distributions depicted in Fig. 12.

4 Conclusion and Outlook

A finite element model to deal with the problem of surface loads with moving load fronts, as it occurs e.g. for the case of dynamic pull-out of reinforcement, is motivated. Exemplarily, a (semi-infinite) rod is chosen at whose surface a constant longitudinal load is applied. Starting from the rod's free end, the front of this load travels with a constant speed. Thus, an overall load increasing with time is acting on the rod. A standard FE method is used to perform the calculations. In order to validate the numerical model an analytical investigation of the given task by means of the method of characteristics is performed as well. The latter investigations take place both for the sub- and supersonic case. That is, for the speed c_q of the load front being lower or higher than the speed c_L a longitudinal wave propagates with along the rod's axis. Between both cases a smooth transition is revealed—a result which is hardly possible to be deduced from a purely numerical approach.

The results of both the finite element and the analytical method show good agreement and thus, the numerical model is capable of dealing with more advanced problems in future work.

Acknowledgments The authors Daniel Balzani and Yannick F. Fangye highly appreciate financial support from the institutional strategy "The Synergetic University" at Technische Universität Dresden as part of the DFG Excellence initiative.

References

Alonso-Mallo, I., Portillo, A.M.: Absorbing boundary conditions and geometric integration: a case study for the wave equation. Math. Comput. Simul. **111**, 1–16 (2015)

Azzam, A.: On the mechanical modeling and analysis of the dynamical fiber pullout mechanism taking into account the damage and viscoelasticity of the bond, Ph.D. thesis, Technische Universität Dresden, Institute of Mechanics and Shell Structures, Dresden (2016)

Curbach, M., Jesse, F.: Eigenschaften und Anwendung von Textilbeton [Specifications and Application of Textile Reinforced Concrete (TRC)]. Beton-und Stahlbetonbau **104**(1), 9–16 (2009)

Curosu, I., Mechtcherine, V., Millon, O.: Effect of fiber properties and matrix composition on the tensile behavior of strain-hardening cement-based composites (SHCCs) subject to impact loading, Cement and Concrete Research, vol. 82, 23–35 (2016)

Doyle, J.F.: Wave Propagation in Structures—Spectral Analysis Using Fast Discrete Fourier Transforms, 2nd edn. Springer, Berlin (1997)

Graff, K.F.: Wave Motion in Elastic Solids. Oxford University Press, Oxford (1991)

Gravenkamp, H., Birk, C., Song, C.: Computation of dispersion curves for embedded waveguides using a dashpot boundary condition. J. Acoust. Soc. Am. **135**(3), 1127–1138 (2014)

Krawczuk, M., Grabowska, J., Palacz, M.: Longitudinal wave propagation. Part I—comparison rod theories. J. Sound Vib. **295**(3–5), 461–478 (2006)

Pao, Y.H., Mow, C.C.: Diffraction of Elastic Waves and Dynamic Stress Concentrations. Crane Russak & Co, New York (1971)

Manolis, G.D., Parvanova, S.L., Makra, K., Dineva, P.S.: Seismic response of buried metro tunnels by a hybrid FDM-BEM approach. Bull. Earthq. Eng. **95**(5), 519–526 (2014)

Sridhar, N., Yang, Q.D., Cox, B.N.: Slip, stick, and reverse slip characteristics during dynamic fibre pullout. J. Mech. Phys. Solids **51**(7), 1215–1241 (2003)

Taylor, R.L.: FEAP—A Finite Element Analysis Program, Manual, Department of Civil and Environmental Engineering, University of California at Berkeley, Berkeley (2013)

Weber, W., Zastrau, B.W.: On SH wave scattering in TRC-Part I: concentric elliptical inclusion. Mach. Dyn. Probl. **33**(2), 105–118 (2009)

Weber, W., Zastrau, B.W.: On the influence of asymmetric roving configurations on the wave scattering behaviour of TRC. In: Brameshuber, W. (Ed.), Proceedings of the International RILEM Conference on Material Science—Vol. I, RILEM Publications, 271–281 (2010)

Weber, W., Zastrau, B.W.: Non-plane wave scattering from a single eccentric circular inclusion - Part I: SH waves. J. Theor. Appl. Mech. **49**(4), 1183–1201 (2011)

Weber, W., Curbach, M.: Wellenausbreitung in bewehrten Feinbetonpaneelen [Wave propagation within panels made of FRC]. Beton-und Stahlbetonbau **108**(8), 562–569 (2013)

Weber, W., Zastrau, B.W.: Analytical description of FRC subjected to transient loads. J. Theor. Appl. Mech. **51**(1), 183–194 (2013)

Weber, W.E., Balzani, D., Fangye, Y.F., Zastrau, B.W.: Numerical modeling of a sub-sonic moving load front along a rod's skin. PAMM **15**(1), 535–536 (2015)

Yang, Q.D., Rosakis, A., Cox, B.N.: Dynamic fibre sliding along debonded, frictional interfaces. Proc. R. Soc. A: Math., Phys. Eng. Sci. **462**(2068), 1081–1106 (2006)

Part VII
Generalized Continua, Multi-Scales and Meta-Materials

Application of Multi-scale Approaches to the Investigation of Sealing Surface Deformation for the Improvement of Leak Tightness in Pressure Relief Valves

Ali A. Anwar, Yevgen Gorash and William Dempster

Abstract This chapter is part of a research program to investigate and model the leak tightness of a Pressure Relief Valve (PRV). Presented here is: a literature review; high-temperature numerical study involving the deformation of contact faces for a metal-to-metal seal accounting for micro and macro effects; and also microscopic measurements of surface finishes and how they are modelled over a micro to nanometre scale. Currently, no review of literature exists which attempts to understand the leakage phenomenon of metal-to-metal seal contact PRV for static closed positions as they reach the set pressure point. This work attempts to do just that by drawing on inspiration from other research areas such as metal-to-metal contact and gasket seals. The key topics of interest surrounding the leakage of fluid through a gap are: fluid flow assumptions, surface characteristics and its deformation, and experimental techniques used to quantify leakage. For the numerical study, the valve geometry is simplified to an axisymmetric problem, which comprises a simple geometry consisting of only three components: a cylindrical nozzle, which is in contact with a disc (representing the valve seat on top), which is preloaded by a compressed linear spring. The nozzle-disk pair is made of the austenitic stainless steel AISI type 316N(L) steel. In a previous study, the macro–micro interaction of Fluid Pressure Penetration (FPP) was carried out in an iterative manual procedure at a temperature of 20 °C. This procedure is now automated and implemented through an APDL script, which adjusts the spring force at a macro scale to maintain a consistent seal at elevated temperatures. Finally, using the Alicona Infinite Focus the surface form and waviness is measured, presented and modelled as 1/4 symmetric over a macro to nanometre scale. It is clear the surface form also needs to be accounted for, something which the literature does not focus on.

A.A. Anwar (✉) · Y. Gorash · W. Dempster
Department of Mechanical and Aerospace Engineering, University of Strathclyde, James Weir Building, 75 Montrose Street, Glasgow G1 1XJ, UK
e-mail: ali.anwar@strath.ac.uk

Y. Gorash
e-mail: yevgen.gorash@strath.ac.uk

W. Dempster
e-mail: william.dempster@strath.ac.uk

K. Naumenko and M. Aßmus (eds.), *Advanced Methods of Continuum Mechanics for Materials and Structures*, Advanced Structured Materials 60,
DOI 10.1007/978-981-10-0959-4_27

1 Introduction

1.1 Scope of Research

There are many commercially available valves which perform different functions. Valves can function as isolators, diverters, flow reversal prevention and reduce pressures within a service system. The service characteristics can range from fluid type, fluid characteristics, pressure, temperature, chemical resistance and finally operational and maintenance requirements (Smith and Vivian 1995). Appropriate valve selection is dependent on complete knowledge of the required function and the service characteristics.

When choosing a valve it has been recommended to factor in the leakage since leak tightness has a direct effect on the operational and maintenance requirements (Zappe 2004). Depending on the service, especially if the fluid is hazardous to humans or the environment, the leak tightness is of the highest interest. Regardless, for any service characteristic and function the leak tightness will have a direct effect on the overall pressure of the system over time, therefore the leak tightness is equitable to a direct cost in operations and maintenance.

As the operating pressure within a Pressure Relief Valve (PRV) reaches the set pressure, the sensitivity of the valve opening prior to reaching an equilibrium (set pressure = operating pressure) increases. To be able to seal the valve up to 90 % of the set pressure and higher requires research into the leak tightness of the valve. The PRV of particular interest is the STARFLOW P3 PRV manufactured by Weir Power and Industrial Division.

This research program was initiated to understand leak tightness of a metal-to-metal contact of a PRV and take the early steps to model the leakage using computational solvers. A preliminary work package by Gorash et al. (2014, 2015), predominantly of Finite Element Analysis (FEA), showed that the inner section of the metallic valve seat is subject to deformation due to a cyclic analysis. Gorash et al. (2014, 2015) recommended that to pursue:

1. A 2D axisymmetric cyclic analysis of a PRV using Fluid Pressure Penetration (FPP) at an elevated temperature;
2. Create an ANSYS Advanced Parametric Language Script (APDL) to automatically account for macro and micro FPP;
3. Combine this all with a multi-scale model which accounts for surface roughness deformation and its effect on leak tightness.

This chapter will focus on a literature review and recommendation 1 and 2 from above. Originally, the third point from above in the recommendations was going to be followed, however, one of the main findings of the research program (after completing step 2) was that a vital surface finish quantity needed to be accounted for and had not been highlighted in the literature for metal-to-metal seals. This surface finish quality is known as the surface form and the measurements taken using the

Alicona Infinite Focus are shown in this chapter. Within the literature the surface form is generally associated with flatness and an implicit assumption is made where the contacting surfaces are assumed "perfectly flat".

After researching this further it had been decided to account for surface form profile and waviness deformation using 3D 1/4 symmetry FEA which accounts for a micro- to nanometre geometry. In the future this would then be connected to a Computational Fluid Solver (CFD) to quantify leakage of the valve due to surface form and waviness.

1.2 Background Information

For the benefit of the reader who is unaware of either a PRV, its components or Surface Metrology, this short section will give a brief overview with respect to the research discussed.

1.2.1 Pressure Relief Valve (PRV)—Components and Functionality

Since only PRVs are considered for this project, it is vital to understand the components and functionality of a PRV. A schematic presentation is provided in the form of a classical ASME valve (see Fig. 1a) and DIN valve (see Fig. 1b). These have been chosen since they represent majority of spring-loaded PRVs available worldwide, which utilise metal-to-metal and seat-to-disc contact.

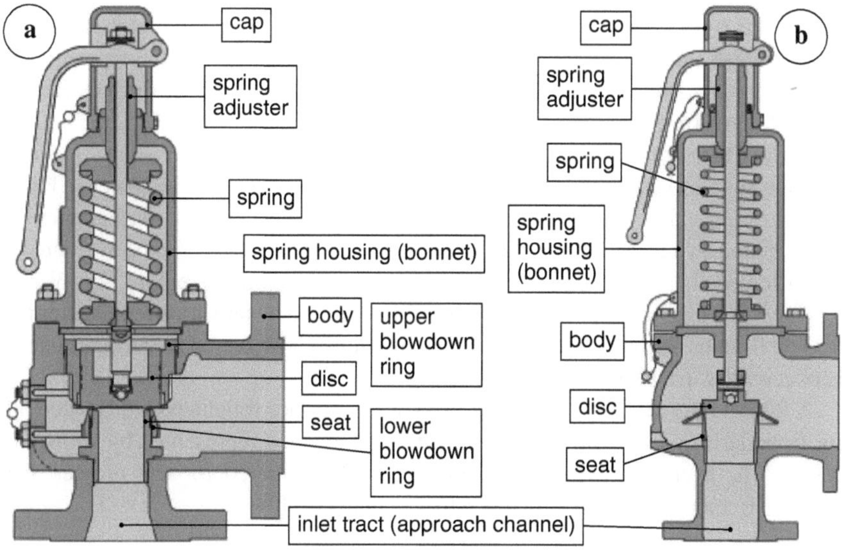

Fig. 1 Valve component schematic for typical **a** ASME and **b** DIN valves (Spirax Sarco 2016)

A PRV functions within a pressured system which is generally connected to either pressurised tanks or pipes. Its purpose is to release overpressure from a pressurised system which has exceeded its design pressure to ensure system safety and allow operations to continue. The valve is designed to release the fluid from the system to relieve pressure, autonomously.

As the system pressure increases the pressure in the inlet tract increases accordingly. The spring stiffness is set using an adjuster (or a compression nut depending on design) which results in a force being applied to the Seat via the Disc. This force will be known as the spring force (F_{sp}). The disc and seat are separate and only have a metal-to-metal contact, effectively acting as the seal. The valve is sealed at a pressure known as the "set pressure".

When the pressure in the system is greater than the set pressure—also known as overpressure—then the force holding the disc down on the seat is outweighed, the disc lifts and the fluid enter into the secondary chamber and releases through the outlet tract. It is also important to note that the effective seat area also increases due to more of the disc being exposed to the inlet pressure when the valve seat is lifted due to the increased pressure.

1.2.2 Metrology: Surface Finish

If a material such as steel is cut and ground down to a smooth finish, it is important to understand what quality of finish is present. To the naked eye it may look to be flat and a smooth finish, however a different story can be told when viewed under magnification. Generally when viewed under magnification a chaotic surface is present. To measure this chaotic surface metrological techniques must be used.

Metrology is a core, long established scientific method of geometrical measurement. This topic is one of the three cores of the project since the understanding of surface finish and how we can quantify it is of great importance. For this research program there are 3 surface metrology concepts to introduce: Average Surface Roughness (R_a), Average Waviness (W_a) and Surface Form, as shown in Fig. 2.

Surface form is the shape that the surface takes when formed. This shape is usually distinguishable visually and then measured and displayed as a profile as seen in Fig. 2. If all the points along the surface form are considered and enclosed via an upper and lower boundary region, encapsulating all the points, the flatness can be attributed to the difference between the highest and lowest points within the region. The average waviness is a magnification of the surface form which accounts for irregularities that deviate from the mean surface represented by sinusoidal waves. The average spacing between the wave period is known W_{sm}.

A further magnification of the waviness shows surface roughness which accounts for irregularities along the waviness profile and again is represented by sinusoidal waves. The average waviness and roughness is considered for this project. It is possible to model the surface using actual scanned profiles.

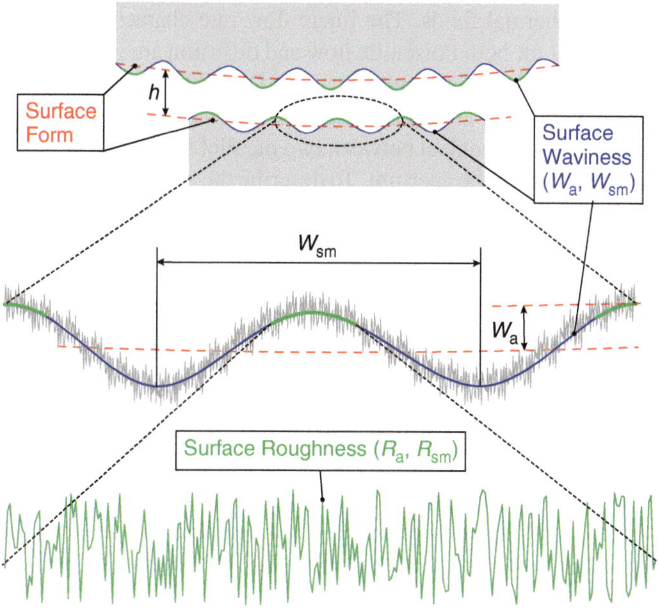

Fig. 2 Surface form, waviness and roughness depicted on two contacting surfaces separated by a gap size of *h*

2 Literature Review

This section identifies the current understanding and technological knowledge of leak tightness of metal-to-metal contact PRVs. Within a PRV the seat and disc would be in contact and would constrain the fluid. Research in this field in direct relation to valves is scarce. However, inspiration can be drawn from relevant fields such as: metal-to-metal contact and gasket seals. When these contacting surfaces come in contact in parallel to each other a finite gap or path is present which is dictated by the surface finish such as: profile, roughness, waviness, flatness, etc. Subsequently, if there is a driving internal pressure the fluid can pass through the path and exit the valve. It has been stated by Burmeister et al. (1967) that the minimum leakage rate which can be obtained for 1 time seal applications is "less that 10^{-8} atmospheric cc/sec of helium".

Previous work suggests that the leakage rate is either viscous laminar flow, molecular flow or, in the intermediate or transitional regime, or a mix of both. Assuming the non-contact area through the gap to be a finite length and the fluid to be isothermal and viscous compressible/incompressible, Poiseuille flow equations are utilised either for a circular cross section (Burmeister et al. 1967) or parallel plates. More recent work shows the development of fluid flow due to diffusion flux based on high to low concentration regions through the contacting surfaces using Fick's law for incompressible isothermal fluids. The lineic flow rate characteristics which determines the gap height for both Poiseuille flow and diffusion are generally put into two characteristics: transmissivity ***K*** and diffusivity ***D*** (BHR Group Ltd. 2000; Mitchell and Rowe 1969; Ledoux et al. 2011; Vallet et al. 2009; Geoffroy and Prat 2004).

The true area of actual contact between two parallel faces is only a small fraction of the nominal area due to the aperture. To describe the path the fluid would take, the chaotic surface must be characterised. Methods to characterise the surface have been created by many authors such as the MOTIF procedure, self-affine fractal analysis and simplistic geometry. Recently, this has been adopted into leak tightness projects to determine its effect on the gap height (Robbe-Valloire et al. 2001; Robbe-Valloire and Prat 2008; Gagnepain and Roques-Carmes 1986; Ganti and Bhushan 1995).

The contact area is dependent upon the aperture of both surfaces in contact and the deformation magnitude is dependent upon the normal load applied and the effective surface hardness of the softer of the two materials. Depending upon the loading the surface aperture can become plastic in areas while the whole structure remains predominantly elastic. Taking this into consideration, the flow path will change depending on the load. Attempts have been made to analytically and computationally describe the surface roughness, elastic (using Herts theory by O'Callaghan and Probert 1987), elastic-perfectly-plastic (refer to Man et al. 2014) and perfectly-plastic and rigid (using theory by Tsukizoe and Hisakado 1965 in Geoffroy and Prat 2004) deformation and understand its effect on the gap height or aperture field.

To verify the leakage rate experimentally British, ASME and API, standards can be used. Recent experimental work by Haruyama et al. (2013) and Geoffroy and Prat (2004) have shown promising methods of detecting and quantifying leakage and its link to the transmissivity ***K*** and diffusivity ***D*** values. More specific research into PRV leakage which detracts from surface finish has been conducted by Ritchie (1989) which examines the effect of misalignment of the valve and its effect on pressure drop. Computationally to model a whole seat and disc of a PRV with the surface roughness and waviness would be possible, but intensive and other techniques such as multi-scale modelling (Thompson 2007b; Jackson and Streator 2006) could be considered instead. Finally, the most recent development by Pérez-Ràfols et al. (2016) is a computational model that is utilised to study the leakage through metal-to-metal seals accounting for both the waviness of the spiral groove and the surface roughness. Correlation between percentage real contact area and actual contact topology and leak rate was confirmed through numerical analysis.

2.1 Leakage: Fluid Flow Assumptions

Depending on the service characteristics the fluid type and characteristics such as pressure, temperature and contaminants will vary. Generally, high performance valves will be capable of withstanding compressible and incompressible Newtonian fluids which range from −196 to −540 °C and pressurised up to 440 bar. Assuming the valve has not opened and the seat and disc is in contact, only the surface deformations at a micro scale create an aperture field through which the fluid can escape out.

In previous studies on gaskets, valves and metal surface contacts, the fluid flux has been assumed to be laminar and either; viscous isothermal compressible/incompressible and equated using Poiseuille law $\boldsymbol{q}_{\mathrm{v}}$ (volumetric flow rate) per unit width in Eq. (1); or diffusive, equated using Fick's law $\boldsymbol{q}_{\mathrm{d}}$ (volumetric flow rate) per unit width in Eq. (2), as shown below:

$$\boldsymbol{q}_{\mathrm{v}} = -\frac{\boldsymbol{K}}{\mu}\,\nabla p \tag{1}$$

$$\boldsymbol{q}_{\mathrm{d}} = -\mathcal{D}\,\boldsymbol{D}\,\nabla c \tag{2}$$

where μ is the viscosity, p is the fluid pressure, $\mathcal{D}$ is the molecular diffusion coefficient and c is the mass fraction of the species. $\boldsymbol{K}$ and $\boldsymbol{D}$ are respectively the transmissivity and diffusivity values. These allow the lineic flow rate characteristics to be described at the scale of the surface. Based on a parallel gap with a height of h these terms can be described as:

$$\boldsymbol{K} = \left\langle \frac{h^3}{12}\left(\boldsymbol{I} + \nabla \boldsymbol{b}\right)\right\rangle, \tag{3}$$

$$\boldsymbol{D} = \left\langle h\left(\boldsymbol{I} + \nabla \boldsymbol{b}\right)\right\rangle, \tag{4}$$

where $\boldsymbol{b}$ is the solution of the closure problem (Vallet et al. 2009).

Majority of authors have adopted a parallel gap (Mitchell and Rowe 1969; Ledoux et al. 2011; Vallet et al. 2009; Geoffroy and Prat 2004) rather than a circular cross section (Burmeister et al. 1967). This generalisation is more appropriate since the diametre of flow path does not need to be known, rather the separation between the contact surfaces is adequate. Also this is only applicable to asperities with local small slopes l, where $h \ll l$ or $l < 10°$ (Vallet et al. 2009). The connection between the transmissivity $\boldsymbol{K}$ and diffusivity $\boldsymbol{D}$ has been theoretically analysed by Geoffroy and Prat (2004) and they conclude that the dependence of the fluid transition in either $\boldsymbol{K}$ or $\boldsymbol{D}$ form is defined by both the gap size and applied load. The caveat with this theory is that a uniform gap height, surface form, waviness and roughness are present with a flat surface deforming it, which is implicitly applied for the latter.

Depending on the rarefaction of the fluid, it could also be in the transition or slip flow regime. This is likely since the surface roughness is at a micro scale and therefore the gap height is also likely to be similar. This can be verified by calculating the Knudsen number and has been accounted for by Gorash et al. (2014) in an analytical model which is an extension of the Poiseuille's Law for fluid flux through a parallel gap written as:

$$q = \frac{h^3}{24\mu RT}\left[P_0^2 - P_1^2 + 12\frac{2-\sigma_a}{\sigma_a}Kn_1 P_1\left(P_0 - P_1\right)\right]\left|\frac{P_0^2 - P_1^2}{P_1}\right|, \tag{5}$$

where μ is the accommodation coefficient, R is the gas constant, T is the temperature and Kn_1 is the Knudsen number at the outlet. The second more subtle assumption here is that the micro fluid flow is based on the outlet not the inlet.

There has been a further modification of this equation to account for chocked mass flow rates at the exit of the microchannel by Chong (2006):

$$\dot{m} = \frac{h^3}{24\mu RT}\left(1 + 12\frac{(2-\mu)}{\mu}\frac{\sigma_a}{P_1 h}\frac{16}{5}\sqrt{\frac{kT}{2\pi m}}\right). \tag{6}$$

where $\pi = 3.14$. This equation is questionable since for chocked conditions for a gas compressibility would have to be considered while Xie's equation does not.

The original form of Eq. (5) was proposed by Arkilic et al. (1997) who also defined the pressure distribution across the gap to be:

$$\frac{P(x)}{P_1} = \left\{\left(6\mathrm{Kn}_1 + \frac{P_0}{P_1}\right)^2 - \left[\left(\frac{P_0^2}{P_1^2} - 1\right) + 12\mathrm{Kn}_1\left(\frac{P_0}{P_1} - 1\right)\right]\left(\frac{x}{L}\right)\right\}^{\frac{1}{2}} - 6\mathrm{Kn}_1, \tag{7}$$

which has shown very good agreement with experimental results for subsonic flows.

2.2 *Representation of Surface Roughness at Micro Scale and Its Effect on Fluid Flow Path*

Micro-scale geometry of a surface profile is chaotic in nature making it important to be able to represent the area in an effective form since this has a direct effect on the gap height, h. There are methods available which make it possible to represent the surface roughness of a model at a micro-scale level. These can be in the form of generic surfaces such as a sinusoidal waves (Geoffroy and Prat 2004; Pérez-Ràfols et al. 2016) or vibrational Eigen modes (Ledoux et al. 2011) or wedges (Mitchell and Rowe 1969). There are more analytical methods based on the surface available such as the sum surface (Robbe-Valloire et al. 2001), MOTIF procedure (Robbe-Valloire and Prat 2008) and fractal analysis (Vallet et al. 2009; Gagnepain and Roques-Carmes 1986; Ganti and Bhushan 1995) which all have benefits and limitations. The benefits and limitations are discussed below and its effect on the gap size.

The sum surface allows the direct analysis of two contacting surfaces. This technique is used by many authors (O'Callaghan and Probert 1987) and based upon a theory created by Tsukizoe and Hisakado (1965), which states, it can be assumed that, "the contact between two rough surfaces can be regarded as the equivalent to the contact between an imaginary rough surface having an appropriate effective topography and a perfectly flat surface" and "the contacting asperities deform in an ideal plastic manner so that, providing no interference from neighbouring asperities occurs

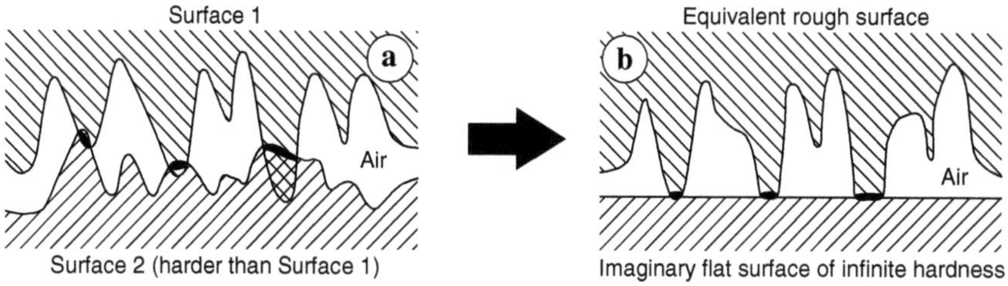

Fig. 3 Illustration of **a** sum surface technique, which is used to generate of **b** equivalent rough surface with *black areas* indicating microcontacts, after O'Callaghan and Probert (1987)

and that work hardening does not take place" as that shown in Fig. 3. Essentially, the flat surface is assumed rigid perfectly-plastic surface. These are accepted and applied assumptions in references (Mitchell and Rowe 1969; Ledoux et al. 2011; Robbe-Valloire et al. 2001). The main limitation with the sum surface technique is that the actual surface has to be measured physically. A further limitation is that representation of the surface is generally in a 2D format and an "effective" gap size, h, has to be calculated and is generally based on an average values as demonstrated by Mitchell and Rowe (1969).

More recently, the surface roughness of both lapped and sand blasted surfaces has been represented as self-affine fractal surfaces combined with the sum surface technique contacting a rigid-perfectly plastic flat surface (Vallet et al. 2009). By using fractal analysis the self-affine fractal surface is based on two parametres and is created using a power law. Using this technique the surface roughness can be modelled in 3D, representing the whole aperture field. 2D representation can be created using this technique but an "effective gap size", h, would have to be employed when calculating leakage rate. Vallet et al. (2009) show good agreement when comparing the fractal surface representation to the "real" surfaces for the lapped surfaces. This shows that for lapped surfaces self-affine fractal surfaces can be used to generically replicate surfaces and represent the aperture field. The main limitation of this method is that the mathematics is intensive and requires the use of computational programs such as MATLAB. By modelling the whole aperture field all possible fluid flow paths can be represented and more accurate gap sizes through the valley and peaks can be calculated.

2.3 *Micro Material Deformation of Rough Surfaces and Its Effect on Fluid Flow Path*

When two surfaces each with their own unique roughness come into contact, the actual contact area is much less than the nominal area. The magnitude of the contact area is dependent on the load applied. So, the accurate deformation of the surface roughness's is of great importance since the voids between the surface roughness's is the gap size (2D) or aperture field (3D) through which the fluid will flow. Also a small change at a micro scale will reverberate in a larger change over a macro scale. Theory by Tsukizoe and Hisakado (1965) essentially assumes the flat surface is rigid and the summed surface is perfectly-plastic and as said these are accepted and applied assumptions in Refs. Mitchell and Rowe (1969), Geoffroy and Prat (2004), Robbe-Valloire et al. (2001).

Using slip line field theory, Mitchell and Rowe (1969) have incorporated the effects of the perfectly-plastic isotropic deformation structural response of two-dimensional wedges to represent the surface roughness in contact with a rigid-perfectly-plastic flat surface which is based on the theory discussed above. The slip line theories main limitation is that it is used to model plastic deformation in plain strain only for a solid represented as a rigid-perfectly-plastic flat surface (University of Cambridge 2004). It is shown that for all contact pressures there is a specific deformed wedge angle found and crucially there is a point at which a maximum leakage rate for specific wedge angle over the seat length. The limitations of the findings are concurrent with the fluid flow assumptions about the gap size and the simplified representation of the surface roughness.

Assuming a simplistic, but effective sinusoidal shape geometry to represent the surface roughness in contact with a flat surface (Geoffroy and Prat 2004; Pérez-Ràfols et al. 2016), it is shown that as the load on the gasket is increased, the incompressible fluid flows from a radial direction, to a circumferential spiral fluid flow through the valleys as shown in Fig. 4. It is also shown that there is a very small region over which the transition from circumferential to radial (or vice versa) occurs and the diffusive and viscous flows are mixed. It is concluded that the radial leakage (which is related to the transmissivity) is of most critical since it is very sensitive to the gap size. However, this theory can only be valid for surfaces which depict a predominantly sinusoidal shape contacting a flat face.

Another analytical technique which accounts for elastic deformation is the Hertz theory (O'Callaghan and Probert 1987). The Hertz elastic deformation theory is only applicable for surfaces with purely spherical aperture contacting shapes. It is employed by Man et al. (2014) for two 3D random rough surfaces in contact.

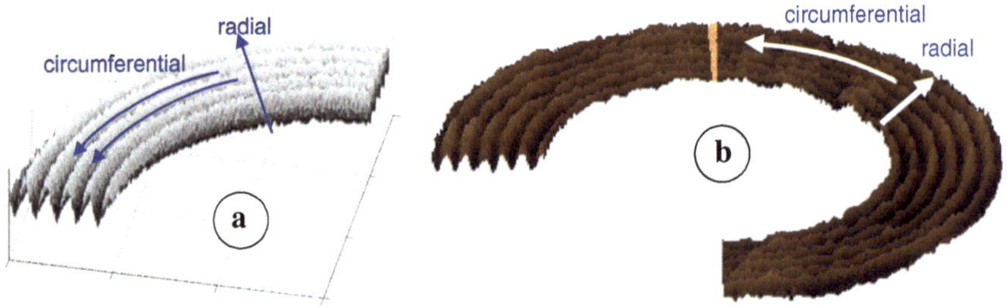

Fig. 4 Schematic illustration of the crest and valley sinusoidal fluctuations indicating radial and circumferential direction used by **a** Geoffroy and Prat (2004) and **b** Pérez-Ràfols et al. (2016)

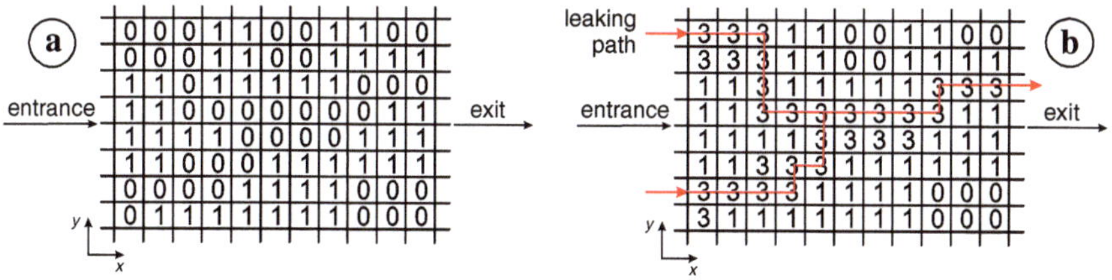

Fig. 5 Recognition algorithm diagram of leak path identification where 0 is no gap and 3 is the fluid flow path: **a** contacting state and **b** leaking path, after (Man et al. 2014)

As the two rough surfaces come into contact and deform due to the loading applied, the leakage path through the aperture field is dictated by the largest gap from one end of the specimen to the other and is calculated using a "recognition algorithm" (see Fig. 5 for details). This fluid flow path recognition is a simplistic, but effective theory to allow one to understand if the microcontact under specific loads will either leak or not for a material.

2.4 *Macro Scale Deformation and Its Effect in Fluid Flow Path*

Using FE-code ANSYS, Gorash et al. (2014, 2015) have shown that at a macro scale the contact length of the seat and disc of a PRV is reduced and is further exasperated due to cyclic opening and closing of the valve. Gorash et al. (2014, 2015) assumed the contact force to be normal to the face of the seat and disc modelled as 2D. Using elements PLANE 183 (for both the seat and disc model), COMBIN14 (spring), CONTA172 (seat contact) and TARGE169 (disc contact) and assuming that for a gas and liquid the fluid flow pressure acting on the seat and disc is parabolic and linear respectively, they have shown that the once the spring preload and the internal set pressure is applied, the contact edges of the disc become plastics for a monotonic material response. This plastic response of the seat is exasperated towards the middle of the disc when cyclic material response is considered. This essentially means the fluid flow path is increased, while effective contact area and length would reduce meaning a higher leakage rate.

2.5 *Experimental Leakage Rate*

One of the earliest collection of advanced studies addressing testing and analysis of PRVs performance was compiled by Singh and Bernstein (1983). The book covers the topics of test facilities design, safety valve experiments, analysis of PRVs performance, and loads on discharge piping. The state-of-art approach to calculate the leakage rate of a PRV is based on recent international standards, e.g. ASME (2014), BSI (2013), API (2014). This method requires the PRV to be set to 90 % of its set pressure using a gas. The outlet of the PRV is closed off with a pipe attached to expel the increase in pressure (i.e. leakage) in some water in the form of bubbles. The leakage has to be less than a specified amount for it to be used in-service.

Based on the average surface roughness, there have been attempts to relate this to the leakage rate through a gasket. Haruyama et al. (2013) created an experiment which quantified the leakage rate of helium through a bolted flange with a new gasket placed between. They concluded that the leakage rate is highest for rough surfaces of $R_a = 3.5\,\mu m$ when a low load force is applied. When a maximum of 400 MPa of force is applied on the flange then leakage rate is the least and is similar for all R_a's being 1.5, 2.5 and 3.5 μm. However, the material properties of the gasket and the flange are not known and so links between the material deformation and leakage is difficult to comment on.

Another method used by Marie and Lasseux (2007) allows quantification of leakage flow of solvents at a micro or nano-scale through a rough metal contact for both viscous and diffusive fluid flow separately for contact pressures up to 700 MPa. Using the leakage results Eqs. (1)–(4), they have managed to find the diffusive and viscous properties from which the effective gap size has been estimated.

2.6 PRV Specific Issues Relating to Leakage

Now that an understanding of the work currently completed on micro deformation and its effect on leakage has been analysed it is important to consider factors which are unique to a PRV which could also cause it to leak.

Depending on the design of the valve, one issue which has been highlighted is that the guide pin can be displaced easily which causes the seat to rotate or displace. This rotation/displacement of the seat causes the valve to leak and subsequently the set pressure decreases. Ritchie (1989) examined this issue and created an analytical model to understand the reduction of seat pressure due to the misalignment angle of the seat. Assuming the valve leaks only when the set pressure has been reached, it was shown that for a disc with radius of 8.47 mm and 155 N of applied force on the seat, the set pressure (100 ψ) decreases by 10 % for a misalignment of 1.225 ° for the seat.

2.7 Summary and Discussion

As it has been shown, previous analytical work accounts for the gap height either in the form of h or h^3 for a parallel gap. This is mainly been accounted for only laminar viscous incompressible/ compressible or diffusive flow through the aperture field, with more recent work in accounting for the rarefaction of the flow. However, the parallel gap assumption is limited to small asperities $h \ll l$ and there has to be an "effective" height used for a 2D surface model which is geometrically represented as simple surfaces. While for 3D surface models computer programs such as MATLAB or equivalent can be used to model lapped surfaces with high accuracy as self-affine fractal surfaces, but are extensive.

In reality Tsukizoe and Hisakado (1965) and the slip line theory is not completely valid for multiple asperity contacts demonstrated by the experiment conducted by Uppal and Probert (1972). They showed that for a multiple asperity surface in contact with a harder flat surface, under a high load the shallower valleys do rise (0.1–0.3 μm) while deep valleys do not, while Tsukizoe and Hisakado (1965) theory considers no movement in the valleys. So, there is an element of elastic behaviour and plastic strain hardening occurring which would have to be considered.

Instead of analytical techniques to describe the deformation of two rough surfaces, FEA programs such as ANSYS could be utilised to describe elastic perfectly plastic deformation of the interface as shown by Megalingam and Mayuram (2012) using actual 3D or 2D scanned surfaces. There is other literature particularly by Thompson (2007a, b, 2011), Thompson and Thompson (2010a, b), which gives further guidance on multi-scale modelling and optical measurements of the surface aperture which is then transferred into ANSYS. The FEA would require some form of verification of the surface deformation accuracy, but if achieved, this will allow a more accurate representation of the fluid flow path.

This could be taken further by modelling the surface as a representative surface using summing technique, self-affine fractal surface or MOTIF procedure, versus, the actual surface in FEA and see the difference in the contact area.

A theory created by Geoffroy and Prat (2004) tries to link the diffusive and transmissive fluid flow to either radial or circumferential, but has not been verified and is only applicable to predominant sinusoidal shaped surfaces in contact with a flat surface, i.e. gaskets. Recently, Pérez-Ràfols et al. (2016) have extended the work of Geoffroy and Prat (2004) by using the summing technique material considerations for a composite elastic modulus (i.e. the effective elastic modulus of the joint) and used a relative size of the whole model, rather than modelling the whole surface. However, in doing so, there is still the implicit assumption made by Pérez-Ràfols et al. (2016) that there is no deviation in the surface form, which may be appropriate for gaskets, but not for this program as will be discussed later in Sect. 4.

Gorash et al. (2014, 2015) have shown that there is a deformation at a macro scale due to the spring force and the pressure of the fluid, which has to be considered to begin. Their work has shown the contact is not uniform between a seat and disc of a PRV rather; it begins and ends within the seat length. So there is a need to understand how the macro and microcontact areas link to give the actual contact area.

The disc rotation on the seat causing leakage maybe more of a design problem rather than a research based problem. The reason for this is that there are clearances prevalent throughout the seat and disc allowing this motion to occur. Also it is understood that the spring force under compression may not be perpendicular to the compression axis of the spring.

Further investigations would be required to understand the effect of grain boundaries on the deformation of surface roughness at a micro scale.

3 Effect of High Temperature on Structural Behaviour of Metal-to-Metal Seal in a Pressure Relief Valve

This section focuses on the structural behaviour of the metal-to-metal contact seal in a PRV when the material, AISI type 316N(L) steel, behaves and reacts as it would when exposed to a higher temperature. In this case the temperature of interest is 538 °C (maximum operating temperature for the investigated valve) and the fluid being a gas, i.e. air. The maximum operating pressure supported by the valve in question is 18.6 MPa. The initial work conducted by Gorash et al. (2014, 2015), will also be discussed in this section.

To study these specific effects of temperature and pressure a finite element analysis is undertaken using ANSYS Parametric Design Language (APDL) script to account for Fluid Pressure Penetration (FPP) through the contact gap. This APDL script allows for pressure between the contact gap to be accounted for sufficiently and automatically adjusts the force retraining the valve. Ultimately, this allows for a cyclic analysis of the valve opening and closing with FPP accounted for. Previously, Gorash et al. (2014, 2015) had to manually account for FPP for every time step until the opened. This meant lengthy solving times.

This research builds upon the work conducted previously (Gorash et al. 2014, 2015). In particular, this research advances in the structural model by analysis conducted at an elevated temperature; with a more accurate material model to consider viscoplastic deformation at high temperature; by creating an APDL script to automatically account for the FPP migration; and adjustment of the spring force to maintain the desired set pressure.

3.1 Seat Contact Configuration

The PRV is simplified into three 2D axisymmetric basic parts: a cylindrical nozzle, which is in contact with a disc (representing the valve seat on top), which is preloaded by a compressed linear spring. This concept is represented in Fig. 6. The effects on the bell-housing and any other parts such as the nozzle ring are not of interest as it is assumed they do not have any effect on the structural behaviour of the seal.

The compression force of the spring is transferred directly to the disc and then to the seat due to the contact. It is this compression force and quality of the contact which maintains the seal up to the set pressure. For this analysis it is assumed that the contact is geometrically uniform. From the study by Gorash et al. (2014, 2015), it has been shown that due to FPP the internal contact between the disc and seat significantly

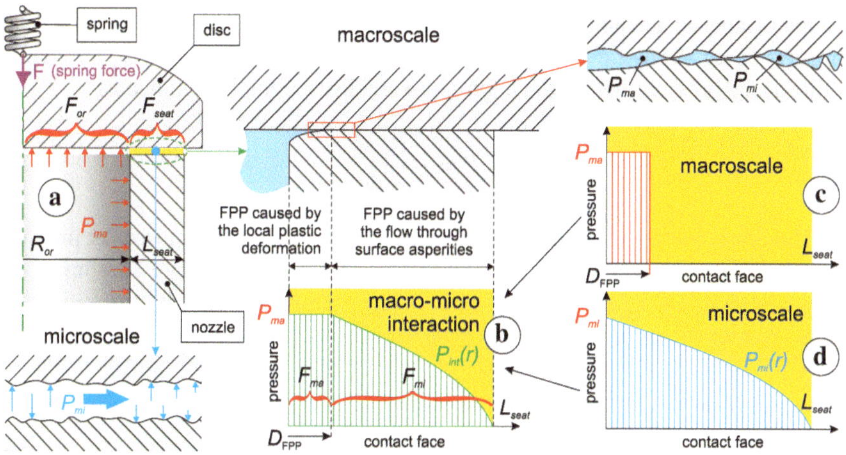

Fig. 6 Concept of seat contact configuration in the contact area of metal-to-metal seal considering FPP (fluid pressure penetration)

deforms due to plastic deformation even at room temperature. The pressure that the FPP attributes to the disc and seat in the contact zone will be known henceforth in this report as the macroscopic pressure distribution.

In reality the contact between the faces of the seat and disc are not uniform, especially at a micro scale, where surface form, waviness and roughness contribute to the leakage. In this case it will be assumed that there is a degree of surface roughness between the contacting faces. To account for this the fluid pressure attributed to the surface roughness of the contacting faces in the gap, Müller and Nau (1998) had shown that the pressure drop across a sealed gap for a compressed fluid can be represented using a power law function dependent on distance as:

$$P(x) = P_1\left[1 - \left(1 - \beta^2\right)\frac{x}{L}\right]^n, \qquad (8)$$

where L is a length of a seal gap; P_1—internal pressure and P_2—external pressure; so the pressure ratio is $\beta = P_2/P_1$ and n is a power law exponent, which is dependent on the type of fluid, e.g. $n = 0.5$ for gas and $n = 1$ for liquid. Henceforth in this report this pressure distribution will be related to as the microscopic pressure distribution. Equation (8) can be extended further by slightly changing the mathematical representation to be aligned with the contact gap in question in a mathematical form as:

$$P(r) = P\left[\frac{r_{\text{out}} - r}{r_{\text{out}} - r_{\text{FPP}}}\right]^n, \qquad (9)$$

where P—internal pressure, r_{out}—outer radius of the contact area, r—inner radius of the contact area, r_{FPP}—radius of FPP. This micro pressure distribution is formulated in this manner to account for the maximum FPP point travelled in the contact zone. Integrating Eq. (9) by r over the length of the valve seat ($L_{\text{seat}} = r_{\text{out}} - r_{\text{in}}$), an average value of the pressure within the micro pressure distribution is obtained as

$$\bar{P} = \frac{P}{1+n}. \tag{10}$$

Therefore, the force required by the spring to maintain the required set pressure between the seat and disc is an accumulation of the operating pressure up to the internal office, macro and micro pressure distribution multiplied by the associated areas. Hence the spring force can be represented in the following form:

$$F_{\text{sp}} = F_{\text{orf}} + F_{\text{macro}} + F_{\text{micro}} \quad \Rightarrow \tag{11}$$

$$F_{\text{sp}} = P\pi r_{in}^2 + P\pi \left(r_{\text{FPP}}^2 - r_{in}^2\right) + P\pi \left(\frac{1}{1+n}\right)\left(r_{\text{out}}^2 - r_{\text{FPP}}^2\right). \tag{12}$$

If $n = 0.5$ for a gas then:

$$F_{\text{sp}} = P\pi \left(\frac{2}{3}r_{\text{out}}^2 - r_{\text{FPP}}^2\right). \tag{13}$$

Alternatively, this can be expressed in terms of spring displacement Δ_{sp}:

$$\Delta_{\text{sp}} = \frac{P\pi}{K}\left(\frac{2}{3}r_{\text{out}}^2 + \frac{1}{3}r_{\text{FPP}}^2\right). \tag{14}$$

3.2 *FEA Modelling and APDL Script*

The numerical study is conducted using academic FE-code ANSYS 16.0. The three components, cylindrical nozzle, disc and spring are all simplified into a 2D axisymmetric, elastic-plastic model, as shown in Fig. 7. The model is set-up as a quasi-static structural analysis allowing the valve to open and close over 100 cycles.

As shown below, the spring is modelled using a COMBIN14 (2-node longitudinal linear spring-damper) finite elements (FEs). A vertical displacement of Δ_{sp} is applied to compress the spring. As previously described the displacement Δ_{sp} is proportional to the sealing set pressure, which as previously elaborated is due to the internal pressure and the macro–micro pressure distribution within the contact area. The spring is connected to the kinematic coupled nodes at the top of the disc. This is to ensure that the vertical force associated with the spring is distributed evenly and vertically down the disc.

The disc and seat are both constructed of PLANE183 (8-node axisymmetric structural solid) FEs with an internal pressure, P, placed on the inside and a fixed boundary constraint placed on the outside near the bottom of the seat (i.e. the connection to the

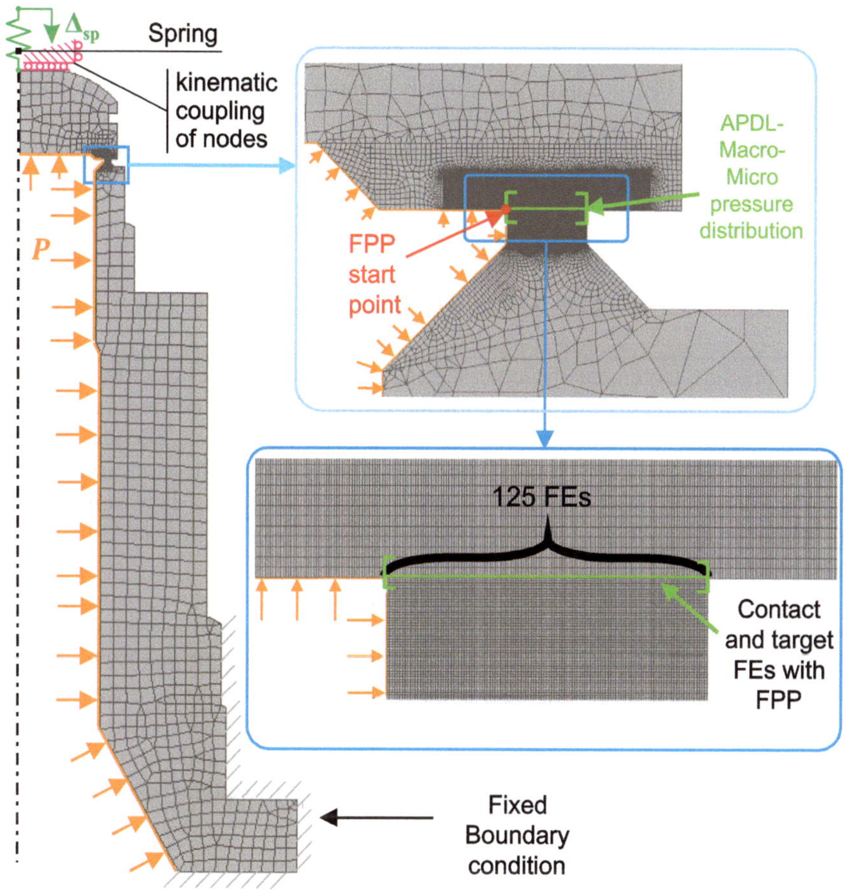

Fig. 7 FE model of the valve seat and disc, with a detailed model of the contact zone, boundary conditions and loadings

outer valve body). The pressure, P, is ramped up with an average pressure change rate of $r_{pr} = 0.744$ MPa/s in an incremental manner until the set pressure of 18.6 MPa is reached and then incrementally reduced to zero. This accounts for 1 cycle of the valve opening and closing.

The contact pair is made up of CONTA172 (2D 3-node surf–surf contact {for disk}) and TARG169 (2D 3-node target {for seat}). The internal contact of the seat and disc is the associated starting point for the FPP feature and is allowed to penetrate along the full contact length. This FPP feature allows pressure to be associated with changing contact conditions, i.e. as the seat and disc deforms a gap is created and it will have the associated pressure applied in it automatically. As the valve ramps up in pressure and cycles open and close, it is expected for the FPP to migrate along the contact length. To account for the micro pressure distribution with respect to the FPP migration point an APDL script is required, so that the macro and micro pressure is applied correctly.

To ensure the FPP is captured accurately a high resolution of contact points are created between the seat and disc (125 FEs in the contact region). In total there are 16,404 elements and 49,939 nodes. For further information specific to element types please refer to ANSYS® Help (2013).

3.3 APDL Script r_{FPP} *Retrieval and Apply*

As the pressure incrementally increases, the FPP through the contact between the disc and seat increases gradually until the pressure is either not high enough to penetrate further or the deformation of the seat/disc reduces not allowing further penetration. The location point of the r_{FPP} is important to allow accurate micro pressure distribution to be applied and hence an accurate spring displacement Δ_{sp}. To capture this migration of the r_{FPP} location across the seat and disc an APDL script is required as shown in Fig. 8.

Figure 8 is a flowchart of the script which activates after the first Load Step (LS) has solved. The script retrieves the FFP contact data produced from the previous LS from the post processor. It then sorts the data out to find the maximum FPP location across the seat and disc. This data point is saved as r_{FPP} and the script instructs the ANSYS program to enter the pre-process (/SOLU) state and adjust the spring displacement and micro pressure distribution using Eqs. (14) and (10) respectively. The micro pressure distribution is also applied only across location r_{FPP} to r_{out}. Once

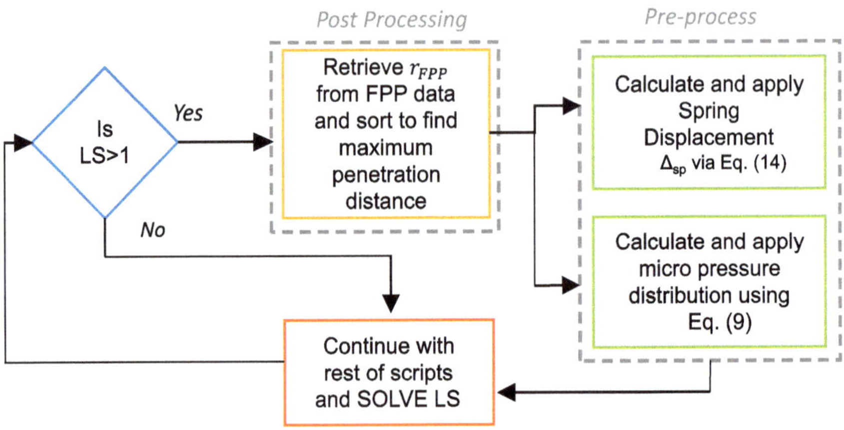

Fig. 8 APDL script flowchart. Beginning at the decision, and then moving though post-processing, pre-process and then returning to solve the problem until the next LS. After which this loop repeats until the last LS

the post-process and pre-process stage is complete the program continues with any other scripts and solves for that LS. This process is restarted at the beginning of the next LS until all the LS's are solved.

3.3.1 APDL Restart Bug

The main challenge to overcome with this script was a bug which is present in ANSYS workbench 16.1. This bug is related to restart controls ("rescontrol"). After creating a new APDL script in the "Outline tree" the user has the option to select which LS the script is to be activated in via "Step selection mode". Due to the nature of the FPP, the script would have to be activated each LS. Naturally, the restart control APDL script should be set to activate for every LS. However, doing this causes the program not to restart with the new r_{FPP} after every LS. Instead it would continue on with solutions found in the first LS. As a workaround solution for this bug, the user has to select "First" LS in "Step Selection mode" to ensure the restart controls work for each LS. This issue has been highlighted with ANSYS technical support.

3.4 *Material Model*

The critical components of the PRV (nozzle and disc) are manufactured of the steel AISI type 316N(L) due to optimally appropriate mechanical properties of this steel grade. AISI type 316N(L) stainless steel has been used within the power generating industry since the early 60s of twentieth century. Commonly used in superheater piping, pressure vessels, heat exchangers and other components exposed to high temperatures of 650 °C as indicated in previous work (Gorash et al. 2014, 2015). The mechanical characteristics of the steel AISI type 316N(L) makes it an optimal material for a valve seat which is expected to undergo high local contact stresses, corrosion-fatigue conditions and possible high-temperature exposure.

Mechanical properties of the steel AISI type 316N(L) in the range of 20–700 °C show significant temperature dependence (Gorash et al. 2012). Available stress–strain experimental data and its fitting by elastic-perfectly-plastic (EPP) and Ramberg–Osgood (R-O) material models were discussed by Gorash et al. (2014, 2015). Compared to martensitic and ferritic steels, austenitic grades including type 316 have lower yield stress σ_{y}, but higher rupture ductility. This complies with experimental observations at room and high temperatures, which confirm that the material behaviour of the steel AISI type 316N(L) is viscous and rate-dependent (Hyde et al. 2010). Thus, an accurate description of the plastic deformations with a unified viscoplasticity model (Chaboche 2008) is essential to address structural integrity and operation issues.

In the previous study (Gorash et al. 2014, 2015), viscoplastic material behaviour of the steel AISI type 316N(L) was simplified to rate-independent plasticity neglecting viscous effects. Available monotonic and cyclic stress-strain curves were fitted by the R-O equation and incorporated through the Multilinear Kinematic Hardening (MLKH) model in ANSYS. Since the dynamic effects have been neglected, the PRV operation was assumed to be quasi-static for FE-simulation in ANSYS.

Consideration of viscoplastic effects for FEA in current work requires the mechanical material properties and Chaboche material constants at a temperature of 538 °C. These values are derived by simple interpolation using the experimental data by Hyde et al. (2010) available for 300, 500, 550 and 600 °C. In order to obtain constants for $T_i = 538$ °C, the constants at lower $T_l = 500$ °C and upper $T_u = 550$ °C neighbouring temperatures are used. The interpolation is done by using the weighting coefficients (lower and upper), which define the proximity of the interpolated temperature to the neighbouring temperatures:

$$w_l = \frac{T_u - T_i}{T_u - T_l} \quad \text{and} \quad w_u = \frac{T_i - T_l}{T_u - T_l}. \tag{15}$$

In this case, the weighting coefficients take the values of $w_l = 0.24$ and $w_u = 0.76$. In general, the following condition should be satisfied for the temperature:

$$T_i = T_l\, w_l + T_u\, w_u. \tag{16}$$

Equation (16) is used to identify all the material constants for AISI type 316N(L) steel at 538 °C using the constants at 500 and 550 °C (Hyde et al. 2010), which are reported in Table 1 and required for implementation of FEA in ANSYS.

In order to conclude about the influence of high temperature on cyclic deformation of the valve seal, the material constants for the Chaboche model are also required for 20 °C. Since the experimental stress-strain curves are available only for a single strain rate value, the rate-independent variant of the Chaboche model is used. The identification of corresponding constants is done using the fitting procedure suggested in previous work (Gorash and MacKenzie 2014). This procedure comprises the initial smoothing of the cyclic stress-strain data by the R-O model and subsequent estimation of the kinematic constants using the Solver add-in of Microsoft Excel. The smoothed cyclic stress-strain data is fitted by the following equation for stress amplitude ($\Delta\sigma/2$) and plastic strain amplitude ($\Delta\varepsilon^p/2$):

Table 1 Material constants for Chaboche model for AISI type 316N(L) steel at 538 °C interpolated from constants at 500 and 550 °C (Hyde et al. 2010) using Eqs. (15) and (16)

T °C	k MPa	E GPa	b	Q MPa	a_1 MPa	C_1	a_2 MPa	C_2	Z MPa · $s^{1/n}$	n
500	32.5	145.54	33.35	30.41	94.6	6472.6	113.3	979.91	175	10
538	*31.36*	*142.29*	*31.56*	*28.43*	*88.29*	*6827.06*	*114.44*	*963.02*	*173.48*	*10*
550	31	141.26	31	27.8	86.3	6939	114.8	957.69	173	10

Table 2 Material constants for Chaboche model for AISI type 316N(L) steel at 20 °C identified using the experimental stress-strain curves and R-O fittings (Gorash et al. 2014, 2015)

k MPa	E GPa	b	Q Mpa	a_1 Mpa	C_1	a_2 Mpa	C_2	a_3 Mpa	C_3	a_4	C_4	a_5 Mpa	C_5
119.1	194	558.3	−119.1	454.4	60.6	134.6	899.1	54.5	14,289	20	267,800	10.5	6,729,430

$$\frac{\Delta\sigma}{2} = \sum_{i=1}^{N} \frac{C_i}{\gamma_i} \tanh\left(\gamma_i \frac{\Delta\varepsilon^{\mathrm{p}}}{2}\right). \tag{17}$$

The number of kinematic back-stresses is increased to five ($N = 5$) in order to incorporate the mixed softening–hardening character of plastic deformation observed during cyclic response. The corresponding values of the kinematic constants (a_1–a_5 and C_1–C_5) for the Chaboche model are reported in Table 2. The next step is the identification of isotropic constants (k, b, Q) for the Chaboche model (see Table 2) through fitting of monotonic stress-strain data by the following equation for stress σ and plastic strain ε^{p}:

$$\sigma = k + Q\left[1 - \exp(-b\varepsilon^{\mathrm{p}})\right] + \sum_{i=1}^{N} \frac{C_i}{\gamma_i}\left[1 - \exp(-\gamma_i\varepsilon^{\mathrm{p}})\right] \tag{18}$$

3.5 *Results and Discussion*

For both the 20 and 538 °C cyclic opening and closing analysis, it is shown that the spring force F_{sp} is required to increase to maintain the desired set pressure as shown in Fig. 9. The percentage increase shown in Fig. 9 is calculated by:

$$\Delta F_{\mathrm{sp}} = \frac{F_{\mathrm{sp-FE}}}{F_{\mathrm{orf}}} \cdot 100. \tag{19}$$

The 538 °C simulation shows roughly a 1.5 point higher set force than at 20 °C at 8.88 %. This is maintained for the first 3 cycles which slightly increases at the 4th to 8.93 %. There is a sharp increase to 9.45 % after which there is an oscillation between 9.3 and 9.45 %. This oscillation lasts until the 15th cycle after which a 9.4 % increase in F_{sp} is maintained for the rest of the 100 cycles.

This transition between 8.88 and 9.45 % can be explained by analysing the plastic deformation across the contacting seat face (see Fig. 10). The whole of the seat face becomes plastically strained with concentration zones at the internal and external edge (with the former being more significant). Of course the graph is only associated with the seat face. The extent of the plastic zone across the seat is shown in Fig. 11 with the two concentration zones being visible again. As can be seen from Fig. 11, the

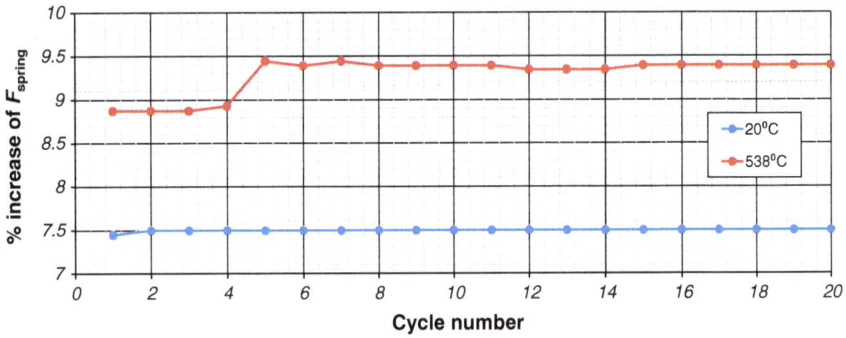

Fig. 9 Spring force (%) increase to account for plastic strain development across contacting faces

disc remains elastic and does not undergo any plastic strain. As the cycles increase the plastic strain at the edges increases significantly—internal edge from 0.1 % (1st cycle) to 0.38 % (100th cycle) and at the external edge—from 0.03 % (1st cycle) to 0.25 % (100th cycle). The internal plastic zone is most prevalent from 0-0.5 mm while the external edge is 1.08–1.25 mm. Since this was an idealised model it is clear that the edges no longer remain square and do deform into radial edges, with a small mound shaped middle section (ranging from 0.4 to 1 mm).

The significant internal plastic zone is due to a combination of the F_{sp}, FPP and internal pressure (based on the macro and micro pressure distribution). As the plastic deformation of the contact face increases, the FPP is allowed to migrate further into the contact zone. This requires an increase in the spring force to maintain the required set pressure.

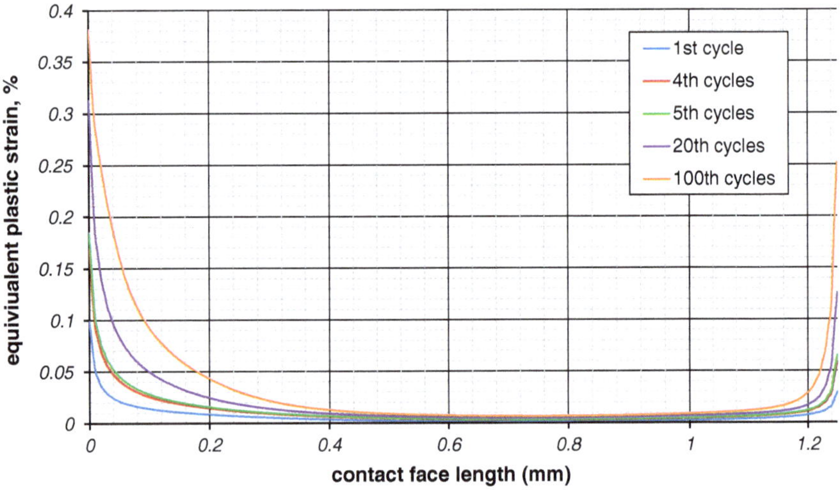

Fig. 10 Equivalent plastic strain (%) across seat length (mm)

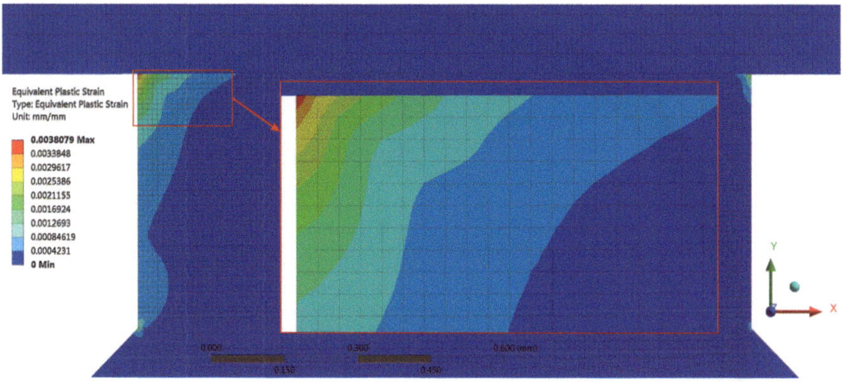

Fig. 11 Equivalent plastic strain of seat and disc at 100 cycles

3.6 Conclusion

Using FE-code ANSYS 16.1, it has been shown that to maintain a required set pressure of 18.6 MPa for a gas at a temperature of 538 °C, the spring force must be increased between 8.88 and 9.45 %. This has been found to be adequate for up to 100 cycles of the valve opening and closing. This increase in spring force is found to be due to the lower yield stress and plastic strain conditions of stainless steel 316N(L) at 538 °C, combined with FPP and the micro pressure distribution representing the surface roughness. This idealised model did show a significant amount of plastic strain at the internal edge of the seat extending from 0 to 0.5 mm in length.

4 Surface Form and Waviness at the Micro and Nanometre Scale

4.1 Surface Metrology Measurements and Investigations

The work so far has focused on 2D axisymmetric FEA with an implicit assumption that the surface finish is "flat", i.e. there is no deviation in the surface form. As mentioned in the Introduction (Sect. 1) recommendation 3—"a multi-scale model accounting for surface roughness deformation"—could be now implemented. However, it was found that the surface form of the discs and seats given by the sponsor varied in magnitude. Therefore, this implicit assumption that the seats are "flat" is no longer valid and recommendation 3 was not the correct course of action.

After some adjustments to the finishing technique used to polish the discs and seats, it was found that the disc surface form dramatically reduced. Importantly the finishing technique used created consistently the same surface finish shape and

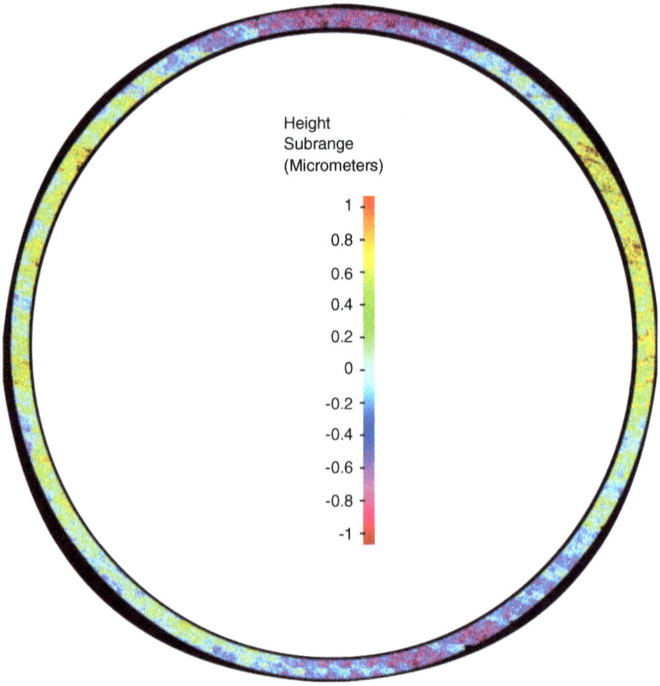

Fig. 12 Surface form of valve seat measured using the Alicona Infinite Focus at 20X objective magnification finding a deviation from the surface form (flatness) of 2.1 μm

deviation of about ±0.5 μm. Using the Alicona Infinite Focus (a confocal and variable focus metrology measurement instrument), it was possible to measure the contacting faces of both the seat and the disc. The surface form measurements of both the valve seat and disc are shown in Figs. 12 and 13.

As can be seen in Fig. 12, the valve seat has a deviation of 2.1 μm in the surface form profile and has almost got a 1/4 symmetric shape to the surface profile. Similarly the valve disc has deviation of 5 μm and has a clear and distinct 1/4 symmetric shape to the surface profile. This profile is termed as a "saddle" shape (Kemet International Limited 2015).

As mentioned in the Literature review (Sect. 2), there is no mention of surface form and its relationship to leakage. Instead preference is given to surface roughness. In an industrial environment, flatness is generally measured using a monochromatic light source. There is a known uncertainty called 2π ambiguity with monochromatic light sources, which induce errors into the measurement.

Since it is a metallic contacting surface, majority of work with respect to leakage is with metal gaskets, where it is assumed the gasket deforms to the point that surface form is no longer a large enough factor to consider with leakage or flatness is only considered. Preference is generally given to surface roughness and waviness rather the surface form shape.

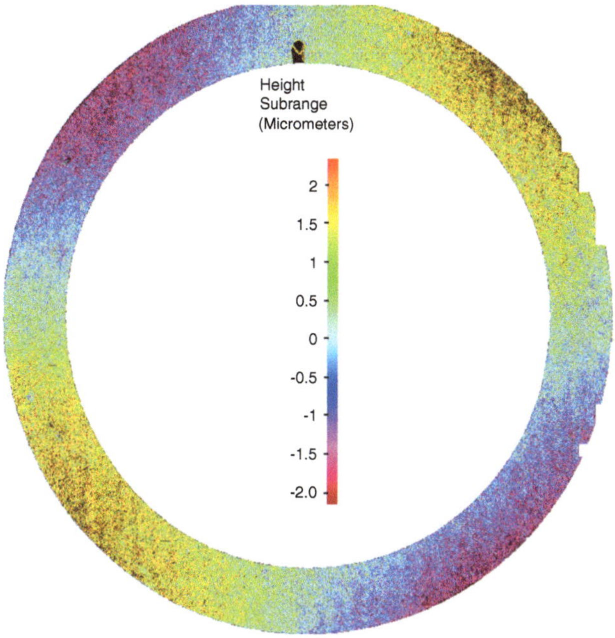

Fig. 13 Surface form of valve seat measured using the Alicona Infinite Focus at 20X objective magnification finding a deviation from the surface form (flatness) of 5 μm

The reality is that there no such thing as a "perfectly flat" surface, which is a term associated with when using a monochromatic light source to measure flatness. Even with the best of finishing techniques there will be some deviation even in the nanometre range.

4.2 Modelling Surface Form and Waviness at the Macro–Nanometre Scale

Taking this surface form shape and waviness into consideration it can be modelled using simplistic surface geometries, in the shape of pyramids, combined with the sum surface technique originally created by Tsukizoe and Hisakado (1965) as shown below in Fig. 14.

As seen earlier the surface form is simplified into a 1/4 symmetric 3D Computer Aided Design (CAD) model. Using the summing technique, the surface form is modelled as a deviation of 7.1 μm and waviness: $W_a = 312$ nm and $W_{sm} = 2.35$ mm. The model is first created over a 2D rectangle, after which the valve seat face is cut out. From this the rest of the valve seat can be modelled.

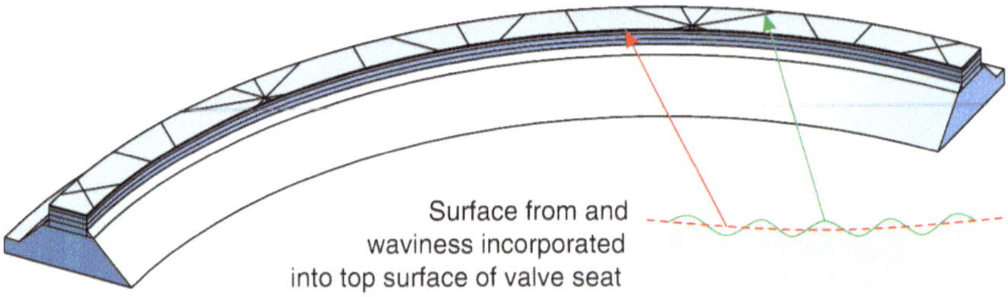

Fig. 14 1/4 symmetry model of valve seat with surface form (7.1 μm deviation) and waviness ($W_a = 312$ nm and $W_{sm} = 2.35$ mm) incorporated into top surface

Using a theoretical approach to modelling of waviness and surface form, radial sinusoidal waves describing waviness can be schematically presented by the following equation:

$$W_r = \cos\left[\left(x^2 + y^2\right)^{1/2}\right], \tag{20}$$

which is shown as contour plot in Fig. 15a and as 3D plot in Fig. 16a. As discussed, the W_a and W_{sm} are respectively found to be in the nanometre and millimetre range. Therefore, the appropriate unit choice for this study for x and y (and for radial W_r and hoop W_h waves), will be micrometres. Circumferential (hoop) sinusoidal waves describing surface form can be schematically described by the following equation:

$$W_h = x \cdot y^3 - y \cdot x^3, \tag{21}$$

which is shown as contour plot in Fig. 15b and as 3D plot in Fig. 16b. A combination of radial and hoop waves, which is usually expected in real contacting surfaces, can be schematically presented by the product of Equations (20) and (21) as $W_c = W_r \cdot W_h$ and shown as contour plot in Fig. 15c and as 3D plot in Fig. 16c.

Sinusoidal waves would be better to model the geometry with, however, it has been found that the CAD software, SolidWorks 2013 is not capable of creating sinusoidal waves due to the W_{sm} length versus the W_a. The W_{sm} versus W_a ratio

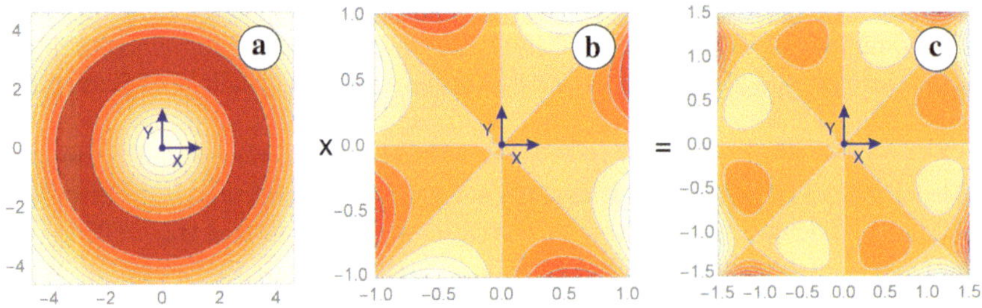

Fig. 15 Schematic contour plots of the sealing surface waves: **a** radial, **b** circumferential (hoop) and **c** their combination

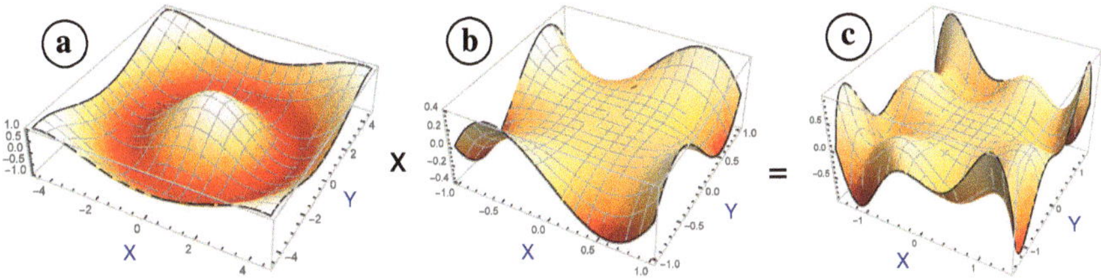

Fig. 16 Schematic 3D plots of the sealing surface waves: **a** radial, **b** circumferential (hoop) and **c** their combination

creates a difficulty within the CAD engine to model due to the line being so close to being linear. It was found that a ratio of about $W_{sm}/W_a < 2500$ can be modelled in SolidWorks 2013 as a sinusoidal wave. This model will be structurally deformed using non-linear FEA considering plasticity in ANSYS and will be reported on later in the year in a GAMM 2016 conference (Anwar et al. 2016a) and in length as part of the thesis.

5 Overall Conclusion

The overall conclusions drawn so far are:

- The literature with respect to PRV leak tightness is few and far between. However, inspiration is drawn from metal-to-metal contacting surface such as gaskets. There is a lot information and research conducted to account for surface roughness and some on waviness and its effect on leakage.
- The elevated temperature effects on the 2D axisymmetric FEA showed that the FPP was capable of migrating further along the contacting faces due to the cyclic softening of the material. This meant that the spring force would have to be readjusted after the first 5 cycles.
- After measuring the surface faces of the seat and the disc, it was found that the surface form was at a micrometre magnitude with a 1/4 symmetry shape and the waviness is exhibited over a nanometre range.

6 Future Direction

With respect to the research conducted so far, the biggest change is taking into account of the surface form and waviness. Therefore, the overall the direction is shown in Fig. 17.

In regards to the 1-way coupling between FEA and CFD this will be reported on in the GAMM 2016 conference (Anwar et al. 2016a). This will focus on the techniques

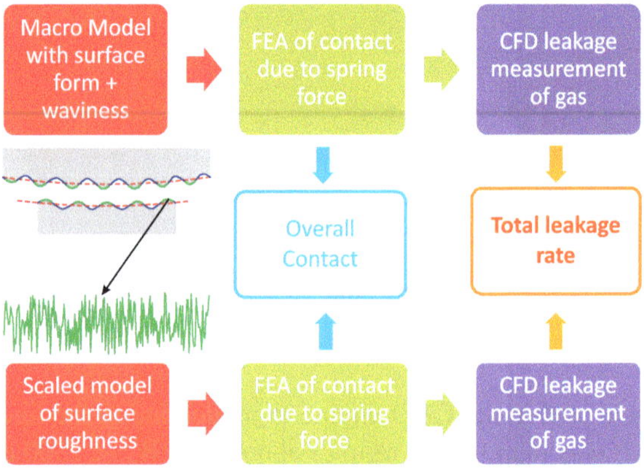

Fig. 17 Overall research direction for PRV leak tightness investigation and model development

used. The initial CFD validation work has been completed and will be reported on at the ASME PVP 2016 conference (Anwar et al. 2016b). The measurements taken using the Alicona Infinite Focus have shown the average roughness (R_a) is about a 10th of the waviness. The surface roughness will be modelled over a small representative surface using real scanned surfaces and then deformed using FEA. This will be reported on in the near future.

Acknowledgments There are many people without whom this research project could have not have been possible. The authors would like to thank:

- The individuals who's research has been mentioned in this paper, since without their great work, this literature review would have not come about;
- Colleagues and supervisors, in particular, Robert Hamilton and David Nash;
- The WEIR group and the WEIR Advanced Research Center (WARC) in particular, Allan Bickley, Allan Stewart, Ian MacQueen, Stéphane Carrier and Fabrice Courdavault;
- Brian Kyte (Sales Director at Alicona UK) for the help in getting access to the Alicona Infinite Focus and addressing technical matters;
- Liza Hall (University of Strathclyde, Advanced Forming and Research Centre, Metrology) for allowing to use the Alicona Infinite Focus and addressing technical matters;
- Rong Su and Wahyudin Syam (University of Nottingham, Institute for Advanced Manufacturing) for allowing to use the Alicona Infinite Focus and addressing technical matters.

References

ANSYS® Help. Mechanical APDL // Element Reference // 7. Element Library // Part I: Element Library. ANSYS, Inc., Canonsburg (PA), USA, Academic Research 14.5.7 edn (2013)

Anwar, A.A., Gorash, Y., Dempster, W., Nash, D.: Deformed gap space using macro-micro FEA model and transferred into a CFD model. In: Proceedings of Joint DMV & GAMM Annual Meeting (GAMM 2016). GAMM e.V., Braunschweig, Germany, 7–11 March (2016a, submitted)

Anwar, A.A., Ritos, K., Gorash, Y., Nash, D., Dempster, W.: Leakage of gas flow through a microchannel in the slip flow regime. In: Proceedings of ASME Pressure Vessels & Piping Conference (PVP 2016). ASME, Vancouver, Canada, 17–21 July 2016 (2016b, submitted)

API. Seat Tightness of Pressure Relief Valves. No. 527 in API Standard. American Petroleum Institute, Washington, USA (2014)

Arkilic, E.B., Schmidt, M., Breuer, K.S., et al.: Gaseous slip flow in long microchannels. J. Microelectromech. Syst. **6**(2), 167–178 (1997). doi:10.1109/84.585795

ASME. Pressure Relief Devices – Performance Test Codes. No. ASME PTC 25-2014 in An American National Standard, The American Society of Mechanical Engineers, New York, USA (2014)

BHR Group Ltd. Valve stem leak – tightness test methodologies. Summary report no. CR1234, European Commission, British Hydromechanics Research Group Ltd., Cranfield, UK (2000)

BSI. Safety devices for protection against excessive pressure. Safety valves. No. BS EN ISO 4126-1:2013 in British Standard, The British Standards Institution, London, UK (2013)

Burmeister, L.C., Loser, J.B., Sneegas, E.C. Advanced valve technology – Revised and enlarged edition. Technology Survey no. NASA SP-5019, Midwest Research Institute, NASA, Washington, D.C., USA (1967)

Chaboche, J.-L.: A review of some plasticity and viscoplasticity constitutive theories. Int. J. Plast. **24**(10), 1642–1693 (2008). doi:10.1016/j.ijplas.2008.03.009

Chong, X.: Subsonic choked flow in the microchannel. Phys. Fluids **18**(12), 127,104–1–127,104–5 (2006). doi:10.1063/1.2408510

Gagnepain, J., Roques-Carmes, C.: Fractal approach to two-dimensional and three-dimensional surface roughness. Wear **109**(1), 119–126 (1986). doi:10.1016/0043-1648(86)90257-7

Ganti, S., Bhushan, B.: Generalized fractal analysis and its applications to engineering surfaces. Wear **180**(1), 17–34 (1995). doi:10.1016/0043-1648(94)06545-4

Geoffroy, S., Prat, M.: On the leak through a spiral-groove metallic static ring gasket. J. Fluids Eng. **126**(1), 48–54 (2004). doi:10.1115/1.1637627

Gorash, Y., MacKenzie, D.: Safe structural design for fatigue and creep using cyclic yield strength. In: Proceedings of 3rd International ECCC Conference – Creep & Fracture in High Temperature Components (ECCC 2014), paper no. ECCC2014-87, Centro Sviluppo Materiali, Rome, Italy, 5–7 May 2014 (2014)

Gorash, Y., Altenbach, H., Lvov, G.: Modelling of high-temperature inelastic behaviour of the austenitic steel AISI type 316 using a continuum damage mechanics approach. J. Strain Anal. **47**(4), 229–243 (2012). doi:10.1177/0309324712440764

Gorash, Y., Dempster, W., Nicholls, W.D., Hamilton, R.: Leak tightness in safety valves: Structural and fluid dynamics analyses, microscopic studies and experimental setup. Report: Project N1a, Weir Advanced Eesearch Centre, University of Strathclyde, Glasgow, UK (2014)

Gorash, Y., Dempster, W., Nicholls, W.D., Hamilton, R.: Modelling of metal-to-metal seals in a pressure relief valve using advanced FE-analysis. In: de Hosson, J.T.M., Hadfield, M., Brebbia, C.A. (eds.) WIT Transactions on Engineering Sciences, vol. 91, pp. 247–258. WIT Press, Southampton (2015). doi:10.2495/SECM150221

Haruyama, S., Nurhadiyanto, D., Choiron, M.A., Kaminishi, K.: Influence of surface roughness on leakage of new metal gasket. Int. J. Press. Vessel. Pip. **111**, 146–154 (2013). doi:10.1016/j.ijpvp.2013.06.004

Hyde, C.J., Sun, W., Leen, S.B.: Cyclic thermo-mechanical material modelling and testing of 316 stainless steel. Int. J. Press. Vess. Pip. **87**(6), 365–372 (2010). doi:10.1016/j.ijpvp.2010.03.007

Jackson, R.L., Streator, J.L.: A multi-scale model for contact between rough surfaces. Wear **261**(11), 1337–1347 (2006). doi:10.1016/j.wear.2006.03.015

Kemet International Limited. How to measure flatness – Technical article. www.kemet.co.uk/blog/lapping/how-to-measure-flatness-technical-article/ (2015). Accessed 11 Jan 2016

Ledoux, Y., Lasseux, D., Favreliere, H., Samper, S., Grandjean, J.: On the dependence of static flat seal efficiency to surface defects. Int. J. Press. Vess. Pip. **88**(11–12), 518–529 (2011). doi:10.1016/j.ijpvp.2011.06.002

Man, J., Zhou, Q., Tao, Z., Zhang, Y., An, Q.: Micro-scale numerical simulation on metal contact seal. Proc. IMechE, Part C: J. Mech. Eng. Sci. **228**(12), 2168–2177 (2014). doi:10.1177/0954406213515644

Marie, C., Lasseux, D.: Experimental leak-rate measurement through a static metal seal. J. Fluids Eng. **129**(6), 799–805 (2007). doi:10.1115/1.2734250

Megalingam, A., Mayuram, M.: Elastic-plastic contact analysis of single layer solid rough surface model using fem. Int. J. Mech. Aerosp. Ind. Mechatron. Manuf. Eng. **6**(1), 133–137 (2012)

Mitchell, L., Rowe, M.: Influence of asperity deformation mode on gas leakage between contacting surfaces. J. Mech. Eng. Sci. **11**(5), 534–549 (1969). doi:10.1243/JMES_JOUR_1969_011_065_02

Müller, H.K., Nau, B.S.: Fluid Sealing Technology: Principles and Applications. Marcel Dekker Inc, New York (1998)

O'Callaghan, P., Probert, S.: Prediction and measurement of true areas of contact between solids. Wear **120**(1), 29–49 (1987). doi:10.1016/0043-1648(87)90131-1

Pérez-Ràfols, F., Larsson, R., Almqvist, A.: Modelling of leakage on metal-to-metal seals. Tribol. Int. **94**, 421–427 (2016). doi:10.1016/j.triboint.2015.10.003

Ritchie, G.: Minimizing pressure relief valve seat leakage through optimization of design parameters. B.Sc. thesis, Dept. of Mechanical Engineering, MIT, Massachusetts, USA (1989)

Robbe-Valloire, F., Prat, M.: A model for face-turned surface microgeometry: application to the analysis of metallic static seals. Wear **264**(11), 980–989 (2008). doi:10.1016/j.wear.2007.08.001

Robbe-Valloire, F., Paffoni, B., Progri, R.: Load transmission by elastic, elasto-plastic or fully plastic deformation of rough interface asperities. Mech. Mater. **33**(11), 617–633 (2001). doi:10.1016/S0167-6636(01)00074-6

Singh, A., Bernstein, M.D. (eds.) Testing and Analysis of Safety/Relief Valve Performance. In: Proceedings of ASME conferences. United Engineering Center, ASME, New York, USA (1983)

Smith, E., Vivian, B.E.: An Introductory Guide to Valve Selection. Introductory Guide Series (REP). Wiley, New York (1995)

Spirax Sarco. Steam engineering tutorials: Introduction to safety valves. www.spiraxsarco.com/Resources/Pages/Steam-Engineering-Tutorials/safety-valves/introduction-to-safety-valves.aspx (2016). Accessed 05 Jan 2016

Thompson, M.K.: Geometric primitive surface roughness in finite element models. In: 2007 Annual Meeting for the Society of Tribologists and Lubrication Engineers, Philadelphia, PA (2007a)

Thompson, M.K.: A multi-scale iterative approach for finite element modeling of thermal contact resistance. Ph.D. thesis, Dept. of Mechanical Engineering, Massachusetts Institute of Technology, Cambridge, USA (2007b)

Thompson, M.K.: A comparison of methods to evaluate the behavior of finite element models with rough surfaces. Scanning **33**(5), 353–369 (2011). doi:10.1002/sca.20252

Thompson, M.K., Thompson, J.M.: Considerations for the incorporation of measured surfaces in finite element models. Scanning **32**(4), 183–198 (2010a). doi:10.1002/sca.20180

Thompson, M.K., Thompson, J.M.: Methods for generating probabilistic rough surfaces in ANSYS. In: Proceedings of 20th Korea ANSYS User's Conference, ANSYS Inc., Gyeongju (Sep. 9–10), South Korea (2010b)

Tsukizoe, T., Hisakado, T.: On the mechanism of contact between metal surfaces - the penetrating depth and the average clearance. J. Basic Eng. **87**(3), 666–672 (1965). doi:10.1115/1.3650635

University of Cambridge. Teaching packages: Slip line field theory. www.doitpoms.ac.uk/tlplib/metal-forming-3/slip_line_field.php (2004). Accessed 13 Oct 2014

Uppal, A., Probert, S.: Deformation of single and multiple asperities on metal surfaces. Wear **20**(3), 381–400 (1972). doi:10.1016/0043-1648(72)90417-6

Vallet, C., Lasseux, D., Sainsot, P., Zahouani, H.: Real versus synthesized fractal surfaces: Contact mechanics and transport properties. Tribol. Int. **42**(2), 250–259 (2009). doi:10.1016/j.triboint.2008.06.005

Zappe, R.W.: Valve Selection Handbook. Gulf Professional Publishing, Oxford (2004)

Against the Fragmentation of Knowledge: The Power of Multidisciplinary Research for the Design of Metamaterials

Francesco dell'Isola, Sara Bucci and Antonio Battista

Abstract The new possibilities arisen in the last years in material manufacturing (3D-printing, electrospinning, roll-to-roll processing, self-assembly, etc.) and the theoretical tools made available by generalized continuum mechanics are still far from achieving their full potential. The main thesis of the present paper is that it is necessary a *multidisciplinary* approach to address the emerging issues in metamaterials' design. Therefore, an improvement in the degree and the depth of the cooperation between scientists from different areas is required. The advancements needed in mechanics and physics of solids and fluids, mathematical and numerical modeling and advanced technology in material construction can be obtained only as a consequence of a synergic effort.

1 Introduction

Technology and hard sciences have always developed in a close and parallel relation. Indeed, a driving force for science to rediscuss the current paradigms, during all History, has been the advancement of new technological possibilities, which allow for new phenomenological evidence to arise. What proves the actual success of the conceptual revolutions connected with the birth of Mechanics, of Thermodynamics

F. dell'Isola (✉) · S. Bucci · A. Battista
MeMoCS, International Research Center for the Mathematics & Mechanics of Complex Systems, Università dell'Aquila, L'Aquila, Italy
e-mail: francesco.dellisola@uniroma1.it

F. dell'Isola
La Sapienza University of Rome, Rome, Italy

S. Bucci
Otto-Von-Guericke-Univerity Magdeburg, Magdeburg, Germany
e-mail: sara.bucci@ovgu.de

A. Battista
LaSIE, FRE-CNRS 3474, Université de La Rochelle, La Rochelle, France
e-mail: antonio.battista@studenti.unina.it

K. Naumenko and M. Aßmus (eds.), *Advanced Methods of Continuum Mechanics for Materials and Structures*, Advanced Structured Materials 60,
DOI 10.1007/978-981-10-0959-4_28

and Electromagnetism, is that those new theories and models, created with the aim of designing and describing new technological advancements, were in the end point embodied in the what has been called *classical physics*. In the opinion of the authors, a similar change of paradigm is going to be experienced by mechanics in the incoming future. However, the new technological possibilities, which permit to produce and control the properties of material both at the micro- and nanoscales, need substantial progresses from a theoretical point of view. It is evident that the new manufacturing techniques, such as electrospinning, 3D-printing, self-assembly, and so on, which have been developed in the recent years, enforce us to reanalyze some of our ideas concerning theoretical mechanics and above all, on the relation existing between it and technology. Indeed, using such techniques, we are able to produce objects which present an exotic and peculiar behavior from a *classical* point of view.

The key challenge, today, is therefore not only to be able to predict the behavior of already existing (possibly advanced) materials, but also to succeed in prescribing the right constitutive and geometric characteristics at the microscale in order to get a certain (even exotic) behavior at the macroscale. Among many paths we can imagine for future developments, the following ones appear critical with respect to the aforementioned technological innovations:

- the development of an improved theoretical framework for generalized continua based on a systematic assumption of the variational approach;
- the development of suitable homogenization techniques allowing for the determination of reliable macroscopic models based on the given micro-/nanostructure and on the physical properties of the considered materials;
- the development of robust and flexible numerical methods to perform effective simulations of the proposed models;
- a sound basis of experimental evidence, to be developed in close connection with the previously mentioned theoretical knowledge and understanding;
- the concrete realization of proofs of concept, constituted by prototypes of new advanced architectured materials.

Therefore the creation of networks,[1] *which allow the interaction of scientists specialized in each of the aforementioned categories, is an unavoidable step if the scientific community wants to successfully meet the challenge.* This idea is not a novelty,[2] of course, but in the opinion of the authors it has to be reexamined in order to promote actual interaction between different fields and not, as often is the case, mere union of the respective results. Indeed, joining different specializations under a unified line of research having a common goal should exploit the highly specialized knowledge currently existing, avoiding that diversification could become fragmentation. In particular, it should prevent the danger of developing just parallel and independent

[1]An example is the recently established M&MoCS International Research Center, see http://memocs.univaq.it/?lang=en.

[2]Indeed the establishment of such networks can be traced back to Hellenistic Science (see Russo et al. 2013). In the *Mouseion* at Alexandria, experts from all disciplines (geometers, physicists, mechanicians, physicians, grammars) were all working together, attacking the same problem from all available point of view.

investigations that, in themselves, are not capable to effectively attack complex problems. We remark that in this connection, it is, in our opinion, also critical the involvement of younger generations of researchers, so as to create, since the very beginning of higher level education, a sort of constructive dependence between students coming from different paths of learning and experienced scientists.

2 Multidisciplinary Nature of Metamaterial Research

When trying to graphically represent the properties of (natural and artificial) materials, one finds that all possible charts matching relevant mechanical and physical properties are made by zones occupied by actually existing materials, as well as by very large empty areas. Some of these areas will remain empty for long because, as far as we understand, the materials there lying would be impossible for fundamental reasons, but others are simply empty even though, in principle, they could be filled by means of a suitable exploitation of the technological possibilities (Ashby 2013). Compared even to few years ago, today the situation has changed so abruptly that the more important and challenging question is related to what can be seen as an *inverse problem*: given a continuum model, which are those mechanical systems that, at a certain length scale, behave as specified by the chosen continuum model? In setting the question in this way, the key point is to understand the microscopic properties of such systems to obtain information on how to realize them (dell'Isola et al. (2015f), Bouaziz et al. (2008), Gibson and Ashby (1997), Brechet and Embury (2013), Fleck et al. (2010)). If the challenge of a multidisciplinary approach is accepted, it will be capable to promote a quantum leap in the exploitation of the technological possibilities already potentially existing, and to drive the advanced manufacturing technology toward the most promising further developments. The new level of ambition in the requirements of peculiar multifunctional properties often create a bottleneck which cannot be overcome by means of traditional materials. Manufacturing materials with the degree of freedom and precision allowed by techniques like 3D printing and electrospinning, associating them, and suitably selecting their geometries, represent an innovative strategy to meet the newly arisen (and yet to come) engineering challenges.

Summarizing the following two points are essential in order to achieve further progresses:

- to establish long-term links among research groups from different areas;
- to avoid the fragmentation of skills and knowledge, i.e., a situation in which the team work is performed without much exchange of information between different specialists.

In other words, multidisciplinarity has to play, in our context, a major role, even more relevant than in the generality of technological and scientific research. Indeed, in the opinion of the authors, all the research directions have to systematically interact:

mathematicians need, in order to supply the rigorous results necessary for numerical computation, to be informed about the conjectures formulated by modelers; physicists and engineers must interact to integrate the capability of understanding phenomena in order to be able to solve more efficiently practical problems; experimentalists must be guided by theoretical knowledge if very unusual phenomena must be discovered and exploited.

Specifically, the networks we have in mind should favour:

- the exchange between physicists and engineers in order to improve the state-of-the-art capability to design application-tailored materials;
- the interaction between mathematicians and more applied scientists in order to direct the theoretical investigation toward more applicable problems and in particular toward the formulation of highly predictive numerical tools;
- the joint work of theoretical mechanicians, physicists, and mathematicians in order to formulate mathematical models capable to drive the design and realization of newly conceived architectured materials, based on complex microstructures and multiphysics/multiscale phenomena;
- the interaction of mathematicians with experienced numerical analysts in order to improve the capability of tailoring macroscopic homogenized models to the description of microscopic complexity (up to nanolevel);
- the collaboration for the study of scientific and technological problems involved in mechanics of natural and artificial tissues and in electromagnetic action on architectured materials together with its possible application for health protection.

We believe that, in this interaction, a special role has to be played by the coordinating power of experienced components and by the capabilities and the willingness of junior components. This will, in our opinion, maximize the chances of developing sharp and ground-breaking solutions in the direction of the concrete realization of new technological application of architectured materials.

3 Some Aspects of the Implementation of a Multidisciplinary Research Work

A long way has been covered by material technology since its beginning (dating back to non-Sapiens hominids). The basic steps of this adventurous travel can be summarized as Brechet and Embury (2013):

1. using the materials available on site (e.g., native metals, bone, or wood);
2. gradual evolution toward the optimization of specific classes of materials on an empirical basis (e.g., development of empirical metallurgy techniques);
3. science-based approaches (e.g., scientific metallurgy and later polymer science etc.);
4. what can be called *hyperchoice of materials*, i.e., the development of scientific methods and tools for comparing and selecting materials coming from different

classes which, individually considered, were already optimized for a specific set of engineering applications;
5. search for multifunctionality of materials, with increasingly ambitious requirements for materials capable of fulfilling conflicting needs.

The importance of theoretical modeling has of course grown very much passing from (1) to (4), and today the demand for multifunctionality is such that the already available theoretical models are not suitable anymore for the full exploitation of the technical possibilities, nor for meeting the needs coming from industry (Bréchet 2000; Fleck et al. 2010). Material scientists have developed an impressive and very specialized body of knowledge, while theoretical mechanicians have pushed the study of classical mechanical models close to the theoretical limit of their potential, and sound nonclassical theoretical frameworks have been developed concerning microstructured/micromorphic media (Green and Naghdi 1995; Masiani et al. 1995; Neff and Forest 2007; Altenbach and Eremeyev 2009; Carcaterra et al. 2015; Federico and Grillo 2012), some of which can by now be considered as *classics* (Eringen 1968; Germain 1973). The developments in numerical analysis in case of micro- and even nanosystems are by now remarkable. However, the interaction between the two aforementioned areas is still not as intensive and fruitful as it can be. The proposed point of view about the modeling of architectured materials and metamaterials by means of suitably reformulated generalized continuum theories is capable, in our opinion, to make it seem even obsolete a clear-cut distinction between the two fields.

Mathematical modeling of materials has been developed in the nineteenth century on the basis of some reasonable and well-grounded assumptions, which are verified by the majority of natural materials and by the great majority of the materials used up to now in engineering. Some natural materials which show sophisticated and often unexpected behaviors are those living tissues produced by natural Darwinian selection whose microstructure:

- is very complex;
- exhibits multiple characteristic length scales;
- involves coupled multiphysics phenomena;
- shows strongly inhomogeneous physical properties at every characteristic length (Dunlop and Fratzl 2013; Tomic et al. 2014).

It is clear that the assumptions used in classical physics for describing mechanical behavior are not anymore suitable when one wants to model living tissues or when one wants to design and build exotic artificial materials tailored to high-performance technological (possibly biomedical) applications, e.g., for the replacement of natural tissues and for providing protection against externally induced damage for living organisms.

It is clear that some well-established *classical* concepts may have to be rethought. For instance, the same concept of stress, strain, local and contact interaction, deformation energy, balance equations, constitutive equations, yield and damage criteria need to be reformulated in some respects, in a context never explored before. To this aim the competences of mathematicians, physicists, numerical analysts, theoretical and applied mechanicians need to be joint to obtain a more performing and predictive Weltanschauung, i.e., a vision of the physical world by means of which physicists and engineers can shape materials and their behaviors for the upcoming future. In other words we need to emancipate from well-established simplification assumptions and modeling hypotheses so as to lead to the development of new design procedures and solutions, as well as, of homogenization techniques under new microstructure assumptions, like high contrast and multiscale implying indeed higher gradient and/or microstructured continuum modeling. Classical mechanics, and in particular continuum mechanics, assumes usually (and rightly!) very simple hypotheses about the mechanical and physical behavior of materials, from which it is naive to expect a universal predictive power. However, a theoretical framework that in our opinion can still have an effective unifying power is the variational one, possibly improved by means of the Hamilton–Reyleigh dissipation mechanism.

4 Short-Term and Long-Term Scientific, Technological, and/or Socioeconomic Impacts

Up to now, one of the strongest features of European science has always been its capability of integrating toward one specific challenge, different competences and capabilities. Notwithstanding the increasing difficulties in supporting multidisciplinary research Europe seems to be able to keep its scientific leadership continuing this long-lasting tradition (which could be tracked up to the achievements of the Hellenistic scientists and in particular to Archimedes: a geometer whose theoretical knowledge produced impressive technological applications). Technology cannot advance without the nourishment supplied by fundamental science, fundamental science finds its ultimate motivation and justification in technological applications. This is why, in our opinion, it is necessary to embody the just-stated principle while confronting a specific challenge: the development of theoretical and experimental tools needed to conceive, optimize, and build novel highly performing architectured materials.

Standard methods, optimization routines, and already existing finite-element analyses could be used to effectively refine a given architectured material once its general constitutive and geometrical characteristics have been chosen. However, what the standard methods are not so good at is a reasonably quick scan of alternative combinations (Ashby 2013), which is of course crucial as the possibilities entailed by the new computer-guided manufacturing techniques are virtually infinite and cannot be tried extensively. As an example, we can consider what are called hybrid materials. The equivalent properties of hybrid materials lie on a trajectory (in the space of possible materials) with end points at the materials that are combined to make them.

A suitable theoretical model should provide a good prevision about the shape of the trajectories. A particular care has to be paid in order to specify the physical meaning of the portions of these curves where small changes of one parameter are associated to great variations of the other parameters: the singularity phenomena arising in these circumstances can indeed be source of very interesting effects, as well as instabilities of even "difficult" types (Luongo 2010; Di Egidio et al. 2007; Luongo 2015). Among the short-term targets, there should be also the concrete realization of prototypes of architectured materials manufactured by means of the technological possibilities displayed by the experimental researchers involved, which should be capable to perform advanced 3D-printing and electrospinning. The focus of the networks should be even more strongly directed toward long-term targets. In this respect, the goal should be to replace the *incremental* character that the advancements in material technology have experienced in the last years with a quicker and sharper, step-like evolution function. We believe that the synergistic effort by theoretical mechanicians, applied mathematicians, numerical analysts, and experimentalists can indeed produce the aforementioned conceptual revolution in continuum mechanics and material science. As for the socioeconomic impact, the achievement of the aforementioned objectives will have an inestimable value for industry and also potentially for environmental issues, as the realization of lighter and not overdimensioned objects will relevantly affect power consumption in manufacturing and shipping. Last, but not least, the materials which we have in mind may play a relevant role in biomechanics and medicine. This feature is to us so valuable that an upper estimate of its potential socioeconomic impact looks simply impossible.

5 A Closer Look at How to Face the Problems Involved

In the present section, we propose a possible way to rationalize the team work. We selected eight main fields of research (FOR), each of which should be involved both for specific tasks and for interacting with the others:

FOR1: Theoretical Continuum Mechanics and Variational Approach

The construction of the general theoretical framework for the description and the prevision of the behavior of advanced architectured materials is probably, as mentioned, the soundest possible ground for exploring the exotic phenomena we have in mind. In particular, they have to extend and generalize the already existing higher gradient and micromorphic models; moreover they have to drive the work of numerical analysts. The researchers should provide assessment about the realization of theoretical models for complex metamaterials showing a good match with both numerical simulations and experimental results. A detailed coverage of the scientific novelties introduced in the modeling have to be clear, offering a general understanding and a unifying perspective relating the proposed theoretical findings with their technological potential and their contribution to the general progress of science.

In order to model architectured materials (i.e., metamaterials in the sense specified in Del Vescovo and Giorgio 2014; a rational collection of recent results is Placidi et al. 2015b), the variational approach is very effective and allows us to obtain well-posed problems with the minimum constitutive assumptions possible (see, e.g., Auffray et al. 2015; dell'Isola et al. 2015a, e, 2016a; dell'Isola and Placidi 2012; Cuomo et al. 2014). Architectured materials can be classified and interpreted as microstructured materials. Particular examples of these materials are for instance micropolar materials (see, e.g., Altenbach and Eremeyev 2009, 2014; Pietraszkiewicz and Eremeyev 2009a, b) but also higher gradient materials (see, e.g., Placidi et al. 2015a; Auffray et al. 2015; dell'Isola et al. 2015a, 2016a) can be seen as a particular case of micromorphic models.

Sometimes the complexity of the medium can be modeled in the framework of mixture theory when different phases of some constituents coexist (see, e.g., Placidi and Hutter 2005, 2006; Andreaus et al. 2014; Giorgio et al. 2015).

A very delicate issue in the study of complex materials is, finally, the damage detection and the characterization of the evolution of cracks inside them (see, e.g., Andreaus and Baragatti 2011, 2009; Placidi 2015, 2014; Thiagarajan and Misra 2004; Rinaldi and Placidi 2014).

FOR2: Advanced Homogenization Techniques

The most suitable continuous (i.e., macroscopic) limit, for a wide class of discrete mechanical system characterized by a high degree of complexity in the microstructure, have to be provided if one wants to fasten the numerical investigation and, as a consequence, the prototyping process. Looking for the correct micro-macro identification, for every conceived microstructured metamaterial, and driving the numerical work toward the most promising models of architectured materials, would be the main concern of researchers in this field. Their objectives have to focus on the description of a suitable continuous limit for:

- (micro) lattices characterized by multiple length scales;
- multilayered materials obtained combining in 3D structures the previous lattices

'Ad hoc' homogenization techniques should be conceived to deal with these particulars micro-macro identifications. Some results in such a direction can be found in Assidi et al. (2011), Carcaterra et al. (2015), dell'Isola et al. (2016b), Dos Reis and Ganghoffer (2010, 2012), Goda et al. (2012), Rahali et al. (2015).

FOR3: Mathematics of Nonlinearity

It is also essential to deepen the understanding of the complex nonlinear phenomena occurring when considering complex multiphysics systems (see, e.g., Javili et al. 2013; Piccardo et al. 2014). This can be done through the study of the instability resulting from the behavior of the considered systems. The ultimate goal should be to assess the degree of accuracy required in order to realistically expect a good fit between models, numerical analyses, and experimental results.

FOR4: Numerical Investigations

New numerical techniques should be developed for the investigation of novel architectured materials. Refined FEM schemes flexible enough to be able to consider even complex geometries and robust enough to work with high degrees of complexity in the microstructure should be also constructed.

The complexity of these newly conceived architectured material gives rise to new numerical issues related to different sources, as for instance the presence of strong heterogeneities resulting from the features of microstructure considered, viscous or independent-rate dissipation, internal resonances, and so forth (see, e.g., Aristodemo and Turco 1994; Turco 2001; Bilotta and Turco 2014; Della Corte et al. 2015; Misra et al. 2007 for some numerical tools on this regard). Besides, in many cases (when the standard Cauchy theory seems to be inadequate) the use of microstructured or higher gradient models may be appropriate to describe a greater variety of unusual behaviors. Recently, very powerful tools in numerical analysis have been developed in the framework of isogeometric analysis (see, e.g., Cazzani et al. 2014b, a, 2015; Greco and Cuomo 2013, 2014, 2016 for more details) in order to deal with the difficulties arisen from this type of models.

The complexity and heterogeneities of the microstructure could be, as already said, source of lack of stability in the materials considered. Typically examples of these phenomena can be observed, e.g., in cellular materials as ceramic or metallic foams as well as honeycomb composites. Usual tools, as perturbation analysis, may encounter problems because they require analytical solution as a reference solution that is often not available. Then in many cases these problems can only be faced with numerical methods (see, e.g., Di Egidio et al. 2007; Luongo and Piccardo 2005; Rizzi et al. 2013; Gabriele et al. 2012).

FOR5: Manufacturing

This research team should be aimed in building prototypes of architectured materials exhibiting strong multifunctionality.

They should extent the technological possibilities in building complex multi-physics systems involving a certain desired coupling between the components and compare the experimental results with the theoretical and numerical previsions in order to refine the models and to adjust the relative parametrization. At the end they should be able to describe the experimental results obtained in addition to the concrete realization of working prototypes and the application of patent for the produced prototypes.

With the spirit of synthesizing the material on the basis of a suitable constitutive assumption on the stored strain energy, some examples of pantographic structures (dell'Isola et al. 2015d, c, 2016b) can be interpreted as a particular carrying out of the fiber sheets described in dell'Isola and Steigmann (2015), Steigmann and Dell'Isola (2015), D'Agostino et al. (2015). Other possibilities that can be explored are those based on a proper modification of an existing material in order to modify its mechanical properties. For instance the addition of some microfillers in a concrete matrix in order to improve the dynamic performances as done in Giorgio and Scerrato (2016), Scerrato et al. (2015, 2014).

FOR6: Mechanics of Discrete Systems

These researchers should try to characterize the discrete mechanical systems which will be selected as the most promising in view of obtaining the desired macroscopic properties and behaviors. They should therefore provide a detailed mechanical description of the micro-level of a large class of architectured materials, including for instance (micro) lattices characterized by multiple length scales, multilayered materials obtained combining in 3D structures the previous lattices and other structures displaying even more complex geometry at the microscale.

FOR7: Coupled Phenomena

The potentiality entailed by the conceivable coupling of the different constitutive elements of the considered multiphysics systems is still far from exhausted. A suitable description of the coupling in a certain class of specific cases is indeed still incomplete. Among the most relevant cases, in the opinion of the authors, there are piezo- and flexo-electromechanical systems thought as elementary components of complex structures, and in general nonlinear electroelastostatics. They will be the main object of the work of FOR5 and will allow for the realization of suitable prototypes of multiphysics architectured materials, exhibiting the desired coupling properties. Materials in which there is a coupling between mechanical and electric states (Lagrangian variables) are also of interest especially in the field of vibration control and noise attenuation. Piezo- or flexoelectric materials can be profitably employed for this purpose (see for some examples Andreaus et al. 2004; Giorgio et al. 2009; Enakoutsa et al. 2015).

FOR8: Engineering and Biomechanical Applications

As already said, finding solutions to already existing problems from engineering and biomedical areas by means of a suitable exploitation of the properties of the developed architectured materials is, in a sense, the main challenge in the field of metamaterials for both its difficulty and usefulness. The involved researchers should start from gathering the requirements from the industrial and biomedical world that can best fit with the results expected and obtained by the network, and communicate efficiently with the industrial sector in order to maximize the possibility of fruitful interaction. We will provide some specific example. The study of wind-excited structures (see, e.g., Pagnini 2010; Piccardo et al. 2015; Luongo and Piccardo 2005) has proven that a high strength–weight ratio under certain shape constraints, as well as targeted anisotropic behaviors and piezo- and flexoelectric induced damping, are of great importance for improving the reliability. It is clear, therefore, that metamaterials can have a great potential impact in this regard. In bone reconstructive surgery, designing (from both mechanical and biological points of view) of suitable implants made of bioresorbable artificial material is an attractive challenge in order to guarantee a proper load-carrying capacity and a fast substitution of biomaterial with newly formed bone for health purposes (see, e.g., Lekszycki and dell'Isola 2012; Ancillao and Andreaus 2013; Giorgio et al. 2016, 2015; Andreaus et al. 2013, 2014). In this context, also the modeling of interaction of artificial material with soft tissue, as,

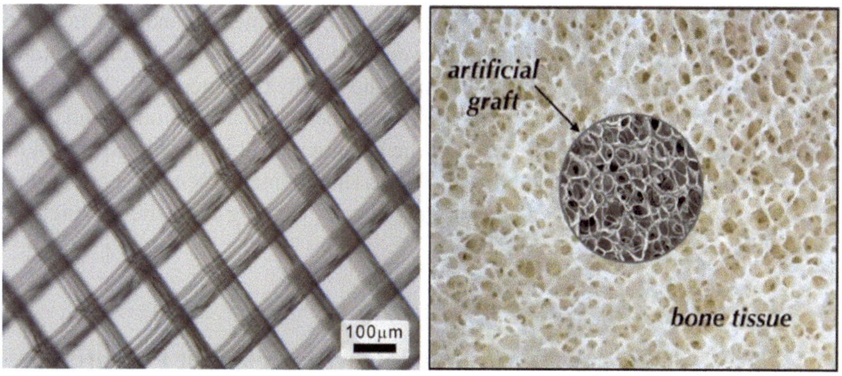

Fig. 1 *Left* an electrospun metamaterial made of polycaprolactone fibers, characterized by extremely slow degradation, and thus suitable for biomedical applications. *Right* the inclusion of a *circular shaped* artificial graft in a bone tissue (scheme). Living bone tissue equipped with an artificial graft can be regarded as a metamaterial whose configuration is described by a displacement field and by an additional kinematical descriptor (e.g., the change of the porosity with respect to the reference configuration)

e.g., cartilage, could be useful (for more details see, e.g., Tomic et al. 2014; Federico and Grillo 2012).

As it is well known that the remodeling process in bone is strictly related to the frequency of the external load. Therefore, the dynamic properties of bones play a key role in the bone functional adaptation. For this reason, a modal analysis should be performed in order to understand how these features changes in the remodeling process and are influenced by external mechanical factors (Fig. 1).

5.1 Risk Assessment and Management

As already mentioned, specific risks are involved in a challenging research activity. In general, the main measure to avoid major problems is closely related with the multidisciplinary approach we propose, since the diversification of the employed approaches makes the team work more robust. For instance, in order to face the possibly enormous difficulties in finding appropriate homogenization methods, the network has to select extremely skilled mathematician which should have, moreover, diversified capabilities and competences within the same theoretical framework constituted by homogenization techniques. In particular, both a *functional* approach (employing gamma- convergence and two-scale convergence methods as their work tools) and a *differential* approach can be carried on, together with the clever employment of formal asymptotic expansion in order to quickly get precious information about the desired continuous limits. To meet the challenge proposed by computational complexity of the considered microstructured materials, different FEMs can be developed and a constant communication between numerical analysts and theoreticians should be in place, which has proven to be so far the best possible insurance against potentially meaningless distortion of the results (Ebinger et al. 2005; Trinh et al. 2012). To avoid the onset of large gaps between theoretical models and experimental results due to a big difference in the accuracy level required, experts in the study of nonlinearity and of instability problems should finally focus their attention to these exact issues, following and developing already existing sophisticated methods (Feyel 2003; Verhoosel et al. 2011).

6 Some Examples of Metamaterials with Relevant Potential and Open Questions

6.1 A Quick Look to Parabolic Pantographs

In this section, we will consider two examples of possible metamaterials which are currently under the attention of several researchers for their potentially advantageous characteristics. The first example can be provided by the just mentioned pantographic structures, i.e., a fabric with two families of orthogonal fibers. Herein we will consider a particular geometry for the fibers, in which they are arranged in parabolic curves. The disposition of the fibers, different from the straight ones, is conceived in order to have a greater rigidity for the same weight, but obviously entails greater difficulties and variety of behaviors at both micro- and macroscale. The reason behind the advantageous strength–weight ratio is probably connected to some kind of arch-like response due to the geometry of the fibers, but the exact mathematical formulation of this qualitative behavior is one of the basic theoretical questions to be addressed. In (Fig. 2) we show straight fibers pantographic structures.

Figure 3 shows the current configuration of a rectangular fabric with aspect ratio 6:1 in a standard bias extension test. As usual the plot of the shear strain relative to the initial fiber axes exhibits three zones in which this deformation is almost unchanged

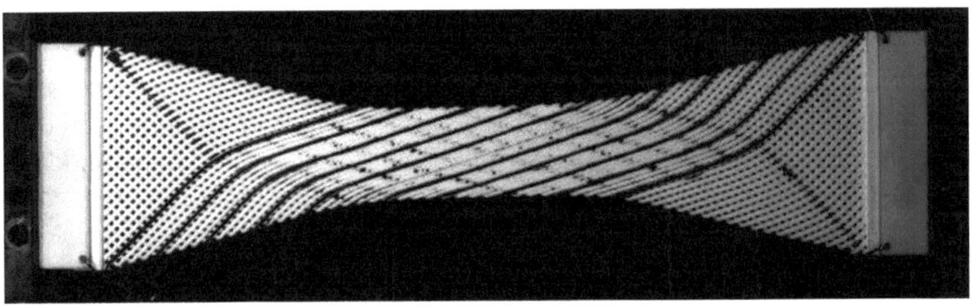

Fig. 2 An architectured material which is lightweight, extremely resistant, and safe in failure. Pantographic sheet under extensional bias test (see dell'Isola et al. 2016b)

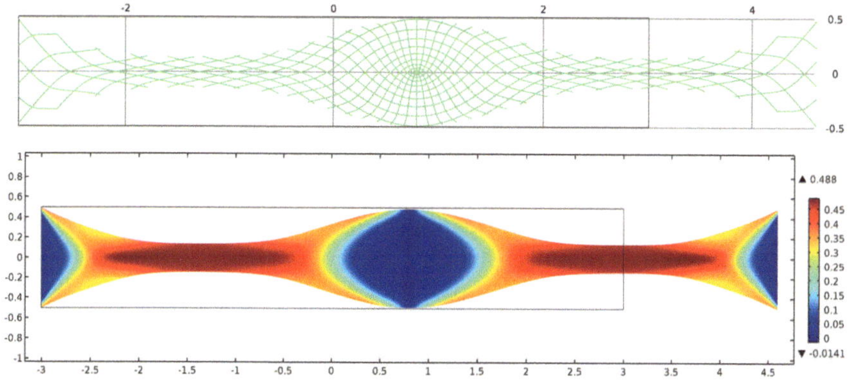

Fig. 3 Bias extension test—aspect ratio of the rectangular lattice 6:1. *Top* Fiber pattern; *bottom* Shear strain relative to the initial fiber axes

with the presence of narrow transition zones due to the presence of a second gradient energy in the model employed to describe bending deformation of fibers (Steigmann and Dell'Isola 2015). Figure 4 shows for a sample of ratio 3:1 a generalized bias test in which there is also an out-of-plane twist deformation imposed on one short edge. The particular arrangement of the fibers, in this case, may result in both out-of-plane and in-plane buckling phenomena if the displacement imposed is beyond a critical value.

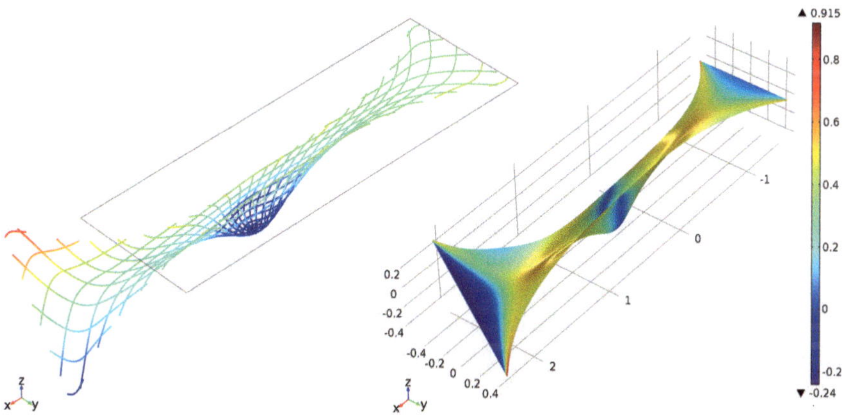

Fig. 4 Test with stretching and twist (45°) (aspect ratio of the rectangular lattice 3:1): Equilibrium shape; the shear strain relative to the initial fiber axes

6.2 Some Preliminary Results on Elastic-Plastic Honeycombs

Another example of metamaterial will be discussed in slightly greater detail, in order to show in a particular case some of the issues discussed before in general terms. Let us, therefore, consider the so-called honeycomb structures.

Honeycomb structures (from now on we will abbreviate the name to honeycombs) are solids with a periodic cellular body which confers them the property of being at the same time very light but still extremely stress resistant. They are produced mainly through two different manufacturing processes: the most used one is the method of expansion, in which the sheets of metal are glued together through binding strips and then expanded, while the less, but still widely, used is the corrugation method (which may be seen as a particularly simple version of roll-to-roll processing), in which the sheets are first deformed in the shape of half-hexagons and then glued together. These structures have been extensively studied and are clearly bio-inspired. The peculiar properties of natural honeycombs were already remarked by Hellenistic scientists (see Russo et al. 2013) and it has been conjectured that one of the first optimization problems was practically solved by bees finding the regular polygon maximizing surface–perimeter ratio. We believe however that there is a lot of room left for technological innovation: inhomogeneities in the microstructure, introduction of active or semi-active components, the addition of composite microscopic substructure are all examples of potentially very fruitful structural modifications which may induce exotic macroscopic behaviors.

Because of their peculiarity, honeycombs are largely used in packaging industries, as well as in computers or electronic components, and among all in aerospace industries or high speed cars. Many authors, such as Gibson and Ashby (1997), Papka and Kyriakides (1998), already attacked the problem of modeling the behavior of such structures with classical balance laws of mechanics, and presented many experimental and numerical results. Here we will show some preliminary results of a wider plan of investigation, approaching the 2D, plane, elastic-plastic problem, trying to focus on the identification of a good representative elementary cell which we will use to perform numerical simulations. From these, we will obtain the stress–strain curves which we will need in future to extrapolate the information necessary to create a mathematical, 'nonclassical', model for the homogenized plastic behavior of such structures. As we will see, we will use a parametrization of the stress state which will allow us to catch in the best way the behavior of honeycombs so as to be able, in future investigations, to extend these results to other periodic structures and cellular solids. We would like to remark the importance of cellular solids in nowadays engineering perspective, since their particular structures enable to minimize the costs of production (a very small amount of material is needed to build them), still giving them the properties required from the final industrial scope. Moreover, studying structures, such as honeycombs, foams, wood, cancellous bones, we will be able to modify them in order to optimize their properties and even to create new ones (thanks to the advanced techniques of 3D printing or electrospinning) with exotic behaviors, matching the demand of industries.

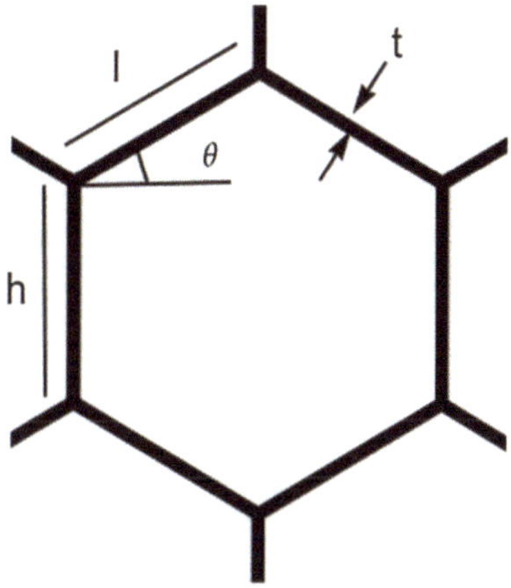

Fig. 5 Hexagonal unitary cell

A preliminary analysis can be performed on the mechanical properties of the core material. Here we will consider Aluminum-5052 (Young's modulus = 68.97 GPa, Poisson's ratio = 0.3, yield strength = 292 MPa). For such a material we consider an elastic-perfectly plastic model. We consider a structure made of regular hexagons with $h = l = 1, t = 0.216$ and $\theta = \pi/6$ (see Fig. 5).

We perform a first simulation on the whole structure, that we fix at the bottom, and to which we impose a strain in the y direction and periodic boundary conditions (so as to have an average strain in x direction = 0). From the numerical result,

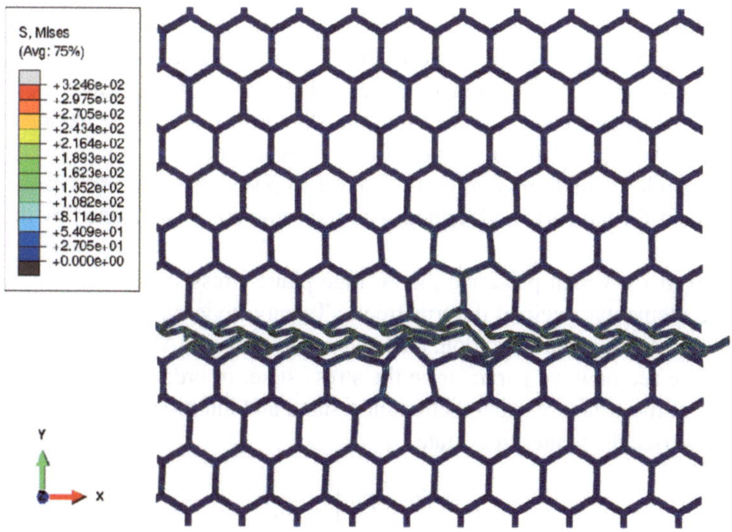

Fig. 6 Localization effect of plastic deformations

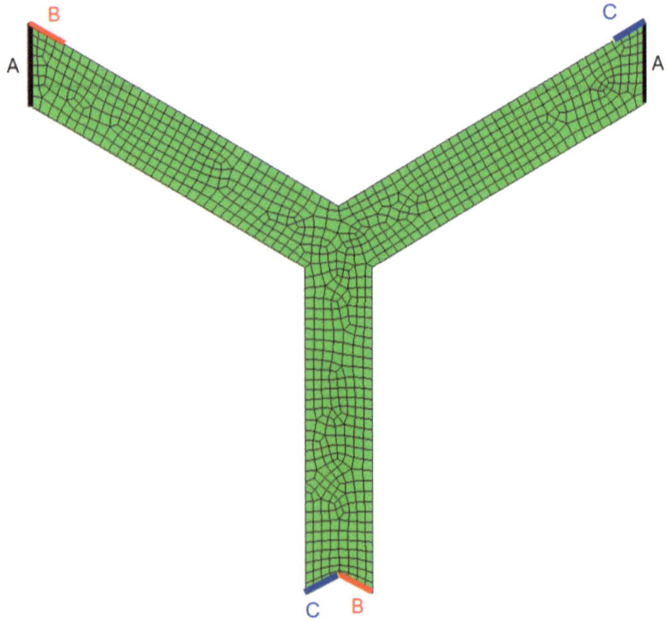

Fig. 7 Elementary cell with highlighted boundary conditions

shown in Fig. 6, we can already see the nontrivial behavior of the honeycombs, in the localization effect of the deformation. Indeed, after the initial elastic and local isotropic (Gibson and Ashby 1997) regime, the plastic deformation (which we initialised through a small concentrated pressure), is localized in rows adjacent one to each other.

This simple result already suggests that a generalized continuum approach is probably suitable in order to model the plastic behavior of the structure.

As an elementary cell, we select the one shown in Fig. 7. Its behavior may be extended to the macrostructure, if periodicity is assumed.

The periodic boundary conditions that we imposed are also shown: the same displacement for the parts highlighted with the same color and named by the same letter is set.

Biaxial loads, which prescribe the average plane stress state, is enforced, so as to let the elementary cell to deform freely. Twenty-node quadratic isoparametric elements, with reduced integration, are used in the program ABAQUS.

Next, we see how we parametrize the stress state, in order to keep in mind the geometry of the structure and to better understand and interpret the obtained results. We parametrize the plane stress state

$$\mathbb{T} = \begin{bmatrix} T_{11} & T_{12} \\ T_{12} & T_{22} \end{bmatrix} \mathbf{e_i} \otimes \mathbf{e_j} \,,$$

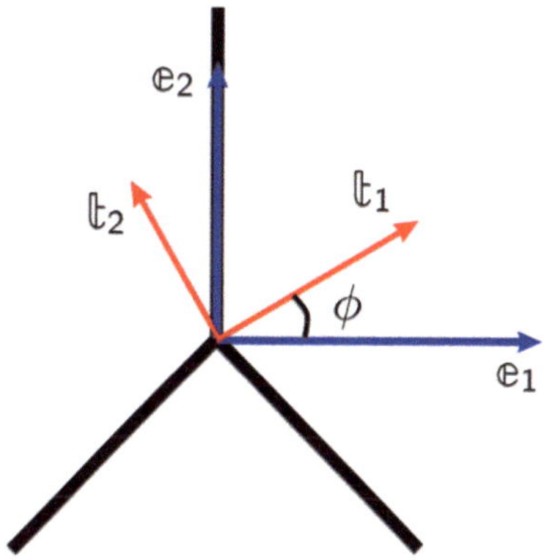

Fig. 8 In black the structure; in *blue* the standard base vectors; in *red* the ones which we use for the parametrization

trough:

- *orientation angle* ϕ
- *magnitude of the stress* $m = \sqrt{\lambda_1^2 + \lambda_2^2}$
- *biaxiality measure* χ s.t. $\lambda_1 = \cos(\chi), \lambda_2 = \sin(\chi)$

Figure 8 may be useful to visualize the parametrization.

For the sake of simplicity we will consider only three kinds of tests:

- compression–compression test, $\lambda_1 < 0, \lambda_2 < 0$
- tension–tension test, $\lambda_1 > 0, \lambda_2 > 0$
- tension–compression test, $\lambda_1 > 0, \lambda_2 < 0$

Moreover, we will show the results for only 21 angles ϕ and 6 angles χ.

The power of using the elementary cell stands behind the fact that the time of computation is drastically reduced and that it allows us to really visualize which are the deformation experienced locally, at the microscale level, by the whole structure as we can see from Fig. 9.

Apart from the visualization argument, what is important are the data that we can extrapolate from each of such simulations. For example, particularly relevant are the stress–strain curves for this basic cases. We can see them in Fig. 10. Only small deformations are considered and on the axes the norms of the linear strain tensor and Cauchy stress tensor are reported. We can clearly distinguish the 6 groups of simulations for each type of test, for the values that χ assumes (notice that $\chi = 0$ is a simple uniaxial test while $\chi = \pi/4$ is a 'isotropic' biaxial test), and the 21 angles per each value of χ, that ϕ assumes. It is already possible to extract some conclusions:

- increasing χ (which means increasing the 'biaxiality' of the tests), increases the stiffness of the response;
- tension–compression tests are, in general, the weakest, tension–tension ones the strongest and compression–compression are in between the former two;
- the isotropy of the elastic regime is visible by the fact that the response for different values of the variables tend to coincide;

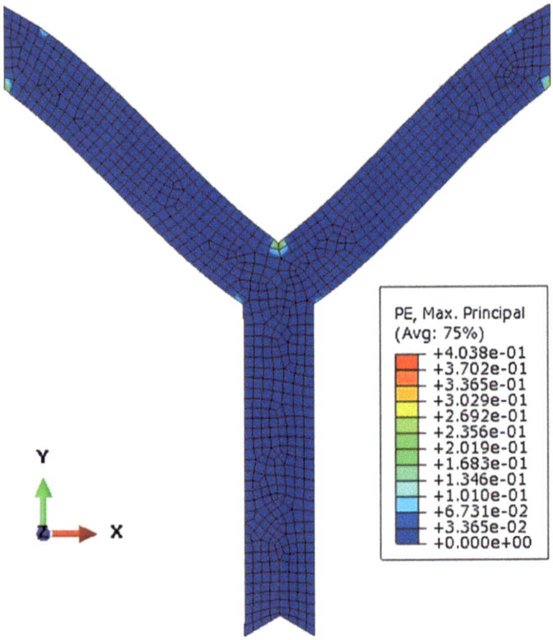

Fig. 9 Final configuration of a compression–compression test

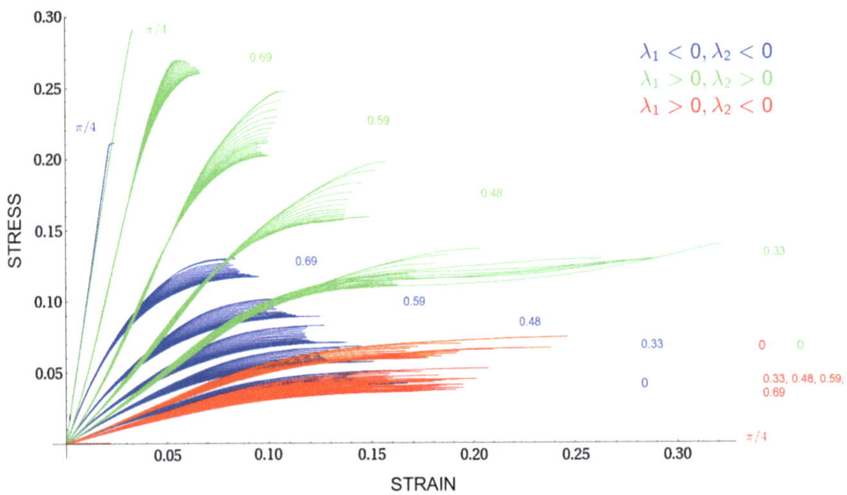

Fig. 10 Stress–strain response of all tests: in *blue* compression–compression; in *green* tension–tension; in *red* tension–compression. Tests reported for different values of χ

- the anisotropy emerges only within the nonlinear elastic part and it is more evident in plasticity

Finally, we would like to point out the unusual nonconvex behavior of such curves, which may be of interest for further investigation with nonlinear theory.

To sum up, we found a candidate elementary cell which should be validated by means of future investigations so as to make sure that it is capable to describe well the behavior of the macrostructure. Obviously these are preliminary results, which will be used in future to investigate the yield limit of such structures (i.e., the limit strength which they can sustain before undergoing plastic deformations) and the following plastic behavior of honeycombs. This last purpose is still challenging if one wants to face it with a nonclassical approach, introducing second gradient theories, which may provide a better description, capable of capturing the effect of localization and therefore predict the collapsing modes of such structures.

7 Conclusions. the Leader's Role: An Eye Kept on Past, Present and Future

Due attention should finally be paid to global research management issues. The researchers should completely agree to work together, toward the achievement of the aforementioned objectives, and a harmonious and fruitful cooperation is simply necessary if good results are expected. As already mentioned, the leading figures in the networks should coordinate and drive the research work, providing the overall strategy. The leaders of such a challenging scientific project have to be fully responsible for the actual realization, ensuring that every collaborating researcher always have a precise idea of the results obtained by the others and of the scientific work that they are expected to do. The network as a whole should always behave as a problem-oriented unity where different competences will coalesce to supply a timely, innovative, and potentially ground-breaking scientific understanding.

We expect that the proposed approach will emerge gradually as the only one really viable in order to make the so needed advancements in metamaterial science. Those researchers and groups that will understand the intrinsically multidisciplinary nature of the required skills, and will be open to enlarge their horizons to new problems ad techniques, will result as the winners in this fascinating competition.

References

Altenbach, H., Eremeyev, V.A.: On the linear theory of micropolar plates. ZAMM-Journal of Applied Mathematics and Mechanics/Zeitschrift für Angewandte Mathematik und Mechanik **89**(4), 242–256 (2009)

Altenbach, H., Eremeyev, V.A.: Vibration analysis of non-linear 6-parameter prestressed shells. Meccanica **49**(8), 1751–1761 (2014)

Ancillao, A., Andreaus, U.: Finite element analysis of the stress state produced by an orthodontic skeletal anchorage system based on miniscrews. J. Cranio-Maxillary Dis. **2**(1), 28 (2013)
Andreaus, U., Baragatti, P.: Fatigue crack growth, free vibrations, and breathing crack detection of aluminium alloy and steel beams. J. Strain Anal. Eng. Des. **44**(7), 595–608 (2009)
Andreaus, U., Baragatti, P.: Cracked beam identification by numerically analysing the nonlinear behaviour of the harmonically forced response. J. Sound Vib. **330**(4), 721–742 (2011)
Andreaus, U., dell'isola, F., Porfiri, M.: Piezoelectric passive distributed controllers for beam flexural vibrations. JVC/J. Vib. Control **10**(5), 625–659 (2004). doi:10.1177/1077546304038224
Andreaus, U., Giorgio, I., Lekszycki, T.: A 2-D continuum model of a mixture of bone tissue and bio-resorbable material for simulating mass density redistribution under load slowly variable in time. Zeitschrift für Angewandte Mathematik und Mechanik **13**, 7 (2013)
Andreaus, U., Giorgio, I., Madeo, A.: Modeling of the interaction between bone tissue and resorbable biomaterial as linear elastic materials with voids. Zeitschrift für angewandte Mathematik und Physik **66**(1), 209–237 (2014)
Aristodemo, M., Turco, E.: Boundary element discretization of plane elasticity and plate bending problems. Int. J. Numer. Methods Eng. **37**(6), 965–987 (1994)
Ashby, M.F.: Cellular Solids: Structure and Properties. Cambridge university press, Cambridge (1997)
Ashby, M.: Designing architectured materials. Scr. Mater. **68**(1), 4–7 (2013)
Assidi, M., Dos Reis, F., Ganghoffer, J.F.: Equivalent mechanical properties of biological membranes from lattice homogenization. J. Mech. Behav. Biomed. Mater. **4**(8), 1833–1845 (2011)
Auffray, N., dell'Isola, F., Eremeyev, V.A., Madeo, A., Rosi, G.: Analytical continuum mechanics à la Hamilton-Piola least action principle for second gradient continua and capillary fluids. Math. Mech. Solids **20**(4), 375–417 (2015)
Bilotta, A., Turco, E.: Numerical sensitivity analysis of corrosion detection. Math. Mech. Solids (2014). doi:10.1177/1081286514560093
Bouaziz, O., Brechet, Y., Embury, J.D.: Heterogeneous and architectured materials: a possible strategy for design of structural materials. Adv. Eng. Mater. **10**(1–2), 24–36 (2008)
Bréchet, Y.: Euromat 99, Microstructures, Mechanical Properties and Processes: Computer Simulation and Modelling. John Wiley & Sons, New York (2000)
Brechet, Y., Embury, J.D.: Architectured materials: expanding materials space. Scr. Mater. **68**(1), 1–3 (2013)
Carcaterra, A., dell'Isola, F., Esposito, R., Pulvirenti, M.: Macroscopic description of microscopically strongly inhomogenous systems: a mathematical basis for the synthesis of higher gradients metamaterials. Archive for Rational Mechanics and Analysis, pp. 1–24 (2015)
Cazzani, A., Malagù, M., Turco, E.: Isogeometric analysis: a powerful numerical tool for the elastic analysis of historical masonry arches. Continuum Mechanics and Thermodynamics, pp. 1–18 (2014a)
Cazzani, A., Malagù, M., Turco, E.: Isogeometric analysis of plane-curved beams. Math. Mech. Solids (2014b). doi:10.1177/1081286514531265
Cazzani, A., Malagù, M., Turco, E., Stochino, F.: Constitutive models for strongly curved beams in the frame of isogeometric analysis. Math. Mech. Solids (2015). doi:10.1177/1081286515577043
Cuomo, M., Contrafatto, L., Greco, L.: A variational model based on isogeometric interpolation for the analysis of cracked bodies. Int. J. Eng. Sci. **80**, 173–188 (2014)
D'Agostino, M.V., Giorgio, I., Greco, L., Madeo, A., Boisse, P.: Continuum and discrete models for structures including (quasi-) inextensible elasticae with a view to the design and modeling of composite reinforcements. Int. J. Solids Struct. **59**, 1–17 (2015)
Del Vescovo, D., Giorgio, I.: Dynamic problems for metamaterials: review of existing models and ideas for further research. Int. J. Eng. Sci. **80**, 153–172 (2014)
Della Corte, A., Battista, A., Dell'Isola, F.: Referential description of the evolution of a 2D swarm of robots interacting with the closer neighbors: perspectives of continuum modeling via higher gradient continua. Int. J. Non-Linear Mech. (2015). doi:10.1016/j.ijnonlinmec.2015.06.016

dell'Isola, F., Placidi, L.: Variational principles are a powerful tool also for formulating field theories. Variational Models and Methods in Solid and Fluid Mechanics. Springer Science & Business Media, New York (2012)

dell'Isola, F., Steigmann, D.: A two-dimensional gradient-elasticity theory for woven fabrics. J. Elast. **118**(1), 113–125 (2015)

dell'Isola, F., Andreaus, U., Placidi, L.: At the origins and in the vanguard of peridynamics, non-local and higher-gradient continuum mechanics: an underestimated and still topical contribution of Gabrio Piola. Math. Mech. Solids **20**(8), 887–928 (2015a)

dell'Isola, F., Della Corte, A., Greco, L., Luongo, A.: Plane bias extension test for a continuum with two inextensible families of fibers: a variational treatment with Lagrange multipliers and a perturbation solution. Int. J. Solids Struct. **81**, 1–12 (2015b)

dell'Isola, F., Giorgio, I., Andreaus, U.: Elastic pantographic 2D lattices: a numerical analysis on static response and wave propagation. Proc. Est. Acad. Sci. **64**, 219–225 (2015c)

dell'Isola, F., Lekszycki, T., Pawlikowski, M., Grygoruk, R., Greco, L.: Designing a light fabric metamaterial being highly macroscopically tough under directional extension: first experimental evidence. Zeitschrift für angewandte Mathematik und Physik **66**(6), 3473–3498 (2015d)

dell'Isola, F., Seppecher, P., Della Corte, A.: The postulations á la D'Alembert and á la Cauchy for higher gradient continuum theories are equivalent: a review of existing results. Proc. R. Soc. A, R. Soc. **471**, 20150415 (2015e)

dell'Isola, F., Steigmann, D., Della Corte, A.: Synthesis of fibrous complex structures: designing microstructure to deliver targeted macroscale response. Appl. Mech. Rev. **67**(6), 060804 (2015f)

dell'Isola, F., Della Corte, A., Giorgio, I.: Higher-gradient continua: the legacy of piola, mindlin, sedov and toupin and some future research perspectives. Math. Mech. Solids **1081** (2016a). doi:10.1177/1081286515616034

dell'Isola, F., Giorgio, I., Pawlikowski, M., Rizzi, N.L.: Large deformations of planar extensible beams and pantographic lattices: heuristic homogenization, experimental and numerical examples of equilibrium. Proc. R. Soc. A, R. Soc. **472**, 20150790 (2016b)

Di Egidio, A., Luongo, A., Paolone, A.: Linear and non-linear interactions between static and dynamic bifurcations of damped planar beams. Int. J. Non-Linear Mech. **42**(1), 88–98 (2007)

Dos Reis, F., Ganghoffer, J.F.: Discrete homogenization of architectured materials: implementation of the method in a simulation tool for the systematic prediction of their effective elastic properties. Technische Mechanik **30**, 85–109 (2010)

Dos Reis, F., Ganghoffer, J.F.: Equivalent mechanical properties of auxetic lattices from discrete homogenization. Comput. Mater. Sci. **51**(1), 314–321 (2012)

Dunlop, J.W.C., Fratzl, P.: Multilevel architectures in natural materials. Scr. Mater. **68**(1), 8–12 (2013)

Ebinger, T., Steeb, H., Diebels, S.: Modeling macroscopic extended continua with the aid of numerical homogenization schemes. Comput. mater. sci. **32**(3), 337–347 (2005)

Enakoutsa, K., Del Vescovo, D., Scerrato, D.: Combined polarization field gradient and strain field gradient effects in elastic flexoelectric materials. Math. Mech. Solids (2015). doi:10.1177/1081286515616048

Eringen, A.C.: Mechanics of Micromorphic Continua. Springer, Berlin (1968)

Federico, S., Grillo, A.: Elasticity and permeability of porous fibre-reinforced materials under large deformations. Mech. Mater. **44**, 58–71 (2012)

Feyel, F.: A multilevel finite element method (fe 2) to describe the response of highly non-linear structures using generalized continua. Comput. Methods Appl. Mech. Eng. **192**(28), 3233–3244 (2003)

Fleck, N.A., Deshpande, V.S., Ashby, M.F.: Micro-architectured materials: past, present and future. Proc. R. Soc. Lond. A: Math., Phys. Eng. Sci., R. Soc. **466**, 2495–2516 (2010)

Gabriele, S., Rizzi, N., Varano, V.: On the imperfection sensitivity of thin-walled frames. Civil-Comp Proceedings 99 (2012)

Germain, P.: The method of virtual power in continuum mechanics. part 2: microstructure. SIAM J. Appl. Math. **25**(3), 556–575 (1973)

Giorgio, I., Scerrato, D.: Multi-scale concrete model with rate-dependent internal friction. Eur. J. Environ. Civil Eng. 1–19 (2016). doi:10.1080/19648189.2016.1144539
Giorgio, I., Culla, A., Del Vescovo, D.: Multimode vibration control using several piezoelectric transducers shunted with a multiterminal network. Arch. Appl. Mech. **79**(9), 859–879 (2009)
Giorgio, I., Andreaus, U., Lekszycki, T., Della Corte, A.: The influence of different geometries of matrix/scaffold on the remodeling process of a bone and bioresorbable material mixture with voids. Math. Mech. Solids (2015). doi:10.1177/1081286515616052
Giorgio, I., Andreaus, U., Madeo, A.: The influence of different loads on the remodeling process of a bone and bioresorbable material mixture with voids. Contin. Mech. Thermodyn. **28**(1–2), 21–40 (2016)
Goda, I., Assidi, M., Belouettar, S., Ganghoffer, J.F.: A micropolar anisotropic constitutive model of cancellous bone from discrete homogenization. J. mech. behav. biomed. mater. **16**, 87–108 (2012)
Greco, L., Cuomo, M.: B-spline interpolation of Kirchhoff-Love space rods. Comput. Methods Appl. Mech. Eng. **256**, 251–269 (2013)
Greco, L., Cuomo, M.: An implicit G1 multi patch B-spline interpolation for Kirchhoff-Love space rod. Comput. Methods Appl. Mech. Eng. **269**, 173–197 (2014)
Greco, L., Cuomo, M.: An isogeometric implicit G1 mixed finite element for Kirchhoff space rods. Comput. Methods Appl. Mech. Eng. **298**, 325–349 (2016)
Green, A.E., Naghdi, P.M.: A unified procedure for construction of theories of deformable media. ii. generalized continua. Proc. R. Soc. Lond. A: Math., Phys. Eng. Sci., R. Soc. **448**, 357–377 (1995)
Javili, A., dell'Isola F, Steinmann P.: Geometrically nonlinear higher-gradient elasticity with energetic boundaries. J. Mech. Phys. Solids **61**(12), 2381–2401 (2013)
Lekszycki, T., Dell'Isola F.: A mixture model with evolving mass densities for describing synthesis and resorption phenomena in bones reconstructed with bio-resorbable materials. Zeitschrift für Angewandte Mathematik und Mechanik **92**(6), 426–444 (2012)
Luongo, A.: A unified perturbation approach to static/dynamic coupled instabilities of nonlinear structures. Thin-Walled Struct. **48**(10), 744–751 (2010)
Luongo, A.: On the use of the multiple scale method in solving "difficult" bifurcation problems. Math. Mech. Solids (2015). doi:10.1177/1081286515616053
Luongo, A., Piccardo, G.: Linear instability mechanisms for coupled translational galloping. J. Sound Vib. **288**(4), 1027–1047 (2005)
Masiani, R., Rizzi, N., Trovalusci, P.: Masonry as structured continuum. Meccanica **30**(6), 673–683 (1995)
Misra, A., Roberts, L.A., Levorson, S.M.: Reliability analysis of drilled shaft behavior using finite difference method and Monte Carlo simulation. Geotech. Geol. Eng. **25**(1), 65–77 (2007)
Neff, P., Forest, S.: A geometrically exact micromorphic model for elastic metallic foams accounting for affine microstructure. modelling, existence of minimizers, identification of moduli and computational results. J. Elast. **87**(2–3), 239–276 (2007)
Pagnini, L.C.: Reliability analysis of wind-excited structures. J. Wind Eng. Ind. Aerodyn. **98**(1), 1–9 (2010)
Papka, S.D., Kyriakides, S.: Experiments and full-scale numerical simulations of in-plane crushing of a honeycomb. Acta mater. **46**(8), 2765–2776 (1998)
Piccardo, G., Tubino, F., Luongo, A.: A shear–shear torsional beam model for nonlinear aeroelastic analysis of tower buildings. Zeitschrift für angewandte Mathematik und Physik, pp 1–19 (2014)
Piccardo, G., Pagnini, L.C., Tubino, F.: Some research perspectives in galloping phenomena: critical conditions and post-critical behavior. Contin. Mech. Thermodyn. **27**(1–2), 261–285 (2015)
Pietraszkiewicz, W., Eremeyev, V.A.: On natural strain measures of the non-linear micropolar continuum. Int. J. Solids Struct. **46**(3), 774–787 (2009a)
Pietraszkiewicz, W., Eremeyev, V.A.: On vectorially parameterized natural strain measures of the non-linear Cosserat continuum. Int. J. Solids Struct. **46**(11), 2477–2480 (2009b)

Placidi, L.: A variational approach for a nonlinear one-dimensional damage-elasto-plastic second-gradient continuum model. Contin. Mech. Thermodyn. (2014). doi:10.1007/s00161-014-0405-2

Placidi, L.: A variational approach for a nonlinear 1-dimensional second gradient continuum damage model. Contin. Mech. Thermodyn. **27**(4), 623–638 (2015)

Placidi, L., Hutter, K.: An anisotropic flow law for incompressible polycrystalline materials. Zeitschrift für angewandte Mathematik und Physik ZAMP **57**(1), 160–181 (2005)

Placidi, L., Hutter, K.: Thermodynamics of polycrystalline materials treated by the theory of mixtures with continuous diversity. Contin. Mech. Thermodyn. **17**(6), 409–451 (2006)

Placidi, L., Andreaus, U., Della Corte, A., Lekszycki, T.: Gedanken experiments for the determination of two-dimensional linear second gradient elasticity coefficients. Zeitschrift für angewandte Mathematik und Physik **66**(6), 3699–3725 (2015a)

Placidi, L., Giorgio, I., Della Corte, A., Scerrato, D.: Euromech 563 cisterna di latina 17–21 march 2014 generalized continua and their applications to the design of composites and metamaterials: a review of presentations and discussions. Math. Mech. Solids (2015b). doi:10.1177/1081286515576948

Rahali, Y., Giorgio, I., Ganghoffer, J.F., dell'Isola F, : Homogenization à la Piola produces second gradient continuum models for linear pantographic lattices. Int. J. Eng. Sci. **97**, 148–172 (2015)

Rinaldi, A., Placidi, L.: A microscale second gradient approximation of the damage parameter of quasi-brittle heterogeneous lattices. ZAMM-Journal of Applied Mathematics and Mechanics/Zeitschrift für Angewandte Mathematik und Mechanik **94**(10), 862–877 (2014)

Rizzi, N.L., Varano, V., Gabriele, S.: Initial postbuckling behavior of thin-walled frames under mode interaction. Thin-Walled Struct. **68**, 124–134 (2013). doi:10.1016/j.tws.2013.03.004

Russo, L., et al.: The Forgotten Revolution: How Science was Born in 300 BC and Why it had to be Reborn. Springer Science & Business Media, Berlin (2013)

Scerrato, D., Giorgio, I., Madeo, A., Limam, A., Darve, F.: A simple non-linear model for internal friction in modified concrete. Int. J. Eng. Sci. **80**, 136–152 (2014)

Scerrato, D., Giorgio, I., Della Corte, A., Madeo, A., Limam, A.: A micro-structural model for dissipation phenomena in the concrete. Int. J. Numer. Anal. Methods Geomech. **39**(18), 2037–2052 (2015)

Steigmann, D.J., Dell'Isola, F.: Mechanical response of fabric sheets to three-dimensional bending, twisting, and stretching. Acta Mech. Sinica **31**(3), 373–382 (2015)

Thiagarajan, G., Misra, A.: Fracture simulation for anisotropic materials using a virtual internal bond model. Int. J. Solids Struct. **41**(11), 2919–2938 (2004)

Tomic, A., Grillo, A., Federico, S.: Poroelastic materials reinforced by statistically oriented fibres—numerical implementation and application to articular cartilage. IMA J. Appl. Math. **79**, 1027–1059 (2014)

Trinh, D.K., Janicke, R., Auffray, N., Diebels, S., Forest, S.: Evaluation of generalized continuum substitution models for heterogeneous materials. Int. J. Multiscale Comput. Eng. **10**(6), (2012)

Turco, E.: An effective algorithm for reconstructing boundary conditions in elastic solids. Comput. Methods Appl. Mech. Eng. **190**(29), 3819–3829 (2001)

Verhoosel, C.V., Scott, M.A., Borden, M.J., Hughes, T.J., de Borst, R.: Discretization of higher-order gradient damage models using isogeometric finite elements. Technical report DTIC Document (2011)

On Equilibrium of a Second-Gradient Fluid Near Edges and Corner Points

Victor A. Eremeyev

Abstract Within the framework of the model of second-gradient fluid we discuss the natural boundary conditions along edges and at corner points. As for any strain gradient model the model of second-gradient fluid demonstrates some peculiarities related with necessity of additional boundary conditions. Here using the Lagrange variational principle we derived the latter boundary conditions for various contact angles.

1 Introduction

Recently, the interest to generalized model of continua grows with respect to modelling of complex material behaviour and for proper description of materials at the micro- and nano-scales. Among these generalized models there are micropolar or Cosserat continua, microstretched and micromorphic media, media with internal variables, gradient elasticity, see Eringen (1999, 2001, 2002), Green and Rivlin (1964), Capriz (1989), Eremeyev et al. (2013), Maugin and Muschik (1994). The actual state of the art may be found in recent collections (Maugin and Metrikine 2010; Altenbach et al. 2011, 2013; Altenbach and Eremeyev 2013 and the references therein). Among the mentioned generalized models of continua the strain gradient elasticity/plasticity is widely used in the mechanics of solids, see e.g., Aifantis (2003), Gurtin (2002), Gao et al. (1999), Huang et al. (2000), Mühlhaus and Aifantis (1991), Forest (2008), Fleck and Hutchinson (1997). For example, the second-gradient models are fruitful for analysis of several types of composites, as shown in dell'Isola and Steigmann (2015), dell'Isola et al. (2015), Rahali et al. (2015).

Considering the gradient elasticity models it is worth to mention second-gradient fluids. The second-gradient fluid called also Korteweg, or Cahn–Hilliard fluid. This model relates to the original works by van der Waals (1893) and Korteweg (1901)

V.A. Eremeyev (✉)
The Faculty of Mechanical Engineering, Rzeszów University of Technology,
al. Powstańców Warszawy 8, 35-959 Rzeszów, Poland
e-mail: veremeyev@prz.edu.pl

K. Naumenko and M. Aßmus (eds.), *Advanced Methods of Continuum Mechanics for Materials and Structures*, Advanced Structured Materials 60,
DOI 10.1007/978-981-10-0959-4_29

and later to Cahn and Hilliard (1958, 1959). Recently, this model found applications on the molecular theory of capillarity (Rowlinson and Widom 2003; de Gennes et al. 2004). It is remarkable that unlike classic fluids second-gradient fluids can exert shear stresses in statics (dell'Isola and Rotoli 1995; Seppecher 1996, 2002, 1989a; dell'Isola et al. 1996, see also Dunn and Serrin 1985; Fried and Gurtin 2006; Brenner 2005; Heida and Málek 2010; Seppecher 1989b, 1993, 1996) where many other properties of the model were discussed.

The paper continues the analysis presented in recent papers (Auffray et al. 2015; dell'Isola et al. 2014; Eremeyev and Altenbach 2014) towards the analysis of the boundary conditions near non-smooth boundaries. First, in Sect. 2 we introduce the strain energy density for a second-gradient fluid which depends on the mass density and norm of its gradient. Then we formulate the Lagrange variational principle in Sect. 3. From this principle we derive the equilibrium equations in Sect. 4 and the natural boundary conditions in Sect. 5.

2 Energy Density of Second-Gradient Fluid

For a second-gradient fluid the energy density W takes the form (Seppecher 1996)

$$W = W(\rho, \nabla\rho), \tag{1}$$

where ρ is the mass density, ∇ is the nabla operator (gradient operator) in the actual configuration and W is the strain energy density per unit volume in the actual configuration. Using the principle of material frame-indifference (Truesdell 1966; Truesdell and Noll 2004) W reduces to the following form:

$$W = W(\rho, \beta), \quad \beta = \nabla\rho \cdot \nabla\rho, \tag{2}$$

where dot stands for the inner (scalar) product.

As an example one may consider the following constitutive equation:

$$W = W_0(\rho) + \frac{1}{2}\lambda (\nabla\rho)^2,$$

where λ is a capillary coefficient and $W_0(\rho)$ is the non-convex function of ρ. The typical form of $W_0(\rho)$ is similar to the van der Waals gas constitutive equations, see Fig. 1. Such constitutive relations may describe phase interfaces of finite thickness, such as the interfacial layer between vapour and fluid, see Rowlinson and Widom (2003), Seppecher (1996), Rosi et al. (2013). This is a simple example of localization phenomenons observed in strain gradient materials.

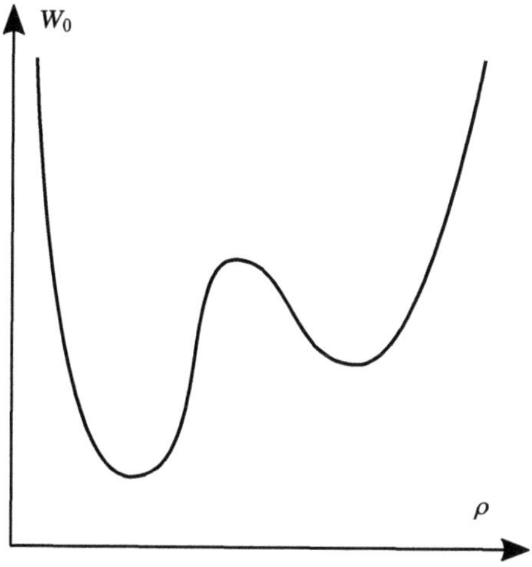

Fig. 1 The free energy of the second-gradient fluid per unit volume versus ρ

3 Lagrange Functional and the Principle of Virtual Work

For derivation of equilibrium equations we use the Lagrange variational principle (Auffray et al. 2015; dell'Isola et al. 2014; Eremeyev and Altenbach 2014). The energy functional is defined as follows:

$$\mathcal{E} = \int_v W(\rho, \beta)\, \mathrm{d}v, \tag{3}$$

where $v \equiv v_f$ is the volume which occupies the fluid in the actual configuration at instant t. From variational equation $\delta\mathcal{E} = 0$, where δ is the variation symbol, we obtain the equilibrium conditions. In order to find the first variation $\delta\mathcal{E}$ we transform $\mathcal{E}$ into Lagrangian form

$$\mathcal{E} = \int_V W(\rho, \beta) J\, \mathrm{d}V, \tag{4}$$

where V the volume of the fluid in the reference configuration and $J = \det \mathbf{F}$, $\mathbf{F} = \mathrm{Grad}\mathbf{x} = (\nabla_X \mathbf{x})^T$ is the deformation gradient, $\mathbf{x}$ is the position vector of the fluid particle in the actual configuration, ∇_X is the nabla operator in the reference configuration and the index T denotes the transposed.

The first variation of $\mathcal{E}$ becomes

$$\delta\mathcal{E} = \int_V \left[J\frac{\partial W}{\partial \rho}\delta\rho + J\frac{\partial W}{\partial \beta}\delta\beta + W\delta J \right] \mathrm{d}V.$$

Introducing the variation of the position vector $\mathbf{v} = \delta\mathbf{x}$ and using the standard formulae

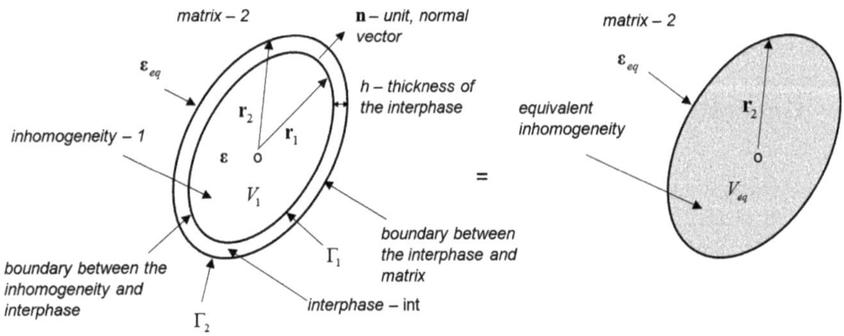

$$\boldsymbol{\Sigma}|_{t+\Delta t} = \Delta \boldsymbol{R} \boldsymbol{\Sigma}|_t \Delta \boldsymbol{R}^T + \Delta \boldsymbol{\Sigma}, \tag{17}$$

$$\Delta \boldsymbol{R} = \left[\boldsymbol{I} - \frac{1}{2} (\Delta \boldsymbol{W} - \Delta \boldsymbol{\Omega}) \right]^{-1} \left[\boldsymbol{I} + \frac{1}{2} (\Delta \boldsymbol{W} - \Delta \boldsymbol{\Omega}) \right], \tag{18}$$

$$\Delta \boldsymbol{W} = \boldsymbol{W} \Delta t, \Delta \boldsymbol{\Omega} = \boldsymbol{\Omega} \Delta t, \tag{19}$$

$$\boldsymbol{\Sigma}|_{t+\Delta t} = \Delta \boldsymbol{R} \boldsymbol{\Sigma}|_t \Delta \boldsymbol{R}^T + \boldsymbol{R} \Delta \boldsymbol{\Sigma} \boldsymbol{R}^T. \tag{20}$$

Figure 11 illustrates the deformed configurations of the longitudinal, middle cross section at the end of the 60th cycle in the analysis. As seen from the figure, the annealed copper plate had larger deflection as a result of the cyclic growth of δ than the as-received copper plate in which the cyclic recovery of δ occurred. Figure 12 shows the hysteresis loops of σ_{xx} and $\varepsilon^{\mathrm{m}}_{xx}$ at point A in the copper layer. Point A is close to the top surface of the copper layer (Fig. 11). Here, $\varepsilon^{\mathrm{m}}_{xx}$ denotes the mechanical part of ε_{xx}, i.e., $\varepsilon^{\mathrm{m}}_{xx} = \varepsilon^{\mathrm{e}}_{xx} + \varepsilon^{\mathrm{p}}_{xx}$. Point A also had the hysteresis loops of σ_{zz} and $\varepsilon^{\mathrm{m}}_{zz}$ that were almost the same as those of σ_{xx} and $\varepsilon^{\mathrm{m}}_{xx}$.

$$\vartheta_{\mathrm{av}}\left(x\right) = \frac{1}{A}\oiint_{A^i_{\mathrm{cro}}} \vartheta\left(x,y,z,t\right)\mathrm{d}y\mathrm{d}z \tag{37}$$

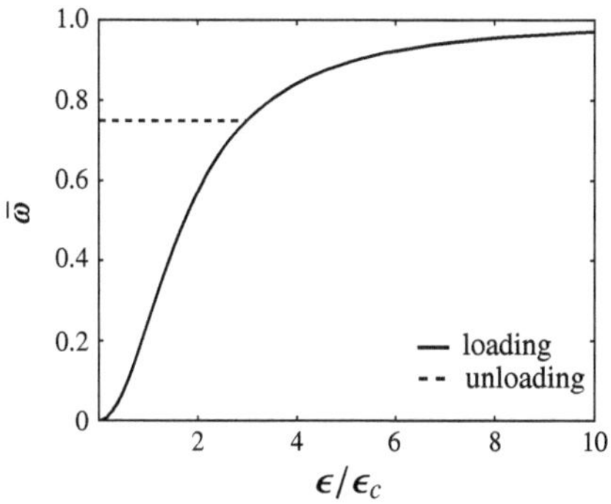

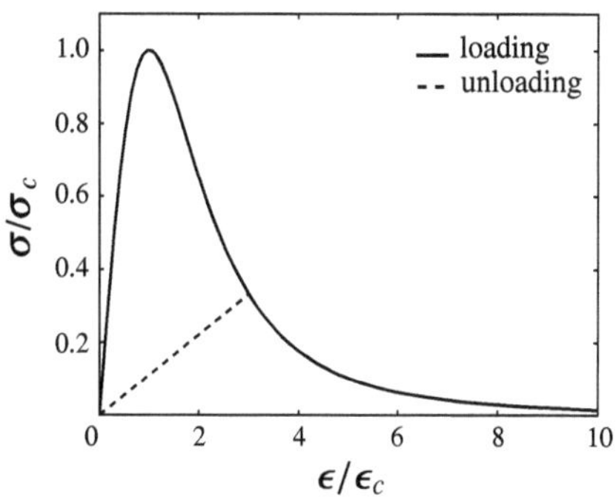

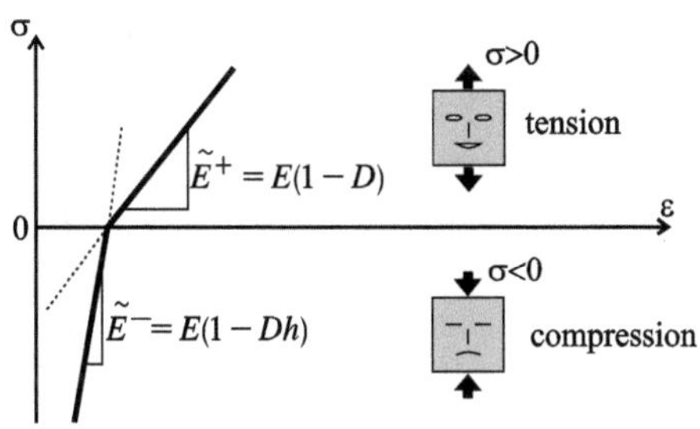

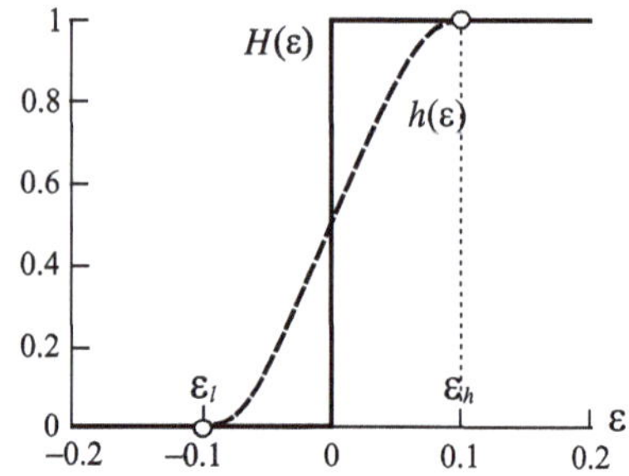

Loading	Stress	Label	Stress relation	$\left(\frac{\sigma_{\mathrm{I}}}{\sigma_+}, \frac{\sigma_{\mathrm{II}}}{\sigma_+}, \frac{\sigma_{\mathrm{III}}}{\sigma_+}\right)$	$\frac{I_1}{\sigma_+}$	$\frac{\sqrt{3I_2'}}{\sigma_+}$	$\frac{3\sqrt[3]{I_3'/2}}{\sigma_+}$	θ	ψ
Tension	σ_+	Z	1	(1, 0, 0)	1	1	1	0	$\frac{\pi}{4}$
Compression	σ_-	D	d	$(-d, 0, 0)$	$-d$	d	d	$\frac{\pi}{3}$	$\frac{3\pi}{4}$
Torsion	τ_*	T	k	$\left(\frac{k}{\sqrt{3}}, -\frac{k}{\sqrt{3}}, 0\right)$	0	k	0	$\frac{\pi}{6}$	$\frac{\pi}{2}$
Biaxial tension[a]	σ_{BZ}	B_{Z}	b_{Z}	$(b_{\mathrm{Z}}, b_{\mathrm{Z}}, 0)$	$2\,b_{\mathrm{Z}}$	b_{Z}	$-b_{\mathrm{Z}}$	$\frac{\pi}{3}$	$0.15\,\pi$
Biaxial compression[a]	σ_{BD}	B_{D}	b_{D}	$(-b_{\mathrm{D}}, -b_{\mathrm{D}}, 0)$	$-2\,b_{\mathrm{D}}$	b_{D}	b_{D}	0	$0.85\,\pi$
Inner pressure[b]	σ_{IZ}	I_{Z}	i_{Z}	$\left(\frac{2}{\sqrt{3}} i_{\mathrm{Z}}, \frac{1}{\sqrt{3}} i_{\mathrm{Z}}, 0\right)$	$\sqrt{3}\,i_{\mathrm{Z}}$	i_{Z}	0	$\frac{\pi}{6}$	$\frac{\pi}{6}$
Outer pressure[b]	σ_{UD}	U_{D}	u_{D}	$\left(-\frac{2}{\sqrt{3}} u_{\mathrm{D}}, -\frac{1}{\sqrt{3}} u_{\mathrm{D}}, 0\right)$	$-\sqrt{3}\,u_{\mathrm{D}}$	u_{D}	0	$\frac{\pi}{6}$	$\frac{2\pi}{3}$
Hydrostatic tension	σ_{AZ}	A_{Z}	a_+^{hyd}	$\left(a_+^{\mathrm{hyd}}, a_+^{\mathrm{hyd}}, a_+^{\mathrm{hyd}}\right)$	$3\,a_+^{\mathrm{hyd}}$	0	0	–	0
Hydrostatic compression	σ_{AD}	A_{D}	a_-^{hyd}	$\left(-a_-^{\mathrm{hyd}}, -a_-^{\mathrm{hyd}}, -a_-^{\mathrm{hyd}}\right)$	$-3\,a_-^{\mathrm{hyd}}$	0	0	–	0

$$\Phi_H = \frac{3^6}{2^6}\,\frac{1}{d_{\rm inc}^3}\,(I_3')^2 - \frac{3^5}{2^5}\,\frac{d_{\rm inc}-1}{d_{\rm inc}^3}\,I_2'\,I_3'\,\sigma_{\rm eq} - \frac{3^4}{2^4}\,\frac{1}{d_{\rm inc}^2}\,(I_2')^2\,\sigma_{\rm eq}^2 - \frac{3^3}{2^3}\,\frac{1-d_{\rm inc}^3}{d_{\rm inc}^3}\,I_3'\,\sigma_{\rm eq}^3 + \frac{3^2}{2^2}\,\frac{1+d_{\rm inc}^2}{d_{\rm inc}^2}\,I_2'\,\sigma_{\rm eq}^4 - \sigma_{\rm eq}^6 \tag{15}$$

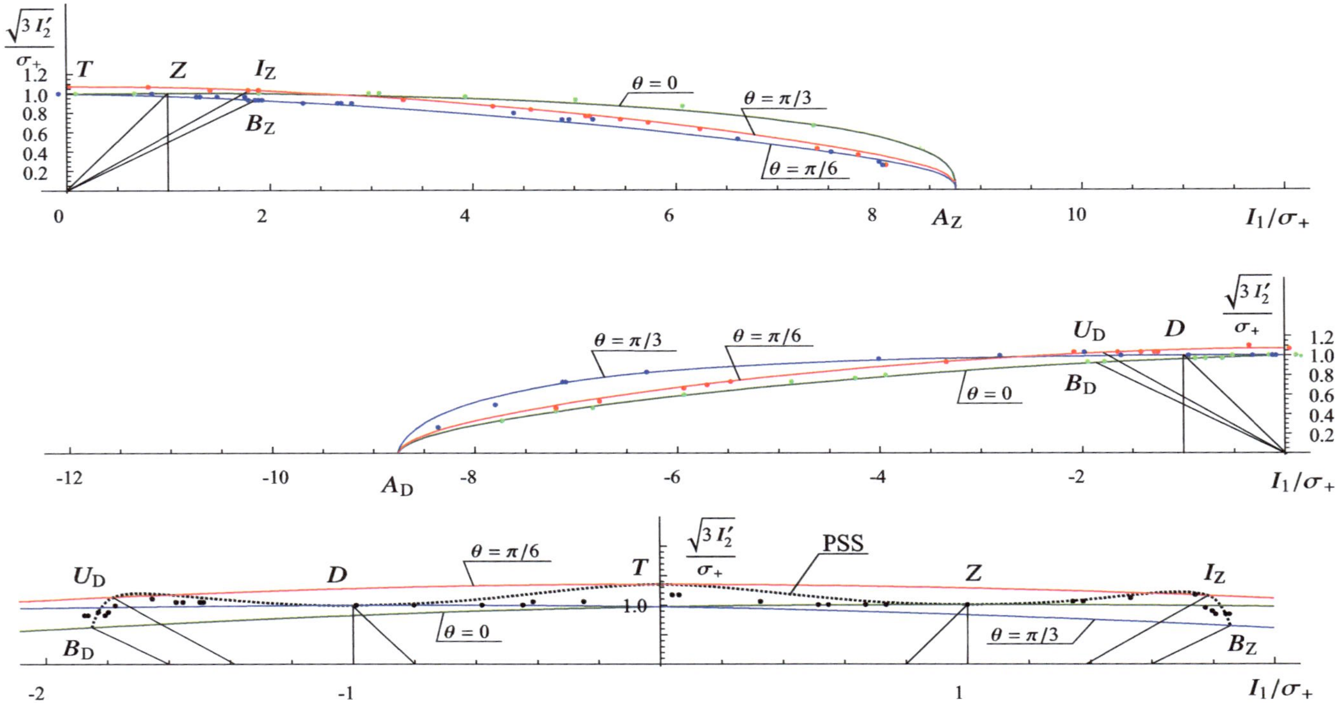

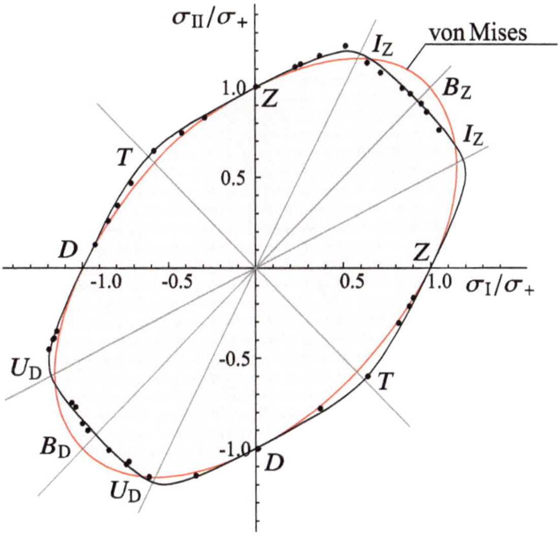

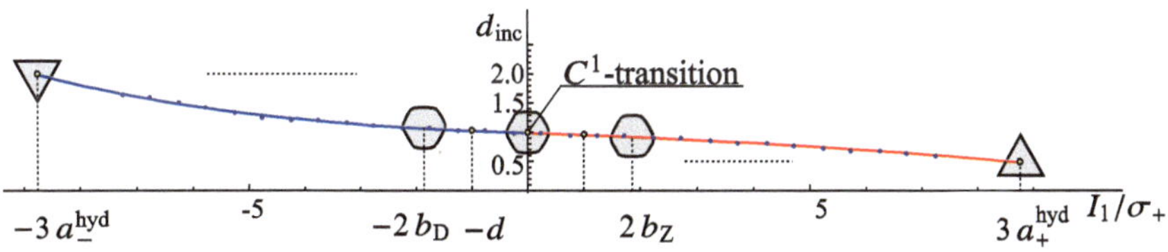

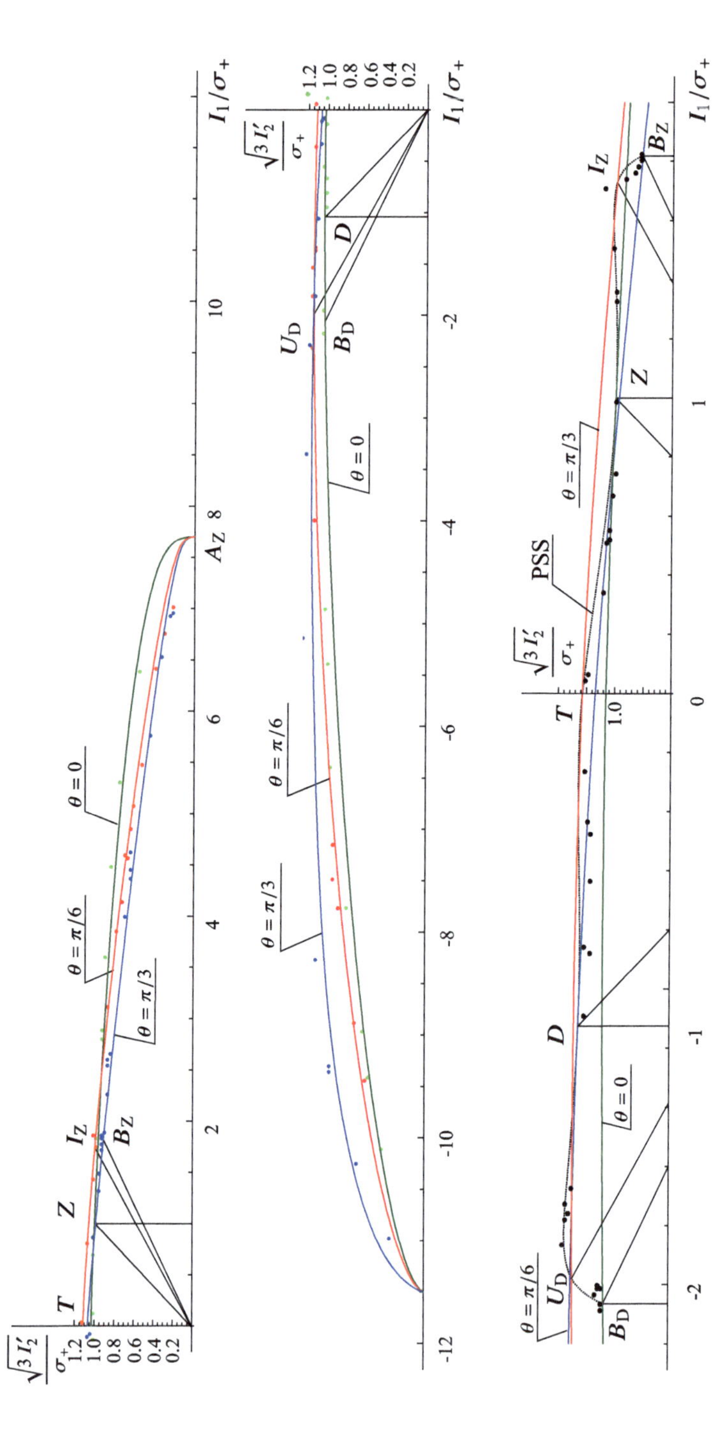

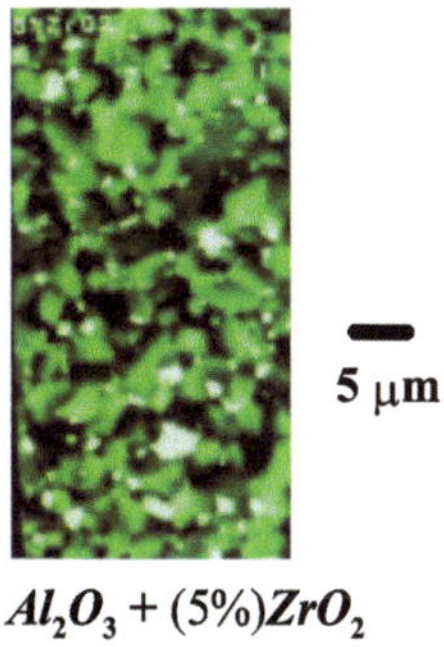

$$\bar{S}^{e}_{ik}(x_j, p_f, p_o, N_c^{(s)}) = (1 - p_f)S^{C1}_{ik} + p_f S^{C2}_{ik}(x_j, p_f) + S^{p_o}_{ik}(x_j, \sigma_r, p_o) + S^{cr}_{ik}(x_j, \sigma_r, p_f, p_o, N_c^{(s)}), \quad (2)$$

$$\bar{S}^{CMCe}_{ik}(x_j, p_f) = (1 - p_f)S^{C1}_{ik} + p_f S^{C2}_{ik}(x_j, p_f). \quad (3)$$

$$\overline{\boldsymbol{D}}_e = -\big(\boldsymbol{d}_j\cdot\boldsymbol{d}_{k,i}-\boldsymbol{d}_j^0\cdot\boldsymbol{d}_{k,i}^0\big)\,\big(\boldsymbol{d}_j^0\otimes\boldsymbol{d}_k^0\big)\times\boldsymbol{g}^i$$
$$= \big(\boldsymbol{d}_{j,i}\cdot\boldsymbol{d}_k-\boldsymbol{d}_{j,i}^0\cdot\boldsymbol{d}_k^0\big)\,\boldsymbol{d}_j^0\otimes\big(\boldsymbol{d}_k^0\times\boldsymbol{g}^i\big). \tag{22}$$

$$\rho J = \rho_0, \quad \mathbf{F}^T \cdot \nabla = \nabla_X, \quad \nabla = \mathbf{F}^{-T} \cdot \nabla_X,$$
$$\delta \mathbf{F}^{-T} = -\mathbf{F}^{-T} \cdot \delta \mathbf{F}^T \cdot \mathbf{F}^{-T}, \qquad \delta \mathbf{F} = (\nabla_X \mathbf{v})^T,$$

where $\mathbf{F}^{-T} = (\mathbf{F}^{-1})^T$ and ρ_0 is the density in the reference configuration, we obtain the following relations (Eremeyev and Altenbach 2014):

$$\delta J = J \nabla \cdot \mathbf{v}, \quad \delta \rho = -\rho \nabla \cdot \mathbf{v}, \quad \delta(\nabla \rho) = -\nabla \mathbf{v} \cdot \nabla \rho - \nabla(\rho \nabla \cdot \mathbf{v}),$$

$$\delta \beta = 2 \nabla \rho \cdot [-\nabla \mathbf{v} \cdot \nabla \rho - \nabla(\rho \nabla \cdot \mathbf{v})].$$

With these formulae we derive $\delta \mathcal{E}$ expressed in terms of $\mathbf{v}$ and its spatial gradient (Eremeyev and Altenbach 2014):

$$\delta \mathcal{E} = \int_v \left\{ \left[W - \rho \frac{\partial W}{\partial \rho} \right] \nabla \cdot \mathbf{v} - 2 \frac{\partial W}{\partial \beta} \nabla \rho \cdot [\nabla \mathbf{v} \cdot \nabla \rho + \nabla(\rho \nabla \cdot \mathbf{v})] \right\} \mathrm{d}v. \tag{5}$$

For further transformations we use the spatial and surface divergence theorems (Lebedev et al. 2010)

$$\int_v \nabla \cdot \mathbf{T} \, \mathrm{d}v = \int_a \mathbf{n} \cdot \mathbf{T} \, \mathrm{d}a, \tag{6}$$

$$\int_a (\nabla_s \cdot \mathbf{T} + 2K \mathbf{n} \cdot \mathbf{T}) \, \mathrm{d}a = \int_{\partial a} \mathbf{m} \cdot \mathbf{T} \, \mathrm{d}s, \tag{7}$$

where $\mathbf{T}$ is a tensor field of any order, $\mathbf{n}$ is the vector of unit normal, $a = \partial v$,

$$\nabla_s = \nabla - \mathbf{n}(\mathbf{n} \cdot \nabla)$$

is the surface nabla operator in the actual configuration, K is the mean curvature of the surface a and $\mathbf{m}$ is the unit vector of the normal to ∂a such that $\mathbf{m} \cdot \mathbf{n} = 0$. Both theorems are fulfilled for smooth regions including ones with edges and corner points, such as shown in Fig. 2. Let us note that for smooth closed surface a set ∂a is empty. Here we keep the last integral in (7) for analysis of boundary conditions along edges.

Finally, we obtain the first variation of the Lagrangian $\mathcal{E}$ in the form

$$\begin{aligned} \delta \mathcal{E} = & \int_v \left\{ -\nabla \left[W - \rho \frac{\partial W}{\partial \rho} + \rho \nabla \cdot \left(2 \frac{\partial W}{\partial \beta} \nabla \rho \right) \right] + \nabla \cdot \left(2 \frac{\partial W}{\partial \beta} \nabla \rho \otimes \nabla \rho \right) \right\} \cdot \mathbf{v} \, \mathrm{d}v \\ & + \int_a \left\{ \left(W - \rho \frac{\partial W}{\partial \rho} + \nabla \cdot \left(2 \frac{\partial W}{\partial \beta} \nabla \rho \right) \rho \right) \mathbf{n} - 2 \frac{\partial W}{\partial \beta} \mathbf{n} \cdot \nabla \rho \otimes \nabla \rho \right. \\ & \left. + 4 \frac{\partial W}{\partial \beta} (\mathbf{n} \cdot \nabla \rho) \rho K \mathbf{n} + \nabla_s \left[2\rho \frac{\partial W}{\partial \beta} (\mathbf{n} \cdot \nabla \rho) \right] \right\} \cdot \mathbf{v} \, \mathrm{d}a \end{aligned}$$

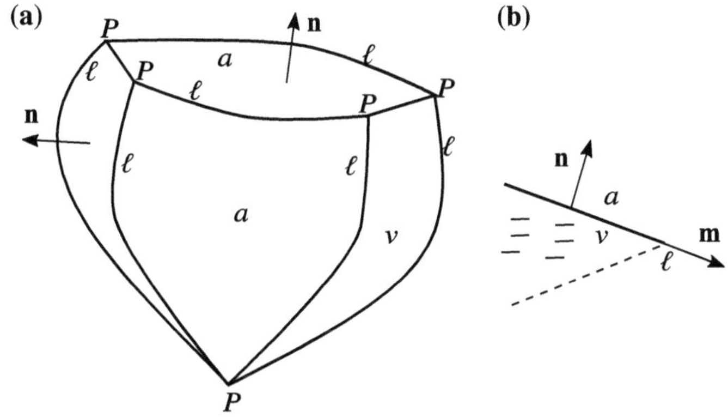

Fig. 2 **a** Volume v with edges ℓ and corner points P and **b** part of the volume near edge

$$
-\int_a 2\frac{\partial W}{\partial \beta}(\mathbf{n}\cdot\nabla\rho)\rho\mathbf{n}\cdot(\mathbf{n}\cdot\nabla)\mathbf{v}\,\mathrm{d}a-\int_{\partial a}2\rho\frac{\partial W}{\partial \beta}(\mathbf{n}\cdot\nabla\rho)\mathbf{m}\cdot\mathbf{v}\,\mathrm{d}s. \tag{8}
$$

The latter formula dictates the possible form of the virtual work functional for the second-gradient fluid. The virtual work is defined as follows:

$$
\delta\mathcal{A}=\int_v \rho\,\mathbf{f}\cdot\mathbf{v}\,\mathrm{d}v+\int_a[\mathbf{t}\cdot\mathbf{v}+c\,\mathbf{n}\cdot(\mathbf{n}\cdot\nabla)\mathbf{v}]\,\mathrm{d}a+\int_\ell \zeta\mathbf{m}\cdot\mathbf{v}\,\mathrm{d}s. \tag{9}
$$

Here $\mathbf{f}$ is the vector of external forces per unit mass, $\mathbf{t}$ is the vector of external surface traction and c and ζ are a surface density of double force and the line density of forces acting along the edges ℓ, respectively.

As a result, we formulate the principle of virtual work for a second-gradient fluid as follows:

The fluid is in equilibrium if and only if the following relation holds true:

$$
\delta\mathcal{E}-\delta\mathcal{A}=0 \tag{10}
$$

for every kinematically admissible virtual displacements $\mathbf{v}$, i.e., for any $\mathbf{v}$ satisfying the kinematic boundary conditions.

For formulations of the principle of virtual power we refer to the landmark papers by Sedov (1968) and Germain (1973a, b), see also Auffray et al. (2015), dell'Isola et al. (2014), dell'Isola and Seppecher (1995, 1997).

4 Equilibrium Equations

From (8), (9) and (10), we obtain the Eulerian equilibrium equation

$$\nabla\left[W-\rho\frac{\partial W}{\partial\rho}+\rho\nabla\cdot\left(2\frac{\partial W}{\partial\beta}\nabla\rho\right)\right]-\nabla\cdot\left(2\frac{\partial W}{\partial\beta}\nabla\rho\otimes\nabla\rho\right)+\rho\mathbf{f}=\mathbf{0}. \quad (11)$$

We transform (11) into the following compact form:

$$\nabla\cdot\mathbf{T}+\rho\mathbf{f}=\mathbf{0}, \quad (12)$$

where the symmetric Cauchy-type stress tensor is defined by

$$\mathbf{T}=-P\,\mathbf{I}-2\frac{\partial W}{\partial\beta}\nabla\rho\otimes\nabla\rho \quad (13)$$

and

$$P=-W+\rho\frac{\partial W}{\partial\rho}-\rho\nabla\cdot\left(2\frac{\partial W}{\partial\beta}\nabla\rho\right).$$

Here $\mathbf{I}$ is the three-dimensional identity tensor, P plays a role of pressure and $\mathbf{T}$ is often called the Korteweg stress tensor (Brenner 2005; Podio-Guidugli and Vianello 2013; Dunn and Serrin 1985).

5 Natural Boundary Conditions

The natural boundary conditions follow from the variational equation

$$\int_a\left\{\left(W-\rho\frac{\partial W}{\partial\rho}+\nabla\cdot\left(2\frac{\partial W}{\partial\beta}\nabla\rho\right)\rho\right)\mathbf{n}-2\frac{\partial W}{\partial\beta}\mathbf{n}\cdot\nabla\rho\otimes\nabla\rho\right. \quad (14)$$

$$\left.+4\frac{\partial W}{\partial\beta}(\mathbf{n}\cdot\nabla\rho)\rho K\mathbf{n}+\nabla_s\left[2\rho\frac{\partial W}{\partial\beta}(\mathbf{n}\cdot\nabla\rho)\right]\right\}\cdot\mathbf{v}\,\mathrm{d}a$$

$$-\int_a 2\frac{\partial W}{\partial\beta}(\mathbf{n}\cdot\nabla\rho)\rho\mathbf{n}\cdot(\mathbf{n}\cdot\nabla)\mathbf{v}\,\mathrm{d}a-\int_{\partial a}2\rho\frac{\partial W}{\partial\beta}(\mathbf{n}\cdot\nabla\rho)\mathbf{m}\cdot\mathbf{v}\,\mathrm{d}s=0. \quad (15)$$

For derivation of natural boundary conditions let us consider two types of kinematically admissible fields $\mathbf{v}$. The first case relates to the free surface. In this case there are no constraints for $\mathbf{v}$. The second type of $\mathbf{v}$ is the case of rigid walls; in other words, we assume that a is fixed, so we have the classic condition $\mathbf{n}\cdot\mathbf{v}=0$ which simply means that the fluid cannot penetrate a solid surface.

5.1 Natural Boundary Conditions for Free Surface

In this case the natural boundary conditions on the surface a take the form

$$\left(W - \rho\frac{\partial W}{\partial \rho} + \nabla\cdot\left(2\frac{\partial W}{\partial \beta}\nabla\rho\right)\rho\right)\mathbf{n} - 2\frac{\partial W}{\partial \beta}\mathbf{n}\cdot\nabla\rho\otimes\nabla\rho + 4\rho K\frac{\partial W}{\partial \beta}(\mathbf{n}\cdot\nabla\rho)\mathbf{n}$$
$$+ \nabla_s\left[2\rho\frac{\partial W}{\partial \beta}(\mathbf{n}\cdot\nabla\rho)\right] = \mathbf{t}, \quad \mathbf{x}\in a, \tag{16}$$
$$- 2\rho\frac{\partial W}{\partial \beta}(\mathbf{n}\cdot\nabla\rho) = c, \quad \mathbf{x}\in a. \tag{17}$$

With $\mathbf{T}$ Eq. (16) transforms into

$$\mathbf{n}\cdot\mathbf{T} + 4\rho K\frac{\partial W}{\partial \beta}(\mathbf{n}\cdot\nabla\rho)\mathbf{n} + \nabla_s\left[2\rho\frac{\partial W}{\partial \beta}(\mathbf{n}\cdot\nabla\rho)\right] = \mathbf{t}, \quad \mathbf{x}\in a. \tag{18}$$

It is obvious that $\mathbf{n}\cdot\mathbf{T}$ does not coincide with the surface traction $\mathbf{t}$ as in the case of the simple materials. Boundary condition (17) plays a role which is similar to additional boundary condition for hyperstresses in the strain gradient elasticity. Let us note that the boundary equations explicitly depend on the mean curvature K of the boundary in the actual configuration.

5.2 *Natural Boundary Conditions for Rigid Surface*

Here $\mathbf{v}$ is tangent vector to the boundary: $\mathbf{n}\cdot\mathbf{v} = 0$. Thus, Eq. (15) takes the form

$$\int_a \left\{-2\frac{\partial W}{\partial \beta}\mathbf{n}\cdot\nabla\rho\otimes\nabla\rho + \nabla_s\left[2\rho\frac{\partial W}{\partial \beta}(\mathbf{n}\cdot\nabla\rho)\right]\right\}\cdot\mathbf{v}\,\mathrm{d}a \tag{19}$$
$$- \int_a 2\frac{\partial W}{\partial \beta}(\mathbf{n}\cdot\nabla\rho)\rho\mathbf{n}\cdot(\mathbf{n}\cdot\nabla)\mathbf{v}\,\mathrm{d}a - \int_{\partial a} 2\rho\frac{\partial W}{\partial \beta}(\mathbf{n}\cdot\nabla\rho)\mathbf{m}\cdot\mathbf{v}\,\mathrm{d}s = 0. \tag{20}$$

Here the admissible surface traction $\mathbf{t}$ is also tangent. From (20) we obtain that

$$- 2\frac{\partial W}{\partial \beta}\mathbf{n}\cdot\nabla\rho\otimes\nabla\rho\cdot\mathbf{A} + \nabla_s\left[2\rho\frac{\partial W}{\partial \beta}(\mathbf{n}\cdot\nabla\rho)\right] = \mathbf{t}, \quad \mathbf{x}\in a, \tag{21}$$
$$- 2\rho\frac{\partial W}{\partial \beta}(\mathbf{n}\cdot\nabla\rho) = c, \quad \mathbf{x}\in a. \tag{22}$$

where $\mathbf{A} = \mathbf{I} - \mathbf{n}\otimes\mathbf{n}$ is the projector operator. The first equation can be also transformed into compact form

$$\mathbf{n}\cdot\mathbf{T}\cdot\mathbf{A} + \nabla_s\left[2\rho\frac{\partial W}{\partial \beta}(\mathbf{n}\cdot\nabla\rho)\right] = \mathbf{t}, \quad \mathbf{x}\in a. \tag{23}$$

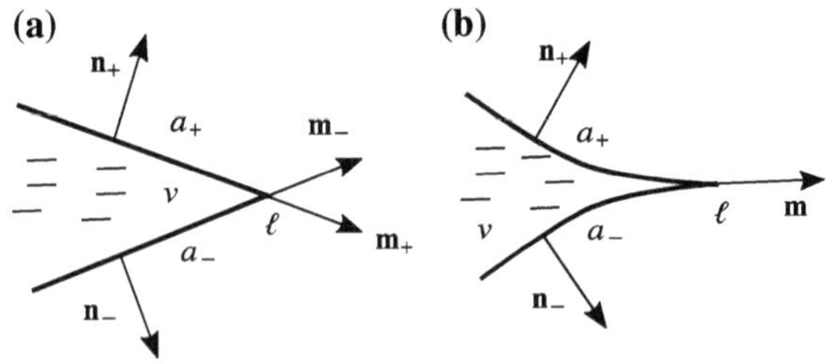

Fig. 3 **a** Edge with nonzero contact angle and **b** edge with zero contact angle (cusp)

5.3 On Edge Conditions

From (8), it also follows the boundary condition along ℓ

$$- 2\rho\frac{\partial W}{\partial \beta}(\mathbf{n}_+ \cdot \nabla\rho)\mathbf{m}_+\Big|_{\ell=\partial a_+} - 2\rho\frac{\partial W}{\partial \beta}(\mathbf{n}_- \cdot \nabla\rho)\mathbf{m}_-\Big|_{\ell=\partial a_-}$$
$$= \zeta_+\Big|_{\ell=\partial a_+}\mathbf{m}_+ + \zeta_-\Big|_{\ell=\partial a_-}\mathbf{m}_-, \quad \mathbf{x} \in \ell. \tag{24}$$

Here $a_\pm$ is the external surfaces of the fluid; $\mathbf{n}_\pm$ and $\mathbf{m}_\pm$ are the unit vectors of normal to $a_\pm$ and $\partial a_\pm$, respectively, see Fig. 3a.

Let us note that the edge condition depends also on the geometrical shape of the contacting surfaces. For example, in the case of cusp, that is when the contact angle is zero, see Fig. 3b, the edge condition transforms into

$$-2\rho\frac{\partial W}{\partial \beta}(\mathbf{n} \cdot \nabla\rho)\mathbf{m}\Big|_{\ell} = \zeta\mathbf{m}\Big|_{\ell}, \quad \mathbf{x} \in \ell. \tag{25}$$

The edge conditions (24) or (25) can be generalized taking into account line tension as in Pietraszkiewicz et al. (2007) as well as for the presence of corner points. For detailed general analysis of edge conditions in strain gradient media we refer to dell'Isola and Seppecher (1995, 1997), dell'Isola et al. (2012).

6 Conclusions

Using the Lagrange variational principle and the principle of virtual power we present the week formulation of the static problem for a second-gradient fluid. As well as for any strain gradient model here the boundary-value problem contains non-classic boundary conditions related with given double forces and line forces. Unlike classical media for a second-gradient fluid, we have to introduce not only surface traction and surface double force density, but also forces acting along edges. The latter may be interpreted as line tension forces.

References

Aifantis, E.C.: Update on a class of gradient theories. Mech. Mater. **35**(3–6), 259–280 (2003)
Altenbach, H., Eremeyev, V.A. (eds.): Generalized Continua: From the Theory to Engineering Applications. CISM Courses and Lectures. Springer, Wien (2013)
Altenbach, H., Maugin, G.A., Erofeev, V. (eds.): Mechanics of Generalized Continua, Advanced Structured Materials, vol. 7. Springer, Berlin (2011)
Altenbach, H., Forest, S., Krivtsov, A. (eds.): Generalized Continua as Models for Materials, Advanced Structured Materials, vol. 22. Springer, Berlin (2013)
Auffray, N., dell'Isola, F., Eremeyev, V.A., Madeo, A., Rosi, G.: Analytical continuum mechanics à la Hamilton-Piola least action principle for second gradient continua and capillary fluids. Math. Mech. Solids **20**(4), 375–417 (2015)
Brenner, H.: Navier-Stokes revisited. Phys. A: Stat. Mech. Appl. **349**(1–2), 60–132 (2005)
Cahn, J.W., Hilliard, J.E.: Free energy of a nonuniform system. I. Interfacial free energy. J. Chem. Phys. **28**, 258 (1958)
Cahn, J.W., Hilliard, J.E.: Free energy of a nonuniform system. III. Nucleation in a two-component incompressible fluid. J. Chem. Phys. **31**, 688 (1959)
Capriz, G.: Continua with Microstructure. Springer, New York (1989)
de Gennes, P.G., Brochard-Wyart, F., Quéré, D.: Capillarity and Wetting Phenomena: Drops, Bubbles, Pearls, Waves. Springer, New York (2004)
dell'Isola, F., Rotoli, G.: Validity of Laplace formula and dependence of surface tension on curvature in second gradient fluids. Mech. Res. Commun. **22**, 485–490 (1995)
dell'Isola, F., Seppecher, P.: The relationship between edge contact forces, double forces and interstitial working allowed by the principle of virtual power. Comptes rendus de l'Académie des sciences Série 2 **321**(8), 303–308 (1995)
dell'Isola, F., Seppecher, P.: Edge contact forces and quasi-balanced power. Meccanica **32**(1), 33–52 (1997)
dell'Isola, F., Steigmann, D.: A two-dimensional gradient-elasticity theory for woven fabrics. J. Elast. **118**(1), 113–125 (2015)
dell'Isola, F., Gouin, H., Rotoli, G.: Nucleation of spherical shell-like interfaces by second gradient theory: numerical simulations. Eur. J. Mech. B/Fluids **15**(4), 545–568 (1996)
dell'Isola, F., Seppecher, P., Madeo, A.: How contact interactions may depend on the shape of Cauchy cuts in *N*th gradient continua: approach "à la d'Alembert". ZAMP **63**, 1119–1141 (2012)
dell'Isola, F., Auffray, N., Eremeyev, V.A., Madeo, A., Placidi, L., Rosi, G.: Least action principle for second gradient continua and capillary fluids: a Lagrangian approach following Piola's point of view. In: The Complete Works of Gabrio Piola, vol. I, pp. 606–694. Springer, Berlin (2014)
dell'Isola, F., Steigmann, D., Della Corte, A.: Synthesis of fibrous complex structures: designing microstructure to deliver targeted macroscale response. Appl. Mech. Rev. **67**(6), 060–804 (2015)
Dunn, J.E., Serrin, J.: On the thermomechanics of interstitial working. Arch. Ration. Mech. Anal. **88**(2), 95–133 (1985)
Eremeyev, V.A., Altenbach, H.: Equilibrium of a second-gradient fluid and an elastic solid with surface stresses. Meccanica **49**(11), 2635–2643 (2014)
Eremeyev, V.A., Lebedev, L.P., Altenbach, H.: Foundations of Micropolar Mechanics. SpringerBriefs in Applied Sciences and Technologies. Springer, Heidelberg (2013)
Eringen, A.C.: Microcontinuum Field Theory I. Foundations and Solids. Springer, New York (1999)
Eringen, A.C.: Microcontinuum Field Theory II. Fluent Media. Springer, New York (2001)
Eringen, A.C.: Nonlocal Continuum Field Theories. Springer, New York (2002)
Fleck, N.A., Hutchinson, J.W.: Strain gradient plasticity. Adv. Appl. Mech. **33**, 295–361 (1997)
Forest, S.: Some links between Cosserat, strain gradient crystal plasticity and the statistical theory of dislocations. Philos. Mag. **88**(30–32), 3549–3563 (2008)
Fried, E., Gurtin, M.E.: Tractions, balances, and boundary conditions for nonsimple materials with application to liquid flow at small-length scales. Arch. Ration. Mech. Anal. **182**(3), 513–554 (2006)

Gao, H., Huang, Y., Nix, W.D., Hutchinson, J.W.: Mechanism-based strain gradient plasticity-I. Theory. J. Mech. Phys. Solids **47**(6), 1239–1263 (1999)
Germain, P.: La méthode des puissances virtuelles en mécanique des milieux continus - première partie, théorie du second gradient. Journal de Mécanique **12**, 235–274 (1973a)
Germain, P.: The method of virtual power in continuum mechanics. part 2: Microstructure. SIAM J. Appl. Math. **25**(3), 556–575 (1973b)
Green, A.E., Rivlin, R.S.: Multipolar continuum mechanics. Arch. Ration. Mech. Anal. **17**, 113–147 (1964)
Gurtin, M.E.: A gradient theory of single-crystal viscoplasticity that accounts for geometrically necessary dislocations. J. Mech. Phys. Solids **50**(1), 5–32 (2002)
Heida, M., Málek, J.: On compressible Korteweg fluid-like materials. Int. J. Eng. Sci. **48**(11), 1313–1324 (2010)
Huang, Y., Gao, H., Nix, W.D., Hutchinson, J.W.: Mechanism-based strain gradient plasticity-II. Analysis. J. Mech. Phys. Solids **48**(1), 99–128 (2000)
Korteweg, D.J.: Sur la forme que prennent les équations des mouvements des fluides si l'on tient compte des forces capillaires par des variations de densité. Archives Néerlandaises des Sciences exactes et naturelles Sér **II**(6), 1–24 (1901)
Lebedev, L.P., Cloud, M.J., Eremeyev, V.A.: Tensor Analysis with Applications in Mechanics. World Scientific, New Jersey (2010)
Maugin, G.A., Metrikine, A.V.: Mechanics of Generalized Continua. One Hundred Years After the Cosserats. Advances in Mechanics and Mathematics, vol. 21. Springer, New York (2010)
Maugin, G.A., Muschik, W.: Thermodynamics with internal variables. Part II. Applications. J. Non-Equilib. Thermodyn. **19**(3), 250–289 (1994)
Mühlhaus, H.B., Aifantis, E.C.: A variational principle for gradient plasticity. Int. J. Solids Struct. **28**(7), 845–857 (1991)
Pietraszkiewicz, W., Eremeyev, V., Konopińska, V.: Extended non-linear relations of elastic shells undergoing phase transitions. ZAMM **87**(2), 150–159 (2007)
Podio-Guidugli, P., Vianello, M.: On a stress-power-based characterization of second-gradient elastic fluids. Contin. Mech. Thermodyn. **25**(2–4), 399–421 (2013)
Rahali, Y., Giorgio, I., Ganghoffer, J., dell'Isola, F.: Homogenization à la piola produces second gradient continuum models for linear pantographic lattices. Int. J. Eng. Sci. **97**, 148–172 (2015)
Rosi, G., Giorgio, I., Eremeyev, V.A.: Propagation of linear compression waves through plane interfacial layers and mass adsorption in second gradient fluids. ZAMM **93**(12), 914–927 (2013)
Rowlinson, J.S., Widom, B.: Molecular Theory of Capillarity. Dover, New York (2003)
Sedov, L.I.: Models of continuous media with internal degrees of freedom. J. Appl. Math. Mech. **32**(5), 803–819 (1968)
Seppecher, P.: Etude des conditions aux limites en théorie du second gradient: cas de la capillarité. Comptes rendus de l'Académie des sciences Série 2, Mécanique, Physique, Chimie, Sciences de l'univers. Sciences de la Terre **309**(6), 497–502 (1989a)
Seppecher, P.: Étude des conditions aux limites en théorie du second gradient: cas de la capillarité. Comptes Rendus de l'Académie des Sciences **309**, 497–502 (1989b)
Seppecher, P.: Equilibrium of a Cahn-Hilliard fluid on a wall—Influence of the wetting properties of the fluid upon the stability of a thin liquid film. Eur. J. Mech. B/Fluids **12**(1), 69–84 (1993)
Seppecher, P.: Les fluides de Cahn-Hilliard. Mémoire d'habilitation à diriger des recherches, Université du Sud Toulon (1996)
Seppecher, P.: Second-gradient theory: application to Cahn-Hilliard fluids. In: Maugin, G.A., et al. (eds.) Continuum Thermomechanics: : The Art and Science of Modeling Matter's Behaviour, pp. 379–388. Springer, Dordrecht (2002)
Truesdell, C.: The Elements of Continuum Mechanics. Springer, New York (1966)
Truesdell, C., Noll, W.: The Non-linear Field Theories of Mechanics, 3rd edn. Springer, Berlin (2004)
van der Waals, J.D.: The thermodynamic theory of capillarity under the hypothesis of a continuous variation of density (English translation by J. S. Rowlinson). J. Stat. Phys. **20**, 200–244 (1893)

Author Index

K. Naumenko and M. Aßmus (eds.), *Advanced Methods of Continuum Mechanics for Materials and Structures*, Advanced Structured Materials 60,
DOI 10.1007/978-981-10-0959-4

MIX
Papier aus verantwortungsvollen Quellen
Paper from responsible sources
FSC® C105338

If you have any concerns about our products,
you can contact us on
ProductSafety@springernature.com

In case Publisher is established outside the EU,
the EU authorized representative is:
Springer Nature Customer Service Center GmbH
Europaplatz 3, 69115 Heidelberg, Germany

Printed by Libri Plureos GmbH
in Hamburg, Germany